Ludwig Reimer

Transmission Electron Microscopy

Physics of Image Formation and Microanalysis

With 264 Figures

Springer-Verlag
Berlin Heidelberg New York Tokyo 1984

Professor Dr. LUDWIG REIMER
Physikalisches Institut, Westfälische Wilhelms-Universität Münster, Domagkstraße 75,
D-4400 Münster, Fed. Rep. of Germany

ISBN 3-540-11794-6 Springer-Verlag Berlin Heidelberg New York Tokyo
ISBN 0-387-11794-6 Springer-Verlag New York Heidelberg Berlin Tokyo

Library of Congress Catolging in Publication Data. Reimer, L. (Ludwig), 1928–. Transmission electron microscopy. (Springer series in optical sciences ; v. 36). 1. Electron microscope, Transmission. I. Title. II. Series. QH212.T7R43 1983 502'.8'2 83-14720

Typesetting and Offset printing: Schwetzinger Verlagsdruckerei GmbH
Bookbinding: J. Schäffer OHG, Grünstadt
2153/3130-543210

Preface

The aim of this book is to outline the physics of image formation, electron-specimen interactions and image interpretation in transmission electron microscopy. The book evolved from lectures delivered at the University of Münster and is a revised version of the first part of my earlier book *Elektronenmikroskopische Untersuchungs- und Präparationsmethoden,* omitting the part which describes specimen-preparation methods.

In the introductory chapter, the different types of electron microscope are compared, the various electron-specimen interactions and their applications are summarized and the most important aspects of high-resolution, analytical and high-voltage electron microscopy are discussed.

The optics of electron lenses is discussed in Chapter 2 in order to bring out electron-lens properties that are important for an understanding of the function of an electron microscope. In Chapter 3, the wave optics of electrons and the phase shifts by electrostatic and magnetic fields are introduced; Fresnel electron diffraction is treated using Huygens' principle. The recognition that the Fraunhofer-diffraction pattern is the Fourier transform of the wave amplitude behind a specimen is important because the influence of the imaging process on the contrast transfer of spatial frequencies can be described by introducing phase shifts and envelopes in the Fourier plane. In Chapter 4, the elements of an electron-optical column are described: the electron gun, the condenser and the imaging system.

A thorough understanding of electron-specimen interactions is essential to explain image contrast. Chapter 5 contains the most important facts about elastic and inelastic scattering and x-ray production. The origin of scattering and phase contrast of non-crystalline specimens is described in Chapter 6. High-resolution image formation using phase contrast may need to be completed by image-reconstruction methods in which the influence of partial spatial and temporal coherence is considered.

Chapter 7 introduces the most important laws about crystals and reciprocal lattices. The kinematical and dynamical theories of electron diffraction are then developed. Electron diffraction is the source of diffraction contrast, which is important for the imaging of lattice structure and defects and is treated in Chapter 8. Extensions of the capabilities of the instrument have awakened great interest in analytical electron microscopy: x-ray microanalysis, electron-energy-loss spectroscopy and electron diffraction, summarized in Chapter 9. The final Chapter 10 contains a brief account of the various specimen-damage processes caused by electron irradiation.

Electron microscopy is an interdisciplinary science with a strong physical background. The full use of all its resources and the interpretation of the results requires familiarity with many branches of knowledge. Physicists are in a favoured situation because they are trained to reduce complex observations to simpler models and to use mathematics for formulating "theories". There is thus a need for a book that expresses the contents of these theories in language accessible to the "normal" electron microscope user. However, so widespread is the use of electron microscopy that there is no such person as a normal user. Biologists will need only a simplified account of the theory of scattering and phase contrast and of analytical methods, whereas electron diffraction and diffraction contrast used by material scientists lose some of their power if they are not presented on a higher mathematical level. Some articles in recent series on electron microscopy have tried to bridge this gap by over-simplification but this can cause misunderstandings of just the kind that the authors wished to avoid. In the face of this dilemma, the author decided to write a book in his own physical language with the hope that it will be a guide to a deeper understanding of the physical background of electron microscopy.

A monograph by a single author has the advantage that technical terms are used consistently and that cross-referencing is straightforward. This is rarely the case in books consisting of review articles written by different specialists. Conversely, the author of a monograph is likely to concentrate, perhaps unconsciously, on some topics at the expense of others but this also occurs in multi-author works and reviews. Not every problem can be treated on a limited number of pages and the art of writing such a book consists of selecting topics to omit. I apologize in advance to any readers whose favourite subjects are not treated in sufficient detail. The number of electron micrographs has been kept to a minimum; the numerous simple line drawings seem better suited to the more theoretical approach adopted here.

There is a tendency for transmission electron microscopy and scanning electron microscopy to diverge, despite their common physical background. Electron microscopy is not divided into these categories in the present book but because transmission and scanning electron microscopy together would increase its size unreasonably, only the physics of the transmission electron microscope is considered – the physics of its scanning counterpart will be examined in a complementary volume.

A special acknowledgement is due to P. W. Hawkes for his cooperation in revising the English text and for many helpful comments. Special thanks go to K. Brinkmann and Mrs. R. Dingerdissen for preparing the figures and to all colleagues who gave me permission to publish their results.

Münster, July 1982 *L. Reimer*

Contents

1. Introduction

1.1 Types of Electron Microscopes

Although the main concern of this book is transmission electron microscopy, the functions and limits of the other types of electron microscopes are also mentioned in this introductory chapter to show the advantages and disadvantages of their various imaging techniques. Several types of electron microscopes and analysing instruments capable of furnishing an "image" can be distinguished. We now examine these briefly, in turn, without considering the historical sequence in which these instruments were developed. In these background sections, references are restricted to review articles and books.

1.1.1 Electron Microscopes for the Direct Imaging of Surfaces of Bulk Specimens

a) Emission Electron Microscopes [1.1, 2, 10]

In an emission electron microscope (Fig. 1.1 a), the cathode that emits the electrons is directly imaged by an electrostatic immersion lens, which accelerates the electrons and produces an intermediate image of the emission intensity distribution at the cathode. This image can be magnified by further electron lenses and is observed on a fluorescent screen or with an image intensifier. The cathode (specimen) has to be plane and its surface should not be too irregular. The electron emission can be stimulated by

a) heating the cathode (thermionic emission), which means that observation is possible only at elevated temperatures and for a limited number of materials; alternatively, the electron-emission temperature need not be raised beyond 500–1000 °C if a thin layer of barium is evaporated on the surface because this lowers the work function;

b) secondary-electron excitation by particle bombardment, by irradiating the cathode surface with an additional high-energy electron beam or an ion beam at grazing incidence;

c) irradiation of the cathode with an ultra-violet light source to excite photoelectrons (photoelectron-emission microscope PhEEM [1.3, 4])

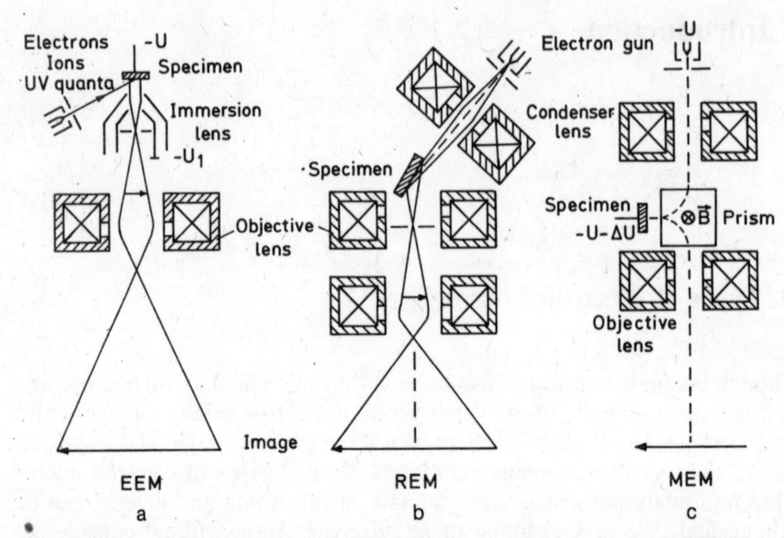

Fig. 1.1 a–c. Schematic ray paths for (**a**) an emission electron microscope (EEM), (**b**) a reflexion electron microscope (REM), (**c**) a mirror electron microscope (MEM)

These instruments have a number of interesting applications but their use is limited to particular specimens; at present, therefore, scanning electron microscopes (Sect. 1.1.2) are the most widely used instruments for imaging bulk specimens, especially because there is no need to limit the roughness of the specimen surface. The final restriction is the limited number of electrons emitted, which restricts the image intensity at high magnification, and the resolution of the immersion lens system is only of the order of 10–30 nm. On the credit side, surfaces can be observed directly in-situ and each of the processes a–c) generates a specific contrast. The photoelectron-emission electron microscope has the advantage of being applicable to nearly any flat specimen surface, including biological specimens [1.5, 6]. The image contrast is caused by differences of the emission intensity (material and crystal orientation contrast) and by angular selection with a diaphragm that intercepts electrons whose trajectories have been deflected by variations of the equipotentials near the surface caused by surface steps (topographic contrast), surface potentials (potential contrast) or magnetic stray fields (magnetic contrast).

b) Reflexion Electron Microscopes [1.7–10]

The electrons that emerge as a result of primary-electron bombardment are either low-energy secondary electrons, which can be imaged in an emission electron microscope (see above) or scanning electron microscope (see below), or primary (backscattered) electrons with large energy losses, which cannot be focused sharply by an electron lens because of chromatic aberration. However, imaging of the surface is possible for grazing electron incidence below 10°, the "reflected" electrons being imaged with an objective lens (Fig. 1.1 b). The energy-loss spectrum of the reflected electrons has a half-width of the order of 100–200 eV. With additional energy selection by means of an electrostatic filter lens, a resolution of 10–20 nm can be attained [1.11]. Because the angle of incidence is so low, small surface steps can be imaged with high contrast. The angular distribution of the reflected electrons at single crystals is a reflexion high-energy electron diffraction (RHEED) pattern with Bragg diffraction spots; images exhibiting crystallographic contrast can be found by selecting individual Bragg spots. A transmission electron microscope can be operated in this mode by tilting the electron gun and condenser-lens system.

c) Mirror Electron Microscopes [1.12, 13]

An electron beam is deflected by a magnetic sector field, and retarded and reflected at a flat specimen surface, which is biased a few volts more negative than the cathode of the electron gun (Fig. 1.1 c). The reflected-electron trajectories are influenced by irregularities of the equipotential surfaces in front of the specimen, which can be caused by surface roughness or by potential differences and specimen charges; magnetic stray fields likewise act on the electron trajectories. An advantage of this method is that the electrons do not strike the specimen; it is the only technique that permits surface charges to be imaged undisturbed. After passing through the magnetic sector field again, the electrons can be selected according to their angular deflection. The lateral resolution of a mirror electron microscope is of the order of 50–100 nm. Single surface steps, 5 nm in height, can produce discernible contrast. Such a mirror electron microscope can be combined with an electron interferometer (Sect. 3.1.4), which offers the possibility of measuring phase shifts caused by the equipotential surfaces or magnetic stray fields with high precision. Methods of scanning mirror electron microscopy [1.14, 15] allow a more quantitative separation of the observed image point and the local beam deflection.

1.1.2 Instruments Using Electron Microprobes

a) Scanning Electron Microscopes (SEM) [1.16–31]

SEM is the most important electron-optical instrument for the investigation of bulk specimens. An electron probe is produced by an one-, two- or three-stage demagnification of the smallest cross-section of the electron beam after acceleration. This electron probe, 5–10 nm in diameter if a thermionic electron gun is used and 0.5–2 nm with a field-emission gun, is scanned in a raster over a region of the specimen (Fig. 1.2). The smallest diameter of the electron probe is limited by the minimum acceptable electron probe current of

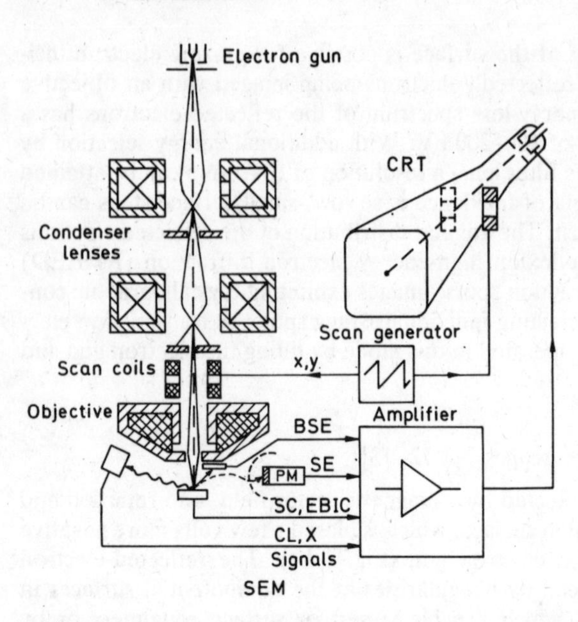

Fig. 1.2. Schematic ray path for a scanning electron microscope (SEM)

10^{-12}–10^{-11} A. It is determined by the need to generate an adequate signal-to-noise ratio, and by the spherical aberration of the final probe-forming lens. The image is displayed on a cathode-ray tube (CRT) rastered in synchronism. The CRT beam intensity can be modulated by any of the different signals that result from the electron-specimen interactions.

The most important signals are produced by secondary electrons with most probable exit energies of 2–5 eV and by backscattered electrons, with energies that range from the energy of the primary electrons to about 50 eV. The secondary-electron yield and the backscattering coefficient depend on the angle of electron incidence (topographic contrast), the crystal orientation (channelling contrast) and electrostatic and magnetic fields near the surface (voltage and magnetic contrast). A signal can also be produced by the specimen current and by electron-beam-induced currents in semiconductors. Analytical information is available from the x-ray spectrum and Auger electrons or from light quanta emitted by cathodoluminescence. The crystallographic structure and orientation can be obtained from electron channelling patterns or electron back-scattering patterns (Sect. 9.3.4) and from x-ray Kossel diagrams.

The resolution of the different modes of operation and types of contrast depends on the information volume that contributes to the signal. Secondary electrons provide the best resolution, because the exit depth is very small, of the order of a few nanometres. The exit volume limits the resolution of field-emission systems with electron-probe diameters smaller than 2 nm. The information depth of backscattered electrons is much greater, of the order of half the electron range, which is as much as 0.1–1 μm, depending on the density of the specimen and the electron energy. The secondary-electron signal also contains a large contribution from the backscattered electrons when these penetrate the surface layer. This is why SEMs mostly operate in the range $E = 10 - 20$ keV. At higher energies, the electron range and the diameter of the electron-diffusion region are greater. Conversely, higher energies are of interest for x-ray microanalysis if the K shells of heavy elements are to be excited. Decreasing the electron energy has the advantage that information can be extracted from a volume nearer to the surface, but the diameter of the electron probe increases owing to the decrease of gun brightness.

Unlike transmission electron microscopy, special specimen-preparation techniques are rarely needed in scanning electron microscopy. Nevertheless, charging effects have to be avoided, by coating the specimen with a thin conductive film for example and organic specimens have to be protected from surface distortions by fixation or cryo-techniques.

b) X-ray and Auger-Electron Microanalysers [1.23–28]

By using a wavelength-dispersive x-ray spectrometer (Bragg reflection at a crystal), we can work with high x-ray excitation rates and electron-probe currents of the order of 10^{-8}–10^{-7} A, though the electron-probe diameter is then larger, about 0.1–1 μm. The main task of an x-ray microanalyser is to analyse the elemental compositions of flat, polished surfaces at normal electron incidence with a high analytical sensitivity. The ray diagram of such an instrument is similar to that of a SEM but two or three crystal spectrometers, which can simultaneously record different characteristic x-ray wavelengths, are attached to the column. The surface can be imaged by one of the SEM modes to select the specimen points to be analysed.

SEM or x-ray microanalyser can be equipped with an Auger-electron spectrometer, of the cylindrical mirror type for example. It is then necessary to work with ultra-high vacuum in the specimen chamber because Auger electrons are extremely sensitive to the state of the surface: a few atomic layers are sufficient to halt them. Special Auger-electron microanalysers have therefore been developed, in which the 1–10 keV electron gun may for example be incorporated in the inner cylinder of a spectrometer. This type of instrument can also work in the scanning mode so that an image of the surface can be formed with secondary electrons or an element-distribution map can be generated using Auger electrons.

1.1.3 Transmission Electron Microscopes

a) Conventional Transmission Electron Microscope [1.32–96]

In a conventional transmission electron microscope (CTEM, or TEM for short) (Figs. 1.3 and 4.17 c, 20), a thin specimen is irradiated with an electron

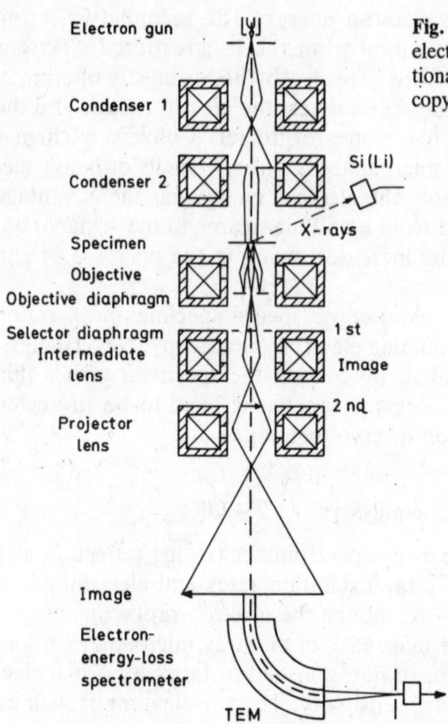

Fig. 1.3. Schematic ray path for a transmission electron microscope (TEM) equipped for additional x-ray and electron energy-loss spectroscopy

beam of uniform current density; the electron energy is in the range 60–150 keV (usually 100 keV) or 200 keV–3 MeV in the case of the high-voltage electron microscope (HVEM).

Electrons are emitted in the electron gun (Sect. 4.1) by thermionic emission from tungsten hairpin cathodes or LaB_6 rods or by field emission from pointed tungsten filaments. The latter are used when high gun brightness is needed. A two-stage condenser-lens system permits variation of the illumination aperture and the area of the specimen illuminated (Sec. 4.2). The electron-intensity distribution behind the specimen is imaged with a three- or four-stage lens system, onto a fluorescent screen (Sect. 4.4). The image can be recorded by direct exposure of a photographic emulsion inside the vacuum (Sect. 4.6).

The lens aberrations of the objective lens are so great that it is necessary to work with very small objective apertures, of the order of 10–25 mrad, to achieve resolution of the order of 0.2–0.5 nm. Bright-field contrast is produced either by absorption of the electrons scattered through angles larger than the objective aperture (scattering contrast) or by interference between the scattered wave and the incident wave at the image point (phase contrast). The phase of the electron waves behind the specimen is modified by the wave aberration of the objective lens. This aberration and the energy spread of the electron gun, which is of the order of 1–2 eV, limits the contrast transfer of high spatial frequencies.

Electrons interact strongly with atoms by elastic and inelastic scattering. The specimen must therefore be very thin, typically of the order of 5 nm–0.5 μm for 100 keV electrons, depending on the density and elemental composition of the object and the resolution desired. Special preparation techniques are needed for this: electropolishing of metal foils and ultramicrotomy of stained and embedded biological tissues for example [1.47, 55, 60, 62, 65, 70, 77, 79]. Thicker specimens can be investigated in a high-voltage electron microscope (Sect. 1.3.3).

TEM can provide high resolution because elastic scattering is an interaction process that is highly localized to the region occupied by the screened Coulomb potential of an atomic nucleus, whereas inelastic scattering is more diffuse; it spreads out over about a nanometre.

A further capability of the modern TEM is the formation of very small electron probes, 2–5 nm in diameter, by means of a three-stage condenser-lens system, the last lens field of which is the objective pre-field in front of the specimen (Figs. 4.17 b, 24). This enables the instrument to operate in a scanning transmission mode (Sect. 4.5.1) with a resolution determined by the electron-probe diameter; this has advantages for imaging thick or crystalline specimens (Sects. 6.1.4 and 8.1.3) and for recording secondary electrons and backscattered electrons, cathodoluminescence and electron-beam-induced currents (Sect. 9.5). The main advantage of equipping a TEM with a STEM attachment is the formation of a very small electron probe, with which elemental analysis and micro-diffraction can be performed on extremely small areas (Sects. 1.3.2, 9). X-ray production in thin foils is confined to the small volume excited by the electron probe, only slightly broadened by multiple scattering. Better spatial resolution is therefore obtainable for segregation effects at crystal interfaces, or precipitates, for example, than in an x-ray microanalyser with bulk specimens, where the spatial resolution is limited to 0.1–1 μm by the diameter of the electron-diffusion cloud.

b) Scanning Transmission Electron Microscope (STEM) [1.97–99]

STEM consists only of a field-emission gun, one probe-forming lens and the electron-detection system, together with an electron spectrometer for electron-energy-loss spectroscopy and for separating the currents of unscattered electrons I_{un}, of elastically scattered electrons I_{el} and of inelastically scattered electrons I_{in} (see Fig. 4.25). This short system can easily be incorporated in an ultra-high-vacuum chamber. Electron-probe diameters of 0.2–0.5 nm can be formed, the spherical aberration of the lens is the limiting factor.

An advantage of STEM instruments is that the contrast can be enhanced by collecting several signals and displaying differences and/or ratios of these by analogue or digital processing. In particular, single atoms on a thin substrate can be imaged with a higher contrast than in the CTEM bright- or dark-field modes (Sect. 6.3.5).

1.2 Electron-Specimen Interactions and Their Applications

Transmission electron microscopes are used not only in the high-resolution mode, in which the contrast is caused by elastic and inelastic scattering, but also in microanalytical modes (Sects. 1.3.2, 9), in which information is collected from the emitted x-rays and the energy losses of inelastically scattered electrons (electron-energy-loss spectroscopy), from cathodoluminescence and electron-beam-induced currents. For every application, a knowledge of the dependence of the electron-specimen interactions on specimen and imaging parameters is essential. The most efficient modes of operation of the instrument can then be selected and the latter can be tuned to give the best possible performance.

1.2.1 Elastic Scattering

Elastic scattering is the most important of all the interactions that create contrast in the electron image. Electrons are elastically scattered at the nuclei of the specimen atoms by the Coulomb force (Fig. 1.4 a). The angular dis-

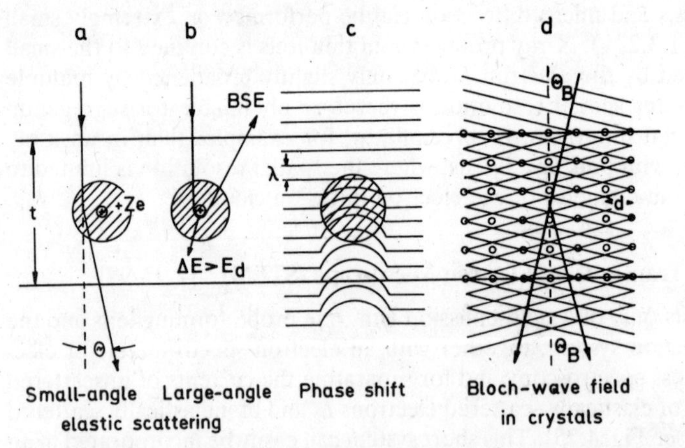

Fig. 1.4 a–d. Elastic electron-specimen interactions: (**a**) small-angle scattering with no energy loss, (**b**) large-angle scattering resulting in backscattered electrons (BSE) and atomic displacement if the energy loss ΔE is larger than the displacement energy E_d of the atoms; (**c**) phase shift of an electron wave by the inner potential; (**d**) Bragg reflection at the lattice planes of a crystal ($2d\sin\theta_B = \lambda$, where d is the lattice-plane spacing, θ_B the Bragg angle and λ the electron wavelength)

tribution of the scattered electrons can be conveniently described in terms of cross-sections (Sect. 5.1.1). The kinetic energies and momenta of colliding particles are conserved in elastic scattering. For small scattering angles θ, a negligible amount of energy is transferred to the nucleus and electrons are scattered with no appreciable energy loss. Considerable energy ΔE can however be transferred from an electron to a nucleus at high electron energies and large scattering angles (Fig. 1.4 b). The energy transferred can exceed the threshold energy E_d, of the order of 10–30 eV, beyond which an atom can be displaced from a crystal position to an interstitial site. (Sect. 5.1.2 and 10.3.2). For small-angle scattering, the term elastic scattering is automatically reserved for interactions in which the primary electrons lose no energy; for large-angle scattering, the primary electrons may lose energy.

The small-angle scattering of electrons governs the scattering contrast, which is caused by interception of scattered electrons by the objective diaphragm (Sects. 4.4.2 and 6.1). For the discussion of high resolution (Sect. 1.3.1), the wave-optical model of electron propagation (Chap. 3) has to be employed. The Coulomb potential of a nucleus or the inner potential of a dense particle shifts the phase of the electron wave with deBroglie wavelength $\lambda = h/mv$ (Fig. 1.4 c). An electron-optical refractive index can be defined (Sect. 3.1.3), which is proportional to the Coulomb energy. Elastic scattering has to be treated quantum-mechanically by solving the Schrödinger equation of the scattering problem (Sect. 5.1.3). In the wave-optical theory of imaging, the phase shifts caused by the specimen and by the wave aberration of the objective lens (Sect. 5.3) are studied and shown to create phase contrast (Sect. 6.2) as an interference effect between the primary and scattered electron waves. Phase shifts produced by the magnetic vector potential in ferromagnetic films can generate phase contrast in an out-of-focus image and are used in Lorentz microscopy (Sect. 6.6) to observe and measure the distribution of magnetization.

1.2.2. Electron Diffraction

In the periodic potential of a crystal lattice, the electron waves propagate as a Bloch-wave field, which exhibits the same periodicity as the lattice. The interaction can also be characterized by the Bragg condition, $2d \sin\theta_B = \lambda$, which relates the angle $2\theta_B$ between a Bragg-diffraction spot and the primary beam to the lattice-plane spacing d and the wavelength λ (Fig. 1.4 d). The kinematical theory of electron diffraction (Sect. 7.2) assumes that the amplitude of a Bragg diffracted wave is small compared to that of the primary wave. This approach is applicable only for very thin foils, less than a few nanometres thick. The dynamical theory (Sect. 7.3), based on the Schrödinger equation, also describes wave propagation in thick crystals and results in a "pendellösung", which means that the amplitudes of the Bragg-diffracted and primary waves oscillate in antiphase as a function of depth and depend sensitively on the tilt of the specimen. The intensity of the primary beam may show anomalous transmission. Though a two-beam approximation (Sect. 7.3.4) is often used to discuss the main effects of dynamical electron diffraction on the beam intensity and image contrast in crystals, a many-beam theory with 20–100 or more diffracted beams has to be used if the observed phenomena are to be explained in detail. The pendellösung causes typical diffraction effects in crystal foils, which can be seen as edge or bend contours, for example (Sect. 8.1). The image intensity depends very sensitively on the strain field of any lattice defects, so that a large variety of defects can be imaged and analysed without resolving the lattice structure (Sects. 8.3–6). A resolution of the order of 1 nm is obtainable if the weak-beam technique, in which a dark-field image is formed with a weakly excited Bragg reflection, is used.

If the objective aperture is so large that the primary beam and one or more Bragg reflections can pass through the diaphragm, the waves interfere in the image plane and form an interference pattern; this can furnish an image of the crystal structure and its faults, if the specimen is thin enough (≤ 10 nm) (Sect. 8.2). Often, however, the image contrast is affected by dynamical diffraction and the phase shift of the electron lens so that a high-resolution image of lattice structures has to be analysed with care; it is usually necessary to compare the results with computer simulations in which the crystal orientation and thickness and the phase shift caused by the spherical aberration and defocus of the electron lens are considered.

1.2.3 Inelastic Scattering and X-Ray Emission

In inelastic scattering (Sect. 5.2) (Fig. 1.5 a), the total energies and the momenta of the colliding particles are conserved, but an electron excitation of the atom or solid is stimulated, and the primary beam loses energy ΔE. Elastic scattering can be considered as an electron-nucleus interaction in which the atomic electron cloud screens only the Coulomb potential, whereas inelastic scattering is an electron-electron interaction.

Typical characteristics of inelastic scattering, such as the more forward-directed angular distribution, and the ratio of inelastic-to-elastic total cross-sections can be explained in terms of a simple theory for which a detailed knowledge of the electron-energy states of the atomic electrons is not required (Sect. 5.2.2). Such information is, however, needed to interpret the energy-loss spectrum. The most important electron excitations that can be observed with energy losses ΔE smaller than 50 eV are the intra- and inter-band and plasmon excitations (Fig. 1.5 b) near the Fermi level E_F (Sect. 5.2.3). The plasmons are longitudinal waves in the electron plasma and decay into either photons or phonons. Ionisation of an inner shell (Fig. 1.5 c) results in an edge-like rise in the energy loss spectrum at $\Delta E = E_I$ where E_I is the ionisation energy of the K, L or M shell; for the K-shell of carbon and aluminium we have $E_K = 285$ eV and 1550 eV, respectively. Elements of higher atomic number contribute in the range $\Delta E = 20$–2000 eV with their L

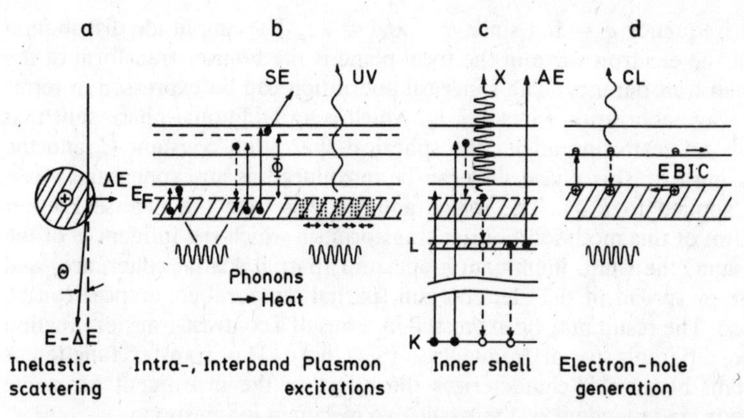

Fig. 1.5 a–d. Inelastic electron-specimen interactions: (**a**) scattering in very small angles with energy loss ΔE; (**b**) intra-, interband and plasmon excitations are shown in the energy-band structure. Excited electrons with energies larger than the work function ϕ can leave as secondary electrons (SE). Plasmons are longitudinal oscillations of the electron plasma and decay in either UV quanta or phonons. (**c**) Inner-shell ionisation results in large energy losses. The vacancy is filled by an electron from a higher state. The excess energy is emitted either as an x-ray quantum or an Auger electron (AE). (**d**) In semiconductors, electron-hole pairs are generated and cause electron-beam-induced currents (EBIC). They recombine by emission of either a light quantum (cathodoluminescence) or a phonon (radiation-less)

and *M* shells. The energy loss spectrum therefore contains a great deal of information about the electronic structure, indicated by the plasmon losses, and about the chemical composition indicated by the ionisation losses.

The vacancy in an ionised shell is filled by an electron from a higher state (Fig. 1.5 c); the energy difference is then emitted either as a x-ray quantum or as an Auger electron (AE). In semiconductors, electron-hole pairs are generated (Fig. 1.5 d) which can result in an electron-beam-induced current (EBIC). Electron-hole pairs recombine radiationless or cause cathodoluminescence (CL).

All excitation energy that does not leave the specimen with the secondary electrons (SE) and photons is converted into phonons and heat (Sect. 10.1).

1.3 Special Applications of Transmission Electron Microscopy

1.3.1 High-Resolution Electron Microscopy

The wave-optical theory of imaging is necessary to discuss high resolution. This can be expressed in terms of a two-stage Fourier transform (Sects. 3.3.2 and 6.2.4). In the focal plane of the objective lens, the diffraction pattern of the specimen is formed; each scattering angle θ corresponds reciprocally to a periodic spacing Λ in the specimen or in other words, is proportional to a spatial frequency $q = 1/\Lambda$ since $\theta = \lambda/\Lambda = \lambda q$. The amplitude distribution $F(q)$ of the electron wave in the focal plane is the Fourier transform of the specimen transparency. The spherical aberration can be expressed in terms of the wave aberration (Sect. 3.3.1), which is an additional phase shift that depends on scattering angle, the spherical-aberration constant C_s and the defocusing Δz. This phase shift can be introduced as an exponential phase factor applied to $F(q)$. The image amplitude is then the inverse Fourier transform of this modified Fourier transform, in which the influences of the diaphragm, the finite illumination aperture (partial spatial coherence) and the energy spread of the electron gun (partial temporal coherence) can be included. The result may be expressed in terms of a contrast-transfer function for the different spatial frequencies (Sect. 6.4). This transfer function is important because it characterizes the effect of the instrument on image formation, independent of the particular specimen in question.

The angular characteristics of inelastically scattered electrons is concentrated within smaller scattering angles than that of elastically scattered electrons. Most of the inelastically scattered electrons normally pass through the objective diaphragm in the bright-field mode. Inelastically scattered electrons do not, however, contribute to high-resolution image details because the inelastic scattering is less localized.

The first electron microscopes had spherical-aberration coefficients C_s of about 5 mm but in present-day microscopes, $C_s = 0.5$–2 mm. The optimum imaging condition in bright field occurs at the Scherzer defocus $\Delta z = (C_s\lambda)^{1/2}$, for which a broad band of spatial frequencies is imaged with positive phase contrast. This band has an upper limit at q_{max}. The value $\delta_{min} = 1/q_{max} = 0.43\,(C_s\lambda^3)^{1/4}$ is often used to define a limit of resolution, though it is not correct to characterize resolution by one number only. For $C_s = 1$ mm and $E = 100$ keV ($\lambda = 3.7$ pm), we find $\Delta z \simeq 60$ nm and $\delta_{min} = 0.2$ nm. Narrow bands of higher spatial frequencies can be imaged if the image is not blurred by imperfect spatial and temporal coherence. These effects limit the resolution of present-day microscopes to 0.2–0.3 nm for aperiodic specimens and to 0.1 nm for lattice-plane imaging.

The efforts of the last few years to increase resolution have been concentrated on

1) Reducing the spherical-aberration coefficient C_s by running the objective lens at a higher excitation. Ideas for compensating the spherical aberration have been proposed but none has so far been successfully applied to routine high-resolution microscopy.
2) Decreasing the wavelength λ by using accelerating voltages greater than 100 kV, or, more exactly, by reducing the product $C_s\lambda^3$, which occurs in the above formula for δ_{min}; a high-voltage electron microscope needs a larger pole-piece system and the value of C_s is correspondingly greater. For crystal-structure imaging, it is an advantage that the Bragg angles decrease as λ since more reflections can contribute to the image with the correct phase unperturbed by spherical aberration.

3) Using a field emission gun to decrease the blurring of the contrast-transfer function at high spatial frequencies caused by partial spatial and temporal coherence. Normal TEMs equipped with thermionic cathodes work with illumination apertures α_i of about 10^{-1} mrad; with a field-emission gun, apertures smaller than 10^{-2} mrad are possible. The energy spread ΔE = 1–2 eV of a thermionic gun can be reduced to ΔE = 0.2–0.5 eV with a field-emission gun but the Boersch effect can increase this somewhat.
4) Improving holography, which was originally devised by Gabor in 1949 in the hope of overcoming the resolution limit imposed mainly by spherical aberration. With the development of the laser, a light source of high coherence, holography rapidly grew into a major branch of light optics. Holography has attracted renewed interest in electron optics with the development of field-emission guns of high brightness and coherence (Sect. 6.5). Though better resolution than in the conventional bright-field mode is still a dream, holography is becoming of increasing interest for quantitative studies of phase shifts.

These limitations on resolution can be partially compensated by a posteriori spatial-frequency filtering in an optical Fraunhofer-diffraction pattern or by digital computation (Sects. 6.4.3 and 6.5.5). These methods are mainly used to reverse the phase in regions in which the sign of the contrast transfer function is wrong and to flatten the transfer function by amplitude filtering. Nevertheless, such methods cannot restore information lost at gaps in the transfer function, unless a more complicated type of processing is applied to a focal series.

High-resolution micrographs of specimens on supporting films are disturbed by a phase-contrast effect that creates defocus-dependent granularity. One way of reducing this granularity is to use hollow-cone illumination (Sect. 6.4.3), which suppresses the granularity of supporting films more than that of specimen structures that contain heavy atoms. Furthermore, the contrast-transfer function does not show sign reversal with this type of illumination.

A further obstacle to obtaining high-resolution images of organic specimens is the radiation damage caused by ionisation and subsequent breakage of chemical bonds and finally by a loss of mass (Sect. 10.2). The radiation damage depends on the electron dose in C cm^{-2} (charge density), incident on the specimen. A dose of 1 C cm^{-2} corresponds to 6×10^4 electrons per nm^2, the value needed to form an image free of statistical noise at high magnification. Most amino-acid molecules are destroyed at doses of 10^{-2} C cm^{-2} and only a few compounds, such as hexabromobenzene and phthalocyanine and related substances can be observed at doses of the order of a few C cm^{-2}. The deterioration and mass loss can be suppressed in various ways.

1) The specimen may be cooled to liquid-helium temperature. However, the ionisation products are only frozen-in and the primary ionisation damage will be the same as at room temperature. Only those secondary-radiation effects that are caused by loss of mass are appreciably reduced.

2) The electron dose may be kept very low, which produces a noisy image. The noise can be decreased by signal averaging, which is straight-forward for periodic structures. Non-periodic structures have to be aligned and superposed by use of correlation techniques (Sect. 6.5.5).
3) The ionisation probability may be decreased by increasing the electron energy. This is discussed in Sect. 1.3.3.

1.3.2 Analytical Electron Microscopy [1.50]

The strength of TEM is that not only can it provide high-resolution images that contain information down to 0.1–0.2 nm but it can also operate in various microanalytical modes with spatial resolution of 1–10 nm.

X-ray microanalysis and electron-energy-loss spectroscopy can be used for the measurement of elemental composition and the energy-loss spectrum also contains information about the electronic structure of the specimen. X-ray microanalysis in TEM (Sect. 9.1) mainly relies on energy-dispersive Si(Li)-detectors, though instruments have been constructed (Sect. 9.1.1) with wavelength dispersive spectrometers, such as are used in x-ray microanalysers (Sect. 1.1.2). The energy-dispersive Si(Li) detector (Sect. 9.1.2), with a resolution $\Delta E_x = 150$–200 eV of x-ray quantum energy $E_x = h\nu$, has the disadvantage that neighbouring characteristic lines are less well separated and the analytical sensitivity is poorer than in a wavelength-dispersive spectrometer; this is counterbalanced by the fact that all lines with quantum energies E_x greater than 1 keV can be recorded simultaneously, even at the low probe currents used in the STEM mode of TEM. Reliable quantitative information concerning elemental composition is provided because the x-rays generated by thin films need only small corrections (Sect. 9.1.3).

A weakness of x-ray microanalysis using an energy-dispersive spectrometer is that elements below Na ($Z = 11$), for which the K-line corresponds to a quantum energy $E_x \simeq 1$ keV, cannot be detected, whereas wavelength-dispersive spectrometers can analyse down to Be ($Z = 3$). This gap can be filled by electron-energy-loss spectroscopy. In the loss spectrum, steep steps are seen at energy losses ΔE above the ionisation energy of a K, L or M shell. Another advantage of electron-energy-loss spectroscopy is that electrons that have been inelastically scattered by ionisation processes are concentrated within small scattering angles and the spectrometer has a large collection efficiency, of the order of 10–50%; the characteristic x-ray quanta, on the other hand, are emitted isotropically and only a small solid angle, of the order of 10^{-2} sr, is collected by an energy-dispersive detector and with a wavelength-dispersive detector, the solid angle is one to two orders of magnitude less. Furthermore, the fluorescence yield ω – the probability of emission of a characteristic x-ray quantum rather than an Auger electron – is very low for light elements. In electron-energy-loss spectroscopy, however, the number of primary ionisation processes is counted. This advantage is lost if the loss spectrum is recorded sequentially by sweeping the spectrum across a detector slit. Systems that provide simultaneous recording by means of multi-detector arrays are therefore of considerable interest.

The fine structure of the inner-shell ionisation steps in the energy-loss spectrum also contain information about the electronic structure; a study of the extended fine structure allows us to measure the nearest-neighbour distances. Further information about the electronic structure can be derived from the plasmon and inter- and intraband excitations at low energy losses ($\Delta E \leq 50$ eV).

X-ray microanalysis and electron-energy-loss spectroscopy can also be exploited for elemental mapping, in which the intensity of a characteristic x-ray line or an energy-loss excitation is displayed directly on the rastered image. This can also be achieved by selective filtering of an electron-microscope image by incorporating a spectrometer in the optical column of the TEM (Sect. 9.2.5). This method is of interest for preventing inelastically scattered electrons from contributing to the image and for investigating the preservation of diffraction contrast of crystals after inelastic scattering.

Information about crystal structure and orientation is provided by the electron-diffraction pattern. The possibility of combining electron diffraction and the various imaging modes is the most powerful feature of TEM for the investigation of the lattice and its defects in crystalline material. With the selected-area electron-diffraction technique (Sect. 9.3.1), it is possible to switch from one mode to another simply by changing the excitation of the diffraction or intermediate lens and to select the diffraction pattern from areas 0.1–1 μm in diameter. Other modes of operation that permit small-area electron diffraction can be used when the instrument is capable of forming an electron probe 1–20 nm in diameter (Sect. 9.3.3). In most cases, crystals are free of defects in such a small area and so convergent-beam diffraction techniques can be applied. In particular, the appearance of high-order Laue zone (HOLZ) Kossel lines in the convergent primary beam provides much additional information about crystal structure (Sect. 9.4.6). A high-order Laue zone pattern of large aperture, of the order of 10°, can be used for three-dimensional reconstruction of the lattice because the Ewald sphere intersects high-order Laue zones in circles of large diameter. Unfortunately, the bores of the lens pole-pieces of TEMs are normally so small that the diffraction pattern cannot be imaged beyond an angle of about 5°.

1.3.3 High-Voltage Electron Microscopy [1.100–105, 4.47–51]

Conventional transmission electron microscopes work with accelerating voltages of 100–120 kV. If the voltage is to be increased to 150–200 kV, a double coaxial high-tension cable and two-stage acceleration are necessary. Beyond this, the high voltage must be generated in a separate tank, typically filled with freon or SF_6 at a pressure of a few bar, which decreases the critical distance for electric breakdown. The high voltage is applied to a cascade of acceleration electrodes, with only 50–100 kV between neighbouring rings.

This structure occupies a considerable space and the column of a HVEM is also larger because the yokes of the electron lenses must be scaled up to avoid magnetic saturation. A building some 10–15 m high is therefore needed to house a HVEM, though more compact successful designs have been built [1.106, 107]. This is one of the reasons why the number of HVEMs in the world is limited to about fifty, despite the numerous advantages of these instruments in materials science and biology. We now summarize these briefly:

a) Increased Useful Specimen Thickness

The investigation of thick amorphous specimens is limited by energy losses because the chromatic aberration of the objective lens blurs image points into image patches of width $C_c\alpha_o\Delta E/E$, where $C_c \simeq 0.5$–2 mm is the chromatic-aberration coefficient, α_o is the objective aperture, ΔE is the full width at half maximum of the electron-energy distribution after passing through the specimen. The mean energy loss ΔE decreases as the electron energy E is raised because the ionisation probability falls. Though the value of C_c is larger for HVEM lenses, the product $C_c\alpha_o$ is of the same order as for 100 keV TEMs and, with the reduction of the ratio $\Delta E/E$, the influence of chromatic aberration is markedly diminished. Biological sections, for example, which can be observed in a 100 keV TEM only if their thicknesses are less than 200 nm, can be studied in a 1 MeV TEM in thicknesses as much as 1 μm. The investigation of whole cells and microorganisms helps to establish the three-dimensional structure and the function of fibrillar and membraneous cell components. A stereo pair of micrographs obtained by tilting the specimen between exposures yields better stereometric information. Large structures can be reconstructed only by analysing serial sections at 100 keV.

Many ceramics and minerals are difficult to prepare in thin-enough layers for 100 keV microscopy but can be studied by HVEM. An increase of useful thickness is also observed for metal foils although, for these, the dependence of dynamical diffraction effects on electron energy and on crystal orientation has to be taken into account (Sect. 8.1.4).

Typical orientations for best transmission are found, depending on the material and on the electron energy. Thanks to this effect and to the reduced chromatic blurring, it is possible to investigate 10 μm of silicon or 2 μm of iron at $E = 1$ MeV, for example. This has two important advantages. At 100 keV, crystalline foils that have been thinned by cleaving, electropolishing or ion-beam etching often possess useful wedge-shaped regions of transmission only in small zones near holes, whereas in HVEM, much larger areas can be examined if nearly the whole prepared specimen disc is a few micrometres thick. Secondly, the thicker parts of the foil are more representative of the bulk material, an important point for dynamical experiments such as mechanical deformation, annealing, in-situ precipitation and environmental experiments (see below). The original dislocation arrangement in deformed metals, for example, cannot be observed at 100 keV because a defect-denuded zone will be present near the surface unless the mobility of the dislocations has been impeded by pre-irradiation with neutrons in a nuclear reactor.

b) Easier Specimen Manipulation

The pole-piece gap of the objective lens is of the order of millimetres in 100 keV instruments and centimetres in HVEM; this extra space makes it a great deal easier to instal complicated specimen stages, for heating, cooling or stretching, or goniometers with additional heating, cooling or stretching facilities. Owing to the increased transmission, higher partial pressures of gases at the specimen can be tolerated by use of a differentially pumped system of diaphragms, for example. This is of interest for in-situ environmental studies of chemical attack and corrosion by gases at high temperature. Similarly, organic specimens can be investigated in the native state with a partial pressure of water (wet cells), though characteristic damage effects arise by radiolysis if the specimen is wet or even frozen (Sect. 10.2.3).

c) Radiation-Damage Experiments

For threshold energies of a few 100 keV, depending on the displacement energy E_d and the mass of the nuclei, energy losses greater than E_d can be transferred to nuclei by elastic large-angle scattering; the nucleus is then knocked from its position in the crystal lattice to an interstitial site, for example. HVEM thus becomes a powerful tool for the in-situ study of irradiation processes and the kinetics of defect agglomeration. By working with a beam of high current density, the damage rate can exceed that in a nuclear reactor of high neutron-flux density, by orders of magnitude. In normal operation, however, the current density is kept low so that the specimen can be investigated for a reasonable time without damage: the cross-sections for elastic scattering and for atomic displacements with $\Delta E \geq E_d$ differ by three to five orders of magnitude.

d) Incorporation of Analytical Modes

At 100 keV electron-energy-loss spectroscopy is restricted to specimen thicknesses less than the mean-free-path length of plasmon losses (10–30 nm) because the steep increase of the energy-loss spectrum at the ionisation energy is blurred by the energy-loss spectrum for low energy losses. Thicker specimens can be investigated in HVEM because the free-path length of plasmon losses increases with energy.

X-ray microanalysis with an energy-dispersive detector is also applicable in HVEM if additional screening against high-energy quanta is installed. The relative contribution of the x-ray continuum decreases owing to the pronounced forward bias of the emission of continuous x-ray quanta at higher energies.

Electron-diffraction analysis can be applied to thicker crystals because the dynamical absorption distance increases as v^2. Many-beam dynamical theory has to be applied even for thin foils, because the radius of the Ewald sphere is now large and many more Bragg reflections are excited simultaneously when a sample is irradiated near a low-indexed zone axis. The critical voltage effect is of special interest (Sect. 7.4.4); this occurs at voltages between 100 kV and 1 MeV for most metals and alloys and can be used for the accurate determination of Fourier coefficients of the lattice potential and of their variation with temperature and composition.

e) High Resolution

The relation between voltage and resolution has already been discussed in Sect. 1.3.1. A notable feature is that many-beam imaging of the crystal structure is used to better advantage. Single atoms can also be resolved and the contrast of single heavy atoms remains adequate at high energies. Most HVEMs have resolutions comparable with that of a 100 keV TEM. A few microscopes have been constructed to take advantage of the reduction of the electron wavelength to increase the resolution but for this, the spherical aberration must also be reduced or at least, prevented from becoming too large. As mentioned in Sect. 1.3.1, radiation damage limits the investigation of organic materials. The decrease of ionisation probability with increasing energy provides a gain of only three between 100 and 1000 keV. The response of photographic emulsions decreases by the same factor. Nevertheless the best images of organic crystals such as phthalocyanine have been obtained with a HVEM in the range 500–700 keV.

2. Particle Optics of Electrons

The acceleration of electrons in the electrostatic field between cathode and anode, the action of magnetic fields with axial symmetry as electron lenses and the application of transverse magnetic and electrostatic fields for electron-beam deflection and electron spectrometry can be analysed by applying the laws of relativistic mechanics and hence calculating electron trajectories. Lens aberrations can likewise be introduced and evaluated by this kind of particle optics. In the case of spherical aberration, however, it will also be necessary to express this error in terms of a phase shift, known as the wave aberration, by using the wave-optical model introduced in the next section.

2.1 Acceleration and Deflection of Electrons

2.1.1 Relativistic Mechanics of Electron Acceleration

The relevant properties of an electron in particle optics are the rest mass m_0 and the charge $Q = -e$ (Table 2.1). In an electric field $\boldsymbol{E}$ and magnetic field $\boldsymbol{B}$, electrons experience the Lorentz force

$$\boldsymbol{F} = Q(\boldsymbol{E} + \boldsymbol{v} \times \boldsymbol{B}) = -e(\boldsymbol{E} + \boldsymbol{v} \times \boldsymbol{B}) \, . \tag{2.1}$$

Inserting (2.1) in Newton's law

$$m\ddot{\boldsymbol{r}} = \boldsymbol{F} \tag{2.2}$$

yields the laws of particle optics.

We start with a discussion of the acceleration of an electron beam in an electron gun. Electrons leave the cathode of the latter as a result of thermionic or field emission (see Sect. 4.1 for details). The cathode is held at a negative potential $\phi_C = -U$ (U: acceleration voltage) relative to the anode which is grounded, $\phi_A = 0$ (Fig. 2.1). The Wehnelt electrode, maintained at a potential $\phi_W = -(U + U_W)$, limits the emission to a small area around the cathode tip. Its action will be discussed in detail in Sect. 4.1.3.

The electrode potentials create an electric field $\boldsymbol{E}$ in the vacuum between cathode and anode, which can also be characterized by equipotentials

ϕ = const (Fig. 2.1). The electric field is the negative gradient of the potential

$$\boldsymbol{E} = -\nabla\phi = -\left(\frac{\partial\phi}{\partial x}, \frac{\partial\phi}{\partial y}, \frac{\partial\phi}{\partial z}\right). \tag{2.3}$$

The existence of a potential implies that the force $\boldsymbol{F} = -e\boldsymbol{E}$ is conservative and that the law of energy conservation

$$E + V = \text{const} \tag{2.4}$$

can be applied, as will be demonstrated by considering the electron acceleration in Fig. 2.1. The kinetic energy at the cathode is $E = 0$, whereas the potential energy V is zero at the anode. The potential energy at the cathode can be obtained from the work W that is needed to move an electron from the anode to the cathode against the force $\boldsymbol{F}$:

$$V = -W = -\int_{A}^{C}\boldsymbol{F}\cdot d\boldsymbol{s} = e\int_{A}^{C}\boldsymbol{E}\cdot d\boldsymbol{s} = -e\int_{A}^{C}\nabla\phi\cdot d\boldsymbol{s} = -e(\phi_C - \phi_A) = eU\,. \tag{2.5}$$

In the reverse direction, the electrons acquire this amount eU of kinetic energy at the anode. This implies that the gain of kinetic energy $E = eU$ of an accelerated electron depends only on the potential difference U, irrespective of the real trajectory between cathode and anode.

Relation (2.5) can also be used to define the potential energy $V(\boldsymbol{r})$ at each point $\boldsymbol{r}$ at which the potential is $\phi(\boldsymbol{r})$:

$$V(\boldsymbol{r}) = Q\phi(\boldsymbol{r}) = -e\phi(\boldsymbol{r})\,. \tag{2.6}$$

However, an arbitrary constant can be added to $V(\boldsymbol{r})$ or $\phi(\boldsymbol{r})$ without changing the electric field $\boldsymbol{E}$ because the gradient of a constant in (2.3) is zero. We arbitrarily assumed $\phi_A = 0$ in the special case discussed above, and the results do not change if we assume that $\phi_C = 0$ and $\phi_A = +U$ for example.

An electron has the kinetic energy $E = 1.602 \times 10^{-19}$ N m if accelerated through a potential difference $U = 1$ V because in SI units:

$$1\,\text{C V} = 1\,\text{A V s} = 1\,\text{W s} = 1\,\text{N m}\,.$$

This energy of 1.602×10^{-19} N m is used as a new unit and is called "one electron volt" (abbreviated: 1 eV). Electrons accelerated through $U = 100$ kV have an energy of $E = 100$ keV.

Relativistic effects have to be considered at these energies especially as acceleration voltages up to some megavolts (MV) are used in high-voltage electron microscopy. Table 2.1 contains, therefore, not only the classical (non-relativistic) formulae but also their relativistic counterparts.

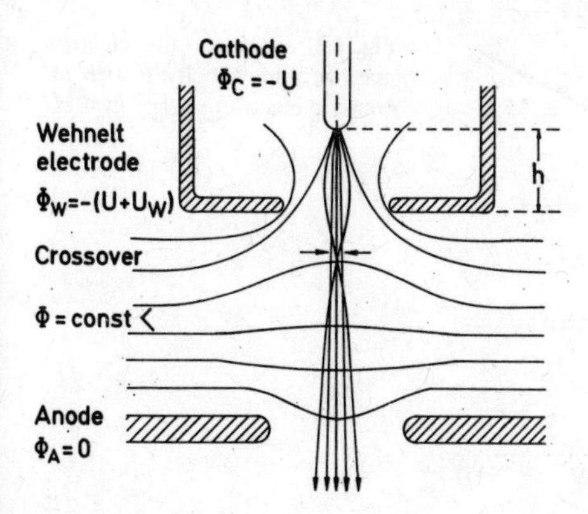

Fig. 2.1. Electron acceleration, trajectories and equipotentials (ϕ = const) in a triode electron gun

Table 2.1. Properties of the electron

Rest mass	$m_0 = 9.1091 \times 10^{-31}$ kg
Charge	$Q = -e = -1.602 \times 10^{-19}$ C
Kinetic energy	$E = eU$, 1 eV $= 1.602 \times 10^{-19}$ N m
Velocity of light	$c = 2.9979 \times 10^{8}$ m s^{-1}
Rest energy	$E_0 = m_0c^2 = 511$ keV $= 0.511$ MeV
Spin	$s = h/4\pi$
Planck's constant	$h = 6.6256 \times 10^{-34}$ N m s

	Non-relativistic ($E \ll E_0$)	Relativistic ($E \gtrsim E_0$)	
Newton's law	$F = \dfrac{dp}{d\tau} = m_0\dfrac{dv}{d\tau}$	$F = \dfrac{dp}{d\tau} = \dfrac{d}{d\tau}(mv)$	(2.7)
Mass	$m = m_0$	$m = m_0(1 - v^2/c^2)^{-1/2}$	(2.8 a)
Energy	$E = eU = \dfrac{1}{2}m_0v^2$	$mc^2 = m_0c^2 + eU = E_0 + E$	(2.9)
		$m = m_0(1 + E/E_0)$	(2.8 b)
Velocity	$v = (2E/m_0)^{1/2}$	$v = c\left[1 - \dfrac{1}{(1 + E/E_0)^2}\right]^{1/2}$	(2.10)
Momentum	$p = m_0v = (2m_0E)^{1/2}$	$p = mv = [2m_0E(1 + E/2E_0)]^{1/2}$ $= \dfrac{1}{c}(2EE_0 + E^2)^{1/2}$	(2.11)
de Broglie wavelength	$\lambda = \dfrac{h}{p} = h(2m_0E)^{-1/2}$	$\lambda = h[2m_0E(1 + E/2E_0)]^{-1/2}$ $= hc(2EE_0 + E^2)^{-1/2}$	(2.12)

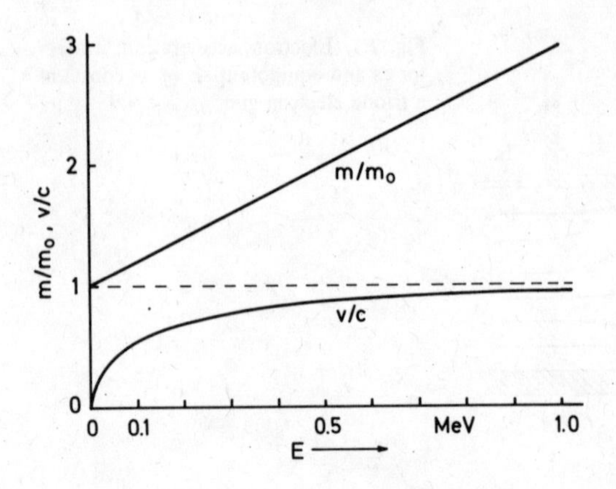

Fig. 2.2. Increase of electron mass m and velocity v with increasing electron energy $E = eU$

The formula (2.8 a) for the increase of the electron mass with increasing velocity v can be obtained from the invariance of the conservation of momentum under a Lorentz transformation, and Newton's law, in the from $\boldsymbol{F} = d\boldsymbol{p}/d\tau$ (2.7), can also be used for relativistic energies.

The most important law of relativistic mechanics is the equivalence of energy and mass: $E = mc^2$. The total energy mc^2 of an accelerated electron is the sum of the rest energy $E_0 = m_0c^2$ and the kinetic energy $E = eU$ (2.9). $E_0 = m_0c^2$ corresponds to an energy of 0.511 MeV. The relativistic increase of the mass m can be formulated not only as in (2.8 a) but also in terms of energy as in (2.8 b) which follows directly from (2.9). The mass, therefore, increases linearly with increasing energy E; it reaches three times the rest mass m_0 at $E = 1$ MeV $\simeq 2E_0$ (Fig. 2.2).

The velocity v (2.10) cannot exceed the velocity of light c (Fig. 2.2) and can be obtained by comparing the right-hand sides of (2.8 a) and (2.8 b). At 100 keV the electron velocity v reaches 1.64×10^8 m s^{-1}, that is more than half of the velocity of light. The momentum $\boldsymbol{p}$ (2.11) is important because the conservation of both energy and momentum has to be considered in electron collisions (Sect. 5.1). The radius r of an electron trajectory in a homogeneous magnetic field $\boldsymbol{B}$ (Sect. 2.1.2) and the de Broglie wavelength λ, (2.12) and Sect. 3.1.1, also depend on the value of the momentum (2.11).

A further property of the electron is its spin (angular momentum) $s = h/4\pi$, and electrons can be polarized by scattering [2.1–3]. Spin polarization does not occur in small-angle scattering, which is responsible for the image contrast in TEM. However, the electron spin does cause deviations from the Rutherford cross-section for large-angle scattering (Sect. 5.1.4) and therefore affects all electron-backscattering effects.

2.1.2 Electron Trajectories in Homogeneous Magnetic Fields

The force generated by the magnetic part of the Lorentz force (2.1) is normal to both the velocity $\boldsymbol{v}$ and the magnetic field $\boldsymbol{B}$ and has a magnitude

$|F| = evB\sin\theta$, θ being the angle between $\boldsymbol{v}$ and $\boldsymbol{B}$. An electron entering a magnetic field with velocity v undergoes an acceleration which is everywhere normal to the local velocity vector. This causes no change in the magnitude of v but does alter its direction. The magnetic field, therefore, is energy conservative.

The continuous change of the direction of $\boldsymbol{v}$ results in a circular trajectory if $\boldsymbol{v} \perp \boldsymbol{B}$ or $\theta = 90°$. On a circular trajectory the centrifugal force $F = mv^2/r$ and the centripetal force $F = evB$ are equal, so that the radius of the circle can be calculated from

$$r = \frac{mv}{eB} = \frac{[2m_0E(1+E/2E_0)]^{1/2}}{eB}$$

$$= 3.37 \times 10^{-6}[U(1+0.9788 \times 10^{-6}U)]^{1/2}B^{-1} \tag{2.13}$$

with r [m], U[V] and B[T] (1 T = 1 tesla = 1 V s m^{-2}).

If θ is not equal to 90°, v can be separated into two components, one, $v_\perp = v\sin\theta$ normal and the other, $v_\parallel = v\cos\theta$, parallel to $\boldsymbol{B}$ (Fig. 2.3). The parallel component $v_\parallel$ results in rectilinear motion in the z direction: $z = v_\parallel t$.

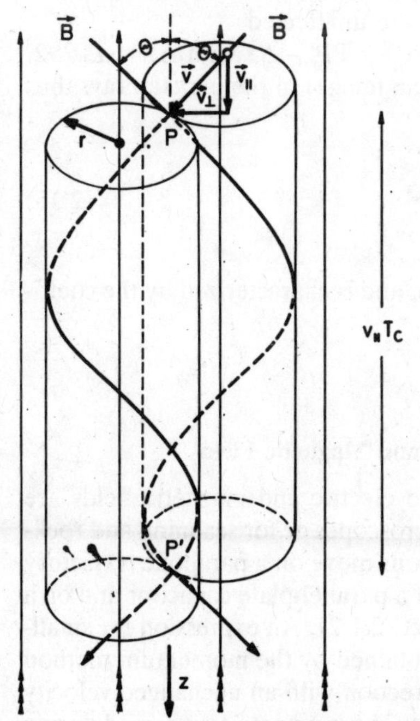

Fig. 2.3. Electron trajectories in a homogeneous magnetic field $\boldsymbol{B}$

The component $v_\perp$ produces motion around a circle of radius $r = mv(eB)^{-1}$. Superposition of the two types of motion results in a helical trajectory. The period T_c or "cyclotron frequency" $\omega_c = 2\pi/T_c$ corresponding to a complete circle is given by

$$T_c = \frac{2\pi r}{v} = \frac{2\pi m}{eB}; \quad \omega_c = \frac{eB}{m}. \tag{2.14}$$

It is thus independent of the velocity for non-relativistic energies.

Electron trajectories passing through a point P (Fig. 2.3) at different angles θ to the z axis will intersect again after travelling a distance

$$\mathrm{PP}' = v_\parallel T_c = vT_c \cos\theta = \frac{2\pi mv}{eB}\left(1 - \frac{1}{2}\theta^2 + \ldots\right) = L_0\left(1 - \frac{1}{2}\theta^2 + \ldots\right) \tag{2.15}$$

with $L_0 = \mathrm{PP}'_0$, which means that P'_0 is an image of P for very small angles θ (paraxial rays). Such helical trajectories also arise in rotationally symmetric inhomogeneous fields inside magnetic electron lenses (Sect. 2.2.1). However, the homogeneous magnetic field is not a lens in the normal sense of the term because all rays parallel to the field are unaffected.

Increasing θ reduces the distance to $\mathrm{PP}' = \mathrm{PP}'_0 - \Delta z$ with $\Delta z = L_0\theta^2/2$. The error disc in the image plane (Gaussian image) of the paraxial rays thus has the diameter

$$\delta = \Delta z \tan\theta \simeq \Delta z\, \theta = \frac{1}{2}L_0\theta^3 = C_s\theta^3. \tag{2.16}$$

This error is known as spherical aberration and is characterized by the coefficient C_s.

2.1.3 Small-Angle Deflections in Electric and Magnetic Fields

Beam deflections produced by transverse electric and magnetic fields are needed for the adjustment of electron microscopes or for scanning and rocking an electron beam (Sect. 4.2.1). Electrons move on a parabolic trajectory in the transverse electric field $|\boldsymbol{E}| = u/d$ of a parallel-plate capacitor and on a circle in a transverse magnetic field $\boldsymbol{B}$ (Sect. 2.1.2). An expression for small-angle deflection ε with $\sin\varepsilon \simeq \varepsilon$ can be obtained by the momentum method (Fig. 2.4). An electron moves in the z direction with an unchanged velocity $v = dz/d\tau$ and with a momentum $p_z = mv$. The momenta transferred during the time of flight $T = L/v$ are as follows:

in an electric field $\boldsymbol{E}$

$$p_x = \int_0^T F\,d\tau = e\int_0^T E\,d\tau = \frac{e}{v}\int_0^L E\,dz = \frac{eEL}{v}\,;$$

in a magnetic field $\boldsymbol{B}$

$$p_x = e\int_0^T vB\,d\tau = e\int_0^L B\,dz = eBL \tag{2.17}$$

and the angles of deflection ε can be obtained from

$$\varepsilon = \frac{p_x}{p_z} = \frac{eEL}{mv^2} = \frac{euL}{2Ed}\,\frac{1+E/E_0}{1+E/2E_0} \tag{2.18 a}$$

$$\varepsilon = \frac{eBL}{mv} = \frac{eBL}{[2m_0E(1+E/2E_0)]^{1/2}} \tag{2.18 b}$$

for the electric and magnetic field, respectively. The formula (2.18 b) for the magnetic deflection will also be obtained in Sect. 3.1.5 by a wave-optical calculation. This formula is important for Lorentz microscopy (Sect. 6.6).

As an example, we calculate field strengths needed to deflect 100 keV electrons through an angle $\varepsilon = 5° \simeq 0.1$ rad in a field of length $L = 1$ cm with a plate or pole-piece separation $d = 2$ mm. The electric field has to be $|\boldsymbol{E}| = 2 \times 10^4$ V cm^{-1}, which implies a voltage u of 4000 V at the plates. The magnetic field $\boldsymbol{B}$ produced by an electromagnet with a slit width d is given approximately by $B = \mu_0 NI/d$ ($\mu_0 = 4\pi \times 10^{-7}$ H m^{-1}, N: number of turns, I: coil current). A deflection ε of 5° requires $B = 10^{-2}$ T a can be achieved with $NI = 20$ A, e.g. 100 turns and $I = 0.2$ A.

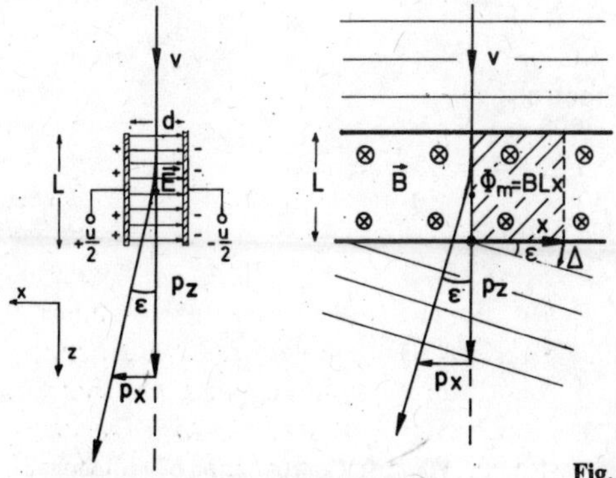

Fig. 2.4 a, b. Small-angle deflections ε in a transverse electric field **(a)** and magnetic field **(b)**

Larger deflection angles (e.g. $\varepsilon = 90°$) are used in electron prisms for energy-loss spectroscopy (Sect. 9.2.1)

2.2 Electron Lenses

2.2.1 Electron Trajectories in a Magnetic Field of Rotational Symmetry

The physical background of electron-lens optics will be described only briefly to give a quantitative understanding of the function of an electron lens (see [2.4–7] for details). Electrostatic lenses will not be discussed because they have lost their importance in electron microscopy since they have larger aberration coefficients and there is a risk of electrical breakdown.

It was shown in Sect. 2.1.2 that a homogeneous magnetic field already acts as an electron lens. Magnetic lenses with short focal lengths are obtained by concentrating the magnetic field by means of magnetic pole-pieces. Figure 2.5 shows the distribution of magnetic field produced by a coil enclosed in an iron shield, apart from an open slit. The magnetic field has rotational symmetry; the distribution on the optic z axis can be represented approximately by Glaser's "Glockenfeld" (bell-shaped field)

$$B_z = \frac{B_0}{1 + (z/a)^2}, \tag{2.19}$$

where B_0 denotes the maximum field in the lens centre and $2a$ the full width at half maximum [2.8]. Other approximations for the field distribution $B_z(z)$

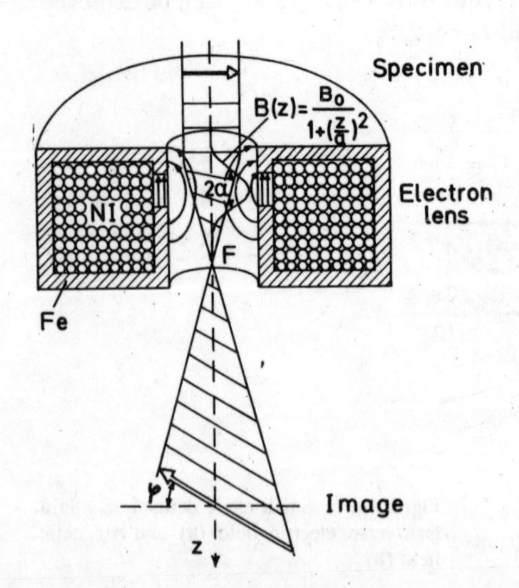

Fig. 2.5. Concentration of a rotationally symmetric magnetic field in the gap of an electron lens (φ: image rotation)

are also in use, but the Glaser field offers the advantage that the most important properties, the positions of foci and principal planes (Sect. 2.2.2) for example, can be calculated straightforwardly. A knowledge of the magnetic field B_z on the axis is sufficient for calculating the paraxial rays because the radial component B_r for off-axis points can be calculated from $B_z(z)$ by using Gauss's law div $\boldsymbol{B} = 0$ (Fig. 2.6):

$$\operatorname{div} \boldsymbol{B} = \pi r^2 B_z(z) - [\pi r^2 B_z(z + dz) + 2\pi r B_r dz] = 0 . \tag{2.20}$$

The first term represents the flux through the bottom and the last terms the flux through the top and side of the cylinder shown in Fig. 2.6. Using the Taylor-series expansion $B_z(z + dz) = B(z) + (\partial B_z/\partial z)\, dz + \dots$ we find

$$B_r = -\frac{r}{2}\frac{\partial B_z}{\partial z} . \tag{2.21}$$

The system of differential equations (Newton's law) for the electron trajectories can be separated in a cylindrical coordinate system r, φ, z:

Radial component: $$m\ddot{r} = F_r + mr\dot{\varphi}^2 \tag{2.22 a}$$

Circular component: $$\frac{d}{dt}(mr^2\dot{\varphi}) = rF_\varphi \tag{2.22 b}$$

Longitudinal component: $$m\ddot{z} = F_z . \tag{2.22 c}$$

The last term in (2.22 a) can be interpretated as the centrifugal force. The second equation (2.22 b) represents the change of angular momentum $\boldsymbol{L}$ caused by the torque $M = rF_\varphi$ ($\dot{\boldsymbol{L}} = \boldsymbol{M}$).

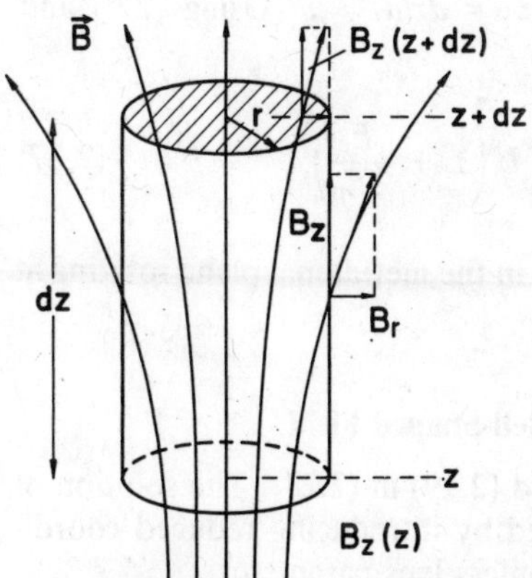

Fig. 2.6. Application of Gauss's law to the magnetic field near the axis

On substituting the Lorentz force $\boldsymbol{F} = -e\boldsymbol{v} \times \boldsymbol{B}$ with $\boldsymbol{v} = (\dot{r}, r\dot{\varphi}, \dot{z})$ and $B_\varphi = 0$ and using (2.21), we obtain

$$m\ddot{r} = -eB_z r\dot{\varphi} + mr\dot{\varphi}^2 \tag{2.23 a}$$

$$\frac{d}{dt}(mr^2\dot{\varphi}) = eB_z r\dot{r} + e\frac{r^2}{2}\dot{z}\frac{\partial B_z}{\partial z} = \frac{d}{dt}\left(\frac{e}{2}r^2 B_z\right) \tag{2.23 b}$$

$$m\ddot{z} = eB_r r\dot{\varphi} . \tag{2.23 c}$$

Integration of (2.23 b) results in

$$mr^2\dot{\varphi} = \frac{e}{2}r^2 B_z + C . \tag{2.24}$$

The constant of integration C becomes zero for meridional rays and only a trajectory $r(z)$ need be considered in a meridional plane rotating at the angular velocity

$$\omega_{\mathrm{L}} = \dot{\varphi} = \frac{e}{2m} B_z . \tag{2.25}$$

This is known as the Larmor frequency, which is half the cyclotron frequency ω_c of (2.14).

For paraxial rays (small values of r), equation (2.23 c) can be approximated by $\ddot{z} = 0$, which implies that v_z is constant. Substitution of (2.25) in (2.23 a) results in

$$m\ddot{r} = -eB_z r\frac{e}{2m} B_z + mr\left(\frac{e}{2m} B_z\right)^2 = -\frac{e^2}{4m} B_z^2 r . \tag{2.26}$$

The time can be eliminated by writing $v_z = dz/d\tau \simeq v$. Using (2.9) and (2.10), we find

$$\frac{d^2r}{dz^2} = -\frac{e}{8m_0U^*} rB_z^2(r) \quad \text{with} \quad U^* = U\left(1 + \frac{E}{2E_0}\right) . \tag{2.27}$$

This is the equation for the trajectory $r(z)$ in the meridional plane rotating at the angular velocity ω_{L}.

2.2.2 Optics of an Electron Lens with a Bell-Shaped Field

Let us now substitute the bell-shaped field (2.19) in (2.27). The solution of the differential equation can be simplified by introducing reduced coordinates $y = r/a$ and $x = z/a$ and a dimensionless lens parameter

$$k^2 = \frac{eB_0^2 a^2}{8 m_0 U^*} \tag{2.28}$$

resulting in

$$\frac{d^2y}{dx^2} = -\frac{k^2}{(1+x^2)^2}\, y \,. \tag{2.29}$$

This equation can be further simplified by the substitutions

$$x = \cot\phi \,; \quad dx = -d\phi/\sin^2\phi \,; \quad 1 + x^2 = \operatorname{cosec}^2\phi \,. \tag{2.30}$$

The meaning of the angle ϕ can be seen from Fig. 2.7. The variable ϕ varies from π for $z = -\infty$ to $\phi = \pi/2$ for $z = 0$ and to $\phi = 0$ for $z = +\infty$. Equation (2.29) then becomes

$$y''(\phi) + 2\cot\phi\, y'(\phi) + k^2 y(\phi) = 0 \,. \tag{2.31}$$

The solution of (2.31) is a linear combination

$$y(\phi) = C_1 u(\phi) + C_2 w(\phi) \tag{2.32}$$

of the two particular integrals

$$\begin{aligned} u(\phi) &= \sin(\omega\phi)/\sin\phi \\ w(\phi) &= \cos(\omega\phi)/\sin\phi \end{aligned} \quad \text{with} \quad \omega = \sqrt{1+k^2} \,. \tag{2.33}$$

The coefficients C_1 and C_2 can be determined from the initial conditions. Thus for a parallel incident ray, the initial conditions are $r = r_0$ for $z = -\infty$ or $y(\pi) = r_0/a$ and $y'(\pi) = 0$ which results in $C_2 = 0$; the radial component of the trajectory becomes

$$y = \frac{r}{a} = -\frac{r_0}{a\omega}\frac{\sin(\omega\phi)}{\sin\phi} \,. \tag{2.34}$$

Such trajectories are plotted in Fig. 2.8 for increasing values of the strength parameter $\omega = \sqrt{1+k^2}$ of the lens.

For a more general discussion, we assume that the ray passes through a point $P_0(y_0, \phi_0)$ in front of the lens. Substituting $y = y_0$ and $\phi = \phi_0$ in (2.32) and solving for C_1 yields

$$C_1 = \frac{y_0 \sin\phi_0}{\sin(\omega\phi_0)} - C_2 \frac{\cos(\omega\phi_0)}{\sin(\omega\phi_0)} \,. \tag{2.35}$$

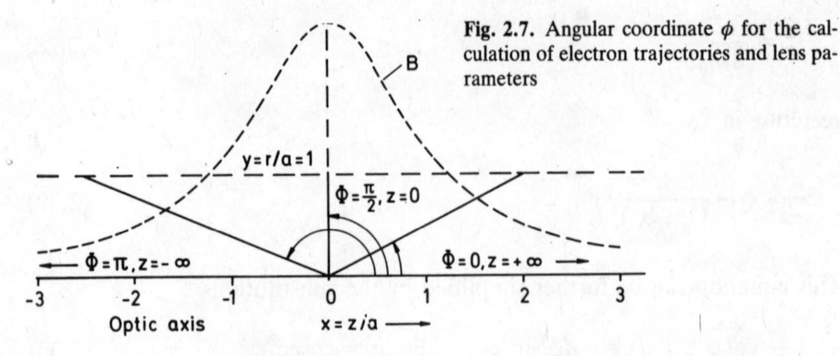

Fig. 2.7. Angular coordinate ϕ for the calculation of electron trajectories and lens parameters

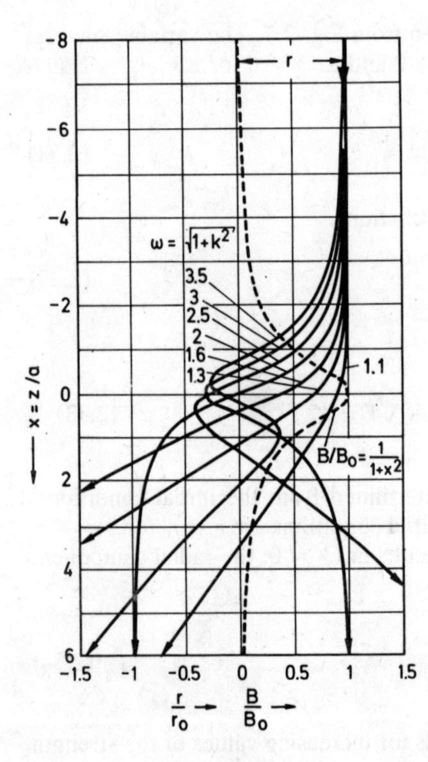

Fig. 2.8. Electron trajectories incident parallel to the axis for increasing values of lens strength $\omega = \sqrt{1 + k^2}$

We substitute (2.35) in (2.32), giving

$$y(\phi) = \frac{\sin(\omega\phi)\sin\phi_0}{\sin(\omega\phi_0)\sin\phi}\, y_0 + \frac{C_2}{\sin\phi}\left[\cos(\omega\phi) - \frac{\cos(\omega\phi_0)}{\sin(\omega\phi_0)}\sin(\omega\phi)\right]. \tag{2.36}$$

The coefficient C_2 can be determined from the direction (slope) of the ray at the point P_0, and different values of C_2 will correspond to different directions. The image point $P_1(y_1, \phi_1)$ corresponding to the object point P_0 can be obtained from the condition that the last square bracket in (2.37) becomes zero, which means that P_1 has the coordinate

$$y_1 = \frac{\sin(\omega\phi_1)\sin\phi_0}{\sin(\omega\phi_0)\sin\phi_1}\, y_0 = My_0 \;; \tag{2.37}$$

independent of C_2 (M: magnification). Multiplying the bracket in (2.36) by $\sin(\omega\phi_0)$ results in the addition theorem for a sine function and the condition for a zero bracket can be written

$$\sin\omega(\phi_1 - \phi_0) = 0 \tag{2.38}$$

which is satisfied by

$$\phi_{1n} = \phi_0 - n\frac{\pi}{\omega}\;; \qquad n = 1, 2, \ldots\,. \tag{2.39}$$

This means that more than one image point can occur in strong lenses. However, $n = 2$ will not be possible until $\omega = \sqrt{1 + k^2} \geq 2$ or $k^2 \geq 3$.

The positions of the object and image points are

$$z_0 = a\cot\phi_0\;; \qquad z_{1n} = a\cot\phi_{1n}\,. \tag{2.40}$$

Substitution of (2.39) in (2.40) gives

$$z_0 = a\cot\left(\phi_{1n} + n\frac{\pi}{\omega}\right) = \frac{a\cot\phi_{1n} - \cot\left(n\frac{\pi}{\omega}\right) - a}{\cot\phi_{1n} + \cot\left(n\frac{\pi}{\omega}\right)}\,. \tag{2.41}$$

This equation may be rewritten in the form

$$\left[z_0 - a\cot\left(n\frac{\pi}{\omega}\right)\right]\left[z_{1n} + a\cot\left(n\frac{\pi}{\omega}\right)\right] = -\,a^2\operatorname{cosec}^2\left(n\frac{\pi}{\omega}\right), \tag{2.42}$$

which is equivalent to Newton's lens equation of light optics

$$Z_0Z_1 = f_0f_1\,, \tag{2.43}$$

where f_0 and f_1 denote the focal lengths and the distances

$$Z_0 = z_0 - z(F_0) ; \qquad Z_1 = z_1 - z(F_1) \tag{2.44}$$

separate the object and image points from the corresponding foci F_0 and F_1. Comparison of (2.42) and (2.43) shows that

$$f_0 = -f_1 = a \operatorname{cosec}\left(n\frac{\pi}{\omega}\right) ; \qquad z(F_0) = -z(F_1) = a \cot\left(n\frac{\pi}{\omega}\right) . \tag{2.45}$$

The focal lengths f are not the same as the distances z (F) of the foci from the lens centre $z = 0$. This means that electron lenses cannot be treated as thin lenses. Principal planes can be introduced, as in light optics, to construct the position of the corresponding image. The positions of the principal planes are, for $n = 1$,

$$z(H_0) = z(F_0) + f_0 = a\frac{\cos\left(\frac{\pi}{\omega}\right) + 1}{\sin\left(\frac{\pi}{\omega}\right)} = a \cot\left(\frac{\pi}{2\omega}\right) = -z(H_1) . \tag{2.46}$$

The positions z (F) of the foci and z(H) of the principal planes are plotted in Fig. 2.9 a as a function of the lens parameters k^2 (2.28). Fig. 2.9 b also shows how the image point can be geometrically constructed for the particular case $k^2 = 1.6$. A ray parallel to the axis is refracted at H_1 and continued as a straight line through the focus F_1; a ray through F_0 is refracted at H_0, continuing parallel to the axis. The intersection of these two lines is the image point. Unlike light-optical lenses, corresponding foci and principal planes are situated on different sides of the lens centre (H_0 and H_1 are interchanged).

The magnification M in (2.37) can be written in terms of f and Z by substituting $\phi = \phi_{1n}$ from (2.39) and using (2.42–45):

$$M = f_0/Z_0 = Z_1/f_1 . \tag{2.47}$$

In reality, the trajectories are curved, and the coordinate system rotates with the angular velocity $\dot{\varphi}$ of (2.25). The total rotation angle φ between image and object (Fig. 2.5) can be calculated by using the substitution $dz = v\,dt$ and (2.28, 30 and 39):

$$\varphi = \frac{e}{2m}\int_{t_0}^{t_1} B_z dt = \frac{e}{2mv}\int_{z_0}^{z_1} B_z dz = \sqrt{\frac{e}{8m_0 U^*}}\int_{z_0}^{z_1}\frac{B_0 dz}{1 + (z/a)^2}$$

$$= -\sqrt{\frac{e}{8m_0 U^*}}\, aB_0 \int_{\phi_0}^{\phi_1} d\phi = k(\phi_0 - \phi_1) = k\frac{\pi}{\omega} = k\frac{\pi}{\sqrt{1+k^2}} . \tag{2.48}$$

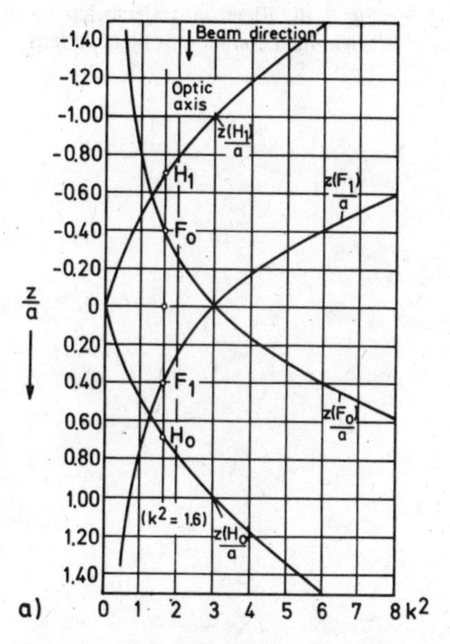

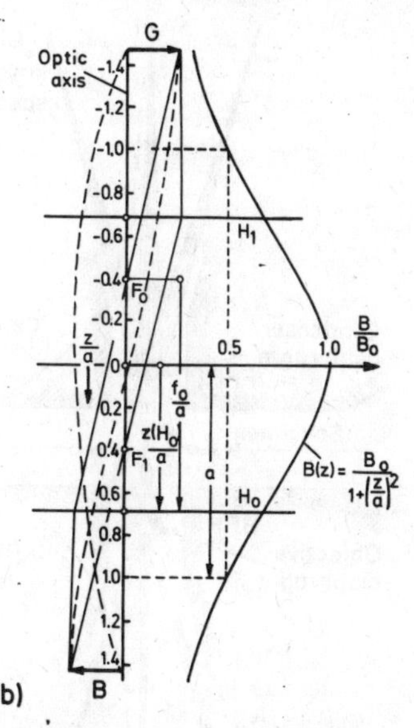

Fig. 2.9. (**a**) Positions of the foci F_1, F_2 and principal planes H_1, H_2 as the lens parameter k^2 is increased and (**b**) example of a geometrical construction for $k^2 = 1.6$ [2.4]

The values of the focal lengths and the positions (2.45) of the foci do not depend on the direction of B_z, whereas the image-rotation angle φ is reversed when B_z or the lens current is reversed.

The image rotation in an electron microscope can be partially compensated, therefore, by changing the sign of the currents in different lenses. The image rotation does not influence the quality of the image but its magnitude has to be known if directions in the image have to be correlated with corresponding directions in the specimen or in an electron-diffraction pattern.

The above formulae are for lenses with symmetric pole-pieces. Lenses with asymmetric pole-piece diameters are often used in practice. If the larger diameter is on the specimen side, more space is available for specimen translation with top-entry specimen stages. These lenses can be treated in a similar way by approximating the lens field on the axis by two Glaser fields (2.19) with different parameters a_1 and a_2 on the two sides [2.9].

2.2.3 Special Electron Lenses

a) Objective Lenses with $k^2 \geq 3$

Objective lenses normally operate with k^2 between 2 and 3. A lens with an excitation $k^2 = 3$ will be optimal in the sense that the focal length is shortest [2.10, 11] and the spherical-aberration coefficient C_s is low (Sect. 2.3.2).

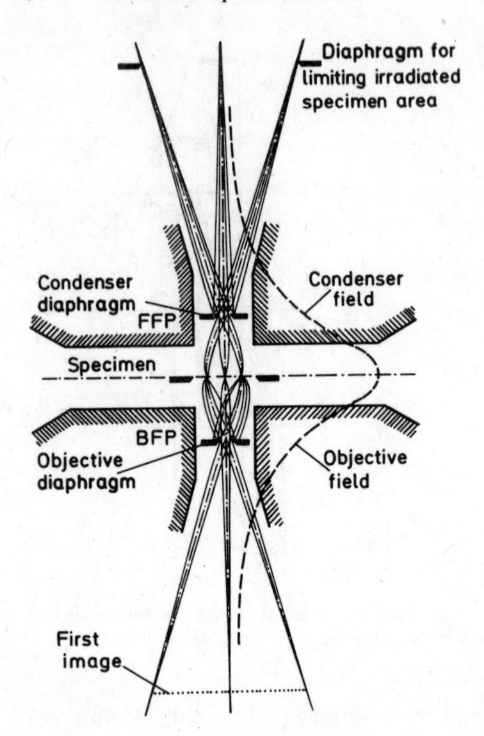

Fig. 2.10. Electron trajectories in a single-field condenser-objective lens

Figures 2.8, 9 show that the focus of such a lens is in the centre of the lens field at $z = 0$. The specimen position is near the lens centre and the pre-field of the lens acts as a condenser lens. Figure 2.10 shows the electron trajectories in such a condenser-objective lens and Fig. 4.17 the corresponding ray diagram with straight lines and two separate lenses representing the pre- and post-field. The front focal plane (FFP) and back focal plane (BFP) are conjugate. A parallel beam in the FFP is focused at the specimen and is again parallel in the BFP. The lens is thus operating in the "telefocal condition". The specimen area illuminated can be limited by placing a diaphragm in a plane conjugate to the specimen plane.

The specimen position is shifted beyond the lens centre in a second-zone lens with $k^2 > 3$ [1.12–14]. A parallel beam entering the lens crosses the lens centre in front of the specimen and is again a parallel beam at the specimen.

b) Superconducting Lenses

The strength of a given magnetic lens with an iron core cannot be increased indefinitely owing to saturation of the magnetization M_s to about 2.1 T ($B = \mu_0 H + M$); strong lenses require an increase of size and power supply. Superconducting hollow cylinders or rings have the property of screening the inner space from external magnetic fields and can trap a magnetic flux that penetrated the ring in the normal conducting state. The critical magnetic field that destroys superconductivity is very high ($B_c \simeq 5$–10 T) in type II superconductors (e.g., Nb-Zr, Nb-Ti, Nb_3Sn). Superconducting lenses can be designed in three different ways [2.15–18]:

1) The lens still has ferromagnetic pole-pieces; these may be of dysprosium or holmium, for which $M_s = 3 - 3.4$ T at low temperatures, excited by a superconducting coil.
2) Superconductors are introduced into the bore of a conventional magnetic lens in the form of hollow cylinders, thus confining the magnetic flux to a smaller space by screening.
3) Here, the flux trapped in superconducting rings or discs is exploited.

Superconducting lenses are of special interest to reduce the weight and size of lenses in high-voltage electron microscopes. Liquid-helium-cooled parts of the column will act as additional cryopumps and with shielding lens the specimen can easily be cooled to 4.2 K to decrease the radiation damage (Sect. 10.2.3).

c) Minilenses

Any decrease of the size of magnetic lenses will have the advantage of decreasing the length of the electron optical column, thus reducing the influence of mechanical vibrations and ac magnetic stray fields. Small lenses (minilenses) are also useful in front of an objective lens, to decrease and control the electron-probe diameter. One way of reducing the size is to use superconducting lenses; alternatively a stronger excitation may be employed with a more efficient watercooling system [2.19–22].

Figure 2.11 shows a single-pole-piece lens (snorkel lens). This system can be used with a small bore diameter to reduce the focal length and the spherical-aberration coefficient; furthermore, the limitations on specimen manipulation normally imposed by the small gaps in conventional lenses are avoided. Projector-lens systems can be designed consisting of two gaps with opposite sign of $B(z)$ to compensate for the image rotation φ (2.48) and offer various other advantages [2.22].

d) Multipole Lenses

A quadrupole lens can be constructed from four pole-pieces of opposite polarity (Fig. 2.12). Because the magnetic field is normal to the electron beam, stronger Lorentz force is exerted. A point object is focused as a line image, as with a cylindrical lens in light optics. In electron microscopes, quadrupole lenses are used as stigmators for compensating the axial astigmatism (Sect. 2.3.7) or to correct the focusing distance of an electron prism for energy analysis (Sect. 9.2.1).

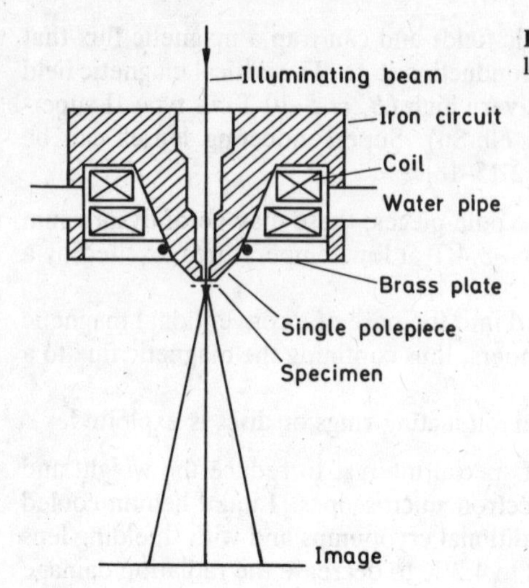

Fig. 2.11. Example of a minilens (snorkel lens) [2.21]

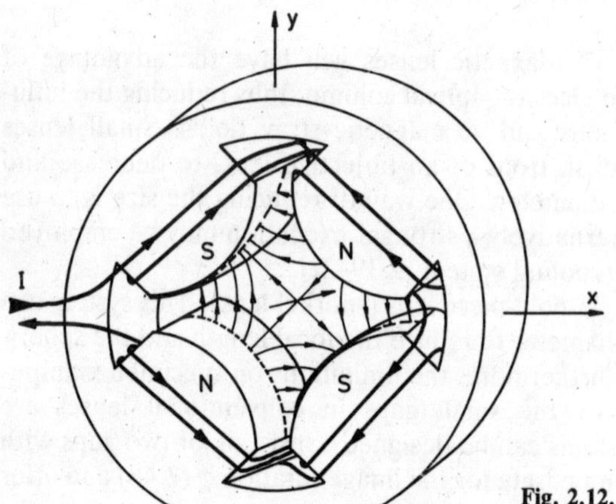

Fig. 2.12. Construction of a quadrupole lens

Octopole lenses consist of eight pole-pieces with alternating polarities. Combinations of octopole and quadrupole lenses can in principle be used to correct lens aberrations (Sect. 2.3.7).

2.3 Lens Aberrations

2.3.1 Classification of Lens Aberrations

There are five possible isotropic aberrations of third order in lenses with rotational symmetry, as in light optics:

1) Spherical aberration (Sect. 2.3.2)
2) Astigmatism (Sect. 2.3.3)
3) Field curvature (Sect. 2.3.3)
4) Distortion (Sect. 2.3.4)
5) Coma (Sect. 2.3.4)

and three further anisotropic errors (Sect. 2.3.5)

6) Anisotropic coma
7) Anisotropic astigmatism
8) Anisotropic distortion.

If the electron beam is not monochromatic, owing to
a) insufficient stabilization of the acceleration voltage,
b) the velocity distribution associated with the thermionic or field emission and
c) energy losses in the specimen,

9) Chromatic aberration (Sect. 2.3.6)

also has to be considered. Departure of the magnetic-lens field from exact rotational symmetry causes an

10) Axial astigmatism (Sect. 2.3.3).

The spherical aberration, a distortion caused by this aberration, the axial astigmatism and the chromatic aberration are the most important aberrations for electron microscopy, and only these on-axis errors will be discussed in detail. The other aberrations can normally be neglected because the electron beam necessarily remains close to the optical axis and small lens apertures are needed for high resolution.

The so-called diffraction error is not caused by the lens itself but is a consequence of the presence of diaphragms; this error will therefore be discussed not in this section but where a wave-optical theory of image formation is formulated in Sections 3.3.2 and 6.2.

2.3.2 Spherical Aberration

The spherical aberration has the effect of reducing the focal length for electron rays passing through outer zones of the lens (Fig. 2.13). Electrons crossing the optic axis at different angles θ or scattered in the specimen through angles θ will intersect the Gaussian image plane at a distance

$$r_s' = C_s\theta^3 M \tag{2.49a}$$

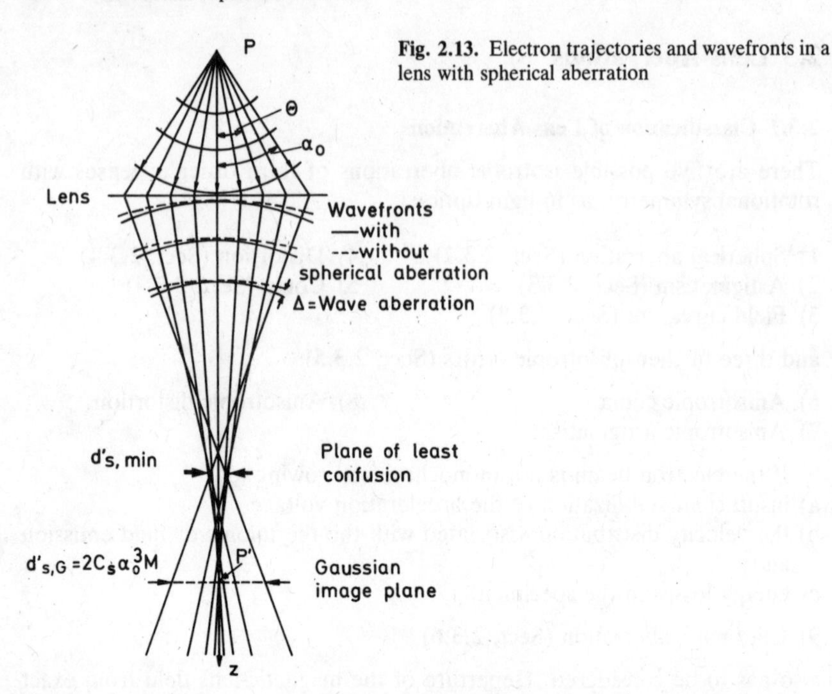

Fig. 2.13. Electron trajectories and wavefronts in a lens with spherical aberration

from the paraxial image point. The Gaussian-image plane is the position of the image when very small apertures are used (paraxial rays). C_s is the spherical-aberration coefficient and M the magnification. We use coordinates x, y or r in the specimen plane and the corresponding coordinates $x' = -Mx$ and y', r', respectively, in the image plane.

A conical electron beam with angular aperture α_o defined by the objective diaphragm does not produce a sharp image point but the beam diameter passes through a minimum, $d'_{s,min}$, in a plane of least confusion; in the Gaussian-image plane, the diameter is $d'_{s,G} = 2\,C_s\alpha_o^3M$. The corresponding diameters in the specimen plane are $d_{s,G} = d'_{s,G}/M$ and $d_{s,min} = d'_{s,min}/M$. The smallest diameter $d_{s,min}$ can be estimated [2.23] from

$$d_{s,min} = 0.5\;C_s\alpha_o^3\;. \tag{2.49 b}$$

The spherical-aberration coefficients of objective lenses are normally of the order of 0.5–4 mm. The values also depend on the magnification (lens excitation) (for experiments, see [2.24]; for theory, see [2.4, 25]).

Calculated values of the spherical-aberration coefficient C_s of magnetic lenses are plotted in Fig. 2.14 as a function of the lens parameter k^2 (see also [2.20, 26, 27]). C_s decreases with increasing lens strength. The minimum focal length occurs at $k^2 = 3$ and C_s shows a flat minimum around $k^2 = 7$. The

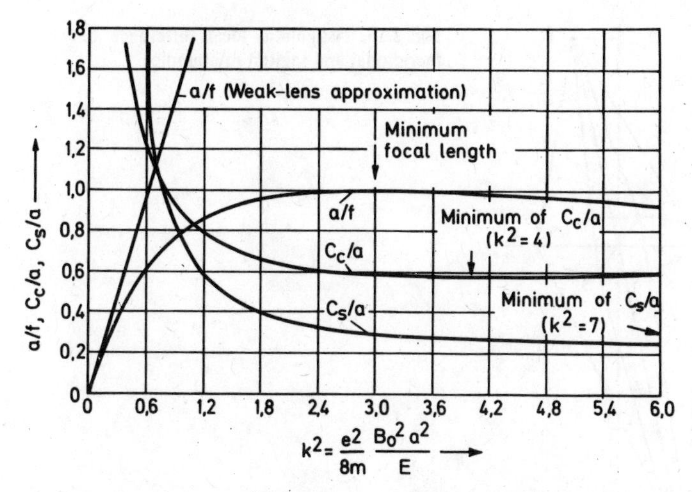

Fig. 2.14. Dependence of reciprocal focal length $1/f$, spherical-aberration constant C_s and chromatic-aberration constant C_c on the lens parameter k^2 [2.4]

spherical aberration of electrostatic lenses is 4–10 times larger than that of comparable magnetic lenses.

The spherical aberration of the objective lens not only influences the resolution but can also be observed when imaging crystalline specimens. The diffracted beams produce shifted twin images if the objective-aperture diaphragm is removed or if the primary and diffracted beam can pass through the diaphragm. The bright bend contours observable in dark field are shifted relative to the corresponding dark contours in bright field. This effect can be used for the measurement of C_s [2.28, 29]. Other methods for measuring C_s arc reported in [2.24, 30, 31]. The same effect limits the useful area in selected-area electron diffraction (Sect. 9.3.1).

A wave-optical formulation of the effect of spherical aberration, which is important for the discussion of phase contrast, will be presented in Sect. 3.3.1.

2.3.3 Astigmatism and Field Curvature

A cone of rays of semi-angle θ from a specimen point distant x from the axis is focused in the Gaussian-image plane as an ellipse with its centre at the Gaussian-image point x'. The principal axes of the ellipse are parallel to x' and y' and their lengths are proportional to x^2 and θ.

Rays passing through points around a circle of radius R in the lens and the corresponding points on the ellipse form an astigmatic bundle of rays, which collapses to perpendicular focal lines F_s and F_m for rays in the sagittal and meridional planes. These foci lie on the curved sagittal and meridional image

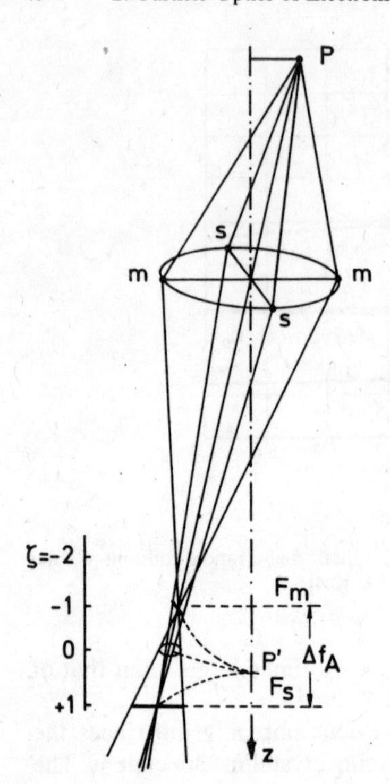

Fig. 2.15. Astigmatic focal difference between meridional and sagittal ray bundles

planes shown in Fig. 2.15. A circle of least confusion is formed in the curved mean image plane. This error disappears for on-axis specimen points ($x = 0$), and this type of astigmatism can be neglected because small apertures are used to decrease the influence of spherical aberration and because the electron beam is necessarily adjusted on-axis to decrease the influence of chromatic aberration.

However, astigmatism can be observed even for points on axis if the lens field is not exactly rotationally symmetric, owing to inhomogenity of the magnetisation of the pole-piece, ellipticity of the pole-piece bores or electric charging of aperture diaphragms. This error is therefore called axial astigmatism. In consequence, a pair of diametrically opposite zones of a circular specimen will be focused sharply at one focal point F_s, and the two other diametrically opposite zones, 90° from the first, will be focused at the other focal point F_m. The difference Δf_A of the focal lengths (Fig. 2.15) will be small and is only of the order of 0.1 to 1 μm. Nevertheless, the resolution can be reduced, as is shown by the following estimate.

The diameter of the error disc at the specimen plane will be

$$d_A = \Delta f_A \alpha_o \ . \tag{2.50}$$

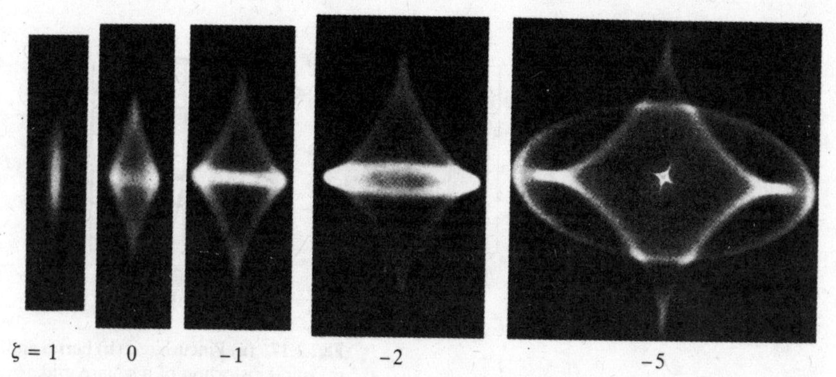

Fig. 2.16. Cross-sections through the caustic at different values of the coordinate ζ of Fig. 2.15 [2.32]

If a resolution $\delta = 0.5$ nm is wanted for an aperture α_o of 10 mrad, d_A should be smaller than δ and, therefore, $\Delta f_A < \delta/\alpha_o = 50$ nm. If we assume that the pole-piece bore is elliptical with semi-axes $b_0 \pm \Delta b$, the relative focal difference becomes

$$\frac{\Delta f_A}{f} = 2\frac{\Delta b}{b_0} \tag{2.51}$$

because the focal length is of the order of the diameter b_0. It follows that Δb must be less than 25 nm with the estimated value of Δf_A. It is very difficult to obtain such high precision in the diameter of the bore.

The simple drawing of Fig. 2.15 is not adequate for calculating the cross-section of the electron beam in an astigmatic image. If all rays, including those not in the sagittal or meridional plane, are considered, a complicated intensity distribution results in the neighbourhood of the focus, the so-called caustic. Figure 2.16 shows observed intensity distributions [2.32, 33] corresponding to cross-sections through the caustic at the positions ζ indicated in Fig. 2.15. The orthogonal focal lines have the coordinates $\zeta = \pm 1$.

2.3.4 Distortion and Coma

Distortion causes a displacement

$$\Delta r' = -C_E r'^3 \tag{2.52}$$

in the Gaussian-image plane for off-axis points. This results in a geometrical distortion of a square, which is known as pincushion distortion for $C_E > 0$ and barrel distortion for $C_E < 0$ (Fig. 2.17 a, b).

The spherical aberration also causes an image distortion in intermediate and projector lenses operating at low magnifications. These large-bore lenses

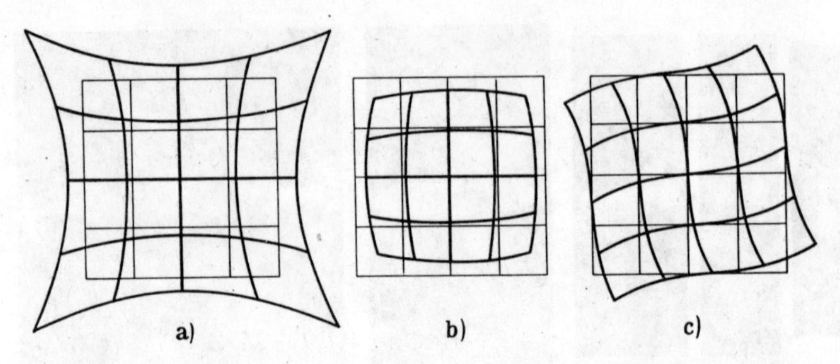

Fig. 2.17. (a) Pincushion, (b) barrel and (c) spiral distortion of a square grid

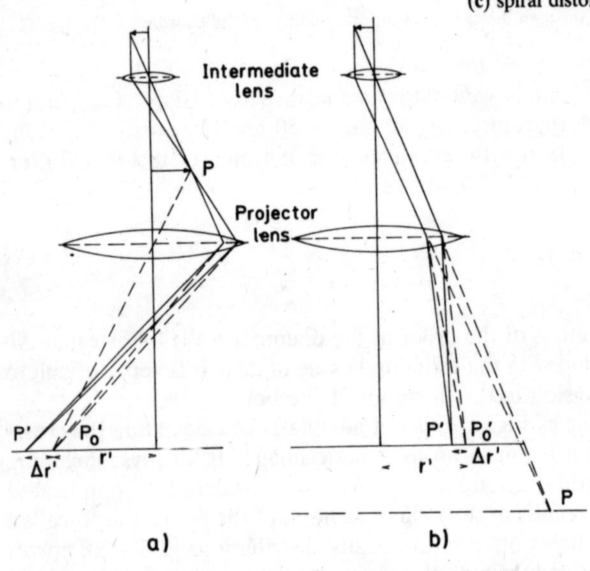

Fig. 2.18 a, b. Examples of distortion caused by the spherical aberration of a projector lens

magnify an intermediate image, in which the angular aperture at any image point is smaller by a factor $1/M$ than the objective aperture α_o. No further decrease of image resolution by the spherical aberration is expected, therefore. However, Fig. 2.18 shows how a distortion of the image can be generated indirectly by the spherical aberration. A conical beam coming from a point P in the intermediate image of Fig. 2.18a converges to an image point P_0' in the absence of spherical aberration but to a point P′ if it is present. The deviation $\Delta r'$ on the image screen increases with r' as r'^3 resulting in a pincushion distortion of a square specimen area. The opposite situation is observed when the intermediate image lies beyond the second lens (Fig. 2.18b); the deviation $\Delta r' \propto r'^3$ is now directed towards the optical axis,

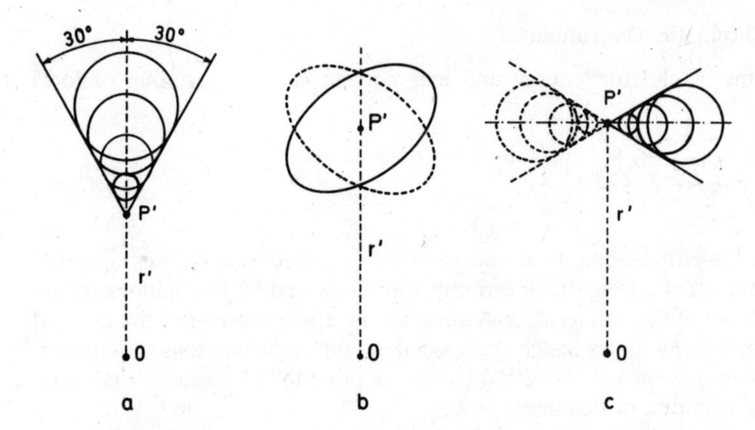

Fig. 2.19. (**a**) The effect of coma on the image P′ of a specimen point P distant $r = -r'/M$ from the axis for increasing angular apertures. (**b**) Anisotropic astigmatism and (**c**) anisotropic coma [(---) magnetic field of the lens reversed]

resulting in barrel distortion. It is possible to compensate this type of distortion by suitable excitation of the lens system, and a pincushion distortion can be compensated by a barrel distortion in another intermediate image step [2.22, 34]. This compensation of distortion is very important for preliminary exploration of the specimen at low magnification.

Coma causes a cone of rays with angles θ to the axis through the specimen point P at an off-axis distance r to be imaged as a circle with radius proportional to θ^2 and to r. The centre of circle does not coincide with the Gaussian-image point $r' = -rM$ but is shifted in the radial direction by twice the radius. Circles corresponding to different angles θ therefore lie within a sector of 60° (Fig. 2.19 a).

2.3.5 Anisotropic Aberrations

The anisotropic distortion is caused by the dependence of the image rotation on the off-axis distance r of the object point; the latter is imaged with an additional rotation angle φ proportional to r'^2. Straight lines in the specimen plane become cubic parabolae in the image plane (Fig. 2.17 c). Reversal of the lens current changes the sense of this distortion.

The anisotropic astigmatism together with the astigmatism discussed in Sect. 2.3.2 results in an ellipse, the principal axes of which are not parallel to the x' and y' axes (Fig. 2.19 b).

The anisotropic coma differs from the coma (Sect. 2.3.4) in the direction of the coma sector, which is now perpendicular to the radius vector r' (Fig. 2.19 c).

2.3.6 Chromatic Aberration

Variations of electron energy and lens current cause a variation of focal length

$$\frac{\Delta f_c}{f} = \frac{\Delta E}{E} - 2\frac{\Delta I}{I} \tag{2.53}$$

because f is proportional to E and B^{-2} or I^{-2}, respectively (I: lens current). This means that chromatic aberration can be caused by fluctuations of the high tension of the accelerating system, by the energy spread of the emitted beam, by energy losses inside the specimen and by fluctuations of lens current. An energy spread of width ΔE causes a point to be imaged as a chromatic-aberration disc of diameter

$$d_c' = d_c M = \frac{1}{2}\, C_c \frac{\Delta E}{E} \frac{1 + E/E_0}{1 + E/2E_0}\, \alpha_o M\ . \tag{2.54}$$

The chromatic-aberration coefficient C_c is of the order of the focal length f for weak lenses and decreases to a minimum of about $0.6\,f$ for stronger lenses (Fig. 2.14) [2.35].

The chromatic aberration of the objective lens influences the electron microscope image in the following manner:

a) The energy spread ΔE of the electron beam limits the resolution. If a resolution $\delta = 0.5$ nm $< d_c = 2r_c$ is wanted for $\alpha_0 = 10$ mrad and $C_c \simeq f = 2$ mm, we must ensure that $\Delta E/E < 1 \times 10^{-5}$. Owing to the Boersch effect (Sect. 4.1.2), the half-width of the electron energy distribution from a thermionic cathode is of the order of 1 eV. If the focusing corresponds to the maximum of this energy distribution only half of this value should be used for ΔE in (2.54). This means that hexaboride or field-emission cathodes with $\Delta E < 1$ eV must be used for high-resolution work, that is, not worse than 0.2–0.3 nm. Furthermore, the high tension and the lens currents have to be stabilized to better than 10^{-5}. The influence of chromatic aberration on contrast transfer will be discussed in detail in Sect 6.4.2.
b) The energy-loss spectrum of transmitted electrons (Sect. 5.2.3) shows energy losses in the range 5–20 eV. For high resolution, the proportion of the electron beam scattered inelastically should be very much smaller than that scattered elastically or unscattered. Though inelastically scattered electrons can be coherent and cause phase contrast at higher magnifications, the inelastic scattering process is less localized, which results in a loss of resolution, irrespective of the chromatic error (Sect. 6.3.2).
c) The number of unscattered and elastically scattered electrons is strongly reduced in thick films, and the energy-loss spectrum is broadened by multiple energy losses (Fig. 5.25 b, c). An operator will focus on the most

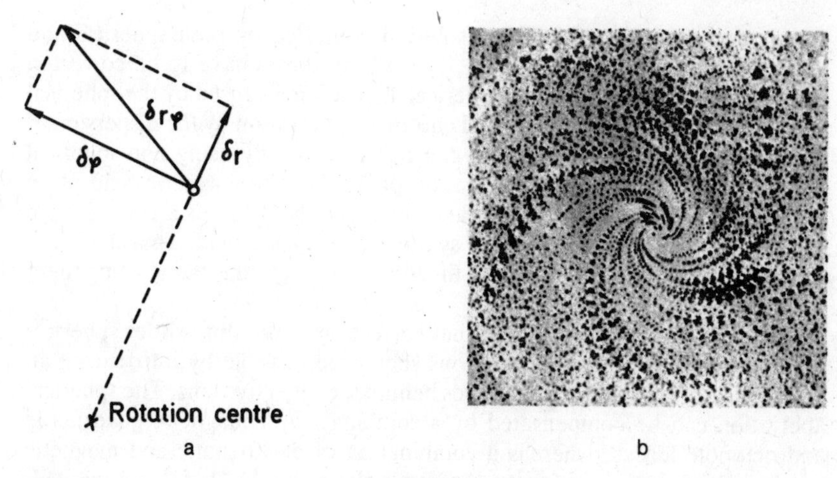

Fig. 2.20. (**a**) Chromatic-error streak and (**b**) demonstration by a superimposed focal series of polystyrene spheres

probable energy (maximum of the energy-loss spectrum). The resolution will be limited by the half-width of the energy-loss spectrum [2.36]. Measurements of the chromatic aberration associated with film thickness are obtained from the blurring of sharp edges [2.37, 38].

Equation (2.54) describes the axial chromatic aberration, which is still present for electron beams entering the objective lens from the axial point of the specimen. When the electron beam passes the specimen at a distance r from the axis a chromatic-error streak $\delta_{r\varphi}$ is observed, which has two components: a radial component δ_r, due to changes of magnification as a function of electron energy, and an azimuthal component δ_φ, due to variation of the image rotation angle φ (2.48). Together, these give (Fig. 2.20 a)

$$\delta_{r\varphi} = C_{r\varphi} r \frac{\Delta E}{E} \quad \text{with} \quad C_{r\varphi}^2 = C_r^2 + C_\varphi^2 \,. \tag{2.55}$$

The constant C_r remains positive whatever the lens excitation whereas C_φ can change sign [2.39]. This chromatic-error streak is illustrated in Fig. 2.20 b, where several exposures of polystyrene spheres, corresponding to different values of the lens current are superimposed.

2.3.7 Corrections of Lens Aberrations

Because the third-order lens aberrations are observable only for non-paraxial rays, aberration correction will be necessary only for lenses working with larger apertures $\alpha > 1$ mrad, such as probe-forming condenser lenses or objective lenses. In the intermediate or projector lenses, the angular aperture is decreased to α/M. The resolution is limited by both spherical and chromatic aberrations (Sect. 6.4.1, 2), so both errors have to be corrected simultaneously. In light optics, spherical aberration is caused by the spherical shape of the glass-lens surfaces and chromatic aberration by the dispersion of the refractive index. Both errors can be corrected by using non-spherical surfaces and/or a suitable combination of lenses. The magnetic field of an electron lens cannot be polished, and spherical aberration is a consequence of the structure of the rotationally symmetric magnetic field. Thus in (2.20), for example, we used the relation $\mathrm{div}\,\boldsymbol{B} = 0$ to obtain the radial component B_r from the axial component B_z.

Scherzer [2.40] demonstrated that correction of the third-order spherical and first-order chromatic aberrations should be possible by introducing an additional system of multipole lenses behind the objective lens. The spherical aberration can be compensated by a combination of magnetic quadrupole and octopole lenses, whereas a combination of electrostatic and magnetic quadrupoles is necessary for the chromatic aberration [2.41–43]. *Koops* et al. [2.44, 45] showed experimentally that such a system works and that the sensitivity to misalignment can be decreased by additional trim coils. However, further experiments will be necessary before such a correction system can be used in a 100 kV high-resolution TEM. Meanwhile we must continue to live with these aberrations. Today, therefore, highly excited objective lenses with small bores are used to decrease the values of C_s and C_c.

Axial astigmatism can be compensated by placing a simple stigmator in the pole-piece bore of a lens. The function of this correction element can be understood from a light-optical analogue [2.46] (Fig. 2.21). Axial astigmatism can be simulated by adding a cylindrical lens C1 to the rotationally symmetric lens L1 (Fig. 2.21 a). The lens C1 acts only on the sagittal bundle,

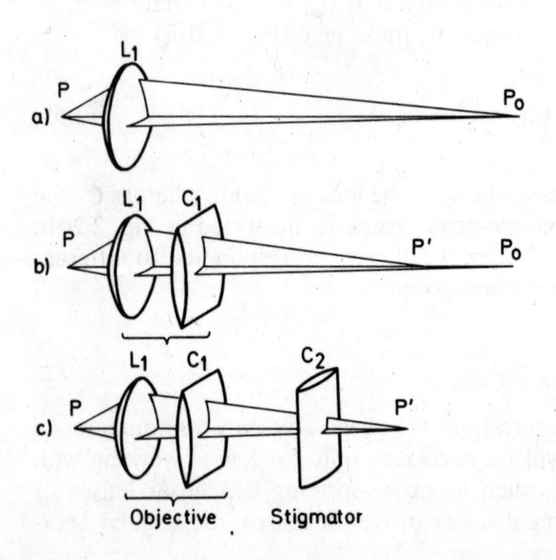

Fig. 2.21 a–c. Light-optical analogue of the action of astigmatism (L1 + C1) and a stigmator C2 [2.46]

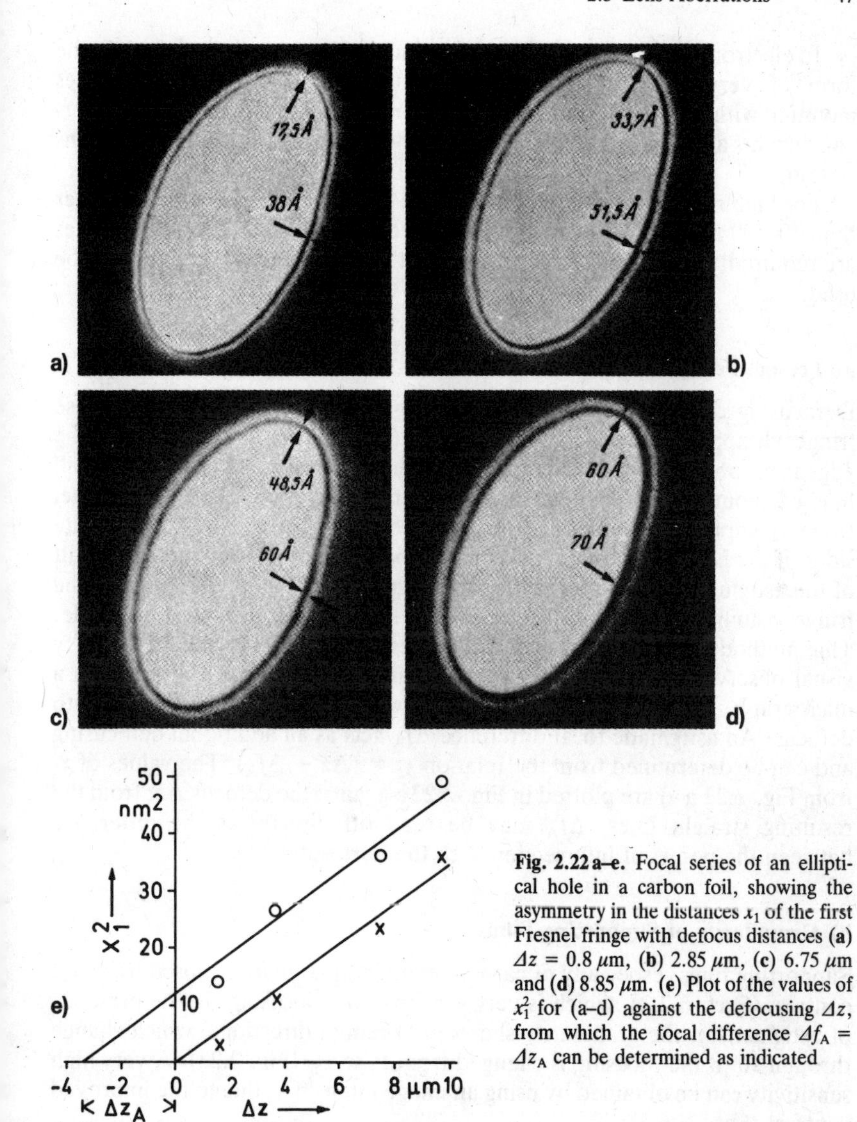

Fig. 2.22 a–e. Focal series of an elliptical hole in a carbon foil, showing the asymmetry in the distances x_1 of the first Fresnel fringe with defocus distances (**a**) $\Delta z = 0.8\ \mu$m, (**b**) 2.85 μm, (**c**) 6.75 μm and (**d**) 8.85 μm. (**e**) Plot of the values of x_1^2 for (a–d) against the defocusing Δz, from which the focal difference $\Delta f_A = \Delta z_A$ can be determined as indicated

resulting in a shorter focal length (Fig. 2.21 b). The stigmator consists of a cylindrical lens C2, rotated through 90° relative to C1. It acts only on the meridional bundle (Fig. 2.21 c), so that P'_s and P'_m coincide in P'. This means that the lens astigmatism is compensated by a perpendicular astigmatism of the same magnitude. The orientation and strength of C2 therefore have to be adjustable.

In electron optics, toric rather than cylindrical lenses are employed, in the form of very weak quadrupole lenses (Sect. 2.2.3). Two quadrupoles mounted with a relative rotation of 45° around the axis and excited by different currents allow us to vary the direction and strength of the quadrupole lens system.

For high resolution, the astigmatic focal difference Δf_A should be smaller than 10–100 nm. Sensitive methods of detecting such small focal differences are required, to adjust the stigmator correctly. The following methods can be used.

a) Fresnel-Fringe Method

Defocusing causes Fresnel-diffraction fringes at edges (Sect. 3.2.2). These fringes disappear in focus. The distance x_1 (3.38) of the first fringe from the edge is proportional to the square root of the defocusing $\Delta z = R_0$. If a small hole of about 0.1 μm diameter in a supporting film is observed, the Fresnel fringes completely disappear in the presence of astigmatism only on opposite sides of the hole. They remain visible in a perpendicular direction as a result of the astigmatic focal difference. The astigmatism is compensated when the fringe visibility for small defocusing is the same around the edge of a hole. This method is capable of revealing values of Δf_A greater than 0.1 μm by visual observation of the viewing screen and about half of this value on a micrograph. Figure 2.22 demonstrates how the distance x_1 increases with defocus. An astigmatic focal difference Δf_A acts as an additional defocusing and can be determined from the relation $x_1^2 \propto (\Delta z + \Delta f_A)$. The values of x_1^2 from Fig. 2.22 a–d are plotted in Fig. 2.22 e against the defocus Δz; from the resulting straight lines, Δf_A may be read off directly as the difference between the points of intersection with the horizontal axis.

b) Granularity of Supporting Films

Supporting films, especially of carbon, exhibit a granularity caused by phase contrast (Sect. 6.2.2), which is very sensitive to defocusing. In the presence of astigmatism, the granularity shows preferential directions, which change through 90° if the focusing is changed from F_s to F_m (Fig. 2.23). A very high sensitivity can be obtained by using an image intensifier, due to the improved contrast (Sect. 4.6.3).

c) Fraunhofer Diffraction

The spatial-frequency spectrum of the granularity can be observed by light-optical Fraunhofer diffraction on developed micrographs (Sect. 6.4.6) or by digital Fourier analysis. Spherical aberration and defocusing lead to gaps in the transfer of the spatial frequencies, which can be seen as a ring pattern in Fraunhofer diffractograms (Fig. 6.27). Astigmatism deforms the rings to

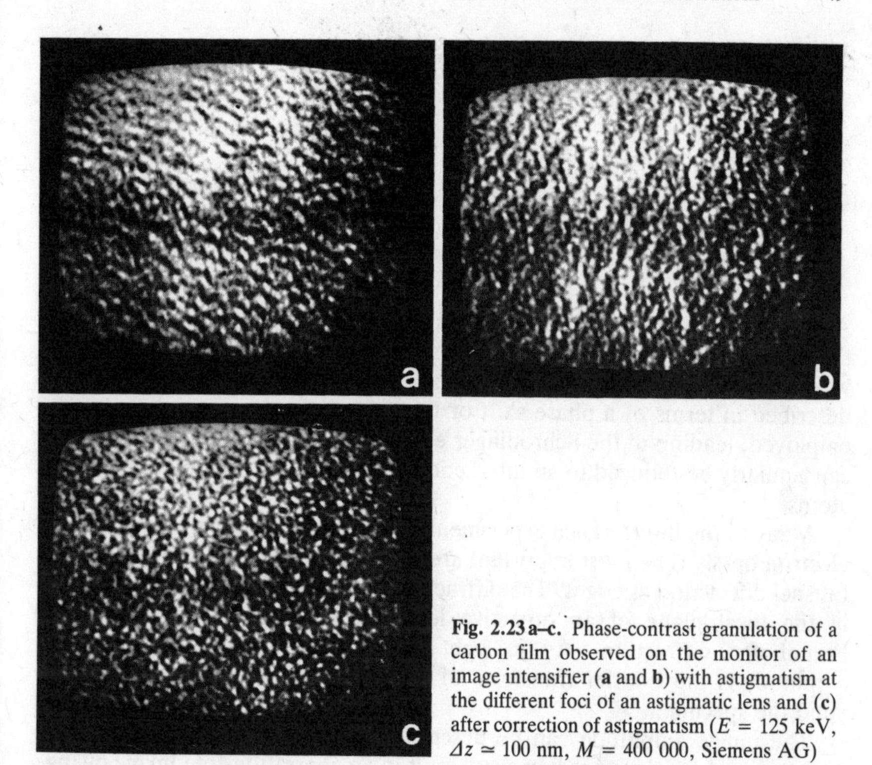

Fig. 2.23 a–c. Phase-contrast granulation of a carbon film observed on the monitor of an image intensifier (**a** and **b**) with astigmatism at the different foci of an astigmatic lens and (**c**) after correction of astigmatism ($E = 125$ keV, $\Delta z \simeq 100$ nm, $M = 400\,000$, Siemens AG)

ellipses or hyperbolae. A disadvantage of this method is that the detection of astigmatism is possible only a posteriori on developed micrographs. However, Fraunhofer diffractograms can also be obtained on-line by special recording techniques (Sect. 6.4.6).

Methods 2 and 3 can detect values of Δf_A greater than 10 nm, which is sufficient for high resolution. All three methods are based on phase-contrast effects caused by defocusing. It is necessary to work with a nearly coherent electron beam to prevent blurring of the fringes and of the granularity. With an illumination aperture α_i of 1 mrad, for example, only one Fresnel fringe can be resolved (Fig. 2.22), whereas with a very coherent beam produced by a field-emission gun, hundreds of fringes may be seen (Fig. 3.10).

3. Wave Optics of Electrons

A de Broglie wavelength can be attributed to each accelerated particle and the propagation of electrons can be described by means of the concept of a wave packet. The interaction with magnetic and electrostatic fields can be described in terms of a phase shift or the notion of refractive index can be employed, leading to the Schrödinger equation. The interaction with matter can similarly be reduced to an interaction with the Coulomb potential of the atoms.

Many of the interference experiments of light optics can be transferred to electron optics. The most important are the Fresnel biprism experiment and Fresnel diffraction at edges. The diffraction pattern far from the specimen or in the focal plane of an objective lens can be described by means of Fraunhofer diffraction. As in light optics, the Fraunhofer-diffraction amplitude is the Fourier transform of the wave-amplitude distribution beyond the specimen.

The image amplitude can be described in terms of an inverse Fourier transform, which does not however result in an aberration-free image owing to the phase shifts introduced by the electron lens and to the use of a diaphragm in the focal plane.

3.1 Electron Waves and Phase Shifts

3.1.1 de Broglie Waves

In 1924, de Broglie showed that an electron can be treated as a quantum of an electron wave, and that the relation $E = h\nu$ for light quanta also should be valid for electrons. As a consequence he stated that the momentum $\boldsymbol{p} = m\boldsymbol{v}$ is also related by $\boldsymbol{p} = h\boldsymbol{k}$ to the wave vector $\boldsymbol{k}$, the magnitude of which (the wave number) may be written $|\boldsymbol{k}| = 1/\lambda$ (λ: wavelength); this is analogous to $p = h\nu/c = hk$ for light quanta. This implies that $\lambda = h/p$ with the relativistic momentum p given by (2.11) – see (2.12) in Table 2.1 and the numerical values in Table 3.1. Substitution of the constants in (2.12) results in the formula

$$\lambda = \frac{h}{mv} = \frac{1.226}{[U(1 + 0.9788 \times 10^{-6}\,U)]^{1/2}} \tag{3.1}$$

with λ [nm] and U [V].

A stationary plane wave that propagates in the z direction can be described by a wave function ψ, that depends on space and time τ:

$$\psi = \psi_0 \cos[2\pi(kz - \nu\tau)] = \psi_0 \cos\left(\frac{2\pi}{\lambda}z - 2\pi\nu\tau\right) = \psi_0 \cos\varphi\ , \tag{3.2}$$

where ψ_0 is called the amplitude and φ the phase of the wave. The phase changes by 2π for τ = const if the difference between two positions $(z_2 - z_1)$ is equal to λ (Fig. 3.1 a).

However, there are important differences between light quanta and electrons when the relations $E = h\nu$ and $p = hk$ are used. The energy E of a light quantum means the energy necessary to generate a quantum or which is transferred when a quantum is absorbed. This energy and the frequency are clearly defined. For electrons, it is incorrect to use the kinetic energy or the relativistic energy $mc^2 = E_0 + E$ in the relation $E = h\nu$. In the presence of an electrostatic field, both quantities and also the momentum p depend on the position. Therefore, k and ν do likewise. The expression for a stationary plane wave will contain values of k and ν that depend on z and can be described by

$$\psi = \psi_0 \cos\left\{2\pi\left[\int_{z_0}^{z} k(z)\,dz - \nu(z)\,\tau\right]\right\}. \tag{3.3}$$

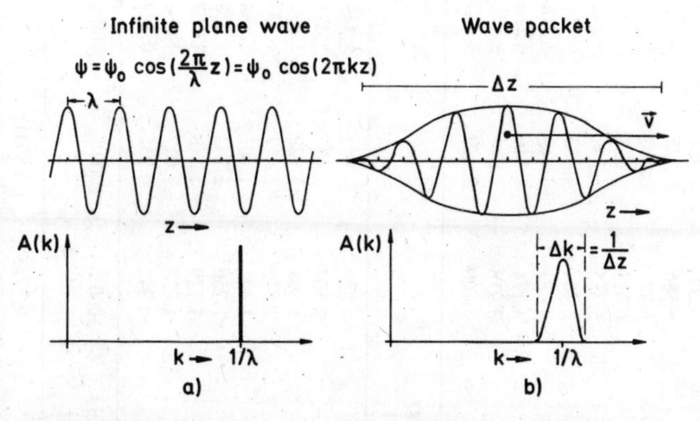

Fig. 3.1. (a) Infinite sine wave with a discrete k-spectrum and **(b)** a wave packet of lateral width Δz and a broadened k-spectrum of width $\Delta k = 1/\Delta z$

Table 3.1. Electron wavelengths λ in picometre (1 pm = 10^{-3} nm)
(e.g., for E = 95 keV, take the value λ = 3.805 pm in the line 90 keV and the row 5 keV)

Energy [keV]	0	1	2	3	4	5	6	7	8	9
0	–	38.751	27.387	22.351	19.347	17.296	15.781	14.604	13.654	12.867
10	12.200	11.627	11.127	10.685	10.291	9.938	9.617	9.326	9.059	8.813
20	8.586	8.375	8.178	7.995	7.822	7.661	7.508	7.365	7.228	7.099
30	6.977	6.860	6.749	6.642	6.541	6.444	6.351	6.261	6.175	6.093
40	6.013	5.937	5.863	5.792	5.723	5.656	5.592	5.529	5.469	5.410
50	5.353	5.298	5.245	5.192	5.142	5.092	5.044	4.998	4.952	4.908
60	4.864	4.822	4.781	4.741	4.701	4.663	4.625	4.588	4.552	4.517
70	4.483	4.449	4.416	4.384	4.352	4.321	4.290	4.261	4.231	4.202
80	4.174	4.146	4.119	4.093	4.066	4.040	4.015	3.990	3.966	3.941
90	3.918	3.894	3.871	3.849	3.827	3.805	3.783	3.762	3.741	3.720
100	3.700	3.680	3.660	3.641	3.622	3.603	3.584	3.566	3.548	3.530

Energy [keV]	0	10	20	30	40	50	60	70	80	90
100	3.700	3.512	3.348	3.203	3.073	2.956	2.850	2.753	2.665	2.583
200	2.507	2.437	2.371	2.310	2.252	2.198	2.147	2.098	2.053	2.009
300	1.968	1.929	1.891	1.855	1.821	1.788	1.757	1.727	1.698	1.670
400	1.643	1.617	1.593	1.568	1.545	1.523	1.501	1.480	1.460	1.440
500	1.421	1.402	1.384	1.366	1.349	1.333	1.317	1.301	1.286	1.271
600	1.256	1.242	1.228	1.215	1.202	1.189	1.176	1.164	1.152	1.140
700	1.129	1.118	1.107	1.096	1.085	1.075	1.065	1.055	1.045	1.036
800	1.027	1.017	1.008	1.000	0.991	0.982	0.974	0.966	0.958	0.950
900	0.942	0.935	0.927	0.920	0.913	0.905	0.898	0.892	0.885	0.878

Energy [keV]	1000	2000	3000	4000	5000	6000	7000	8000	9000	10000
	0.8715	0.5041	0.3568	0.2765	0.2258	0.1909	0.1654	0.1459	0.1305	0.1180

As explained above, the wavelength λ will be the distance $z_2 - z_1$ over which the phase changes by 2π for τ = const. If the changes of k and ν are small over this distance, the phase change becomes

$$\varphi_2 - \varphi_1 = 2\pi = 2\pi\left(k - \frac{d\nu}{dz}\tau\right)(z_2 - z_1) . \tag{3.4}$$

This gives us a time-dependent wavelength $\lambda = z_2 - z_1 = [k - \tau(d\nu/dz)]^{-1}$ at a fixed position, which is in contradiction with the assumption of a stationary wave. This dilemma can be solved only by substituting the sum E' of the kinetic and the potential energy in $E = h\nu$, i.e.,

$$E' = mc^2 + V = mc^2 - e\phi = h\nu . \tag{3.5}$$

This, too, poses a problem because an arbitrary constant can be added to the potential ϕ or, the potential energy V, as shown in (2.6). The frequency of an electron wave is therefore not a clearly defined quantity. This does not matter, because it is not an observable quantity.

A similar contradiction is also found for the wave vector $\boldsymbol{k}$. The quantities $E' = mc^2 - e\phi$ and $p = mv$ do not form a relativistic four-vector, whereas that is a necessary condition if the physical laws are to satisfy the invariance requirements of relativity. An electrostatic field is time dependent if it is seen from a frame of reference moving with a velocity $\boldsymbol{v}$ relative to the original frame. It is, therefore, associated with a magnetic field via Maxwell's equations. The correct value $\boldsymbol{p}'$ that must be used for $\boldsymbol{p}$ in $\boldsymbol{p} = h\boldsymbol{k}$ – the canonical momentum – is

$$\boldsymbol{p}' = m\boldsymbol{v} - e\boldsymbol{A} = h\boldsymbol{k} . \tag{3.6}$$

The magnetic vector potential $\boldsymbol{A}$ is related to the magnetic field by $\boldsymbol{B} = \nabla \times \boldsymbol{A}$. The vector $(p'_x, p'_y, p'_z, E'/c)$ thus becomes a relativistic four-vector.

In the relation (3.6), the vector potential $\boldsymbol{A}$ is not uniquely defined because an arbitrary vector field $\boldsymbol{A}'$ that satisfies the condition $\nabla \times \boldsymbol{A}' = 0$ can be added to $\boldsymbol{A}$ without affecting the value of $\boldsymbol{B}$ because $\boldsymbol{B} = \nabla \times (\boldsymbol{A} + \boldsymbol{A}') = \nabla \times \boldsymbol{A}$. The arbitrary field $\boldsymbol{A}'$, therefore has only to be curl-free. Just like the frequency, then, the wave number k and the wavelength λ are not uniquely defined quantities for electrons and should not be observable quantities.

This is a very strange conclusion for an electron microscopist, who daily sees electron-diffraction patterns and uses (3.1), but it transpires that electron-interference effects can be observed even though the wavelength of an electron is not a clearly defined quantity. Consider the following hypothetical experiment, which can serve as a model for all interference and diffraction experiments. A wave from a source Q (Fig. 3.2) is scattered at the two

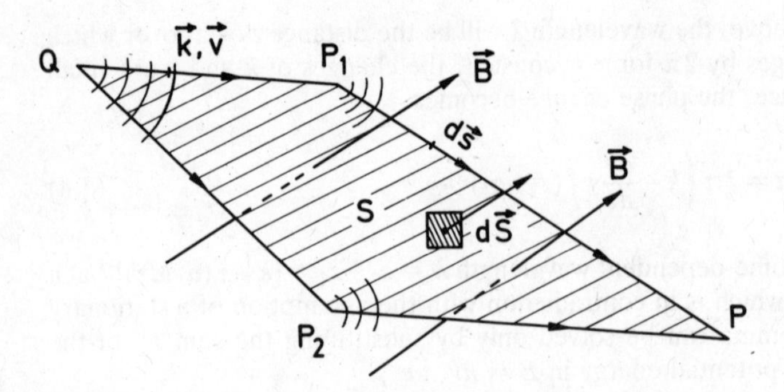

Fig. 3.2. Calculation of the phase difference between to waves scattered at P_1 and P_2

scattering centres P_1 and P_2, beyond which the scattered waves overlap at P. There will be constructive interference if the difference between the phases is an integral multiple of 2π or an even integral multiple of π and destructive interference if the phase is an odd integral multiple of π, for $\tau =$ const.

In the presence of a magnetic and electrostatic field, the phase difference between QP_1P and QP_2P becomes

$$\varphi_2 - \varphi_1 = 2\pi\left(\int_{Q}^{P_2} \boldsymbol{k} \cdot d\boldsymbol{s} + \int_{P_2}^{P} \boldsymbol{k} \cdot d\boldsymbol{s} - \int_{Q}^{P_1} \boldsymbol{k} \cdot d\boldsymbol{s} - \int_{P_1}^{P} \boldsymbol{k} \cdot d\boldsymbol{s}\right)$$
$$= \frac{2\pi}{h} \oint (m\boldsymbol{v} - e\boldsymbol{A}) \cdot d\boldsymbol{s} \,. \tag{3.7}$$

We assume that no additional phase shifts occur in the scattering event. The signs of the last integral have been changed by interchanging the lower and upper integration limits, and the four integrals have thus reduced to an integral over the closed loop QP_2PP_1Q, in which (3.6) for $\boldsymbol{k}$ has been used. Stokes's law can be applied to the integral involving $\boldsymbol{A}$ and the result may be expressed in terms of the magnetic flux ϕ_m enclosed within the loop of area S

$$\varphi_2 - \varphi_1 = \frac{2\pi}{h} \oint m\boldsymbol{v} \cdot d\boldsymbol{s} - \frac{2\pi e}{h} \int_S (\nabla \times \boldsymbol{A}) \cdot d\boldsymbol{S} = \frac{2\pi}{h} \oint m\boldsymbol{v} \cdot d\boldsymbol{s} - \frac{2\pi e}{h} \phi_m \,. \tag{3.8}$$

An arbitrary additional vector field $\boldsymbol{A}'$, such that $\nabla \times \boldsymbol{A}' = 0$, which caused the trouble in the definition of the wavelength, does not influence the difference between the phases in (3.8). Equation (3.8) therefore shows that the relation $k = 1/\lambda = mv/h$ can be used to calculate phase or wave number differences for interference and diffraction experiments in the absence of a magnetic field ($\phi_m = 0$), and that if magnetic flux does pass through the loop

($\phi_m \neq 0$), it causes an additional phase shift. Experiments to verify the existence and magnitude of this latter phase shift are described in Sect. 3.1.5.

Because only the first term of the phase in (3.2) is important for interference experiments, we reduce the wave function (3.2) of a plane wave to

$$\psi = \psi_0 \cos(2\pi kz) = \psi_0 \cos(2\pi \boldsymbol{k} \cdot \boldsymbol{r}) \quad \text{or} \quad \psi = \psi_0 e^{i\varphi} = \psi_0 e^{2\pi i \boldsymbol{k} \cdot \boldsymbol{r}} \tag{3.9}$$

in complex notation, which simplifies the mathematics. For many applications, it is also of interest to use spherical waves

$$\frac{e^{2\pi ikr}}{\quad} \tag{3.10}$$

A_Q is a measure of the magnitude of the source Q, and r denotes the distance from the source. The plane-wave function (3.9) and the spherical-wave function (3.10) are special solutions of the time-independent wave equation (3.22).

The widely used terms "wavefront" or "wave surface" can be defined as surfaces of constant phase φ. The wavefronts of a plane wave are planes normal to the direction of propagation. In the case of a spherical wave, they are concentric spheres. The rays of particle optics are trajectories normal to the wavefronts.

3.1.2 Probability Density and Wave Packets

A parallel electron beam with N electrons per unit volume and velocity v represents a current density

$$j = eNv \tag{3.11}$$

in A cm^{-2}, where Nv is the flux of particles, that is, the number of electrons traversing unit area per unit time. In electron microscopy, we can measure the current or current density, proportional to Nv, or we can count the number of electrons N by single-particle detection. To combine these possibilities of measuring with the wave concept, we use the quantum-mechanical formula for a flux of particles

$$j = e\frac{i\hbar}{2m}(\psi\nabla\psi^* - \psi^*\nabla\psi) \ . \tag{3.12}$$

When this formula is applied to a plane wave (3.9), the operator ∇ becomes $\partial/\partial z$, and substitution of (3.9) in (3.12) results in

$$j = e\frac{\hbar}{m} 2\pi k|\psi_0|^2 = ev|\psi_0|^2 \tag{3.13}$$

by use of the relation $k = mv/h$. Comparison of (3.13) with (3.11) shows that $|\psi_0|^2 = N$, which corresponds to the interpretation of $|\psi|^2 = \psi\psi^*$ as a probability density or of $\psi\psi^* d\tau = N\, d\tau$ as the probability of finding an electron in the volume element $d\tau$. We shall call the quantity

$$I = |\psi|^2 = \psi\psi^* \tag{3.14}$$

an intensity, which can be used to relate the wave amplitude to measurable quantities.

We have to be careful when substituting $|\psi_0| = N^{1/2}$ because we can describe only one electron by a de Broglie wave ($N = 1$); interference effects between electron waves can occur only within the wave field of one electron (see also discussion in Sect. 3.1.4).

Because $\psi\psi^*$ means the probability of finding an electron, its integral over all space should be unity for one electron:

$$\int \psi\psi^* d^3\boldsymbol{r} = 1\,. \tag{3.15}$$

An infinite sine wave such as (3.9) cannot be normalized by means of (3.15). The concept of a wave packet is therefore introduced, to combine the motion of a particle of velocity $\boldsymbol{v}$ with the concept of a wave. A monochromatic wave with a discrete wavelength λ or wave number $k = 1/\lambda$ represents a sine wave with an infinite extension (Fig. 3.1 a). A limited wave packet moving with the particle velocity $\boldsymbol{v}$ (Fig. 3.1 b) can be obtained by superposing a broad spectrum $A(k)$ of wavelengths or wave numbers:

$$\psi = \int_{-\infty}^{+\infty} A(k)\, e^{2\pi ikz} dk\,. \tag{3.16}$$

The amplitudes in front of and behind the wave packet vanish by destructive interference. The amplitudes are summed up by constructive interference with the correct phase only inside the wave packet. The width Δk of the wave-number spectrum $A(k)$ and the spatial width Δz of the wave packet are related by the Heisenberg uncertainty principle $\Delta p \Delta z = h$ or $\Delta k \Delta z = 1$ (compare the Fourier transform of a finite sine wave in Table 3.3).

In practice, it is inconvenient to use a broad spectrum $A(k)$, which corresponds to the superposition (3.16) of many partial waves. Therefore, we continue to use the expression (3.9). The results obtained for the centre of the k spectrum are not appreciably different from the behaviour of the wave packet, provided that $\Delta k \ll k$.

3.1.3 Electron-Optical Refractive Index and the Schrödinger Equation

An electron-optical refractive index n can be introduced as in light optics, where it is defined as the ratio of the velocity c in vacuum to the velocity c_m in matter or by the corresponding ratio of the wavelengths: $n = c/c_m = \lambda/\lambda_m$. The velocity of electrons in matter is influenced by the Coulomb attractive force

$$\boldsymbol{F} = -\frac{e^2 Z}{4\pi\varepsilon_0 r^2}\boldsymbol{u}_r \tag{3.17}$$

between the electron of charge $-e$ and the atomic nucleus of charge Ze ($\boldsymbol{u}_r$: unit vector in radial direction). It is more convenient to use the potential energy

$$V(r) = -\frac{e^2 Z_{\mathrm{eff}}}{4\pi\varepsilon_0 r} \tag{3.18}$$

of an electron in the Coulomb field. The effective atomic number Z_{eff} takes into account the increased screening of the nuclear charge by the atomic electrons with increased r (Sect. 5.1.3). We have only to replace the energy in (2.12) by $E - V(r)$ to obtain the dependence of electron wavelength on r. The refractive index in the absence of a magnetic field becomes

$$n(r) = \frac{\lambda}{\lambda_m} = \frac{p_m}{p} = \left[\frac{2(E-V)E_0 + (E-V)^2}{2EE_0 + E^2}\right]^{1/2}. \tag{3.19}$$

This formula can be simplified if it is assumed that $V(r) \ll E$ and E_0,

$$n(r) = 1 - \frac{V(r)}{E}\frac{E_0 + E}{2E_0 + E} + \dots; \tag{3.20}$$

$n \geq 1$ because $V(r)$ in matter is negative.

Figure 3.3 shows schematically the potential energy $V(r)$ along a row of atoms. The mean value $V_i = -eU_i$, that is the constant term of a Fourier expansion, is called the inner potential (Table 3.2). This inner potential causes a phase shift relative to a wave travelling in vacuum. [The large local variations of $V(r)$ in the specimen produce elastic scattering of electrons at the nuclei, Sect. 5.1.3.]

An optical path difference $\Delta = (n-1)t$ and hence a phase shift φ that corresponds to a layer of thickness t can be introduced by writing

$$\varphi = \frac{2\pi}{\lambda}\Delta = \frac{2\pi}{\lambda}(n-1)t = \frac{2\pi}{\lambda}\frac{eU_i}{E}\frac{E_0 + E}{2E_0 + E}t. \tag{3.21}$$

Thus for carbon films, for example, we have $U_i = 8$ V, giving $n - 1 = 4 \times 10^{-5}$ for 100 keV electrons, for which $\lambda = 3.7$ pm. A film thickness t of 23 nm will be needed to obtain a phase shift φ of $\pi/2$.

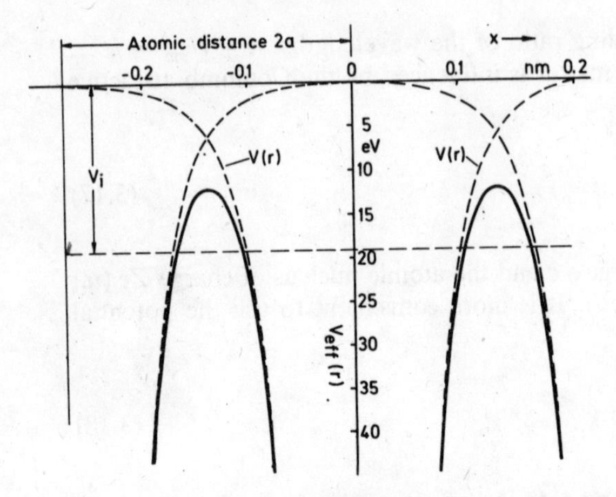

Fig. 3.3. Potential $V(r)$ of a crystal lattice along a row of Ge atoms with interatomic spacing $2a$ and definition of the inner potential $V_i = eU_i$

Table 3.2. Values of the inner potential U_i [V] of various elements

Be	7.8 ± 0.4	[3.1]	Au	21.1 ± 2	[3.3]
C	7.8 ± 0.6	[3.2]		22.1–27.0	[3.2]
Al	13.0 ± 0.4	[3.2]	Si	11.5	[3.5]
	12.4 ± 1	[3.3]	Ge	15.6 ± 0.8	[3.4]
	11.9 ± 0.7	[3.4]	W	23.4	[3.5]
Cu	23.5 ± 0.6	[3.2]	ZnS	10.2 ± 1	[3.3]
	20.1 ± 1.0	[3.4]			
Ag	20.7 ± 2	[3.3]			
	17.0 ± 21.8	[3.2]			

The electron-optical refractive index or the inner potential U_i can be determined from the shift of single-crystal diffraction spots or Kikuchi lines in electron-diffraction patterns with oblique incidence (RHEED, Sect. 9.3.4) [3.5, 6]. An interference effect due to double refraction can be observed in small polyhedric crystals (e.g., MgO smoke); however, this can be explained completely only by the dynamic theory of electron diffraction [3.7, 8]. The phase shift also causes modifications of the Fresnel fringes at the edge of transparent foils (Sect. 3.2.2). However, the most accurate method of measuring U_i involves the use of electron interferometry (Sect. 3.1.4).

Substitution of the wave number $k_m = n(r)\,k$ in the time-independent wave equation

$$\nabla^2\psi + 4\pi^2k^2\psi = 0\,, \tag{3.22}$$

which is also valid for electromagnetic waves and light quanta, yields the wave-mechanical Schrödinger equation in relativistically correct form,

$$\nabla^2\psi + 4\pi^2 n(r)^2 k^2 \psi = \nabla^2\psi + 4\pi^2 \left(1 - \frac{V}{E}\frac{2E_0 + 2E}{2E_0 + E}\right)\frac{2EE_0 + E^2}{h^2c^2}\psi = 0\,, \tag{3.23}$$

or with $\hbar = h/2\pi$,

$$\nabla^2\psi + \frac{2m_0}{\hbar^2}\left[E\left(1 + \frac{E}{2E_0}\right) - V(r)\left(1 + \frac{E}{E_0}\right)\right]\psi = 0\,, \tag{3.24}$$

or

$$\left[\frac{\hbar^2}{2m}\nabla^2 + E^* - V(r)\right]\psi = 0 \quad \text{with} \quad E^* = E\frac{2E_0 + E}{2(E_0 + E)}\,. \tag{3.25}$$

3.1.4 Electron Interferometry and Coherence

In one type of light-optical interferometer, a wave is split into two coherent waves. After propagating along different paths, the latter are recombined with a corresponding phase shift and generate an interference pattern. Two-beam interference with electron waves was first observed from holes in thin single-crystal films [3.9, 10]. *Marton* et al. [3.11] tried to create a Mach-Zehnder interferometer by electron diffraction at three superposed single-crystal films. However, the most satisfactory method relies on the biprism interferometer by *Möllenstedt* and coworkers [3.12–15].

Figure 3.4 a shows a light-optical analogy using a Fresnel biprism. The refracted waves behind the prism are generated by the virtual sources Q_1 and Q_2. The optical phase difference $\varphi_2 - \varphi_1 = 2\pi\Delta/\lambda$ can be calculated as a function of the coordinate x in the viewing plane from the path difference Δ given by (3.21). From Fig. 3.4 a, we see that

$$\Delta = \left[L^2 + \left(x + \frac{a}{2}\right)^2\right]^{1/2} - \left[L^2 + \left(x - \frac{a}{2}\right)^2\right]^{1/2} \simeq \frac{ax}{L} \quad \text{if} \quad x, a \ll L\,. \tag{3.26}$$

Constructive-interference maxima are obtained if $\Delta = n\lambda$ or $\varphi_2 - \varphi_1 = 2\pi n$, n being an integer. The distance between the maxima becomes

$$\Delta x = \frac{L}{a}\lambda = \frac{\lambda}{2\beta}\,. \tag{3.27}$$

The electric field between a thin wire ($\simeq 1$ μm ∅) and earthed plates can form such a biprism for electrons (Fig. 3.4 b). A shadow of the wire is seen at

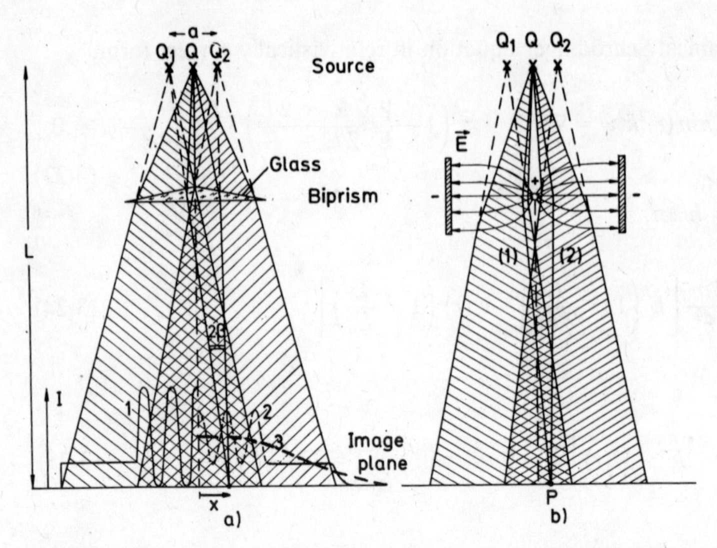

Fig. 3.4. (**a**) Fresnel-biprism experiment of light optics. Curve 1 (*left*): intensity distribution of interference fringes for coherent illumination and curves 2 and 3 (*right*) for partially coherent and incoherent illumination. (**b**) Electron-optical realisation of a biprism experiment with a positively biased wire

the viewing plane if the wire is earthed. As the positive bias of the wire is increased, the two waves with wave vectors $\boldsymbol{k}_1$ and $\boldsymbol{k}_2$ can overlap, resulting in an amplitude distribution

$$\psi = \psi_0 \left[\exp(2\pi i \boldsymbol{k}_1 \cdot \boldsymbol{r}) + \exp(2\pi i \boldsymbol{k}_2 \cdot \boldsymbol{r})\right] ,$$

$$\psi = \psi_0 \left[\exp(\pi i(\boldsymbol{k}_1 - \boldsymbol{k}_2) \cdot \boldsymbol{r}) + \exp(-\pi i(\boldsymbol{k}_1 - \boldsymbol{k}_2) \cdot \boldsymbol{r})\right] \exp(\pi i(\boldsymbol{k}_1 + \boldsymbol{k}_2) \cdot \boldsymbol{r}) ,$$

$$\psi = 2\psi_0 \cos\left[\pi(\boldsymbol{k}_1 - \boldsymbol{k}_2) \cdot \boldsymbol{r}\right] \exp(2\pi i k z) . \tag{3.28}$$

The intensity distribution $I(x) = \psi\psi^*$ becomes

$$I(x) = 4 I_0 \cos^2(2\pi\beta x/\lambda) \tag{3.29}$$

in which we have written $I_0 = |\psi_0|^2$ and $(\boldsymbol{k}_1 - \boldsymbol{k}_2) \cdot \boldsymbol{r} = 2kx \sin\beta \simeq 2\beta x/\lambda$ (see Curve 1 in Fig. 3.4 a).

If an extended source is used rather than a point source, as assumed above, the probability (3.29) of observing an electron at any point x will not be changed. However, electrons from other points of the extended source will produce shifted interference patterns. The maxima and minima are totally blurred if the extension Δa of the source is larger than the distance Δx between the maxima. The illumination is then said to be "incoherent" (see

Curve 3 in Fig. 3.4 a). In the centre of the overlap, the intensity becomes $2I_0$, the value expected if no interference effects occur. Partially coherent illumination, with $\Delta a < \Delta x$, leads to decrease of the maxima and the minima no longer fall to zero (Curve 2 in Fig. 3.4 a).

When the source size is sufficiently small, $\Delta a \ll \Delta x = L\lambda/a$, the radiation is said to be spatially coherent. The angle, $\alpha_i = \Delta a/2L$, can be interpreted as the illumination aperture, that is the cone angle of the rays from different points of the source at the point P in the observation plane; the condition $\Delta a \ll \Delta x$ for spatial coherence is thus equivalent to $a\alpha_i \ll \lambda/2$, which is also used as a coherence condition in light optics.

Another coherence condition for temporal coherence results from the finite coherence length l_c of a wave packet. The path difference Δ between two interfering waves has to be much smaller than l_c. The value of $l_c = v\Delta\tau$ is related to the emission time $\Delta\tau$. In light emission, a normal dipole transition has an emission time $\Delta\tau \simeq 10^{-8}$ s so that with $v = c$, we have $l_c \simeq 3$ m $\simeq 6 \times 10^5 \lambda$ for $\lambda = 0.5$ µm. For electrons, $\Delta\tau$ can be estimated from the Heisenberg uncertainty relation $\Delta E \Delta\tau \simeq h$ where $\Delta E \simeq 1$ eV is the energy spread at the electron gun; this gives $\Delta\tau \simeq 4 \times 10^{-15}$ s. Thus 100 keV electrons, for which $v = 1.64 \times 10^8$ m s^{-1}, have a coherence length $l_c = v\Delta\tau = 600$ nm $\simeq 2 \times 10^5 \lambda$. *Möllenstedt* and *Wohland* [3.16] produced path differences Δ of the order of l_c for 2.5 keV electrons, using a biprism combined with a Wien filter (Sect. 9.2.1) and confirmed that the biprism interference pattern decreases in amplitude if $\Delta \simeq 0.5\, l_c$.

The influence of spatial and temporal coherence on phase contrast is discussed in Sect. 6.4.2. For further discussion of coherence and the introduction of coherence functions, see [3.17, 18].

The biprism experiment sheds light on another important aspect of wave optics. In particle optics, the concept of a trajectory is used. In our example, the particle can pass either side of the wire. In wave optics, the wave of a single electron passes on both sides of the wire simultaneously and we can observe only the probability of detecting the electron at some position x. It is, therefore, nonsense to ask on which side the electron has passed. If we put a detector on one side of the wire, half of the total number of electrons will be detected, but we thereby suppress all wave amplitudes on this side and will observe no interference pattern.

Introduction of a thin foil on one side of the wire causes a phase shift given by the inner potential U_i (Sect. 3.1.3). The phase shift can be measured more accurately by using an interference microscope, which images both the specimen and an interference pattern. The biprism can be placed either in front of [3.3, 19] or behind the objective lens [3.20, 21]. The first arrangement is particularly suitable for magnifying the interference fringes, the latter, the specimen structure. A shadow of the wire is obtained with an earthed wire. The two parts of the image on either side of the shadow overlap with increasing positive bias and interference fringes appear (Fig. 3.5). The fringe separation decreases with increasing overlap (3.27). It is advisable to use a

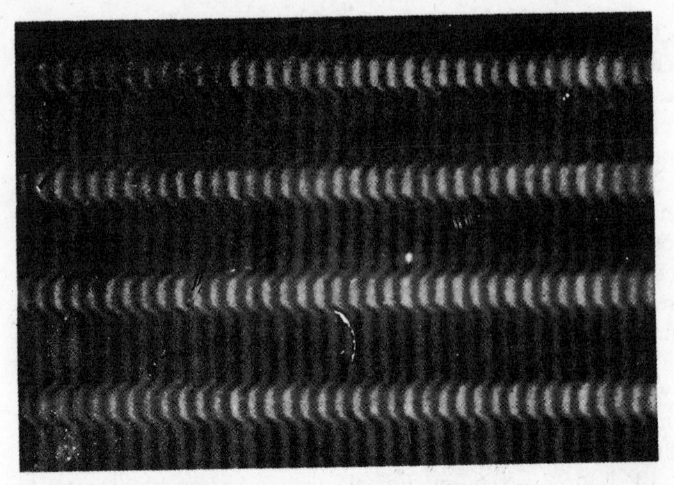

Fig. 3.5. Part of an electron interferogram of a specimen with evaporated beryllium steps, showing the phase shift by the inner potential [3.1]

simple specimen geometry for measuring the inner potential, such as evaporated strips [3.1] (Fig. 3.5) or circular areas [3.3].

Biprism interference fringes in a commercial electron microscope can be obtained also by setting a spider's thread ($\simeq$ 0.3 μm ∅) across 2 mm diaphragm in the plane of the selected-area diffraction diaphragm [3.22]. The fibre becomes positively biased by electron bombardment.

Möllenstedt and *Lichte* [3.23, 24] created a surface interference microscope by splitting the electron beam with a first biprism in front of a magnetic prism, which deflects the beam as in Fig. 9.10 onto the specimen surface, which is biased a few volts more negative than the cathode. The reflected electrons again pass through the magnetic prism and the two separate beams are reunited by a second biprism to form an interference pattern. This device can be used to detect very small phase differences caused by deformations of the last equipotential surface in front of the specimen at which the electron beam is reflected. Such deformations may be produced by surface steps and potential differences, which can be measured with high accuracy, of the order of the electron wavelength λ.

3.1.5 Phase Shift by Magnetic Fields

The action of a magnetic field can also be described in terms of a refractive index. In Sect. 3.1.3, we introduced n for the electric field $\boldsymbol{E}$ by using the electric potential ϕ. These two quantities are related by $\boldsymbol{E} = -\nabla\phi$. The action of $\boldsymbol{B}$ can be analogously described by the magnetic vector potential $\boldsymbol{A}$ ($\boldsymbol{B} = \nabla \times \boldsymbol{A}$). The momentum $\boldsymbol{p} = m\boldsymbol{v}$ has to be replaced by the canonical momentum $\boldsymbol{p}' = m\boldsymbol{v} - e\boldsymbol{A}$ (Sect. 3.1.1). For a constant electric potential ϕ, the following expression for the relative refractive index is obtained (neglecting an arbitrary additive constant)

$$n = \frac{p'}{p} = 1 - \frac{e\lambda}{h}\boldsymbol{A}\cdot\boldsymbol{u}_t\,. \tag{3.30}$$

where $\boldsymbol{u}_t$ is the unit vector in the direction of the trajectory normal to the wavefront. The presence of the scalar product $\boldsymbol{A}\cdot\boldsymbol{u}_t$ in (3.30) means that the refractive index also depends on the direction.

The phase shift expected from (3.30) can be measured by means of a biprism. Consider two optical rays (Fig. 3.4 b) from the source Q to the point P of the interference pattern. The optical phase difference $\varphi_2 - \varphi_1$ was calculated in (3.8) and depends on only the enclosed magnetic flux ϕ_m. A phase shift occurs even if there is no magnetic field at the trajectory, and hence if no magnetic term of the Lorentz force acts on the electrons. For the phase shift, only the magnetic flux through the enclosed area is important. A magnetic flux $\phi_m = h/e = 4.135 \times 10^{-15}$ Vs is sufficient to cause a phase shift $\varphi = 2\pi$ corresponding to a path difference $\Delta = \lambda$ and to a shift of the interference pattern by one fringe distance. Such a small flux can be created by an iron whisker with a cross section of 2000 nm^2 and a saturation magnetization $B_s = 2.1$ T [3.25, 26] or by a 25 nm permalloy film evaporated on the biprism wire [3.27]. The theoretical value of the phase shift was confirmed from two exposures of the fringe system obtained with B_s in opposite directions. By use of three biprism wires, a larger spatial separation of the electron rays can be realized and the flux of a coil 20 μm diameter can be enclosed [3.28, 29].

Boersch and *Lischke* [3.30, 31] verified the quantization of the enclosed flux in superconductors, which is a multiple of the flux quantum (fluxon) $h/2e$. One fluxon corresponds to a shift of the fringe pattern by one half of the fringe distance.

The deflection of electrons in a transverse magnetic field of length L through an angle ε (Sect. 2.1.3 and Fig. 2.4 b) means, in wave-optical terms, that the incident wave is tilted through an angle ε after passing the magnetic field. With an arbitrary origin $x = 0$ in Fig. 2.4 b, rays at a distance x enclose a magnetic flux $\phi_m = BLx$. Equation (3.8) gives the same value for the deflection angle $\varepsilon = \Delta/x$, with $\Delta = \lambda(\varphi_2 - \varphi_1)/2\pi$, as that given by (2.18 b), obtained by use of classical mechanics.

3.2 Fresnel and Fraunhofer Diffraction

3.2.1 Huygens Principle and Fresnel Diffraction

In wave optics, all other wavefronts can be calculated once the shape of one of them is known, by use of the Kirchhoff diffraction theory based on the wave equation (3.22). However, the simpler treatment offered by Huygens principle can also be used in electron optics; this states that each surface element dS of a wavefront generates a secondary spherical wave with amplitude

$$d\psi = \frac{A(\theta)}{i\lambda}\psi\frac{e^{2\pi ikR}}{R}dS\,; \qquad A(\theta) = (1+\cos\theta)/2\,, \tag{3.31}$$

where ψ denotes the amplitude of the incident wave at dS, and θ the angle of emission to the normal of the wavefront. A new wavefront is generated by the superposition of all of the secondary waves (Fig. 3.6). At a point P in front of the wavefront, all of the amplitudes of the secondary waves have to be summed, considering their phase shifts. The factor $A(\theta)$ is unity in the direction of the propagating wave and decreases with increasing θ. For the reverse direction A is zero. The exact form of $A(\theta)$ is not important for the following calculation. The factor $1/i = \exp(-i\pi/2)$ in (3.31) represents a phase shift of $-\pi/2$ relative to the incident wave.

We apply Huygens principle to the propagation of a spherical wave (Fig. 3.6). The known wavefront is thus spherical with radius r and amplitude ψ (3.10). The surface element dS in (3.31) becomes $dS = rd\chi \cdot 2\pi r \sin\chi$, by use of the spherical polar coordinates r and χ. The distance R to the point P can be calculated from $R^2 = r^2 + (r+R_0)^2 - 2r(r+R_0)\cos\chi$, from which we obtain $2R\,dR = 2r\,(r+R_0)\,\sin\chi d\chi$ and it follows that $dS = 2\pi \cdot [r/(r+R_0)]R\,dR$. The amplitude ψ_P at P is obtained by integration over all the secondary waves,

$$\psi_P = \int_S \frac{A(\theta)}{i\lambda}A_Q\frac{e^{2\pi ikr}}{r}\frac{e^{2\pi ikR}}{R}dS = \frac{2\pi A_Q e^{2\pi ikr}}{i\lambda(r+R_0)}\int_{R_0}^{R_{max}} A(\theta)e^{2\pi ikR}\,dR\,. \tag{3.32}$$

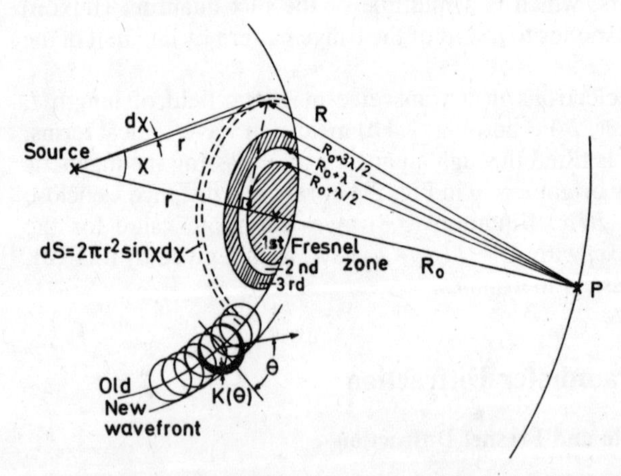

Fig. 3.6. Illustration of Huygens' principle and Fresnel zones showing how the wave amplitude at the point P is obtained by summing the amplitudes of the Huygens wavelets from a spherical wavefront of radius r

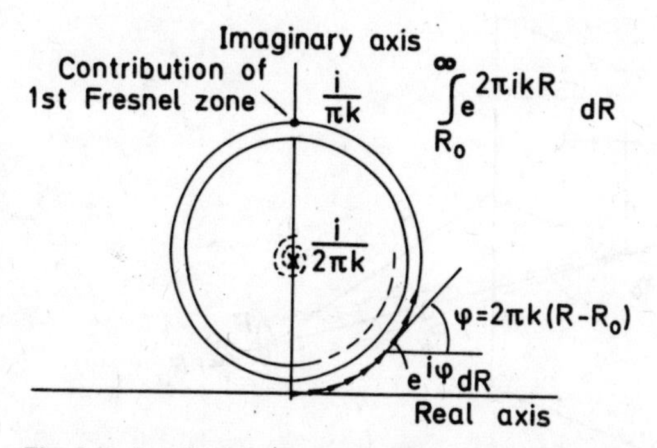

Fig. 3.7. Amplitude-phase diagram for the integral in (3.32)

The result can be established by means of an amplitude-phase diagram (APD) (Fig. 3.7). The term $\exp(i\varphi)dR$ can be represented in the complex number plane by a line element dR inclined at an angle φ to the real axis. The integration in (3.32) means adding infinitesimal line elements dR with increasing $\varphi = 2\pi kR$, resulting in a circle. The radius of the circle decreases because of the decrease of $A(\theta)$ with increasing θ. The result is a spiral, which converges to the centre of the circle. Starting from the lower limit of integration $R = R_0$, the integral reaches its greatest value when $\varphi = 2\pi k(R - R_0) = \pi$ or $R - R_0 = \lambda/2$. The value of the integral then decreases again because the phase shift of the secondary wave becomes greater than π. This is the basic idea of the Fresnel-zone construction. If a circle of radius $R = R_0 + \lambda/2$ is drawn, as in Fig. 3.6, the first Fresnel zone is obtained, as indicated by the hatched area on the wavefront, which contributes to the amplitude ψ_P with a positive value. The second Fresnel zone between the corresponding values $R_0 + \lambda/2$ and $R_0 + \lambda$ results in a negative contribution, the next Fresnel zone again gives a positive contribution, and so on with alternating signs. The convergence of the APD to the centre of the circle means that the integral in (3.32) becomes only half of the value ψ_P of the first Fresnel zone. This results in the following value for the integral in (3.32)

$$\int_{R_0}^{R_{max}} A(\theta)e^{2\pi ikR}\,dR = \frac{1}{2}\int_{R_0}^{R_0+\lambda/2} e^{2\pi ikR}\,dR = \frac{1}{2}\frac{1}{2\pi ik}e^{2\pi ikR_0}\underbrace{(e^{i\pi}-1)}_{-2} . \qquad (3.33)$$

Substituting this value in (3.32) gives

$$\psi_P = A_Q \exp[2\pi ik(r + R_0)]/(r + R_0) . \qquad (3.34)$$

This is the expected formula for the wavefront at a distance $r + R_0$ from the point source. This simple example demonstrates the power of Huygens principle.

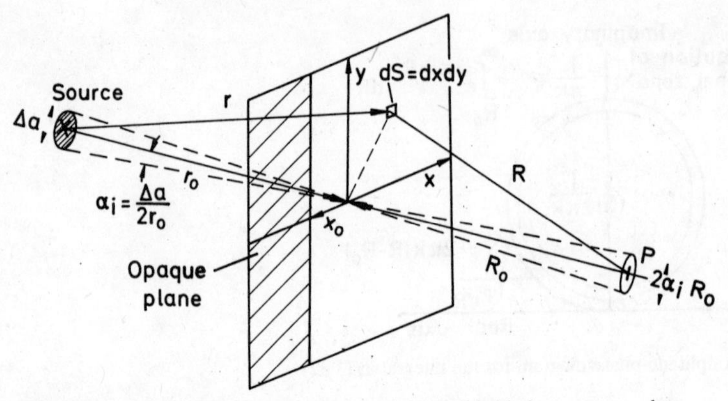

Fig. 3.8. Calculation of wave amplitude at a point P behind an opaque edge

We now use Huygens principle with another choice of coordinates, which directly yields Fresnel diffraction at an edge (Sect. 3.2.2). The surface element dS is placed in an x-y plane normal to the line from the source Q to the point P (Fig. 3.8). The distance r becomes

$$r = (r_0^2 + x^2 + y^2)^{1/2} = r_0\left(1 + \frac{x^2+y^2}{r_0^2}\right)^{1/2} = r_0\left(1 + \frac{x^2+y^2}{2r_0^2} + \dots\right), \quad (3.35)$$

with a corresponding formula for R if r_0 is replaced by R_0. Substitution of (3.35) into (3.32) results in the following formula; $A(\theta)$ has been omitted because the integral already converges for small values of x and y, for which $A(\theta) = 1$:

$$\psi_P = \frac{A_Q \exp[2\pi ik(r_0+R_0)]}{i\lambda r_0 R_0} \iint \exp\left(2\pi ikx^2 \frac{r_0+R_0}{2r_0R_0}\right) \exp\left(2\pi iky^2 \frac{r_0+R_0}{2r_0R_0}\right) dx\, dy \,. \quad (3.36)$$

The substitutions $u = x\left(\frac{2(r_0+R_0)}{\lambda r_0 R_0}\right)^{1/2}$ and $v = y\left(\frac{2(r_0+R_0)}{\lambda r_0 R_0}\right)^{1/2}$ give

$$\psi_P = \frac{A_Q \exp[2\pi ik(r_0+R_0)]}{2i(r_0+R_0)} \int_{-\infty}^{+\infty} \exp(i\pi u^2/2)\, du \int_{-\infty}^{+\infty} \exp(i\pi v^2/2)\, dv$$

$$= \frac{A_Q \exp[2\pi ik(r_0+R_0)]}{i(r_0+R_0)} \frac{1}{2} [C(u) + iS(u)]_{-\infty}^{+\infty} [C(v) + iS(v)]_{-\infty}^{+\infty} , \quad (3.37)$$

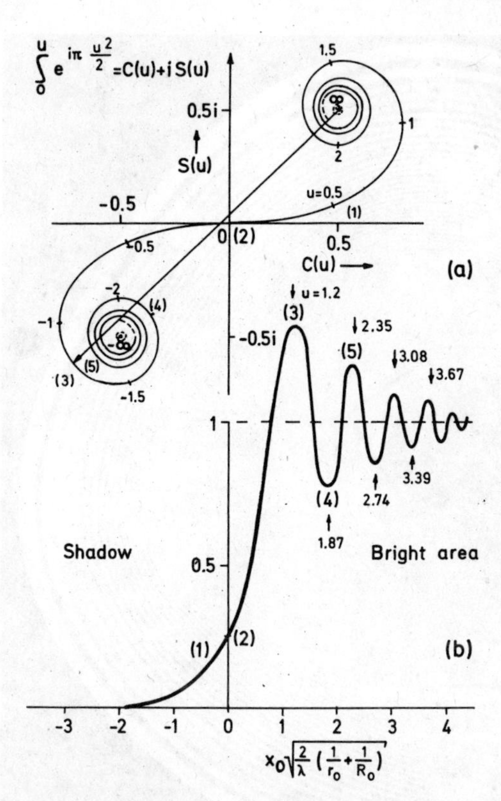

Fig. 3.9. (**a**) Amplitude-phase diagram (*Cornu spiral*) for one of the integrals in (3.37). (**b**) Intensity distribution of Fresnel fringes in the Fresnel diffraction pattern of an opaque edge

where $C(u)$ and $S(u)$ are the tabulated Fresnel integrals. $C(u) + iS(u)$ produces the Cornu spiral in the APD of Fig. 3.9a. The more-complicated shape of the APD as compared to Fig. 3.7 results only from the different choice of coordinate system. The point of convergence is $-0.5 - 0.5i$ for the limit of integration $u \to -\infty$ and $0.5 + 0.5i$ for $u \to +\infty$. The total amplitude is obtained by connecting the two points of convergence and is hence $1 + i$. Because we have to consider the product of two integrations, in the x and the y directions, the two quantities in square brackets in (3.37) give $(1+i)^2 = 2i$, and again we obtain the expected value ψ_P of the wave excitation at P at a distance $r_0 + R_0$ from the source Q.

3.2.2 Fresnel Fringes

The last section has shown that the wave amplitude at a point in front of a wavefront and the wave propagation can both be described by Huygens elementary waves and Fresnel integrals. This formalism will now be applied to electron-opaque obstacles. One important example is the appearance of Fresnel-diffraction fringes at opaque half-planes (Figs. 2.22 [illegible] 10).

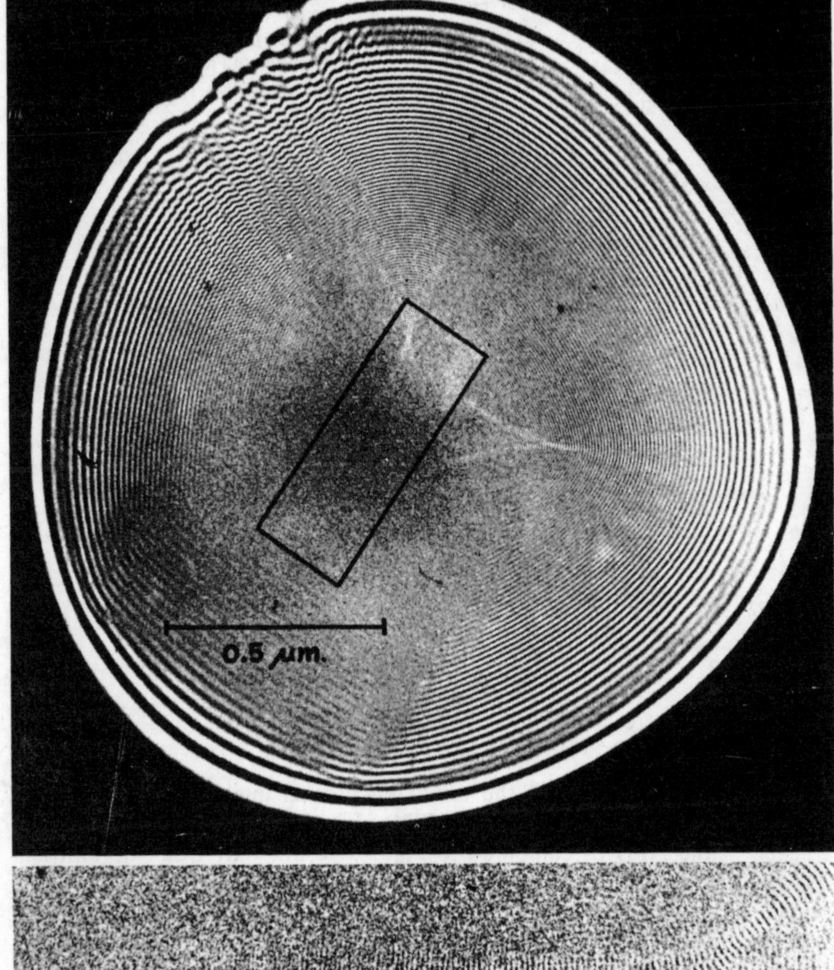

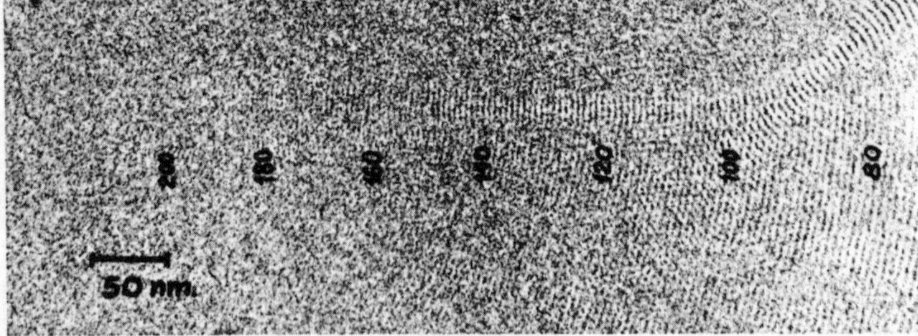

Fig. 3.10. Numerous Fresnel fringes around a hole in a carbon foil obtained with a highly coherent field-emission gun (*Siemens AG*)

At a distance x_0 from the shadow of a half-plane (Fig. 3.8), the intensity is obtained by integration in the x direction from $-x_0$ to $+\infty$. In the y direction, we again consider strips of width dy from $-\infty$ to $+\infty$, and the integral in the y direction has the value $1 + \mathrm{i}$, as before. The amplitude contribution from the x direction is obtained by connecting the point of convergence $0.5 + 0.5\mathrm{i}$ for $x_0 = u = +\infty$ to the corresponding point $u = -x_0[2(r_0 + R_0)/\lambda r_0 R_0]^{1/2}$ on the Cornu spiral. The coordinate u is the arc length along the Cornu spiral. At the point (3) of the spiral farthest from the positive-convergence point, we obtain a maximum of amplitude and intensity. In Fig. 3.9, further corresponding points on the Cornu spiral and in the intensity distribution are numbered. In all practical cases $r_0 \gg R_0$, and a relatively accurate formula can be obtained for the positions of the maxima u_n or x_n from the condition that the phase in the integral of (3.37) will be $\pi u^2/2 = \pi(2n - 5/4)$ for $n = 1, 2, \ldots$ or, in other words, that the tangent to the Cornu spiral is inclined at an angle of $-45°$ to the real axis. This condition results in

$$u_n = \sqrt{(8n-5)/2}\ ; \qquad x_n = \sqrt{\lambda R_0(8n-5)/4}\ . \tag{3.38}$$

The intensity distribution of these Fresnel fringes in a plane at a distance R_0 below an edge can be imaged by defocusing the objective lens [3.32]. Observation of these Fresnel fringes is important for the recognition and correction of astigmatism (Sect. 2.3.7). The number of Fresnel fringes visible is a measure of the spatial coherence of the electron beam. Figure 3.10 shows several Fresnel fringes around a hole in a supporting film illuminated with a field-emission gun. The influence of spherical aberration of the objective lens on the intensity and position of Fresnel fringes has to be considered [3.33]; for transparent supporting films, the decrease of wave amplitude caused by scattering and the phase shift due to the inner potential U_i also have an effect [3.34–37].

If the source has a finite size Δa, the specimen is irradiated with an angular aperture $\alpha_\mathrm{i} = \Delta a/2r_0$ (Fig. 3.8), which causes blurring of the Fresnel diffraction pattern proportional to $2\alpha_\mathrm{i}R_0$. The intensity distribution with such partially coherent illumination is given by the convolution of the coherent distribution with the geometric shadow distribution of the source at a distance R_0. Because the distances between the diffraction maxima $x_{n+1} - x_n$ (3.38) decrease with increasing n, the diffraction maxima of high-order disappear first, and only one Fresnel maximum can be observed when the larger illumination aperture necessary for visual observation of the viewing screen at high magnification is used (Fig. 2.22). All of the maxima are blurred when the angular aperture α_i is very large, as in the case of incoherent illumination; the intensity distribution is then the same as that of the purely geometric shadow of the edge, thrown by an extended source.

3.2.3 Fraunhofer Diffraction

Fresnel diffraction goes over into Fraunhofer diffraction if a plane incident wave is used and if the diffraction pattern is observed at an infinite distance. Alternatively, this pattern can be observed in the focal (diffraction) plane of a lens (Fig. 3.11). A parallel beam inclined at a small angle θ to the optical axis converges to a point in this plane at a distance $f\theta$ from the optic axis. No further phase shifts occur if the lens is aberration-free. The amplitude of the incident wave can be described by $\psi = \psi_0 \exp(2\pi ikz)$ and is modified by the specimen to

$$\psi = \psi_0 a_s(\boldsymbol{r}) \exp[i\varphi_s(\boldsymbol{r})] \exp(2\pi ikz) = \psi_s(\boldsymbol{r}) \exp(2\pi ikz) \tag{3.39}$$

($\boldsymbol{r}$: radius vector in the specimen plane from the origin on the optic axis, $a_s(\boldsymbol{r}) \leq 1$: local decrease of amplitude (absorption) and $\varphi_s(\boldsymbol{r})$: phase shift caused by the specimen).

Furthermore, a geometrical phase shift Δ_g has to be included, which results from the deflection $\Delta_g = -\theta x$ of the plane wave front in the direction θ. Figure 3.12 shows that two points O and P separated a distance $\boldsymbol{r}$ cause an optical-path difference

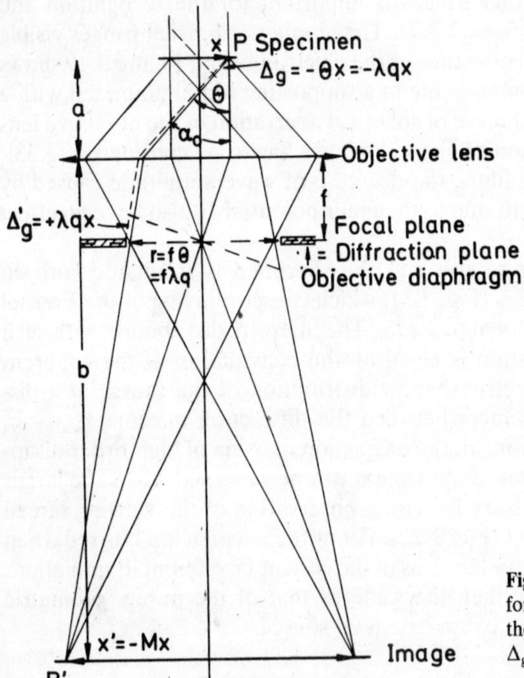

Fig. 3.11. Ray diagram for image formation by an objective lens; from the wavefronts (---), the phase shifts Δ_g and Δ'_g can be derived

$$\Delta_g = \boldsymbol{u}_0 \cdot \boldsymbol{r} - \boldsymbol{u} \cdot \boldsymbol{r} = -\lambda(\boldsymbol{k} - \boldsymbol{k}_0) \cdot \boldsymbol{r} = -\lambda \boldsymbol{q} \cdot \boldsymbol{r} , \tag{3.40}$$

($\boldsymbol{u}_0 = \lambda \boldsymbol{k}_0$ and $\boldsymbol{u} = \lambda \boldsymbol{k}$ are unit vectors in the direction of the incident and scattered waves, respectively.) The phase difference is given by

$$\varphi_g = \frac{2\pi}{\lambda} \Delta_g = -2\pi(\boldsymbol{k} - \boldsymbol{k}_0) \cdot \boldsymbol{r} = -2\pi \boldsymbol{q} \cdot \boldsymbol{r} . \tag{3.41}$$

Figure 3.12 shows that

$$|\boldsymbol{k} - \boldsymbol{k}_0| = |\boldsymbol{q}| = 2k \sin\frac{\theta}{2} \simeq \frac{\theta}{\lambda} . \tag{3.42}$$

A diffraction grating with period Λ (lattice spacing) generates a diffraction maximum at an angle $\sin\theta \simeq \theta = \lambda/\Lambda$. This implies that q in (3.40–42) is equal Λ^{-1}; it is known as the spatial frequency by analogy with the relation $\nu = T^{-1}$ between the temporal frequency ν and the period T. From now on we shall use $\boldsymbol{q}$ as a coordinate in the diffraction plane.

The amplitude $F(\boldsymbol{q})$ in the diffraction plane can be obtained by integration over all of the surface elements $dS = d^2\boldsymbol{r}$ of the specimen plane,

$$F(\boldsymbol{q}) = \int_S \psi_s(\boldsymbol{r}) \exp(\mathrm{i}\varphi_g)\, dS = \int_S \psi_s(\boldsymbol{r}) \exp(-2\pi \mathrm{i} \boldsymbol{q} \cdot \boldsymbol{r})\, d^2\boldsymbol{r} . \tag{3.43}$$

This shows that $F(\boldsymbol{q})$ is the Fourier transform of $\psi_s(\boldsymbol{r})$.

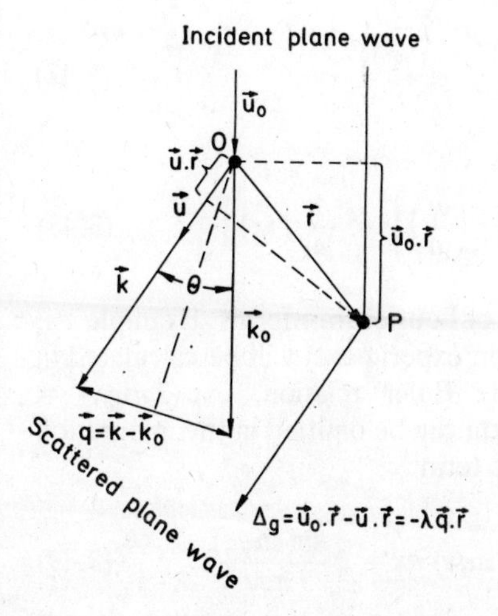

Fig. 3.12. Demonstration of the path-length difference Δ_g caused by scattering at two points O and P

3.2.4 Mathematics of Fourier Transforms

This section contains a short review of the mathematics of Fourier transforms, because Fourier transforms are important not only in the description of Fraunhofer diffraction and electron diffraction at crystal lattices but also in the electron-optical theory of image formation. For simplicity, we normally discuss one-dimensional functions $f(x)$. There is no difficulty in extending this to two- and three-dimensional Fourier transforms (see examples in Table 3.3).

Let $f(x)$ be a real or complex function of the real variable x. The Fourier transform of $f(x)$ is defined by the mathematical operation (abbreviated $\mathcal{T}$)

$$\mathcal{T}\{f(x)\} = F(q) = \int_{-\infty}^{+\infty} f(x)\,\mathrm{e}^{-2\pi i q x}\,dx\,. \tag{3.44}$$

$f(x)$ can be obtained from $F(q)$ by the inverse Fourier transform $\mathcal{T}^{-1}$, which has the opposite sign in the exponent

$$\mathcal{T}^{-1}\{F(q)\} = f(x) = \int_{-\infty}^{+\infty} F(q)\,\mathrm{e}^{+2\pi i q x}\,dq\,. \tag{3.45}$$

The following relations can be obtained from the definition (3.44) of a Fourier transform:

1) *Linearity*

$$\mathcal{T}\{af(x) + \mathrm{b}g(x)\} = aF(q) + bG(q)\,. \tag{3.46}$$

2) *Translation theorem*

$$\mathcal{T}\{f(x-x')\} = F(q)\,\mathrm{e}^{-2\pi i q x'} \quad \text{or} \quad \mathcal{T}^{-1}\{F(q-q')\} = f(x)\,\mathrm{e}^{+2\pi i q' x}\,. \tag{3.47}$$

3) *Scale change*

$$\mathcal{T}\{f(ax)\} = \frac{1}{|a|}F\left(\frac{q}{a}\right) \quad \text{or} \quad \mathcal{T}^{-1}\left\{F\left(\frac{q}{a}\right)\right\} = \frac{1}{|a|}f(x)\,. \tag{3.48}$$

Table 3.3 contains concrete examples of Fourier transforms. Example 1 a, a rectangular function (slit in a diffraction experiment) will be calculated in detail as an example. Because of the Euler relation, $\exp(2\pi i q x) = \cos(2\pi q x) + \mathrm{i}\sin(2\pi q x)$ the last sine term can be omitted in the integration of (3.44) owing to the asymmetry of this term

$$F_1(q) = \int_{-\infty}^{+\infty} f_1(x)\,\mathrm{e}^{-2\pi i q x}\,dx = \int_{-a/2}^{+a/2} \cos(2\pi q x)\,dx = a\,\frac{\sin(\pi q a)}{\pi q a}\,. \tag{3.49}$$

Table 3.3 Examples of Fourier transforms

Specimen function $f(x)$	Fourier transform $\mathcal{T}\{f(x)\} = F(q)$
1a) One-dimensional slit $f_1(x) = \begin{cases} 1 & \text{if } \lvert x\rvert < a/2 \\ 0 & \text{if } \lvert x\rvert > a/2 \end{cases}$ 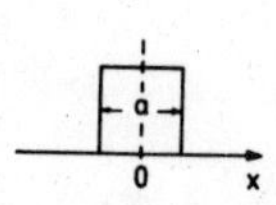	$F_1(q) = a \dfrac{\sin \pi a q}{\pi a q}$ 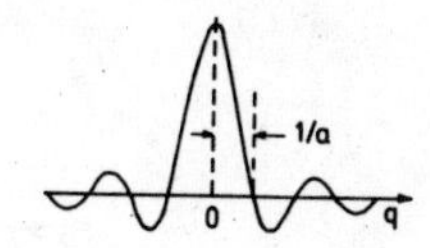
1b) Point source ($a \to 0$) one-: $f_1(x) = \delta(x)$, $\int \delta(x)dx = 1$ two-dimensional: $f_1(\boldsymbol{r}) = \delta(\boldsymbol{r})$, $\iint \delta(\boldsymbol{r})d^2\boldsymbol{r} = 1$	$F_1(q) = 1$ (isotropic scattering) 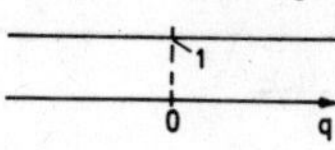
1c) $f_1(x) = \delta(x - b)$	$F_1(q) = \exp(2\pi i b q)$; $\lvert F_1(q)\rvert = 1$ 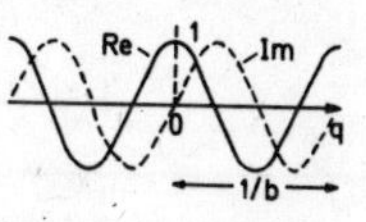
2) Transparent rectangle $f_2(\boldsymbol{r})$ 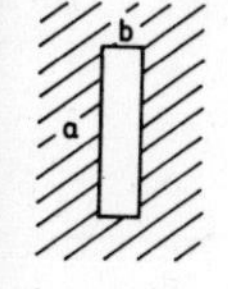	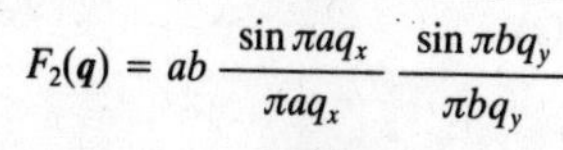$F_2(\boldsymbol{q}) = ab \dfrac{\sin \pi a q_x}{\pi a q_x} \dfrac{\sin \pi b q_y}{\pi b q_y}$ 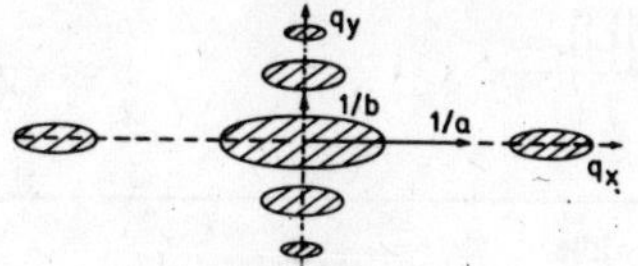
3) Three-dimensional parallelepiped $f_3(\boldsymbol{r})$ 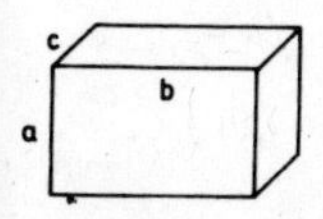	$F_3(\boldsymbol{q}) = abc \dfrac{\sin \pi a q_x}{\pi a q_x} \dfrac{\sin \pi b q_y}{\pi b q_y} \dfrac{\sin \pi c q_z}{\pi c q_z}$
4) Circular hole $f_4(\boldsymbol{r}) = \begin{cases} 1 & \text{if } \lvert \boldsymbol{r}\rvert < R \\ 0 & \text{if } \lvert \boldsymbol{r}\rvert > R \end{cases}$ 	$F_4(\boldsymbol{q}) = \pi R^2 \dfrac{J_1(2\pi q R)}{2\pi q R}$ (Airy-distribution) 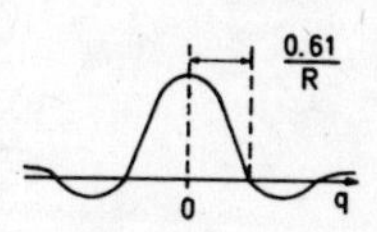

Table 3.3 (continued)

Specimen function $f(x)$	Fourier transform $\mathcal{T}\{f(x)\} = F(q)$
5) Sphere of radius R: $f_5(\mathbf{r})$	$F_5(\mathbf{q}) = \frac{4\pi}{3} R^3 \, 3 \, \frac{\sin u - u \cos u}{u^3}$; $u = 2\pi q R$
6) Gaussian function $f_6(x) = \exp[-(x/a)^2]$	$F_6(q) = \sqrt{\pi}\, a \exp[-(\pi q a)^2]$
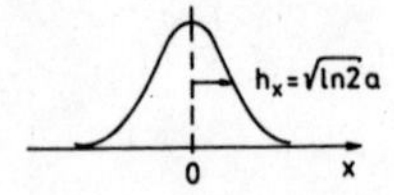	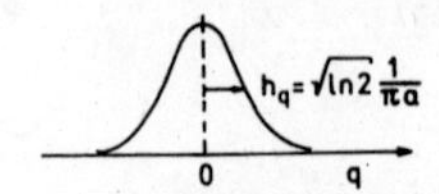
7) One-dimensional point lattice $f_7(x) = \sum_{n=1}^{N} \delta(x - x_n)$	$F_7(q) = \frac{\sin \pi q N d}{\sin \pi q d}$
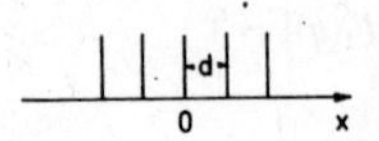	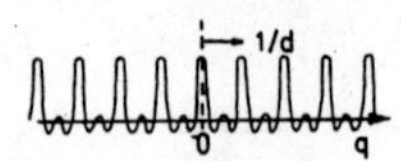
8) N slits of width a $f_8(x) = f_1(x) \otimes f_7(x)$	$F_8(q) = F_1(q) \cdot F_7(q) = a \frac{\sin \pi a q}{\pi a q} \frac{\sin \pi q N d}{\sin \pi q d}$
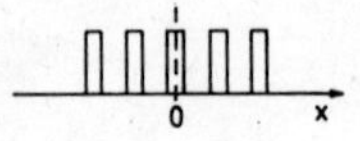	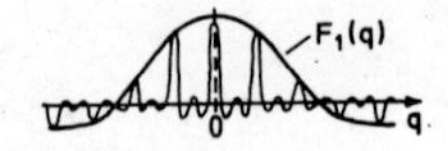
9) Infinite wave $f_9(x) = \cos(2\pi x/\Lambda)$ $= \frac{1}{2}(e^{2\pi i x/\Lambda} + e^{-2\pi i x/\Lambda})$	$F_9(q) = \frac{1}{2}\delta\left(q \pm \frac{1}{\Lambda}\right)$
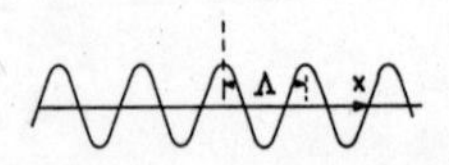	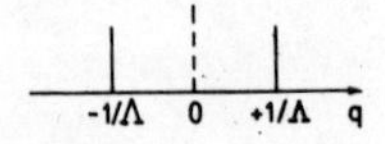
10) Wave packet of width a $f_{10}(x) = f_1(x) \cdot f_9(x)$	$F_{10}(q) = F_1(q) \otimes F_9(q)$
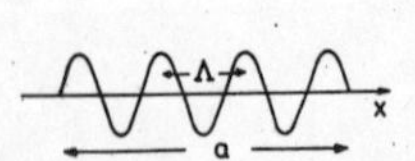	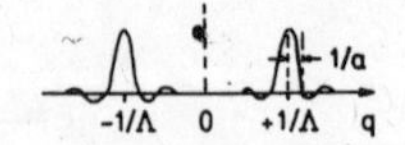

Table 3.3 (continued)

Specimen function $f(x)$	Fourier transform $\mathcal{T}\{f(x)\} = F(q)$
11) Infinite point row $N \rightarrow \infty$ $f_{11}(x) = \sum_{n=-\infty}^{+\infty} \delta(x - nd)$ 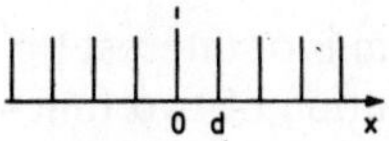	$F_{11}(q) = \sum_{n=-\infty}^{+\infty} \delta\left(q - \frac{n}{d}\right) = \sum_{n=-\infty}^{+\infty} \delta(q - q_n)$ 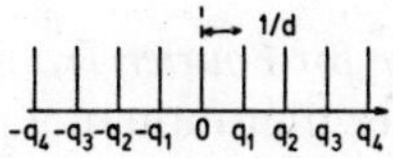
12) Infinite periodic function $f_{12}(x) = f_{11}(x) \otimes f_1(x)$ 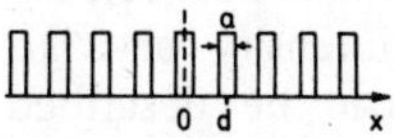	$F_{12}(q) = F_{11}(q) \cdot F_1(q) = \sum_{n=-\infty}^{+\infty} F_n \delta(q - q_n)$ $F_n = F_1(q_n)$ 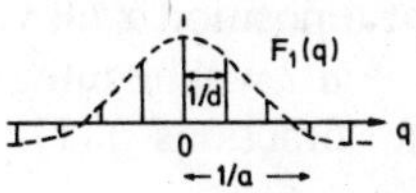
Fourier sum: $f(x) = \sum_{n=-\infty}^{+\infty} F_n \cos(2\pi q_n x) = \frac{1}{2} + \frac{2}{\pi}\left[\cos(2\pi q_1 x) - \frac{1}{3}\cos(2\pi q_3 x) + \frac{1}{5}\cos(2\pi q_5 x) + \ldots\right]$	
13) Three-dimensional point lattice $f_{13}(\boldsymbol{r}) = \sum_{l=-\infty}^{+\infty} \sum_{m} \sum_{n} (\boldsymbol{r} - l\boldsymbol{a}_1 + m\boldsymbol{a}_2 + n\boldsymbol{a}_3)$	$F_{13}(\boldsymbol{q}) = \sum_h \sum_k \sum_l \delta(\boldsymbol{q} - (h\boldsymbol{a}_1^* + k\boldsymbol{a}_2^* + l\boldsymbol{a}_3^*))$ $\boldsymbol{a}_i^* = \frac{\boldsymbol{a}_j \times \boldsymbol{a}_k}{\boldsymbol{a}_i \cdot (\boldsymbol{a}_j \times \boldsymbol{a}_k)}$
14) Limited lattice of extended objects e.g. $f_{14}(\boldsymbol{r}) = [f_{13}(\boldsymbol{r}) \otimes f_5(\boldsymbol{r})] \cdot f_3(\boldsymbol{r})$	$F_{14}(\boldsymbol{q}) = [F_{13}(\boldsymbol{q}) \cdot F_5(\boldsymbol{q})] \otimes F_3(\boldsymbol{q})$ Lattice-structure amplitude

If the width of the slit a tends to zero a δ-function results (point source in Example 1b). The width of the diffraction maximum in $F_1(q)$ then goes to infinity. This means that the diffraction amplitude $F(q)$ of a point source is isotropic in all directions q. The Fourier transform of a δ-function at the position b relative to the origin (Example 1c) is obtained by using the translation theorem (3.47).

Further examples are shown in Table 3.3: a rectangular slit (2), a parallelepiped (3), a circular diaphragm (4) and a sphere (5), illustrating Fourier transforms of functions in more than one dimension.

A Gaussian function (Example 6) is again a Gaussian function after a Fourier transform but with the reciprocal half-width.

The Fourier transform $F_7(q)$ of a one-dimensional point lattice with N points distance d apart shows principal maxima with spacing $1/d$ if the numerator and denominator of $F_7(q)$ simultaneously become zero. There are $N-1$ zeros between these principal maxima, where only the numerator is zero. The widths of the principal maxima decrease in proportion to $1/Nd$ and the amplitudes of the subsidiary maxima are further decreased. Example 7 becomes example 11 for $N \to \infty$.

Convolution Theorem for Fourier Transforms. This theorem is of interest for calculating the Fourier transform of a product or a convolution of two functions each of whose Fourier transforms is known. We first introduce the concept of convolution of two functions $f(x)$ and $g(x)$. Let us consider, for example, measurement of the intensity distribution $f(x)$ of a photographic emulsion with a densitometer. Let $g(x)$ be the slit function $f_1(x)$ introduced in Table 3.3 describing the transmission of the slit. The slit moves across the function $f(x)$ and at a position x all values of the function $f(x)$ between the limits $x - a/2$ and $x + a/2$ will be integrated. The resulting intensity curve is a convolution of the functions $f(x)$ and $g(x)$, which can be described mathematically by

$$C(x) = \int_{-\infty}^{+\infty} f(\xi)\, g(x-\xi)\, d\xi = f(x) \otimes g(x) . \tag{3.50}$$

If $F(q)$ and $G(q)$ are the Fourier transforms of $f(x)$ and $g(x)$, respectively, then the convolution theorem states that

$$\begin{aligned} &\text{a) } \mathcal{T}\{f \otimes g\} = F(q) \cdot G(q) \\ &\text{b) } \mathcal{T}\{f \cdot g\} = F(q) \otimes G(q) . \end{aligned} \tag{3.51}$$

The proof of (3.51 a) passes through the following stages: reversal of the order of integration, use of the translation theorem (3.47) and withdrawal of $G(q)$ from the integral because it no longer depends on ξ:

$$\begin{aligned} \mathcal{T}\{f(x) \otimes g(x)\} &= \int_{-\infty}^{+\infty} \left[\int_{-\infty}^{+\infty} f(\xi)\, g(x-\xi)\, d\xi \right] e^{-2\pi i q x}\, dx \\ &= \int_{-\infty}^{+\infty} \left[\int_{-\infty}^{+\infty} g(x-\xi)\, e^{-2\pi i q x}\, dx \right] f(\xi)\, d\xi \\ &= \int_{-\infty}^{+\infty} G(q)\, e^{-2\pi i q \xi} f(\xi)\, d\xi = F(q) \cdot G(q) . \end{aligned}$$

Some applications of this convolution theorem will now be discussed in detail. The diffraction grating consisting of N slits with spacing a (Example 8) can be described as the convolution of a discrete point function $f_7(x)$ that coincides with the centres of the slits, with the function $f_1(x)$ of a single slit. By use of (3.51 a), the Fourier transform $F_8(q)$ is equal to the product of the Fourier transforms $F_1(q)$ of the single slit and $F_7(q)$ of the lattice function. The slit function $F_1(q)$, therefore, acts as an envelope, modulating the amplitudes of the principal maxima.

Examples 9 and 10 contain an application of (3.51 b) to a wave of finite extent, as already discussed in Sect. 3.1.2 and illustrated in Fig. 3.1 b. The Fourier transform $F_9(q)$ of the infinite sine wave $f_9(x)$ has non-zero values only for $q = \pm 1/\Lambda$. A finite wave can be described by the product $f_{10}(x) = f_1(x) \cdot f_9(x)$, which is zero for $x < -a/2$ and $x > a/2$. The Fourier transform is obtained from (3.51 b) by a convolution of the Fourier transforms of the individual functions: $F_{10}(q) = F_9(q) \otimes F_1(q)$. This means that the δ-functions at the positions $q = \pm 1/\Lambda$ will be broadened by the function $F_1(q)$. If the sine wave does not decrease abruptly to zero at $x = \pm a/2$ but is multiplied by a Gaussian function $f_6(x)$, the example shown schematically in Fig. 3.1 b is obtained.

An infinite row of points (Example 11), with spacing d, has a Fourier transform consisting of an infinite number of δ-functions at the positions $q_n = n/d$ (n integer), whereas only a first-order maximum appears for the function $f_9(x)$. Each δ-function at q_n corresponds to a function $f(x) = \exp(2\pi i q_n x)$ (Example 1 c). An infinite periodic function $f_P(x)$ with the period d (Example 12) can be described by a convolution of the infinite point row (Example 11) with a function $f(x)$ defined in the interval $-d/2 < x < +d/2$. The Fourier transform of $f(x)$ is, therefore, the envelope of the maxima of $\delta(q - q_n)$. At the positions q_n, the Fourier amplitudes become

$$F_n = \frac{1}{d} \int_{-\infty}^{+\infty} f(x) \exp(-2\pi i q_n x)\, dx \,. \tag{3.52}$$

The inverse Fourier transform of $F_P(x) = \sum_{-\infty}^{+\infty} F_n \delta(q - q_n)$ gives the description of a periodic function in terms of a sum of sine and cosine terms (Fourier sum):

$$f_P(x) = \sum_{n=-\infty}^{+\infty} F_n \exp(-2\pi i q_n x) = \sum_{n=-\infty}^{+\infty} [a_n \cos(2\pi q_n x) + b_n \sin(2\pi q_n x)] \,, \tag{3.53}$$

where the coefficients a_n and b_n can be calculated from

$$a_n = \frac{1}{d} \int_{-d/2}^{+d/2} f(x) \cos\left(2\pi \frac{n}{d} x\right) dx \,; \qquad b_n = \frac{1}{d} \int_{-d/2}^{+d/2} f(x) \sin\left(2\pi \frac{n}{d} x\right) dx \,. \tag{3.54}$$

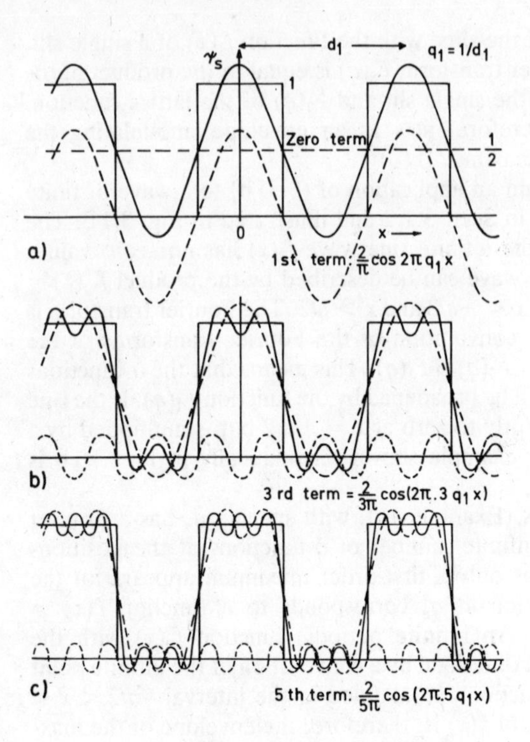

Fig. 3.13. Approximation of a step function by an increasing number of terms of a Fourier series

If this formula is applied to a periodic rectangular function, a diffraction grating of slit width $a/2$, for example, all of the b_n become zero because $f(x)$ is a symmetric function and

$$a_0 = \frac{1}{2}; \qquad a_n = \frac{1}{\pi n} \sin \frac{n\pi}{2},$$

$$\text{which means} \quad a_n = \begin{cases} 0 & \text{for } n \text{ even} \\ \dfrac{1}{\pi n}(-1)^{(n+1)/2} & \text{for } n \text{ odd.} \end{cases} \tag{3.55}$$

Figure 3.13 shows how the rectangular function can be approximated successively by an increasing number of sine functions of the Fourier sum (3.53) with the coefficients (3.54) up to the fifth order. The last curve in Fig. 3.13 can be observed experimentally as the intensity distribution of a grid in the image plane if all diffraction maxima with $n > 5$ are removed by a diaphragm in the focal plane of the objective. A pure sine wave will be observed when only the first maximum is transmitted by the diaphragm (first curve in Fig. 3.13). The image will then contain only the information that there is a periodicity d in the specimen but no information about the detailed form of the periodic function.

Fourier transforms of electron-microscope images can be obtained by means of an optical diffractometer (Sect. 6.4.6) or by digital computation. In the latter case, the Fourier transform or its inverse will be calculated using a limited number of discrete image points. This number should be kept as small as possible without loss of information. The necessary number can be deduced from the sampling theorem of Shannon. The density distribution $\sigma(x, y)$ of the specimen can be obtained from a digital scan over a square of side L, either by measuring a photographic record with a microdensitometer or by direct recording of the image intensity. The smallest spatial frequency present will be $q = 1/L$. All higher spatial frequencies are multiples of this frequency; spectral points with the coordinates $q_{xn} = n/L$ and $q_{ym} = m/L$ (m, n integer). The Fourier transform $S(q)$ of $\sigma(x, y)$ inside the square of area L^2 can be calculated from the sum

$$S(q) = \mathcal{T}\{\sigma(x, y)\} = \sum_{n=-N}^{+N} \sum_{m=-N}^{+N} F_{nm} \frac{\sin[\pi L(q_{xn} - q)]}{\pi L(q_{xn} - q)} \frac{\sin[\pi L(q_{ym} - q)]}{\pi L(q_{ym} - q)} \tag{3.56}$$

with the coefficients

$$F_{nm} = \frac{1}{N^2} \sum_{i=0}^{N-1} \sum_{j=0}^{N-1} \sigma(x_i, y_j) \exp[-2\pi i(q_{xn}x_i + q_{ym}y_j)] . \tag{3.57}$$

If δ denotes the resolution limit of the electron-microscope image then $q_{max} = 1/\delta = N/L$ will be the upper limit of the useful spectral points and of the number N^2 of sampling points. The area of the densitometer slit or of the electron detector should be of the order of δ^2 to ensure good averaging and to reduce the noise.

3.3. Wave-Optical Formulation of Imaging

3.3.1 Wave Aberration of an Electron Lens

The spherical aberration can be treated in wave optics in the following manner. An object point emits a spherical, scattered wave with concentric wavefronts of equal phase (Fig. 2.13). An ideal lens would introduce the phase shifts necessary to create a spherical wave beyond the lens, converging onto the image point. The rays of geometric optics are trajectories orthogonal to the wavefronts, and the wave amplitudes scattered into different angles θ of the cone with the aperture α are summed in the image point with equal phase. A radial decrease is observed in the intensity distribution of the

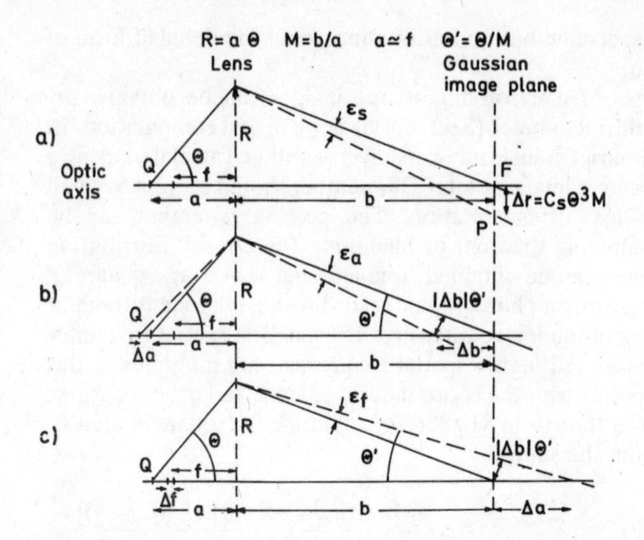

Fig. 3.14 a–c. Ray diagram for evaluating the angular deviations ε caused by (**a**) spherical aberration, (**b**) change Δa of specimen position and (**c**) change Δf of focal length

blurred image point only because of the finite aperture (Airy disc, see Fig. 3.17b). The spherical aberration reduces the focal length for rays at larger scattering angles θ. Because the rays and wavefronts are orthogonal, the wavefronts beyond a lens with spherical aberration are more strongly curved in the outer zones of the lens; there is a difference Δ of optical path relative to the spherical wavefronts of an ideal lens (Fig. 2.13). The wave amplitudes are therefore not all in phase at the Gaussian image point. The smallest diameter of the intensity distribution, similar to an Airy disc, will be observed in front of the Gaussian image plane (Fig. 3.17c). A phase shift by defocusing of the lens has also to be considered; this can be generated either by a displacement Δa of the specimen or by a change Δf of focal length.

Figure 3.14 will be used to calculate the dependence of the phase shift $W(\theta) = 2\pi\Delta/\lambda$, which is known as the wave aberration, on the scattering angle θ. First, however, a comment on its sign should be added, because different conventions are found in the literature. By using $\exp(2\pi ikz)$ instead of $\exp(-2\pi ikz)$ for a plane wave in Sect. 3.1.1, the convention is made that the phase increases with increasing z, that is, the direction of wave propagation. Because the optical path length along a trajectory in Fig. 2.13 is decreased by Δ, the phase shift $W(\theta)$ is also decreased. This phase shift has to be represented, therefore, by a phase factor $\exp[-iW(\theta)]$ with a negative sign in the exponent (see also comments [3.38, 39]).

A ray that leaves the specimen point Q at a scattering angle θ reaches the lens at a distance $R \simeq a\theta$ ($\theta \simeq$ a few 10 mrad) from the optic axis (Fig. 3.14a). The ray intersects the optic axis in the Gaussian image plane at F, if there is no spherical aberration, and at P, a distance $\Delta r = C_s\theta^3 M$ from F in this plane, if the spherical aberration does not vanish, see (2.49 a). This causes a small angular deviation

$$\varepsilon_s \simeq \Delta r/b = C_s\theta^3 M/b \ . \tag{3.58}$$

By use of the relations $\theta = R/a$, $M = b/a$ and $a \simeq f$, (3.58) becomes

$$\varepsilon_s = C_s R^3/f^4 \ . \tag{3.59}$$

We now assume that there is no spherical aberration and that the specimen distance a is increased by Δa (Fig. 3.14 b). The focal length f of the lens is unchanged. The variation Δb of the image distance can be calculated from the well-known lens equation

$$\frac{1}{f} = \frac{1}{a + \Delta a} + \frac{1}{b + \Delta b} = \frac{1}{a}\left(1 - \frac{\Delta a}{a} + \dots\right) + \frac{1}{b}\left(1 - \frac{\Delta b}{b} + \dots\right) . \tag{3.60}$$

Solving for Δb and using $1/f = 1/a + 1/b$, we obtain

$$\Delta b = -\Delta a \, b^2/a^2 \ . \tag{3.61}$$

The corresponding angular deviation is obtained from

$$\varepsilon_a = |\Delta b|\theta'/b = \Delta a \, R/f^2 \tag{3.62}$$

by use of $\theta' \simeq R/b$.

A third case (Fig. 3.14 c), where the focal length is changed to $f + \Delta f$, can be treated in a similar way. The lens equation $1/(f + \Delta f) = 1/a + 1/(b + \Delta b)$ gives $\Delta b = \Delta f \, b^2/f^2$ and so

$$\varepsilon_f = -|\Delta b|\theta'/b = -\Delta f \, R/f^2 \ . \tag{3.63}$$

Adding the three angular deviations of the geometric optical trajectories, we obtain the total angular deviation

$$\varepsilon = \varepsilon_s + \varepsilon_a + \varepsilon_f = C_s(R^3/f^4) - (\Delta f - \Delta a) \, R/f^2 \ . \tag{3.64}$$

Figure 3.15 shows an enlargement of part of the lens between two trajectories, and their orthogonal wavefronts, which reach the lens at distances R and $R + dR$ from the optical axis. The angular deviation causes an optical-path difference $ds = \varepsilon dR$ between the two trajectories. These path differences ds have to be summed (integrated) to get the total path difference Δ or the phase shift $W(\theta)$ relative to the optical axis:

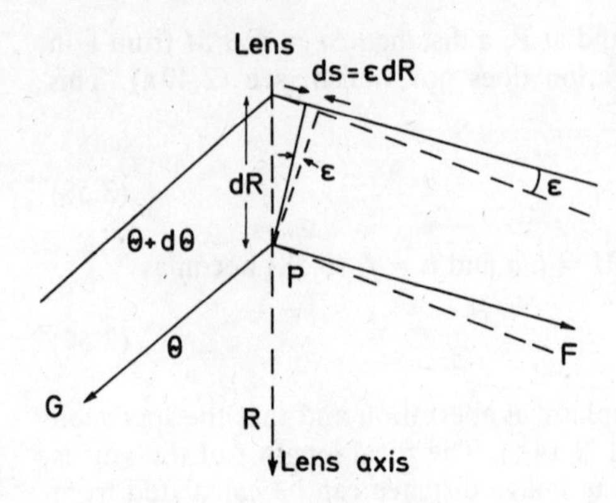

Fig. 3.15. Part of the outer zone of a lens at a distance R from the optic axis, showing the relation between the angular deviation ε and the optical path difference $ds = \varepsilon dR$

$$W(\theta) = \frac{2\pi}{\lambda}\Delta = \frac{2\pi}{\lambda}\int_0^R ds = \frac{2\pi}{\lambda}\int_0^R \varepsilon\, dR = \frac{2\pi}{\lambda}\left[\frac{1}{4}C_s\frac{R^4}{f^4} - \frac{1}{2}(\Delta f - \Delta a)\frac{R^2}{f^2}\right] \tag{3.65}$$

With $R/f \simeq \theta$ and the defocusing $\Delta z = \Delta f - \Delta a$, the so-called Scherzer formula [3.40] is obtained

$$W(\theta) = \frac{\pi}{2\lambda}(C_s\theta^4 - 2\Delta z\theta^2)\,, \tag{3.66}$$

or by introducing the spatial frequency $q = \theta/\lambda$ (3.42)

$$W(q) = \frac{\pi}{2}(C_s\lambda^3 q^4 - 2\Delta z\lambda q^2)\,. \tag{3.67}$$

In more accurate calculations, the change of the positions of the principal planes of the lens and the variation of C_s when the lens excitation is changed also have to be considered [2.25]. Axial astigmatism (Sect. 2.3.3) can be included in (3.66) by introducing an additional term

$$W_A = \frac{\pi}{2\lambda}\Delta f_A \sin[2(\chi - \chi_0)]\;; \tag{3.68}$$

which depend on an azimuthal angle χ.

The relation (3.66) is important for the discussion of phase contrast and for the study, in wave-optical terms, of the formation of a small electron probe for scanning transmission electron microscopy. Because the wave aberration depends on the two parameters, C_s and λ, it is convenient to discuss the wave aberration in terms of reduced coordinates [3.41, 42]

$$\theta^* = \theta(C_s/\lambda)^{1/4} \quad \text{and} \quad \Delta z^* = \Delta z(C_s\lambda)^{-1/2}\,. \tag{3.69}$$

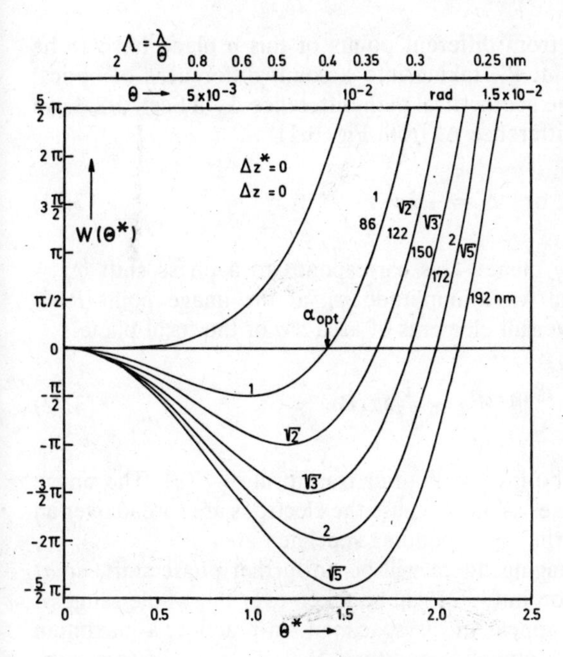

Fig. 3.16. Wave aberration $W(\theta^*)$ as a function of the reduced scattering angle θ^* for various reduced defocusing distances Δz^*. The upper scale shows the values of $\Lambda = \lambda/\theta$ for the special case $E = 100$ keV and $C_s = 2$ mm

This results in the reduced wave aberration

$$\frac{W(\theta^*)}{2\pi} = \frac{\theta^{*4}}{4} - \frac{\theta^{*2}}{2}\Delta z^* \,. \tag{3.70}$$

Figure 3.16 shows $W(\theta^*)$ for different values of reduced defocus $\Delta z^* = \sqrt{n}$ (n an integer), for which the minima of $W(\theta^*)$ are $-n\pi/2$.

Defocusing with positive Δz is called underfocusing and with negative Δz, overfocusing. Scattered electron waves are shifted with a phase $\pi/2$ relative to the unscattered wave (Sect. 6.2.1). For this reason, a reduced defocusing of $\Delta z^* = 1$ (Scherzer focus) is advantageous for phase contrast because $W(\theta)$ has the value $-\pi/2$ over a relatively broad range of scattering angles or spatial frequencies in the vicinity of the minimum for $\Delta z^* = 1$ in Fig. 3.16.

3.3.2 Wave-Optical Theory of Imaging

The rays from an object point P are reunited by the lens at the image point P′ (Fig. 3.11), a distance $x' = -Mx$ from the optic axis, where x is the off-axis distance of P ($M = b/a$: magnification). Rays with equal scattering angles from different points of the specimen intersect in the focal plane of the lens. It has already been shown in Sect. 3.2.3 that the wave amplitude $F(\boldsymbol{q})$ in this plane is obtained from the amplitude distribution $\psi_s(\boldsymbol{r})$ behind the specimen by a Fourier transform.

The wave amplitudes from different points of this $\boldsymbol{q}$ plane have to be summed at the image point P′, taking into account differences of optical path. Just as we defined the geometrical path difference Δ_g in Sect. 3.2.3, we now formulate the path difference Δ_g' from Fig. 3.11

$$\Delta_g' = +\lambda q x = -\Delta_g \quad \text{or} \quad \Delta_g' = +\lambda \boldsymbol{q} \cdot \boldsymbol{r} \tag{3.71}$$

for the two-dimensional $\boldsymbol{q}$ plane. This corresponds to a phase shift $\varphi_g' = 2\pi \boldsymbol{q} \cdot \boldsymbol{r}$. As in (3.43), the wave amplitude ψ_m at the image point P' is obtained by integrating over all elements of area $d^2\boldsymbol{q}$ of the focal plane

$$\psi_m(\boldsymbol{r}') = \frac{1}{M} \iint F(\boldsymbol{q})\, e^{+2\pi i \boldsymbol{q}\cdot\boldsymbol{r}}\, d^2\boldsymbol{q} = \frac{1}{M} \psi_s(\boldsymbol{r}) \,. \tag{3.72}$$

Thus, ψ_m is obtained as the inverse Fourier transform of $F(\boldsymbol{q})$. The image intensity $I = \psi_m \psi_m^*$ decreases as M^{-2} because the electrons are spread over an area M^2 times as large as the corresponding specimen area.

For aberration-free imaging, there will be no further phase shift, apart from φ_g', and the integration in (3.72) will be taken over the whole range of spatial frequencies $\boldsymbol{q}$ that appear in the specimen. In practice, a maximum scattering angle $\theta_{max} = \alpha_o$ (objective aperture) that corresponds to a maximum spatial frequency q_{max} is used. This limitation on spatial frequencies q by an objective diaphragm can be expressed in terms of a multiplicative masking function $M(q)$ which would have the values $M(q) = 1$ for $|\boldsymbol{q}| < q_{max}$ and $M(q) = 0$ for $|\boldsymbol{q}| > q_{max}$ in the normal bright-field mode. Because the wave aberration $W(q)$ in (3.67) due to spherical aberration and defocusing depends only on q, the action of this contribution can be represented by a multiplication of the amplitudes at the focal plane by the phase factor $\exp[-iW(q)]$. Equation (3.72) has therefore to be modified to

$$\psi_m(\boldsymbol{r}') = \frac{1}{M} \iint F(q) \underbrace{\{e^{-iW(q)}\, M(q)\}}_{H(q)}\, e^{2\pi i \boldsymbol{q}\cdot\boldsymbol{r}}\, d^2\boldsymbol{q} \,. \tag{3.73}$$

$H(q)$ is known as the pupil function. The convolution theorem (3.51 b) can be applied to (3.73)

$$\psi_m(\boldsymbol{r}') = \frac{1}{M} \psi_s(\boldsymbol{r}) \otimes h(\boldsymbol{r}) = \frac{1}{M} \iint \psi_s(\boldsymbol{r}_1)\, h(\boldsymbol{r}-\boldsymbol{r}_1)\, d^2\boldsymbol{r}_1 \,, \tag{3.74}$$

where $h(\boldsymbol{r}) = \mathcal{F}^{-1}\{H(\boldsymbol{q})\}$ is the inverse Fourier transform of the pupil function $H(\boldsymbol{q})$. This means that sharp image points will not be obtained. Instead, each image point will be blurred (convoluted) with the point-spread function $h(\boldsymbol{r})$. The image of a point source that scatters in all scattering angles with

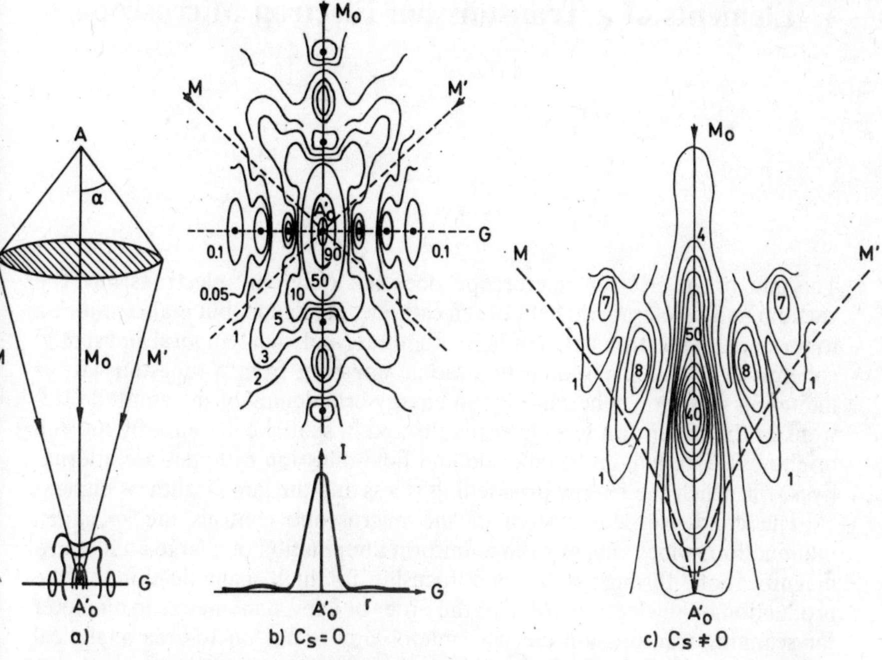

Fig. 3.17. (**a**) The limiting rays M and M′, which produce an image of the point A at the image point A_0' in the Gaussian-image plane G. (**b**) and (**c**) show the enlarged intensity distribution near A_0'; Lines of equal density for an aberration-free lens (b) and for a lens with spherical aberration (c). The curve at the bottom of (b) represents the cross section $I(r)$ through the Gaussian focus (*Airy disc*)

equal amplitudes $[F(q) = \text{const and } \psi_s(r) = \delta(0)]$ would be the function $h(r)/M$.

We consider now the case $W(q) = 0$ (no spherical aberration or defocusing). The image amplitude of a point source can be calculated as the Fourier transform of $M(q)$ (see Example 4 in Table 3.3)

$$\psi_m(r') \propto h(r) \propto \frac{J_1(x)}{x} \quad \text{with} \quad x = \frac{2\pi}{\lambda}\alpha_o r \,. \tag{3.75}$$

This amplitude distribution, which corresponds to the intensity distribution $I(x) \propto [J_1(x)/x]^2$, of the blurring function of a point source with an aberration-free but aperture-limited objective lens is called the Airy distribution (Fig. 3.17 b). The intensity distribution for $\Delta z \neq 0$ is symmetrical in Δz, and is plotted in Fig. 3.17 b as lines of equal intensitity. The same situation is shown in Fig. 3.17 c with spherical aberration. The distribution is asymmetric in Δz; the smallest error disc occurs at underfocus, in agreement with the geometrical-optical construction of Fig. 2.13. The dotted line is the caustic of geometrical optics. Further discussion of the wave-optical imaging theory will be found in Sect. 6.2.

4. Elements of a Transmission Electron Microscope

The gun of an electron microscope does not only emit electrons into the vacuum and accelerate them between cathode and anode, but is also required to produce an electron beam of high brightness and high temporal and spatial coherence. The conventional thermionic emission from a tungsten wire is limited in temporal coherence by an energy broadening of the emitted electrons of the order of a few electronvolts and in spatial coherence by the gun brightness. Lanthanum hexaboride and field-emission cathodes are alternatives, for which the energy broadening is less and the gun brightness higher.

The condenser-lens system of the microscope controls the specimen illumination, which ranges from a uniform illumination of a large area at low magnification, through a stronger focusing for high magnification, to the production of an electron probe of the order of a few nanometres in diameter for scanning transmission electron microscopy or for micro-area analytical methods.

The useful specimen thickness depends on the operation mode used and the information desired. Specimen manipulation methods are of increasing interest but are restricted by the size of the specimen and by the free space inside the pole-piece system of the objective lens.

The different imaging modes of a TEM can be described by ray diagrams, as in light optics, which can also be used to evaluate the depth of focus or to establish a theorem of reciprocity between conventional and scanning transmission electron microscopy.

Observation of the image on a fluorescent screen and image recording on photographic emulsions can be replaced by techniques that allow electron currents to be measured quantitatively.

4.1 Electron Guns

4.1.1 Physics of Electron Emission

The conduction electrons in metals or compounds have to overcome the work function ϕ if they are to be emitted from the cathode into the vacuum. Figure 4.1 shows the dependence of potential energy on a coordinate z normal to the surface. The Coulomb energy $V(z)$ of an electron in front of a

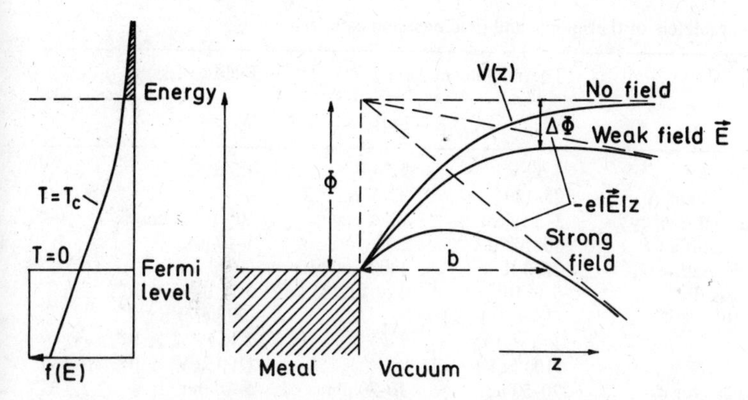

Fig. 4.1. Fermi distribution $f(E)$ and potential energy of electrons at the metal-vacuum boundary

conducting surface can be calculated by considering the effect of a mirror charge with opposite sign behind the surface; with an external electric field $\boldsymbol{E}$, the potential energy $V = -e|\boldsymbol{E}|z$ is superposed on that of the mirror charge, giving

$$V(z) = -\frac{e^2}{16\pi\varepsilon_0}\frac{1}{z} - e|\boldsymbol{E}|z \, . \tag{4.1}$$

As a result, the work function ϕ is decreased by $\Delta\phi$ (Schottky effect):

$$\phi_{\text{eff}} = \phi - \Delta\phi = \phi - e\sqrt{\frac{e|\boldsymbol{E}|}{4\pi\varepsilon_0}} \, . \tag{4.2}$$

This decrease can, however, be neglected in normal thermionic cathodes.

Increasing the cathode temperature leads to a broadening of the Fermi distribution $f(E)$ at the Fermi level (Fig. 4.1), and for high temperatures, electrons in the tail of the Fermi distribution acquire enough kinetic energy to overcome the work function ϕ. The current density j_c [A cm^{-2}] of the cathode emission can be estimated from Richardson's law

$$j_c = AT_C^2 \exp(-\phi/kT) \, , \tag{4.3}$$

where $k = 1.38 \times 10^{-23}$ J K^{-1} is Boltzmann's constant, T_C is the cathode temperature, $A \simeq 120$ A K^{-2} cm^{-2} is a constant that depends on the cathode material. The Schottky effect causes an observable increase of the current emitted only for high field strengths $|\boldsymbol{E}| > 10^5$ V cm^{-1}.

Most metals melt before they reach a sufficiently high temperature for thermionic emission. An exception is tungsten, which is widely used at a

Table 4.1. Parameters of thermionic and field emission cathodes

	Thermionic emission		Field emission
	W	LaB_6	W
Work function ϕ	4.5 eV	2.7 eV	4.5 eV
Richardson constant A	75–120	30 A cm^{-2} K^{-2}	–
Emission current density j_c	1–3 A cm^{-2}	25 A cm^{-2}	10^4–10^6 A cm^{-2}
Total current emitted I	10–100 μA		1–10 μA
Working temperature T_C	2800 K	1400–2000 K	(1000 K)
Gun brightness β [A cm^{-2} sr^{-1}]	5×10^4 (E = 10 keV) 1–5×10^5 (100 keV)	3×10^5 (15 keV)	5×10^7–5×10^8 (20 keV) 2×10^8–2×10^9 (100 keV)
Crossover diameter d_c	20–50 μm (hairpin) 10–30 μm (pointed cath.)	10–20 μm	5–10 nm
Energy width ΔE	1–2 eV	0.5–2 eV	0.2–0.4 eV
Lifetime	25 h	150–200 h	–
Vacuum	10^{-2}–10^{-3} Pa	10^{-3}–10^{-4} Pa	10^{-7}–10^{-8} Pa
	(1 Pa = 1 N m^{-2} = 10^{-2} mbar)		

working temperature T_C of 2500–3000 K (melting point T_m = 3650 K); lanthanum hexaboride (LaB_6) cathodes (or other borides) with T_C = 1400–2000 K are also employed because their work function is lower (Table 4.1). The tungsten metal evaporates continuously during operation, limiting the lifetime t of the filament. This can be seen from the following values [4.1]:

T_C [K]	2500	2600	2700	2800	2900
t [h]	200	90	35	12	5

The width b of the potential barrier at the metal-vacuum boundary decreases with increasing electric field $\boldsymbol{E}$; for $|\boldsymbol{E}| > 10^7$ V cm^{-1} the width b becomes less than 10 nm (Fig. 4.1) and electrons at the Fermi level can penetrate the potential barrier by the wave-mechanical tunnelling effect. High values of the field $|\boldsymbol{E}| \sim U_1/r$ can be achieved by making the radius r of the tungsten tip small, $r \simeq 0.1$–1 μm (U_1 denotes the voltage applied between tip and the first anode). The current density of field emission can be estimated from the Fowler-Nordheim formula (see also [4.2])

$$j = \frac{k_1 |\boldsymbol{E}|^2}{\phi} \exp\left(-\frac{k_2 \phi^{3/2}}{|\boldsymbol{E}|}\right). \tag{4.4}$$

The constants k_1 and k_2 depend only weakly on $|\boldsymbol{E}|$. Whereas current densities j of only 1–2 A cm^{-2} can be obtained with thermionic emission, values of 10^4–10^6 A cm^{-2} may be reached with field emission. However, the emitting area is so small that the total current is not greater. For field emission the current is of the order of 1–10 μA, compared to 5–50 μA for thermionic cathodes.

4.1.2 Energy Spread and Gun Brightness

The tail of the Fermi distribution $f(E)$ (shaded area in Fig. 4.1) results in a Maxwell-Boltzmann distribution of the exit momenta p or energies $E = p^2/2\,m$

$$f(p) \propto \exp(-E/kT_C)\ . \tag{4.5}$$

Electrons are emitted in all directions within the half-space; the electron motion is characterized by the tangential (t) and normal (n) components of $\boldsymbol{p}$, so that (4.5) has to be multiplied by the volume element (density of states) $2\pi p^2 dp$ of the momentum space to get the number of electrons with momenta between p and $p + dp$ or energies between $E = (p_t^2 + p_n^2)/2\,m$ and $E + dE$. This yields the normalized total energy distribution (Fig. 4.2)

$$N(E)\,dE = \frac{E}{(kT_C)^2}\exp(-E/kT_C)\,dE\ , \tag{4.6}$$

with a most probable energy $E_p = kT_C$,
a mean energy $\langle E \rangle = 2kT_C$
and a half-width $\Delta E = 2.45\ kT_C$.

Thus, for a cathode temperature T_C of 2500 K, the half-width ΔE will be 0.5 eV. This energy spread is superposed on the accelerating energy $E = eU$. This theoretical value will occur only when the cathode is operated in the saturation mode with low current density. In the normal-operation mode with a triode gun (Sect. 4.1.3), an anomalous energy spread is observed (Boersch effect [4.3]), with the result that $\Delta E \simeq$ 1–2 eV. This can be explained by Coulomb interaction of the electrons in the crossover [4.4–6].

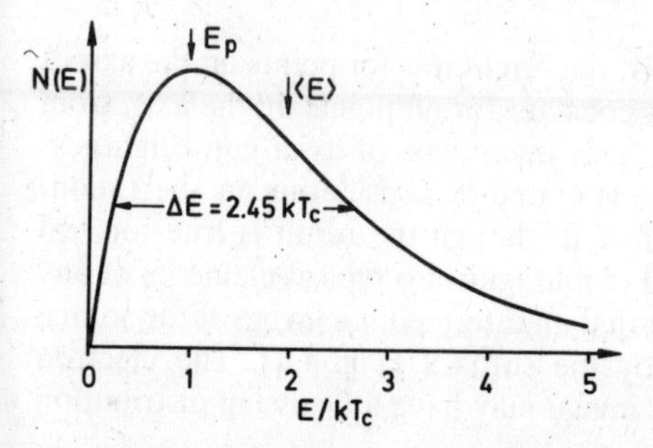

Fig. 4.2. Maxwellian distribution of electron energies emitted from a thermionic cathode ($\langle E \rangle$: mean energy, E_p: most-probable energy, ΔE: energy width)

The energy spread of field-emission guns is of the order of $\Delta E = 0.2$–0.4 eV [4.7].

The components p_t of the initial exit momenta tangential to the exit surface result in an angular spread of the electron beam and limit the value of the *gun brightness* β. This quantity is defined as the current density $j = \Delta I/\Delta S$ per solid angle $\Delta\Omega = \pi\alpha^2$ where α denotes the half-aperture of the cone of electrons that pass through the surface element ΔS

$$\beta = \frac{\Delta I}{\Delta S \Delta \Omega} = \frac{j}{\pi \alpha^2} \ . \tag{4.7}$$

The maximum possible value β_{max} for a thermionic cathode can be estimated from the following simplified model (Fig. 4.3) (see [4.8–10] for details). The components p_t and p_n are each described by a Maxwell-Boltzmann distribution (4.5), with mean square values

$$\langle p_t^2 \rangle = \langle p_n^2 \rangle = 2\, m_0 k T_C \ . \tag{4.8}$$

The electron acceleration contributes an additional kinetic energy $E = eU$ so that, in all, using (2.11), we find

$$\langle p_n^2 \rangle = 2\, m_0 k T_C + 2\, m_0 E (1 + E/2\,E_0) \ . \tag{4.9}$$

The angular aperture α of a virtual electron source behind the cathode surface can be obtained from the vector sum of $\boldsymbol{p}_t$ and $\boldsymbol{p}_n$ (Fig. 4.3): $\alpha = p_t/p_n$ or $\langle \alpha^2 \rangle = \langle p_t^2 \rangle / \langle p_n^2 \rangle$. Substituting (4.8, 9) in (4.7) gives

$$\beta_{max} = \frac{j_c}{\pi} \left[1 + \frac{E}{kT_C} (1 + E/2\,E_0) \right] \ . \tag{4.10}$$

This formula is valid even for non-uniform fields in front of the cathode.

Numerical values of the gun brightness are listed in Table 4.1. The maximum value β_{max} can be attained with thermionic cathodes by using optimum operation conditions (Sect. 4.1.3). Otherwise lower values are attained from 0.1 to 0.5 β_{max}.

The axial gun brightness β, that is, the brightness for points on the axis of an electron-optical column, remains constant for all points on the axis, from the cathode tip to the final image. This invariance of axial gun brightness along the optic axis will now be demonstrated by considering an aberration-free lens with a diaphragm in front of it, though the result is true for real lenses with aberrations. Lenses and diaphragms are typical elements of any electron-optical system. We assume that an intermediate image of the source is formed in the plane indicated by the suffix 1 (Fig. 4.4). The electron current density in this intermediate image may have a Gaussian distribution

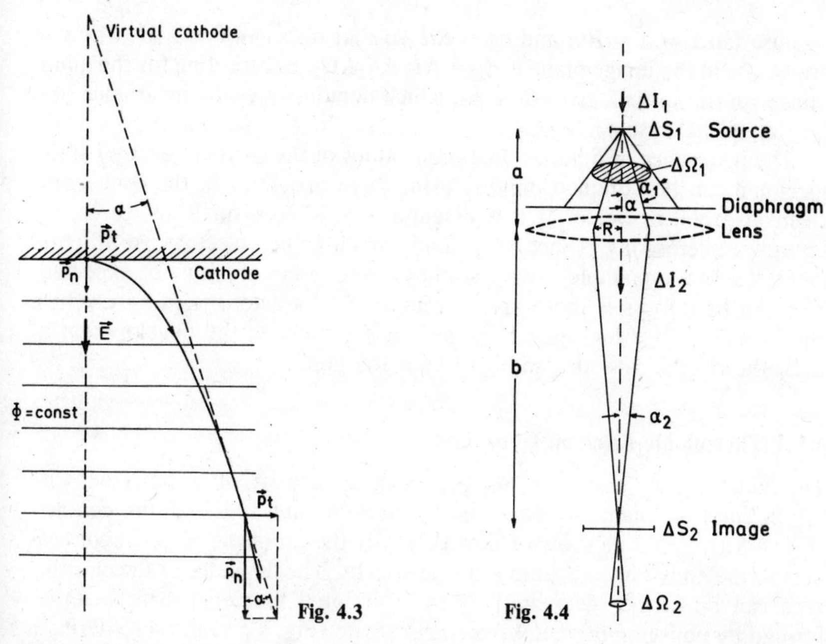

Fig. 4.3 Fig. 4.4

Fig. 4.3. Evaluation of the angular spread α of electrons emitted; with a transverse exit momentum p_t and a uniform electric field in front of the cathode, the trajectories are parabolic

Fig. 4.4. Demonstration of the conservation of gun brightness on the axis of an electron optical system in the presence of apertures and lenses

(4.13). We consider only the centre of this distribution because we are interested only in the axial brightness. A fraction ΔI_1 of the total current passes the area ΔS_1 with an angular aperature α_1.corresponding to a solid angle $\Delta\Omega_1 = \pi\alpha_1^2$. The gun brightness in this plane is

$$\beta_1 = \frac{\Delta I_1}{\Delta S_1 \Delta\Omega_1} = \frac{\Delta I_1}{\Delta S_1 \pi\alpha_1^2} . \tag{4.11}$$

The diaphragm in front of the lens cuts off a fraction of the current ΔI_1 and only a fraction

$$\Delta I_2 = \Delta I_1 \frac{\pi\alpha^2}{\pi\alpha_1^2} \tag{4.12}$$

will pass through the diaphragm. This current is concentrated in an image area $\Delta S_2 = \Delta S_1 M^2$ where $M = b/a$ is the magnification, which can be smaller than unity if the lens is demagnifying. The aperture is decreased to $\alpha_2 = \alpha/M$

because $\tan\alpha \simeq \alpha = R/a$ and $\alpha_2 \simeq R/b$ so that $\alpha_2/\alpha = a/b = 1/M$. The gun brightness in the image plane is $\beta_2 = \Delta I_2/\Delta S_2 \Delta\Omega_2$. Substituting for the quantities with the suffix 2 gives $\beta_1 = \beta_2$, which demonstrates the invariance of β for this special case.

The invariance of β means that high values of the current density j at the specimen can be obtained only by using large apertures of the convergent electron probe or beam. If it is essential to use very small apertures, for Lorentz microscopy (Sect. 6.5) and small-angle electron diffraction (Sect. 9.3.4) for example, correspondingly low values of j must be expected. The gun brightness is therefore an important characteristic of an electron gun. The need for high gun brightness has stimulated the development of LaB_6 thermionic cathodes and field-emission guns.

4.1.3 Thermionic Electron Guns

The most widely used thermionic cathodes consist of a tungsten wire 0.1–0.2 mm in diameter bent like a hairpin and soldered on contacts (Fig. 4.5 a). It will be shown below that only the tip of the filament contributes to the emission current in a triode system. The diameter of the emitting area can be further decreased by using pointed filaments, which can be formed by polishing the wire to a lancet shape (Fig. 4.5 b) or by soldering a straight wire on a hairpin cathode and electropolishing the tip to a point with a small radius of curvature (Fig. 4.5 c). Both types of pointed filaments are sensitive to high-voltage breakdown and positive-ion bombardment, which can blunt the tip. A good vacuum and slow increase of high tension are necessary when working with cathodes of this type [4.11].

LaB_6 cathodes consist of small pointed crystals [4.12–14]. They require indirect heating because their electrical resistance is too high for direct current heating. The heating power can be decreased by supporting a small crystal between carbon rods or fibres (Fig. 4.5 d) [4.15, 16] or binding them

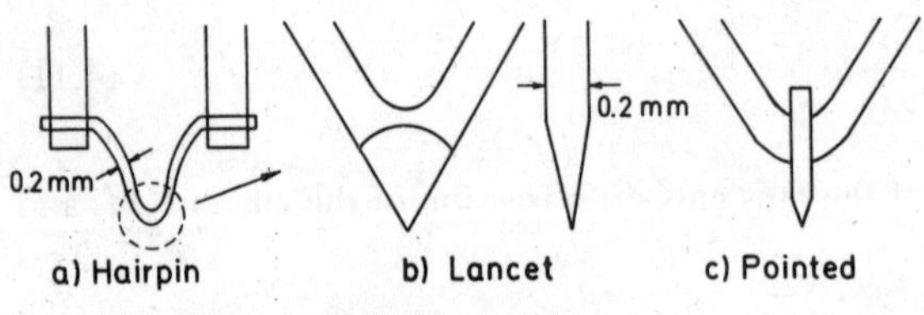

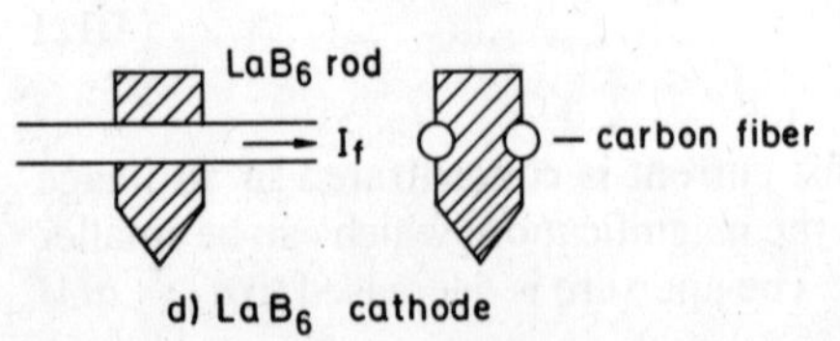

Fig. 4.5 a–d. Types of cathodes: **(a)** tungsten-hairpin cathode, **(b)** polished, lancet cathode, **(c)** electrolytically etched pointed filament, **(d)** LaB_6 cathode

to refractory metals (rhenium or tantalum) that have a low rate of reaction with LaB_6. Different types of LaB_6 mounting techniques are compared by *Crawford* [4.17]. These LaB_6 cathodes can be used instead of tungsten hairpin cathodes but they do however need a better vacuum than the latter, to reduce the damage caused by positive-ion bombardment. Interest in these cathodes is increasing owing to their higher gun brightness and the lower value of their energy spread ΔE (Table 4.1). The emission current is greatest for (110) oriented tips, ten times higher than for the (510) orientation [4.18].

An electron gun consists of three electrodes (triode system):

1) the heated filament, which forms the cathode, at the potential $\phi_C = -U$,
2) the Wehnelt electrode, at a potential ϕ_W some hundreds of volts more negative than the cathode and
3) the earthed anode ($\phi_A = 0$).

Figure 2.1 shows the equipotentials ϕ = const in a cross-section through a triode gun and Fig. 4.6 those near the cathode tip. In Fig. 4.6a, the negative bias of the Wehnelt electrode is not great enough to decrease $|E|$ at the cathode surface. The zero equipotential intersects the tip around a circle. All of the electrons emitted from a large cathode area (non-shaded) are accelerated. Beyond the circle, the electric-field strength is of opposite sign, and no electrons can leave the shaded area. In Fig. 4.6b, the negative bias is further increased and the area of emission is thus reduced. In Fig. 4.6c, the zero

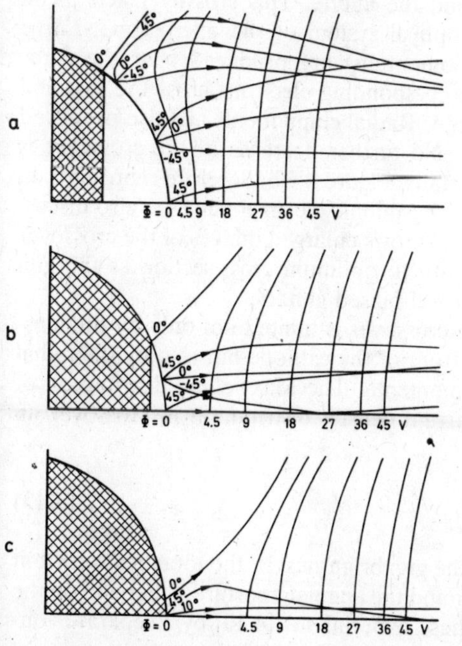

Fig. 4.6a–c. Equipotentials ϕ = const in front of the cathode tip for (from **a** to **c**) increasing negative bias $-U_W$ of the Wehnelt electrode; electron trajectories are shown with an exit energy of 0.3 eV and various angles of emission [4.19]

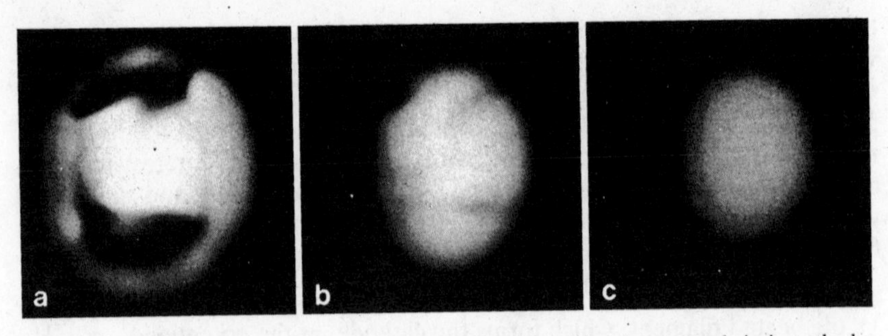

Fig. 4.7 a–c. Enlarged images of the crossover of an auto-biased tungsten-hairpin cathode. From (**a** to **c**) the heating current of the gun filament is increased

equipotential reaches the tip of the cathode. No electrons will leave the cathode if the negative Wehnelt bias is increased further. Figure 4.8 shows this cut-off voltage as a steep decrease of the total beam current I_c to zero for different heights h (Fig. 2.1) of the cathode tip.

Figure 4.6 also shows some electron trajectories, with an initial exit energy of 0.3 eV and different angles of emission. In Fig. 4.6 b, the electrons enter a more or less uniform electric field in the vicinity of the cathode, which exerts small additive radial force components on the electron trajectories; the cross-section of the electron beam passes through a minimum, known as a *crossover*, between the cathode and the anode. This crossover acts as an electron source for the electron optical system of the microscope. Large radial components of velocity (momentum) are produced near the zero-equipotentials in Fig. 4.6 a. The corresponding electrons cross the axis and result in a *hollow-beam* cross-section. Radial components are also produced in Fig. 4.6 c near the cut-off bias. No further decrease of the crossover is observed, but the emission current falls. Figure 4.6 b therefore represents an optimum condition for operating a thermionic electron gun (see also discussion of brightness below). Figure 4.7 shows enlarged images of the crossover and the transition from *hollow-beam* to optimum cross-section as the gun filament current is increased in an self-biased gun.

The minimum diameter of the crossover is limited not only by the lens-like action of the electric field in front of the cathode but also by the radial components of the electron exit momenta. The Maxwellian distribution of exit velocities gives the radial current-density distribution in crossover an approximately Gaussian shape:

$$j(r) = j_0 \exp[-(r/r_0)^2] . \tag{4.13}$$

Figure 4.8 demonstrates how the gun brightness β, the total beam current I_c, the diameter d_c of the crossover and the angular aperture α_c depend on the Wehnelt bias U_W (generated in these experiments [4.19] by a separate volt-

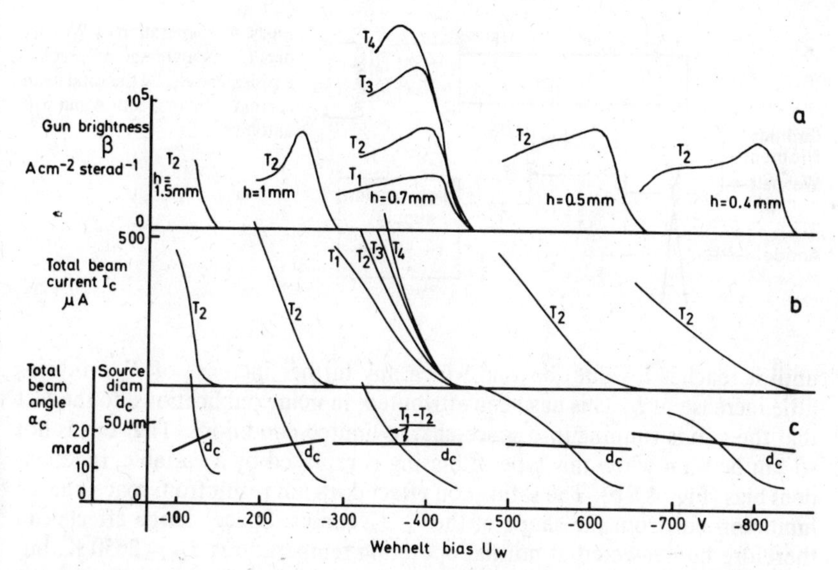

Fig. 4.8. (**a**) Gun brightness β, (**b**) total beam current I_c and (**c**) beam aperture α_c as a function of Wehnelt bias U_W for different heights h of the gun filament behind the Wehnelt bore [4.19]

age supply) for different heights h (Fig. 2.1) of the filament tip behind the Wehnelt bore. The dependence of I_c on U_W (Fig. 4.8 b) shows that the negative cut-off bias on the right-hand side of the I_c-U_W plots increases with decreasing h. The maximum of the gun brightness β occurs just below the corresponding cut-off bias (Fig. 4.8 a). This is caused by an increase of beam aperture α_c with decreasing bias (Fig. 4.8 c), which decreases the gun brightness. The maximum attainable value of β for constant cathode temperature does not vary much with the height h of the cathode tip. The diameter of the crossover d_c in Fig. 4.8 c is not strongly influenced by the biasing.

In practice, the Wehnelt electrode is biased not by a separate voltage supply but by the voltage drop $U_W = I_c R_W$ across the resistor R_W in the high-tension supply line (Fig. 4.9), produced by the emission current I_c. The resistance R_W can be altered by means of a mechanical potentiometer or a vacuum diode, the filament heating of which is varied. It will now be shown that this system is self-biasing. By use of an independent voltage supply for U_W, a dependence of the emission current I_c on U_W as shown in Figs. 4.8 b and 4.10 a will be observed as the filament current I_f or cathode temperature T_C is increased. By use of Wehnelt biasing produced by the voltage drop across R_W, the working points shown in Fig. 4.10 a are obtained; these are the points of intersection of the straight lines $I_c = U_W/R_W$, plotted for different values of R_W (1–12 $M\Omega$) with the temperature curves. From this diagram, the dependence of I_c on I_f or T_C for constant R_W (Fig. 4.10 b) can be constructed. This plot shows a *saturation current* I_c which increases as T_C is raised

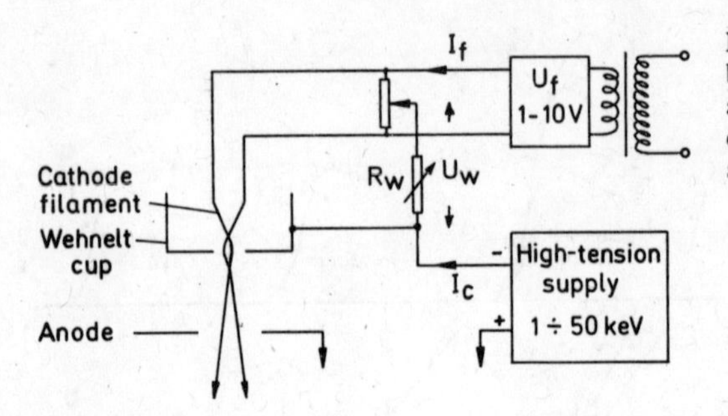

Fig. 4.9. Generation of Wehnelt bias U_W as a voltage drop across a resistance R_W by the total beam current I_c in an electron gun with autobias

until it reaches a value, beyond which any further increase of T_C produces little increase of I_c. This has been attributed, in many publications, to the fact that the gun is running into space-charge-limited conditions. That this is not so can be seen when this type of biasing is replaced by a variable, independent bias (Fig. 4.8b). The saturation effect does not result from space-charge limitation but from the shape of the I_c-U_W curves. Space-charge effects can therefore be neglected at normal operating temperatures $T_C \simeq 2650$ K, but can occur at high values of T_C, where they cause a decrease of gun brightness [4.19].

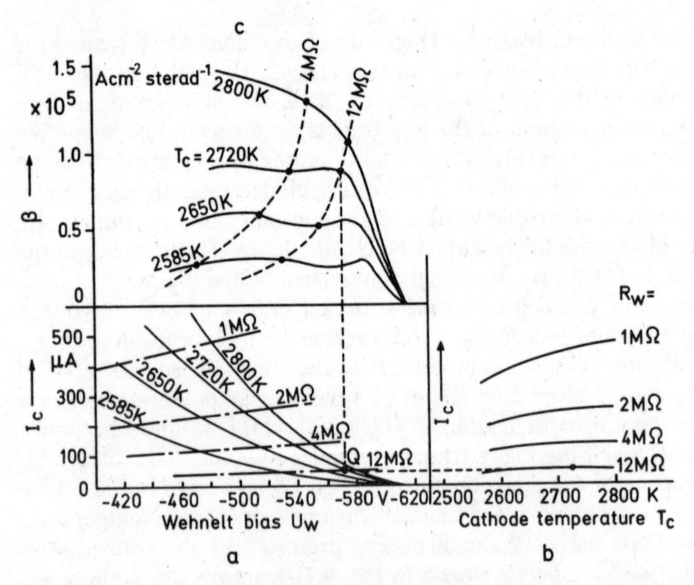

Fig. 4.10. (c) Dependence of gun brightness β on Wehnelt bias U_W for different cathode temperatures T_C, (---) working points for constant R_W. (a) The same dependence for the total beam current, (-·-·-) $I_c = U_W/R_W$ for different R_W. (b) Construction of the dependence of I_c on cathode temperature T_C from the intersection of the dash-dotted line in (a) with the I_c-U_W curves [4.19]

The optimum value of R_W for a given cathode geometry and temperature can be constructed by dropping a vertical line from the maximum of β in Fig. 4.10c onto the corresponding I_c-U_W curve. The slope of the straight line from the point of intersection Q to the origin of Fig. 4.10a determines the optimum resistance R_w. These optimum points are situated just at the onset (*knee*) of the saturation of the emission current in Fig. 4.10b. Decreasing R_W and increasing T_C produce higher brightness and larger saturation currents, but shorten the lifetime of the cathode.

4.1.4 Field-Emission Guns

Field-emission guns consist of a pointed cathode tip and at least two anodes (Fig. 4.11 and 25). Tungsten is normally used as the tip material because etching is easy, but it has the disadvantage of a high work function ϕ and sensitivity to adsorption layers. Wires of 0.1 mm diameter are spotwelded on a tungsten-hairpin cathode and electrolytically etched to a radius of curvature of about 100 nm (Fig. 4.5c). The hairpin can be heated to eliminate adsorbed gas atoms from the tip, or to work at a higher temperature, of the

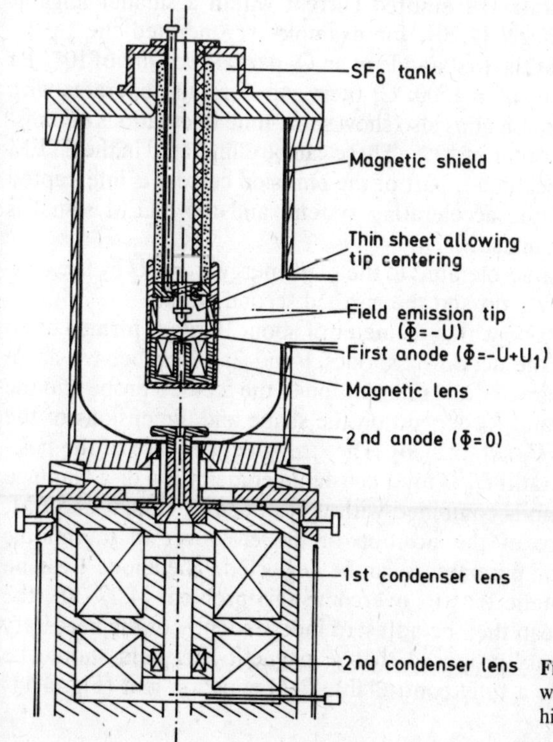

Fig. 4.11. Field-emission gun with a magnetic focusing lens behind the first anode [4.32]

order of 1500 K or to raise the temperature when the tip requires remoulding.

Drawn tungsten wires usually exhibit a crystal orientation with [110] along the wire axis. The current density emitted at the tip has a dark spot in the centre due to the high work function of the (110) planes. For this reason, (100), (310) or (111) oriented tips are prepared; they emit with a bright central spot. Normally the cathode tip emits into a large cone of semi-angle 1 rad. Use of a (100) tip concentrates the emission to about 0.1 rad.

The positive voltage U_1 of a few kV at anode 1 in Figs. 4.11 and 4.25 generates a field strength $|\boldsymbol{E}| \simeq U_1/r$ of about 5×10^7 V cm^{-1} at the cathode tip; this produces a field-emission current of the order of 1–10 μA. The field emission current (4.4) depends on the work function ϕ and on $|\boldsymbol{E}|$. Both quantities vary during operation of the gun. The work function changes due to diffusion of impurities from within the tip material or to surface reactions or the adsorption of gases. The electric-field strength changes as a result of damage to the tip by ion bombardment. This damage is unacceptable unless an ultrahigh vacuum $\leq 5 \times 10^{-8}$ Pa is maintained in the field-emission system. Even with a constant emission current, these factors can alter the solid angle of emission. The current emitted by a field-emission gun therefore drifts over long periods and the tip has to be reactivated and remoulded from time to time to concentrate the emitted current within a smaller angular cone. *Veneklasen* and *Siegel* [4.20], for example, regenerated the tip by reversing the voltage U_1 at the first anode in an O_2 partial pressure of 10^{-7} Pa at a cathode temperature $T_C \simeq 1500$ °C; their gun normally operates with $T_C = 1000$ °C. Field-emission guns also show short-time fluctuations of emission, which can be of the order of 10%. These can be eliminated in the STEM mode by a signal-ratio technique: part of the emission current is intercepted by a diaphragm behind the accelerating system, and the STEM signal is divided by this reference signal.

The electrons are post-accelerated to the final energy $E = eU$ by the voltage U between the cathode tip and the earthed second anode.

A focused electron probe with a diameter of about 10 nm is formed as an image of the source by the action of anodes 1 and 2, which behave as an electrostatic lens. The diameter and the position of the focused probe and the aberration constants C_s and C_c depend on the shape and dimensions of the anodes and on the ratio U_1/U [4.21–26]. The strong dependence of the position of the probe on the ratio U_1/U for a constant geometry is a disadvantage when a field-emission gun is combined with the condenser system of TEM, whereas the dependence of the position of the crossover of thermionic cathodes on the operating parameters can be neglected. The anode 1 can be replaced by an electrostatic lens to overcome this problem [4.27–30]; the electron probe position can then be adjusted independently of the necessary voltage U_1. Other authors proposed that a magnetic-lens field should be superimposed to provide a fully controllable field-emission gun (Fig. 4.11) [4.31, 32].

Field-emission guns have the advantage of a high gun brightness and low energy spread (Table 4.1). They are of interest in all work that needs high coherence, which means low beam apertures and high current densities: high-resolution phase contrast, electron holography and interferometry, Lorentz microscopy and STEM. The high coherence of a field-emission gun is demonstrated in Fig. 3.10 by the large number of resolvable Fresnel fringes. If it is necessary to work with larger probe currents, for x-ray microanalysis for example, thermionic cathodes give better performance (Sect. 4.2.2)

4.2 The Illumination System of a TEM

4.2.1 Two-Lens Condenser System

The condenser lens system of TEM (Fig. 4.12) performs the following tasks:

1) Focusing of the electron beam on the specimen in such a way that sufficient image intensity is obtainable even at high magnification.
2) Irradiation of a specimen area that corresponds as closely as possible to the viewing screen, whatever the magnification, thereby reducing specimen drift by heating and limiting the radiation damage and contamination in non-irradiated areas.
3) Variation of the illumination aperture α_i which is in the order of 1 mrad for medium magnifications, must be ≤ 0.1 mrad for high resolution and application of phase contrast and $\leq 10^{-2}$ mrad for Lorentz microscopy, small-angle electron diffraction and holographic experiments.
4) Production of a small electron probe (2–100 nm diameter) for x-ray microanalysis, microbeam electron-diffraction techniques and the scanning mode.

TEMs are equipped with two condenser lenses to satisfy these requirements, and the prefield of a strongly excited objective lens can act as an additional condenser lens especially for point 4 (Sect. 4.2.3).

Figure 4.12 shows the most important modes of operation of a two-lens condenser system. In the cases a–c, only the condenser lens C2 is excited. When focusing (b), the familiar lens formula $1/f_2 = 1/s_2 + 1/s_2'$ can be applied, and the crossover is demagnified by the factor $d_s/d_c = s_2'/s_2$. The current density j_s at the specimen and the illumination aperture α_i reach a maximum and the diameter d_s of the irradiated specimen area a minimum (Fig. 4.13). For underfocus (a) and overfocus (c) j_s and α_i decrease and d_s increases. A condenser diaphragm (100–200 μm diameter) near the centre of the condenser lens selects only the centre of the beam. In focus, d_s has the same value as with no diaphragm because the crossover is imaged in both cases; the maximum current density j_s in the centre of the beam and the illumination

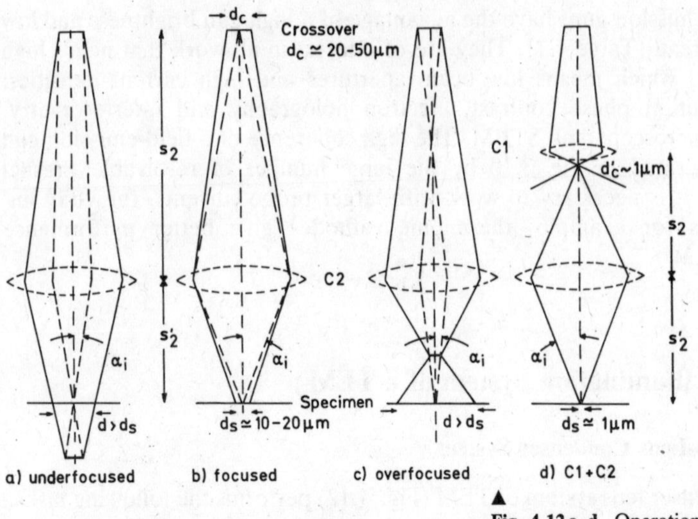

▲ **Fig. 4.12 a–d.** Operation of a two-lens condenser system for illuminating the specimen. (**a**) Under-focus and (**c**) over-focus and (**b**) in-focus operation with condenser lens C2. (**d**) Additional use of condenser lens C1 for demagnifying the crossover and focusing the demagnified image on the specimen with condenser lens C2

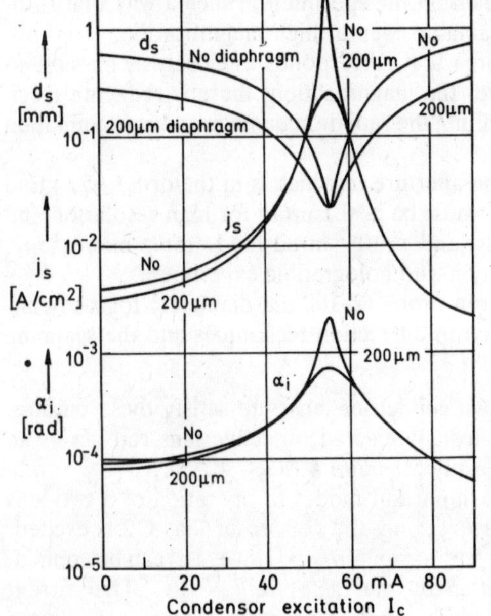

◀ **Fig. 4.13.** Dependence of the diameter d_s of the irradiated area, the current density j_s and the illumination aperture α_i in the specimen plane on the excitation of condenser lens C2 with no diaphragm and with a 200 μm diaphragm

aperture α_i are, however, decreased as the diaphragm is made smaller. The current density and the aperture are related via the gun brightness (4.7) $\beta = j_s/\pi\alpha_i^2$.

The size of the final fluorescent screen corresponds to a specimen area of 1 μm at $M = 100\,000$. It is sufficient, therefore, to illuminate specimen areas as small as this. This can be achieved by fully exciting the condenser lens C1 (Fig. 4.12d); the strongly demagnified intermediate image of the crossover with a diameter $d_c' \simeq 1$ μm can then be imaged on the specimen by condenser C2 with $M = s_2'/s_2$, resulting in $d_s \simeq 0.5$–1 μm. The diameter of the irradiated area can be varied by over- or underfocusing C2.

Very small values of α_i (for Lorentz microscopy or small-angle electron diffraction, for example) can be obtained by exciting condenser lens C1 and using the demagnified image of the crossover an an electron source (C2 switched off). The smallest obtainable illumination aperture will be $\alpha_i = r_c'/(s_2 + s_2') \simeq 10^{-2}$ mrad. In view of (4.7), this operation mode only works with very small current densities j_s. These results for the case of thermionic cathodes can be improved by use of a field-emission gun, which has a higher gun brightness and an electron crossover of $\simeq 10$ nm diameter in front of the gun system (Sect. 4.1.4).

The condenser lens C1 works with a relative large entrance aperture and is equipped, therefore, with a stigmator to compensate the astigmatism and to decrease the diameter d_c' of the crossover image.

The electron-gun system of a microscope can be adjusted onto the axis of the condenser lens system by tilting and shifting the gun system. Further adjustments are necessary to bring the electron beam onto the axis of the objective lens and magnifying lens system. The specimen structures spiral around the image rotation centre if the high voltage, or preferably the lens current of the objective lens is varied periodically (wobbled). For this alignment, holey formvar films or polystyrene spheres on supporting films can be used as specimens for easy observation of this spiral movement (Fig. 2.20b). The distance of the image rotation centre from the point of intersection of the objective axis can be calibrated by shifting the condenser lens system relative to the objective lens. The point of intersection of the objective field axis with the final screen does not necessarily coincide with the centre of the final screen [4.33–35]. Its position can be determined by reversing the objective lens current. Specimen structures at the point of intersection will remain stationary.

The alignment procedure needed to bring the electron beam on axis involves a mechanical shift and tilt of the condenser lens system or electromagnetic deflection of the electron beam by pairs of alignment coils (Fig. 4.14a, b). Such coils also can be used to generate the dark-field mode (Sect. 4.4.2 and 6.1.2). The incident electron beam is tilted (Fig. 4.14b) so that a cone of scattered electrons or Bragg-reflected electrons is on axis and can pass the objective diaphragm. The transition from the bright- to the dark-field mode and back can easily be achieved by switching the alignment coils off and on, respectively. These coils can also be used for irradiation with a rocking beam at low and medium magnifications as an additional aid for focusing (Sect. 4.4.4) or, together with the objective-lens pre-field, for scanning and rocking (Fig. 4.14c, d) the electron probe in a scanning mode (Sect. 4.2.3) or for special diffraction techniques (Sect. 9.3).

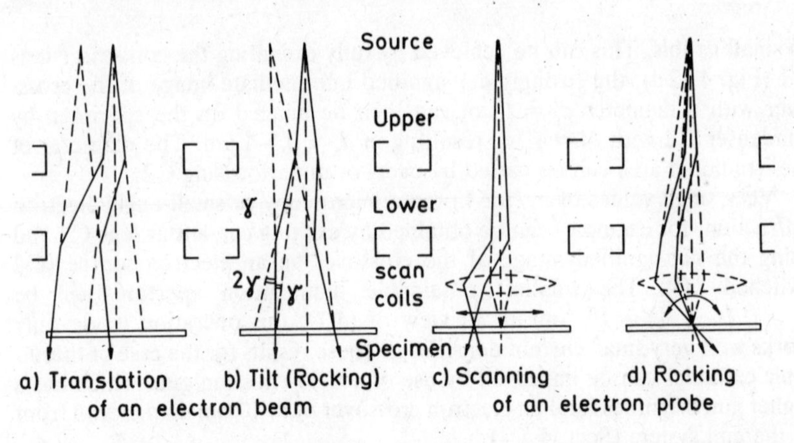

Fig. 4.14. (**a**) Shift and (**b**) tilt of an electron beam by a double deflection coil system or (a) scanning and (b) rocking when exciting the coils by a saw-tooth current. (**c**) Scanning and (**d**) rocking when working with the pre-field of an objective lens as an additional condenser lens

A dark-field mode with hollow-cone illumination (Sect. 6.1.2) can be created by replacing the circular diaphragm in the condenser lens C2 by an annular diaphragm. Alternatively, the beam may be deflected around a circle (Sect. 6.4.3).

4.2.2 Electron-Probe Formation

The illumination of specimens with small electron probes of diameter ≤ 0.1 μm is important for the x-ray microanalysis and energy-loss spectroscopy of small specimen areas and for microbeam-diffraction methods in TEM and for scanning transmission electron microscopy (STEM). An electron probe is formed by a one- to three-stage demagnification of the electron-gun crossover. If the geometrical diameter is d_0, the total probe current will be given by

$$I_p = \frac{\pi}{4} d_0^2 j_0 \, . \tag{4.14}$$

In reality, the intensity is distributed more as a Gaussian distribution (4.13). This, however, will only change the results of our simple estimation by correction factors of the order of unity. The conservation of gun brightness (4.7) on the optical axis implies $j_0 = \pi\beta\alpha_p^2$ (α_p is the electron-probe aperture). Substitution in (4.14) gives

$$I_p = \frac{\pi^2}{4} \beta d_0^2 \alpha_p^2 \, . \tag{4.15}$$

Solving for d_0, we find

$$d_0 = \left(\frac{4 I_p}{\pi^2 \beta}\right)^{1/2} \frac{1}{\alpha_p} = \frac{C_0}{\alpha_p}\,, \tag{4.16}$$

which shows that for a given probe current I_p small values of d_0 can be obtained only for large values of the gun brightness β and probe aperture α_p.

The geometrical diameter d_0 is blurred by the action of lens aberrations; chromatic aberration produces an error disc of diameter d_c (2.54) and spherical aberration of diameter d_s (2.49). The aperture limitation α_p causes a diffraction-error disc of diameter $d_d = 0.6\,\lambda/\alpha_p$, that is, the half-width of the Airy distribution in Fig. 3.17b.

To estimate the final probe size d_p, this blurring can be treated as a quadratic superposition of the error-disc diameters [4.36], though this is strictly valid only when the error discs are all of Gaussian shape

$$d_p^2 = d_0^2 + d_d^2 + d_s^2 + d_c^2 \tag{4.17}$$

$$d_p^2 = [C_0^2 + (0.6\,\lambda)^2]\frac{1}{\alpha_p^2} + \frac{1}{4}\,C_s^2\,\alpha_p^6 + \left(C_c\frac{\Delta E}{E}\right)^2 \alpha_p^2\,. \tag{4.18}$$

The constant C_0 (4.16) will be much greater than the wavelength for a thermionic cathode and the diffraction- and chromatic-error terms in (4.18) can be neglected.

Figure 4.15 shows how the diameters d_0 and d_s superpose and produce a minimum probe diameter d_{min} at an optimum aperture α_{opt} for a constant probe current I_p. The optimum aperture is obtained by writing $\partial d_p/\partial\alpha_p = 0$, i.e.,

$$\alpha_{opt} = (4/3)^{1/8}\,(C_0/C_s)^{1/4}\,, \tag{4.19}$$

and substitution in (4.18) gives

$$d_{min} = (4/3)^{3/8}\,(C_0^3 C_s)^{1/4}\,. \tag{4.20}$$

The constant C_0 will be much smaller than the wavelength for a field-emission gun, and the energy spread ΔE is also smaller. A superposition of the largest terms in (4.18), now d_d and d_s, again yields a minimum (Fig. 4.15). For this case, a wave-optical calculation should strictly be used (see e.g. Fig. 3.17c) but this has no influence on the position of the minimum. The only difference is that the increase due to the spherical aberration for $\alpha > \alpha_{opt}$ does not occur so strongly.

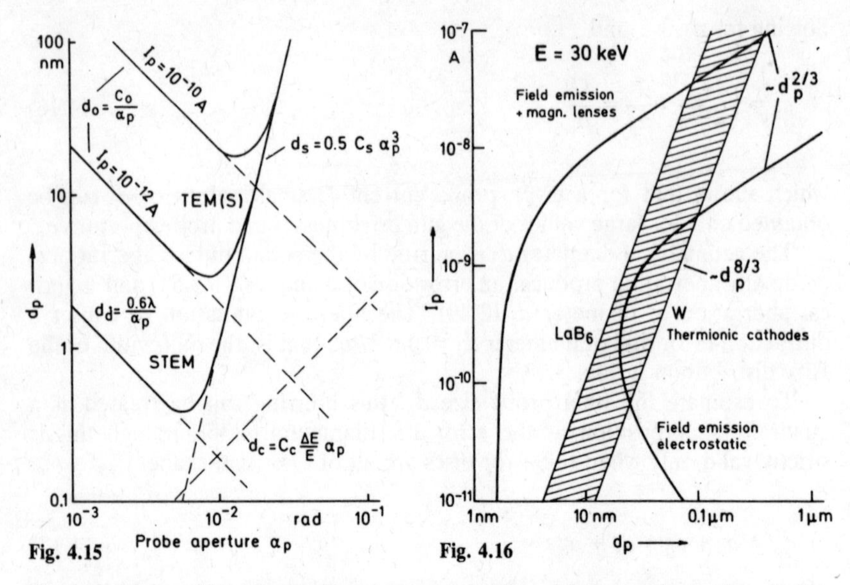

Fig. 4.15 **Fig. 4.16**

Fig. 4.15. Superposition of error discs shown as a function of probe aperture α_p in a double-logarithmic plot. Upper two curves: superposition of d_o and d_s for $I_p = 10^{-10}$ and 10^{-12} A for the scanning transmission mode in a TEM. Lower curve: superposition of d_d and d_s for a field-emission STEM. $E = 100$ keV, $\Delta E = 1$ eV, $C_s = C_c = 2$ mm, $\beta = 10^5$ A cm^{-2} sr^{-1}

Fig. 4.16. Double-logarithmic plot of maximum probe current I_p versus probe diameter d_p for LaB_6 and W thermionic cathodes and field-emission guns

It is of practical interest to express the maximum probe current I_p as a function of the probe diameter d_p (Fig. 4.16). For thermionic cathodes, (4.20) can be solved for I_p which is included in C_0 (4.16)

$$I_p = (3\pi^2/16)\beta C_s^{-2/3} d_p^{8/3} . \tag{4.21}$$

To obtain a similar formula for field emission [4.37–39], it can be assumed that the source diameter is so small that the lens system of the field-emission gun produces an error disc of diameter

$$d_{s1} = 0.5\ C_{s1}\alpha_1^3 \tag{4.22}$$

caused only by the sperical aberration of this system. The corresponding error disc due to the spherical aberration C_{s2} of the objective lens becomes

$$d_{s2} = 0.5\ C_{s2}\alpha_2^3 . \tag{4.23}$$

The angles α_1 and α_2 are related by $\alpha_2 = \alpha_1/M$ where $M < 1$ is the demagnification. The total diameter of the electron probe is again given by quadratic superposition

$$d_p^2 = d_{s1}^2 M^2 + d_{s2}^2 = \frac{1}{4}\,\alpha_1^6 (C_{s1}^2 M^2 + C_{s2}^2 M^{-6}) \,. \tag{4.24}$$

With J as the current per solid angle, the probe current becomes $I_p = \pi\alpha_1^2 J$ and substitution of α_1 in (4.24) gives

$$d_p^2 = \frac{1}{4}\left(\frac{I_p}{\pi J}\right)^3 (C_{s1}^2 M^2 + C_{s2}^2 M^{-6}) \,. \tag{4.25}$$

It was shown in Sect. 4.1.4 that in a field-emission gun, the position of the electron probe – the image of the emission area – can be varied. This also changes the demagnification of the probe-forming lens. The optimum demagnification can be obtained from $\partial d_p/\partial M = 0$. Substituting this value in (4.25) and solving for I_p yields

$$I_p = \frac{1.3\,\pi J}{C_{s1}^{1/2}\,C_{s2}^{1/6}}\,d_p^{2/3} \,. \tag{4.26}$$

The probe current I_p increases only as $d_p^{2/3}$ (Fig. 4.16) and reaches saturation at a relatively small value of d_p because I_p cannot become larger than the emission current. If larger probe currents are necessary, LaB_6 and tungsten thermionic cathodes perform better for large probe diameters. Though these calculations are somewhat oversimplified, they show that the field-emission gun can have disadvantages if large currents are needed, for x-ray microanalysis, for example.

For production of electron probes smaller than 0.1 μm, the field-emission gun has the advantage of providing larger beam currents for constant probe diameter, which is important for increasing the signal-to-noise ratio in STEM.

The importance of the probe current I_p for achieving a good signal-to-noise ratio is shown by the following estimate. A signal S is produced by a number

$$n = f I_p \tau / e \tag{4.27}$$

of electrons, where f denotes the fraction of electrons recorded by the detector ($f < 1$) or the number of recorded x-ray quanta per electron ($f \ll 1$), and τ denotes the recording time for one image point, that is, the frame time (1/20–1000 s) divided by the number of image points (10^4–10^6). As a result of statistical shot noise, the noise signal is $N = n^{1/2}$. The signal-to-noise ratio must be larger than a value $\varkappa$, which should be on the order 3–5 to detect a signal in a noisy record. If a signal difference ΔS on a background signal S is to be detected, then

$$\frac{\Delta S}{S} \geq \varkappa \frac{N}{S} = \varkappa \frac{n^{1/2}}{n} = \frac{\varkappa}{(fI_p\tau/e)^{1/2}} \tag{4.28}$$

and I_p has to satisfy the inequality

$$I_p \geq \left(\frac{\varkappa}{\Delta S/S}\right)^2 \frac{e}{f\tau} \, . \tag{4.29}$$

Numerical example: For $\varkappa = 3$, $\Delta S/S = 5\%$, $f = 0.1$, $\tau = 1$ ms (scanning of a frame with 10^6 image points in 1000 s) and $e = 1.6 \times 10^{-19}$C, we find $I_p \geq 3.6$ pA.

4.2.3 Illumination with an Objective Pre-Field Lens

A condenser-objective lens (Fig. 2.10 and 4.17) with an excitation $k^2 = 3$ not only has the advantage of a lower spherical-aberration coefficient C_s but also simplifies the transition from the extended illumination needed for the TEM bright- and dark-field modes (Fig. 4.17 a) to the illumination required to form a small electron probe for the scanning transmission mode and for x-ray and energy-loss analysis and electron diffraction of small specimen areas (Fig. 4.17 b)

As discussed in Sect. 2.2.3, this type of lens operates in the telefocal condition with the specimen in the lens centre. The action of the pre-field condenser and post-field objective field can be represented in a ray diagram by two separate lenses. The optimum working conditions for illumination with an extended electron beam (Fig. 4.17 a) will be achieved by fully exciting condenser lens C 1 and focusing the crossover on the front focal plane (FFP) with condenser lens C 2. This can be checked by imaging the back focal plane (BFP) on the final viewing screen, the BFP being conjugate to the FFP. Furthermore, the specimen plane and the plane of the condenser C 2 diaphragm of diameter d_2 are conjugate, and the diameter of the irradiated area is thus $d_s = Md_2$ with the demagnification $M = f_0/s_2'$. The illumination aperture $\alpha_i = d_c'/2f_0$ is limited by the diameter d_c' of the crossover image in the FFP. Thus for $f_0 = 1$ mm, $s_2' = 200$ mm, $d_2 = 100$ µm and $d_c' = 0.5$ µm, we find $d_s = 0.5$ µm and $\alpha_i = 0.25$ mrad which are optimum operation conditions for high resolution.

A very small spot diameter d_s can be obtained for the scanning mode (Fig. 4.17 b) by switching off C2 and using a small C2 diaphragm. The geometrical diameter d_0 of the electron probe in the specimen plane can be estimated with the demagnification factor $M = f_0/s_1' \simeq 1/250$; if the diameter

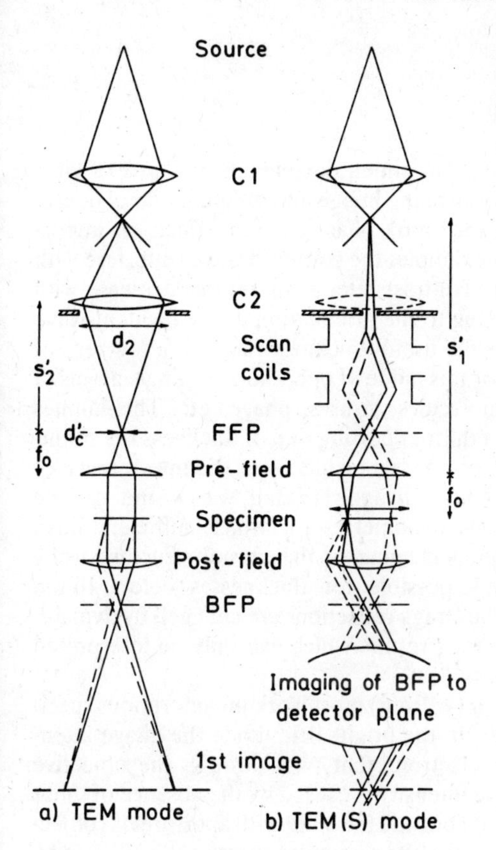

Fig. 4.17 a, b. Specimen illumination by an objective pre-field: (**a**) large-area illumination for TEM, (**b**) electron-probe forming and scanning for the scanning transmission mode of a TEM

of the crossover image in the focal plane of C1 is 0.5 μm, we obtain $d_0 \simeq 2$ nm. The aperture α_p of the electron probe is determined by the projected diameter of the C2 diaphragm in the FFP. The aperture will be of the order of 5 mrad and, therefore, one order of magnitude larger than the illumination aperture α_i in the extended-beam-producing mode of Fig. 4.17 a. As shown in Sect. 4.2.2, a large probe aperture α_p will be necessary to produce a small probe diameter d_p.

The electron probe can be scanned across the specimen plane by means of pairs of scanning coils, as in Fig. 4.17 b, which rock the incident electron beam. If the pivot point of this beam rocking is at the FFP, the pivot point of the rays behind the specimen will be at the BFP because these planes are conjugate. This means that the position of the first diffraction pattern in the BFP is stationary, and the BFP can be imaged on the detector plane in the scanning transmission mode (Sect. 4.5.1).

4.3 Specimens

4.3.1 Useful Specimen Thickness

The maximum useful thickness of a specimen depends on the type of electron-specimen interaction used to form the image and on the mode of operation. In the high-resolution mode (≤ 1 nm), phase-contrast effects are important; in the bright-field mode, for example, the scattered waves interfere with the primary incident wave. Phase-contrast effects, therefore, decrease with increasing specimen thickness owing to the attenuation of the incident-wave amplitude (Sect. 6.2). This limits the useful thickness range to a few tens of nanometres. Typical specimens for this mode of operation are single atoms of heavy elements, organic macromolecules, viruses, phages etc. This implies that the specimen is mounted on a thin supporting film of thickness $t \leq 5$ nm.

The imaging of lattice planes of crystals results from the interference of the primary beam and one or more Bragg-reflected waves and can be observed for thicknesses of a few nanometres for which sufficient wave amplitude remains. High-resolution imaging of the crystal structure using several Bragg reflections is only possible for thicknesses below 10 nm because the wave amplitudes of the Bragg reflections are changed by dynamical electron diffraction; false contrast results which can only be interpreted by computer model calculations.

For medium and low resolution (≥ 1 nm) most work on amorphous specimens relies on scattering contrast. In the bright-field mode the image intensity depends on the number of electrons that pass through the objective diaphragm. The decrease of image intensity is caused by the absence of those electrons which have been scattered outside the cone with aperture α_o (objective aperture). In biological specimens, the scattering contrast is increased by staining the tissue or a thin section with heavy atoms. Numerical examples of scattering contrast are reported in Sect. 6.1.3. Another example is the negative staining technique where microorganisms or macromolecules are embedded in a layer of a heavy metal compound, such as phosphotungstic acid.

The energy lost by electrons during inelastic scattering and the chromatic aberration of the objective lens limit the maximum useful specimen thickness to 0.1–0.3 μm for 100 kV and about 1 μm for 1 MV. The chromatic aberration can be avoided in the STEM mode of TEM. However, the resolution is limited by the broadening of the electron probe due to multiple scattering (Sect. 5.3.3).

By using the primary beam in bright field or a Bragg-reflected beam in dark field, lattice defects in crystalline specimens can be imaged. The maximum thickness is limited by the intensity of the primary or Bragg-reflected beam and by the chromatic error for thick specimens. The intensities of the beams depend on the crystal orientation and better penetration is observed in the case of anomalous transmission near a Bragg condition. At 100 keV, the useful thickness of metal foils and other crystalline material is of the order of

50–200 nm. The increase in the useful thickness when the accelerating voltage is increased from 100 kV to 1 MV is only of the order of three to five times (see also Sect. 8.1.4). However, a large number of specimens, electropolished metal foils, for example, are some 200–500 nm thick over most of the specimen area and in many cases, the only areas that can be used at 100 keV are the edges of holes in the centre of electropolished discs.

4.3.2 Specimen Mounting

Metals and other materials can be used directly as thin discs of 3 mm diameter and $\simeq 0.1$ mm thickness if they can be thinned in the centre by electropolishing or chemical or ion etching.

Other specimens for TEM (crystal flakes, surface replicas, evaporated films, biological sections) are mounted on copper grids with 100–200 μm meshes. Grids of 3 mm diameter are commercially available with different mesh sizes and orientation marks.

Small particles, microorganisms, viruses, macromolecules and single molecules need a supporting film possessing the following properties:

1) Low atomic number to reduce scattering
2) High mechanical strength
3) Resistance to electron irradiation (and heating)
4) Low granularity (caused by phase contrast) for high resolution
5) Easy preparation.

For medium magnifications, formvar films of 10–20 nm are in use, which are produced by dipping a glass slide in a 0.3% solution of formvar in chloroform and floating the dried film on a water surface. A higher mechanical strength is obtained by evaporating an additional thin film of carbon ($\simeq 5$ nm) on a formvar or collodion film. Pure carbon films are more brittle but can be used as 3–5 nm films on plastic supporting films with holes.

For high resolution, the granularity of carbon films is useful for investigating the contrast-transfer function of TEM (Sect. 6.4.6) but it obscures the image of small particles, macromolecules and single atoms. Numerous attempts have therefore been made to prepare supporting films with less phase-contrast granularity: amorphous aluminium oxide [4.40], boron [4.41], single-crystal films of graphite [4.42] or vermiculite [4.43].

The specimen grids or discs are mounted in a specimen cartridge, which can be transferred through an airlock system either into the upper bore of the pole-piece of the objective lens (top entry) or, mounted on a rod, into the pole-piece gap (side entry). The specimen position is near the centre of the bell-shaped lens field for a strongly excited objective lens with $k^2 \simeq 3$. The pole-piece gap also contains the objective diaphragms and the anti-contamination blades or *cold finger* (Sect. 10.4.2), which decrease the partial pressure of organic molecules near the specimen. This decreases the space for special specimen manipulations. The gap is only of the order of a few millimetres for 100 kV-TEM and 1–2 cm in a HVEM.

4.3.3 Specimen Manipulation

The principal methods of specimen manipulation are summarized in Table 4.2. Only the most important points will be discussed here. For details, see the summaries [4.44–46] and the proceedings of the HVEM symposia [4.47–51].

Specimen rotation about an axis parallel to the electron beam can be used to bring specimen structures into a convenient orientation in the final image. Tilting devices with one axis normal to the electron beam can produce stereo pairs for quantitative measurement and stereoscopic observation of the three-dimensional specimen structure. A specimen goniometer can tilt the specimen with high precision in any desired direction. Side-entry goniometers are available, which cause a specimen shift less than 1 μm when tilting the specimen ± 30°. The second degree of freedom for angular adjustment is often exploited as a specimen rotation about an axis normal to the specimen plane. Top-entry goniometers can often tilt the specimen and move the specimen normal over a cone around the optic axis. Goniometer stages can be useful to biological-tissue sections, to bring lamellae systems or other structures into favourable orientations, for example. Crystalline specimens have to be tilted in a goniometer for

a) observation of lattice fringes,
b) observation of diffraction contrast of lattice defects with distinct Bragg reflections or known orientation,
c) determination of the Burgers vector of lattice defects,
d) determination of crystal orientation by electron diffraction.

A large variety of specimen tilting, heating and stretching cartridges have been developed for 100 kV TEMs. The problem arises whether the phenomena of deformation, annealing and precipitation are the same in the thin-specimen areas that can be studied at 100 kV as in the bulk material. Such specimen manipulations are therefore of particular interest in HVEM in which there is more space in the pole-piece gap and the specimens will be more similar to bulk material, because greater thicknesses can be penetrated.

Specimen-cooling devices operating at temperatures below – 150 °C can be used to reduce the contamination of inorganic material because of radiation-induced etching of carbon in the presence of oxygen molecules (Sect. 10.4.2). Such devices must not be confused with the cooled anti-contamination blades mentioned above. Specimens that melt at room temperature or due to electron-beam heating or which sublimate in the vacuum may be observable if the specimen is cooled. A special application is the direct observation of cryosections. The sections have to be transfered from the

Table 4.2. Specimen manipulations

Procedure	Application
1. Specimen rotation	
Rotation on an axis parallel to the electron beam	Orientation of specimen structures or diffraction patterns relative to the edges of the final screen
2. Specimen tilt	
a) Tilt (± 5–10°) about an axis in the specimen plane	Stereopairs
b) Tilt about an axis normal to the beam and rotation about an axis parallel to the beam	Lattice defects Determination of orientation
c) Double tilt (± 25–50°) about two perpendicular axes normal to the beam	Favourable orientation of biological sections
d) Specimen goniometer (± 25–50°) Small specimen drift by adjustment of tilt axis in height and position (accuracy: ± 0.1°)	Three-dimensional reconstruction
3. Straining devices	
Straining of the specimen by movement of two clamps by mechanics or piezo-electric effect	Straining of metals and high polymers
4. Specimen heating	
Direct heating of a grid	Recovery and recrystallisation
Indirect heating (≃ 10 W for 1000 °C)	Precipitation and transition phenomena
5. Specimen cooling	
a) Cooling to between –100° and –150 °C with liquid nitrogen	Temperature-sensitive specimens Decrease of specimen contamination Direct observation of cryosection
b) Cooling with liquid helium (4–10 K)	Structure of condensed gases Decrease of radiation damage Superconducting states Magnetic structure in low Curie-point ferromagnetics
6. Environmental cells	
Gas pressure between diaphragms covered with foils or separated from the microscope vacuum by additional pumping stages Spraying the specimen with a gas jet	Gas reactions on the specimen Corrosion tests Biological specimens in wet atmosphere
7. Other in-situ methods	
Evaporation in the specimen chamber	Investigation of film growth
Particle bombardment by an ion source or the beam of HVEM	Radiation-damage experiments
Magnetization of the specimen by additional coils (Lorentz microscopy)	Direct observation of movement of ferromagnetic domains
Specimen cartridge with a Faraday cage in front of the specimen for measuring the backscattering coefficient	Determination of specimen thickness

cryo-microtome to the cooled specimen cartridge of the microscope via a cooling chain.

Specimen-cooling devices operating at liquid-helium temperature need very careful design and construction [4.52, 53]. The specimen and an additional storage tube for liquid helium have to be shielded against heat-radiation losses by surrounding them in a liquid-nitrogen-cooled trap. Specimen or physical effects that are normally present only at very low temperatures can be observed: the crystal structure of condensed gases, for example, magnetic fields around superconducting domains and ferromagnetic films of low Curie temperature. The mobility of radiation-induced lattice defects decreases at low temperatures. These defects can be generated directly in a cooled specimen by bombardment with α-particles or high-energy electrons beyond the threshold energy (Sect. 5.1.2); the coagulation of dislocation loops or stacking faults can then be observed when the temperature is raised. The suppression of the radiation damage of organic specimens is another application of liquid-helium-cooled stages. However, specimen cooling obviously only retards secondary radiation effects such as distortion of the crystal lattice, which leads to fading of the electron-diffraction pattern, but cannot prevent primary damage of the individual organic molecules (Sect. 10.2).

Environmental cells in which partial pressures of inert or reactive gases up to atmospheric pressure is maintained allow us to observe in-situ reactive processes between a gas and the specimen; with a partial pressure of water, hydrated biological specimens can be observed [4.54]. Such studies are limited by electron scattering at the gas molecules. The large gas pressure in the specimen area can be obtained either by using a differentially pumped system of diaphragms or by confirming the gas between diaphragms covered with thin supporting films. High-voltage electron microscopy is more suitable for environmental experiments because much more space is available in the pole-piece gap and the scattering in the gas atmosphere is less severe. Table 4.2 contains some further examples of so-called in-situ experiments. A Faraday cage in a top-entry cartridge can be used to measure the backscattering coefficient for determination of specimen thickness (Sect. 5.3.4).

4.4 The Imaging System of a TEM

4.4.1 Objective-Lens System

Though the first intermediate image formed by the objective lens has a magnification M of only 20–30 times, it is from this lens that the highest performance will be demanded. The mechanical tolerances necessary have already been discussed in Sect. 2.3.3. The initial astigmatism of an objective lens must be so small that the main task of the stigmator is to compensate the

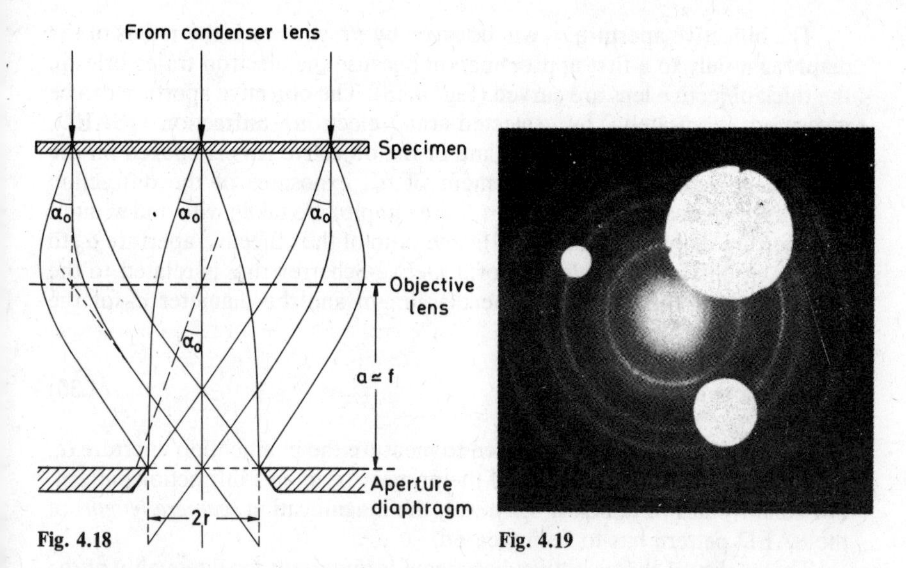

Fig. 4.18 **Fig. 4.19**

Fig. 4.18. Action of the objective diaphragm in the focal plane of the objective lens as an angular-selective diaphragm

Fig. 4.19. Measurement of objective apertures by superposition of the shadows of three objective diaphragms in a selected-area electron diffraction pattern of an evaporated gold film

astigmatism caused by contamination of the diaphragm and other perturbing effects. The resolution-limiting errors such as spherical and chromatic aberrations are only important for the objective lens because a magnification M decreases the apertures for the following lenses to $\alpha = \alpha_0/M$. The diameter of the spherical-aberration disc is proportional to α^3 (2.49). Even for a modest magnification M of 20–50 times at the first intermediate image, the aperture is so small that the spherical aberration of the intermediate and subsequent lenses can be neglected, even though these lenses normally have larger values of C_s and C_c than the objective lens, their pole-piece diameters being larger. The distortions (pincushion or barrel) associated with the spherical aberration in projector lenses (Sect. 2.3.4) causes only an image distortion but no loss of resolution.

Any one of three or four diaphragms of 20–200 μm diameter can be inserted in the focal plane of the objective lens, thus permitting the objective aperture α_0 to be changed (Fig. 4.18). The trajectories in Fig. 4.18 show that a diaphragm introduced in this plane selects all electrons that are scattered through angles $\theta \geq \alpha_0$. Decreasing the objective aperture increases the scattering contrast (Sect. 6.1). For high resolution, an aperture as large as possible is used so that high spatial frequencies can contribute to the image (Sect. 6.2), and so that the contamination and charging of the diaphragm do not disturb the image.

The objective aperture α_o will be given by r/f, where r is the radius of the diaphragm only to a first approximation because the electron trajectories in the thick objective lens are curved (Fig. 4.18). The objective aperture can be measured accurately by selected-area electron diffraction (SAED, Sect. 9.3.1), in which the focal plane of the objective lens is imaged on the final image screen. For measurement of α_o, exposures of the diffraction pattern (of an evaporated gold film, for example) are taken with and without the aperture diaphragm (Fig. 4.19). The ratio of the objective aperture α_o to the Bragg diffraction angle $2\theta_B$ of a Debye-Scherrer ring is related to the diameter d_0 of the shadow of the diaphragm and the diameter d_B of the Debye-Scherrer rings by

$$\alpha_o/2\theta_B = d_0/d_B \,. \tag{4.30}$$

The same procedure can be used to measure the illumination aperture α_i, which corresponds to the radius of the primary beam in a diffraction pattern. For small values of α_i below 0.1 mrad, the magnification (*camera length*) of the SAED pattern has to be increased.

The quality of the objective diaphragm is important for the quality of the image. The diaphragm has to be of a heat-resistant material (Pt, Pt-Ir, Mo or Ta) capable of tolerating the largest possible current density in the focal plane; this may reach 10 A cm^{-2}. Dust, fragments of the specimen and contamination in general can cause local charging, which generates an additional astigmatism, especially if small apertures are used. Charging effects can be postponed by using thin-foil diaphragms that consist of thin metal foils (1–2 μm) with circular holes [4.55–57].

4.4.2 Imaging Modes of a TEM

The imaging system of a TEM consists of at least three lenses (Fig. 4.20): objective lens, the intermediate lens and the projector lens. The intermediate lens can magnify the first intermediate image, which is formed just in front of this lens (Fig. 4.20 a), or the first diffraction pattern, which is formed in the focal plane of the objective lens (Fig. 4.20 b), by reducing the excitation (selected-area electron diffraction, Sect. 9.3.1). In many microscopes, an additional diffraction lens is inserted between the objective and intermediate lenses to image the diffraction pattern and to enable the magnification to be varied in the range 10^2 to 10^6.

The *bright-field mode* (BF) (Figs. 4.20 a and 4.21 a) with a centred objective diaphragm is the typical TEM mode, with which scattering contrast (Sect. 6.1.1) and diffraction contrast (Sect. 7.1.1) can be produced with objective apertures α_o between 5 and 20 mrad. For high-resolution phase contrast (Sect. 6.2), the aperture should be larger ($\alpha_o \simeq 20$ mrad) to transfer high spatial frequencies. The only purpose of the diaphragm in this mode is to decrease the background by absorbing electrons scattered at very large

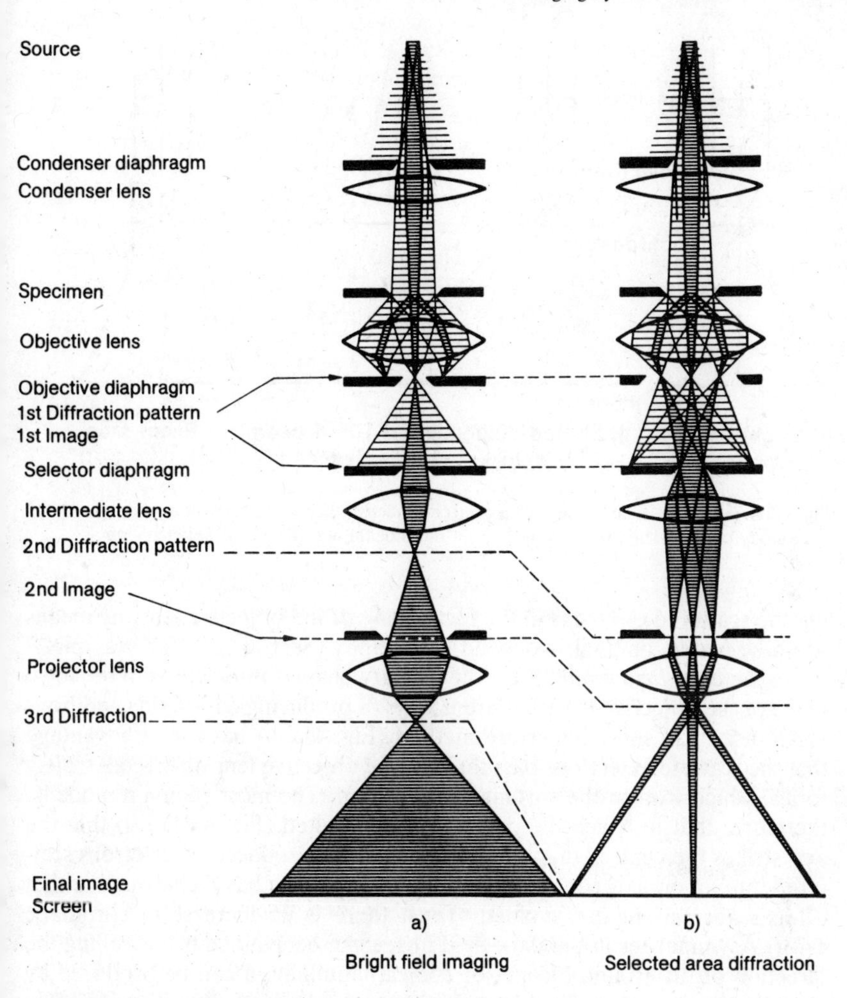

Fig. 4.20 a, b. Ray diagram for a transmission electron microscope in (**a**) the bright-field mode and (**b**) selected-area electron diffraction (SAED) mode

angles. The resolution is limited by the attenuation of the contrast-transfer function (CTF) caused by chromatic aberration (Sect. 6.4.2) and not by the objective aperture α_o. Normally, the specimen is irradiated with small illumination apertures $\alpha_i \leq 1$ mrad. For high resolution, an even smaller aperture, $\alpha_i \leq 0.1$ mrad, is necessary to avoid additional blurring of the CTF (Sect. 6.4.2). When unconventional types of contrast transfer are desired, it is often necessary to change the illumination condition, by tilting the beam or using hollow-cone illumination, for example, or by selecting and phase shift-

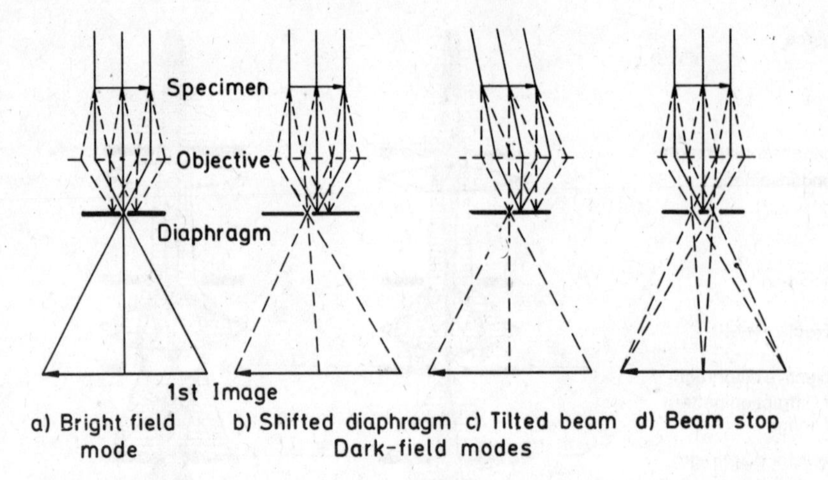

Fig. 4.21. (**a**) Bright-field mode with a centred-objective diaphragm and production of a dark-field mode by (**b**) a shifted diaphragm, (**c**) a tilted beam and (**d**) a central beam stop

ing the scattered electrons in the focal plane of the objective lens, by means of phase plates, or single-sideband holography (Sect. 6.5.2), for example.

In the *dark-field mode* (DF), the primary beam is intercepted in the focal plane of the objective lens. Different ways of producing dark-field conditions are in use. The shifted-aperture method (Fig. 4.21 b) has the disadvantage that the scattered electrons pass through the objective lens on off-axis trajectories, which worsen the chromatic aberration. The most common mode is, therefore, that in which the primary beam is tilted (Fig. 4.21 c) so that the axis strikes the centred diaphragm. The image is produced by electrons scattered into an on-axis cone of aperture α_o. This mode has the advantage that off-axis aberrations are avoided. Thus, there is no increase of chromatic error. Asymmetries in the dark-field image can be avoided by swivelling the direction of tilt around a cone, or conical illumination can be produced by introducing an annular diaphragm in the condenser lens. Another possibility is to use a central beam stop that intercepts the primary beam in the BFP; for this, a thin wire stretched across a circular diaphragm may be employed (Fig. 4.21 d). DF micrographs need a longer exposure time, because there are fewer scattered electrons. It is also possible to image single heavy atoms in the DF mode (Sect. 6.3.2) but the CTF of this mode is nonlinear, whereas the CTF of the BF mode is linear for weak phase specimens and allows image-correction procedures to be applied.

The DF mode can also be employed to image crystalline specimens with selected Bragg-diffraction spots. Increasing the objective aperture allows us to transfer the primary and one Bragg-reflected beam through the objective diaphragm. These beams can interfere in the final image. The fringe pattern is then an image of the crystal-lattice planes (Sect. 8.2.1). Optimum results are obtained for this *lattice-plane imaging mode* when the primary beam is tilted by the Bragg angle $+\theta_B$. The Bragg-reflected beam which is deflected by $2\theta_B$, passes through the objective lens with an angle $-\theta_B$ relative to the axis.

In a *many-beam imaging mode,* more than one Bragg reflection and the primary beam form a lattice image that consists of crossed lattice fringes, or an image of the lattice and its unit cells if a large number of Bragg reflections are used (Sect. 8.2.2). This mode is restricted to large unit cells, which produce diffraction spots of low Bragg angle Θ_B so that the phase shifts of the many beams produced by spherical aberration are not sufficiently different to cause imaging artifacts.

In a *multi-beam imaging mode* (MBI) (Sect 8.1.3) used at low magnification without resolution of the lattice structure, the DF images of the diffracted beams exactly overlap the BF images of the primary beam only if the spherical aberration is low enough. This situation can arise in HVEM or in the scanning transmission mode. The advantages of this mode are that the "pendellösung" or stacking-fault fringes are cancelled. Oscillating contrast is increased along dislocations if DF and BF are anti-complementary, whereas the fringe contrast is cancelled if DF and BF are complementary. These cases can occur at the top and bottom of a foil respectively. This MBI mode contains only contrast caused by differences of absorption of Bloch waves and also preserves the contrast of lattice defects.

Further operating modes of a TEM are described in other sections: *Scanning transmission mode* (Sect. 4.5), *electron-diffraction modes* (Sect. 9.3), *Lorentz microscopy* (Sect. 6.6) and *analytical modes* (x-ray microanalysis, Sect. 9.1; electron-energy-loss spectroscopy, Sect 9.2).

4.4.3 Magnification and Calibration

If structures as small as 0.2 nm are to be resolved, the instrument must be capable of magnifying this distance until it is larger than the resolution of the photographic emulsion (20–50 μm); this requires a magnification M of at least 250 000 times, for which more than two imaging lenses are needed.

The magnification depends on the excitation of the objective lens. If the magnification is to be constant to within about $\pm 1\%$, the following precautions have to be taken:

1) The height of the specimen in the specimen cartridge must be reproducible. Depending on the microsocpe used, a variation of the vertical position of ± 50 μm results in a variation of ± 2–5% in the magnification [4.58, 59].
2) The lens current and acceleration voltage must be highly stable. The lens current is related to the height of the specimen. Differences of specimen height can therefore be compensated by reading the lens current necessary for focusing and using a calibration curve relating lens current and

Fig. 4.22. Surface replica of a diffraction grating with 2160 lines per mm for the calibration of medium magnifications

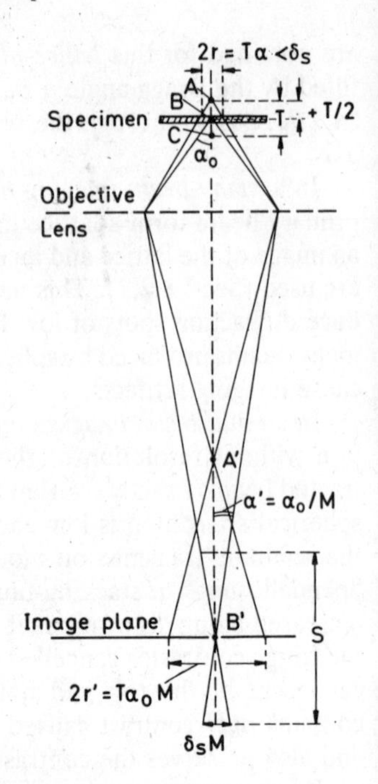

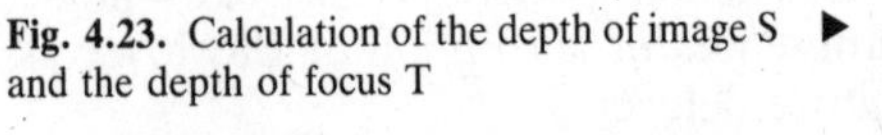

Fig. 4.23. Calculation of the depth of image S and the depth of focus T ▶

magnification. However, accuracies of the lens currents of the order of ±1% are needed, and a compensating circuit has to be used to measure the lens current [4.60].

3) Hysteresis effects in the iron part of the objective lens must be avoided [4.59]. This can be achieved by setting the lens excitation at its maximum value and then reducing the lens current down to (but not below) that needed for focusing. This cycle of maximum excitation and focusing has to be repeated two or three times.

Up to values of about 40 000 times, the magnification can be calibrated by means of surface replicas of metal gratings (Fig. 4.22) which are commercially available. For magnifications of the order of 100 000 times, images of lattice planes (Sect. 8.2.1) can be used, provided that the lattice constants are not altered by radiation damage. Dowell [4.61] discussed the procedure and the possible errors (±2%) if the lattice constant is calibrated with a TlCl standard by electron diffraction; these errors can mainly be attributed to the distortion of electron-diffraction patterns by the projector lens. For biological specimens, catalase crystals can be employed. The measured values of the lattice constants are 8.8 ± 0.3 nm [4.62] and 8.6 ± 0.2 nm [4.63]. Polystyrene spheres should be avoided for magnification calibration because their diameters are affected by the preparation, by radiation damage and by contamination [4.64].

4.4.4 Depth of Image and Depth of Focus

The depth of image is defined in Fig. 4.23. A blurring of the image $\delta_s M$ will be observed at a distance $\pm S/2$ from the final image on the viewing screen, where δ_s and S are related as follows

$$\delta_s M = \alpha' S \; ; \qquad S = \frac{\delta_s M}{\alpha'} = \frac{\delta_s M^2}{\alpha_o} . \tag{4.31}$$

Here, $\alpha' = \alpha_o/M$ denotes the angular aperture in the final image. Numerical example: For $M = 10\,000$, $\alpha_o = 10$ mrad, $\delta_s = 5$ nm, we find that $S > 50$ cm. Because of this large depth of image, a focused image will be obtained on the photographic plate even though the latter is some centimetres below the viewing screen, which is normally inclined for focusing.

Another property of the instrument is the depth of focus (Fig. 4.23). This is the axial distance $\pm T/2$ within which specimen details on the axis will be focused with a resolution δ_s. From Fig. 4.23, we have

$$T < \delta_s/\alpha . \tag{4.32}$$

The values of T listed in Table 4.3 can be expected when $\delta_s M$ is set equal to 50 μm, corresponding to the resolution of a photographic emulsion. Whether the objective aperture α_o or the illumination aperture α_i has to be used for α depends on the specimen. This is important because they vary by orders of magnitude: 5 mrad $\leq \alpha_o \leq$ 20 mrad and 10^{-2} mrad $\leq \alpha_i \leq$ 1 mrad. In thin specimens, which scatter electrons only weakly, the majority of the electrons are unscattered and emerge from the specimen with the illumination aperture α_i. In thick specimens, the electrons are strongly scattered and leave the specimen in a broad cone from which those within the aperture α_o are selected. Focusing at low magnifications sometimes becomes difficult owing to the large depth of focus. For thick specimens, a larger aperture can be used for focusing, which decreases the depth of focus T; for thin specimens, the illumination aperture can be artificially increased by rocking the electron beam [4.65].

Table 4.3. Depth of focus T for $\delta_s M$ = 50 μm (resolution of a photographic emulsion)

M	δ_s [nm]	T for the apertures: α = 20	5	1 mrad
200	250	12.5 μm	50 μm	250 μm
2 000	25	1.2 μm	5 μm	25 μm
10 000	5	0.25 μm	1 μm	5 μm
40 000	1	50 nm	0.2 μm	1 μm
160 000	0.5–1	25–50 nm	0.1–0.2 μm	0.5–1 μm

Light microscope: M = 1000, δ = 0.2 μm, α = 1 : T = 0.2 μm

These geometrical estimates of the depth of focus cannot be used for wave-optical phase contrast at high resolution because defocusing differences Δz as small as some tens of nm already change the image-intensity distribution. If the specimen contains a periodicity Λ or a spatial frequency $q = 1/\Lambda$, a diffraction maximum will be formed at $\sin\theta \simeq \theta = \lambda/\Lambda$. The maxima and minima of the specimen periodicity will be reversed in contrast when the second term of the wave aberration $W(\theta)$ in (3.66) caused by the defocusing Δz changes the phase of the diffracted beam by π. Setting $W(\theta) = \pi\Delta z\theta^2/\lambda \le \pi$ results in

$$\Delta z < \frac{\lambda}{\theta^2} = \frac{\Lambda}{\theta}. \qquad (4.33)$$

With $\Delta z \leftrightarrow T, \Lambda \leftrightarrow \delta_s, \theta \leftrightarrow \alpha_o$, this formula corresponds to (4.32). Numerical example: For $\lambda = 3.7$ pm (100 keV), $\alpha_o = 10$ mrad, $\Lambda = 0.37$ nm, we find $\Delta z < 37$ nm.

4.5 Scanning Transmission Electron Microscopy (STEM)

4.5.1 Scanning Transmission Mode of TEM

Unlike the conventional transmission mode of a TEM, in which the whole imaged specimen area is illuminated simultaneously, the specimen is scanned in a raster point-by-point with a small electron probe in the scanning transmission mode of a TEM.

The prefield of the objective lens is used as an additional condenser lens (Sect. 4.2.3) to form a small electron probe at the specimen when operating in the STEM mode (Fig. 4.17 b) [4.66]. The objective lens works near $k^2 = 3$ (condenser-objective single lens, Sect. 2.2.3). An electron-probe diameter of the order of 2–5 nm can be produced with thermionic cathodes. No further lenses are needed below the objective lens. Nevertheless, these lenses may be excited to image the first diffraction pattern in the back focal plane (BFP) through the small pole-piece bores of the subsequent lenses onto the electron-detector plane, above or below the final image plane. The generator that produces the saw-tooth currents for the deflection coils simultaneously deflects, in synchronism, the electron beam of a cathode-ray tube (CRT). The intensity of the CRT beam can be modulated by any of the signals that can be obtained from the electron-specimen interactions. The transmitted electrons can be recorded in the bright- and dark-field modes with a semiconductor detector or a scintillator-photomultiplier combination. These modes can be selected by placing large circular or sector diaphragms in front of the detector; we recall that an enlarged far-field diffraction pattern is produced in the detector plane by each object element in turn, and does not move during scanning if the pivot point of the primary-beam rocking is at the FFP (Fig. 4.17b).

In the conventional TEM, small illumination apertures α_i are used in the bright- and dark-field modes (Sect. 4.4.2). In the STEM mode, a small electron probe can be obtained only with a large value of $\alpha_i \simeq 10$ mrad (Sect. 4.2.2). The detector aperture α_d has to be matched to this illumination condition. Thus, in the BF mode it will be necessary to use a detector aperture $\alpha_d \simeq \alpha_i$. Otherwise a large part of the unscattered electrons would not be recorded and the signal-to-noise ratio would be correspondingly decreased. Details of contrast mechanisms and differences between STEM and the conventional TEM modes are discussed in Sect. 6.1.4 for amorphous specimens and in Sect. 8.1.3 for crystalline specimens.

An annular semiconductor detector or scintillator can be used below the specimen to record the forward-scattered electrons (FSE) scattered through angles $\theta \geq 10°$ [4.67]. Another can be placed above the specimen to record the backscattered electrons (BSE). Secondary electrons (SE) with exit energies ≤ 50 eV will move round the axis in spiral trajectories owing to the strong axial magnetic field and can be detected by a scintillator-photomultiplier combination situated between the objective and condenser lenses (Fig. 4.24). This SE mode can be used to image the surface structure of the specimen.

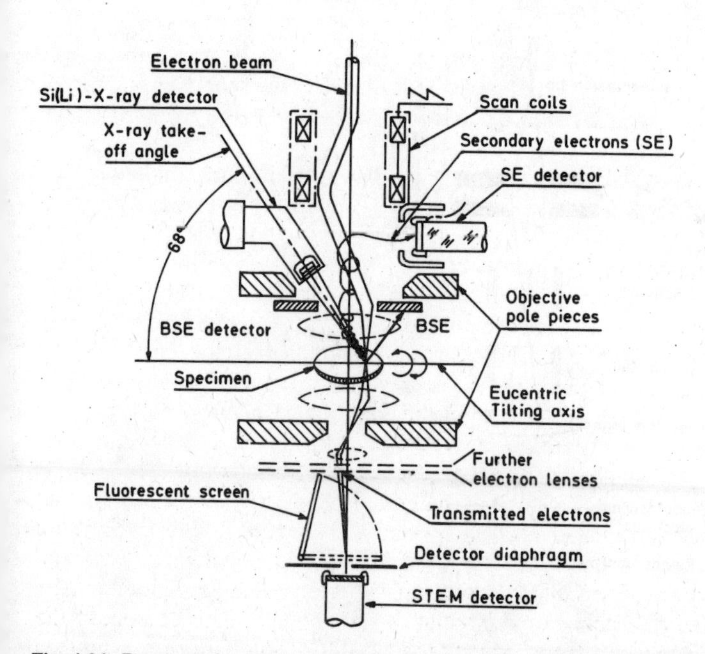

Fig. 4.24. Detectors for x-rays, secondary electrons (SE), backscattered electrons (BSE) and transmitted electrons (TE) in the scanning transmission mode of a TEM

The effect of chromatic aberration of the objective lens can be avoided in the STEM mode. This is of interest for thick specimens. However, the gain of resolution will be limited by the top-bottom effect caused by multiple scattering (Sect. 5.3.3). The main advantages of the STEM mode are production and positioning of small electron probes ≤ 0.1 μm for the micro-beam electron diffraction and convergent-beam diffraction techniques (Sect. 9.3.3) and for x-ray analysis of small specimen areas (Sect. 9.1). The STEM mode can also be used to generate other signals, which are in common use in surface-scanning electron microscopy; electron-beam-induced current (EBIC) in semiconductors or cathodoluminescence (Sect. 9.5), for example. These can be combined with the high-resolution capabilities of a TEM. Another advantage is the possibility of analogue and digital manipulation of the signal.

4.5.2 Field-Emission STEM

This type of electron microscope is designed to work only in the scanning transmission mode. Figure 4.25 schematically shows a version introduced by *Crewe* and coworkers [4.68–71]. A field-emission gun is used to produce a very small electron probe of of 3–10 nm behind the gun system (Sect. 4.1.4).

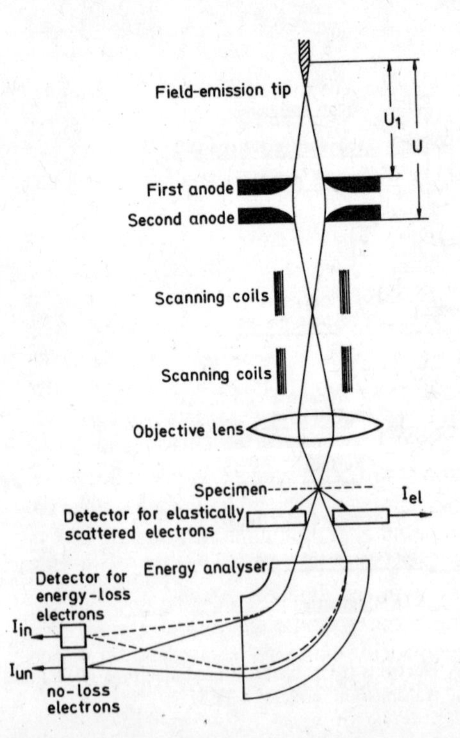

Fig. 4.25. Field-emission STEM with an electron-energy-loss spectrometer

Only one magnetic lens with a short focal length, low spherical aberration and equipped with a stigmator is needed to demagnify this probe to 0.2–0.5 nm on the specimen. The scanning coils are arranged in front of the lens. A signal I_{el} of large-angle elastically-scattered electrons can be detected by an annular detector. The cone of small-angle scattered electrons, which enters an electron spectrometer, can be separated into unscattered (I_{un}) and inelastically (I_{in}) scattered signals. These three signals will be discussed in detail in Sect. 6.3.5 and can be combined by analogue procedures. This offers new possibilities for contrast enhancement and noise filtering that are not available in a conventional TEM. The spectrometer can also be used for electron-energy-loss spectroscopy (Sect. 9.2) of a selected area or for energy-selecting microscopy (Sect. 9.2.5).

The field-emission gun, lens and spectrometer occupy little space and the whole STEM column can be mounted in an ultrahigh-vacuum vessel that operates at 10^{-8}–10^{-7} Pa. This allows the gun to operate satisfactorily and drastically reduces specimen contamination.

4.5.3 Theorem of Reciprocity

The reciprocity theorem was first discussed by Helmholtz (1860) in light optics. In geometrical optics, it is known as the reciprocity of ray diagrams. However, in wave optics it also implies that the excitation of a wave at a point P by a wave from a source Q is the same as that detected at Q with the source at P.

The ray diagram of STEM is the reciprocal of that of TEM [4.72, 73]. This will be demonstrated by the schematic ray diagram of Fig. 4.26. The *source* in the ray diagram of TEM in Fig. 4.26 a is already a demagnified image of the crossover produced by the condenser lenses. The intermediate image can be further enlarged by the subsequent lenses, not shown in the diagram. The specimen is illuminated with an illumination aperture α_i of the order of 0.1–1 mrad, which is much smaller than the objective aperture α_o = 5–20 mrad. The ray diagram of STEM (Fig. 4.26 b) has to be read in the reverse direction. The objective lens now demagnifies a source point on the specimen. A large probe aperture $\alpha_p \simeq \alpha_o$ is necessary to obtain the smallest possible spot size (Sect. 4.2.2). A fraction of the incident cone of the electron probe and scattered electrons are collected by the detector aperture α_d. If $\alpha_d = \alpha_i \ll \alpha_o = \alpha_p$ the same image contrast is obtained as in TEM (Sect. 6.1.4). A scanning unit between source and objective lens deflects the electron probe in a raster across the specimen. Projected backwards, the rays scan over a virtual source, which corresponds to the image plane of a TEM in Fig. 4.26 a. We can argue that in STEM, the CRT is needed to *image* this virtual plane by modulating the CRT with the detector signal.

By enlarging the ray diagram near the specimen in Fig. 4.27, we can demonstrate that the theorem of reciprocity can also be applied to wave-optical imaging, Fresnel fringes and phase contrast, for example. We con-

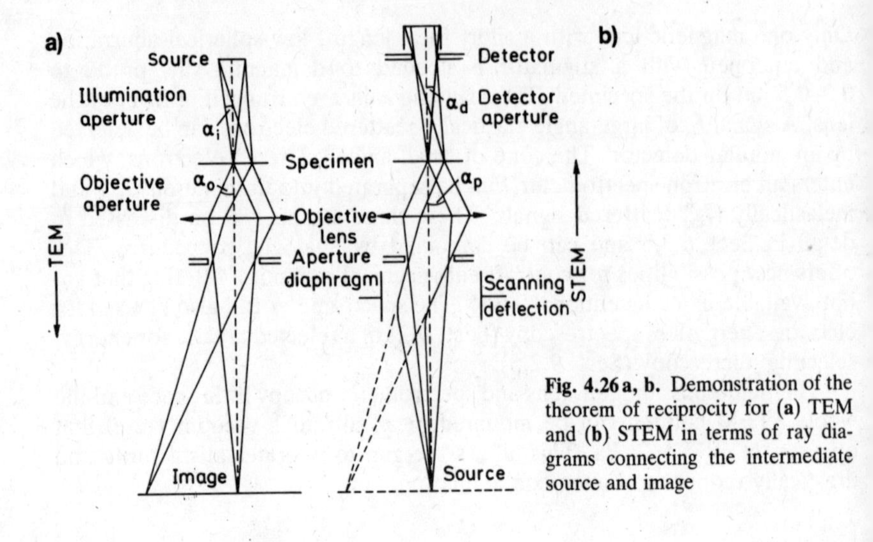

Fig. 4.26 a, b. Demonstration of the theorem of reciprocity for (**a**) TEM and (**b**) STEM in terms of ray diagrams connecting the intermediate source and image

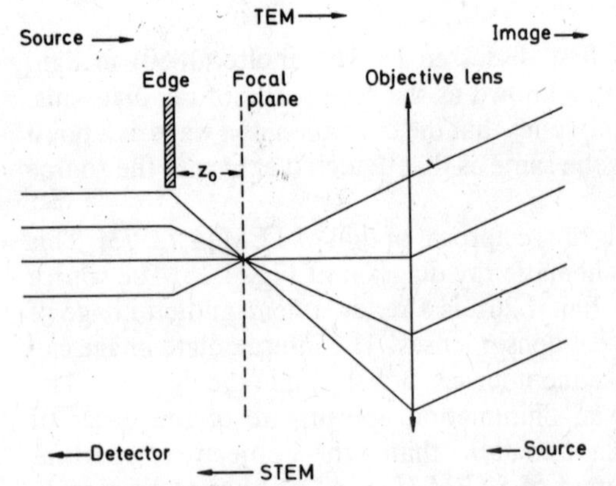

Fig. 4.27. Demonstration of the theorem of reciprocity of TEM and STEM for phase contrast (Fresnel fringes at an edge) (z_0 = defocusing)

sider the case of Fresnel fringes. Source and detector are assumed to be very distant so that the incident and exit waves can be regarded as plane waves. The objective lens in TEM enlarges the intensity distribution in the plane at a distance $\Delta z = z_0$ (defocusing) behind the specimen. At one point of this plane, the Huygens elementary wavelets from each point of the specimen edge overlap with their corresponding geometric phase shifts and form the Fresnel fringes of an edge (Fig. 3.9). When the diagram is reversed for discussion of the STEM mode, an electron probe is formed in the focal plane at a distance Δz in front of the specimen edge. The same geometric phase shifts as in TEM will occur during the wave propagation to the detector. It should be mentioned that the distance of the first Fresnel fringe from the edge increases as $\Delta z^{1/2}$ with increasing defocusing Δz [4.74] whereas a fringe distance that increased as Δz was observed [4.75], which can be explained by refraction at the wedge-shaped edge.

The phase shifts caused by the spherical aberration of the lens also act in the same manner in TEM and STEM. It was shown in Sect 3.2.2 that an increase of the illumination aperture α_i causes a blurring of the Fresnel fringes by $\pm \alpha_i \Delta z$, thus decreasing the number of observable fringes. The same effect would be obtained when recording a TEM image with a slit of width $2\alpha_i \Delta z$. An analogous blurring is observed in STEM if the detector area or the detector aperture α_d is increased. Therefore, if phase-contrast effects are to be observed in STEM, a small detector aperture has to be used ($\alpha_d \ll \alpha_p$). It will be shown in Sect 6.1.4 that this is an unfavourable operating condition. Exposure of the specimen to damaging radiation has to be kept low and all of the unscattered electrons have to be collected in order to image single atoms. This means that α_d should be approximately equal to α_p. This corresponds to extremely incoherent illumination in TEM. Single atoms are imaged in STEM only by their scattering contrast. In TEM, the optimum condition for imaging atoms corresponds to phase-contrast operation, for which $\alpha_i \ll \alpha_o$.

4.6 Image Recording and Electron Detection

4.6.1 Fluorescent Screens

The final image of a TEM can be observed on a fluorescent screen consisting of ZnS or ZnS/CdS powder, which is excited by cathodoluminescence. The colour can be varied by adding small concentrations of activator atoms, such as Cu or Mn. The maximum emission is normally in the green (550 nm), where the sensitivity of the human eye is maximum.

A fluorescent screen should provide a high light output L and a good resolution δ. The eye needs less intensity when the image is well resolved whereas more intensity is required to focus a blurred image. The quantity L/δ^2 can be used as a measure of the quality of a screen. Table 4.4 contains some data about various fluorescent screens. Although CdS and ZnS single crystals have high quality, they are not used, because the high resolution cannot be fully utilized in the recording step since photographic emulsions have a poorer resolution, of the order of 20–50 μm.

The light intensity of a fluorescent screen is proportional to the incident electron current density j, usually measured in A cm^{-2}. For constant j, the intensity might be expected to increase in proportion to the electron energy, because more light quanta are generated by high-energy electrons. Actually, a slower rate of increase is observed due to the increasing depth of generation and the subsequent absorption and scattering of the light quanta. The light-generating efficiency likewise decreases when the electron range exceeds the

Table 4.4. Characteristics of fluorescent screens irradiated with 60 keV electrons [4.76]

Fluorescent material	Luminescene L (relative)	Resolution (absolute) [μm]	(relative)	L/δ^2 (relative)
ZnS/CdS powder	1	50	1	1
ZnO powder	0.15	25	0.5	0.6
Uranium glass	0.003	5	0.1	0.3
ZnS single crystal	0.01–0.1	5	0.1	1–10
CdS single crystal				
(red luminescence)	0.15	5	0.1	15
cooled (green lum.)	0.2	7		
ZnS(Mn) evaporated layer (yellow-orange)	0.2	5	0.1	4

thickness of the fluorescent layer; this can be a problem in high-voltage electron microscopy [4.77].

The decay of intensity with time proceeds in two stages: a fast decrease with a time constant of the order of 10^{-5}–10^{-3} s is followed by an afterglow of the order of seconds. For a faster response, in STEM, for example, fluorescent materials with time constants less than 1 μs are needed (Sect. 4.6.6).

4.6.2 Photographic Emulsions

Photographic emulsions supported on glass plates or flexible films are directly exposed to the electrons inside the microscope vacuum. The gelatin of the photoemulsion contains a considerable amount of water, and it is necessary to dehydrate the photographic material in a desiccator at 1 Pa and to load the microscope camera as quickly as possible [4.78]. When using a transparent fluorescent screen in contact with a glass fibre-optic plate, images can be recorded in air by direct contact with the outer side of the fibre-optic plate [4.79].

The basic processes that occur in the exposure of photographic emulsions to electrons will now be discussed; for more details see [4.80–85]. The ionisation probability of electrons is so large that each silver halide particle penetrated is rendered developable and can be reduced to a silver grain. High-energy electrons in the MeV range can probably penetrate some grains without ionisation. For light, on the contrary, several quanta have to be absorbed in one grain for it to be made developable. Unlike light exposure, therefore, there is no illumination threshold for exposure to electrons.

The following law for the photographic density D can be derived with this exposure mechanism. The density D of a developed photographic emulsion is defined as decimal logarithm of the ratio of the light transmission L_0 of an unexposed part of the plate and that of an exposed region (L):

$$D = \log_{10}(L_0/L) . \tag{4.34}$$

A saturation density D_{max} results when all of the grains are developed. Owing to the statistical nature of silver-grain production, the density D of an unsaturated plate exposed to a charge density J will be given by

$$D = D_{max}(1 - e^{-cJ}) , \tag{4.35}$$

where

$$J = j\tau = en \tag{4.36}$$

in C cm^{-2}; j denotes the current density in A cm^{-2}, τ is the exposure time and n the number of incident electrons per unit area. Equation (4.35) has been confirmed for low and medium densities. The validity of this law means that a long exposure with low j produces the same density as a short exposure with high j if the product $j\tau$ is constant. This law of reciprocity is not true for light. For the latter, the relation can be expressed in terms of the Schwarzschild exponent $\varkappa$, different from unity, equal densities being obtained for $j\tau^{\varkappa}$ = const. All experiments show that $\varkappa = 1$ for exposure to electrons. However, for some emulsions the results depend on the delay between exposure and development [4.86].

Equation (4.35) leads to the proportionality

$$D = cD_{max}J = \varepsilon J , \tag{4.37}$$

where the sensitivity ε is defined by

$$\varepsilon = (dD/dJ)_{J\to 0} = cD_{max} . \tag{4.38}$$

This is valid for $D \leq 0.2\ D_{max}$, which means, in practice, $D \leq 0.6$–1.5 (Fig. 4.28 a).

If N grains are developed per unit area with a mean projected area $\bar{a}$, the density D can be written for small J as

$$D = \frac{1}{2.3} N\bar{a} = \frac{1}{2.3}\frac{pJ\bar{a}}{e} , \tag{4.39}$$

with ln 10 = 2.3 and p the mean number of grains exposed by one electron. This gives for the sensitivity

$$\varepsilon = \frac{1}{2.3}\frac{p\bar{a}}{e} . \tag{4.40}$$

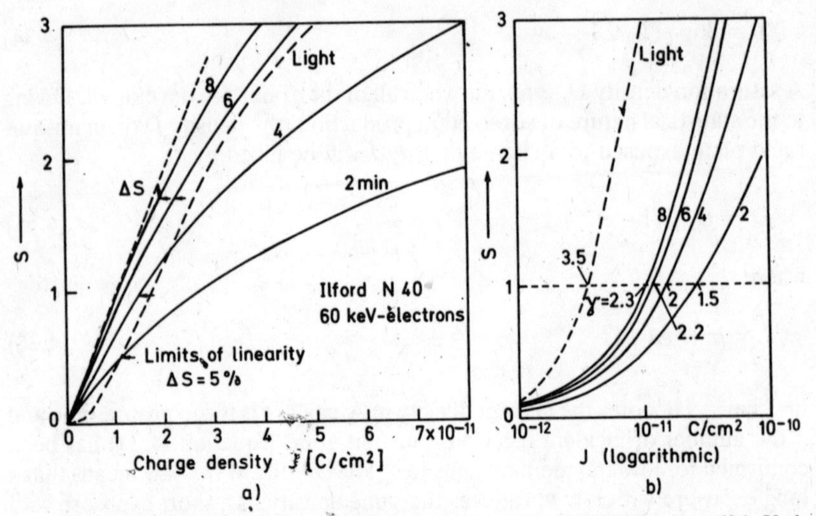

Fig. 4.28. (**a**) Density curves of a photographic emulsion Ilford N 40 exposed to 60 keV electrons and to light for the indicated developing times. (**b**) Double-logarithmic plot with values of $\gamma = dB/d(\log J)$ at $D = 1$

The mean number of grains exposed depends on the electron energy and the following parameters of the emulsion: quantity of silver per unit area (0.4–0.6 mg cm^{-2}), mean density $\varrho = 1$–2 g cm^{-3}, thickness of emulsion $t = 1$–50 µm and grain diameter (0.5–2 µm). The electron energy and the mean density determine the electron range R, which lies between 75 and 120 µm for 100 keV electrons. The sensitivity increases as E increases if $R < t$ and decreases if $R > t$ because the ionisation probability per unit path length of an electron trajectory decreases with increasing energy. Photoemulsions, therefore, exhibit decreasing sensitivity with increasing energy in HVEM, which can be partly compensated by using thicker emulsions [4.77, 4.87]. The sensitivity can also be increased by placing a metal foil in front of the emulsion, which decreases the electron energy and increases the ionisation probability [4.88].

It is usual to plot D versus $\log J$ (Fig. 4.28 b). This curve has a straight region with slope γ for medium densities. This slope is used to characterise the photographic emulsion because high values of γ correspond to high contrast recording. A relative variation of current density $\Delta j/j$ or of charge density $\Delta J/J$ produces a relative variation of light transmission $\Delta L/L = -\gamma\Delta J/J$. During exposure to light, the value of γ can be high even at low density owing to the existence of a threshold (see exposure to light in Fig. 4.28 a). Because the density curve for electron exposure does not show a threshold, γ cannot increase beyond a certain limit. The proportionality (4.37) can be written as

$$D = \varepsilon J = \varepsilon\, 10^{\log J} , \tag{4.41}$$

and the maximum possible slope γ is given by

$$\gamma_{\max} = \frac{dD}{d(\log J)} = \varepsilon \ln 10 \cdot 10^{\ln J} = 2.3\ D . \tag{4.42}$$

With electron exposures, therefore, it is impossible to obtain a value of γ greater than 2.3 for a density of unity. γ can increase as long as the density increases with J. No further increase is observed when D approaches the saturation value $D_{\max}$. A further increase of contrast can be obtained by a suitable choice of the photographic material used for printing the micrograph.

The resolution of an emulsion is limited by two effects: the diameter of the electron-diffusion cloud and the granularity of the emulsion. When exposed to light, a halo is formed by scattering at the silver halide grains, the radius of which depends on the grain size. The diffusion halo in electron exposure depends only on electron energy and the mean density of the emulsion. If a slit of width d is illuminated with unit intensity, a density distribution (edge spread function)

$$S(x) = \frac{2.3\ d}{x_k}\ 10^{-2|x|/x_k} \tag{4.43}$$

($d \ll x_k$) is obtained [4.81]. The quantity x_k, typically 30–50 μm, is the width over which the intensity falls to 10% of the central value. Figure 4.29 a shows the intensity recorded by an emulsion for which $x_k = 50$ μm exposed to a slit of width $d = 10$ μm.

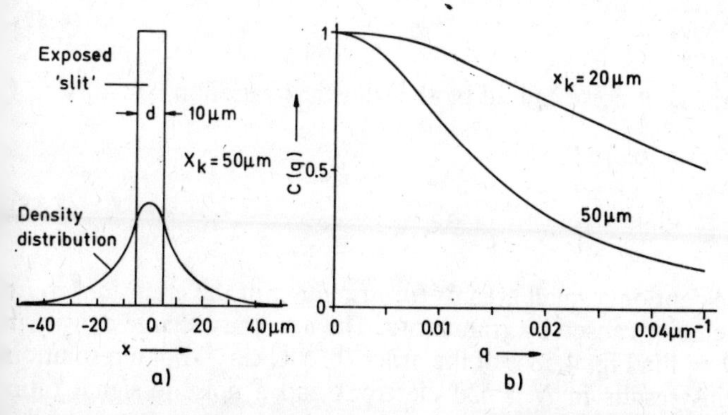

Fig. 4.29. (**a**) Edge-spread function and (**b**) contrast-transfer function $C(q)$ caused by electron diffusion in the photographic emulsion

Suppose now that the density varies periodically with a spacing corresponding to a spatial frequency $q = 1/\Lambda : D = D_0 + \Delta D \cos(2\pi q x)$. This function has to be convoluted with $S(x)$, which results in a decrease of the density amplitude from ΔD to $\Delta D'$. The contrast-transfer function,

$$C(q) = \frac{\Delta D'}{\Delta D} = \frac{1}{1 + \left(\frac{\pi q x_k}{\ln 10}\right)^2} = \frac{1}{1 + (1.36\, q x_k)^2}, \tag{4.44}$$

is plotted in Fig. 4.29 b.

The granularity can be considered in the following manner. The number p of neighbouring grains exposed is greater when an electron passes through the emulsion than when it is stopped by it. Depending on the grain size, the thickness of the emulsion and the electron energy, p lies between 6 and 50. For light ($p = 1$), the mean-square deviation of the density with a photometer slit of area A is

$$\overline{\Delta D_{\mathrm{L}}^2} = \frac{1}{2.3} \frac{\bar{a}}{A} D. \tag{4.45}$$

An emulsion exposed to a homogeneous current density j appears more granular than one exposed to light because, during electron exposure, more neighbouring silver grains are produced in clusters. The observed mean-square deviation $\overline{\Delta D_{\mathrm{E}}^2}$ will lie between the limits

$$\overline{\Delta D_{\mathrm{L}}^2} < \overline{\Delta D_{\mathrm{E}}^2} < (p+1)\, \overline{\Delta D_{\mathrm{L}}^2}. \tag{4.46}$$

In order to detect a periodicity, the amplitude $\Delta D'$, already decreased by electron diffusion, must be approximately five times greater than the noise

$$\Delta D' \geq 5 \sqrt{\overline{\Delta D_{\mathrm{E}}^2}}. \tag{4.47}$$

Furthermore, the shot noise caused by the statistical variation $\Delta N = N^{1/2}$ of the number

$$N = n\delta^2 = \frac{j\tau}{e} \delta^2 \tag{4.48}$$

of electrons incident on a small area δ^2 (δ: resolution of the emulsion) must be less than the noise caused by granularity. The necessary charge density $j\tau$ for a density $D = 1$ in Fig. 4.28 is of the order 10^{-11} C cm^{-2}. For a resolution δ of 30 µm, this results in $N = 530$ electrons and a noise-to-signal ratio $\Delta N/N = N^{-1/2} = 4\%$. The human eye can detect relative intensity variations of the order of 5%. This means the sensitivity of photographic emulsions to electron exposure is of just the right order, and photographic emulsions are optimum for the recording of electron micrographs. A film size of $A = 6 \times 9\ \mathrm{cm}^2$ contains $A/\delta^2 = 6 \times 10^6$ image points which corresponds to a very high storage capability.

4.6.3 Image Intensification

Difficulties in correcting the astigmatism and in focusing arise at high resolution because the qualities of fluorescent screens are not perfectly matched to the resolution and sensitivity of photographic emulsions. In particular, a larger radiation dose is needed for visual observation than is necessary for photographic recording. This problem can be partially solved by the use of an image intensifier. When the sensitivity is increased to the point where single electrons can be recorded as bright spots on a TV screen, the limit of particle noise is reached, and a satisfactory compromise between noise and detail recognition or contrast can be found. Furthermore, TV recording has the advantage that the background can be subtracted electronically and the contrast is increased. Dynamical effects can be recorded on videotape.

A transmission fluorescent screen may be coupled to a TV camera tube by means of a glass fibre-optic plate. A plumbicon valve [4.89] or a SEC valve with a KCl layer of high secondary-electron yield can be used. If an image intensifier tube is inserted between the glass fibre-optic plate and the

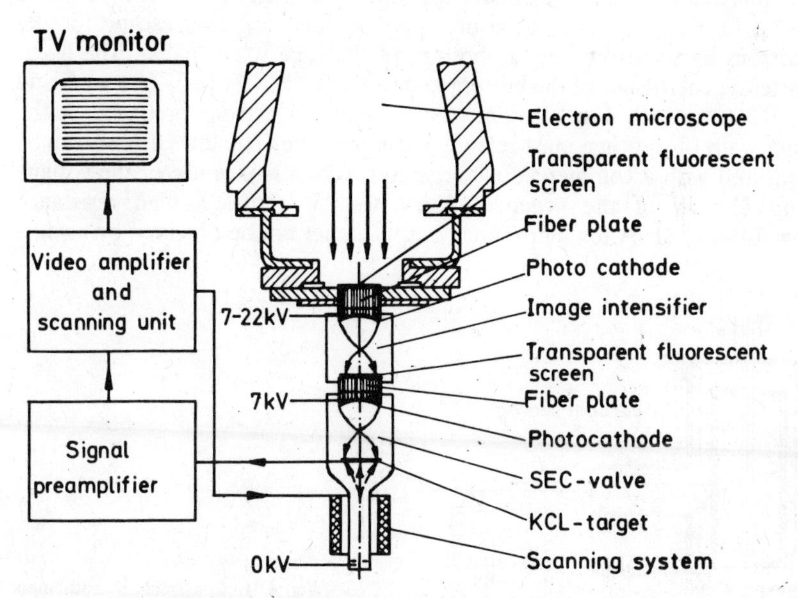

Fig. 4.30. Image intensifier consisting of fluorescent layer, fibre-optic plate, image-intensifier tube and TV camera

TV camera (Fig. 4.30), it is possible to record single electrons [4.90]. As with photographic emulsions, images can be stored and integrated over a longer time than the normal TV frame time. If a signal-to-noise ratio of 10 is required, the current density in the image plane need be only 10^{-12} A cm^{-2} for a frame frequency of 25 s^{-1} and 10^{-14} A cm^{-2} for a storage time of 4 s. For comparison, photographic emulsions need 10^{-12} A cm^{-2} for a 10 s exposure time and ZnS fluorescent screens need 10^{-11}–10^{-10} A cm^{-2} for observation at high resolution.

Channel plates consist of parallel tubes, 40–50 μm in diameter, coated with a material of high secondary-electron yield. A bias of 1000 V between the front and reverse side of the plate produces a voltage drop along the inner sides of the tubes and the latter act as secondary-electron multipliers. Gains of 10^3–10^4 can be obtained [4.91–93]. However, only 10% of 100 keV electrons produce a primary secondary electron; moreover, the channel tubes contaminate and the gain is decreased in the normal vacuum of an electron microscope, at the high count rates needed.

4.6.4 Faraday Cages

Direct measurement of electron currents is of interest for determination of electron current densities and electron-beam currents. Quantitative measurement needs a Faraday cage. The latter consists of an earthed shield that contains a hole somewhat smaller than the inner cage (Fig. 4.31). The hole has to be small enough to ensure that the solid angle of escape for the electrons backscattered at the bottom of the cage is negligible. The backscattering coefficient of the bottom material must be low (η = 6% for C and η = 13% for Al). Furthermore, the secondary electrons produced at the inner walls of the cage must remain inside the cage. The low currents can be measured with a commercial electrometer, which makes use of the voltage drop $U = RI$ of the order of 1 mV to 1 V across a high resistance $R = 10^6$–10^{10} Ω. A low impedance output signal can be obtained by using a

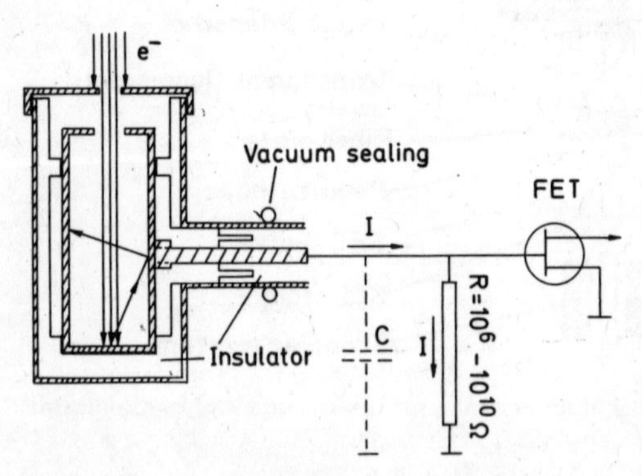

Fig. 4.31. Construction and input circuit of a Faraday cage for measuring electron currents

field-effect transistor (FET) or a vibrating reed electrometer, which amplifies the ac signal of a capacitor with one vibrating plate. The high resistance R and the by-no-means-negligible capacitance C of the cage, the cables and the electrometer input result in a time constant $\tau_0 = RC$ which may reach a few seconds for very small currents. A Faraday cage, therefore, cannot be used to record fast variations of low electron currents.

4.6.5 Semiconductor Detectors

A semiconductor detector consists of a p-n junction diode below a surface layer. High-energy electrons of energy E create $n = E/\overline{E}_i$ electron-hole pairs. The mean energy $\overline{E}_i$ for creating one pair is 3.6 eV in silicon at room temperature. The electron-hole pairs created in a zone of depth $z = R$ (R: electron range) are separated in the sensitive volume of the depletion layer of the junction and can partially diffuse to the junction before recombination. This produces a detector current I_D with a gain g_D

$$I_D = fI_p g_D \quad \text{with} \quad g_D = f_c \frac{E - E_{min}}{\overline{E}_i} , \tag{4.49}$$

where f is the fraction of the incident probe current I_p that impinges on the active area of the detector and f_c the collection efficiency of the junction for electron-hole pairs created; $E_{min} = 5$–10 keV is a minimum threshold energy of the incident electrons, due to absorption in an evaporated-gold contact layer and/or to an increased surface recombination rate (dead layer).

Because of the relatively large capacitance C of the depletion layer, a low-impedance current amplifier has to be used to convert I_D to a video voltage of a few 100 meV. The time constant $\tau_0 = RC$ decreases with decreasing area of the depletion layer and increasing current fI_p. The capacitance can be further decreased by employing reverse biasing of the p-n junction, but a zero bias is normally used. Currents of 10^{-11} A can be recorded in about 10^{-5} s which corresponds to a cut-off frequency of 100 kHz of the video signal. By decreasing C, it is possible to observe backscattered electrons at TV scan rates [4.94, 95].

4.6.6 Scintillation Detectors

Scintillator materials emit light quanta (photons) under electron bombardment. ZnS, which is used for fluorescent screens in TEM, has a high efficiency but its light-intensity decay time is of the order of 10^{-3} s and the afterglow persists for several seconds; it cannot be used, therefore, for fast recording. Plastic scintillators (NE 102 A of Nuclear Enterprise Ltd., for example) and P-47 powder (yttrium silicate doped with 1% cerium) have become standard scintillator materials for SEM and STEM, because their time constants are of the order of 10^{-8} s and their efficiency is not worse than one tenth that of ZnS.

Plastic scintillators are evaporated with a conductive and light absorbing Al coating about 100 nm thick. The light emission decreases with increasing irradiation time owing to radiation damage of the organic material. However, the thin damaged layer can be removed by polishing. P-47 powder layers exhibit a much larger radiation resistance. Methods of preparing P-47 layers with optimum thickness are reported in the references [4.96, 97].

The photons emitted are collected by a light pipe, in front of the photomultiplier, which reflects the light by total reflection with a transmittance T. The photons are converted to photo-electrons at the photocathode of the multiplier with a quantum efficiency q_c between 5 and 20%. The photoelectrons are accelerated by a potential difference of + 100 V to an electrode of high secondary-electron yield $\delta_{PM} = 8$–15. The total gain of the photomultiplier is obtained by successive acceleration and secondary-electron emission at $n = 8$–10 electrodes, resulting in a total gain $g_{PM} = \delta_{PM}^n$. The pulse of g_{PM} electrons or the current induced by a higher rate of incident electrons causes a voltage drop U across a resistor $R = 100$ kΩ, which can be amplified by operational amplifiers. For an incident probe current I_p and a detector collection efficiency f, which depends on the signal generated (transmitted, secondary or backscattered electrons) and on the solid angle of collection, the signal is

$$U = fI_p \frac{E}{\bar{E}_{ph}} Tq_c \delta_{PM}^n R , \tag{4.50}$$

where $\bar{E}_{ph}$ denotes the energy needed to produce one photon in the scintillator. Such a scintillator-photomultiplier combination can be operated with a large bandwidth Δf, up to some MHz and low noise. It is possible to achieve a rms noise amplitude that is only a factor 1–2 larger than the shot noise. The latter is the noise amplitude

$$I_{rms} = (2eI_p\Delta f)^{1/2} \tag{4.51}$$

associated with an electron current I_p, caused by statistical fluctuations of the number of electrons incident during equal sampling times.

5. Electron-Specimen Interactions

The elastic scattering of electrons by the Coulomb potential of a nucleus is the most important of the interactions that contribute to the image contrast. Cross-sections and mean-free-path lengths are used to describe quantitatively the scattering process. A knowledge of the screening of the Coulomb potential of the nuclei by the atomic electrons is important when calculating the cross-sections at small scattering angles.

The inelastic scattering is concentrated within smaller scattering angles and the excitation of energy states results in energy losses. The most important mechanisms are plasmon and interband excitations and inner-shell ionisations. The inelastic scattering process is less localized than elastic scattering and cannot contribute to high resolution but the analytical modes of energy-loss spectroscopy become of greater interest. The inner-shell ionisation also results in the subsequent emission of characteristic x-ray quanta or Auger electrons, when the electrons return into the initial states.

Even quite thin specimen layers, of the order of a few nanometres, do not show the angular or energy-loss distribution corresponding to a single scattering process. Multiple scattering effects have to be considered as the specimen thickness is increased and this can also result in electron-probe broadening.

5.1 Elastic Scattering

5.1.1 Cross-Section and Mean Free Path

The most convenient quantity for characterizing the angular distribution of scattered particles is the differential cross-section, which is introduced in Fig. 5.1 a, using the model of Coulomb scattering of an electron by a nucleus. The electrons travel on hyperbolic trajectories due to the attractive Coulomb force (3.17) between electron and nucleus. If there were no interaction, the electron would travel straight past the nucleus; the shortest distance between them, the impact parameter, is denoted by a. Increasing a decreases the scattering angle θ. Electrons that pass through an element of area $d\sigma$ of the parallel incident beam will be scattered into a cone of solid angle $d\Omega$. The ratio $d\sigma/d\Omega$ is known as the differential cross-section and is a function of the scattering angle θ.

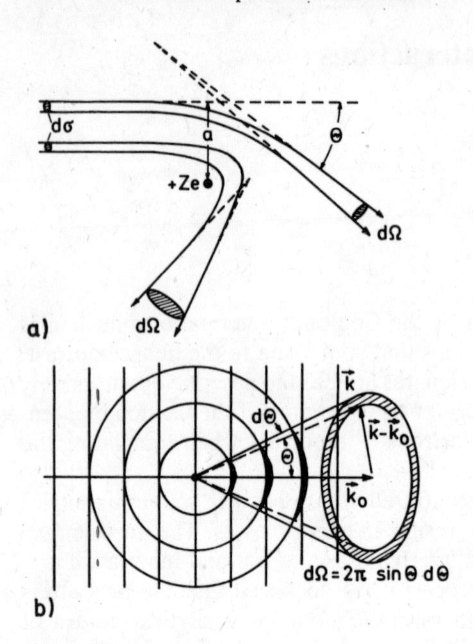

Fig. 5.1. (**a**) Elastic electron scattering in the particle model and explanation of the differential cross-section $d\sigma/d\Omega$ (p: impact parameter), (**b**) Scattering in the wave model with the superposition of a plane incident wave of wave number $\boldsymbol{k}_o$ and a spherical scattered wave of amplitude $f(\theta)$, depending on the scattering angle θ

This cross-section $d\sigma/d\Omega$ cannot be calculated exactly from this classical particle model; wave mechanics has to be used (Sect. 5.1.3). Far from the nucleus, the total wave field can be expressed as the superposition of the undisturbed plane incident wave of amplitude $\psi = \psi_0 \exp(2\pi i k_0 z)$ and a spherical scattered wave of amplitude

$$\psi_{sc} = \psi_0 f(\theta) \frac{1}{r} e^{2\pi ikr} \tag{5.1}$$

depending on the scattering angle θ (Fig. 5.1b).

The current density $j_0 = eNv$ of a parallel beam has been introduced in (3.11); Nv is the flux of particles that pass through unit area per unit time. Scattering into the solid angle $d\Omega$ is observed when the electron hits the fraction $d\sigma$ of the unit area. The scattered current dI_{sc} that passes through the area $dS = r^2 d\Omega$ will be

$$dI_{sc} = j_{sc} r^2 d\Omega = j_0 d\sigma \quad \text{which implies} \quad j_{sc} = (j_0/r^2)(d\sigma/d\Omega) . \tag{5.2}$$

Substituting the scattered-wave amplitude ψ_{sc} (5.1) in the quantum-mechanical expression for the current density (3.12) yields

$$j_{sc} = ev \frac{|f(\theta)|^2}{r^2} |\psi_0|^2 = j_0 \frac{|f(\theta)|^2}{r^2} , \tag{5.3}$$

with $\nabla = \partial/\partial r$. The current that passes through the area $dS = r^2 d\Omega$ then becomes

$$dI_{sc} = j_{sc} r^2 d\Omega = j_0 |f(\theta)|^2 d\Omega \,. \tag{5.4}$$

Comparing this with (5.2), we find

$$\frac{d\sigma}{d\Omega} = |f(\theta)|^2 \,. \tag{5.5}$$

Suppose that n electrons are incident on a solid film per unit area and that the film has a mass-thickness $dx = \varrho dz$ in units g cm^{-2}. There will be $N_s = N\varrho\, dz$ atoms per unit area in a layer of thickness dz with $N = N_A/A$ atoms per gram (N_A is Avogadro's number and A is the atomic weight). Scattering into the solid angle $d\Omega$ occurs when the electrons hit a small area $d\sigma$ in the vicinity of each atom. A scattering event will be recorded when the electrons strike a fraction $f = N_s d\sigma = N_A \varrho\, dz\, d\sigma/A$ of the unit area; a fraction $f = dn/n$ of the incident electrons will be scattered into a solid angle $d\Omega$

$$\frac{dn}{n} = -N d\sigma\, dx = -N \frac{d\sigma}{d\Omega}\, dx\, d\Omega \,. \tag{5.6}$$

The negative sign indicates that n is decreased by scattering.

The number of electrons scattered through angles $\theta \geq \alpha$ can be calculated by dividing the corresponding solid angle into small segments $d\Omega = 2\pi \sin\theta\, d\theta$ (Fig. 5.1 b), and integrating over θ from α to π. This gives the partial cross-section $\sigma(\alpha)$:

$$\sigma(\alpha) = \int_{\alpha}^{\pi} \frac{d\sigma}{d\Omega}\, 2\pi \sin\theta\, d\theta \,. \tag{5.7}$$

The *total elastic cross-section* is obtained if the lower limit of integration is zero, i.e.,

$$\sigma_{el} = \int_{0}^{\pi} \frac{d\sigma}{d\Omega}\, 2\pi \sin\theta\, d\theta \,. \tag{5.8}$$

This quantity can be used to calculate the number of unscattered electrons, for example. Whether or not scattering occurs is determined by the total (elastic and inelastic) cross-section $\sigma_t = \sigma_{el} + \sigma_{inel}$ (σ_{inel} is the total inelastic cross-section, Sect. 5.2.2).

As in (5.6), the decrease dn of the number of unscattered electrons will be

$$\frac{dn}{n} = -N\sigma_t dx \,. \tag{5.9}$$

Integrating, we find

$$\ln n = -N\sigma_t x + \ln n_0 \,, \tag{5.10}$$

where the constant of integration, $\ln n_0$, is determined by the initial number $n = n_0$ of incident electrons per unit area at $x = 0$. This shows that the number of unscattered electrons decreases exponentially with increasing mass-thickness x

$$n = n_0 \exp(-N\sigma_t x) = n_0 \exp(-x/\Lambda_t) \,. \tag{5.11}$$

The length $\Lambda_t = 1/N\sigma_t$ is known as the total mean free path (in units g cm^{-2}) between scattering events.

5.1.2 Energy Transfer in an Electron-Nucleus Collision

An elastic collision is defined as a collision in which the total kinetic energy and the momentum are conserved. The laws of conservation of energy and momentum before and after the collision can be written without any detailed knowledge of the interaction process between the particles. We characterize quantities after the collision by a dash and those of the nucleus by the suffix n. From Fig. 5.2, the conservation of momentum can be expressed as:

$$\boldsymbol{p} = \boldsymbol{p}' + \boldsymbol{p}'_n \quad \begin{cases} p = p' \cos\theta + p'_n \cos\psi & (5.12) \\ 0 = p' \sin\theta - p'_n \sin\psi & (5.13) \end{cases}$$

and the conservation of kinetic energy

$$E = E' + E'_n \,. \tag{5.14}$$

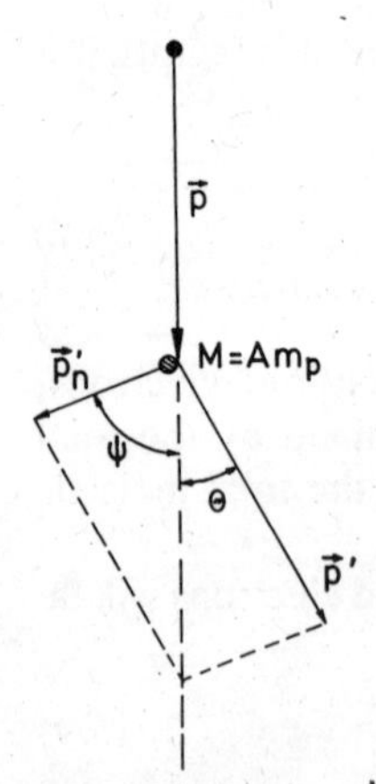

Fig. 5.2. Conservation of momentum in elastic scattering

Equation (2.11) has to be used for the relativistic momentum of the electron, whereas the non-relativistic formula

$$p'_{\rm n} = (2ME'_{\rm n})^{1/2} \tag{5.15}$$

can be used for the momentum of the nucleus, because its rest mass $M = Am_{\rm p}$ is very large ($m_{\rm p}$: atomic mass unit). Solving (5.13) for $\sin\psi$, and (5.14) for E' and substituting these quantities in (5.12), we obtain

$$\frac{1}{c}[E(E+2E_0)]^{1/2} = \frac{1}{c}[(E-E'_{\rm n})(E-E'_{\rm n}+2E_0)]^{1/2}\cos\theta$$

$$+\left[2ME'_{\rm n}\left(1-\frac{(E-E'_{\rm n})(E-E'_{\rm n}+2E_0)}{2Mc^2E'_{\rm n}}\sin^2\theta\right)\right]^{1/2}. \tag{5.16}$$

The energy transfer $E'_{\rm n}$ to the nucleus will be small compared to E, so that $E - E'_{\rm n} \simeq E$. Transfering the first term on the right-hand side of (5.16) to the left-hand side, squaring the equation and using the relation $1-\cos\theta = 2\sin^2(\theta/2)$, we find

$$E'_{\rm n} = \frac{2E(E+2E_0)}{Mc^2}\sin^2\frac{\theta}{2} = \frac{E(E+1.02)}{496A}\sin^2\frac{\theta}{2}, \tag{5.17}$$

with $E'_{\rm n}$, E and E_0 in MeV.

From the conservation of energy (5.14), this energy $E'_{\rm n}$ transferred to the nucleus must be equal to the energy lost by the electron, $\Delta E = E - E'$. Table 5.1 shows typical values of $E'_{\rm n}$. This energy loss is negligible for small scattering angles, owing to the presence of the factor $\sin^2(\theta/2)$. Therefore, in elastic electron-nucleus small-angle scattering we can say that there is effectively no energy loss of the primary electron.

However, the energy losses are not negligible for greater electron energies and scattering angles. If the energy transfer $E'_{\rm n}$ is greater than the dis-

Table 5.1. Energy transfer $E'_{\rm n}$ to a nucleus, which is equal to the energy loss ΔE of the primary electron in an elastic scattering process with a scattering angle θ for 100 keV and 1 MeV electrons

	E = 100 keV			E = 1 MeV		
	C	Cu	Au	C	Cu	Au
θ	(A = 12)	(A = 63.5)	(A = 197)			
0.5°	0.5 meV	0.1 meV	0.03 meV	9 meV	1.7 meV	0.54 meV
10°	0.15 eV	29 meV	9 meV	2.7 eV	0.5 eV	0.17 eV
90°	10 eV	1.9 eV	0.6 eV	179 eV	34 eV	11 eV
180°	20 eV	3.8 eV	1.2 eV	359 eV	68 eV	22 eV

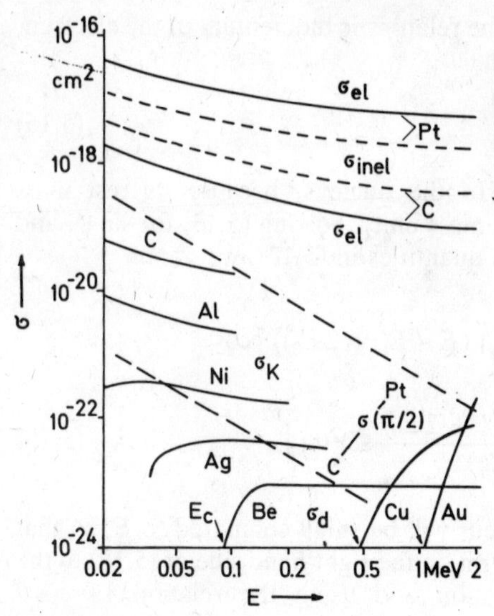

Fig. 5.3. Comparison total cross sections σ for elastic scattering (σ_{el}), for inelastic scattering (σ_{inel}), for K-shell ionisation (σ_K), for back-scattering into angles $\theta \geq \pi/2$ ($\sigma_{\pi/2}$) and for an atomic displacement with a displacement energy $E_d = 20$ eV (σ_d) as functions of electron energy E

placement energy $E_d \simeq 10 - 30$ eV, nuclei can be displaced from their lattice points to interstitial sites, resulting in radiation damage, which has to be considered in high-voltage electron microscopy (Sect. 10.3.2). Carbon atoms can also be knocked out of organic compounds by this direct transfer of momentum. However, the cross-section σ_d for this knock-on process is smaller by orders of magnitude than the cross-sections σ_{el} and σ_{inel} (Sect. 5.1.5 and 5.2.2) for elastic and inelastic scattering (Fig. 5.3). The displacement cross-section σ_d for an energy transfer $\Delta E = E_n' \geq E_d$ can be calculated by first determining the minimum scattering angle θ_{min} for which $E_n' = E_d$ by use of (5.17). We expect $E_n' \geq E_d$ for all $\theta \geq \theta_{min}$ and

$$\sigma_d = \int_{\theta_{min}}^{\pi} \left(\frac{d\sigma}{d\Omega}\right)_M 2\pi \sin\theta \, d\theta \tag{5.18}$$

in which we use the differential Mott cross-section $(d\sigma/d\Omega)_M$, appropriate for large scattering angles and relativistic energies (Sect. 5.1.4). The threshold energy E_c for transfer of the minimum energy E_d to a nucleus can be obtained by setting $E_n' = E_d$ and $\theta = 180°$ in (5.17).

5.1.3 Elastic Differential Cross-Section for Small-Angle Scattering

The elastic cross-section $d\sigma/d\Omega$ or the scattering amplitude $f(\theta)$ can be calculated from the wave-mechanical Schrödinger equation (3.24). The asymptotic solution far from the nucleus can be represented by a plane, unscattered wave and a scattered, spherical wave (5.1) with an amplitude $f(\theta)$ depending on the scattering angle θ (Fig. 5.1b)

$$\psi_s = \psi_0[\exp(2\pi i \boldsymbol{k}_0 \cdot \boldsymbol{r}) + i f(\theta) r^{-1} \exp(2\pi i k r)] \ . \tag{5.19}$$

The scattering amplitude

$$f(\theta) = |f(\theta)| \, e^{i\eta(\theta)} \tag{5.20}$$

is complex. This means that there is not only a phase shift of 90° between the scattered and incident wave [factor i = $\exp(i\pi/2)$ in (5.19)] but also a smaller additional phase shift $\eta(\theta)$, which depends on the scattering angle. The important phase shift of 90° between the scattered and incident wave, which also occurs when light is scattered, can be understood from the following model. An incident wave $\psi = \psi_0 \sin(2\pi kz)$ is phase shifted by $\varphi = (2\pi/\lambda)(n-1)t$, see (3.21), when passing through a thin specimen structure of local thickness t; the amplitude behind the specimen thus becomes

$$\psi_s = \psi_0 \sin(2\pi kz + \varphi) = \psi_0 \sin(2\pi kz)\cos\varphi + \psi_0 \cos(2\pi kz)\sin\varphi \ . \tag{5.21}$$

For small φ, $\sin\varphi \simeq \varphi$ and $\cos\varphi \simeq 1$, and (5.21) may be written

$$\begin{aligned} \psi_s &= \psi_0 \sin(2\pi kz) + \psi_0\varphi \sin(2\pi kz + \pi/2) \ ; \\ \psi_s &= \psi \qquad\qquad\quad\ + i\psi_{sc} \ . \end{aligned} \tag{5.22}$$

This shows that ψ_s can be separated into the incident wave ψ and a 90° phase-shifted wave ψ_{sc}. At large distances, the scattered wave is spherical; the amplitude $f(\theta)$ depends on the nature of the specimen. The factor i before ψ_{sc} expresses the 90° phase shift between ψ and ψ_{sc} in complex notation.

The scattering amplitude $f(\theta)$ is the amplitude of the spherical wave far from the scattering event (Fig. 5.1b) and is therefore identical with the diffraction amplitude in Fraunhofer diffraction. We need to calculate the wavefront behind the atom (Fig. 5.4) in the form of (3.39). We can assume that $a_s(\boldsymbol{r}) = 1$ because there is no absorption of electrons, and the phase shift $\varphi_s(\boldsymbol{r})$ can be obtained from the optical path difference Δ relative to the wavefront in vacuum

$$\varphi_s(\boldsymbol{r}) = \frac{2\pi}{\lambda}\Delta = \frac{2\pi}{\lambda}\int_{-\infty}^{+\infty} [n(\boldsymbol{r}) - 1)] \, dz = -\frac{2\pi}{\lambda E}\frac{E + E_0}{E + 2E_0}\int_{-\infty}^{+\infty} V(\boldsymbol{r}) \, dz \ . \tag{5.23}$$

Equation (3.20) has been used for the electron-optical refractive index $n(\boldsymbol{r})$; this contains the potential energy, which involves not only the Coulomb potential of the nucleus but also that of the atomic electrons. The latter cause a screening of the nuclear charge $+Ze$. The charge distribution $\varrho(\boldsymbol{r}_j)$ inside an atom can be described by a δ-function at the nucleus ($r = 0$), together with the charge density $-e\varrho_e(\boldsymbol{r}_j)$ of the electron cloud with probability density $\varrho_e(\boldsymbol{r}_j)$; ψ_{0s} denotes the atomic-wave amplitude at the position $\boldsymbol{r}_j$

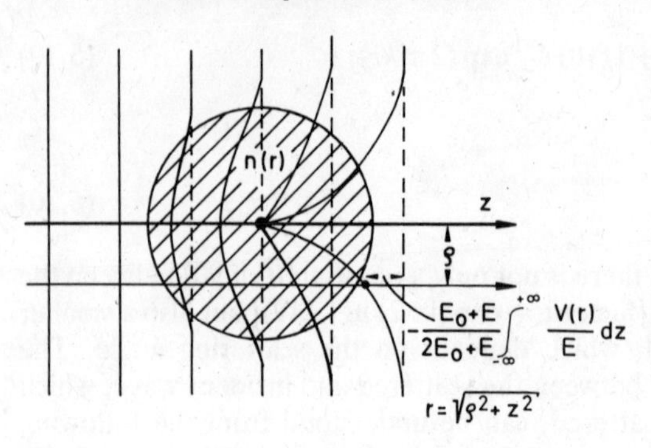

Fig. 5.4. Phase shift φ of a plane incident wave front passing the Coulomb potential $V(r)$ of an atom

$$e\varrho(\boldsymbol{r}_j) = eZ\delta(0) - e\varrho_e(\boldsymbol{r}_j) = eZ\delta(0) - e\sum_{s=1}^{Z}\psi_{0s}\psi_{0s}^* . \tag{5.24}$$

The probability density $\varrho_e(\boldsymbol{r}_j)$ can be calculated from the Thomas-Fermi model or by the Hartree-Fock method. An element of volume $d^3\boldsymbol{r}_j$ at a distance $\boldsymbol{r}_j$ from the nucleus contributes $-e^2\varrho(\boldsymbol{r}_j)\ (4\pi\varepsilon_0|\boldsymbol{r}_i - \boldsymbol{r}_j|)^{-1}$ to the Coulomb energy of a beam electron at a distance $\boldsymbol{r}_i$. The total Coulomb energy becomes

$$V(\boldsymbol{r}_i) = -\frac{e^2}{4\pi\varepsilon_0}\int\frac{\varrho(\boldsymbol{r}_j)}{|\boldsymbol{r}_i - \boldsymbol{r}_j|}d^3\boldsymbol{r}_j . \tag{5.25}$$

If $\varrho_e(\boldsymbol{r}_j)$ is assumed to be rotationally symmetric, (5.25) simplifies to

$$V(r_i) = -\frac{e^2 Z_{\text{eff}}}{4\pi\varepsilon_0 r_i} \quad \text{with} \quad Z_{\text{eff}} = Z - \int_0^{r_i}\varrho_e(r_j)\,4\pi r_j^2\,dr_j . \tag{5.26}$$

The final term in Z_{eff} is a measure of the electron charge inside a sphere of radius r_i. Its contribution to the Coulomb potential $V(r_i)$ can be evaluated by concentrating this charge at the centre of the sphere, whereas charges outside the sphere with $r_j \geq r_i$ do not contribute (elementary law of electrodynamics). It is advantageous in the calculations that follow to approximate the screening action in (5.26) in various ways:

a) one exponential term (Wentzel atom model)

$$V(r) = -\frac{e^2 Z}{4\pi\varepsilon_0 r}\,\mathrm{e}^{-r/R} \quad \text{with} \quad R = a_{\mathrm{H}}Z^{-1/3} \tag{5.27 a}$$

$a_{\mathrm{H}} = 0.0529$ nm is the Bohr radius

b) a sum of exponentials [5.1–3]

$$V(r) = -\frac{e^2 Z}{4\pi\varepsilon_0 r}\sum_{i=1}^{k} b_i \exp(-a_i r)\ ; \qquad \sum_{i=1}^{k} b_i = 1\ . \tag{5.27b}$$

c) These Coulomb potentials of single, free atoms have to be modified if the atoms are densely packed in a solid. To a first approximation, the overlap of Coulomb potentials (Fig. 3.3) can be expressed as a sum

$$\begin{aligned} V_{\text{eff}}(r) &= V(r) + V(2a-r) - 2V(a) \quad &\text{for} \quad r \le a \\ &= 0 \quad &\text{for} \quad r \ge a, \end{aligned} \tag{5.27c}$$

where $2a$ is the distance between neighbouring atoms or the diameter of a sphere with the same volume as the Wigner-Seitz cell of the lattice [5.4, 5]. This is known as the *muffin-tin model*. Calculated elastic cross-sections agree with experimental results (Sect. 6.1.1). It would be difficult to test further approximations, in which the changes of the wave functions of atomic electrons due to packing effects and electron band structure are considered, for example.

We now take the Fourier transform (3.43) of ψ_s. But first, it is useful to rewrite (3.39) in the form

$$\psi_s(\boldsymbol{r}) = \psi_0 + \psi_0\{\exp[i\varphi_s(\boldsymbol{r})] - 1\}\ . \tag{5.28}$$

If (5.28) is substituted in (3.43), using $\boldsymbol{q}\cdot\boldsymbol{r} = qr\cos\chi$ (r, χ are polar coordinates in the specimen plane) and $d^2\boldsymbol{r} = r dr d\chi$, the Fourier transform becomes

$$F(q) = \int_0^\infty \int_0^{2\pi} \psi_s(r)\, e^{-2\pi i q r \cos\chi}\, r\, dr\, d\chi\ . \tag{5.29}$$

The Fourier transform of the first term in (5.28) results in a δ-function at $\boldsymbol{q} = 0$ and represents the incident, unscattered wave, which is concentrated at the focal point of the diffraction plane. The second term of (5.28) corresponds to the scattered amplitude. The integral over χ for constant r yields the Bessel function J_0, giving

$$F(q) = \psi_0\left[\delta(0) + 2\pi\int_0^\infty \{\exp[i\varphi_s(r)] - 1\}\, J_0(2\pi q r)\, r\, dr\right]. \tag{5.30}$$

This can be related to the scattering amplitude $f(\theta)$ by rewriting (5.19) in the form

$$F(q) = \psi_0[\delta(0) + i|f(\theta)|e^{i\eta(\theta)}]\ . \tag{5.31}$$

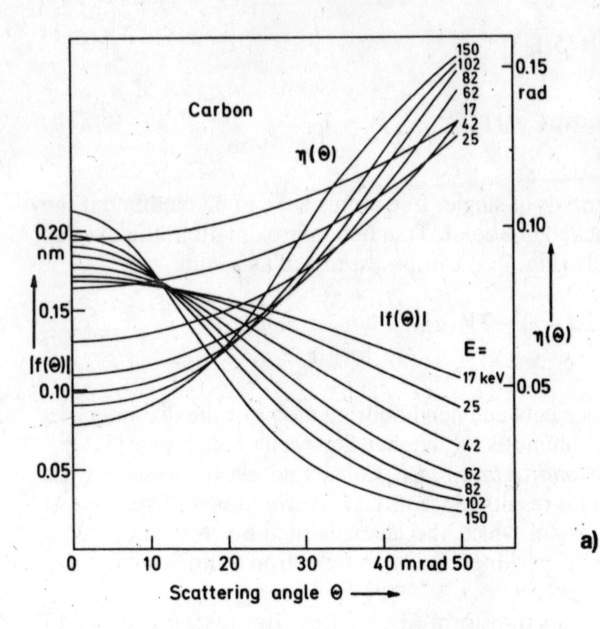

Fig. 5.5 a, b. Values of the scattering amplitude $|f(\theta)|$ and phase shift $\eta(\theta)$ of the complex scattering amplitude $f(\theta) = |f(\theta)| \exp[i\eta(\theta)]$ calculated by the WKB method using a muffin-tin model for **(a)** carbon and **(b)** platinum

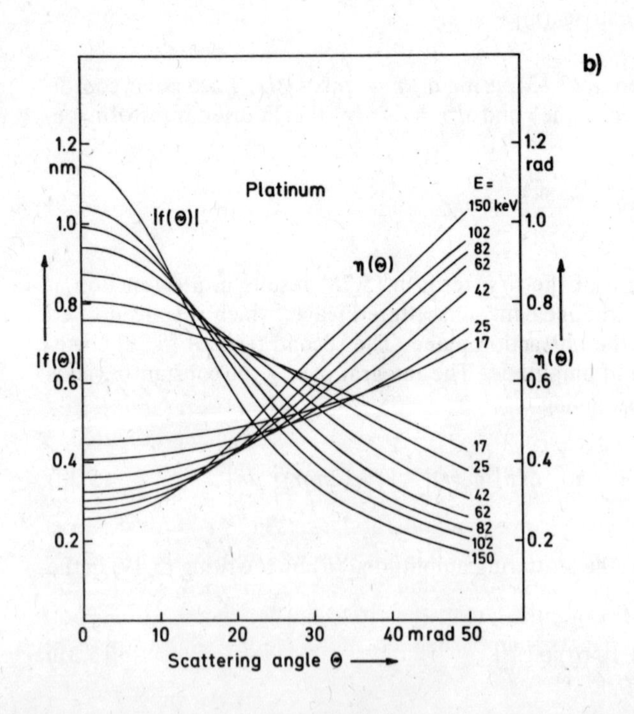

This is the so-called WKB method (Wentzel, Kramer, Brillouin) in the small-angle approximation of *Molière* [5.6], also associated with the name of *Glauber* [5.7].

Results obtained by this method agree with those given by the exact partial-wave method to an accuracy of 1% up to scattering angles of 10° [5.8]. Figure 5.5 shows calculated scattering amplitudes $f(\theta)$ for C and Pt atoms obtained with the muffin-tin model [5.5]. The value of $f(\theta)$ increases with increasing electron energy for small θ, but decreases for large θ. In consequence, the total elastic cross-section σ_{el} decreases with increasing energy (see Fig. 5.3 and Sect. 5.1.5). The additional phase shift $\eta(\theta)$ that is superimposed on the normal shift of $\pi/2$ between incident and scattered wave is very much less for C than for Pt. Complex scattering amplitudes have also been reported in [5.4, 5.9].

For so-called *weak-phase specimens*, for which $\varphi_s(\mathbf{r}) \ll 1$, the Taylor series $\exp[i\varphi_s(\mathbf{r})] = 1 + i\varphi_s(\mathbf{r}) + \dots$ can be used. Substituting in (5.30), we obtain the, so-called, *Born approximation*

$$F(\mathbf{q}) = \psi_0[\delta(0) + \underbrace{i2\pi\int\varphi_s(\mathbf{r})\, e^{-2\pi i\mathbf{q}\cdot\mathbf{r}}\, d^2\mathbf{r}}_{f(\mathbf{q})}] . \tag{5.32}$$

By use of $\varphi_s(\mathbf{r}) = (2\pi/\lambda)\int(n-1)\,dz$ and the refractive index n from (3.20), $f(\mathbf{q})$ in (5.32) becomes

$$f(\mathbf{q}) = -\frac{2\pi}{\lambda E}\cdot\frac{E+E_0}{E+2E_0}\int V(\mathbf{r})\, e^{-2\pi i\mathbf{q}\cdot\mathbf{r}}\, d^3\mathbf{r} . \tag{5.33}$$

$f(\mathbf{q})$ can be transformed to $f(\theta)$ by multiplying with the factor λ^{-1}. The factor in front of the integral in (5.33) is written in various ways in the literature, which are connected by the identities:

$$\frac{\pi e^2}{\varepsilon_0\lambda^2 E}\,\frac{E+E_0}{E+2E_0} \equiv \frac{e^2 m_0(1+E/E_0)}{4\pi\varepsilon_0\hbar^2} \equiv \frac{1+E/E_0}{a_H} , \tag{5.34}$$

where $a_H = 0.0569$ nm is the Bohr radius.

The scattering amplitude $f(\theta)$ is a real quantity in the Born approximation and the additional phase shift $\eta(\theta)$ is zero. The Born approximation cannot, therefore, be used for atoms of high atomic number because these are never "weak-phase specimens". The resulting difference between the WKB method and the Born approximation is shown in Fig. 6.3 for the total elastic cross-section σ_{el} or $x_a = A/(N_A\sigma_{el})$ (see discussion in Sect. 6.1.1).

The Born approximation has the advantage that an analytical solution can be obtained for simple potential models, and that the dependence of $f(\theta)$ on different parameters can be comprehended more readily. Substitution of

(5.25) in (5.33), with the coordinate $\boldsymbol{r}_j$ describing the atomic charge density and $\boldsymbol{r}_i$ the beam electron gives

$$f(q) = \frac{2\pi}{\lambda E}\,\frac{E+E_0}{E+2E_0}\,\frac{e^2}{4\pi\varepsilon_0}\underbrace{\int_0^\infty \varrho(\boldsymbol{r}_j)\,\mathrm{e}^{-2\pi i\boldsymbol{q}\cdot\boldsymbol{r}_j}\,d^3\boldsymbol{r}_j}_{Z-f_x}\cdot\underbrace{\int_0^\infty \frac{\exp[-2\pi i\boldsymbol{q}\cdot(\boldsymbol{r}_i-\boldsymbol{r}_j)]}{|\boldsymbol{r}_i-\boldsymbol{r}_j|}\,d^3\boldsymbol{r}_i}_{1/\pi q^2}\,. \tag{5.35}$$

This, in turn, implies

$$f(\theta) = \frac{\lambda^2(1+E/E_0)}{8\pi^2 a_\mathrm{H}}\,(Z-f_x)\,\frac{1}{\sin^2\dfrac{\theta}{2}}\,, \tag{5.36}$$

in which we have used (5.34) and (3.42); writing $\sin(\theta/2) \simeq \theta/2$, we finally obtain

$$\frac{d\sigma_\mathrm{el}}{d\Omega} = |f(\theta)|^2 = \frac{\lambda^4(1+E/E_0)^2}{4\pi^4 a_\mathrm{H}^2}\,\frac{(Z-f_x)^2}{\theta^4}\,, \tag{5.37}$$

where f_x is the scattering amplitude for x-rays. This quantity is dimensionless, whereas $f(\theta)$ for electrons has the dimension of a length. The differential cross-section $d\sigma/d\Omega = |f(\theta)|^2$ is the Rutherford formula for a screened Coulomb potential, see also (5.41).

The x-ray scattering amplitude f_x will be zero if only the scattering at the nucleus is considered and the influence of screening by the electron cloud is neglected. In this case, the Rutherford cross-section for the unscreened nucleus of charge $+Ze$ is obtained. When screening is considered, the x-ray amplitude f_x goes to zero for large θ, and the cross-section should become independent of the distribution $\varrho_\mathrm{e}(r_j)$ of the electron cloud for large-angle scattering. Screening effects influence only scattering through small angles, where $f_x \neq 0$. The unscreened Rutherford cross-section has a singularity at $\theta = 0$ because $\sin^4(\theta/2) \to 0$ as $\theta \to 0$ in the denominator of (5.36). However, the numerator also goes to zero as $\theta \to 0$ because f_x tends to $Z = \int \varrho_\mathrm{e}(r_j)\,d^3\boldsymbol{r}_j$ as θ or q tends to zero; $f(0)$ therefore takes a finite value, which is sensitive to the choice of screening model.

This influence of screening on $f(0)$ can be better understood if we substitute (5.27 a) for $V(r)$ in (5.33) and consider small scattering angles, for which $\sin\theta \simeq \theta$ [5.10, 11]. We have

$$f_x = \frac{Z}{1+4\pi^2 q^2 R^2}\,. \tag{5.38}$$

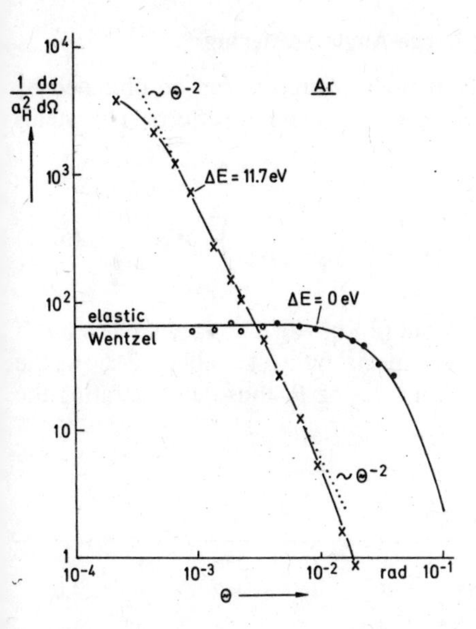

Fig. 5.6. Angular dependence of elastic and inelastic differential cross-sections $d\sigma/d\Omega$ (resonance energy loss at $\Delta E = 11.7$ eV) for 25 keV electrons scattered on an argon gas target ($a_H = 0.0529$ nm: Bohr radius) [5.17]

Substitution in (5.36) gives

$$f(\theta) = \frac{2ZR^2(1 + E/E_0)}{a_H[1 + (\theta/\theta_0)^2]} \quad \text{with} \quad \theta_0 = \frac{\lambda}{2\pi R} \tag{5.39}$$

or

$$\frac{d\sigma_{el}}{d\Omega} = \frac{4Z^2R^4(1 + E/E_0)^2}{a_H^2} \frac{1}{[1 + (\theta/\theta_0)^2]^2}. \tag{5.40}$$

The scattering amplitude (5.39) falls to $f(\theta_0) = 0.5 f(\theta)$ at the *characteristic screening angle* θ_0. Calculations of $f(\theta)$ in the Born approximation using electron density distributions $\varrho_e(r_j)$ given by relativistic Hartree-Fock calculations have been published in [5.12–16].

Differential scattering cross-sections can only be measured on gas targets if the concentration of atoms is so low that multiple scattering does not occur. Figure 5.6 shows measurements of the elastic and inelastic differential cross-sections of argon atoms, which confirm the dependence (5.40) of $d\sigma_{el}/d\Omega$ on the scattering angle θ predicted by the Wentzel model (5.27 a). The angular intensity distribution resulting from scattering in thin films is influenced by multiple scattering even when the films are very thin (Sect. 5.3.1). The results can be compared with calculated values of $d\sigma/d\Omega$ only after applying a deconvolution procedure to the measured data; alternatively the transmission $T(\alpha)$ through a thin film into a cone of aperture α (Sect. 6.1.1) may be compared with calculated cross-sections $\sigma(\alpha)$ as defined in (5.7).

5.1.4 Differential Cross-Section for Large-Angle Scattering

The differential cross-section for large-angle scattering can be obtained by setting $f_x = 0$ in (5.37), which means neglecting screening effects. This yields the Rutherford cross-section

$$\frac{d\sigma_R}{d\Omega} = \left(\frac{Ze^2}{8\pi\varepsilon_0 E}\right)^2 \left(\frac{E+E_0}{E+2E_0}\right)^2 \frac{1}{\sin^4(\theta/2)} = \left(\frac{Ze^2}{8\pi\varepsilon_0 mv^2}\right)^2 \operatorname{cosec}^4\frac{\theta}{2} \,. \quad (5.41)$$

The final expression can be obtained from (2.8 b, 10). The dependence on Z^2 and E^{-2} for non-relativistic energies is of special interest. Table 5.2 shows the decrease of the term $\operatorname{cosec}^4(\theta/2)$ with increasing θ, thus demonstrating the rapid decrease with increasing θ.

Table 5.2 Decrease of $\operatorname{cosec}^4(\theta/2)$ with increasing θ

θ	10°	20°	30°	40°	50°	60°	70°	80°	90°
$\operatorname{cosec}^4(\theta/2)$	17300	1100	223	73	31	16	9.2	5.9	4.0
θ	100°	110°	120°	130°	140°	150°	160°	170°	180°
$\operatorname{cosec}^4(\theta/2)$	2.9	2.2	1.8	1.5	1.3	1.15	1.06	1.02	1.0

Figure 5.3 shows the cross-section σ_B for backscattering into angles $\theta \geq \pi/2$, which is an order of magnitude less than the total elastic cross-section σ_{el}.

However, the Rutherford formula (5.41) is an approximation, because it neglects not only the screening but also the influence of electron spin. In order to include the latter, the Dirac equation [5.18] has to be solved instead of the Schrödinger equation. The difference between the resulting Mott cross-section and the Rutherford cross-section (5.41) can be expressed in terms of the ratio

$$r = \frac{d\sigma_{Mott}/d\Omega}{d\sigma_R/d\Omega} \,. \quad (5.42)$$

The Mott scattering cross-section of an unscreened nucleus can be written in the form of an infinite series in powers of $\alpha = Z/137$; neglecting terms of the order α^3 [5.19]

$$r = 1 - \beta^2 \sin^2(\theta/2) + \pi\alpha\beta \sin(\theta/2)\,[1 - \sin(\theta/2)] \quad (5.43)$$

with $\beta = v/c$. However, this formula can be used only for elements for which $Z \leq 20$. For heavy elements, terms in α^3 and α^4 can no longer be neglected and numerical methods have to be used; results are available in [5.19–22].

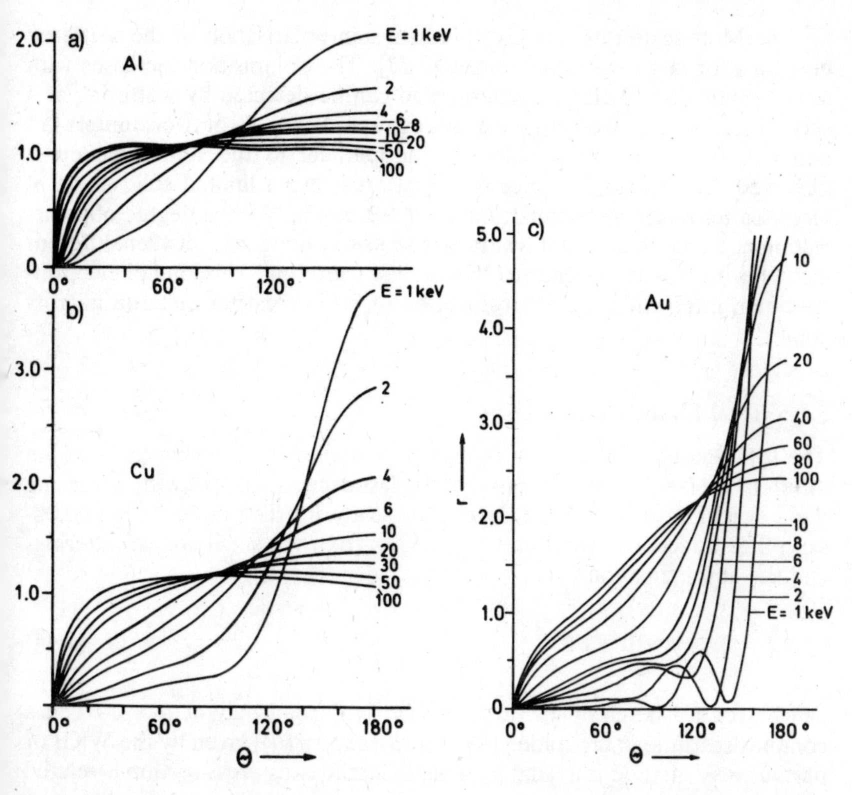

Fig. 5.7 a–c. Ratio r of Mott scattering cross-section $(d\sigma/d\Omega)_M$ and Rutherford cross-section $(d\sigma/d\Omega)_R$ for (**a**) Al, (**b**) Cu and (**c**) Au [5.26]

For more-exact calculations, a screened Coulomb potential has to be employed; the equations can be solved with the partial-wave method [5.23, 24]. Figure 5.7 shows the variation of r with θ for Al, Ge and Au. In this calculation [5.26], the muffin-tin model (Eq. (5.27c) and Fig. 5.3) was used to represent, to a first-order approximation, the packing effects of the atoms in solids. The effect of screening at low scattering angles can be seen in the decrease of r with decreasing θ. The numerical values below $\theta = 10°$ agree with WKB calculations [5.5] using the same $V(r)$. If the results for higher scattering angles are compared with those of [5.20], rather good agreement is found for the shape of $r(\theta)$. The Mott scattering cross-sections calculated [5.24] have been confirmed by scattering experiments on mercury-gas targets, for example [5.25]. The use of Mott scattering cross-sections in Monte-Carlo calculations [5.26] likewise gives better agreement with experimental results for electron backscattering at solid films and bulk material than the use of the Rutherford formula.

The Mott scattering law also predicts spin polarisation of the scattered electrons for large scattering angles [5.27]. The polarisation increases with atomic number and electron energy and can be detected by scattering at a second foil; an anisotropy of the counting rates N_1 and N_2 of two counters C 1 and C 2 in a direction parallel and antiparallel to the spin direction is observed. The number of electrons scattered into a limited solid angle at large scattering angles is small; furthermore, not only is the degree of polarisation not a parameter that would permit us to distinguish between different elements but it is also decreased by multiple scattering. This kind of electron-specimen interaction will, therefore, be of no interest for electron microscopy.

5.1.5 Total Elastic Cross-Section

The total elastic cross-section σ_{el} can be obtained by applying (5.8) to the differential cross-section. Because of the fast decrease of $f(\theta)$ with increasing θ, an exact knowledge of the large-angle scattering distribution is not necessary. The largest contribution to σ_{el} comes from the small-angle scattering, and the approximation

$$\sigma_{el} = 2\pi \int_0^\infty |f(\theta)|^2 \theta \, d\theta \tag{5.44}$$

can be used. The complex value of $f(\theta)$ is needed only at $\theta = 0$ when a complex scattering amplitude $f(\theta) = |f(\theta)| \exp[i\eta(\theta)]$ given by the WKB or partial-wave method is available. The total scattering cross-section is related to the imaginary part by the optical theorem of quantum-mechanical scattering theory:

$$\sigma_{el} = 2\lambda \operatorname{Im}\{f(0)\} = 2\lambda |f(0)| \sin\eta(0) \,. \tag{5.45}$$

Substituting the expression $d\sigma_{el}/d\Omega$ given by (5.40) and $R = a_H Z^{-1/3}$ in (5.44), we obtain

$$\sigma_{el} = \frac{Z^2 R^2 \lambda^2 (1 + E/E_0)^2}{\pi a_H^2} = \frac{h^2 Z^{4/3}}{\pi E_0^2 \beta^2} \,. \tag{5.46}$$

The absolute value of σ_{el} or its reciprocal $x_a = A/(N_A \sigma_{el})$ does not agree well with experiment (Fig. 6.3) because the Wentzel screening model (5.27 a) is too simple. However, the cross-section σ_{el} is observed to be proportional to β^{-2} for low-Z material and the predicted saturation for energies larger than 1 MeV is found for all elements (Figs. 5.3 and 6.3). Also, the proportionality with $Z^{4/3}$ is a typical result of the Wentzel model and of the Born approximation. Calculations of the total elastic cross-section for Hartree-Fock-Slater or Dirac-Slater atoms [5.28] can be approximated by

$$\sigma_{el} = \frac{1.5 \times 10^{-6}}{\beta^2} Z^{3/2}\left(1 - 0.23\frac{Z}{137\beta}\right) \quad \text{for} \quad \frac{Z}{137\beta} < 1.2 , \tag{5.47}$$

where σ_{el} is measured in nm.

5.2 Inelastic Scattering

5.2.1 Electron-Specimen Interactions with Energy Loss

Whereas an elastic collision conserves kinetic energy and momentum, an inelastic collision conserves the total energy and momentum, and part of the kinetic energy is converted to atom-electron excitation. The primary electron is observed to lose energy even at small scattering angles. The following excitation mechanisms may be distinguished:

1) Excitation of oscillations in molecules [5.29] and phonon excitations in solids [5.30, 31]. These energy losses (Fig. 5.8) are of the order of 20 meV – 1 eV and can be observed only after monochromatization of the primary electron beam, which has an energy width of the order of 1 eV when a thermionic electron gun is used. These interaction processes are also excited by the infrared part of the electromagnetic spectrum. The observed energy losses are of considerable interest for molecular and solid-state physics but, owing to the low intensity of a monochromatized beam, high spatial resolution is impossible. At the moment, therefore, these processes are of little interest in electron microscopy.

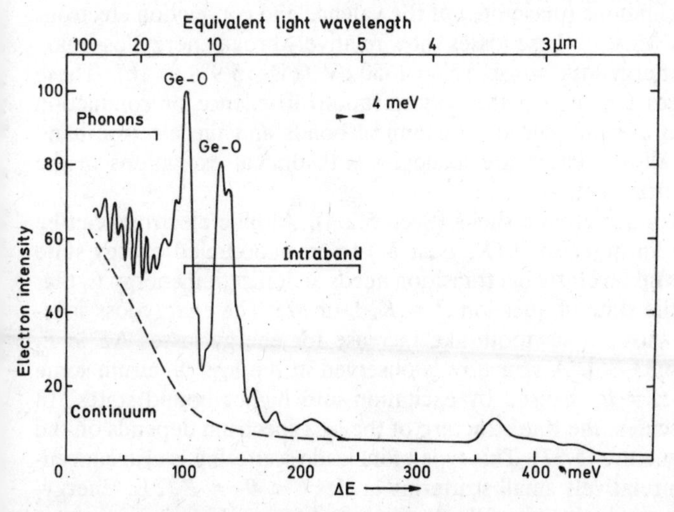

Fig. 5.8. Energy-loss spectrum of an evaporated Ge film due to phonon excitation, excitation of the GeO bonding and to intraband transitions for energy losses $\Delta E \leq 500$ meV [5.31]

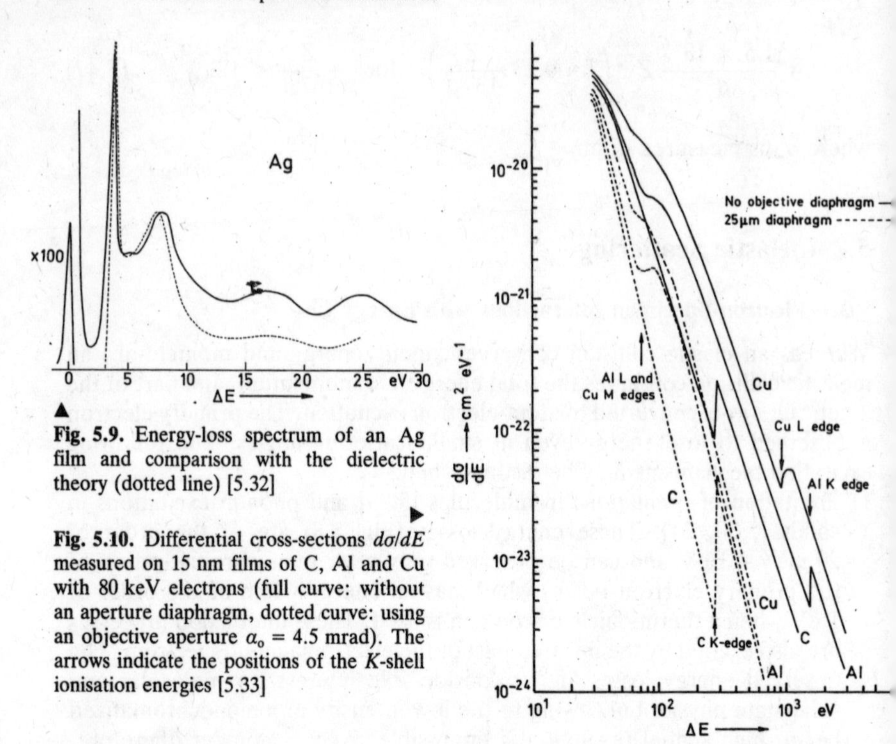

▲ **Fig. 5.9.** Energy-loss spectrum of an Ag film and comparison with the dielectric theory (dotted line) [5.32]

▶ **Fig. 5.10.** Differential cross-sections $d\sigma/dE$ measured on 15 nm films of C, Al and Cu with 80 keV electrons (full curve: without an aperture diaphragm, dotted curve: using an objective aperture $\alpha_o = 4.5$ mrad). The arrows indicate the positions of the *K*-shell ionisation energies [5.33]

2) Intra- and interband excitation of the outer atomic electrons, excitation of collective oscillations (plasmons) of the valence and conduction electrons (Sect. 5.2.3). Most of these losses show relatively broad energy-loss maxima in the energy-loss range $\Delta E = 1$–50 eV (Figs. 5.9 and 16). These plasmon losses depend on the concentration of valence or conduction electrons and are influenced by chemical bonds and the electron-band structure in alloys. There are analogies with optical excitations in the visible and ultraviolet.
3) Ionisation of inner atomic shells (Sect. 5.2.4). Atomic electrons can be excited from an inner shell (*K*, *L* or *M*) to an unoccupied energy state above the Fermi level; such a transition needs an ionization energy E_I that depends on the state in question (I = *K*, *L* or *M*). The energy-loss spectrum $d\sigma/dE$ shows a sawtooth-like increase for energy losses $\Delta E \geq E_I$ (Fig. 5.10) [5.33–36]. A structure is observed in the loss spectrum some eV above $\Delta E = E_I$, caused by excitation into higher bound states. In organic molecules, the fine structure of the loss spectrum depends on the molecular structure [5.37]. This type of inelastic scattering is also concentrated within relatively small scattering angles $\theta < \theta_E = E_I/2E$. Energy-loss spectroscopy is therefore the best method for analyzing elements of low atomic number (e.g., C, N, O) in thin films with thicknesses smaller

than or comparable to the mean-free-path for inelastic scattering. When the electron gap in the inner shell is filled by an electron from an outer shell, the excess energy is emitted as an x-ray quantum or transferred to another atomic electron, which is emitted as an Auger electron (Sect. 5.4.2).

The inelastic interactions therefore form the basis for several analytical methods in electron microscopy (Sect. 9.2). However, inelastic scattering is not favourable for high resolution because the inelastic-scattering process is less localized than the elastic one. In a classical particle model, an electron can be inelastically scattered even when passing the atom at a distance of a few 0.1 nm. This is also illustrated by the fact that inelastic scattering is concentrated at smaller scattering angles than elastic scattering (Fig. 5.6). In order to resolve a specimen periodicity of Λ, scattering amplitudes are needed out to an angle $\theta = \lambda/\Lambda$; there are too few inelastically scattered electrons at large θ for the imaging of small spacings Λ or high spatial frequencies $q = 1/\Lambda$.

The total inelastic cross-section σ_{inel} is larger than the elastic cross-section σ_{el} for elements of low atomic number and smaller for high Z. The energy losses in thick specimens decrease the resolution due to the chromatic aberration.

The largest part of excitation energy is converted to heat (phonons) (Sect. 10.1). Excitations and ionizations in organic specimens cause bond ruptures and irreversible radiation damage (Sect. 10.2). Colour centres and other point defects and clusters are generated in ionic crystals (Sect. 10.3).

Inelastic scattering can be described by a cross-section $d^2\sigma_{\text{inel}}/d(\Delta E)\,d\Omega$, depending on the scattering angle θ and the energy loss ΔE. It becomes difficult to establish such a two-dimensional cross-section from theory or experiment for several reasons: the complexity of the energy-loss spectrum, its dependence on specimen structure and foil thickness (e.g., surface-plasmon losses) and the occurence of multiple elastic and inelastic scattering. In both theory and experiment, therefore, $d\sigma_{\text{inel}}/d\Omega$ is obtained for individual energy losses ΔE (e.g. plasmon losses or ionization of inner shells). In addition, it is of interest to derive approximate expressions for $d\sigma_{\text{inel}}/d\Omega$, independent of ΔE, and for the mean energy loss dE_{m}/dx per element of path length $dx = \varrho\, dt$.

5.2.2 Inelastic Scattering at a Single Atom

During an inelastic-scattering event, an excitation energy $\Delta E = E_m - E_n$ may be transfered to an electron of an atom that is excited from the ground state (n) with energy E_n and wave function ψ_{ns} ($s = 1, \ldots, Z$) to the excited state (m) with energy E_m and wave function ψ_{ms}. Selection rules, such as $\Delta l = \pm 1$, govern the allowed excitations, similar to those for optical excitation.

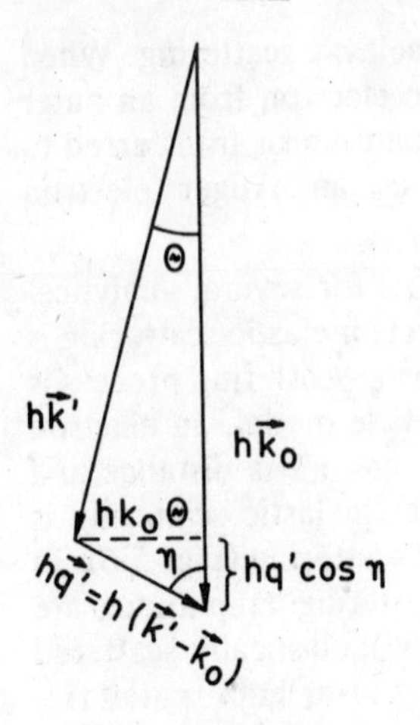

Fig. 5.11. Conservation of momentum in inelastic scattering

The total energy and the momentum remain the same before and after the collision. We introduce the parameter $\boldsymbol{q}' = \boldsymbol{k} - \boldsymbol{k}_0$ (Fig. 5.11) instead of $\boldsymbol{q} = \boldsymbol{k} - \boldsymbol{k}_0$ used for elastic scattering. Using the relations $p = hk$ and $E = p^2/2\,m$ (non-relativistic), the conservation of momentum

$$k'^2 = k_0^2 + q'^2 - 2\,k_0 q' \cos\eta \tag{5.48}$$

and the conservation of energy

$$\Delta E = E - E' = \frac{h^2}{2m}(k_0^2 - k'^2)\,, \tag{5.49}$$

we obtain the relation

$$\Delta E = \frac{h^2 k_0 q'}{m} \cos\eta\,. \tag{5.50}$$

Furthermore, from Fig. 5.11 the relation

$$q'^2 = (k_0\theta)^2 + (q'\cos\eta)^2 = k_0^2(\theta^2 + \theta_E^2) \tag{5.51}$$

can be read off, in which $\theta_E = \Delta E/2\,E$ or in relativistic form

$$\theta_E = \frac{\Delta E}{E}\,\frac{E + E_0}{E + 2\,E_0} = \frac{\Delta E}{mv^2}\,. \tag{5.52}$$

As in (5.35), we obtain the first-order Born approximation for the inelastic-scattering cross-section by replacing q by q' (5.51) and the electron-density distribution (5.24)

$$\varrho_e(r_j) = \sum_{s=1}^{Z} \psi_{ns}\psi_{ns}^* \quad \text{by} \quad \sum_{s=1}^{Z} \psi_{ns}\psi_{ms}^* ;$$

this gives

$$\frac{d\sigma_{\text{inel}}}{d\Omega} = \frac{(1+E/E_0)^2}{4\pi^4 a_H^2 q'^4} \underbrace{\frac{k'}{k}}_{\simeq 1} \sum_{m \neq n} \left| \sum_{s=1}^{Z} \int \psi_{ns}\psi_{ms}^* \exp(-2\pi i \boldsymbol{q}' \cdot \boldsymbol{r}_j)\, d^3\boldsymbol{r}_j \right|^2 . \quad (5.53)$$

In the inner summation, products of terms containing different values of the suffix $s = 1, \ldots, Z$ (exchange terms) can be neglected. The summations over m and s can thus be interchanged, giving

$$\sum_{s=1}^{Z} \sum_{m \neq n} \left| \int \psi_{ns}\psi_{ms}^* \exp(-2\pi i \boldsymbol{q}' \cdot \boldsymbol{r}_j)\, d^3\boldsymbol{r}_j \right|^2$$

$$= \sum_{s=1}^{Z} \Bigg\{ \underbrace{\sum_{m} \left| \int \psi_{ns}\psi_{ms}^* \exp(-2\pi i \boldsymbol{q}' \cdot \boldsymbol{r}_j)\, d^3\boldsymbol{r}_j \right|^2}_{1} - \underbrace{\left| \int \psi_{ns}\psi_{ns}^* \exp(-2\pi i \boldsymbol{q}' \cdot \boldsymbol{r}_j)\, d^3\boldsymbol{r}_j \right|^2}_{f_x^2/Z^2} \Bigg\}$$

$$= Z - f_x^2/Z . \quad (5.54)$$

The first term in the square brackets is equal to unity, a consequence of the summation rule for oscillator strengths. The last term is the contribution of *one* electron to the x-ray scattering amplitude f_x.

The approximation in which this is set equal to f_x/Z is strictly valid only if the electron-density distributions of the atomic electrons are equal (H and He atoms only). When (5.54) is substituted in (5.53) for other atoms, the resulting approximation permits us to obtain analytical formulae using the Wentzel model (5.27 a). The quantity q' in (5.53) contains the energy losses ΔE. *Koppe* [5.38] suggested substituting the mean value $\Delta E = J/2$ where J is the mean ionisation energy of the atom $\simeq 13.5\, Z$ in eV. By use of f_x from (5.38), this gives [5.10]

$$\frac{d\sigma_{\text{inel}}}{d\Omega} = \frac{(1+E/E_0)^2}{4\pi^4 a_H^2 q'^4} Z \left[1 - \frac{1}{(1+4\pi^2 q^2 R^2)^2} \right] , \quad (5.55)$$

or with $q'^2 = (\theta^2 + \theta_E^2)/\lambda^2$,

$$\frac{d\sigma_{\text{inel}}}{d\Omega} = \frac{\lambda^4 (1+E/E_0)^2}{4\pi^4 a_H^2} \frac{Z\left\{1 - \dfrac{1}{[1+(\theta/\theta_0)^2]^2}\right\}}{(\theta^2+\theta_E^2)^2} . \quad (5.56)$$

This formula for the inelastic differential cross-section may be compared with its elastic counterpart (5.40). The characteristic angle θ_0, which is responsible for the decrease of the elastic differential cross-section $d\sigma_{el}/d\Omega$, is of the order 10 mrad and the angle θ_E, responsible for the decrease of $d\sigma_{inel}/d\Omega$ with increasing θ, of the order 0.1 mrad. This confirms that inelastic scattering is concentrated within much smaller angles than elastic scattering (Fig. 5.6). For very large scattering angles, $\theta \gg \theta_0$ and $\gg \theta_E = J/4E$, the ratio

$$\frac{d\sigma_{inel}/d\Omega}{d\sigma_{el}/d\Omega} = \frac{1}{Z} \tag{5.57}$$

depends on only the atomic number, whereas for small θ, $d\sigma_{inel}/d\Omega > d\sigma_{el}/d\Omega$ for all elements (Fig. 5.6).

A total inelastic cross-section σ_{inel} can be defined in the same way as the elastic one σ_{el}, using (5.8) and (5.44). Integration of (5.56) gives [5.10]

$$\nu = \frac{\sigma_{inel}}{\sigma_{el}} = \frac{4}{Z}\ln\left(\frac{h}{\pi m_0 J R \lambda}\right) \simeq \frac{26}{Z}\,. \tag{5.58}$$

Table 5.3 shows theoretical and experimental values of the ratio ν.

Table 5.3. Values of the ratio $\nu = \sigma_{inel}/\sigma_{el}$

E [keV]	C	Ge	Pt		
20–60	1.67 – 1.89	0.52–0.58	–	[5.39]	
40	2.6 ± 0.3	0.5	0.2	[5.40]	
100	3.0	0.8	–	[5.41]	
100	4.2	0.95	0.39	[5.10]	theory
100	2.2	–	–	[5.42]	theory
25	1.6	–	–	[5.43]	
80	2.9 ± 0.6	–	–	[5.44]	
80	Be: 4.5, Al: 1.8, As: 0,95, Sn: 0,65			[5.45]	

5.2.3 Energy Losses in Solids

Only the most important theoretical and experimental results concerning energy losses in solids will be discussed. Extensive reviews have been published [5.32, 46–49].

A number of interaction processes have to be considered to explain the characteristic energy losses of a material. An atomic electron can be excited to a higher energy state by an electron-electron collision. Indeed, energy losses are found in scattering experiments on gases that can be explained as the energy differences between spectroscopic terms. Thus, a 7.6 eV loss in

Hg vapour corresponds to the optical resonance line [5.50]. The electrons in the outer atomic shell of a solid are broadened in energy bands. Excitations from one band to another (interband excitation) must be distinguished from those inside one band (intraband excitation). Non-vertical interband and intraband transitions can also be observed [5.51–53]. Exact information about the band structure above the Fermi level is not available for most materials and the electron energy-loss spectrum is therefore compared with the light-optical constants in the visible and ultraviolet spectrum by means of the so-called *dielectric theory*. In this theory, plasma oscillations are also considered as longitudinal density oscillations of the electron gas [5.54].

The underlying idea of the dielectric theory can be understood in the following manner. The optical constants of a solid can be described either by a complex refractive index $n + i\varkappa$, where $\varkappa$ is the absorption coefficient or by a complex permittivity $\varepsilon = \varepsilon_1 + i\varepsilon_2 = (n + i\varkappa)^2$. In electrodynamics, the rate of energy dissipation per unit volume is $dW/dt = \boldsymbol{E} \cdot \boldsymbol{D}$. Electrons that penetrate into the crystal with velocity v represent a moving point charge and can be described by a δ-function $\varrho = e\delta(x - v\tau)$ which produces a $\boldsymbol{D}$ field related to ϱ by $\operatorname{div}\boldsymbol{D} = -\varrho$. If we assume a periodic electric field $\boldsymbol{E} = \boldsymbol{E}_0 \exp(-i\omega t)$ and use the relation $\boldsymbol{D} = \varepsilon\varepsilon_0\boldsymbol{E}$, the mean energy dissipation will be

$$\overline{\frac{dW}{d\tau}} = \varepsilon_0\varepsilon_2 \frac{\omega|\boldsymbol{E}_0|^2}{2} = \frac{1}{\varepsilon_0}\frac{\varepsilon_2}{|\varepsilon|^2}\omega\frac{D_0^2}{2} = \frac{1}{\varepsilon_0}\omega\frac{D_0^2}{2}\left(-\operatorname{Im}\left\{\frac{1}{\varepsilon}\right\}\right) \tag{5.59}$$

because

$$\frac{1}{\varepsilon} = \frac{1}{\varepsilon_1 + i\varepsilon_2} = \frac{\varepsilon_1 - i\varepsilon_2}{|\varepsilon|^2}\,. \tag{5.60}$$

The permittivity $\varepsilon(\omega) = \varepsilon_1(\omega) + i\varepsilon_2(\omega)$ is a function of ω. The short local excitation of the $\boldsymbol{D}$ field corresponds to a broad spectrum of ω values in a Fourier integral. The differential cross-section is obtained by perturbation theory

$$\frac{d^2\sigma}{d(\Delta E)\,d\Omega} = \frac{1}{\pi^2 a_H EN}\frac{E + E_0}{E + 2E_0}\frac{-\operatorname{Im}\{1/\varepsilon(\Delta E, \theta)\}}{\theta^2 + \theta_E^2}$$

$$= \frac{1}{\pi^2 a_H m v^2 N}\frac{-Im\left\{\frac{1}{\varepsilon}\right\}}{\theta^2 + \theta_E^2}\,. \tag{5.61}$$

For a free (f) electron gas with N electrons per unit volume, the conduction electrons of a metal or the valence electrons of a semiconductor, for example, the dependence of $\varepsilon^{(f)}(\omega)$ on frequency or the energy loss $\Delta E = \hbar\omega$ may be calculated as follows: The alternating electric field $\boldsymbol{E} = \boldsymbol{E}_0 \exp(-i\omega\tau)$

exerts a force on an electron given by Newton's law

$$m^*\ddot{\boldsymbol{x}} + \gamma\dot{\boldsymbol{x}} = -e\boldsymbol{E} \ . \tag{5.62}$$

This expression contains a friction term proportional to $\dot{x}$, which represents the deceleration due to energy dissipation; m^* denotes the effective mass of the conduction electrons. The solution of this equation has the form

$$\boldsymbol{x} = \frac{e}{m^*\omega^2}\frac{\omega^2 - i\omega\gamma}{\omega^2 + \gamma^2}\boldsymbol{E} \ . \tag{5.63}$$

The displacement $\boldsymbol{x}$ of the charge $-e$ causes a polarization $\boldsymbol{P} = -eN\boldsymbol{x} = \varepsilon_0\chi_e\boldsymbol{E}$ where χ_e is the dielectric susceptibility, $\varepsilon = \varepsilon_0(1 + \chi_e)$. Substituting for $\boldsymbol{x}$ from (5.63), we obtain

$$\varepsilon_1^{(f)} = \varepsilon_0\left(1 - \frac{\omega_p^2}{\omega^2}\frac{1}{1 + (\gamma/\omega)^2}\right) ; \qquad \varepsilon_2^{(f)} = \varepsilon_0\frac{\gamma}{\omega}\frac{\omega_p^2}{\omega^2}\frac{1}{1 + (\gamma/\omega)^2} \tag{5.64}$$

in which ω_p is the *plasmon frequency*

$$\omega_p = \left(\frac{Ne^2}{\varepsilon_0 m^*}\right)^{1/2} \tag{5.65}$$

The dependence of $\varepsilon_1^{(f)}$ and $\varepsilon_2^{(f)}$ on frequency is shown in Fig. 5.12. The factor $-\mathrm{Im}\{1/\varepsilon\} = \varepsilon_2/|\varepsilon|^2$ that appears in (5.61) passes through a sharp maximum when the denominator reaches a minimum, which means that $\varepsilon_1^{(f)} = 0$ at $\omega = \omega_p$ for small values of the damping constant γ. This plasmon loss $\Delta E_p = \hbar\omega_p$ excites longitudinal charge-density oscillations in the electron gas, which are quantized and are known as *plasmons* [5.55]. For a large number of materials, the observed energy loss ΔE_p agrees with that predicted by (5.65) (see Table 5.4 in which ν is the number of conduction or valence electrons per atom).

However, the position of the plasmon losses can be considerably influenced by interband excitations (bond states: b). In the $\varepsilon_1^{(b)}$-curve, the bond states show a typical anomalous dispersion near the resonance frequency. In the special case of Fig. 5.12 with $\omega_b < \omega_p$ the superposition of $\varepsilon_1^{(b)}$ and $\varepsilon_1^{(f)}$ causes a shift of the plasmon loss to a higher frequency ω_p', for which $\varepsilon_1^{(b)} + \varepsilon_1^{(f)} = 0$. When $\omega_b > \omega_p$, the plasmon loss may be shifted to a lower frequency. Thus, agreement between the calculated and measured values of the 15 eV plasmon loss in Al (Fig. 3.17) in Table 5.4 is accidental. The optical constants indicate that $\varepsilon_1^{(f)} = 0$ for $\Delta E_p = 12.7$ eV. An oscillator contribution at 1.5 eV (interband excitation) shifts the loss to 15.2 eV. For silver, there is a transition of $4f$ electrons to the Fermi level at 3.9 eV and further interband excitations occur at $\Delta E > 9$ eV. The latter shift the energy

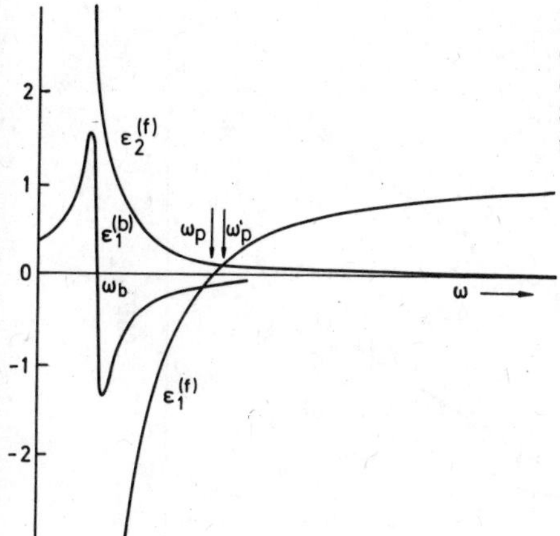

Fig. 5.12. Components of the complex dielectric permittivity $\varepsilon_1 + i\varepsilon_2$ for the free electron gas (f) and for a bound state (b). Shift of the plasmon frequency ω_p at $\varepsilon_1^{(f)} = 0$ to ω_p'

Table 5.4. Comparison of experimental and theoretical values of plasmon losses ΔE_p and the half-width $(\Delta E)_{1/2}$ in eV, the constant a of the dispersion law (5.66) and the cut-off angle θ_c (ν is the number of electrons per atom)

	ν	ΔE_p exp.	theor.	$(\Delta E)_{1/2}$	a exp.	theor.	θ_c (40 keV) [mrad] exp.	theor.
Al	3	15.0	15.8	0.6	0.40 ± 0.01	0.44	15	13
Be	2	18.9	18.4	5.0	0.42 ± 0.04	0.42	–	–
Mg	2	10.5	10.9	0.7	0.39 ± 0.01	0.37	12	11
Si	4	16.9	16.6	3.2	0.41	0.45	–	–
Ge	4	16.0	15.6	3.3	0.38	0.44	–	–
Sb	5	15.3	15.1	3.3	0.37 ± 0.03	0.38	–	–
Na	1	3.7	5.9	0.4	0.29 ± 0.02	0.25	10	9

at which $\varepsilon_1 = 0$ to 3.75 eV. This intensive energy loss (Fig. 5.9) is, therefore, a plasmon loss, which cannot be separated from the 3.9 eV interband loss. Figure 5.9 shows, as an example, a comparison of a measured energy-loss spectrum (full curve) with one calculated by use of the dielectric theory with optical values of ε (dotted curve). Two further maxima are also in agreement with the optical data.

The dependence of $\Delta E_p = \hbar\omega_p \propto N^{1/2}$ on the electron density N (5.65) results in a weak decrease of ΔE_p with increasing temperature because of the thermal expansion [5.56, 57]. Changes of ΔE_p in alloys, due to the change of electron concentration, are of special interest because these shifts can be used for local analysis of the composition of an alloy by energy-loss spectroscopy (Sect. 9.2.3).

Plasmon losses show a dispersion, in the sense that the magnitude of ΔE_p depends on the momentum transferred and, therefore, on the scattering angle θ

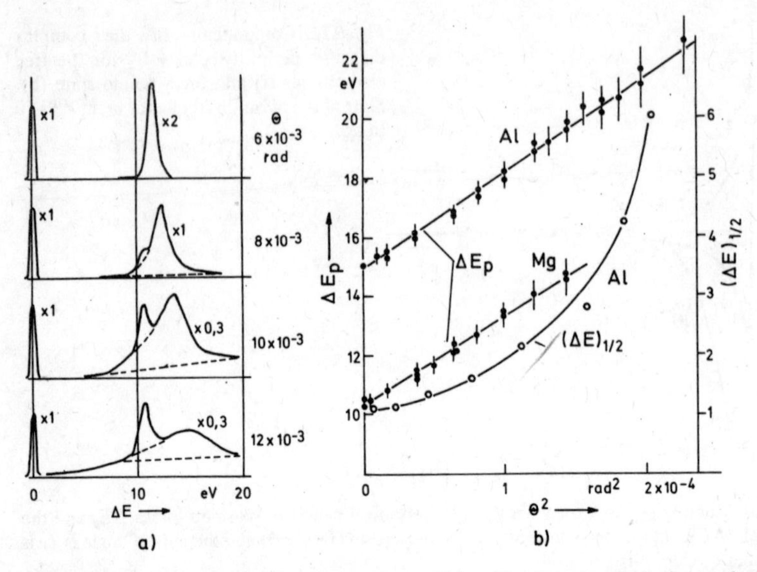

Fig. 5.13. (**a**) Dispersion of the plasmon loss of Mg with increasing scattering angle θ and the unshifted line due to elastic large angle scattering and inelastic small-angle scattering. (**b**) Verification of the dispersion relation (5.66) by plotting ΔE_p versus θ^2 for the Al- and Mg-plasmon losses [5.64] and the broadening $(\Delta E)_{1/2}$ of the plasmon loss of Al with increasing θ [5.59, 60]

$$\Delta E_p(\theta) = \Delta E_p(0) + a\theta^2 \quad \text{with} \quad a = \frac{6}{5}\frac{EE_F}{\Delta E_p(0)}, \tag{5.66}$$

where E_F is the Fermi energy. In addition, the half-width $(\Delta E)_{1/2}$ of the loss maxima increases with increasing θ (Fig. 5.13b) [5.58–65]. The plot of ΔE_p versus θ^2 in Fig. 5.13b demonstrates the validity of this dispersion law. The constant a_{exp} in (5.66) (Table 5.4) can be obtained from the slope of the curve. Interband excitations show no dispersion and normally have a larger half-width. However, a dispersion with $a = 0.15$ has been observed for the 13.6 loss of LiF which is an exciton excitation [5.66]. If the energy loss is observed at a scattering angle θ, not only is the shifted value $\Delta E_p(\theta)$ observed but also the unshifted $\Delta E_p(0)$, which results either from primary small-angle plasmon scattering and secondary elastic large-angle scattering or vice versa (Fig. 5.13a). This effect can also be seen in photographic records of loss spectra (Fig. 9.9). The values of the plasmon losses and their dispersion are anisotropic in anisotropic crystals such as graphite [5.67, 68] and also in cubic crystals (Al, for example) for large scattering angles [5.69].

Integration of (5.61) over ΔE gives the contribution of a plasmon loss to the differential cross-section:

$$\frac{d\sigma}{d\Omega} = \frac{\Delta E_p}{2\pi a_H N E} \frac{E + E_0}{E + 2E_0} \frac{1}{\theta^2 + \theta_E^2} G(\theta, \theta_c) . \tag{5.67}$$

The cross-section decreases as θ^{-2} for medium scattering angles, between $\theta_E < \theta < \theta_c$ (Fig. 5.14). The correction function $G(\theta, \theta_c)$ introduced by *Ferrell* [5.71] takes into account the fact that $d\sigma/d\Omega$ has to become zero at a cut-off angle θ_c, which implies that plasmon wavelengths shorter than the mean distance between valence electrons are damped more strongly. There is thus a maximum momentum that can be transferred in the inelastic collision, and scattering angles greater than θ_c are not possible.

Integration over the solid angle Ω in (5.67) and using the approximation $G(\theta, \theta_c) = 1$ for $\theta < \theta_c$ and $G(\theta, \theta_c) = 0$ for $\theta \geq \theta_c$ yields the total cross-section for plasmon excitation

$$\sigma_p = \frac{E_p}{2 a_H m v^2} \ln(\theta_c/\theta_E) \tag{5.68}$$

and the corresponding mean free path $\Lambda_p = 1/N\sigma_p$ (Sect. 5.1.1). Figure 5.15 shows calculated and measured values of Λ_p in the range $E = 100 - 1000$ keV.

Multiple inelastic scattering is observed in thick specimens, which means that multiples of the plasmon losses also appear in the loss spectrum. This can be seen, in particular, in the loss spectrum of Al, which shows one sharp plasmon loss at 15.2 eV (Fig. 5.16). The probability $P_n(t)$ for the appearance of an energy loss $\Delta E = n\Delta E_p$ in a specimen layer of thickness t can be

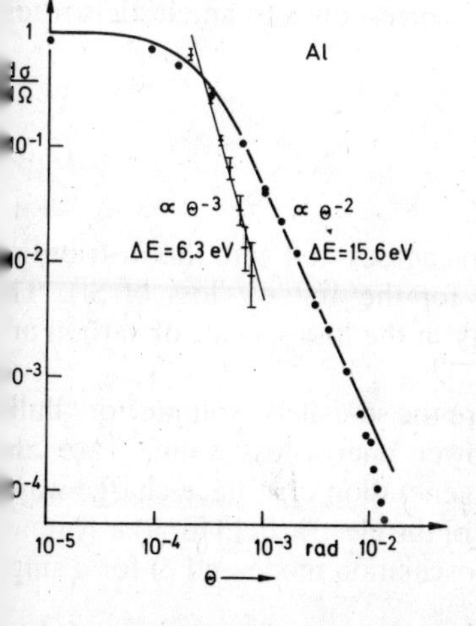

Fig. 5.14. Angular dependence of the volume-plasmon-loss cross-section for $\Delta E = 15.6$ eV [5.64] and of the surface-plasmon loss ($\Delta E = 6.3$ eV) of Al [5.59, 60]

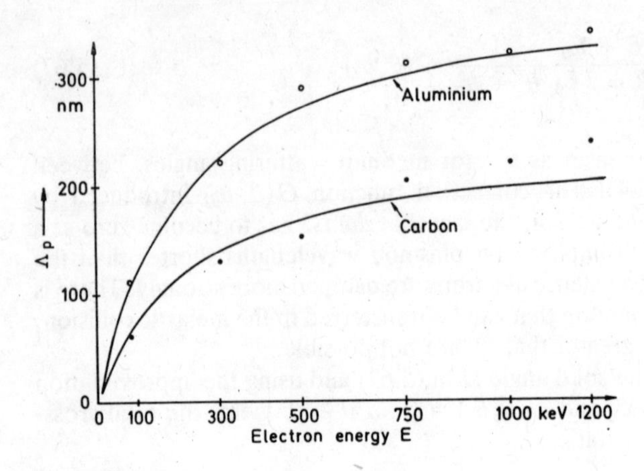

Fig. 5.15. Mean-free-path length Λ_p for Al and carbon plasmon losses as functions of electron energy [5.70]

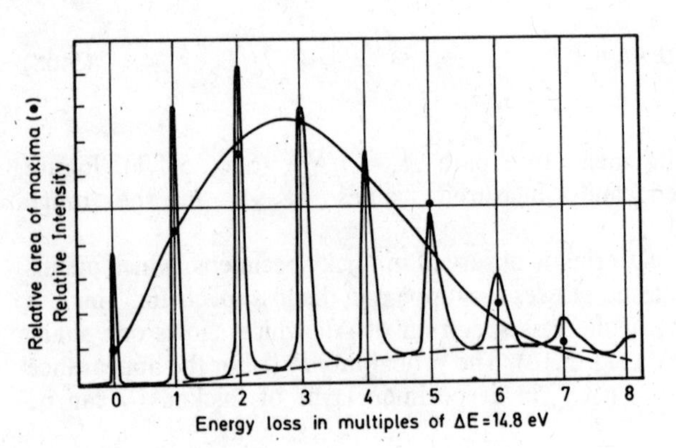

Fig. 5.16. Multiple characteristic plasmon losses of 20 keV electrons passing through a 208 nm Al film and comparison of the areas of the loss maxima (●) with a theoretical Poisson distribution [5.69]

described by a Poisson distribution ($n = 0$ corresponds to an elastic-scattering event with no energy loss)

$$P_n(t) = \left(\frac{t}{\Lambda_p}\right)^n \frac{\exp(-t/\Lambda_p)}{n!} . \quad (5.69)$$

The integrated intensity of the loss maxima agrees well with this distribution (see Fig. 5.16 for Al [5.72], and [5.73] for the 16.9 eV loss of Si). The convolution with the background intensity in the loss spectra of carbon and aluminium films is considered in [5.74, 75].

The plasmon losses discussed above are the so-called "volume" or "bulk" losses. Surface-plasmon losses, with lower energy-loss values, are also observed; these can be explained by the generation of surface-charge waves [5.76]. Figure 5.17 shows the distribution of the electric field for a) a symmetric ω^- and b) an asymmetric ω^+ surface-oscillation mode and c) for a sing

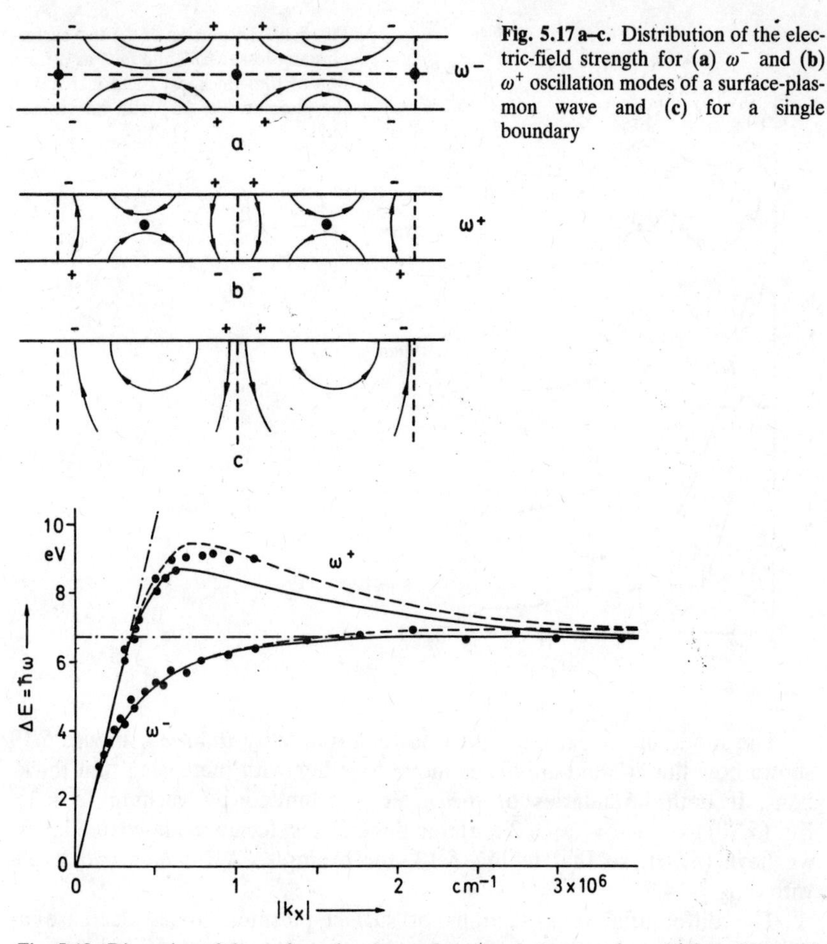

Fig. 5.17 a–c. Distribution of the electric-field strength for (a) ω^- and (b) ω^+ oscillation modes of a surface-plasmon wave and (c) for a single boundary

Fig. 5.18. Dispersion of the surface-plasmon modes ω^+ and ω^- versus $k_x = \theta/\lambda$ for a 16 nm Al film covered on each side with a 4 nm oxide film. The dashed curve corresponds to ε for amorphous Al_2O_3 and the full curve for α-Al_2O_3 [5.78]

boundary in a thicker layer. Both oscillation modes show strong dispersion [5.64, 77–79], which depends not only on the wave number $k_x = \theta/\lambda$ of the surface waves but also on the specimen thickness t (see example in Fig. 5.18). The saturation values for large θ and t are

$$\omega^{\pm} = \frac{\omega_p}{\sqrt{1+\varepsilon}}, \tag{5.70}$$

where ε is the relative permittivity of the neighbouring medium, typically vacuum, oxide or supporting film.

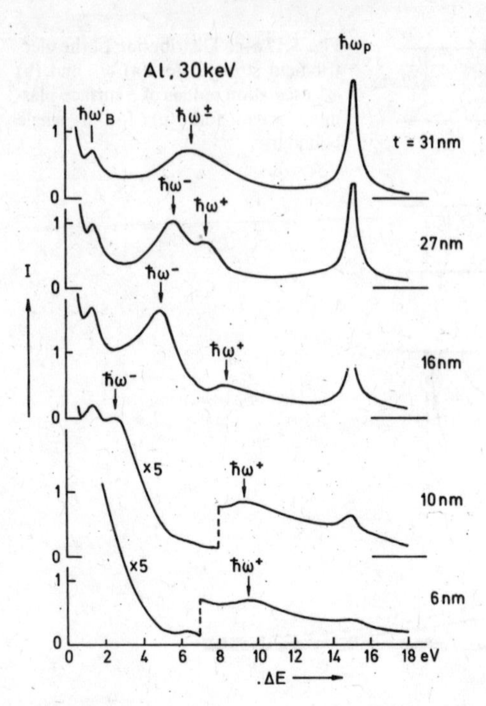

Fig. 5.19. Dispersion of the two surface plasmon losses $\hbar\omega^+$ and $\hbar\omega^-$ as a function of film thickness t and increase of the plasmon loss $\hbar\omega_p$ with increasing t [5.79]

The mode ω^+ is excited with a lower probability than ω^-. Figure 5.19 shows how the ω^+ and ω^- losses move together with increasing film thickness. If both boundaries of the layer are limited by vacuum ($\varepsilon = 1$), Eq. (5.70) gives $\omega^\pm = \omega_p/\sqrt{2}$ for large thicknesses; for an oxide-coated layer, we have (5.70), so that in Fig. 5.18, for example, $\Delta E = \hbar\omega^\pm = 6.25$ eV with $\varepsilon_{\text{oxide}} = 4.7$.

The differential cross-sections of surface-plasmon losses decrease as $\theta\,\theta_E/(\theta^2 + \theta_E^2)^2$ for increasing θ, and hence as θ^{-3} for $\theta > \theta_E$ (Fig. 5.14). At non-normal incidence of the primary electrons, the excitation of surface-plasmon losses shows an asymmetric angular distribution [5.76, 78–81].

If the electron velocity v is greater than the velocity of light in the specimen layer (e.g., for Si), energy losses $\Delta E = 3.4$ eV are observed due to the generation of Čerenkov radiation [5.82, 83]. Guided-light modes can be excited in thin dielectric films, such as graphite [5.68].

The plasmons decay into phonons after a short lifetime of 10^{-15}–10^{-14} s and contribute to specimen heating (Sect. 10.1). At non-normal electron incidence, a light quantum with energy $\hbar\omega_p$ may also be emitted [5.84–88]. This radiation is part of the *transition radiation* and of the long wave-length tail of the continuous x-ray spectrum [5.89, 90].

5.2.4 Energy Losses Caused by Inner-Shell Ionisation

Ionisation of an atom in an inner shell I = K, L or M needs an ionisation energy E_I, which is the difference between the K, L or M energy level and the upper unoccupied energy state, the Fermi level in solids (zero energy level in Fig. 5.35).

The ionisation of hydrogen-like levels produces a saw-tooth-shaped peak in the energy-loss spectrum with a step increase of the cross-section $d\sigma/d(\Delta E)$ for energy losses $\Delta E \geq E_I$ and a smooth decrease for greater ΔE (Fig. 5.10). In practice, this simple shape of the energy loss spectrum is modified by several other influences [5.91]:

1) Neighbouring sharp peaks can be observed, caused by the energy separation of inner shells due to spin-orbit coupling, e.g. subshells L_1, L_2 and L_3.

2) Energy losses ΔE in excess of E_I can excite the atomic electron from the inner level to an unoccupied state above the Fermi level, where the selection rule $\Delta l = \pm 1$ has to be considered. This creates a fine structure behind the edge, depending on the nature of the chemical bonding and the density of states in the band structure of a solid. Differences result for example in the K-ionisation loss of amorphous carbon and graphite (Fig. 5.20) [5.36]. Fine structure depending on the molecular configuration can be observed in organic compounds (Fig. 5.21) [5.37].

3) Electrons that leave the atom are influenced by not only the Coulomb potential $-Ze^2/4\pi\varepsilon_0 r$ but also by a centrifugal repulsive barrier that corresponds to an effective potential $l(l+1)\hbar^2/2mr$ [5.92] if transitions with $l \geq 2$ are observed. This displaces the maxima of the hydrogen-like saw-tooth profile.

4) Fine structure consisting of weak maxima 10–100 eV above the ionisation peak may be observed (Fig. 5.22) [5.91, 93, 94]; this is also observed in x-ray absorption spectra and is called EXAFS (extended x-ray absorption fine structure). It is caused by oscillations of the photoelectron cross-sections due to backscattering by the nearest neighbouring atoms. These losses furnish a sensitive tool for extracting information about the arrangement of nearest-neighbour atoms and the Fourier transform of the EXAFS spectrum provides a radial structure function [5.95]. Thus, the main maxima indicated by arrows in Fig. 5.22 and split by the influence of second-nearest neighbours give the values $r_1 = 0.28 \pm 0.01$ nm for Al and $r_1 = 0.2 \pm 0.01$ nm for the oxide, in agreement with crystallographic x-ray data [5.91].

Calculation of the excitation of a discrete atomic level in Sect. 5.2.2 and of plasmons in Sect. 5.2.3 showed that $d\sigma/d\Omega$ is greatest in the forward direction ($\theta = 0$) and decreases with a half-width of $\theta_E = E_I/2E$ as θ increases. The same law is valid for energy losses somewhat greater than the ionisation energy E_I, and the angular distribution also shows that forward scattering is dominant. The angle θ_E is a measure of the aperture of the scattering cone within which electrons with losses $\Delta E = E_I$ are to be

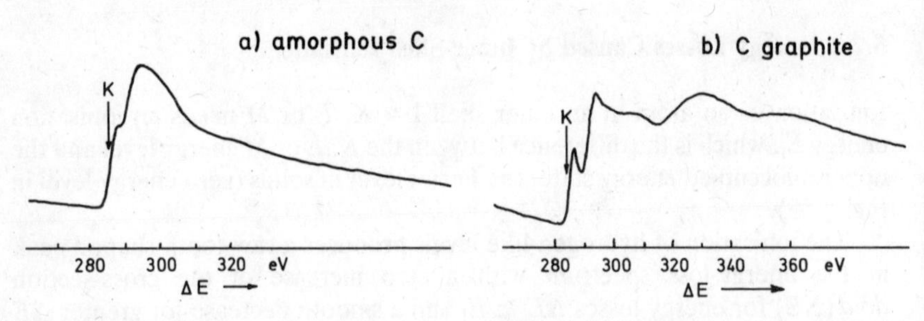

Fig. 5.20 a, b. Differences in the energy-loss spectrum at the *K* ionisation edge for carbon in (**a**) amorphous carbon films and (**b**) graphite films [5.36]

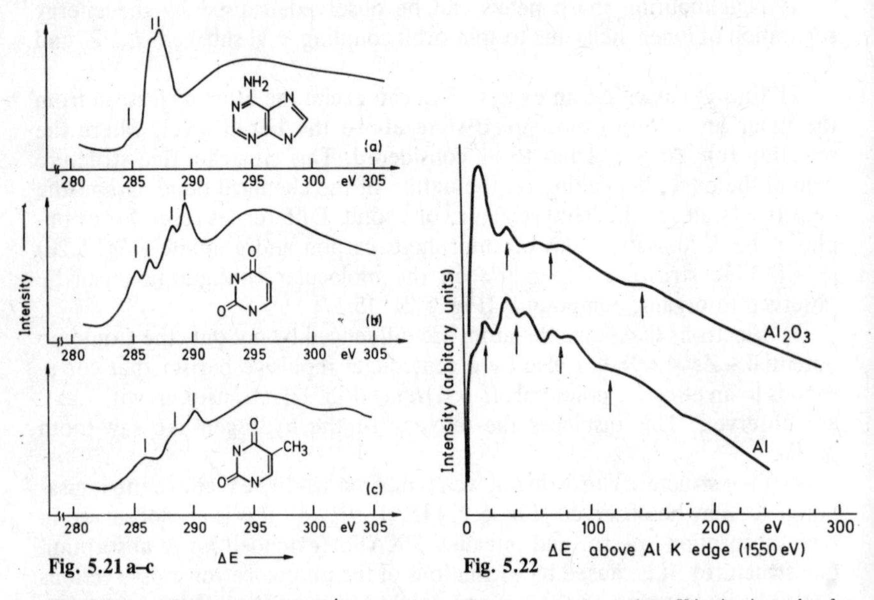

Fig. 5.21 a–c

Fig. 5.22

Fig. 5.21 a–c. Fine structure in the energy loss spectrum near the carbon *K* ionisation edge for (**a**) adenine, (**b**) uracil and (**c**) thymine [5.37]

Fig. 5.22. Energy-loss maxima (extended fine structure) of Al and Al_2O_3 films above the *K* ionisation edge of Al due to interactions with the neighbouring atoms [5.93]

expected. Even for $\Delta E = 1$ keV, a value $\theta_E = E_I/2E = 5$ mrad is found; this is still less than the characteristic angle θ_0 (5.39) for elastic scattering.

When $\Delta E \gg E_I$, the atomic electrons can be treated approximately as free electrons. The binding energy is negligible compared to the kinetic energy of the excited electrons. The laws of conservation of momentum and energy (5.12–14) for the electron-nucleus collision can be transferred to the electron-electron collision. The scattering angle θ_f and the energy loss ΔE are related by

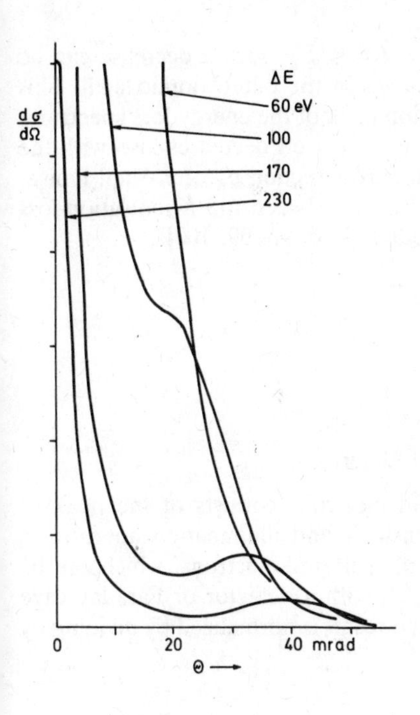

Fig. 5.23. Angular dependence of the differential cross-section of 80 keV electrons scattered by a carbon film for different energy losses ΔE below the k ionisation loss at $\Delta E = 285$ eV and generation of a scattering maximum for large ΔE and θ [5.96]

$$\theta_{\mathrm{f}} = (\Delta E/E)^{1/2} \tag{5.71}$$

for non-relativistic energies $E_{\mathrm{I}} \ll \Delta E < E < E_0$. A sharp maximum of the angular distribution of scattered electrons with energy losses $\Delta E \gg E_{\mathrm{I}}$ can therefore be expected at θ_{f}; the peak in the forward direction associated with energies $\Delta E \gg \mathrm{E_I}$ is no longer found. Figure 5.23 shows measurements of the angular distribution of $d\sigma/d\Omega$ for different energy losses ΔE below the K-ionisation energy of carbon ($E_{\mathrm{I}} = 285$ eV). These show a continuous transition from forward scattering to a maximum at large scattering angles with increasing energy loss [5.96]. The broad maximum can be explained by the following modified model. Application of the modified equations (5.12–14) to electron-electron collisions supposes that the atomic electron is at rest before the collision. A still-classical theory [5.97] considers that the atomic electron has a velocity distribution on its orbit. This theory predicts a broadening of the maxima at the scattering angle θ_{f} of the correct order of magnitude and also predicts the observed shift of the maximum to smaller values than θ_{f}. Application of quantum-mechanical reasoning to this problem [5.98] results in a decrease of $d\sigma/d(\Delta E)$ proportional to $\Delta E^{-4.5}$ with increasing $\Delta E \gg E_{\mathrm{I}}$, if the angle of collection is large enough to collect all electrons with energy losses ΔE. A decrease of $d\sigma/d(\Delta E)$ proportional to E^{-n} with exponents n between 3.5 and 4.5 is found experimentally, as is shown by double-logarithmic plots (Fig. 5.10).

In front of the K-ionisation loss, for $\Delta E < E_K$, such a decrease can be observed due to the excitation of electrons in the outer atomic shell. This decrease continues below the K-ionisation peak of the energy-loss spectrum, and the contribution of the K-shell ionisation itself decreases also with the same law for $\Delta E \gg E_K$ [5.96]. The total cross-section σ_K of K-shell ionisations can be determined from the difference between the K-ionisation-loss spectrum and the extrapolated background [5.36, 96, 99, 100].

5.3 Multiple-Scattering Effects

5.3.1 Angular Distribution of Scattered Electrons

The angular distribution of transmitted electrons consists of the peak of unscattered, primary electrons of intensity I and illumination aperture α_i together with the angular distribution of scattered electrons, which can be measured by recording the current $\Delta I(\theta)$ with a detector or Faraday cage having a solid angle $\Delta\Omega$ of collection; the result is normalized by dividing by the incident current I_0. This quantity

$$\frac{1}{I_0}\frac{\Delta I(\theta)}{\Delta\Omega} = \frac{N_A \varrho t}{A}\frac{d\sigma}{d\Omega} = \frac{x}{\Lambda_t} S_1(\theta) \tag{5.72}$$

is proportional to the mass-thickness $x = \varrho t$ only for very small values of x; Λ_t denotes the total free-path length of (5.11).

The intensity I of the unscattered, primary beam decreases exponentially with increasing mass-thickness x according to (5.11), i.e., $I/I_0 = \exp(-x/\Lambda_t)$. It will be shown in Sect. 6.1.1 that the elastic mean free path Λ_{el} is identical with the value x_a in Table 6.1 and Λ_t can be calculated by use of (5.11):

$$\frac{1}{\Lambda_t} = \frac{1}{\Lambda_{el}} + \frac{1}{\Lambda_{inel}} = \frac{N_A}{A}(\sigma_{el} + \sigma_{inel}) = \frac{N_A \sigma_{el}}{A}(1+\nu) = \frac{1+\nu}{x_a}, \tag{5.73}$$

where the ratio $\nu = \sigma_{inel}/\sigma_{el}$ is defined in (5.58) and listed in Table 5.3.

For $E = 100$ keV, a value $\Lambda_t = 12\ \mu g\ cm^{-2}$ or $t = 120$ nm is found for organic material of density $\varrho = 1\ g\ cm^{-3}$, but for evaporated Ni and Fe films the same mass-thickness is obtained, or $t \simeq 15$ nm [5.41]. This decrease of primary-beam intensity is important for the visibility of phase-contrast effects, which are generated by interference between the primary and scattered electron waves. In Lorentz microscopy (Sect. 6.6), the domain contrast is created by using a primary beam of very small illumination aperture, $\alpha_i \lesssim 10^{-2}$ mrad; elastic and inelastic-scattered electrons cause a blurring of the domain contrast in the Fresnel mode.

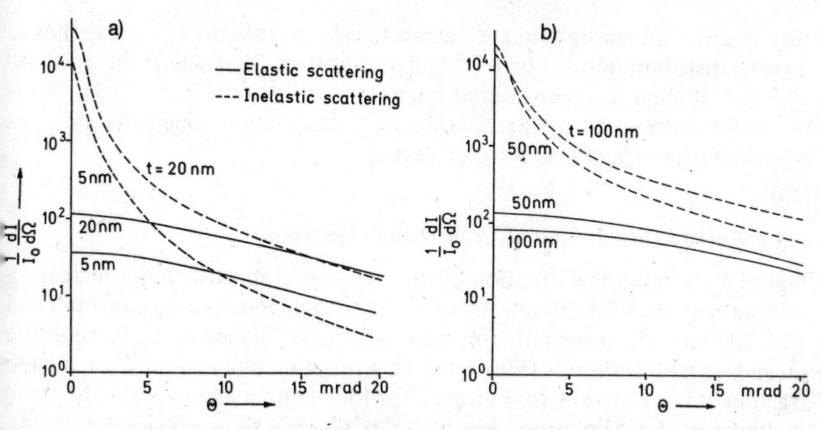

Fig. 5.24 a, b. Angular distribution of the elastic and inelastic scattering intensities of 100 keV electrons in carbon films of increasing thickness t = 5–100 nm [5.102]

The angular distribution for multiple scattering can be obtained by evaluating a multiple-scattering integral [5.10] or by superposition of multiple-scattering functions $S_n(\theta)$, which are calculated by an n-fold convolution of $S_1(\theta)$ defined in (5.72)

$$S_n(\theta) = S_{n-1}(\theta) \otimes S_1(\theta) . \tag{5.74}$$

These are then weighted with the coefficients of a Poisson distribution [5.42]

$$\frac{1}{I_0}\frac{\Delta I(\theta)}{\Delta\Omega} = \exp\left(-\frac{x}{\Lambda_t}\right) \sum_{n=1} \left(\frac{x}{\Lambda_t}\right)^n \frac{S_n(\theta)}{n!} . \tag{5.75}$$

A procedure is described whereby the two-dimensional integration necessary for the convolution in (5.74) is reduced to a one-dimensional integration, by use of projected distributions [5.101].

Figure 5.24 shows the contribution of elastic and inelastic scattering calculated from (5.75) for different carbon-film thicknesses [5.102]. For thin films (5 nm), the angular distribution can be assumed to be approximately proportional to the differential cross-section $d\sigma/d\Omega$ of single atoms. The intensity distribution is modified by multiple scattering as the thickness is increased. The elastic contribution at small scattering angles θ increases up to 50 nm but decreases for greater thicknesses, due to elastic multiple scattering into larger angles and to inelastic scattering.

These calculations neglect all interference effects. In crystalline specimens, destructive interference decreases the scattered intensity between the primary beam and the Bragg diffraction spots; the scattered intensity is caused by thermal diffuse scattering (electron-phonon scattering) and inelastic scattering. In amorphous specimens, the short-range order corresponds to a radial distribution function of neighbouring atoms that causes diffuse maxima and minima in the scattered intensity distribution (Sect. 7.5.1). However, this distribution oscillates around the distributions calculated here, in which interference effects were neglected.

5.3.2 Energy Distribution of Transmitted Electrons

Figure 5.25 shows the variation of the energy-loss spectrum with increasing thicknesses for 1.2 MeV electrons [5.70]. Analogous results are obtained with 100 keV electrons only for correspondingly thinner films, because the mean free path is shorter (Fig. 5.15). In a very thin film (Fig. 5.25 a), a large fraction of the electrons pass through the film without energy loss. The three multiples of the Al plasmon loss at $\Delta E = 15.2$ eV show a Poisson distribution (5.69). The intensity of higher energy losses is very low, and an increase by L-shell ionisation appears at $\Delta E = 80$ eV. At medium thicknesses (Fig. 5.25 b), the zero-loss peak is strongly reduced, and seven plasmon losses can be detected. The plasmon losses are superimposed on a broad maximum, due to overlapping of the L-ionisation peak and the plasmon losses. In similar experiments [5.103], the plasmon losses were not resolved, but two maxima were observed 80 eV apart. The plasmon losses disappear in very thick specimens (Fig. 5.25 c) and only a broad energy distribution with a most-probable energy E_p and a full-width at half maximum ΔE_{FWHM} or ΔE_H is observed.

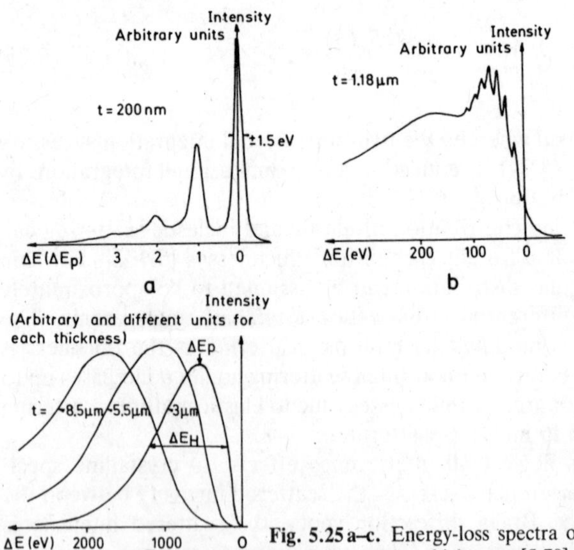

Fig. 5.25 a–c. Energy-loss spectra of 1200 keV electrons in Al foils of increasing thickness t [5.70]

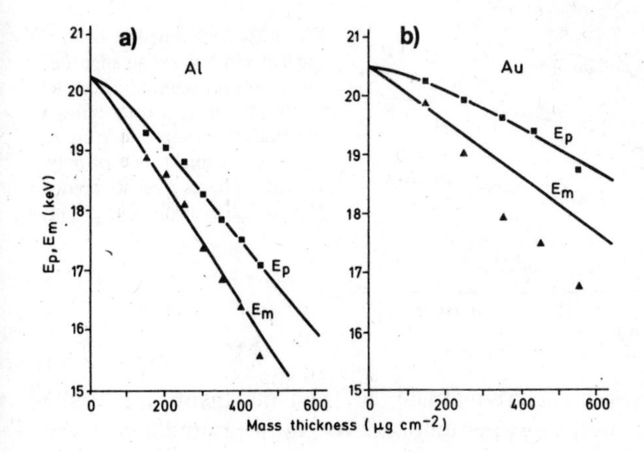

Fig. 5.26 a, b. Most-probable energy E_p and mean energy E_m of 20 keV electrons in (**a**) Al and (**b**) Au films of increasing mass thickness $x = \varrho t$ calculated with (5.76) and (5.79) and measured values [5.108]

For the value of the most-probable energy loss, a theory of Landau [5.104] can be used, which considers the atomic structure only in terms of a mean ionisation energy $J \simeq 13.5\ Z$ (see also modifications of this theory in [5.105, 106]:

$$E - E_p = \frac{N_A e^4 Zx}{8\pi\varepsilon_0^2 A E_0 \beta^2}\left[\ln\left(\frac{N_A e^4 Zx}{4\pi\varepsilon_0^2 J^2 A(1-\beta^2)}\right) - \beta^2 + 0.198\right] \tag{5.76}$$

($E_0 = m_0c^2$, $\beta = v/c$ and $x = \varrho t$ is the mass-thickness). The validity of this formula for 100 keV electrons has been confirmed for Al [5.70] and MgO [5.107]. Other experiments for Al [5.103] agree less closely. Measurements of E_p at E = 20–40 keV on Al- and Au-films also show agreement with (5.76) (Fig. 5.26) [5.108]. However, the observed values of the full-width at half maximum ΔE_{FWHM} of the energy distribution are greater than those given by the following expression, which can be derived from the Landau theory

$$\Delta E_{\text{FWHM}} = 4.02\,\frac{N_A e^4 Zx}{8\pi\varepsilon_0^2 A E_0 \beta^2}\,. \tag{5.77}$$

For many applications, it is not sufficient to characterize the energy distribution by the most probable energy E_p and ΔE_{FWHM}; it is also of interest to know the mean energy, which can be calculated from

$$\text{a)}\quad E_m = \frac{\int_0^E N(E)\,E\,dE}{\int_0^E N(E)\,dE} \qquad \text{or} \qquad \text{b)}\quad E_m = E + \int_0^x \left(\frac{dE_m}{dx}\right)_B dx \tag{5.78}$$

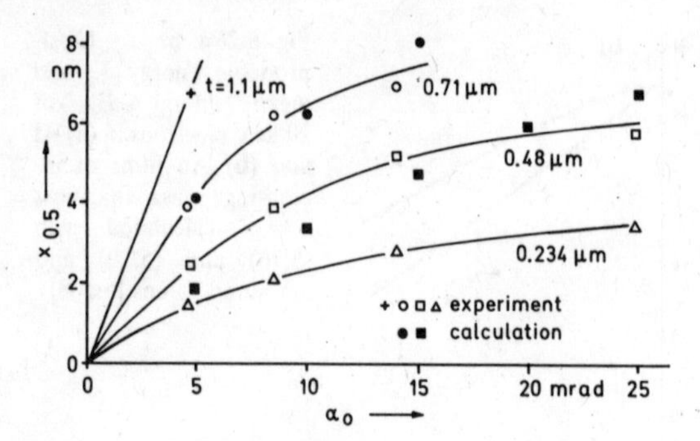

Fig. 5.27. Experimental value of the half-width $x_{0.5}$ of an edge (see Fig. 5.30 a for definition) caused by the chromatic aberration as a function of objective aperture α_o for different thicknesses of polystyrene spheres. Solid points: Monte-Carlo calculations [5.111]

by use of the energy-loss spectrum $N(E)$, or from the theoretical Bethe formula for the mean energy loss per unit path length, measured in terms of mass-thickness [5.109]

$$\left(\frac{dE_m}{dx}\right)_B = -\frac{e^4 N_A Z}{4\pi\varepsilon_0^2 A E_0 \beta^2} \ln\left(\frac{E_0\beta^2}{2J}\right). \tag{5.79}$$

This formula can also be used to calculate the specimen heating (Sect. 10.1) and the radiation damage by ionisation (Sect. 10.2). Calculations of the mean energy E_m based on (5.78 b and 79) agree with the experimental values obtained from measured energy distributions $N(E)$ by applying (5.78 a) (Fig. 5.26). This agreement was not found in [5.110]; there the measured values of E_m were too low. The stronger decrease of the experimental values for large mass-thicknesses of gold can be attributed to an increase of the effective path length, caused by multiple scattering.

The energy losses impair the resolution, as a result of chromatic aberration (2.54). Measurements of the width $x_{0.5}$ of the blurred intensity spread (Figs. 5.29 c and d) at the edges of indium crystals placed below polystyrene spheres of different thicknesses (see Fig. 5.30 a for a definition of $x_{0.5}$) are plotted in Fig. 5.27 as a function of objective aperture. The reason why the measured values of $x_{0.5}$ do not increase as α_0 (2.54) is that the step intensity distribution consists of a steep central and a flat outer part. Monte-Carlo calculations, which take into account the plural scattering of the carbon plasmon-loss spectrum, from which $x_{0.5}$ can be obtained by the same method, predict the same dependence on aperture and agree with the experimental results. Fig. 5.29 c and d at E = 100 keV and 200 keV show that the effect of chromatic aberration decreases with increasing E.

5.3.3 Electron-Probe Broadening by Multiple Scattering

The angular distribution of scattered electrons in thick films (0.1–1 µm) produces a spatial distribution that, in turn, broadens the incident electron probe normal to the beam direction. This effect limits the resolution of the scanning

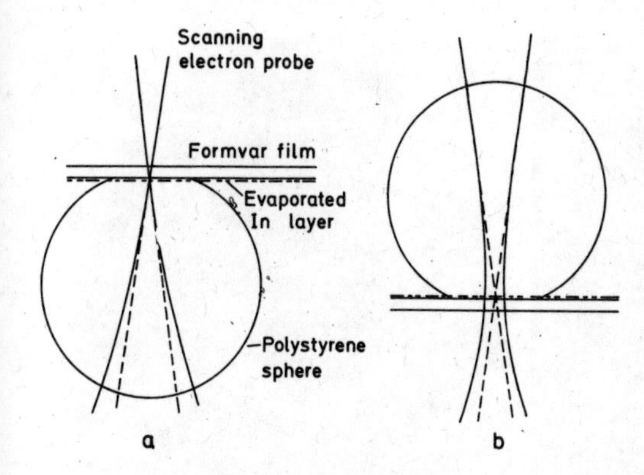

Fig. 5.28 a, b. Specimen structure and electron-beam broadening in the two cases in which the polystyrene spheres are (**a**) below and (**b**) above the evaporated indium layer

transmission mode and although the chromatic aberration of the conventional transmission mode shown in Fig. 5.27 is avoided; it likewise limits the resolution of x-ray microanalysis of thick specimens.

This scattering can be observed as a *top-bottom* effect in the scanning transmission mode and is illustrated in Figs. 5.28 and 29. The specimen consists of a thin formvar supporting film onto which indium, which condenses as small crystals on the substrate, has been evaporated. This specimen is coated with polystyrene spheres of 1 μm diameter to simulate a thick specimen of known thickness. The indium layer is scanned by an unbroadened electron probe with the polystyrene sphere below the layer (Figs. 5.28 a and 29 a). The image of the indium crystals is sharp, and the subsequent scattering of electrons in the polystyrene sphere and the broadening of the electron beam merely decrease the intensity recorded with the STEM detector without affecting the resolution. With the polystyrene spheres uppermost, the indium layer is scanned by a broadened probe, the edges of the indium crystals are blurred and the resolution is reduced (Figs. 5.28 b and 29 b).

A resolution parameter can be obtained by measuring the intensity distribution across the edge of the indium crystals with a densitometer and measuring the width $x_{0.5}$ between the points at which the step reaches 0.25 and 0.75 of its total intensity (Fig. 5.30). Measured values of $x_{0.5}$ behind polystyrene spheres of thickness t are plotted in Fig. 5.30 for different electron energies.

A value of $x_{0.5} \simeq 10$ nm is found for $E = 100$ keV and $t = 1$ μm. The order of magnitude is the same for the chromatic aberration for objective apertures $\alpha_0 \geq 10$ mrad (Fig. 5.27). Whereas the blurring of specimen structures by chromatic aberration is approximately the same over the whole specimen thickness, structures at the top of a 1 μm layer are imaged in the scanning mode with a better resolution. It was therefore suggested that the chromatic aberration in the conventional TEM mode could be avoided in this way [5.113]. However, the top-bottom effect sets a limit on the improvement

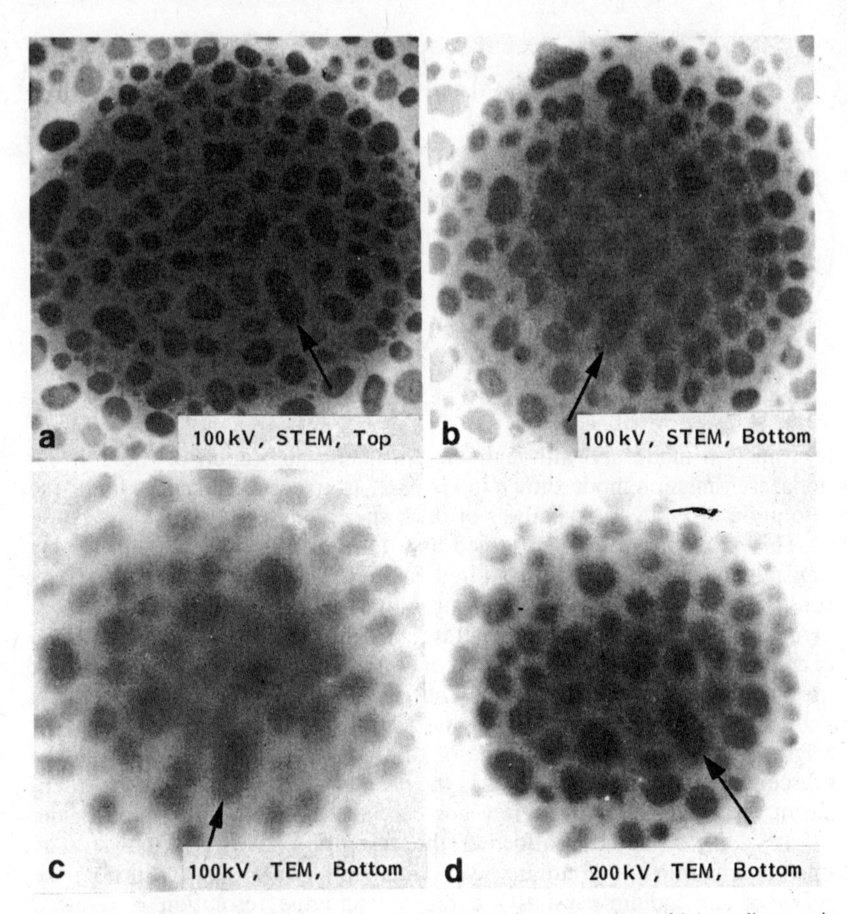

Fig. 5.29 a–d. Images of indium crystals and the same polystyrene sphere of 1.1 μm diameter in the 100 keV STEM mode with the polystyrene sphere (**a**) below and (**b**) above the indium layer to demonstrate the top-bottom effect, (**c**) and (**d**) normal TEM images of the same area at E = 100 and 200 keV respectively, blurred by the chromatic aberration and the energy losses in the polystyrene sphere

that can be achieved [5.112]. The advantages of the scanning mode can be seen in other applications (Sect. 4.5.1).

The effect of chromatic aberration decreases with increasing energy and, in high-voltage electron microscopy, multiple scattering can also cause a top-bottom effect in the conventional bright-field TEM mode; here, however, structures at the bottom of the specimen are imaged with a better resolution [5.111, 114].

This spatial broadening of the electron probe can be calculated from the differential cross-section $d\sigma/d\Omega$ by applying a multiple-scattering integral [5.115], by solving the Boltzmann transport equation [5.116–118] or by

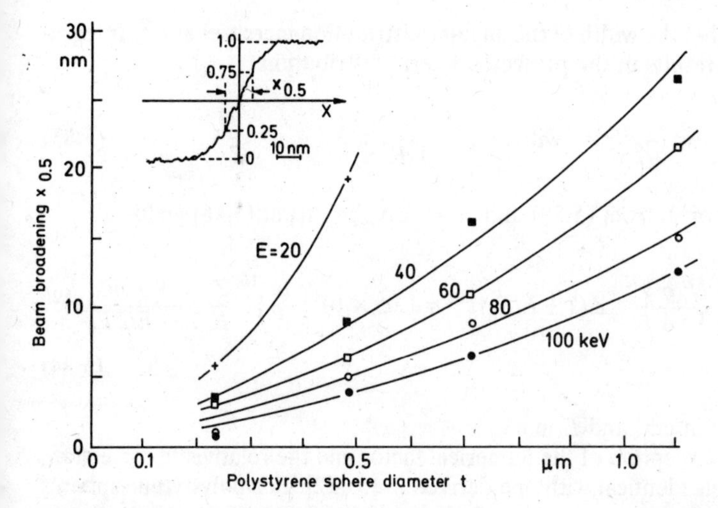

Fig. 5.30. (a) Densitometer recording across an edge of an indium crystal and definition of $x_{0.5}$. (b) Measurements of $x_{0.5}$ in the STEM mode for specimens below polystyrene spheres of thickness t [5.112]

Monte-Carlo calculations [5.119]. A disadvantage of all of these methods is that they do not lead to analytic formulae.

An approximate formula for estimating the beam broadening can be obtained if we return to a multiple-scattering theory proposed by *Bothe* [5.120] (see also [5.121]). The differential cross-section (5.40) $d\sigma_{el}/d\Omega \propto [1 + (\theta/\theta_0)^2]^{-2}$ can be approximated by a two-dimensional Gaussian function of the form $\exp[-(\theta_x^2 + \theta_y^2)/\theta_0^2]$. In both formulae $\overline{\theta^2} = \theta_0^2$. This function has the advantage that convolutions can be evaluated straightforwardly, thanks to the following property of Gaussians

$$\exp(-x^2/a^2) \otimes \exp(-x^2/b^2) \propto \exp[-x^2/(a^2 + b^2)] \,. \tag{5.80}$$

Bothe obtained the projected probability function which expresses the likelihood of finding an electron at a depth z, a distance x from the axis, and with a projected scattering angle θ_x (Λ: mean-free-path length)

$$P(z, x, \theta_x) = \exp\left[-\frac{4}{\theta_0^2}\left(\frac{\theta_x^2}{z} - \frac{3x\theta_x}{z^2} + \frac{3x^2}{z^3}\right)\right] \,. \tag{5.81}$$

Integration over x gives the projected angular distribution for a film of thickness $z = t$

$$f(t, \theta_x) \propto \exp(-\theta_x^2/\overline{\theta^2}) \quad \text{with} \quad \overline{\theta^2} = \theta_0^2 t/\Lambda \,. \tag{5.82}$$

This means that the width of the angular distribution increases as $t^{1/2}$. Integration over θ_x results in the projected lateral distribution

$$I(t,x) = \exp(-x^2/x_0^2) \quad \text{with} \quad x_0^2 = \frac{\theta_0^2}{3\Lambda} t^3 . \tag{5.83}$$

Substitution of θ_0 from (5.39) and $\Lambda = A/N_A \varrho \sigma_{el}$ from (5.46) yields

$$x_0 = \frac{\lambda^2}{2\pi a_H} \left(\frac{N_A \varrho}{3\pi A}\right)^{1/2} Z(1+E/E_0)\, t^{3/2} = 1.05 \times 10^5 \left(\frac{\varrho}{A}\right)^{1/2} \frac{Z}{E} \frac{1+E/E_0}{1+E/2E_0} t^{3/2} \tag{5.84}$$

with x_0 and t in cm, and E in eV.

With the exception of the numerical factor and the relativistic correction, this formula is identical with one derived in [5.122]. For polystyrene spheres ($\varrho = 1.05$ g cm^{-3}), $t = 1$ µm and $E = 100$ keV Eq. (5.84) gives $x_{0.5} = 0.96\, x_0 = 20$ nm, which is larger than the measured value of 10 nm (Fig. 5.30). However, the blurred image of an edge in the scanning mode is produced only by small-angle scattering $\theta \leq \alpha_d$, which reduces the beam broadening actually observed.

5.3.4 Electron Diffusion, Backscattering and Secondary-Electron Emission

Electron diffusion in bulk material is more important for scanning electron microscopy. In TEM, the specimens normally have to be thin enough to avoid the multiple-scattering effects of thick films. However, a knowledge of electron interactions with solids is necessary for the discussion of recording by photographic emulsions, scintillators and semiconductors. Walls and diaphragms in the microscope column are struck by electrons, and backscattered electrons (BSE) and secondary electrons (SE) from the specimen can be used as signals in the scanning mode. The most important facts about electron diffusion and BSE and SE emission from thin films will therefore be summarized here.

Electron trajectories in a solid are curved by large-angle elastic-scattering processes (*Mott scattering*). The mean electron energy decreases along the trajectories as a result of energy losses. This decrease can be described by the Bethe formula (5.79). Integration of (5.79) yields $E_m(x)$ (Fig. 5.31). Setting $E_m = 0$ gives the *Bethe range* R_B, which increases with increasing atomic number Z. However, the trajectories become more strongly curved with increasing Z, owing to the presence of the factor Z^2 in the Rutherford cross-section (5.41). In practice, the range R is approximately independent of Z if it is measured in units of mass-thickness (e.g., g cm^{-2}) and is of the order of R_B only for small Z. The range can be estimated from the empirical formula

$$R \simeq \frac{20}{3} E^{5/3} , \tag{5.85}$$

in the energy range $10 \leq E \leq 100$ keV, with R in μg cm^{-2} and E in keV.

The depth distribution $Q(z)$ of energy dissipation by ionisation describes the probability of producing electron-hole pairs in semiconductors, or photons in scintillators, and the generation of heat. In Fig. 5.31, $Q(z)$ curves are plotted for C and Au. These have a maximum below the surface and also demonstrate the existence of a range R approximately independent of Z.

A fraction η of the incident electrons can leave the specimen as BSE with energies reduced by inelastic scattering; η is known as the backscattering coefficient. Integration of the Rutherford cross-section (5.41) from $\theta = \pi/2$ to $\theta = \pi$ (backscattering) gives

$$\eta = \frac{e^4 Z^2}{16\pi\varepsilon_0^2 E^2} \left(\frac{E + E_0}{E + 2E_0}\right)^2 Nt , \tag{5.86}$$

where $N = N_A \varrho / A$ is the number of atoms per unit volume. Plots of η / NZ^2 against the film thickness t are indeed approximately independent of the material (Fig. 5.32). The linear increase of η with increasing film thickness can be used to measure the latter by placing a small Faraday cage in front of the specimen [5.123–125]. For a more-accurate comparison of theory and experiment, it is necessary to consider the Mott cross-sections (Fig. 5.7) and the contribution of electron-electron scattering with large energy losses. Monte-Carlo calculations are suitable for these calculations [5.26, 126]. The backscattering coefficient may be influenced by channelling effects and also depends on the orientation of the crystal foil relative to the electron beam. These effects can be used to record channelling patterns with BSE by rocking the incident electron beam (Sect. 9.3.3).

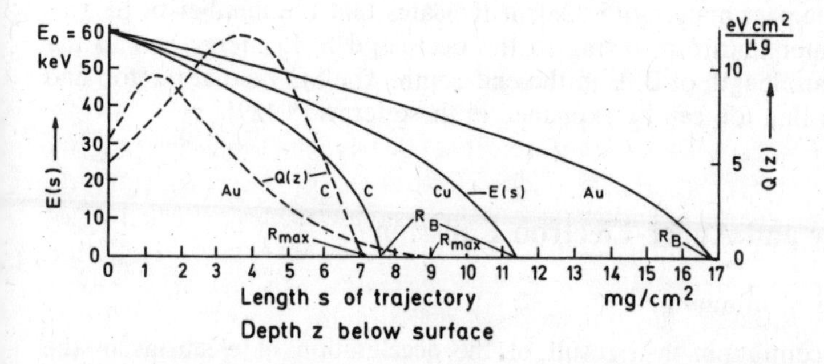

Fig. 5.31. Decrease of mean energy E_m along the electron trajectories for different elements and definition of the Bethe range R_B. Depth distribution of ionisation density $Q(z)$ (dashed curves) and practical range R_{max}

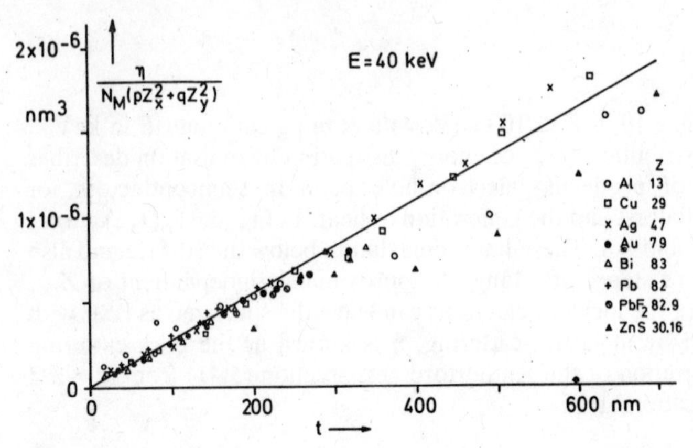

Fig. 5.32. Increase of the backscattering coefficient η with increasing film thickness t plotted as η/NZ^2 versus thickness ($N = N_A\varrho/A$: number of atoms per unit volume) [5.124]

Electrons excited by inelastic collisions with an energy sufficiently far above the Fermi level to overcome the work function can leave the specimen as secondaryelectrons (SE); by convention, these have an energy $E_{SE} \leq 50$ eV and emerge from a small exit depth of the order of $t_{SE} = 1$–10 nm [5.127]. The secondary-electron yield δ is proportional to the Bethe loss $|dE_m/dx|$ (5.79) in the surface layer and to the path length $t_{SE} \sec\phi$ inside the exit depth; ϕ is the angle between the incident direction and the surface normal. The total SE yield is the sum of the SE generated by the primary beam (δ_{PE}) and by the backscattered electrons or the transmitted electrons on the bottom surface (δ_{BSE})

$$\delta = \delta_{PE} + \delta_{BSE} = \delta_{PE}(1 + \beta\eta) . \tag{5.87}$$

The fraction β is greater than unity and can increase to values of two or three for compact material [5.128]; it indicates that the number of SE per BSE is greater than δ_{PE}, owing to the decreased BSE energy and to the increased path length of BSE in the exit depth. The SE yield at the top and bottom of a thin foil can be explained in these terms [5.129].

5.4 X-Ray and Auger-Electron Emission

5.4.1 X-Ray Continuum

The x-ray continuum is a result of the acceleration of electrons in the Coulomb field of the nucleus. It is well known from electrodynamics that an accelerated charge can emit an electromagnetic wave. A periodic acceleration $a(t)$ produces a monochromatic wave (dipole radiation). Because the

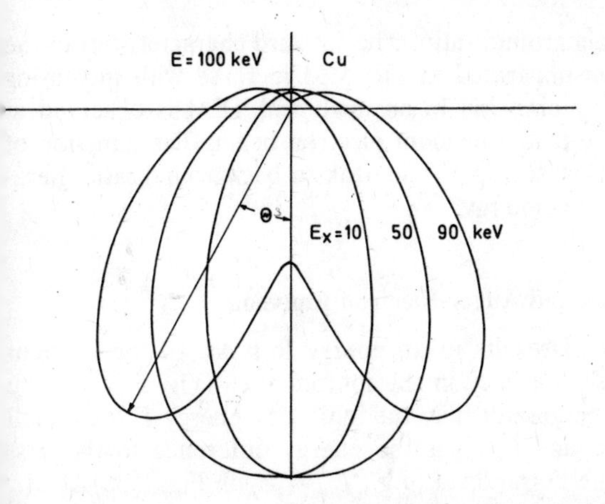

Fig. 5.33. Angular characteristics (polar diagram of the emitted x-ray continuum) of a thin Cu foil irradiated with $E = 100$ keV electrons for three quantum energies $E_x = 90$, 50 and 10 keV. The curves are normalized to the same maximum value

spectral distribution emitted is the Fourier transform of $a(t)$ and the interaction time is very short, the resulting spectrum is very broad, and the x-ray quanta can be emitted with energies in the range $0 \leq E_x \leq E = eU$.

The angular distribution of the x-ray continuum is very anisotropic (Fig. 5.33). The number dN_x of quanta emitted per unit time, with energies between E_x and $E_x + dE_x$, into a solid angle $d\Omega$ at an angle θ relative to the forward direction of electron incidence, inside a film of thickness t and density ϱ (no multiple scattering) with $N_A \varrho t/A$ atoms per unit area, can be described by

$$dN_x = \frac{I(\theta)}{E_x} \frac{N_A \varrho t}{A} \frac{I_p}{e} dE_x d\Omega \ , \tag{5.88}$$

where I_p is the probe current and

$$I(\theta) = I_x \frac{\sin^2\theta}{(1-\beta\cos\theta)^4} + I_y\left(1 + \frac{\cos^2\theta}{(1-\beta\cos\theta)^4}\right) \tag{5.89}$$

where β denotes v/c for the incident electrons [5.143]. Analytical formulae for I_x and I_y are published [5.130], which approximate the exact values with an accuracy of 2%. To a first-order approximation, I_x and I_y are proportional to Z^2/E.

The x-ray continuum can be used to calibrate the film thickness in the microanalysis of biological sections, for example (Sect. 9.1.4). It also contributes to the background below the characteristic x-ray peaks thereby decreasing the peak-to-background ratio. The forward characteristics of the continuous x-ray emission illustrated in Fig. 5.33 increase with increasing electron energy. The x-ray emission in an analytical TEM is observed at angle $\theta = 90–135°$ relative to the incident electron beam. The emission of characteristic x-ray quanta is isotropic. The peak-to-background ratio therefore increases with increasing energy.

5.4.2 Characteristic X-Ray and Auger-Electron Emission

Ionisation of an inner shell results in an energy loss ΔE of the incident electron (Sect. 5.2.4) and a vacancy in the ionized shell (Fig. 5.34 a). The electron energy E has to be greater than the ionisation energy E_{nl} of a shell with quantum numbers n and l. E_{nl} is the energy difference to the first unoccupied state above the Fermi level (e.g., $E_{nl} = E_K$ in Fig. 5.35).

The ionisation cross-section can be calculated from a formula given in [5.131]

$$\sigma_{nl} = \frac{e^4 Z_{nl}}{16\pi\varepsilon_0^2 E E_{nl}} b_{nl} \ln\left(\frac{4E}{E_{nl}}\right), \tag{5.90}$$

where b_{nl} and B_{nl} are numerical constants, e.g. $b_K = 0.35$, $B_K = 1.65\,E_K$ for the K shell ($n = 1$); Z_{nl} denotes number of electrons with quantum numbers n, l. The ratio $u = E/E_{nl} \geq 1$ is called the *overvoltage ratio*. An empirical formula $B_K = [1.65 + 2.35\exp(1-u)]\,E_K$ is proposed [5.132], which gives $B_K = 1.65\,E_K$ for large u and $B_K = 4\,E_K$ for u near to unity. This leads to the following formula for K-shell ionisation ($Z_{nl} = 2$),

$$\sigma_K = \frac{e^4}{8\pi\varepsilon_0^2 E E_K} b_K \ln\left(\frac{4E}{B_K}\right) = \frac{e^4}{8\pi\varepsilon_0^2 u E_K^2} b_K \ln\left(\frac{4uE_K}{B_K}\right). \tag{5.91}$$

The product $\sigma_K E_K^2$ should therefore depend on only u. This is confirmed for low and medium atomic numbers (Fig. 5.36). For high atomic numbers, σ_K becomes larger [5.133]. Cross-sections σ_K for some elements are also plotted in Fig. 5.3 for comparison with other important cross-sections of electron-specimen interactions. A review concerning the theoretical and experimental cross-sections of the K and L shells has been prepared [5.134, 135].

The vacancy in the inner shell is filled by electrons from outer shells (Fig. 5.34 b). The energy difference, $E_L - E_K$ for example can be emitted as an x-ray quantum of discrete energy $E_x = h\nu = E_L - E_K$. Because of the quantum-mechanical selection rules, $\Delta l = \pm 1$ and $\Delta j = 0, \pm 1$, for example, the *Kα lines* that result from the transition of an L ($n = 2$) electron to the K shell ($n = 1$) consists of only a doublet ($K\alpha_1$ and $K\alpha_2$ in Fig. 5.35).

To a first approximation, in which the sub-shells are disregarded the quantum energies $E_{x,K}$ of the K series can be estimated from modified energy

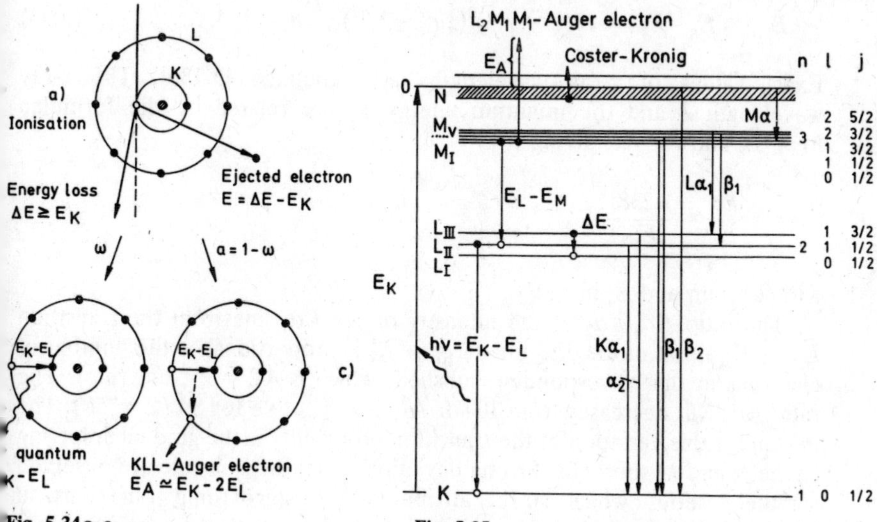

Fig. 5.34 a–c

Fig. 5.35

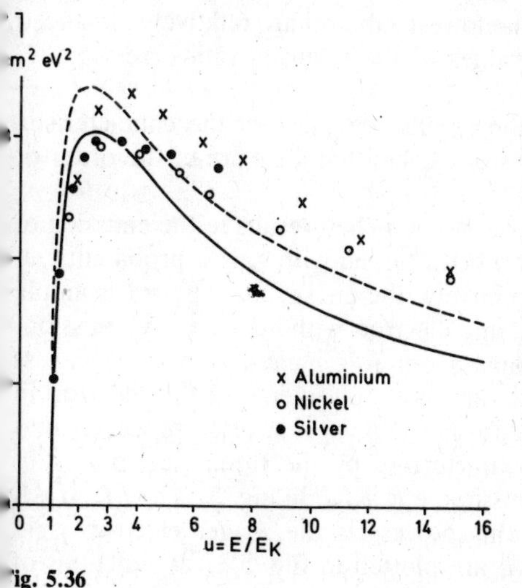

Fig. 5.36

Fig. 5.34 a–c. Schematic representation of (**a**) the ionisation process, (**b**) x-ray emission and (**c**) Auger-electron emission

Fig. 5.35. Energy level of the atomic subshells with quantum numbers n, l, j and ionisation of the K shell; possible transitions of electrons to fill the vacancies in inner shells and nomenclature of emitted x-ray lines are shown. Example of the emission of a KLL Auger electron and a Coster-Kronig transition

Fig. 5.36. Plot of σ_K/E_K^2 (σ_K: ionisation cross-section of the K shell, E_K: ionisation energy) versus the over-voltage ratio $u = E/E_K$ [5.134]

erms of the Bohr model ($E_1 = E_K$)

$$E_{x,K} = E_n - E_1 = -R(Z-1)^2(1/n^2 - 1/1^2)\ ; \quad n = 2\text{: } K\alpha, n = 3\text{: } K\beta\ , \qquad (5.92)$$

vhere $R = 13.6$ eV denotes the ionisation energy of the hydrogen atom. The eduction of Z by 1 represents the screening of the nuclear charge Ze by the emaining electron in the K shell. Likewise, for the L series ($E_2 = E_L$):

$$E_{x,L} = E_n - E_2 = -R\,(Z - 7.4)^2(1/n^2 - 1/2^2)\ ; \quad n = 3, 4, \ldots \tag{5.93}$$

Exact values of x-ray wavelengths are tabulated [5.136]. The x-ray wavelength λ and the quantum energy E_x are related by the formulae $h\nu = E_x$ and $\nu\lambda = c$, giving

$$\lambda = \frac{hc}{E_x} = \frac{1.238}{E_x} \tag{5.94}$$

with λ in nm and E_x in keV.

The ratio $K\alpha_1/K\alpha_2$ of the intensity of the $K\alpha_1$ line from the transition $L_{II} \to K$ to that of the $K\alpha_2$ line ($L_{III} \to K$) is proportional to the number of electrons in the corresponding subshell, which is $4/2 = 2$ (sum rule). The ratio $K\alpha_1/K\beta_1$ decreases from 10 for Al ($Z = 13$) to 3 for Sn ($Z = 50$). The reason for this variation of the transition probability is the gradual emptying of the N and M subshells. Strong deviations from the sum rule are observed for the L series, which can be attributed to Coster-Kronig transitions, in which a vacancy in a L_I or L_{II} subshell is filled by an electron from another subshell (L_{III}). The energy is transferred to an electron near the Fermi level (Fig. 5.35). The lines with L_{III} as the lowest sublevel are relatively enhanced by this effect. For experimental values of the intensity ratios, see [5.137, 9.12, 14].

Unlike the x-ray continuum, the angular emission of the characteristic quanta is isotropic. The half-widths of the emitted lines are of the order of 1–10 eV.

However, not every ionisation of an inner shell results in the emission of an x-ray quantum. This process can be observed with only a probability ω, the *x-ray fluorescence yield*. Alternatively, the energy $E_L - E_K$ for example may be transferred to another atomic electron without an x-ray emission. The latter electron leaves the atoms as an *Auger electron* with an excess energy $E_A = (E_L - E_K) - E_I$; E_I is the ionisation energy of this electron in the presence of a vacancy in one sub-shell, taking relaxation processes into account. An Auger electron is characterized by the three electronic sub-shells which are involved in the emission, e.g. KLL in Fig. 5.34 c or $L_2M_1M_1$ in Fig. 5.35. The probability of this process is the Auger electron yield $a = 1 - \omega$. The quantities ω and a are plotted in Fig. 5.37 as functions of atomic number Z [5.138–140].

Because the value of ω is very low for light elements, detection of light elements by x-ray microanalysis is very inefficient. Electron energy-loss spectroscopy becomes more attractive (Sect. 9.3). The other alternative, Auger-electron spectroscopy, cannot be used in TEM on account of the strong magnetic field at the specimen. It is employed only in SEM and in special Auger-electron microprobe instruments (Sect. 1.1).

The continuous and characteristic x-ray quanta emitted interact with a solid by three processes:

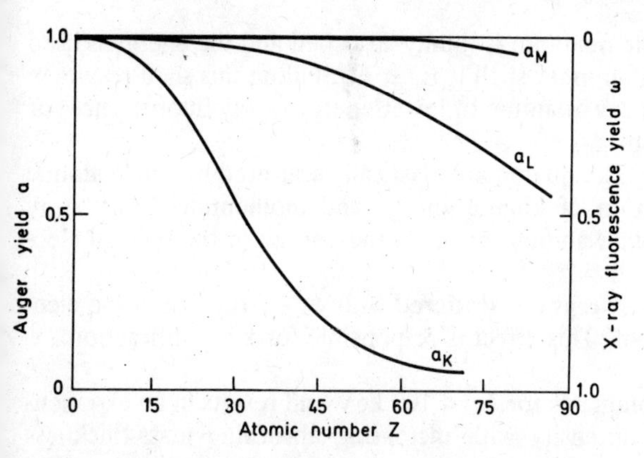

Fig. 5.37. Auger-electron yield a and fluorescence yield $\omega = 1 - a$ as a function of atomic number Z [5.138]

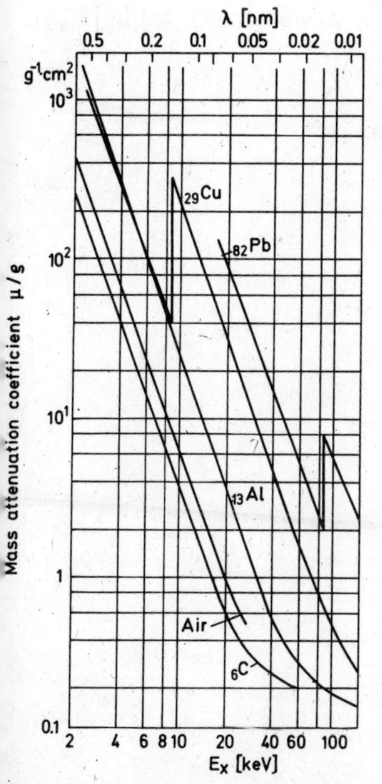

Fig. 5.38. Dependence of mass attenuation coefficient μ/ϱ on quantum energy E_x (*lower scale*) and x-ray wavelength λ (*upper scale*)

1) *Photo-ionisation.* The quantum is totally absorbed and the energy is used to ionise an atom in an inner shell if $E_x \geq E_{nl}$. Filling this shell results in the emission of an x-ray quantum of lower energy (x-ray fluorescence) or of an Auger electron.
2) *The Compton effect.* The quanta are elastically scattered by single atomic electrons (conservation of kinetic energy and momentum). The x-ray energy decreases by an amount equal to the energy of the ejected electron.
3) *Thomson scattering.* X-rays are scattered without energy loss at the electron shell of the atom. This effect is responsible for x-ray diffraction.

The first process dominates for $E_x < 100$ keV and results in an exponential decrease of x-ray intensity with increasing absorbing mass-thickness $x = \varrho t$:

$$N = N_0 \exp(-\mu t) = N_0 \exp[-(\mu/\varrho)\varrho t] \, . \tag{5.95}$$

The *mass attenuation coefficient* $\mu/\varrho = N_A \sigma_x / A$ [g^{-1} cm^2] is related to a single *fluorescence cross-section* σ_x and decreases as E_x^{-n} ($n = 2.5$–3.5) with increasing E_x, with a sharp increase if $E_x \geq E_{nl}$ of an atomic shell (Fig. 5.38). Numerical values and formulae for μ/ϱ are to be found in *International Tables for X-Ray Crystallography*, Vol. 3 (Kynoch Press, Birmingham) and in [5.141, 142].

6. Scattering and Phase Contrast for Amorphous Specimens

Elastic scattering through angles larger than the objective aperture causes absorption of the electrons at the objective diaphragm and a decrease of transmitted intensity. This scattering contrast can be explained by particle optics. The exponential decrease of transmission with increasing specimen thickness can be used for quantitative determination of mass-thickness or of the total mass of an amorphous particle, for example.

The superposition of the electron waves at the image plane results in interference effects and causes phase contrast, which depends on defocusing and spherical aberration, on the objective aperture and also on the particular illumination conditions.

The phase contrast of heavy atoms is sufficient for imaging single atoms. The image of a single atom will be a diffraction disc surrounded by diffraction fringes, which can result in image artifacts when fringes from neighbouring atoms overlap, for example.

It is possible to characterize the imaging process independently of the specimen structure by introducing the contrast-transfer function, which describes how individual spatial frequencies of the Fourier spectrum are modified by the imaging process. The contrast-transfer function of the normal bright-field mode alternates in sign and decreases at high spatial frequencies owing to the imperfect temporal and spatial coherence. Methods of suppressing the change in sign, by hollow-cone illumination for example, have been proposed.

The idea of holography as an image restoration method as originally proposed by Gabor for electron microscopy is handicapped by the lack of coherence of the electron beam. The use of field-emission guns now allows holography to be applied in electron microscopy for the quantitative measurement of phase shifts.

The phase shift caused by the magnetic fields inside ferromagnetic domains can be exploited to image the magnetic structure of thin films or small particles; this is known as Lorentz microscopy.

6.1 Scattering Contrast

6.1.1 Transmission in the Bright-Field Mode

We assume in this section that the electrons move as particles through the imaging system. This means that we shall be considering the intensity and not the wave amplitude, even in the focal plane of the objective lens, and that we shall sum intensities and not wave amplitudes in the image plane. In the purely wave-optical theory of imaging (Sect. 3.3.2), we always sum over wave amplitudes and obtain the image intensity by squaring the wave amplitude in the final image plane. The resulting *phase contrast* will be considered in Sect. 6.2. (It will be shown in Sect. 6.3.3, formulae (6.27–30) that the scattering contrast can be treated by the more-general phase-contrast theory if complex scattering amplitudes are used.) The scattering contrast, therefore describes the image intensity for low and medium magnifications, where phase-contrast effects do not normally have to be considered unless a highly coherent electron beam and large defocusing are employed.

In the bright-field mode, the diaphragm in the focal plane of the objective lens acts as a stop (Fig. 4.18) that absorbs all electrons scattered through angles $\theta \geq \alpha_o$ (objective aperture). Only electrons scattered through $\theta < \alpha_o$ can pass through the diaphragm. We can thus define a transmission $T(\alpha_o)$ which depends on the objective aperture α_o, and also on the electron energy E, the mass-thickness $x = \varrho t$ (ϱ: density, t: thickness) and the material composition (atomic weight A and atomic number Z). We assume that the illumination aperture α_i of the incident electron beam is appreciably smaller than $\alpha_o (\alpha_i \ll \alpha_o)$, which is usually the case in normal TEM work, whereas in STEM the two apertures are normally comparable (Sect. 6.1.4).

Scattering contrast is typically observed with amorphous specimens, surface replicas or biological sections (see quantitative examples in Sect. 6.1.3). Even for amorphous specimens, the assumption that the waves scattered at single atoms add incoherently is not fully justified because the angular distribution of the scattered intensity (diffraction pattern) shows diffuse maxima (Sect. 7.5.1). However, the total number of electrons scattered into a cone of half-angle α_o is very insensitive to such diffuse maxima in the diffraction pattern, because the angular distribution oscillates about that, corresponding to completely independent scattering (Fig. 7.22). Polycrystalline films with very small crystals, platinum for example, can also be treated by the theory of scattering contrast [6.1]. Evaporated films that contain larger crystals, Ag or Au for example, may deviate from this simple theory, owing to dynamical diffraction effects, and their mean transmission averaged over a large film area cannot be described exactly by the formulae of scattering contrast.

Equation (5.7) can be used to calculate the number of electrons scattered through scattering angles $\theta \geq \alpha_o$ and intercepted by the diaphragm in the focal plane of the objective lens. Substituting the small-angle approximation (5.40) for elastic scattering, which also allows us to set the upper limit of the integral equal to infinity, gives [6.2]

$$\sigma_{el}(\alpha_o) = 2\pi \int_{\alpha_o}^{\infty} |f(\theta)|^2 \, \theta \, d\theta = 2\pi \frac{4Z^2R^4(1+E/E_0)^2}{a_H^2} \int_{\alpha_o}^{\infty} \frac{\theta}{[1+(\theta/\theta_0)^2]^2} \, d\theta$$

$$= \frac{Z^2R^2\lambda^2(1+E/E_0)^2}{\pi a_H^2} \frac{1}{1+(\alpha_o/\theta_0)^2} . \tag{6.1}$$

The total elastic cross-section σ_{el} (5.46) is obtained by setting $\alpha_o = 0$. According to (5.11), the mean free mass-thickness $x_a = \varrho\Lambda_{el}$ between elastic scattering events becomes

$$x_a = \varrho\Lambda_{el} = \frac{\varrho}{N\sigma_{el}} = \frac{\pi A a_H^2}{N_A Z^2 R^2 \lambda^2 (1+E/E_0)^2} , \tag{6.2}$$

where $N = N_A\varrho/A$ denotes the number of atoms per unit volume. If complex scattering amplitudes $f(\theta)$ are available, from WKB calculations say, σ_{el} can be obtained from the optical theorem (5.45); subtracting the number of electrons passing through the diaphragm then gives

$$\sigma_{el}(\alpha_o) = \underbrace{2\lambda \,\mathrm{Im}\,\{f(0)\}}_{\sigma_{el}} - 2\pi \int_0^{\alpha_o} |f(\theta)|^2 \theta \, d\theta . \tag{6.3}$$

This means that the complex scattering amplitude need be known only in the region $0 \le \theta \le \alpha_o$.

A formula for the cross-section $\sigma_{inel}(\alpha_o)$, analogous to (6.1), can be calculated by using the differential inelastic cross-section (5.56). The term θ_E that contains the mean ionisation energy J can be neglected in q', because it will be important only for very small scattering angles. We find

$$\sigma_{inel}(\alpha_o) = 2\pi \int_{\alpha_o}^{\infty} \frac{d\sigma_{inel}}{d\Omega} \theta \, d\theta$$

$$= 2\pi \frac{\lambda^4 Z(1+E/E_0)^2}{4\pi^4 a_H^2} \int_{\alpha_o}^{\infty} \frac{1}{\theta^4} \left(1 - \frac{1}{[1+(\theta/\theta_0)^2]^2}\right) \theta \, d\theta \tag{6.4}$$

$$= \frac{4ZR^2\lambda^2(1+E/E_0)^2}{\pi a_H^2} \left[\frac{-1}{4[1+(\alpha_o/\theta_0)^2]} + \ln \sqrt{1+(\theta_0/\alpha_o)^2}\right] .$$

The decrease of transmission $T(\alpha_o)$ through the aperture α_o with increasing mass-thickness $x = \varrho t$ can be obtained as in (5.9)

$$\frac{dn}{n} = -\frac{N_A}{A} [\sigma_{el}(\alpha_o) + \sigma_{inel}(\alpha_o)] \, dx . \tag{6.5}$$

Integration gives

$$T(\alpha_o) = n/n_0 = \exp[-x/x_k(\alpha_o)] \tag{6.6}$$

where the *contrast thickness* $x_k(\alpha_o)$ is given by

$$\frac{1}{x_k(\alpha_o)} = \frac{4}{Zx_a}\left[\frac{Z-1}{4[1+(\alpha_o/\theta_0)^2]} + \ln\sqrt{1+(\theta_0/\alpha_o)^2}\right] \tag{6.7}$$

and x_a is defined in (6.2).

The exponential decrease (6.6) of transmission can be checked by plotting the quantity

$$K = \log_{10}\frac{1}{T} = 0.4343\frac{x}{x_k} \tag{6.8}$$

against the mass-thickness x. The expected linear increase of K is observed for small mass-thicknesses (Fig. 6.1) [6.3–8]. The agreement is less good for larger mass-thicknesses, owing to multiple scattering. The increase of K becomes less or, in other words, a higher transmission is observed than that predicted by (6.6), because electrons first scattered through large angles can be scattered back towards the incident direction and can hence pass through the objective diaphragm. For high energies and large apertures, the situation can be reversed; K then shows a larger increase than expected from the value

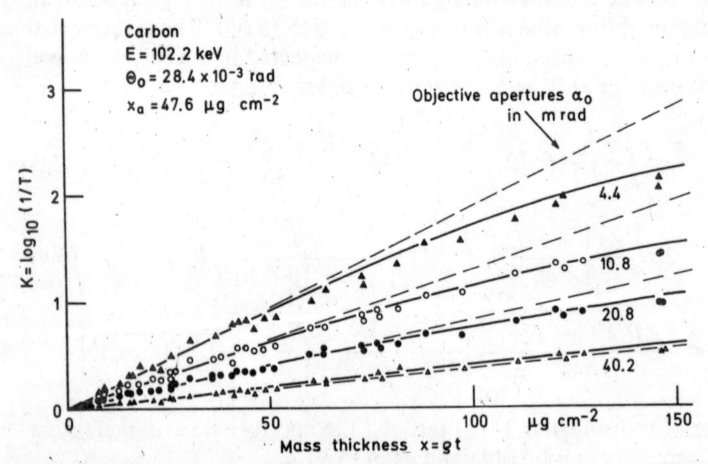

Fig. 6.1. Contrast $K = \log_{10}(1/T)$ of carbon films as a function of mass-thickness $x = \varrho t$ for different objective apertures α_o ($E = 100$ keV). The full curves were calculated using a multiple-scattering integral [6.2] with the constants $x_a = 47.6\ \mu g\ cm^{-2}$ and $\theta_0 = 28.4$ mrad, obtained from a best fit of the initial slopes at small x [6.8]

of x_k because electrons are scattered out of the cone with aperture α_o by multiple scattering. The full curves in Fig. 6.1 were calculated on the basis of a multiple-scattering integral [6.2] and show good agreement with the experimental results. The curves were calculated with the values of x_a and θ_0 of Table 6.1, which were obtained by fitting the initial slopes of the $K(x)$ curves. The limits of linearity of these curves are discussed in [6.8–10]. For very large mass-thicknesses ($x \geq 100$ µg cm^{-2} in Fig. 6.1), the transmission T is proportional to the solid angle $\pi\alpha_o^2$ of electrons passing through the objective aperture α_o. This is a consequence of the broadened angular distribution of the scattered electrons, which decreases slowly with increasing θ for the ranges of apertures used. This thickness range is of no interest for TEM and STEM, however, because of the large energy losses and probe broadening due to multiple scattering.

The contrast thickness $x_k(\alpha_o)$ can be obtained, using (6.8), from the initial slope of $K(x)$ in Fig. 6.1. Figure 6.2 shows measured values for different apertures α_o and electron energies; for comparison, calculated values using (6.3) and complex scattering amplitudes $f(\theta)$ given by the WKB method (Sect. 5.1.3) (pure elastic scattering), modified to take account of the inelastic contribution [6.11] are also plotted. The calculation of the scattering amplitudes $f(\theta)$ assumed dense atomic packing, represented by the muffin-tin model (5.27 c). The measured values of $x_k(\alpha_o)$ can be approximated by the parameters x_a and θ_0, which appear in (6.7). Values of these quantities are tabulated in Table 6.1. The mean free mass-thickness x_a for elastic scattering (6.2) and the characteristic angle $\theta_0 = \lambda/2\pi R$ (5.39) then depend on only one parameter, the screening radius R of the Wentzel model (5.27 a). For different electron energies E or wavelengths λ, the following scaling rules have been derived [6.12, 13]:

$$x_a\lambda^2/(1 + E/E_0)^2 = \text{const} ; \qquad \theta_0/\lambda = \text{const} . \tag{6.9}$$

Table 6.1. Experimental values [6.8] of mean-free-path length x_a and characteristic angle θ_0. Mean-free-path length $\Lambda = 10\, x_a/\varrho$ with Λ [nm], x_a [µg cm^{-2}] and ϱ [g cm^{-3}]

	C		Ge		Pt	
E [keV]	x_a [µg cm^{-2}]	θ_0 [mrad]	x_a [µg cm^{-2}]	θ_0 [mrad]	x_a [µg cm^{-2}]	θ_0 [mrad]
17.3	10.1	92.4	–	–	6.5	53.8
25.2	14.4	69.9	6.8	50.6	8.1	52.4
41.5	22.4	46.6	10.6	42.6	11.6	50.8
62.1	31.8	37.8	14.4	38.2	14.1	43.2
81.8	39.7	32.4	17.8	34.4	16.8	40.2
102.2	47.6	28.4	21.0	30.8	19.2	38.4
150	70.6	21.6	28.0	23.4	23.4	25.8
300	114.0	17.8	42.0	19.0	31.6	16.2
750	139.2	10.2	58.7	11.5	50.7	13.2
1200	168.0	6.5	62.1	6.8	46.8	8.0

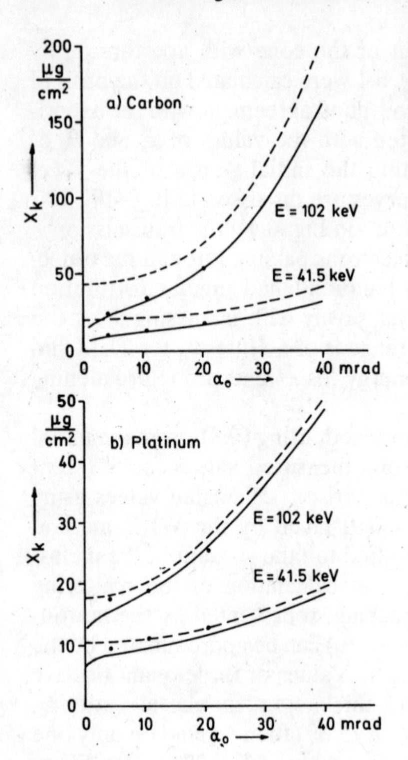

Fig. 6.2 a, b. Contrast thickness x_k of (**a**) carbon and (**b**) platinum for $E = 40$ and 100 keV [6.8] (---) theoretical values considering elastic scattering only, (—) considering both elastic and inelastic scattering [6.11]

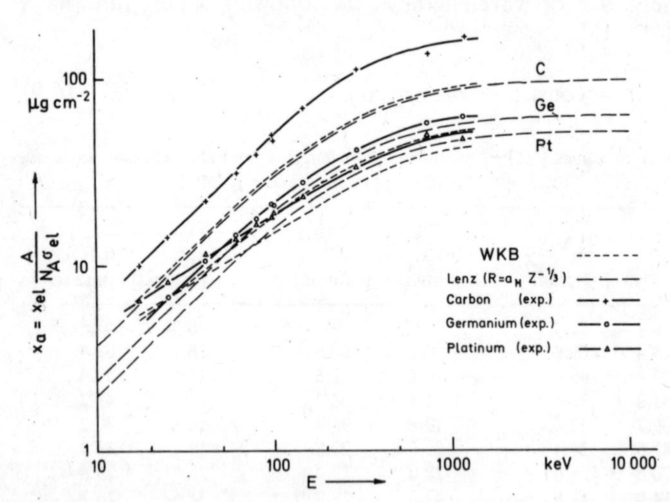

Fig. 6.3. Variation of electron mean free mass-thickness x_a for C, Ge and Pt films with electron energy E. (– – –) calculations based on the Lenz theory [6.2] (Born approximation); (---) calculations by the WKB method [6.8]

Thus x_a and θ_0 can be calculated for any energy, once they are known for one energy. These scaling rules are, however, valid only for films of low atomic number, such as carbon. This limitation is a typical consequence of using the Born approximation. Measured values of x_a are plotted versus electron energy in Fig. 6.3. The values for carbon differ from those given by (6.2) by only a constant vertical shift in the logarithmic scale of Fig. 6.3, which means a constant factor. The theory is thus confirmed, so far as the dependence on electron energy is considered, apart from the constant factor that is determined by the scattering potential $V(r)$ of the atoms. However, values of x_a and θ_0 for Ge and Pt (Fig. 6.3) do not satisfy the scaling relation (6.9) and should rather be compared with those given by WKB calculations (Fig. 6.2); x_a is then not necessarily identical with the mean free mass-thickness for elastic scattering.

For all elements, x_a attains a saturation value at high electron energies (Fig. 6.3), whereas the contrast thickness $x_k(\alpha_o)$ continues to increase for a fixed value of α_o (Fig. 6.4) (see also [6.14]). The increase can be understood from the fact that, with increased energy, the electrons are scattered through smaller angles (Fig. 5.5). For this reason, smaller apertures are normally used in high-voltage electron microscopy.

An empirical law [6.7, 8]

$$K = \log_{10}\frac{1}{T} = b\,\frac{Z^a}{A}x \tag{6.10}$$

can be used to describe the dependence of K on the atomic number Z for a constant α_o and electron energy E; a and b are aperture- and energy-dependent constants. A double logarithmic plot of KA/x versus Z gives a straight line of slope a (Fig. 6.5). Measurements on gases, which are ideal examples

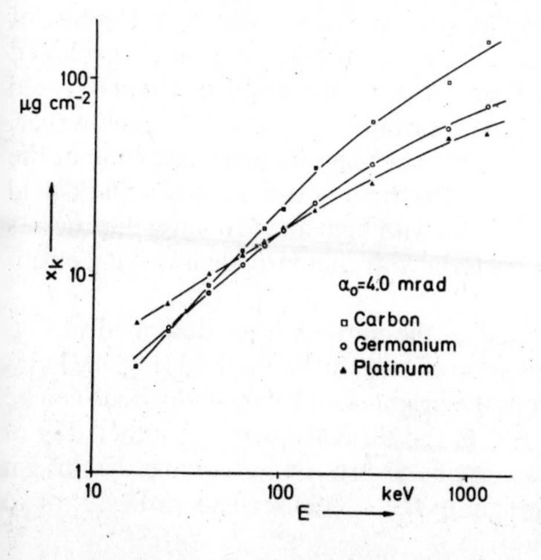

Fig. 6.4. Dependence of the contrast thickness x_k of carbon films in the electron energy for different objective apertures α_o

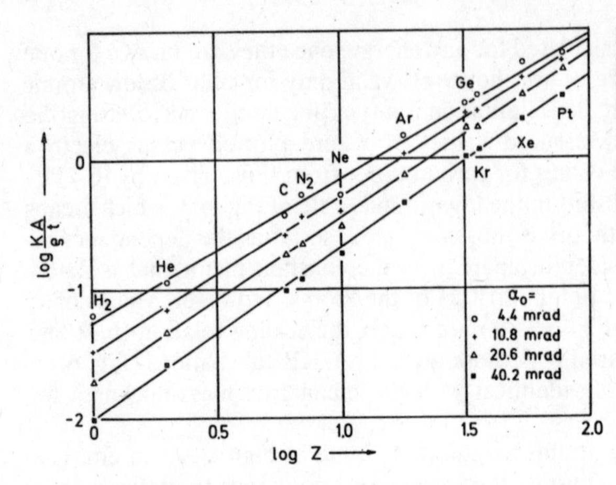

Fig. 6.5. Double logarithmic plot of $KA/\varrho x$ versus atomic number $Z\,[K = \log_{10}(1/T)]$ for different objective apertures α_o (E = 100 keV) demonstrating the validity of an empirical power law $K\alpha Z^a$

of amorphous specimens, can also be approximated by a straight line. The Wentzel atomic model (5.27 a) with $R = a_H Z^{-1/3}$ leads to $K \propto Z^{4/3}/A$ for purely elastic scattering, if (6.2 and 8) are used. Rutherford scattering would give $K \propto Z^2/A$. In reality, none of these exponents of Z is valid. The case in which E = 60 keV and α_o = 4 mrad is of special interest, because a = 1.1 for these values and the slow decrease of Z/A with increasing Z is thus compensated. In consequence, the value of K is nearly constant for equal mass-thicknesses x of different elements; this is of interest for the determination of mass-thickness from measurements of K or the transmission T (Sect. 6.1.5).

6.1.2 Dark-Field Mode

The bright-field mode is not convenient for specimens with very small mass-thicknesses such as DNA molecules or virus particles, because a decrease of transmission of at least 5% is needed for visual detection on a developed photographic emulsion. Better contrast can be expected in the dark-field mode, if a thin supporting film is used (see example in Sect. 6.1.3). However, the requisite electron charge density in C cm^{-2} and the exposure time of the photographic emulsion are greater than for the bright-field mode. Dark-field imaging is also advantageous if structures with high and low mass-thicknesses are to be imaged simultaneously; bacteria with cilia provide a striking example [6.15].

Dark-field images can be formed in the various ways described in Fig. 4.21. To decrease the effect of lens aberrations, the tilt method (Fig. 4.21 c) is widely used, and the transition from the bright- to the dark-field mode can be effected by switching on the current in the tilt coils [6.16]. Another way of distributing the intensity of the primary beam around the circular diaphragm is to work with an annular diaphragm in the condenser lens [6.17–19] or to

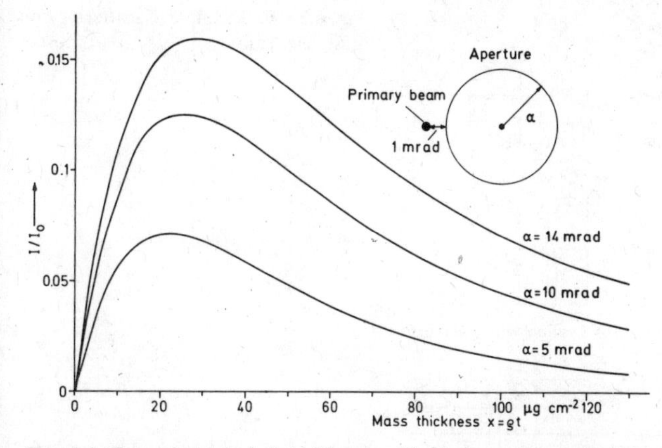

Fig. 6.6. Example of dark-field intensity I/I_0 [I_0: intensity of the incident electron beam] as a function of carbon film mass-thickness $x = \varrho t$ for different objective apertures α_o in the tilted-beam mode; the distance of the primary beam from the periphery of the centred objective diaphragm is 1 mrad ($E = 100$ keV)

deflect the electron beam electronically on a cone by the tilt coils between condenser and objective lenses [6.20].

The dark-field intensity I/I_0 is plotted against mass-thickness x in Fig. 6.6 for the tilted-beam mode and various centred apertures; the primary beam is at a distance of 1 mrad from the periphery of the centred diaphragm. The intensity I/I_0 passes through a maximum because the number of electrons scattered through the dark-field aperture first increases as the mass-thickness and subsequently decreases with increasing mass-thickness, as a result of multiple scattering.

6.1.3 Examples of Scattering Contrast

The following quantitative examples of scattering contrast (Fig. 6.7) illustrate how the scattering contrast affects different imaging problems and how this contrast can be calculated with the aid of various theories and experimental data; they also indicate how a measured transmission can be quantitatively evaluated. The x_k values used have been calculated from (6.7), by use of the experimental x_a and θ_0 values of Table 6.1.

a) Shadow-Casting Film (Fig. 6.7 a)

Shadowing of surface replicas with evaporated films of heavy metals increases the contrast and resolution (Fig. 8.8). A shadow such, as is shown in Fig. 6.7 a is clearly recognizable. Denoting the image intensity without a

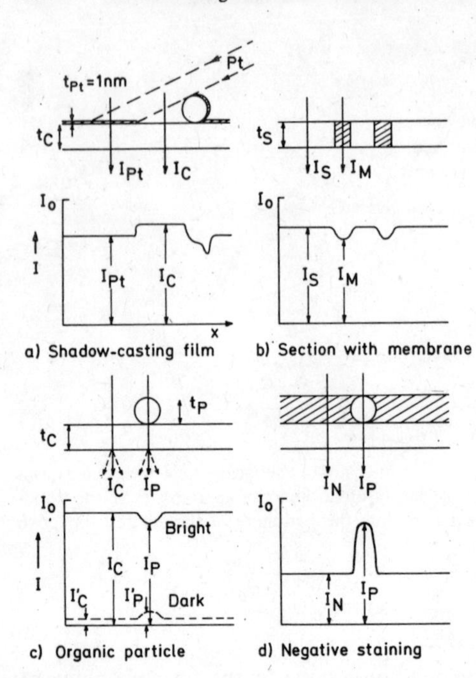

Fig. 6.7 a–d. Example of scattering-contrast calculations

specimen by I_0, the intensity with the carbon supporting film by I_C and with the evaporated platinum film by I_{Pt}, the following relations are found:

$$I_C = I_0 \exp\left(-\frac{\varrho_C t_C}{x_{k,C}}\right) ; \quad I_{Pt} = I_0 \exp\left[-\left(\frac{\varrho_C t_C}{x_{k,C}} + \frac{\varrho_{Pt} t_{Pt}}{x_{k,Pt}}\right)\right] . \tag{6.11}$$

The ratio of the platinum and carbon intensities

$$\frac{I_{Pt}}{I_C} = \exp\left(-\frac{\varrho_{Pt} t_{PC}}{x_{k,Pt}}\right) \tag{6.12}$$

is observed in the image. If t_{Pt} is small, so that the exponential law of transmission (6.6) is obeyed, the thickness of the carbon supporting film has no effect on the ratio I_{Pt}/I_C in (6.12). For a given value of I_{Pt}/I_C, the thickness of the shadowing film must be at least

$$t_{Pt} = \frac{x_{k,Pt}}{\varrho_{Pt}} \ln \frac{I_C}{I_{Pt}} . \tag{6.13}$$

Numerical example: For $E = 80$ keV, $\alpha_o = 4$ mrad, $x_{k,Pt} = 17.5\ \mu g\ cm^{-2}$ and $\varrho_{Pt} = 21\ g\ cm^{-3}$, we find $t_{Pt} = 0.9$ nm for $I_{Pt}/I_C = 0.9$.

b) Stained Membrane in a Biological Section (Fig. 6.7b)

Measurements at $E = 60$ keV and $\alpha_o = 5$ mrad of the transmission of a thin section of an OsO_4-stained mitochondrial membrane embedded in Vestopal result in mean values of $T_S = I_S/I_0 = 0.765$ for the embedding medium and $T_M = I_M/I_0 = 0.67$ at a membrane. The thickness of the section can be calculated from the first value by assuming that the main contribution to the contrast comes from carbon ($x_{k,C} = 14.6\ \mu g\ cm^{-2}$); this gives $x_S = x_{k,C} \ln(1/T_S) = 3.9\ \mu g\ cm^{-2}$ so that with $\varrho_S = 1.1\ g\ cm^{-3}$, the section thickness $t_S = 35.5$ nm. For more-accurate quantitative measurements, the mass loss by radiation damage (Sect. 10.2) has to be considered.

Assuming $x_{k,Os} \simeq x_{k,Pt} = 13.0\ \mu g\ cm^{-2}$, the second value T_M implies $x_{Os} = x_{k,Os} \ln(T_S/T_M) = 1.7\ \mu g\ cm^{-2}$ for the equivalent mass-thickness of the incorporated osmium. The relative fraction of *Os* atoms becomes

$$\frac{\text{number of C atoms}}{\text{number of Os atoms}} = \frac{x_S}{x_{Os}} \frac{A_{Os}}{A_C} = 36 .$$

The same ratio $T_M/T_S = 0.88$ would be observed at $E = 1$ MeV and $\alpha_o = 1.5$ mrad for a membrane in a section of thickness $t_S = 120$ nm. For a section as thick as this, the resolution is already reduced at $E = 60$ keV by the effect of chromatic aberration (Sect. 5.3.2).

c) Organic Particle on a Supporting Film (Fig. 6.7c)

This case is described by a formula similar to (6.12)

$$\frac{I_P}{I_C} = \exp\left(-\frac{\varrho_P t_P}{x_{k,C}}\right) . \tag{6.14}$$

In bright field, an unstained particle with $t_P = 10$ nm and $\varrho_P = 1\ g\ cm^{-3}$ generates an intensity ratio $I_P/I_S = 0.97$ for $E = 100$ keV, $\alpha_o = 10$ mrad and $x_{k,C} = 32\ \mu g\ cm^{-2}$, which is beyond the limit of visibility. However, such a particle can be seen in phase contrast at optimum defocusing (Sect. 6.2).

If the same particle of 10 nm diameter ($x_P = 1\ \mu g\ cm^{-2}$) on a carbon support film of $x_C = 1\ \mu g\ cm^{-2}$ ($t_C = 5$ nm) is observed in the dark-field mode, the ratio I'_P/I'_S increases to 2 because the dark-field intensities are proportional to x for small mass-thicknesses. From Fig. 6.6, the ratio I'_S/I_0 can be seen to be 0.01 for $x_S = 1\ \mu g\ cm^{-2}$. A 10–30 fold longer exposure time therefore is needed than for a bright-field image.

d) Negatively Stained Particle (Fig. 6.7d)

The same particle, 10 nm in diameter, is now negatively stained by embedding it in a thin layer of phosphotungstic acid PWO_4 ($\varrho_N = 4\ g\ cm^{-3}$). For the same imaging condition as in c), the contrast thickness for PWO_4 will be approximately the same as that of Pt: $x_{k,N} \simeq x_{k,Pt} = 19\ \mu g\ cm^{-2}$. Where the particle is situated, an increase of the transmitted intensity ratio

$$\frac{I_P}{I_N} = \exp\left[\left(\frac{\varrho_N}{x_{k,N}} - \frac{\varrho_P}{x_{k,C}}\right) t_P\right] = 1.19 \tag{6.15}$$

can be expected, which means a considerable gain of contrast in comparison with the decrease $I_P/I_S = 0.97$ for an unstained particle in the bright-field mode.

6.1.4 Scattering Contrast in the STEM Mode

It is a characteristic of the bright-field transmission mode in TEM that the illumination aperture α_i is very much smaller than the objective aperture α_o (Fig. 6.8 a). In consequence, the transmission $T = I/I_0$ depends on only the objective aperture α_o. Small shifts of the objective aperture or small inclinations of the incident beam hardly alter the intensity I that goes through the diaphragm. It was shown in Sect. 4.2.2 that the smallest possible spot size of an electron probe for STEM can be obtained only with a relatively large probe aperture $\alpha_p \simeq 10$ mrad. The theorem of reciprocity (Sect. 4.5.3) indicates that the same transmission can be expected if the electron-probe aperture α_p is approximately equal to α_o, whereas the detector aperture is small: $\alpha_d \simeq \alpha_i$ (Fig. 6.8 b). However, a lower intensity I_0 is recorded in the absence of a specimen, because I_0 is a fraction α_d^2/α_p^2 of the intensity of the incident electron probe with aperture α_p. The intensity I with a specimen present is determined by the decrease of the intensity I_0 due to scattering through larger angles together with the increase due to scattering from the other directions of incidence back into the detector aperture. The same ratio $T = I/I_0$ can

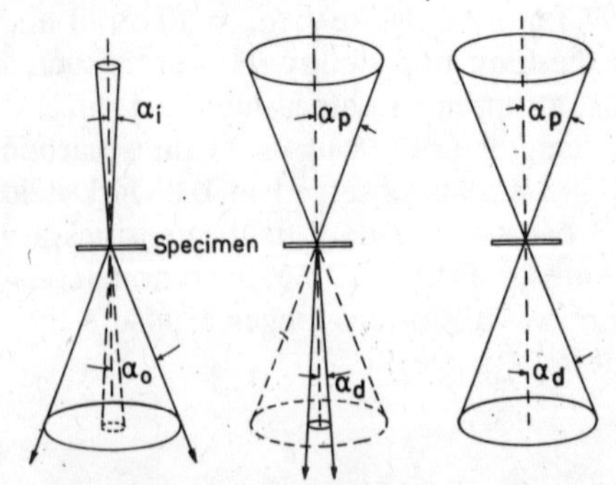

Fig. 6.8 a–c. Apertures in the **(a)** TEM mode, **(b)** reciprocal apertures in the STEM mode, **(c)** optimum STEM mode with $\alpha_p \simeq \alpha_d$

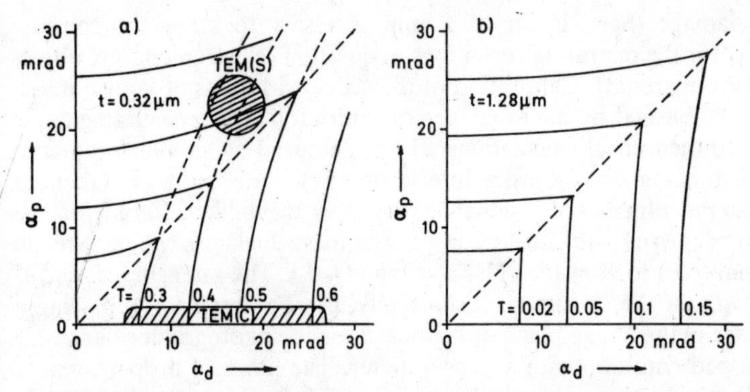

Fig. 6.9 a, b. Lines of equal transmission T in an α_p-α_d diagram [(α_p) electron-probe aperture, (α_d) detector aperture] for (**a**) $t = 320$ nm and (**b**) $t = 1.28$ μm carbon films ($E = 100$ keV). The hatched areas indicate the ranges of the conventional (C) and scanning transmission (S) modes of TEM

therefore be expected as in the TEM mode if we normalize with respect to the intensity I_0 that actually passes through the detector aperture. In practice, however, the electron irradiation must be minimized and the signal-to-noise ratio must be made as high as possible; it thus becomes more convenient to work with $\alpha_d \simeq \alpha_p$ in the STEM mode (Fig. 6.8c), so as to collect all of the electrons of the incident beam when no specimen is present.

Figure 6.9a contains calculated lines of equal transmission $T = I/I_0$ for a relatively thin carbon film ($t = 320$ nm) in an α_p-α_d field. The full curves are those for which I_0 represents the intensity going through the detector aperture. The dashed lines for $\alpha_p > \alpha_d$ are those for which I_0 is the total current in the electron probe. When thicker films are used ($t = 1.28$ μm in Fig. 6.9b), the angular width of the electron-scattering distribution becomes broader than the apertures used. The transmission then becomes less dependent on the aperture and is determined by only the larger of the two apertures α_p and α_d [6.21, 22].

If α_d is increased while α_p is kept constant (corresponding to motion along a line parallel to the abcissa of Fig. 6.9a), the transmission decreases for small mass-thicknesses x. The decrease of T with increasing mass-thickness is still exponential but the value of x_k for $\alpha_d \simeq \alpha_p$ is larger in the STEM mode than it is for $\alpha_i \ll \alpha_o = \alpha_d$ in the TEM mode.

6.1.5 Measurement of Mass-Thickness and Total Mass

The exponential law of transmission (6.6) in the conventional TEM bright-field mode and the STEM mode can be used for quantitative determination of the mass-thickness of amorphous specimens, such as supporting films, biological sections and microorganisms (see examples in Sect. 6.1.3) [6.23–25]. The method can also be used to measure the loss of mass by radiation damage (Sect. 10.2). It is only necessary to know the contrast thickness x_k for the operating conditions in question (electron energy, objective aperture, material). Calibration of this value with films of known mass-thickness, established by microbalance or interferometric measurements, is preferable to theoretical calculations. If t is measured by an interferometric method (two-beam or Tolansky interferometry), care must be taken to ensure that the film has the same density ϱ as the bulk material for the calculation of the mass-thickness $x = \varrho t$. The mass-thickness will be directly proportional (6.8) to $K = \log_{10}(1/T) = \log_{10}(I_0/I_s)$. The intensities I_s and I_0 with and without the specimen, respectively, can be obtained by placing a Faraday cage in the image plane or by measuring the photographic density D of a developed emulsion with a densitometer. The limits of proportionality between D and the charge density $J = j\tau$ [C cm^{-2}] should be checked [4.86].

In the STEM mode, the signal provided by a scintillator-photomultiplier combination is directly proportional to the intensity. A signal proportional K and $x = \varrho t$ can be obtained on-line by means of a logarithmic amplifier [6.26] and can be displayed as a Y-modulation trace on the cathode-ray tube (CRT). This method can also be used to plot lines of equal transmission (mass-thickness) directly and these can be superimposed on the CRT image. Isodensity curves can be produced from photographic records by special reproduction techniques [6.27].

These methods yield the local mass-thickness of a specimen. The total mass of a particular particle can be evaluated by numerical integration over the projected area, which is straightforward with digital or analogue integration of a logarithmic STEM signal.

The total mass can be determined from photographic records (negatives) by a special photometric method [6.28, 29] and has been applied to such specimens as erythrocytes, spermatozoa, phages and virus particles. A large area A of the film that includes the mass (particle) to be investigated provides the transmitted light intensity L_p of the photographic negative, whereas the same area in a region that does not contain the particle (pure supporting film) furnishes the light transmission L_f of the latter; the transmission of an unexposed area of the emulsion is denoted by L_0. Using the relations $T = I/I_0 = \exp(-x/x_k)$ and $D = \log_{10}(L_0/L) = \varepsilon j\tau$ (4.37) for the electron transmission T and density D, the total mass can be calculated from

$$m = \int_A \varrho t \, dA = \frac{A}{c' \log(L_0/L_f)} \frac{L_p - L_f}{L_f} . \tag{6.16}$$

The constant c' can be calibrated by using latex spheres or carbon films of known thickness. *Brakenhoff* et al. [6.19] proposed a similar technique for the dark-field mode.

These methods for the quantitative measurement of mass-thickness are applicable only to amorphous specimens; in the crystalline state, a film of the same mass-thickness will show a decrease of the diffuse scattering depending on specimen temperature (thermal diffuse scattering, Sect. 7.4.1), and the intensities of the Bragg reflections depend strongly on the specimen thickness and orientation. Polycrystalline films with large crystals (Cu, Ag and Au evaporated films, for example) show an averaged transmission that can be twice the value found with an amorphous film. For films with very small crystals (such as Al, Ni, Pt) however, transmission is of the same order as that of amorphous films of equal mass-thickness, provided that the crystals are so small that their diffraction intensity is within the limits of kinematical diffraction theory [6.1].

6.2 Phase Contrast

6.2.1 The Origin of Phase Contrast

We have shown that there is a phase shift of 90° ($\pi/2$ radians) between the unscattered and scattered wave (Sect. 5.1.3). The complex scattering amplitude of the atom creates an additional phase shift $\eta(\theta)$, which can be neglected for low-Z material. If ψ_i is the amplitude of the incident wave in the final image ($I_0 = \psi_i \psi_i^* = |\psi_i|^2$) and ψ_{sc} that of the scattered spherical wave that passes through the objective diaphragm, there will be a phase shift of $\pi/2$, if we assume that the imaging lens introduces no additional phase shift. We examine the 90° phase shift further by plotting $\psi_i + i\psi_{sc}$ as a complex amplitude. Figure 6.10a shows that for $\psi_{sc} \ll \psi_i$, the resulting amplitude $\psi_i + i\psi_{sc}$ has approximately the same absolute value as ψ_i, so that $I = |\psi_i + i\psi_{sc}|^2$ does not differ significantly from $I_0 = |\psi_i|^2$; this means that the phase object is invisible. If the phase of the scattered wave could be shifted by a further 90° (Fig. 6.10b), the superposition would become $\psi_i - \psi_{sc}$ and hence $I = |\psi_i - \psi_{sc}|^2 = \psi_i^2 - 2\psi_i\psi_{sc} + \ldots < I_0$. This is called *positive phase contrast.* If ψ_{sc} were shifted by $3\pi/2$ or $-\pi/2$, the superposition would be $\psi_i + \psi_{sc}$ (Fig. 6.10c) so that $I > I_0$; this is called *negative phase contrast.* In light microscopy, these phase shifts can be produced by inserting

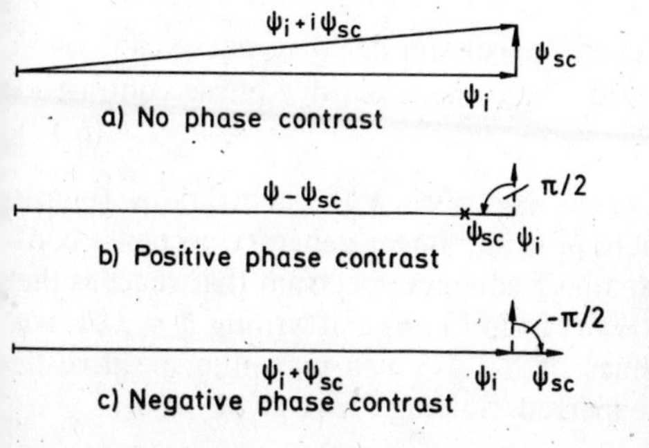

Fig. 6.10. (**a**) Vector addition of the image amplitude ψ_i and the scattered amplitude ψ_{sc} phase shifted by $\pi/2$ or 90°, (**b**) positive phase contrast produced by an additional phase shift of $+\pi/2$, (**c**) negative phase contrast produced by an additional phase shift of $-\pi/2$ or $+3\pi/2$

a Zernike phase plate in the focal plane of the objective lens; such a plate shifts the scattered wave by an optical-path-length difference of $\lambda/4$ and has a central hole through which the primary beam passes unmodified. A path difference $\lambda/4$, which corresponds to a phase shift of $\pi/2$, can be produced by passing 100 keV electrons through a 23 nm thick carbon foil with inner potential U_i (Sect. 3.1.3). This possibility has been investigated in attempts to create the desired phase shift by means of a carbon foil with a central perforation (Sect. 6.4.5). However, practical difficulties arise with such phase plates because in continuous operation, the foil becomes charged and contaminated by electron irradiation.

The effect of spherical aberration and defocusing may be expressed in terms of the wave aberration $W(\theta)$ (Fig. 3.16). The shape of the wave-aberration curve for different values of defocus shows that a phase shift of $\pm\pi/2$ cannot be obtained simultaneously for all scattering angles; only for a limited range of scattering angles or their corresponding spatial frequencies q will $W(\theta)$ produce the desired phase shift. Values of defocus for which $W(\theta)$ takes a minimum value of $-\pi/2$ are particularly favourable ($\Delta z^* = 1$ in Fig. 3.16).

6.2.2 Defocusing Phase Contrast of Supporting Films

Supporting films (especially of carbon) show a characteristic granular structure at high resolution, the appearance of which changes with the defocus (Figs. 2.23 and 6.12); this granularity was first reported by *Sjöstrand* [6.30] and discussed as phase contrast by *von Borries* and *Lenz* [6.31]. Carbon films show statistical fluctuations of local mass-thickness and, therefore, of the electron-optical phase shift. The two-dimensional Fourier transform of the phase shift contains a wide range of spatial frequencies with approximately equal probability (white noise). For this reason, carbon films are ideal test specimens for investigating the transfer characteristic of an electron-optical imaging system for different spatial frequencies.

A single spatial frequency q that corresponds to a spacing or periodicity $\Lambda = 1/q$, creates a diffraction maximum at a scattering angle $\theta = \lambda/\Lambda = \lambda q$. Those spatial frequencies for which the wave aberration is an odd multiple of $\pi/2$, thus:

$$W(\theta) = (2m-1)\frac{\pi}{2} \qquad \begin{cases} m = \text{even:} & \text{maximum negative phase contrast} \\ m = \text{odd:} & \text{maximum positive phase contrast} \end{cases} \tag{6.17}$$

will be imaged with maximum phase contrast. Wave aberrations (phase shifts) for which $W(\theta) = m\pi$ where m is an integer generate no phase contrast and thus leave gaps in the spatial frequency spectrum that reaches the image. Substituting in (6.17) for $W(\theta)$ from (3.66) and writing $\theta = \lambda/\Lambda$, we obtain an equation for those values of Λ for which maximum positive or negative phase contrast is to be expected. Solving for Λ gives

$$\Lambda = \lambda \left[\frac{\Delta z}{C_s} \pm \left(\frac{\Delta z^2}{C_s^2} + \frac{(2m-1)\lambda}{C_s} \right)^{1/2} \right]^{-1/2} . \tag{6.18}$$

This formula of *Thon* [6.32, 33] is an extension of an earlier formula of *Lenz* and *Scheffels* [6.34], in which only those terms of $W(\theta)$ caused by defocusing were considered: $W(\theta) = \pi \Delta z \theta^2 / \lambda = \pi/2$ and for defocusing only, we find

$$\Lambda = \sqrt{2 \Delta z \lambda} , \tag{6.19}$$

which is valid for large Λ and defocusing Δz and relates a specimen periodicity Λ to the optimum defocusing Δz, at which the periodicity will be imaged with optimum phase contrast.

The periodicities Λ that are imaged with maximum positive or negative phase contrast can be measured in light-optical Fraunhofer diffraction patterns of the developed photographic emulsion (Sect. 6.4.6); typical curves are plotted in Fig. 6.11 as functions of defocusing Δz. The full curves were calculated from (6.18) and show excellent agreement. Even in focus ($\Delta z = 0$), the granularity of the carbon film does not disappear, owing to the term in $W(\theta)$ that contains the spherical aberration. The resolution is limited (horizontal dashed line in Fig. 6.11) by the attenuation of the contrast transfer caused by the chromatic aberration and the finite illumination aperture (Sect. 6.4.2).

Crystalline areas with periodic structures have been observed in carbon foils [6.35] by using a tilted primary beam, as used for the imaging of lattice

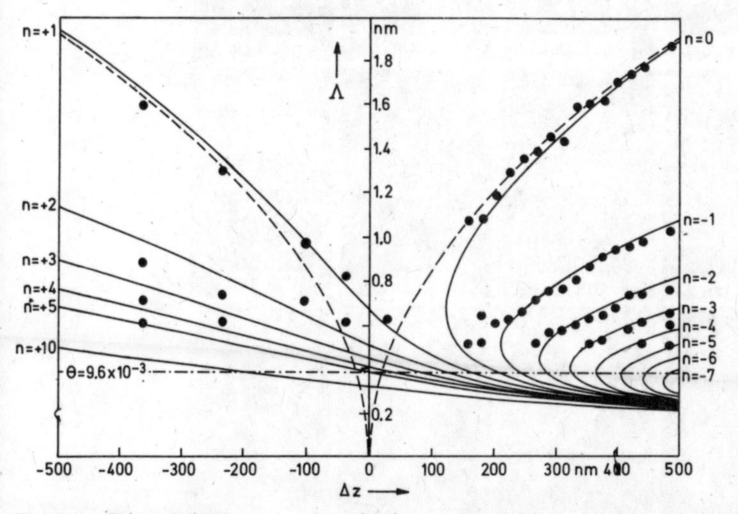

Fig. 6.11. Comparison of measured spatial frequencies q with maximum positive and negative phase contrast obtained by laser diffraction on micrographs of carbon foils, and theoretical curves based on (6.18) [6.33]

planes (Sect. 8.2.1). However, when such structures are seen in amorphous specimens with this mode of imaging, they may be caused by selective filtering of spatial frequencies. This selective filtering results from modification of the contrast-transfer function by tilted-beam illumination (Sect. 6.4.3). Bright spots of 0.3–0.5 nm diameter were observed in dark-field imaging with an annular aperture [6.36], which were most intense at an overfocus of 275 nm. These spots were attributed to Bragg reflections on small crystallites (see also [6.37]).

As we have seen, the granularity of carbon foils is very useful for investigating the contrast transfer of TEM but it disturbs the images of small particles, macromolecules and single atoms. Numerous attempts have been made therefore, to prepare supporting films with less granularity in phase contrast (Sect. 4.3.2).

An electron-optical method of decreasing the phase contrast of the supporting film relative to the contrast of single atoms and structures with stronger phase shifts involves using hollow-cone illumination (Sect. 6.4.3).

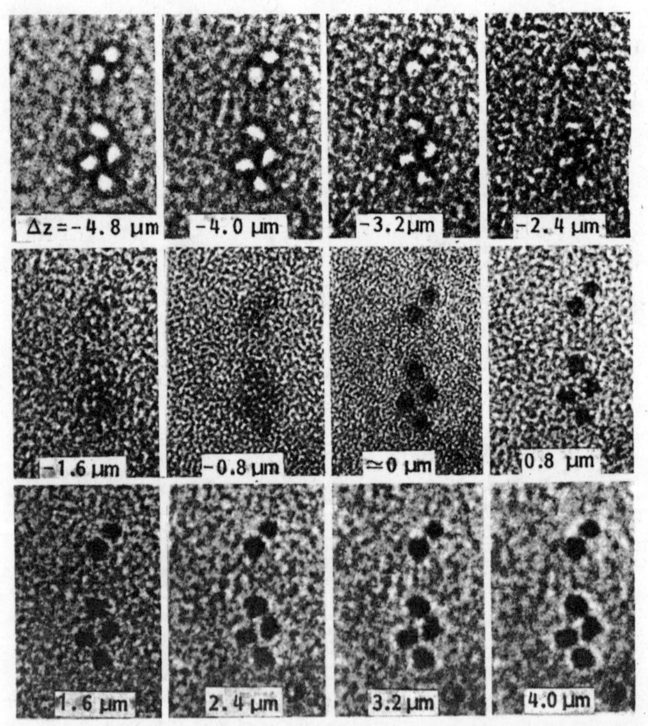

Fig. 6.12. Defocus series of ferritin molecules on a 5 nm carbon supporting film (E = 100 keV)

6.2.3 Examples of Phase Contrast of Small Particles and Lamellar Systems

Figure 6.12 shows a through-focus series of ferritin molecules on a carbon supporting film 5 nm thick. In focus (third image in second row) the molecules show weak scattering contrast, due to the iron-rich core of the molecule ($\simeq$ 5 nm diameter). This contrast is caused by the loss of electrons that have been scattered through large angles and intercepted by the objective diaphragm. The part of the electron wave that passes through the objective aperture is phase shifted 90°. An increase of contrast by phase contrast can be observed for underfocus ($\Delta z > 0$). The image of the molecules becomes darker in the centre. Normally, the operator instinctively focuses for maximum contrast, which means underfocusing. In overfocus ($\Delta z < 0$), the phase shift $W(\theta)$ becomes positive and the molecules appear bright in the centre. For a quantitative interpretation of the dependence of image intensity in the centre on defocusing, see [6.38].

Reversed phase contrast may occur in some specimens, incorporated molecules of o-phenanthroline in electrodeposited nickel films for example [6.39]. The molecules are imaged as bright spots in underfocus and as dark spots in overfocus. This confirms that they are really vacancies in the nickel film ($\simeq$ 1 nm in diameter) that contain the organic molecules. Because of the lower inner potential U_i of the vacancies, the wavefront behind the inclusions will exhibit an opposite phase shift. Phase contrast can also be observed in defocused images of crystal foils with vacancy clusters [6.40].

In phase contrast, the number of electrons that pass through the objective diaphragm will be constant and all will reach the image. This means that if the intensity at some point of the image is increased by summing the amplitudes with favourable phase shifts, the intensities at neighbouring image points will be decreased, so that the mean value of the intensity is only reduced by scattering contrast. If the image of a particle is darker in the centre as a result of positive phase contrast, it will be surrounded by a bright rim and vice versa (Fig. 6.12). Beyond this bright ring, further rings follow with decreasing amplitudes. In complex structures, and especially in periodic structures, these bright and dark fringes can interfere and cause artifacts. Figure 6.13 demonstrates such an effect for myelin lamellae. The contrast of the membranes can be reversed by overfocusing ($\Delta z < 0$). The width of the dark stripes increases with increasing overfocus but at $\Delta z = -4.8$ μm, twice the number of dark lines can be seen. In underfocus, an increase of the dark contrast of the membranes can again be observed.

The two examples of Figs. 6.12 and 13 show that the phase-contrast effects in a defocus series can be interpreted when the specimen structure is known from a focused image or from the method of specimen preparation. For structures smaller than 1 nm however, this becomes difficult, because the spherical-aberration term of $W(\theta)$ also has to be considered. In this case, more-complicated image-reconstruction methods have to be used (Sect. 6.5) to extract information about the specimen from one micrograph or a series.

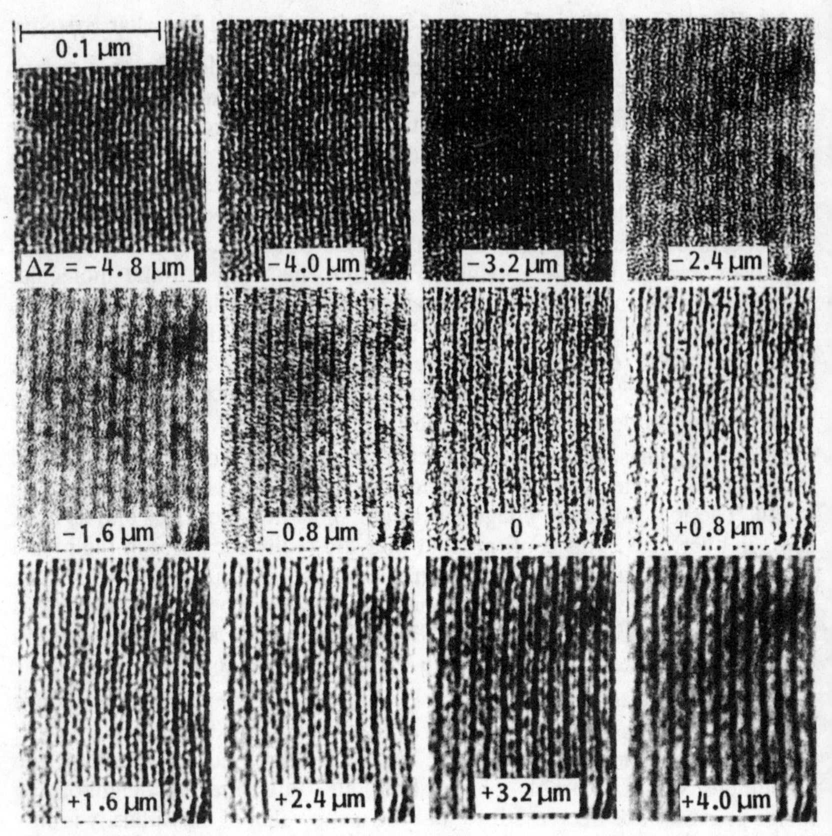

Fig. 6.13. Defocus series of an ultramicrotome section through myelin lamellae (stained with OsO_4 and embedded in Vestopal)

6.2.4 Theoretical Methods for Calculating Phase Contrast

The wave-optical theory of imaging has already been described in Sect. 3.3.2. We set out from formula (3.39) for the modified plane wave behind the specimen. The amplitude ψ_0 will be normalized to unity. The local amplitude modulation $a_s(\boldsymbol{r})$ is assumed to differ little from one: $a_s(\boldsymbol{r}) = 1 - \varepsilon_s(\boldsymbol{r})$ where $\varepsilon_s(\boldsymbol{r})$ is small. If the phase shift $\varphi_s(\boldsymbol{r})$ is also much less than one, then the exponential term in (3.39) can be expanded in a Taylor series

$$\psi_s(\boldsymbol{r}) = 1 - \varepsilon_s(\boldsymbol{r}) + \mathrm{i}\varphi_s(\boldsymbol{r}) + \dots \,. \tag{6.20}$$

With this approximation, the specimen is said to be a weak-amplitude, weak-phase object. In practice, electron-microscope specimens thinner than 10 nm and of low atomic number do behave as weak-phase objects. The amplitude modulation $\varepsilon_s(\boldsymbol{r})$ can be neglected. When the phase contrast of particles with high atomic number is calculated, colloidal gold particles for example, the decrease $\varepsilon_s(\boldsymbol{r})$ of amplitude must however be considered [6.38, 41].

Equations (3.73, 74) contain the complete mathematical treatment of phase contrast. Depending on the information required and the nature of the phase contrast, the following procedures can be used:

1) If the scattering amplitude $F(q)$ of a specimen is known, the image amplitude $\psi_m(\boldsymbol{r}')$ is given approximately by (3.73). For high resolution, $F(\boldsymbol{q})$ is related to the scattering amplitude $f(\theta)$ of a single atom (Sect. 5.1.3) by $F(q) = \lambda f(\theta)$. Examples are discussed in Sect. 6.3.2.
2) For constant conditions and variations of the specimen structure, it will be advantageous to use the convolution (3.74) of the object function $\psi_s(\boldsymbol{r})$ with the Fourier transform $h(\boldsymbol{r})$ of the pupil function $H(\boldsymbol{q})$ because in this case, intermediate calculation of $F(\boldsymbol{q})$ would be a waste of computation time.
3) If more-general information is wanted about the contrast transfer – what spatial frequencies q are imaged with positive or negative phase contrast for a given electron lens or how the contrast transfer is influenced by a finite illumination aperture or by the energy spread of the incident electron beam, for example – then the pupil function or the contrast-transfer function can be used (Sect. 6.4).

6.3 Imaging of Single Atoms

6.3.1 Imaging of a Point Source

We first consider an idealized point specimen, which scatters isotropically into all scattering angles or which is the source of a spherical wave of amplitude $f(\theta)$, independent of the scattering angle θ. As shown in Sect. 3.3.2, the amplitude-blurring or point-spread function $h(r, r')$ is obtained as the image. The scattering amplitude of a single atom decreases with increasing θ. However, to a first approximation the values of the objective aperture normally used, $\alpha_o = 10$–20 mrad, cut off the wave before the scattering amplitude $f(\theta)$ begins to decrease appreciably (Fig. 5.5) so that an isotropic point source with $f(\theta) = f(0) = \text{const}$ is a good approximation and describes the imaging of single atoms satisfactorily. In high-voltage electron microscopy, $f(\theta)$ begins to decrease for scattering angles smaller than the optimum objective aperture (Sect. 6.3.4).

We introduce polar coordinates r' and χ in the image plane and normalize the magnification to unity ($M = 1$). The scalar product in (3.73) becomes $\boldsymbol{q} \cdot \boldsymbol{r} = qr' \cos\chi = \theta r' \cos\chi/\lambda$; we have $d^2\boldsymbol{q} = \theta d\theta/\lambda^2$ and $F(q) = \lambda f(\theta)$. For the bright- and dark-field modes (1 and 0, respectively, for the first term), we obtain

$$\psi_m(r') = \frac{1}{0} + \frac{i}{\lambda} \int_0^{\alpha_0} \int_0^{2\pi} f(\theta) e^{-iW(\theta)} \exp\left(\frac{2\pi i}{\lambda} \theta r' \cos\chi\right) \theta \, d\theta d\chi \,. \quad (6.21)$$

The difference between bright and dark field is that, in the former, the primary incident wave (normalized in amplitude to unity) contributes to the image amplitude, whereas in the dark-field mode, it will be absorbed by a central beam stop or by a diaphragm. The factor i indicates that there is a phase shift of 90° between the primary and scattered waves.

If the specimen and the scattering amplitudes are assumed to be rotationally symmetric, the integration over χ in (6.21) gives the Bessel function J_0. The term involving $W(\theta)$ can be rewritten, using the Euler formula,

$$\psi_m(r') = \frac{1}{0} + \frac{2\pi i}{\lambda} \int_0^{\alpha_0} f(\theta)[\cos W(\theta) - i \sin W(\theta)] \, J_0\left(\frac{2\pi}{\lambda}\theta r'\right) \theta \, d\theta$$

$$= \frac{1}{0} + \varepsilon_m(r') + i\varphi_m(r') \,. \quad (6.22)$$

In the absence of the wave aberration $[W(\theta) = 0]$, the real part $\varepsilon_m(r')$ of (6.22) becomes zero and the same result is obtained as in Fig. 6.10 a, namely that the 90° phase-shifted imaginary part $\varphi_m(r')$ makes no contribution to bright-field image contrast because $\varphi_m(r') \ll 1$. With non-vanishing wave aberration $W(\theta)$, the real part of (6.22) that contains $\sin W(\theta)$ is non-zero: $\varepsilon_m(r') \neq 0$. The image intensity is obtained by squaring the absolute amplitude i.e.,

$$I(r') = \psi_m(r')\psi_m^*(r') = \left[\frac{1}{0} + \varepsilon_m(r')^2\right]^2 + \varphi_m^2(r')$$

$$= \begin{cases} 1 + 2\varepsilon_m + \varepsilon_m^2 + \varphi_m^2 \simeq 1 + 2\varepsilon_m(r') + \dots & \text{bright field} \quad (6.23\,\text{a}) \\ \varepsilon_m^2 + \varphi_m^2 & \text{dark field}\,. \quad (6.23\,\text{b}) \end{cases}$$

Because both ε_m and φ_m are very much smaller than unity, the quadratic terms can be neglected in the bright-field mode. If we consider the intensity variation

$$\Delta I(r') = I(r') - I_0 = 2\varepsilon_m(r') = \frac{4\pi}{\lambda} \int_0^{\alpha_0} f(\theta) \sin W(\theta) \, J_0\left(\frac{2\pi}{\lambda}\theta r'\right) \theta \, d\theta \quad (6.24)$$

for the bright-field mode relative to the background $I_0 = 1$, the integrand in (6.24) can be split into three factors, which are plotted in Fig. 6.14. The

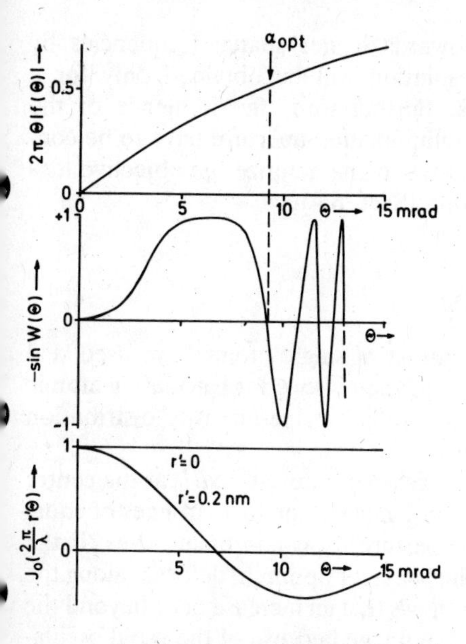

Fig. 6.14 a–c. Plot of the three factors (**a**) $\theta f(\theta)$, (**b**) $-\sin W(\theta)$ and (**c**) $J_0(2\pi\theta r'/\lambda)$ in the integrand of (6.24)

factor $\theta f(\theta)$ expresses the fact that an annular element $2\pi\theta\, d\theta$ increases as θ. The factor $\sin W(\theta)$ passes through a maximum when the maximum of the wave aberration $W(\theta)$ in Fig. 3.16 is at $W(\theta) = -\pi/2$; this occurs for a reduced defocusing of $\Delta z^* = 1$. The Bessel function J_0 is unity for $r' = 0$ at the centre of the atom. The reduced optimum aperture will be that for which $W(\theta)$ is again zero, that is, $\alpha^*_{opt} = \sqrt{2}$ for $\Delta z^* = 1$. The subsequent rapid oscillations of $\sin W(\theta)$ with increasing θ will give no further significant contribution to the integral in (6.24). By use of (3.69), the values

$$\Delta z_{opt} = (C_s\lambda)^{1/2}\,; \quad \alpha_{opt} = 1.41\,(\lambda/C_s)^{1/4} \tag{6.25}$$

are obtained for the so-called *Scherzer focus* with maximum positive phase contrast [6.42] and the corresponding optimum aperture.

When the distance r' from the centre of the atom increases, the oscillations of the Bessel function J_0 in (6.24) are shifted to smaller values of θ (Fig. 6.14 c), which decreases the value of the integral and can even change its sign. The Airy distribution (Fig. 3.17 b) is obtained for $F(\theta)$ = const and for $\sin W(\theta) = -1$ at all scattering angles. At Scherzer focus, the half-width of the image-intensity distribution passes through a minimum:

$$\delta_{min} = 0.43\,(C_s\lambda^3)^{1/4}\,. \tag{6.26}$$

This quantity δ_{min} can be used to define the *resolution* of TEM. However, a single number proves to be insufficient to characterize the resolution. Thus, specimen details closer together than δ_{min} can be imaged by shifting the minimum of the wave aberration towards higher spatial frequencies by defocusing. However, this better resolution will be obtained only for a limited range of spatial frequencies. Furthermore, the influence of the chromatic aberration and of the finite illumination aperture have to be considered. It is, therefore, more informative to characterize the objective lens of TEM by its contrast-transfer function (Sect. 6.4).

6.3.2 Imaging of Single Atoms in TEM

Most of the calculations of image contrast of single atoms have used real values of $f(\theta)$ in (6.21) [6.43–50]. One of the reasons for calculating atomic images is to study the behaviour of the radial image-intensity distribution when different parameters are varied.

Figure 6.15 shows the calculated decrease of intensity $\Delta I/I_0$ at the centre of a platinum atom ($E = 100$ keV, $C_s = 1$ mm) in the form of lines of equal $\Delta I/I_0$ with defocus Δz and objective aperture α_o as coordinate axes. If the objective aperture is varied through the Scherzer optimum defocus, along the line BB′, the upper curve of Fig. 6.15 shows that an increase of α_o beyond the optimum aperture does not improve the image because of the rapid oscillations of $W(\theta)$ (see also the contrast-transfer function in Sect. 6.4.1). In practice, therefore, large diaphragms should be used because a smaller diaphragm that corresponds to the optimum aperture at Scherzer focus can become charged around the periphery, thereby causing additional phase shifts. If the defocus is varied through the Scherzer focus along the line AA′, the left curve in Fig. 6.15 shows that the phase contrast oscillates with increasing defocus Δz. The atom is alternatively imaged in positive and negative phase contrast. Positive phase contrast is observed not only at the Scherzer focus, at $\Delta z^* = 1$, but again at $\Delta z^* = \sqrt{5}$ where a broad interval of spatial frequencies is transfered with the phase shift $W(\theta) = -5\pi/2$ (Figs. 3.16 and 6.19c). The inset in the left top corner of Fig. 6.15 contains the radial variation of $I(r')$ for a platinum atom at the Scherzer focus. Once again, a bright annular ring is observed around the central darker region (Sect. 6.2.2), which reconciles the larger decrease of the intensity in the central region with the fact that the number of electrons transmitted is constant.

Figure 6.16 shows the influence of some parameters on the radial intensity distribution $I(r')$ of a single bromine atom [6.51]. A decrease of the spherical-aberration constant C_s (full curves in Fig. 6.16a) increases the positive phase contrast at the centre ($r' = 0$), strengthens the bright annular ring and reduces the half-width. For $\Delta z^* = \sqrt{5}$ (dotted line), stronger oscillations are observed at large distances r'. If the sign is changed (phase shift of 180°) in the region in which $-\sin W(\theta)$ is negative, as in an optical-reconstruction scheme proposed by *Marechal* and *Hahn* (Sect. 6.5.4) a result is obtained for $C_s = 2$ mm that is comparable with that for $C_s = 0.5$ mm at the

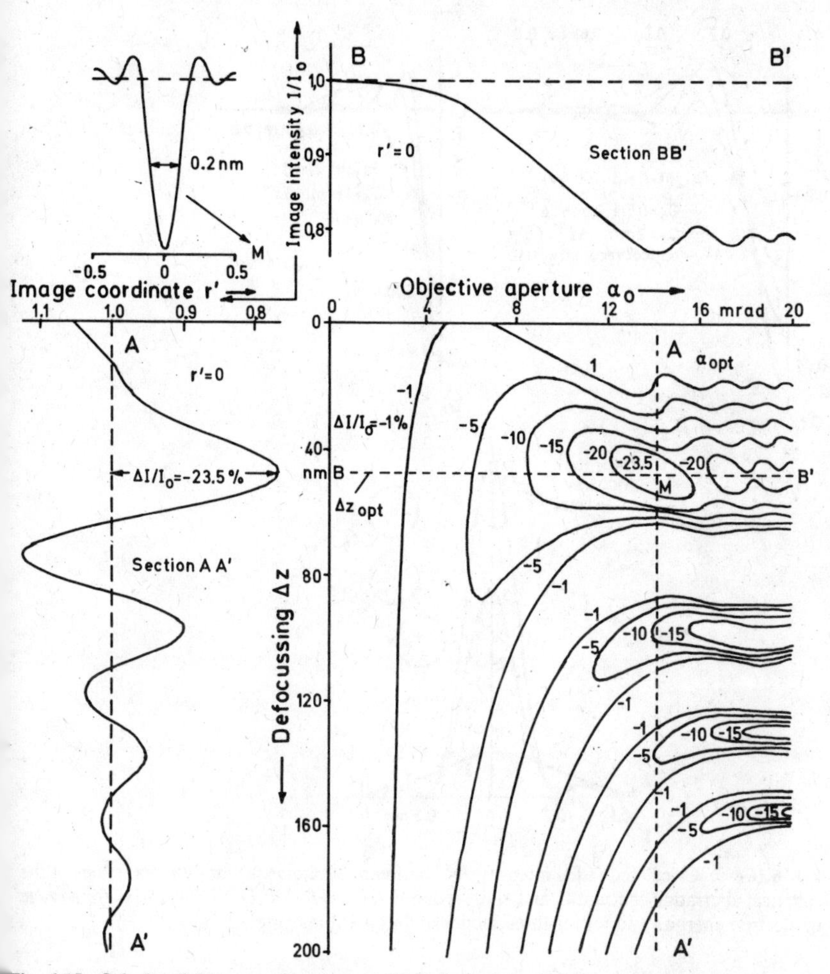

Fig. 6.15. Calculated decrease of intensity $\Delta I/I_0$ in the centre of platinum atoms as a function of defocus Δz and objective aperture α_o for $E = 100$ keV and $C_s = 0.5$ mm. Sections along the lines AA′ and BB′ are shown at the left and top respectively. The radial intensity distribution of a platinum atom at the Scherzer focus M for Δz_{opt} and α_{opt} is seen at the left top corner

Scherzer optimum defocus at $\Delta z^* = 1$ (dashed curve in Fig. 6.16 a). The influence of electron energy (Fig. 6.16 b) is discussed in Sect. 6.3.4.

Figure 6.16 c contains calculated dark-field intensity distributions for three different modes of dark-field imaging (Sect. 6.1.2). The advantage of using a central beam stop (1) or tilted illumination with a central diaphragm (2) rather than a shifted diaphragm (3) can be seen clearly (see also [6.52, 53]). In modes (2) and (3), the radial intensity distributions are somewhat

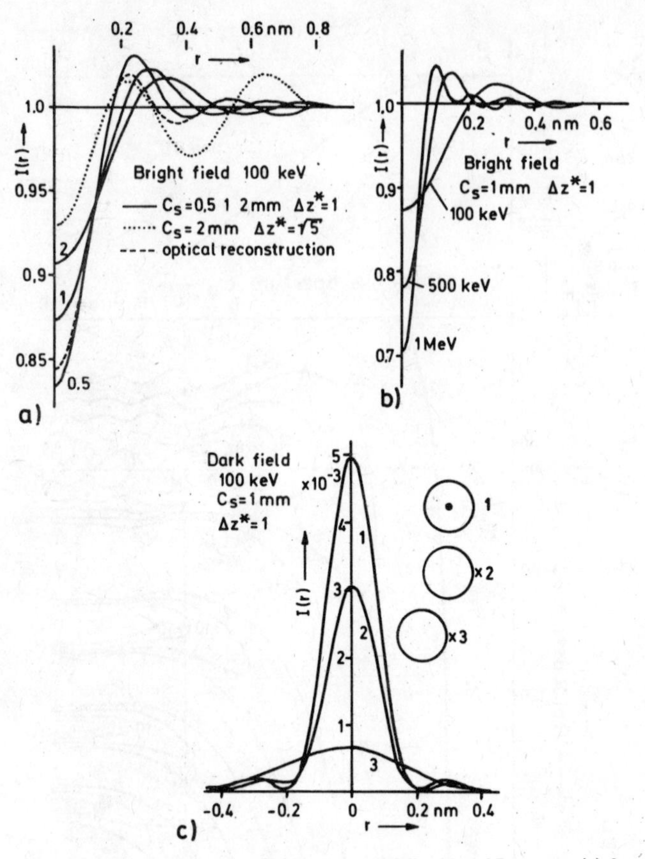

Fig. 6.16 a–c. Calculated radial intensity distribution of Br atoms (**a**) for different values of the spherical-aberration constant C_s and of the reduced defocus Δz^* at $E = 100$ keV, (**b**) for a range of electron energies and (**c**) for three modes of dark-field imaging

asymmetrical. Illumination with a hollow cone (annular diaphragm in the condenser lens) (Sect. 6.4.3) corresponds to an incoherent superposition of images obtained with mode (2); this averages the weak asymmetry of the image discs (see also [6.54, 55]).

The intensity of dark-field images is much lower and is smaller than the decrease of intensity in the bright-field mode. The dark-field mode has the advantage of higher contrast. Single atoms appear as relatively bright dots against the weak background intensity of a supporting film [6.56–59]. However, longer exposure times are needed than in the bright-field mode. Furthermore, the contrast transfer of the dark-field mode is non-linear [6.54, 55, 60, 61]. In (6.23) $\varepsilon_m(r')$ appears as a linear term in the bright-field mode but as a quadratic term in dark field. The Fourier spectrum of a specimen periodicity Λ with spatial frequency $q = 1/\Lambda$ consists of the central beam and two diffracted beams of order ± 1. Removal of the central beam in dark field will result in twice the spatial frequency between the two diffracted waves so that a specimen periodicity of $\Lambda/2$ will be observed in dark field.

The presence of neighbouring atoms leads to a superposition of the image amplitudes of the individual atoms, which can produce parasitic structures in bright- and dark-field imaging. Consider for example the dotted curve of Fig. 6.16 a, which corresponds to the image of a Br atom at a defocus $\Delta z^* = \sqrt{5}$; if two neighbouring atoms are separated by a distance of 0.4 nm, the second minimum of $I(r')$ for one atom will coincide with the central decrease of the other, thus causing an increase in the contrast of both. If, however, they are separated by 0.8 nm, the secondary minima at $r' = 0.4$ nm will increase; a third atom will apparently be seen, though this, in fact, will be an image artifact.

Hitherto, we have discussed the contribution of elastic scattering to phase contrast. The image amplitudes of the inelastic scattered electrons also have to be considered. However, inelastic scattering is concentrated within smaller scattering angles than elastic scattering and already decreases strongly inside the objective aperture. We know that the image amplitude is the Fourier transform of the scattering amplitude $f(\theta)$. A narrow scattering distribution results in a broader image disc [6.50, 62]. It can also be argued that there are no inelastically scattered electrons at high spatial frequencies, where they would be necessary for high resolution. In a classical model of scattering, we can say that inelastic scattering is less-localized than elastic scattering. An electron that passes far from an atom can nevertheless excite an atomic electron by Coulomb interaction. Inelastic scattering is, therefore, useless for obtaining high-resolution information.

Images of single atoms have been observed in molecules of known structure: triangles of heavy atoms separated by distances of the order of 1 nm in triacetomercuryaurin [6.63], uranium-stained mellitic acid [6.64], monolayers of thorium-hexafluoracetylacetonate [6.65] and single W atoms and clusters [6.66]. These confirm that the calculated contrast and resolution in the bright-field mode are of the right order of magnitude. In the dark-field mode, single atom images have been obtained for U, Os, Ir, Pd [6.16], Th [6.56], Rh [6.67] and, at high voltages (200 and 3000 keV), for U, Ba, Sr, Fe ($Z = 26$) [6.68, 69]. The dark-field mode with conical illumination (hollow-beam) was employed for observing Hg [6.59] and for U and Ba atoms [6.68].

At present, these experiments merely show that single atoms can indeed be imaged in principle; they also clearly demonstrate that high resolution is limited not by lack of contrast but by the background noise of the supporting film or organic matrix and by radiation damage (Sect. 10.2).

6.3.3 Complex Scattering Amplitudes and Scattering Contrast

For more-accurate calculation of the contrast of single atoms, the above-mentioned decrease of $f(\theta)$ with increasing θ as well as the phase shift $\eta(\theta)$ of the complex scattering amplitude (5.20) have to be included in (6.22). The phase shift $\eta(\theta)$ has to be added to the existing phase shift of 90° between primary and scattered wave; it causes a decrease of the amplitude in Fig. 6.10 a even if the lens introduces no additional phase shift. The use of complex scattering amplitudes expands the theory of phase contrast into a more general form that also contains the scattering contrast [6.38, 47, 48]. To demonstrate this, we assume that $W(\theta) = 0$ and replace $f(\theta)$ by $|f(\theta)| \exp[i\eta(\theta)]$ in (6.22) for the bright-field mode. When the Euler formula is applied to $\exp[i\eta(\theta)]$, Eq. (6.22) becomes

$$\psi_m(r') = 1 + \frac{2\pi i}{\lambda} \int_0^{\alpha_o} |f(\theta)|[\cos\eta(\theta) + i\sin\eta(\theta)]\, J_0\left(\frac{2\pi}{\lambda}\theta r'\right)\theta\, d\theta\,. \tag{6.27}$$

If we assume that $|f(\theta)| \sin\eta(\theta)| \simeq |f(0)| \sin\eta(0) = \text{const}$ for all scattering angles $\theta \leq \alpha_o$, the relation $\int x J_0(x)\, dx = x J_1(x)$ can be used and (6.27) becomes

$$\begin{aligned}\psi_m(r') &= 1 - |f(0)| \sin\eta(0) \frac{\alpha_o}{r'} J_1\left(\frac{2\pi}{\lambda}\alpha_o r'\right) \\ &\quad + i|f(0)| \cos\eta(0) \frac{\alpha_o}{r'} J_1\left(\frac{2\pi}{\lambda}\alpha_o r'\right) \\ &= 1 - \varepsilon_m(r') + i\varphi_m(r')\,.\end{aligned} \tag{6.28}$$

The radial variation of the decrease of intensity $\Delta I(r')$ is obtained as in (6.24) by use of all of the terms of (6.23 a)

$$\Delta I(r') = 2\varepsilon_m + \varepsilon_m^2 + \varphi_m^2\,. \tag{6.29}$$

The dominant first term is negative, which means that a decrease of intensity is observed in bright field. Integrating the intensity variation $\Delta I(r')$ over the whole image disc, we obtain

$$\begin{aligned}2\pi \int_0^\infty \Delta I(r') r' dr' &= -\, 4\pi |f(0)| \sin\eta(0)\, \alpha_o \int_0^\infty J_1\left(\frac{2\pi}{\lambda}\alpha_o r'\right) dr' \\ &\quad + 2\pi |f(0)|^2 [\sin^2\eta(0) + \cos^2\eta(0)]\, \alpha_o^2 \int_0^\infty \frac{J_1^2\left(\frac{2\pi}{\lambda}\alpha_o r'\right)}{r'} dr' \\ &= -2\lambda |f(0)| \sin\eta(0) + \pi\alpha_o^2 |f(0)|^2\,.\end{aligned} \tag{6.30}$$

The first term is identical with σ_{el}, as the optical theorem (5.45) shows. The last term is the elastically scattered intensity that goes through the objective aperture. The whole integral is equal to $-\sigma_{el}(\alpha_o)$, see (6.3). This is none other than the contribution of one atom to the decrease of intensity by scattering contrast. Formula (6.5) for the decrease of intensity by a layer of atoms is obtained by multiplying $\sigma_{el}(\alpha_o)$ by the number of atoms $N_A dx/A$ per unit area of a film of mass-thickness dx. If the individual atoms cannot be resolved, an average over the intensity decrease of all atoms is observed, as in (6.30).

6.3.4 Dependence of Phase Contrast on Electron Energy

The decrease of intensity $\Delta I/I_0$ of single atoms in a positive phase contrast increases with increasing electron energy (Fig. 6.16 b) as $f(\theta)$ becomes concentrated towards lower scattering angles (Fig. 5.5). However, this is only the case for the imaging of single atoms. For more-coarse structures, the phase contrast decreases with increasing energy [6.38].

We assume a weak-phase object for which the Born approximation (5.33) can be used unhesitatingly. The integral in (5.33) will be independent of energy for a given potential distribution $V(r)$, the energy dependence that appears in the factor in (5.33)

$$\frac{E+E_0}{\lambda E(E+2E_0)} \propto m\lambda \propto v^{-1} \propto \frac{E+E_0}{(2EE_0+E^2)^{1/2}} . \tag{6.31}$$

When the image amplitude is calculated with the aid of (6.21, 22), this factor does not change if the inverse Fourier transform is applied with respect to q rather than θ. If the specimen contains only one spatial frequency q or if we consider a single spatial frequency of a carbon film, optimum phase contrast with $W(\theta) = -\pi/2$ can be obtained by selecting a suitable value of the defocus. The quantity $\varepsilon_m(r')$ in (6.22, 23) is, therefore, proportional to v^{-1}. The same reasoning will be valid for small particles or specimen structures for which the distribution of scattering amplitudes $F(q)$ is concentrated within angles smaller than the objective aperture. An optimum defocus value can be found with the same $W(q)$ curve for different energies.

For single atoms, we first reconsider the case of an approximately constant scattering amplitude $F(q)$ within the optimum objective aperture α_{opt} (6.25), which can be written as $q_{opt} = \alpha_{opt}/\lambda = 1.41\,(C_s\lambda^3)^{-1/4}$. For the central point $r' = 0$, Eq. (6.24) gives

$$\varepsilon_m(0) = 2\pi \int_0^{q_{opt}} F(q) \sin W(q)\, q\, dq \propto v^{-1} \int_0^{q_{opt}} \sin W(q)\, q\, dq \tag{6.32}$$

or, with the reduced coordinate $\theta^* = (C_s\lambda^3)^{1/4} q$,

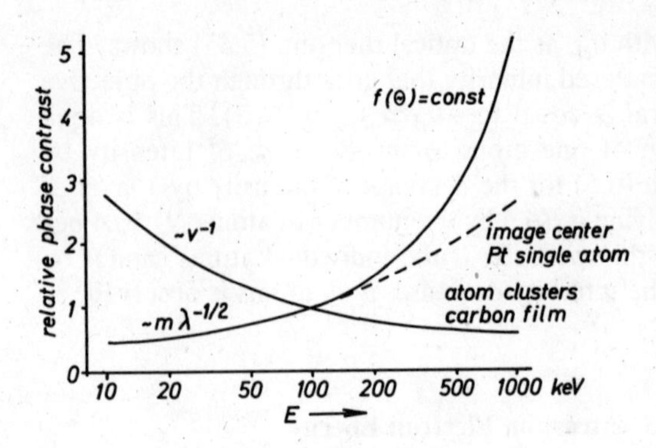

Fig. 6.17. Energy variation of phase-contrast effects (normalized to unity at $E = 100\,\mathrm{keV}$) for large atom clusters or small particles, with $f(\theta)$ decreasing inside the optimum aperature; for single atoms, showing that the decrease of $f(\theta)$ inside the aperture (dashed curve) has an effect for $E \geq 150\,\mathrm{keV}$; and assuming $f(\theta) = \mathrm{const}$ inside the aperture

$$\varepsilon_\mathrm{m}(0) \propto v^{-1}(C_\mathrm{s}\lambda^3)^{-1/2}\int_0^{1.41} \sin W(\theta^*)\,\theta^* d\theta^* \propto m\lambda^{-1/2}$$
$$\propto (E_0 + E)(2EE_0 + E^2)^{1/4}. \tag{6.33}$$

This last quantity increases with energy. Figure 6.17 shows the factors (6.31 and 33) normalized to unity at $E = 100$ keV. Calculations of the optimum positive phase contrast of single atoms show, however, a slower increase with increasing energy, due to the decrease of $f(\theta)$ inside the aperture α_opt. The concentration of $f(\theta)$ around small scattering angles in Fig. 5.5 is greater than the decrease of α_opt with increasing E. Furthermore, it is assumed in the calculation that C_s is constant, whereas HVEM objective lenses normally have larger values of C_s; (6.33) shows that the phase contrast has to be multiplied by $(C_{\mathrm{s},100}/C_{\mathrm{s},E})^{1/2}$. Nevertheless, it can be concluded that the ratio of ΔI for a single atom and of the granularity of the supporting film (Sect. 6.2.2) improves at high electron energies.

6.3.5 Imaging of Single Atoms in the STEM Mode

It was shown in Sect. 4.2.2 that small electron probes can be obtained only with large probe apertures α_p. The theorem of reciprocity (Sect. 4.5.3) indicates that phase-contrast effects can be observed also with $\alpha_\mathrm{p} \gg \alpha_\mathrm{d}$. In the normal STEM mode with $\alpha_\mathrm{p} \simeq \alpha_\mathrm{d} \simeq 10$ mrad, therefore, the illumination is incoherent, which blurs phase-contrast effects (see also Sect. 6.4.4). However, the contrast of atoms can be increased by using the following three signals, all of which can be obtained with a field-emission STEM equipped with an electron spectrometer (Fig. 4.25):

1) The signal I_el generated by the elastically scattered electrons. All electrons scattered through angles larger than the detector aperture α_d, which is of the same order as the probe aperture α_p, are collected. This signal contains a few inelastically scattered electrons but these can be neglected, because inelastic scattering is concentrated at small scattering angles. Similarly, some

of the elastically scattered electrons remain inside the detector cone and pass into the spectrometer where they contribute to the unscattered electron signal I_{un}. We can suppose that 50% of the electrons are scattered inside a cone of aperture θ_0. This characteristic angle θ_0 is tabulated in Table 6.1 and α_d should be appreciably smaller than θ_0. For calculation of I_{el}, the total elastic scattering cross-section σ_{el} (e.g. the approximation (5.47)) can be used.

2) The signal $I_{un} = I_p - (I_{el} + I_{in})$ that corresponds to the unscattered electrons, which pass through the specimen and spectrometer with no energy loss (I_p: probe current)

3) The signal I_{in} is generated by all of the inelastically scattered electrons, with the exception of those scattered through angles larger than α_d; the approximation (5.58) can be used for the total inelastic scattering cross-section σ_{in}.

A homogeneous supporting film (suffix s) contains $N = N_A \varrho / A$ atoms per unit volume and a thin film of thickness t will produce the signals

$$I_{el,s} = \sigma_{el,s} N t I_p \; ; \qquad I_{in,s} = \sigma_{in,s} N t I_p \; . \tag{6.34}$$

The probe current I_p is concentrated within the probe diameter d_p. The current density is therefore $j_p \simeq I_p / d_p^2$. The image of a single heavy atom (suffix a) will also take the form of an error disc of diameter d_p. At its centre, a signal contribution

$$I_{el,a} = \sigma_{el,a} j_p \simeq \sigma_{el,a} I_p / d_p^2 \tag{6.35}$$

will be observed. This relation was verified experimentally for single U, Hg and Ag atoms [6.70]. The contrast of single atoms can be increased by exploiting the fact that only this signal contributes to the high-resolution information. The inelastic scattering of a heavy atom is distributed over a much larger area (Sect. 6.3.2 and [6.71]) owing to the non-localization of inelastic scattering. The signal I_{el} (Fig. 6.18 b) from a supporting film with varying mass-thickness together with isolated individual heavy atoms (Fig. 6.18 a) contains a long-range contribution that depends on the local mass-thickness, in which the contributions of the single atoms of the supporting film overlap and their images are not resolved. In addition, there exists a short-range fluctuation associated with the higher spatial frequencies of the supporting film and with the local increase of elastic scattering at the individual heavy atoms. The inelastic signal I_{in} (Fig. 6.18 c) also contains the long-range variation of mass-thickness but the image of the higher spatial frequencies of the supporting film and that of the single atoms are blurred on account of non-localization. The contrast of single atoms can be increased and filtered by combining the various signals on-line, by use of the following analogue techniques [6.72, 73]:

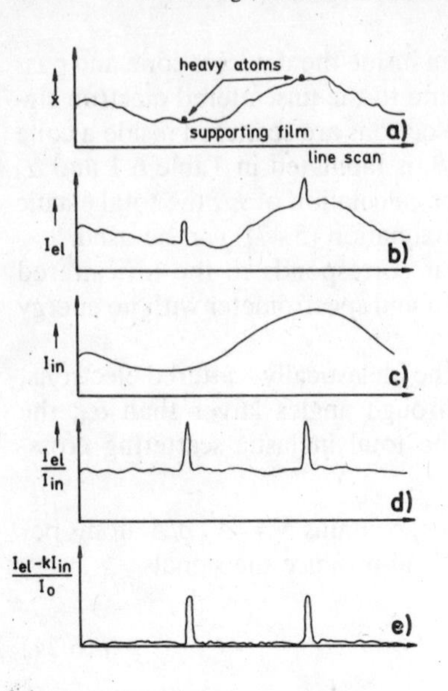

Fig. 6.18. (**a**) Schematic variation of STEM signals of single heavy atoms on a supporting film, (**b**) elastic signal I_{el}, (**c**) inelastic signal I_{in}, (**d**) ratio I_{el}/I_{in} and (**e**) difference signal $(I_{el} - kI_{in})/I_0$

1) The ratio I_{el}/I_{in} renders the contrast, due to the long-range variations of the mass-thickness of the supporting film, uniform (Fig. 6.18d).
2) The difference signal $(I_{el} - kI_{in})/I_0$ can also be used to suppress the long-range variations of mass-thickness (Fig. 6.18e). Divison by the emission current I_0 of the field emission gun eliminates effects due to fluctuations of this current.
3) If two annular detectors are used, the scattering angle between the two detectors can be chosen in such a way that heavy atoms scatter mainly on the outer annular detector. It is now possible to eliminate the short-range fluctuations of mass-thickness from the supporting film.

The following quantitative values for Hg atoms ($Z = 80$) on a carbon ($Z = 6$) substrate [6.72] give an idea of the number of electrons per unit area needed to record a high-resolution STEM micrograph at $E = 40$ keV. The supporting film ($t = 2$ nm, $\varrho = 2$ g cm^{-3}) contains $N_A \varrho t/A = 200$ nm^{-2} carbon atoms, and the probe area is taken to be $d_p^2 = 0.05$ nm^2. This gives

$$I_{el,s}/I_p = 2.9 \times 10^{-2} \qquad I_{el,a}/I_p = 0.13$$

$$I_{in,s}/I_p = 4.4 \times 10^{-2} \qquad I_{in,a}/I_p \simeq 0$$

and the ratio signals become

$$I_{el,s}/I_{in,s} = 0.65 \qquad I_{el,a}/I_{in,s} = 3.3\ .$$

The first ratio will be observed for the pure supporting film and the second when a Hg atom is present. The ratio of these two ratios is the increase of the signal inside the image disc of a Hg atom relative to the background of the supporting film: $I_{el,a}/I_{el,s} = 4.6$. It will be necessary to record about ten electrons per atom in order to form an image disc that can be separated from the background, for which about two electrons are needed per the same area. This implies that $n = 10\, I_p/I_{el} d_p^2 = 1.5 \times 10^3$ electrons nm^{-2} or a charge density of $J = j\tau = ne = 2.5 \times 10^{-2}$ C cm^{-2}. This charge density is already high enough to cause severe damage to organic material; most organic molecules will be destroyed at such charge densities by irreversible radiation damage (Sect. 10.2).

The positions of atoms on carbon substrates are seen to change in a sequence of micrographs [6.74–76] whereas clusters of two or more atoms remain stationary. Examination of biological molecules stained with heavy atoms will be possible in STEM only if the atoms stay at their reaction sites.

6.4 Contrast-Transfer Function (CTF)

6.4.1 CTF for Amplitude and Phase Specimens

The method whereby the imaging properties of an objective lens are described by a contrast-transfer function, independent of any particular specimen structure, was first developed in light microscopy and subsequently applied to electron microscopy by *Hanszen* and coworkers [6.60, 77–79]; for reviews and a full list of the early papers see [6.80–83].

For a specimen with a single spatial frequency q, (6.20) can be written as

$$\psi_s(x) = 1 - \varepsilon_q \cos(2\pi q x) + i\varphi_q \cos(2\pi q x) + \dots \,. \tag{6.36}$$

The Fourier transform $F(q)$ of $\psi_s(x)$ consists of two diffraction maxima of order ± 1:

$$F(\pm q) = \frac{1}{2}(-\varepsilon_q + i\varphi_q)\,. \tag{6.37}$$

Equation (6.21) simplifies to a sum over the amplitudes of the primary beam and the two diffracted beams:

$$\begin{aligned}\psi_m(x') &= 1 + \sum_{\pm q} \frac{1}{2}(-\varepsilon_q + i\varphi_q)\, e^{-iW(q)} e^{2\pi i q x'} \\ &= 1 + (-\varepsilon_q + i\varphi_q)\, e^{-iW(q)} \cos(2\pi q x')\,.\end{aligned} \tag{6.38}$$

The image intensity becomes

$$
\begin{aligned}
I(x') &= |\psi_m(x')|^2 \\
&= 1 - 2\cos W(q)\,\varepsilon_q \cos(2\pi q x') + 2\sin W(q)\;\varphi_q \cos(2\pi q x') + \dots \\
&= 1 - D(q)\qquad \varepsilon_q \cos(2\pi q x') - B(q)\qquad \varphi_q \cos(2\pi q x')\,.
\end{aligned} \tag{6.39}
$$

The factor of the term in ε_q is the CTF of the amplitude structure of the specimen

$$
D(q) = 2\cos W(q)\,. \tag{6.40}
$$

Similarly, the factor of the term containing φ_q is the CTF of the phase structure

$$
B(q) = -2\sin W(q) = -2\sin\left[\frac{\pi}{2}\,(C_s\lambda^3 q^4 - 2\,\Delta z \lambda q^2)\right]. \tag{6.41}
$$

The sign of $B(q)$ is chosen so that $B(q) > 0$ for positive phase contrast. Equation (6.41) can be written in terms of the reduced coordinates (3.69) and (3.70)

$$
B(\theta^*) = -2\sin W(\theta^*) = -2\sin\left(\frac{\theta^{*4}}{4} - \frac{\theta^{*2}}{2}\,\Delta z^*\right). \tag{6.42}
$$

We discuss only the more important case of CTF for phase structures. Figure 6.19 shows the CTF $B(\theta^*)$ for three values of the reduced defocus $\Delta z^* = 1$, $\sqrt{3}$, $\sqrt{5}$ and for the neighbouring values of Δz^* indicated in Fig. 6.19 as a function of the reduced angular coordinate θ^* (3.69). The ideal CTF would take the value $B(q) = 2$ for all q. The real CTFs shown in Fig. 6.19 pass through zero at certain points, around which there are transfer gaps; for the corresponding values of θ^* or q, no specimen information reaches the image. Further spatial-frequency-transfer intervals are obtained with negative values of $B(q)$, which means imaging with negative phase contrast for the corresponding range of q. With negative phase contrast, the maxima and minima in the image of a periodic structure are interchanged compared to those seen with positive phase contrast. Broad bands of spatial frequencies (main transfer bands) with the same sign of CTF are expected when the minimum of $W(q)$ in Fig. 3.16 is an odd multiple of $-\pi/2$. The main transfer bands are indicated in Fig. 6.19 by arrows. These transfer bands become somewhat broader if the underfocus is increased slightly beyond the values $\Delta z^* = \sqrt{n}$ (see CTFs with a central dip in Fig. 6.19). In focus ($\Delta z^* = 0$), there is no main transfer band. This is also the case for overfocus ($\Delta z^* < 0$), for which the oscillations of CTF are more frequent.

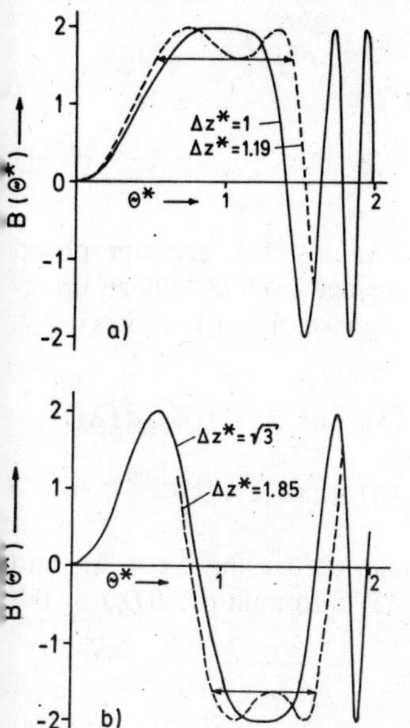

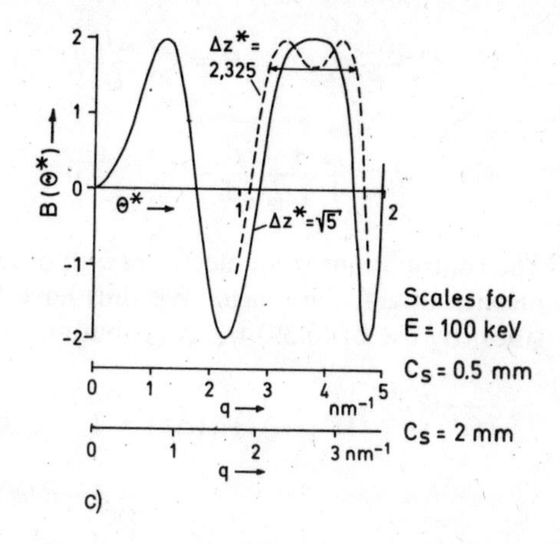

Fig. 6.19 a–c. Contrast-transfer functions $B(\theta^*) = -2\sin W(\theta^*)$ for weak-phase specimens in reduced coordinates $\theta^* = (C_s/\lambda)^{1/4}\,\theta$ for various values of reduced defocus $\Delta z^* = (C_s\lambda)^{-1/2}\Delta z$. The arrows indicate the main transfer intervals

.4.2 Influence of Energy Spread and Illumination Aperture

Ve assumed in Sect. 6.4.1 that a) the electron beam is monochromatic (tem-oral coherence) and b) the incident wave is plane or spherical (point source spatial coherence). In reality, the electron-emission process gives a beam ith an energy width of $\Delta E = 1$–2 eV for thermionic and 0.2–0.3 eV for eld-emission guns (Sect. 4.1.2), and the electron source (crossover) has a nite size, corresponding to an illumination aperture α_i. So long as $\alpha_i \ll \alpha_o$, e illumination is said to be (spatially) partially coherent; when α_i and α_o are f the same order, the illumination becomes incoherent. The variations of lectron energy ΔE as well as those of the acceleration voltage and the lens urrents ΔU and ΔI respectively result in variations Δf of the defocusing 2.53). The influence on CTF has been investigated in [6.84]. The energy read can be approximated by a Gaussian distribution

$$j(\Delta f) = \frac{2\sqrt{\ln 2}}{\sqrt{\pi}H}\exp\left[-\ln 2\left(\frac{\Delta f}{H/2}\right)^2\right], \tag{6.43}$$

hich is normalized so that $\int_{-\infty}^{+\infty} j(\Delta f)\,d(\Delta f)$ is equal to unity and has the full idths at half-maximum:

$$H = C_c \frac{\Delta E}{E} f_r\,; \quad H = C_c \frac{\Delta U}{U} f_r \quad \text{or} \quad H = 2\,C_c \frac{\Delta I}{I} f_r\,,$$

$$\text{where} \quad f_r = \frac{1 + E/E_0}{1 + E/2E_0}\,. \tag{6.44}$$

The contributions from electrons with different values of Δf are superposed incoherently at the image. We thus have to average over the image intensities. By use of (6.39) the contribution from a phase object becomes

$$\begin{aligned}\overline{I(x')} &= \int_{-\infty}^{+\infty} I(x')\,j(\Delta f)\,d(\Delta f) = 1 - \varphi_q \cos(2\pi q x') \int_{-\infty}^{+\infty} B(q)\,j(\Delta f)\,d(\Delta f)\\ &= 1 - B(q)\,K_c(q)\,\varphi_q \cos(2\pi q x')\,. \end{aligned}\tag{6.45}$$

$B(q) = -2\sin W(q)$ in (6.41) contains the mean defocusing $\Delta z = \Delta f$, and the result of the integration (averaging) in (6.45) is to multiply $B(q)$ by the envelope function

$$K_c(q) = \exp\left[-\left(\frac{\pi\lambda q^2 H}{4\sqrt{\ln 2}}\right)^2\right], \tag{6.46}$$

which depends only on q and not on $B(q)$. The function $K_c(q)$ therefore acts as an envelope function; it damps the CTF oscillations for increasing q (Fig. 6.20). The contrast transfer of low spatial frequencies will not be affected because the spatial frequency appears in the exponent of (6.46) to the power 4. We can define a limiting spatial frequency $q_{max} = 1/\Lambda_{min}$ for which $K_c(q) = 1/e = 37\%$. The exponent in (6.46) then becomes unity. Solving for Λ_{min} gives

$$\Lambda_{min} = \left(\frac{\pi\lambda H}{4\sqrt{\ln 2}}\right)^{1/2}. \tag{6.47}$$

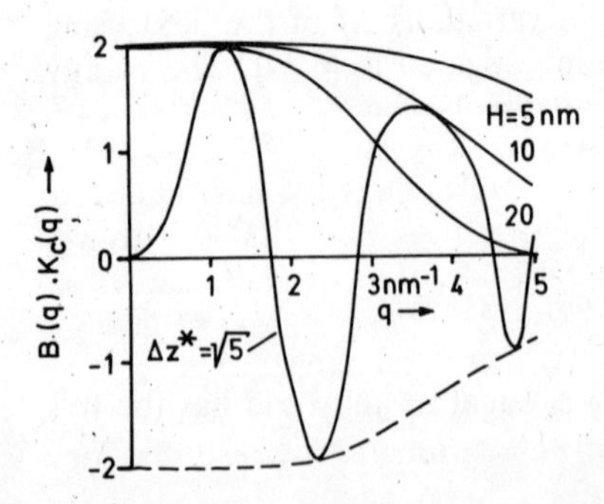

Fig. 6.20. Envelope $K_c(\theta^*)$ of the contrast transfer function $B(\theta^*)$ for different values of the parameter H. Values of $H = 5$, 10, 20 nm correspond to $\Delta E = 1, 2$ and 4 eV, respectively for $E = 100$ keV, $C_s = C_c = 0.5$ mm

Numerical example: For $E = 100$ keV, $C_c = 1$ mm, $\Delta E = 1$ eV or $\Delta I/I = 5 \times 10^{-6}$, we find $\Lambda_{min} = 0.2$ nm ($q_{max} = 5\ \mathrm{nm}^{-1}$). To obtain this resolution, the half-width of the energy spread, of the accelerating voltage or of the objective lens current must not exceed these values. A Gaussian distribution is only an approximation to the true energy-spread distribution. In reality an asymmetric distribution similar to a Maxwellian distribution should be used for the energy-spread function. Calculations show that this asymmetry has little effect [6.84].

If a finite electron-source size and hence a finite illumination aperture is used, many of the electrons in a supposedly parallel beam in fact travel at an oblique angle to the optic axis; this angle is characterized by an angular coordinate $s = \theta/\lambda$. The action on CTF is discussed in [6.85–87]. Each spatial frequency q produces diffraction maxima of order ± 1 on either side of the primary beam. The diffraction maxima with angular coordinates $\boldsymbol{q} + \boldsymbol{s}$ will pass through the objective lens with different phase shifts from those for a central beam, for which $s = 0$. For small values of s, the phase shift can be described by the first term of a Taylor series

$$W(\boldsymbol{q} \pm \boldsymbol{s}) = W(\boldsymbol{q}) \pm \nabla W(\boldsymbol{q}) \cdot \boldsymbol{s} + \ldots \qquad \text{where } \nabla \text{ is the gradient.}$$

Equation (6.39) now becomes

$$\begin{aligned} I(x') &= 1 - \varphi_q \cos(2\pi q x')[\sin W(\boldsymbol{q} + \boldsymbol{s}) + \sin W(\boldsymbol{q} - \boldsymbol{s})] \\ &= 1 - \varphi_q \cos(2\pi q x')\{2 \sin W(\boldsymbol{q}) \cos[\nabla W(\boldsymbol{q}) \cdot \boldsymbol{s}]\} . \end{aligned} \tag{6.48}$$

If a two-dimensional Gaussian distribution is assumed for the s values, so that

$$j(s) = \frac{\ln 2}{\pi H^2} \exp\left[-\left(\frac{s}{H}\right)^2 \ln 2\right] ; \quad H = \frac{\alpha_i}{\lambda} ; \quad 2\pi \int_{-\infty}^{+\infty} j(s)\, s\, ds = 1 \tag{6.49}$$

hen averaging over all s as in (6.45) again results in an envelope function

$$K_s(q) = \exp\left(-\frac{[\nabla W(q)]^2 H^2}{4 \ln 2}\right) = \exp\left(-\frac{(\pi C_s \lambda^2 q^3 - \pi \Delta z q)^2 \alpha_i^2}{\ln 2}\right) . \tag{6.50}$$

Jnlike the envelope $K_c(q)$ (6.46), which depends only on q, $K_s(q)$ depends lso on the illumination aperture and defocusing. After first decreasing, $K_s(q)$ passes through a minimum and rises again to unity where $W(q)$ eaches a minimum and hence $\nabla W(q) = 0$. The main transfer bands, in hich $\nabla W(q)$ is small over a wide range of spatial frequencies, will be fluenced least, therefore (Fig. 6.21, curves with $\alpha_i^* = 0.09$). The attenua-on of CTF by the envelope $K_s(q)$ can be confirmed by laser diffraction Sect. 6.4.6) [6.88, 89].

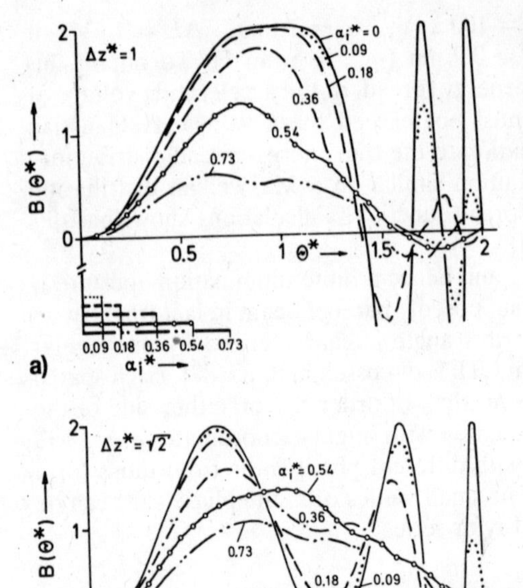

Fig. 6.21 a, b. Contrast-transfer functions for phase contrast at the defocus values (**a**) $\Delta z^* = 1$ and (**b**) $\Delta z^* = \sqrt{2}$ and increasingly large reduced illumination apertures $\alpha_i^* = (C_s/\lambda)^{1/4}\alpha_i$ [6.84]

Under usual conditions of TEM in the presence of both energy spread and finite source size ($\Delta E \leq 2\,\mathrm{eV}$, $\alpha_i \leq 1$ mrad), the effective envelope can be approximately written as a product of envelope functions $K_c(q)$ and $K_s(q)$ which describe the effects of energy spread and illumination spread separately [6.87]. For larger values of α_i or of the reduced aperture $\alpha_i^* = \alpha_i(C_s/\lambda)^{1/4}$, Fig. 6.21 shows the numerical results for $\Delta z^* = 1$ and $\sqrt{2}$ [6.84]. An envelope representation is no longer possible; $B(\theta^*)$ is now damped inside the main transfer intervals, as well. For $\Delta z^* = \sqrt{2}$, for example, and a large illumination aperture, $B(\theta^*)$ has a broad interval of equal sign but with reduced amplitude, of the order 1 instead of 2 for the maxima at low α_i.

6.4.3 CTF for Tilted-Beam and Hollow-Cone Illumination

In the axial illumination mode, each spatial frequency contributes with the diffraction maxima of order ±1 (double-sideband transfer). The superposition of the primary beam and the two sidebands is responsible for the gaps in CTF. A tilted-beam illumination with the primary beam near the centre objective diaphragm cuts off one sideband (single-sideband transfer); in one direction (across the aperture), twice the maximum spatial frequency for axial illumination can be transferred. If this extended transfer is to be achieved in more than one direction, several micrographs must be recorded with different azimuths of the tilted beam. This method is also of interest fo three-dimensional reconstruction (Sect. 6.5.6). The superposition of several exposures with a range of tilted-beam-illumination azimuths is of special interest because this is equivalent to hollow-cone illumination. Single-sideband transfer can also be achieved with axial illumination by using a shifted circular diaphragm or a specially designed half-plane diaphragm (see single-sideband holography in Sect. 6.5.2). A disadvantage of all of these modes is that the primary-beam spot passes near the diaphragm, which can introduce unreproducible phase shifts due to local charging. Tilted-beam or hollow-cone methods that do not require a physical diaphragm or can function with one of larger diameter will therefore be of interest. Some important properties and advantages of these non-standard operation modes will now be discussed in detail.

In the tilted-beam illumination mode, an extended range of spatial frequencies is transfered without a transfer gap but with a variable phase difference between the primary beam and the diffracted beam, caused by the difference between $W(|\boldsymbol{\theta}|)$ for $\boldsymbol{\theta} = \boldsymbol{\alpha}$ corresponding to the direction (tilt) of the primary beam and for $\boldsymbol{\theta} = \boldsymbol{\alpha} + \boldsymbol{q}/\lambda$, corresponding to the diffracted beam. In axial illumination, the images of single atoms and small particles are surrounded by concentric Fresnel fringes of Airy-disc-like intensity distributions (Figs. 3.17 and 6.16). With tilted-beam illumination, the fringe system is asymmetric with bright and dark central intensities depending on defocus and aberrations. The different phase shifts $W(\theta)$ create different lateral shifts of the corresponding specimen periodicities or Fourier components in the image. Thus, particles typically show an asymmetric contrast with bright and dark intensities on opposite sides, which resembles oblique illumination with light (pseudo-topographic contrast).

The contrast transfer of tilted-beam illumination is linear for weak amplitude and phase objects. The appropiate CTFs have been calculated and discussed in [6.82, 85, 90–95], among others. The effect of partial spatial coherence (finite illumination aperture) can be expressed as an envelope function only to a first approximation. For partial temporal coherence (energy spread of the electron gun), an important finding with tilted-beam illumination is the existence of an achromatic circle [6.96]. Whereas the envelope $K_c(q)$ for axial illumination shows a rapid decrease at the resolution limit, the envelope function for tilted-beam illumination increases again to a maximum for q values around twice the resolution limit for axial illumination. This means that conditions can be found for which the gain of resolution offered by tilted-beam illumination by a factor of about two compared to axial illumination is not counteracted by the deleterious effect of chromatic aberration [6.97].

Hollow-cone illumination can be produced with an annular condenser diaphragm. However, a large fraction of the electron beam is absorbed and it is better to move a beam of low aperture around a cone by exciting a two-stage deflection system [6.20]. This can, in practice, be reduced to superposition of a limited number of exposures for example, eight different azimuths

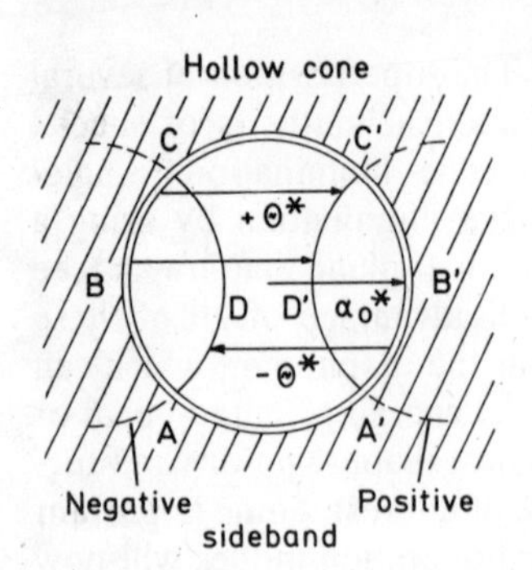

Fig. 6.22. Hollow-cone illumination with a reduced aperture $\alpha_o^* = (C_s/\lambda)^{1/4}\alpha_o$. Electrons scattered through angles θ^* can pass through the diaphragm only within the segments ABC and A′B′C′

around a cone with half-angle (tilt) of 10 mrad for an illumination aperture of $\simeq 0.1$ mrad [6.98, 99]. The asymmetric fringe systems of the tilted-beam illumination are cancelled and the granular contrast of supporting films also decreases, whereas the central contrast of stronger phase objects will be the same for all different azimuths.

The theory of hollow-cone illumination is discussed in more detail in [6.85, 100–103]. Figure 6.23 shows calculated CTFs for phase structures at a reduced defocus $\Delta z^* = 1$ and different values of $\alpha_o^* = \alpha_o (C_s/\lambda)^{1/4}$ [6.85]. The dashed line B_{id} is the CTF for an ideal lens without aberrations, calculated on the assumption that the directions around the hollow cone or the discrete number of tilt angles superpose incoherently. A given spatial frequency with the scattering angle $\theta^* \leq 2\alpha_o^*$ can be transferred only by beams on the arcs ABC and A′B′C′ of the hollow cone (Fig. 6.22), and CTF becomes proportional to the ratio of these arcs to the total arc-length 2π of the hollow beam

$$B_{id}(\theta^*) = \frac{4\arccos(\theta^*/2\alpha_o^*)}{2\pi} = \frac{2}{\pi}\arccos\left(\frac{\theta^*}{2\alpha_o^*}\right). \tag{6.51}$$

In the presence of a wave aberration ($W(\theta^*) \neq 0$), the phase shift $W(\theta^*)$ is not uniform over the arcs ADC and A′D′C′. Because $|\sin W(\theta^*)| \leq 1$, all the CTFs clearly lie below $B_{id}(\theta^*)$. The curve for $\alpha_o^* = 1.49$ in Fig. 5.23 shows that the ideal curve B_{id} is approached very closely for $\Delta z^* = 1$ with $q^*_{max} = 2\alpha_o^* = 2.98$. This CTF may be compared with that for axial illumination in Fig. 6.19 a. CTFs of hollow-cone illumination in Fig. 6.23 do not show any contrast reversals. The only disadvantage is that $B(\theta^*)$ in Fig. 6.23 reaches a maximum value of only 0.8, as compared to 2 for axial illumination.

Figure 6.24, based on a calculation of *Rose* [6.102], shows that hollow-cone illumination is also applicable without an aperture-limiting diaphragm. The defocusing is $\Delta z^* = 1/\sqrt{2}$ and the quantities α_1^* and α_2^* are the inner and outer apertures of the cone of finite width. This optimum condition differs from that of Fig. 6.23 mainly in that the narrow regions of negative sign occur for small and large θ^*, and in the presence of ripple oscillations on CTF. The optimum conditions are those in which the phase contrast reinforces the scattering contrast.

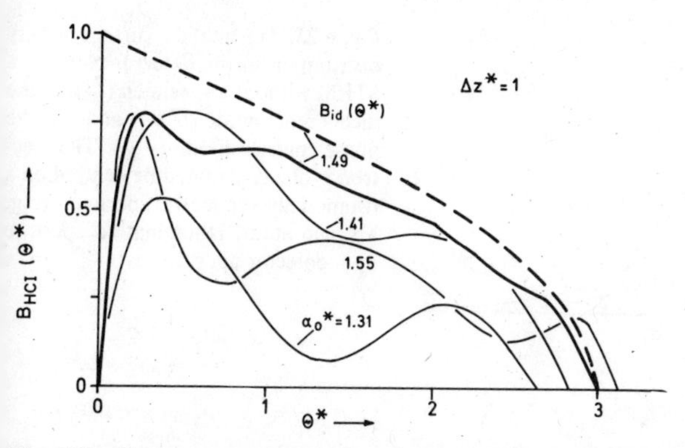

Fig. 6.23. Phase-contrast-transfer function for hollow-cone illumination for different values of α_o^* and a reduced defocus $\Delta z^* = 1$. (---) ideal CTF which is proportional to the length of the segments ABC and A′B′C′ in Fig. 6.22 [6.84]

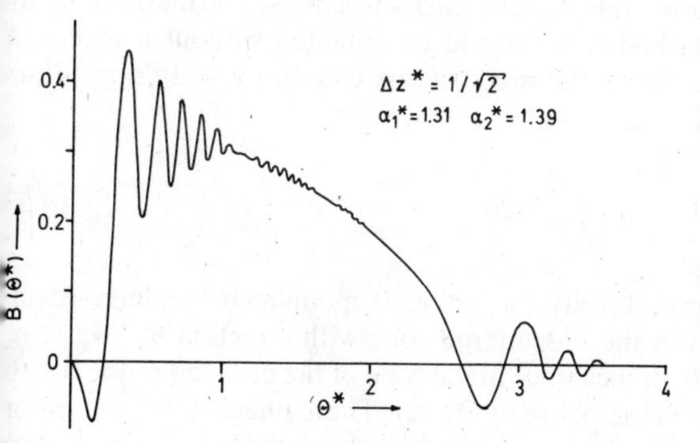

Fig. 6.24. Example of hollow-cone illumination phase CTF with no objective diaphragm, for a reduced defocus $\Delta z^* = 1/\sqrt{2}$ and a broad cone of illumination of inner and outer diameters α_1^* and α_2^* [6.102]

6.4.4 Contrast Transfer in STEM

As discussed in Sect. 4.5.3, the phase-contrast effects in the STEM mode will be the same as in TEM if the corresponding apertures are interchanged: $\alpha_d = \alpha_i \simeq 0.1$ mrad and $\alpha_p = \alpha_o \simeq 10$ mrad (Fig. 6.8 a, b). However, only a small fraction of the incident cone would be collected by the detector if these conditions were employed. There are two collection possibilities: either the electrons inside the cone of aperture $\alpha_d = \alpha_p$ are collected (Fig. 6.8 c) or an annular detector, which collects only the elastically scattered electrons, is used (Sect. 6.3.5). These modes can be described as bright and dark field respectively. The recorded intensities will be complementary:

$$I_B(r) = I_0 - I_D(r) , \tag{6.52}$$

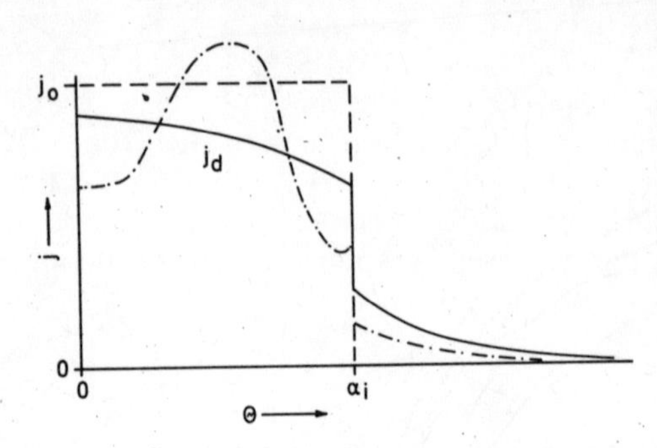

Fig. 6.25. Schematic current-density distribution in the detector plane of a STEM without a specimen (---), in the incoherent mode (—) and in the phase-contrast mode (-·-·-). The electron probe is assumed be centred on a rotationally symmetric specimen (e.g. a single atom) (Illumination aperture α_i = detector aperture α_d)

where I_0 denotes the intensity recorded without a specimen. A detector tha collects the whole illumination cone produces an incoherent bright-fielc image if interference effects between the scattered and unscattered wave need not be considered. This is merely conservation of the number of elec trons. Figure 6.25 shows the various current-density distributions in the detector plane. The dashed curve would be obtained without a specimen whereas the full curve shows the modification caused by scattering. Equa tion (6.52) means that

$$2\pi \int_0^{\alpha_d} (j_0 - j_d)\, \theta \, d\theta = 2\pi \int_{\alpha_d}^{\infty} j_d \theta \, d\theta \tag{6.53}$$

However, the current density for $\theta \leq \alpha_d$ is modulated by phase effects Each point (direction) of the unscattered cone with direction $\boldsymbol{\theta}_u$ ($|\boldsymbol{\theta}_u| < \alpha_d$ corresponds to an angle of incidence in the cone of the electron probe but th probe-forming lens shifts the phase by $W(\theta_u)$. These phase shifts are respor sible for the shape of the electron probe and the deviations from an Airy disc-like probe profile. The intensity at each point of the detector plane is th result of interference between the unscattered wave of direction $\boldsymbol{\theta}_u$ and wave elastically scattered into this direction with a scattering angle $\boldsymbol{\theta}_s$. Th elastically scattered wave experiences a phase shift of $\pi/2$ during the scatte ing process (Sect. 5.1.3) and an additional phase shift relative to the unsca tered wave: $W(|\boldsymbol{\theta}_u - \boldsymbol{\theta}_s|) - W(\theta_u)$. For a fixed direction $\boldsymbol{\theta}_u$, the result superposing all possible scattering angles $\boldsymbol{\theta}_s$ with the condition $|\boldsymbol{\theta}_u - \boldsymbol{\theta}_s| <$ has to be evaluated. This leads to a modulation of the current-density di tribution inside the illumination cone, drawn schematically as the dash-dc ted line in Fig. 6.25 and consisting of zones of decreased and increase intensity.

Phase contrast therefore can be produced by dividing the detector pla into annular zones [6.104], which collect electrons with mainly constructi or mainly destructive interference. A single narrow annular detector is t reciprocal of hollow-cone illumination in TEM. Hence, CTF of this STE mode has a triangular shape, modified by phase-contrast effects similar to those of Fig. 6.23.

Another possible phase-sensitive detector consists of two semicircular discs, separated by a narrow gap, normal to the scan direction [6.102, 105–107]. This is capable of giving differential phase contrast, which represents the gradient of the object parallel to the scan direction. An obvious extension from semicircles to quadrants gives the two components of the gradient [6.108, 109].

6.4.5 Improvement of CTF Inside the Microscope

The transfer gaps and changes of sign caused by spherical aberration render the electron-optical CTF very different from the ideal CTF with $B(q) = +2$; the latter can be attained in the light-optical phase-contrast method in which a Zernike plate shifts the phase by $+\pi/2$ in the focal plane of the objective lens. Similar methods have been proposed for the TEM, too.

Hoppe [6.110] suggested that a plate consisting of rings alternatively transparent and opaque to electrons could be used to suppress spatial frequencies or scattering angles that are transferred with a negative sign in $B(q)$; a related proposal was made by *Lenz* [6.111]. Such a *zone plate* can also be used in the optical filtering of micrographs [6.112]. Calculations show that no significant improvements can be expected, because of the broad gaps in the CTF that correspond to spatial frequencies transferred with the wrong sign and consequently suppressed [6.46]. *Hanszen* [6.81] proposed the use of two complementary zone plates, which would cover the whole CTF without gaps for two different values of defocus.

Phase shifts can be generated by means of a carbon film of uniform thickness with a hole for the primary beam [6.113] or with a contamination spot created by the primary beam [6.114]. In another proposal, a phase shift is generated by allowing a fibre stretched across the diaphragm to become electrically charged [6.115, 116]. Profiled phase plates of variable thickness [6.78] can be produced by electron-beam writing or by growing a contamination layer with the required local thickness on a carbon film supported by the diaphragm [6.117–119]. Calculations showed that electron scattering in the phase plate does not influence the imaging significantly [6.120]. Indeed the transfer gaps in the CTF vanish, as shown by laser diffraction [6.121]. However, no practical examples of image improvement have yet been reported.

All of these interventions in the focal plane, including single-sideband holography (Sect. 6.5.2), have the disadvantage that the diaphragm, some 100 μm in diameter, has to be adjusted precisely on axis in the focal plane, and that, whenever the electron beam strikes the transparent or opaque part of the diaphragm, charging can occur, which influences the phase shift unpredictably. For these reasons, none of these a priori correction methods is in practical use. The present tendency is to apply a posteriori correction methods to the final micrographs (Sects. 6.5.4 and 5).

6.4.6 Control and Measurement of the CTF by Optical Diffraction

For a weak-phase specimen, a specimen periodicity Λ or spatial frequency $q = 1/\Lambda$ is linearly transferred to the image as a periodicity ΛM with an amplitude proportional to $|B(q)|$. The periodicities in the micrograph can be analyzed by applying light-optical Fraunhofer diffraction to the photographic record.

Figure 6.26 shows how an optical bench is organized for recording diffractograms. A He-Ne laser with $\lambda_L = 632.8$ nm is used as the light source for the following reasons. A periodicity ΛM produces its first diffraction maximum at the diffraction angle $\sin\theta \simeq \theta = \lambda_L/\Lambda M$. Photographic emulsions have a resolution of about 30 μm under electron bombardment (Sect. 4.6.2) corresponding to a maximum possible diffraction angle $\theta_{max} = 0.6328$ μm / 30 μm = 20 mrad. The largest periodicity that should be detectable in the diffractogram of the micrograph is of the order of 1 mm, which requires a minimum detectable diffraction angle $\theta_{min} = 0.6$ mrad. The image structures, with periodicities ΛM, have to be matched to this interval between 30 μm and 1 mm by means of a suitable magnification M. The Fraunhofer diffraction of developed emulsions, therefore, shows pronounced forward characteristics. A He-Ne laser emits an approximately parallel light beam with a diameter of about 1 mm and a divergence angle (aperture) of 0.4 mrad, which is sufficient to resolve diffraction maxima at an angle θ_{min} from the primary beam. A parallel beam with such a small aperture can be obtained with a conventional light source only by demagnification of the latter and placing a very small diaphragm in the focal plane of a collimator lens, which wastes much of the light. Exposure times of the order of several minutes are needed, whereas a He-Ne laser needs only a fraction of a second and the diffractogram can be observed directly on a screen.

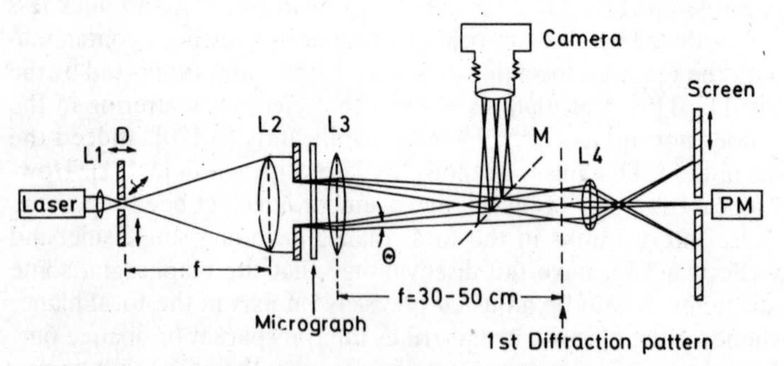

Fig. 6.26. Fraunhofer diffraction arrangement for recording optical diffractograms of electron micrographs

The central mode of the laser is selected by a lens L_1 followed by a 10–20 μm diaphragm. Lens L_2 generates a parallel beam of large diameter, and the lens L_3 produces a Fraunhofer-diffraction pattern in its focal plane. The focal length f is 30–50 cm, to get a diffractogram of large diameter. The diffractogram can be either photographed or enlarged by means of an additional lens L_4 ($f \simeq 5$ cm) on a white-paper screen for visual observation or for measurement of the radial intensity distribution of the diffractogram with a slit and a photomultiplier.

The specimen periodicity Λ or spatial frequency $q = 1/\Lambda$ can be calibrated by diffraction from a photograph of a sheet of graph paper. The micrograph is placed between lenses L_2 and L_3. Most photographic emulsions show not only a density variation but also a variation of optical thickness nt, resulting in an additional phase shift. Thus the emulsion Gevaert Scientia D-56 shows a phase-modulation amplitude $\Delta\varphi = 1.5\ \pi\Delta D$ when exposed to a sine-wave modulation with a density amplitude ΔD. Calculations and experiments show that the contribution from the phase shift results in a 10–20 times greater intensity at the diffractogram than a pure amplitude modulation caused by density variations [6.122]. The influence of the phase shift can be eliminated by immersing the micrograph in a cuvette containing a mixture of benzene and CS_2, with the same refractive index n as the gelatin of the emulsion [6.123]. Alternatively, the emulsion can be bleached, and the diffractogram is then generated by the pure phase modulation of the transmitted light. A sine-wave density modulation only results in diffraction maxima of order ± 1 whereas a combination of density and phase modulation produces higher diffraction orders. For more-quantitative work and for optical-reconstruction experiments (Sect. 6.5.4) phase compensation in a cuvette is essential but for visual observation of the diffractogram, the higher brightness of the diffractogram caused by the phase modulation is advantageous. If the diffracted intensity of the emulsion used for electron exposure is very weak, the variations of the optical thickness being small, the micrograph can be copied onto a emulsion (such as that used for slides, for example) with a stronger phase modulation.

CTF can be controlled and measured with a specimen showing a white noise of spatial frequencies, which implies that the amplitudes φ_q in (6.36) must be independent of q. This is nearly true for thin carbon supporting films, as already shown in Sect. 6.2.2. The variation of the image intensity $I(x')$ is then proportional to $|B(q)|\,\varphi_q$ (6.39). The amplitude of the light transmitted by the micrograph is proportional to the density D (Sect. 4.6.2), which is in turn proportional to $I(x')$. The intensity in the diffractogram, therefore, will be proportional to $|B(q)|^2$. A typical diffractogram (Fig. 6.27 a) shows the gaps (zero points) of $B(q)$. However, diffraction maxima that belong to regions of $B(q)$ of different sign cannot be distinguished. This requires comparison with formulae such as (6.41).

The following information can be obtained from a diffractogram:

1) The q values that correspond to the maxima in the diffractogram can be measured and plotted for a defocus series, as in Fig. 6.11. A diffractogram thus contains information about the defocusing Δz and the spherical-aberration constant C_s. The diffractogram shows maxima of $|B(q)|^2$ if $W(q) = n\pi/2$ if n is odd. The gaps of constant transfer result if n is even. By use of (3.67) for $W(q)$, the relation $W(q) = n\pi/2$ can be transformed to

$$C_s \lambda^3 q^2 - 2\Delta z \lambda = n/q^2 . \tag{6.54}$$

Plotting n/q^2 versus q^2 results in a straight line if a correct numbering of n is selected. C_s can be read from the slope and Δz from the intersection of the straight line [6.124]. A tilted carbon film contains a whole range of defocus values and it is possible to deduce the dependence on Δz in Fig. 6.11 from a single micrograph [6.125]. It is also possible to determine defocusing distances Δz as large as a few millimetres if a small illumination aperture $\alpha_i \leq 10^{-2}$ mrad is used, as in Lorentz microscopy (Sect. 6.6) [6.126, 127]. In this case, the term that contains the spherical aberration in (6.41) can be neglected and the relation $\Delta z = |2m - 1|/(2\lambda q^2)$, where $m = 1, 2, \ldots$, can be used to determine Δz from the q-values of the maxima in the diffractogram.

2) Astigmatism can be considered in the wave aberration $W(q)$ or CTF $B(q) = -2\sin W(q)$ by adding a term (3.68) to (3.66). This results in an elliptical distortion of the diffraction rings for small astigmatism (Fig. 6.27 a) and a hyperbolic distortion for stronger astigmatism (Fig. 6.27 b).

3) A continuous drift of the image during exposure of the emulsion results in a blurring of the diffractogram parallel to the direction of drift (Fig. 6.27 c); a sudden jump of specimen position during exposure duplicates the entire structure, so that a pattern of interference fringes is superimposed on the main pattern (6.27 d) (see also the next Item 4)

4) The envelopes $K_c(q)$ (6.46) and $K_s(q)$ (6.50) can be determined from the decrease of the diffraction-maxima amplitude for large q. The largest spatial frequency transferred is inversely proportional to the resolution limit and can be read from a diffractogram of two superposed micrographs by the following procedure [6.128, 129]. When the two micrographs are shifted through a small distance d, every resolved structure appears twice in the transmitted-light amplitude, which means that each structure is convoluted with a double source that consists of two points a distance d apart. Using the Fourier convolution theorem (3.51), we see that the diffractogram of a single micrograph will be multiplied by the Fourier transform of a double-point source; the intensity in the diffractogram is hence multiplied by $\cos^2(\pi q d)$. The diffractogram of the superposed micrographs is therefore overprinted with a pattern of interference fringes with a spacing $\Delta q = 1/d$ as in Fig. 6.27 d. The limit of contrast transfer can be seen from the limit of recognizable fringes. It is important to use two successive micrographs and not two copies of one micrograph. In the latter case, fringes can also be produced by

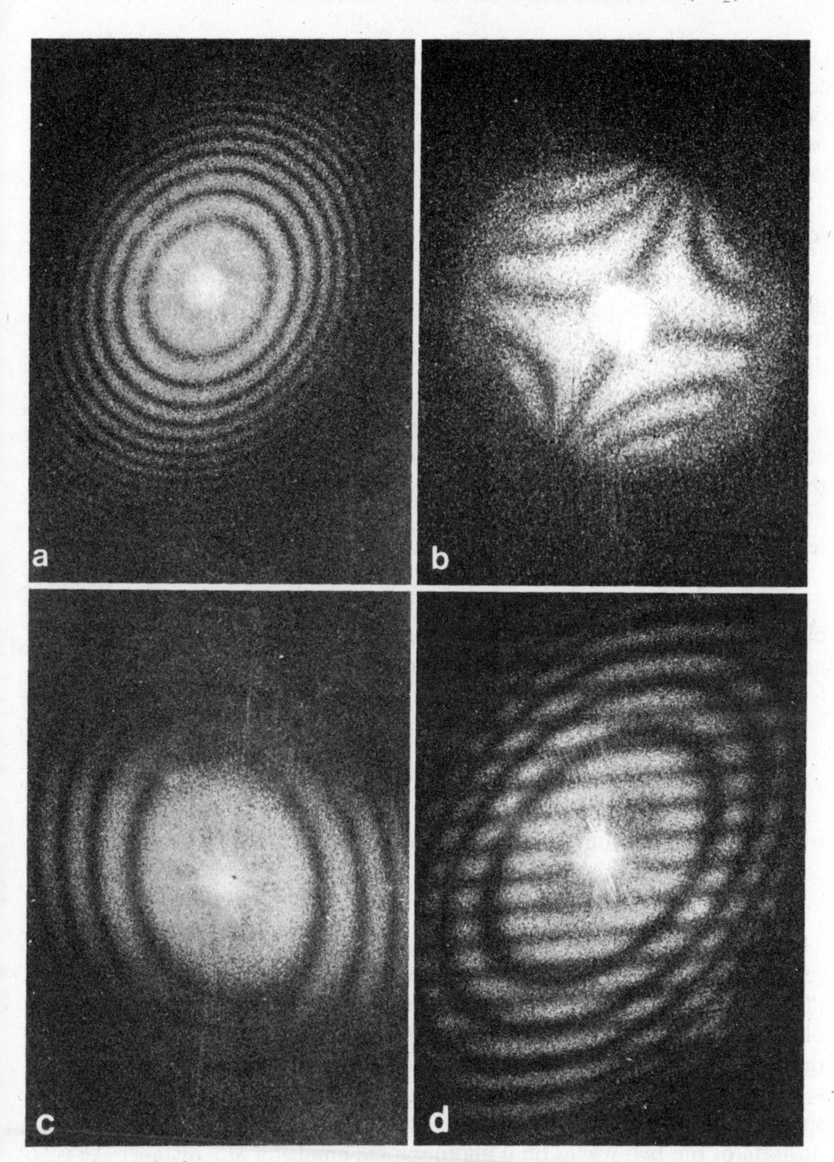

Fig. 6.27 a–d. Diffractograms of micrographs of carbon foils, showing (**a**) the gaps of the contrast transfer and a weak astigmatism (elliptical diffraction rings), (**b**) stronger astigmatism which results in hyperbolic diffraction fringes, (**c**) detection of a continuous specimen drift and (**d**) of a sudden jump in the specimen position during the exposure; the fringe pattern is caused by the doubling of the image structure

clusters of silver grains generated by a single electron and reproduced in both copies. In SEM and STEM, it is necessary to shift the image on the cathode-ray tube between the two exposures by about 1 cm. Otherwise, the fringe pattern may be caused by the granularity of the CRT screen, continuing out to larger spatial frequencies [6.130].

5) The correction methods of the CTF discussed in Sect. 6.4.5 can be controlled by studying diffractograms.

These examples of the application of laser diffraction show the importance of this technique for the control of the imaging process and of the final micrographs. No high-resolution work should be done without using this diffraction technique.

It is a drawback that the diffraction can be done only a posteriori and not a priori, that is, before recording and developing the micrograph. Some attempts have therefore been made to record the electron current density of the final image on a special screen, the optical properties of which change under electron bombardment and from which a diffractogram can be recorded directly by laser diffraction. A thermoplastic film is used [6.131], which converts the current distribution of the image into a thickness profile, for this purpose. Light-optical diffractograms are then generated by the phase differences associated with the varying optical-path lengths. The stored image can be erased by heating the thermoplastic film. The Pockels effect in ferroelectric crystals incorporated in a commercial *Titus tube* has likewise been used [6.132, 133].

6.5 Electron Holography and Image Processing

6.5.1 Fresnel and Fraunhofer In-Line Holography

The idea of holography was first introduced by *Gabor* [6.134] to improve the resolution of the electron microscope by (a posteriori) light-optical processing of micrographs to cancel the effect of spherical aberration. The recording of a hologram and the reconstruction of the wavefront will be described for the case of in-line holography, proposed originally by *Gabor*. The unscattered part of an incident plane wave acts as a reference wave (Fig. 6.28 a). A specimen (object) point P produces a spherical, scattered wave. The superposition of the two waves on a photographic emulsion at a distance Δz is an interference pattern which consists of concentric fringes. The same pattern will be seen if we insert a magnifying-lens system between the object and the micrograph and record with a defocusing Δz. If the lens is not free of aberrations, the interference pattern will be modified by the additional phase shifts.

In the reconstruction (Fig. 6.28 b), the micrograph is irradiated with a plane wave, and acts as a diffraction grating or Fresnel-zone lens. The fringe distance decreases with increasing distance from the centre and the corre-

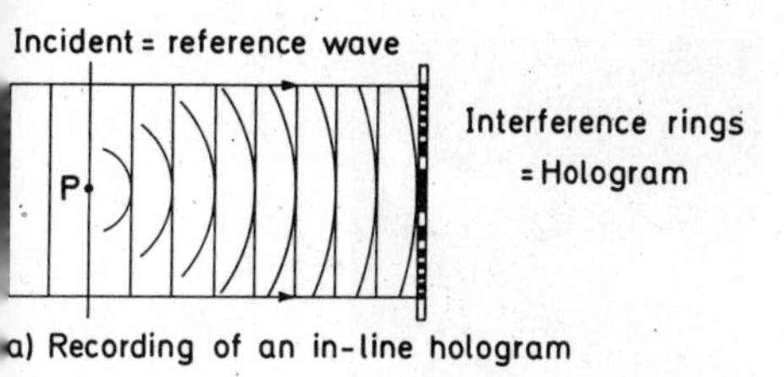

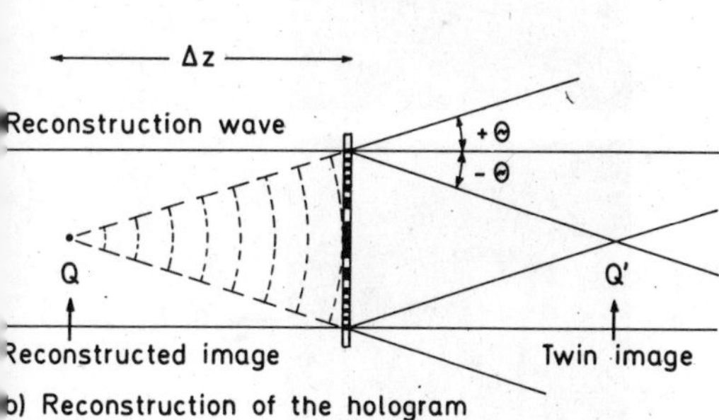

Fig. 6.28. (**a**) Recording and (**b**) reconstruction of an in-line hologram

sponding diffraction angle ($\pm\theta$) increases. The two diffracted waves sidebands) form spherical waves centred on Q and Q′. We see that we econstruct the spherical wave from P = Q behind the hologram. However, ve see, too, that in-line holography has the disadvantage of producing a twin mage at Q′. If we are looking from the right, the distance between Q and Q′ as to be large enough so that if one of the twin images is in focus, the other is lurred. The latter is a Fresnel diffraction pattern with a defocusing $2\Delta z$ elative to the focused twin image, and a single object point is imaged as a veak concentric-ring system with a large inner radius $r_1 \simeq \sqrt{\Delta z \lambda}$. A specinen structure smaller than this radius and situated in a large structure-free rea can be reconstructed without any disturbance from the twin image.

Off-axis points will create an asymmetric fringe system and points in front r behind the object plane behave like Fresnel-zone lenses with smaller or arger fringe diameters, respectively. Each of these fringe systems recontructs a point source in the correct position relative to P. A hologram can herefore produce a full-three-dimensional image of the specimen.

In-line holograms can be further classified into Fresnel, Fraunhofer and ingle-sideband holograms (Sect. 6.5.2). The difference is explained in ig. 6.29, in which a periodic object of lattice constant or spatial frequency $= 1/\Lambda$ is used as an example. Figure 6.29 a shows the superposition of the iffracted waves of order ± 1 (sidebands) and the incident plane wave. In resnel holography (small defocusing Δz), the three waves overlap. In the lanes A, B, C, ..., the three-wave field results in zero intensities for defocus alues $\Delta z = n\Lambda^2/\lambda$ (n integer). These defocus values correspond exactly to e zeros of CTF $B(q) = -2\sin W(q)$ if only the defocus term of $W(q)$ is

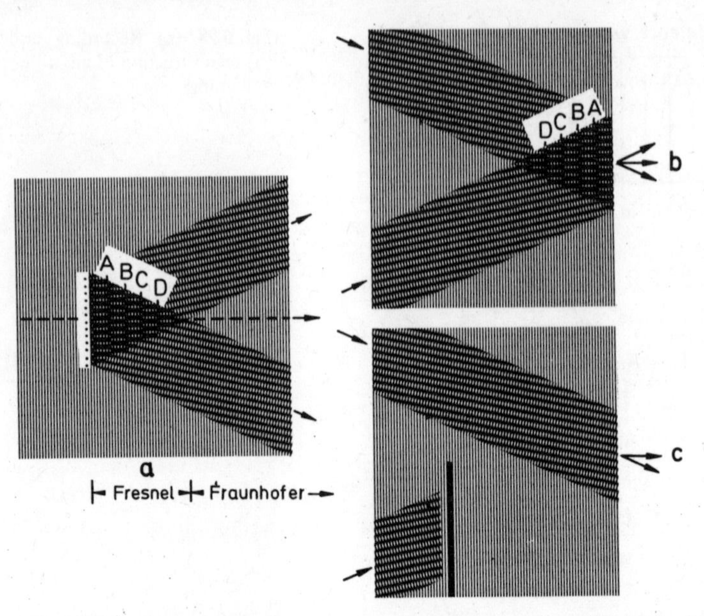

Fig. 6.29. (**a**) Plane incident wave and diffracted waves of order ± 1 in the Fresnel an Fraunhofer regions. (**b**) Overlap of the waves in the image plane. (**c**) Avoidance of the transfe gaps at A, B, C, ... by absorbing one sideband (single-sideband holography)

considered. Near the Gaussian image plane, the three waves overlap to forr a magnified image (Fig. 6.29 b). A magnification M decreases the scatterin angle of the sidebands to $\pm \theta/M$ and increases the fringe (lattice) distance t ΛM. Spherical aberration in the imaging system creates an additional phas shift between the plane incident wave and the two sidebands. The photc graphic emulsion is placed in the Fresnel region for small defocusing. Suc Fresnel in-line holograms were originally proposed by Gabor.

For the light-optical reconstruction of an in-line hologram, an objectiv lens is needed with an appropriately scaled spherical aberration to allow fc the difference between the wavelengths of the electrons used for recordin and that of the light employed for reconstruction (see also Sect. 6.5.4). Th reconstructed image shows a contrast-transfer function $B^2(q)$ because CTF have to be multiplied in such a twofold imaging process. This means that th intervals corresponding to spatial frequencies with a negative sign of $B(q)$ i the original electron-optical image are reproduced with the correct sigı However, the information gaps in CTF cannot be avoided so that if tł photographic emulsion is placed at A, B, ... in Fig. 6.29 b the Fresnel in-lir hologram cannot contain any information about the corresponding spati frequency.

In Fraunhofer in-line holograms, a larger defocusing is used so that the sidebands do not overlap and no transfer gaps occur (Fig. 6.29). This method has been tested in electron microscopy [6.135, 136]. It was found that gold particles smaller than 1 nm can be reconstructed but these are also visible in the normal bright-field mode.

If a specimen area of diameter d_0 is to be recorded and reconstructed, the spatial frequencies present in the spectrum will lie between $q_{min} = 1/d_0$ and q_{max} corresponding to the resolution limit; q_{min} corresponds to a diffraction angle $\theta_{min} = \lambda/d_0$ so that a defocus value

$$\Delta z = \frac{d_0}{\theta_{min}} = \frac{d_0^2}{\lambda} \tag{6.55}$$

will be necessary to separate the primary beam and the sidebands (Fig. 6.29). This defocusing is large even if the diameter of the specimen area d_0 is quite small. A large defocusing causes a blurring of the hologram due to the finite illumination aperture α_i, which sets a limit on the minimum periodicity Λ_{min} or maximum spatial frequency q_{max}

$$\frac{1}{q_{max}} = \Lambda_{min} = \alpha_i \Delta z = \frac{\alpha_i d_0^2}{\lambda} \tag{6.56}$$

The following numerical example, $\Lambda_{min} = 0.2$ nm, $\lambda = 3.7$ pm ($E = 100$ keV), $d_0 = 100$ nm, $\alpha_i = 7.5 \times 10^{-8}$ rad, shows the limitation of in-line holography. Only a small specimen area of diameter d_0 can be imaged and an extremely low aperture α_i is necessary, which can be obtained only with a field-emission gun.

The influence of non-axial aberrations (coma, Seidel astigmatism, field curvature and distortion) render the image formation anisoplanatic [6.137–139]. These aberrations cause a shift and rotation of a transferred specimen sine wave of period Λ. If shifts and rotations of $\Lambda/8$ can be tolerated, the radius of the isoplanatic patch can be estimated to be about 100 nm, which corresponds to a circular area of 4 cm at the final image screen if the magnification M is 4×10^5.

In a light-optical reconstruction of a hologram, the hologram must be either immersed in a cuvette with the same refractive index as the gelatin (pure amplitude object, see also Sect. 6.4.6) or bleached (pure phase object). A light-optical reconstruction of a bleached Fresnel in-line hologram with the conditions (6.62) gives a total CTF $B^2(q)$ whereas a $\lambda/4$ plate for the primary beam will be necessary for an immersed hologram [6.140].

6.5.2 Single-Sideband Holography

Figure 6.29 c shows that a large defocusing to separate the sidebands and hence to avoid the contrast-transfer gaps (transition from Fresnel to

Fraunhofer in-line holography) is not necessary if one of the sidebands is suppressed by a diaphragm [6.141–144]. The best way of doing this is to insert a half-plane diaphragm in the focal plane of the objective lens with a small opening for the primary beam. Charging effects can disturb the image if the primary beam passes too close to the diaphragm [6.145]; this can be reduced either by preparing the diaphragm in a special way [6.146] or by heating it [6.147].

The influence on contrast transfer can be seen from (6.38), retaining only one diffraction order (sideband) instead of both (of order ± 1). The alternative signs in the following formulae correspond to the two possible single-sideband images recorded with complementary half-plane diaphragms; we find

$$\psi_b(x') = 1 - \frac{1}{2}\varepsilon_q \mathrm{e}^{-\mathrm{i}W(q)}\,\mathrm{e}^{\pm 2\pi \mathrm{i}qx'} + \frac{1}{2}\,\mathrm{i}\varphi_q \mathrm{e}^{-\mathrm{i}W(q)}\,\mathrm{e}^{\pm 2\pi \mathrm{i}qx'} \tag{6.57}$$

and hence

$$I_b(x') = \psi_b\psi_b^* = 1 - \varepsilon_q \cos[2\pi qx' \mp W(q)] \mp \varphi_q \sin[2\pi qx' \mp W(q)] + \dots \tag{6.58}$$

This means that there are no transfer gaps, a result that can also be inferred from Fig. 6.29c. The wave aberration produces only a lateral shift of the image-lattice periodicity. The image of the phase component is shifted by $\pi/2$ or $\lambda/4$ even when $W(q) = 0$. The resulting contrast is asymmetric. This is typical of single-sideband imaging. Thus edges appear bright on one side and dark on the other. This asymmetry is reversed when the complementary half-plane diaphragm is used [reversal of the sign of the last term in (6.58)].

The image of the amplitude component in (6.58) remains unshifted. This can be used to separate the amplitude and phase components. The sum of two single-sideband holograms recorded with complementary half-plane diaphragms increases the amplitude component and cancels the phase component, whereas the difference between them cancels the amplitude and increases the phase component. The sum or difference of the holograms can be computed digitally or obtained by superposition of the micrographs by use of a positive and a negative copy for the difference [6.148]. The wave aberration $W(q)$ can be corrected in the light-optical reconstruction by using an objective lens with the appropriate spherical aberration and defocusing (Sect. 6.5.4). The focal plane of this lens again contains two sidebands. With a half-plane diaphragm complementary to that used earlier in the electron-optical imaging, the corrected intensity distributions (6.58) can be obtained free of the lateral shift caused by $W(q)$.

6.5.3 Off-Axis Holography

Off-axis or out-line holography, proposed by *Leith* and *Upatnieks* [6.149], uses a separate reference beam for recording the phase in the interference pattern. The superposition of a reference wave (wave vector $\boldsymbol{k}_1$), which may for example pass through a hole in the support film, and a wave ($\boldsymbol{k}_2$), which transmits the specimen structure off-axis and is modified in amplitude and phase (3.39), can be performed by means of an electrostatic biprism (Sect. 3.1.4) [6.136, 150–152]. With a field-emission source, of the order of a hundred fringes can be recorded. In the narrow region of overlap of the two beams, the wave amplitude is (Fig. 6.30 a)

$$\psi = \psi_0 [a_s(\boldsymbol{r}) \, \mathrm{e}^{\mathrm{i}\varphi_s(\boldsymbol{r})} \, \mathrm{e}^{2\pi \mathrm{i} \boldsymbol{k}_2 \cdot \boldsymbol{r}} + \mathrm{e}^{2\pi \mathrm{i} \boldsymbol{k}_1 \cdot \boldsymbol{r}}] \, . \tag{6.59}$$

The hologram is a record of the bisprism interference fringes, which have an intensity distribution

$$\begin{aligned} I(\boldsymbol{r}) = \psi\psi^* &= I_0 \{1 + a_s^2(\boldsymbol{r}) + 2 a_s(\boldsymbol{r}) \cos[2\pi x/d + \varphi_s(\boldsymbol{r})]\} \\ &= I_0 [1 + a_s^2(\boldsymbol{r}) + a_s(\boldsymbol{r}) \, \mathrm{e}^{\mathrm{i}\varphi_s(\boldsymbol{r})} \, \mathrm{e}^{2\pi \mathrm{i} x/d} + a_s(\boldsymbol{r}) \, \mathrm{e}^{-\mathrm{i}\varphi_s(\boldsymbol{r})} \, \mathrm{e}^{-2\pi \mathrm{i} x/d}] \end{aligned} \tag{6.60}$$

with $x \parallel \boldsymbol{k}_1 - \boldsymbol{k}_2$ and $d = \lambda/2\beta$ is the fringe spacing. The fringes contain the amplitude modulation $a_s(\boldsymbol{r})$ of the specimen wave and the phase shift $\varphi_s(\boldsymbol{r})$ as a local shift of the interference fringes.

The hologram is reconstructed by irradiating it with a coherent light wave (Fig. 6.30 b). The light amplitude behind the hologram can also be represented by (6.60) but the amplitude factors are now modified by the γ-value (Sect. 4.6.2) of the emulsion. Thus, the factor $\exp(2\pi \mathrm{i} x/d)$ in the third term of the last bracket of (6.60) is equivalent to a phase shift by a prism with a deflection angle of $\theta = \lambda_L/d$ (λ_L denotes the wavelength of the reconstruction light wave); for the last term in the bracket, the angle would be $-\theta$. This means that the specimen wave $a_s(\boldsymbol{r}) \exp[\mathrm{i}\varphi_s(\boldsymbol{r})]$ is completely reconstructed

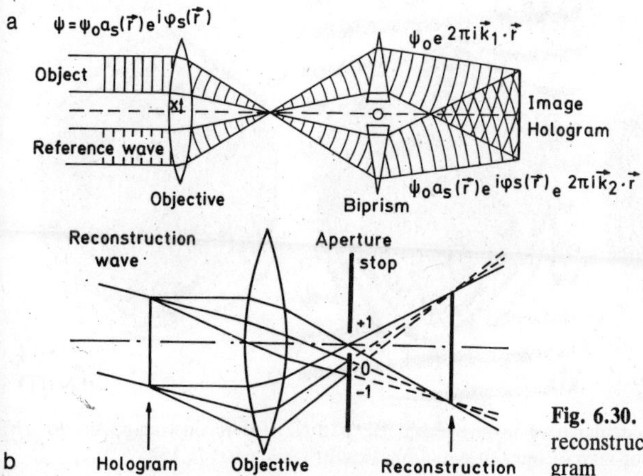

Fig. 6.30. (a) Recording and (b) reconstruction of an off-axis hologram

in the deflected wave (first diffraction order). This first order can be selected by a diaphragm placed in the focal plane of the lens in Fig. 6.30b, thereby suppressing the twin image. Although the wave amplitude is fully restored with the correct phase and its amplitude is modified only by the γ of the recording process, the phase will again be lost in the recorded hologram.

However, the phase can be recovered by splitting the reconstruction wave with a Mach-Zehnder interferometer (see also Fig. 6.32) placed in front of the hologram [6.153, 154]. The two reconstruction waves are inclined at $\pm\theta$ to the axis, and the central beam selected by the diaphragm contains the superposition of diffracted waves of orders ± 1 from the two reconstruction waves; the amplitude is thus

$$\psi \propto \{a_s(\boldsymbol{r}) \exp[i\varphi_s(\boldsymbol{r})] + a_s(\boldsymbol{r}) \exp[-i\varphi_s(\boldsymbol{r})]\} \propto \cos[\varphi_s(\boldsymbol{r})] \tag{6.61}$$

in the reconstructed image. This merely represents an intensity distribution proportional to $\cos^2[\varphi_s(\boldsymbol{r})]$ or fringes of equal phase (Fig. 6.31). These fringes can be lines of equal specimen thickness, produced by the phase shift corresponding to the inner potential U_i (Sect. 3.1.3 and Table 3.2), for example, or lines parallel to $\boldsymbol{B}$ magnetic field lines caused by the phase shift of the magnetic vector potential (Sect. 3.1.5) [6.155].

Hanszen and coworker [6.156–158] have suggested the arrangement shown in Fig. 6.32. If only the hologram H_1 is used, the wave behind H_1 is superposed on a parallel reference beam. The tilt of this beam can be changed by means of the mirror M, its amplitude by a filter F_2 and the phase by a pressure cell. In the reconstruction, this superposition results in a new interference fringe system in which the direction, spacing and phase can be changed. Using a copy H_2 of a hologram H_1 in the second beam, a small shift in the holograms produces a fringe shape that contains information about the

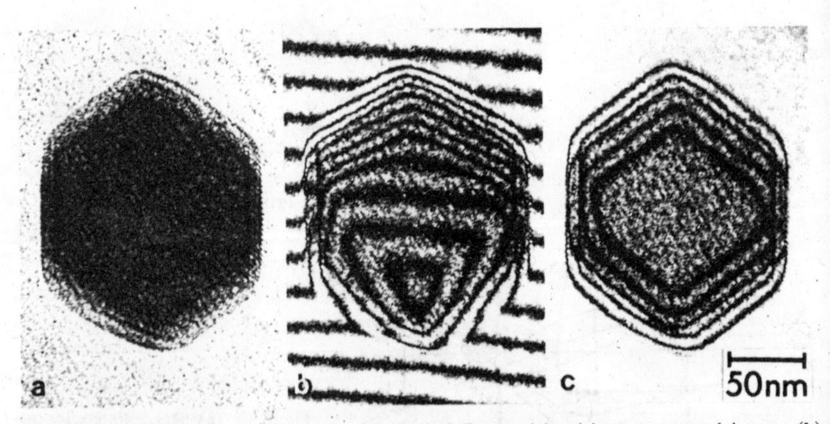

Fig. 6.31 a–c. Interference image of a decahedral Be particle: (**a**) reconstructed image, (**b**) interferogram, (**c**) contour map of lines of equal phase shift (thickness) [6.154]

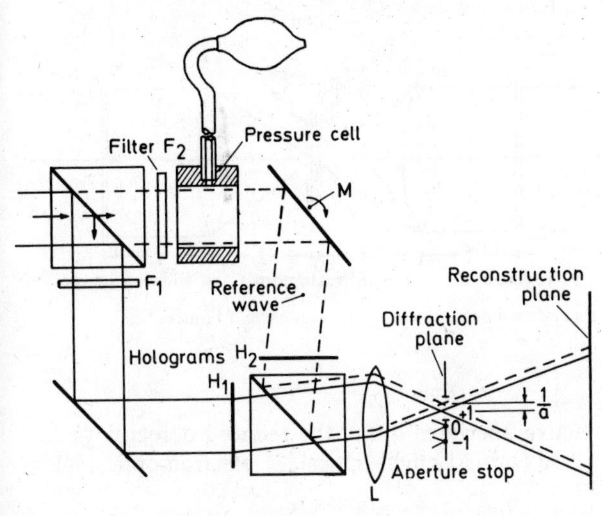

Fig. 6.32. Mach-Zehnder interferometer for recording the phase structure of holograms

slope of the phase distribution; alternatively, the phase value can be doubled if diffraction orders of opposite sign are used for the reconstruction. A further interesting application is the study of phase shifts in crystals depending on the tilt angle relative to the exact Bragg position [6.157, 158]. For a detailed study of electron holography, see [6.159].

6.5.4 Optical Filtering and Image Processing

Figure 6.33 shows the principal ray path of an optical-filtering process. The electron micrograph is irradiated with a parallel laser beam. As for optical diffraction (Sect. 6.4.6), the micrographs can be immersed in a fluid with the same refractive index as the gelatin of the emulsion to compensate the phase shift caused by thickness variations of the gelatin layer. Only the optical density caused by the silver grains will then influence the incident wave. A 1:1 imaging with the lenses L_2 and L_3 can be used to obtain a diffraction pattern in the focal plane of L_2, where the optical Fourier filter has to be situated. The diameter of the diffraction pattern can be increased to a few millimeters by making the focal length of L_2 large (30–50 cm). Filtering processes in the focal plane are easier than in the electron microscope, where the diameter of the diffraction pattern in the back focal plane of the objective lens is of the order of only a few tenths of a millimetre.

If the sole aim is optical filtering in the focal plane of L_2, L_2 and L_3 must have no spherical aberration within the aperture used. If the Gabor reconstruction method is to be used or if the electron-optical transfer is being simulated by light optics [6.160] the following relations between the spatial

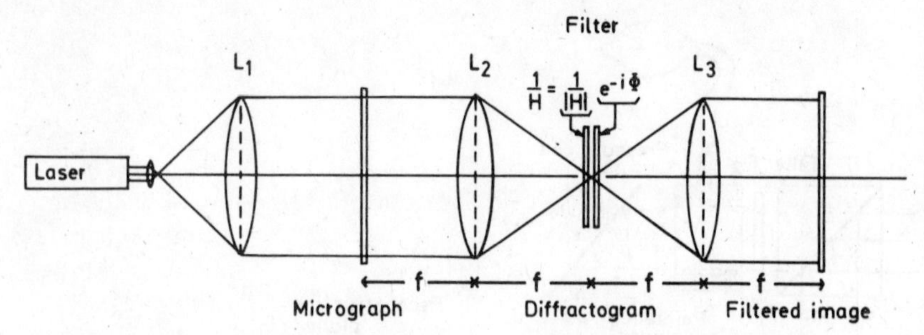

Fig. 6.33. Ray path of a light-optical image-processing system using a Fourier filter

frequencies q, the objective apertures α and the reduced defocusings Δz^* (3.69) have to be respected (suffix L: light-optical, E: electron-optical quantities):

$$\frac{q_E}{q_L} = \left(\frac{\lambda_L^3 C_{s,L}}{\lambda_E^3 C_{s,E}}\right)^{1/4}; \quad \frac{\alpha_L}{\alpha_E} = \left(\frac{\lambda_L C_{s,E}}{\lambda_E C_{s,L}}\right)^{1/4}; \quad \frac{\Delta z_L^*}{\Delta z_E^*} = \left(\frac{\Lambda_L C_{s,L}}{\lambda_E C_{s,E}}\right)^{1/2}. \tag{6.62}$$

Various imaging modes (e.g., bright-field, dark-field and hollow-cone illumination) and their contrast-transfer characteristics can be simulated with these relations [6.80, 140, 160, 161].

The filter theory that will now be discussed applies to weak-phase objects, for which we can set $\psi_s(\boldsymbol{r}) = 1 + i\varphi_s(\boldsymbol{r})$ in (6.20); this will give an indication of the optimum design of a Fourier filter in the focal plane of L_2. The image intensity $I(\boldsymbol{r}')$ can be expressed in terms of the blurring or point-spread function $h(\boldsymbol{r})$ introduced in (3.74), which is essentially the inverse Fourier transform of the pupil function $H(\boldsymbol{q})$

$$I(\boldsymbol{r}') = 1 + 2\varphi_s(\boldsymbol{r}') \otimes h(\boldsymbol{r}') = 1 + 2\int\int \varphi_s(\boldsymbol{r}_1)\, h(\boldsymbol{r}' - \boldsymbol{r}_1)\, d^2\boldsymbol{r}_1 \,. \tag{6.63}$$

Behind the micrograph (negative), the light amplitude for $\gamma = 1$ will be

$$A(\boldsymbol{r}) = 1 - \varphi_s(\boldsymbol{r}) \otimes h(\boldsymbol{r}) \,. \tag{6.64}$$

The diffracted light amplitude in the focal plane of L_2 can be represented by

$$F(\boldsymbol{q}) = \mathcal{T}[A(\boldsymbol{r})] = \delta(\boldsymbol{q}) - F_s(\boldsymbol{q}) \cdot H(\boldsymbol{q}) \,, \tag{6.65}$$

where $\delta(\boldsymbol{q})$ represents the primary beam at $\boldsymbol{q} = 0$ and $F_s(\boldsymbol{q})$ is the Fourier transform of $\varphi_s(\boldsymbol{r})$. The convolution theorem (3.51) maps the convolution in (6.64) into a multiplication in (6.65). If an optical filter $1/H(\boldsymbol{q})$ is present in the focal plane of L_2, the amplitude in the reconstructed image plane will be

$$A(\boldsymbol{r}') = 1 - \varphi_s(\boldsymbol{r}') \otimes \mathcal{F}^{-1}\left[H(\boldsymbol{q})\frac{1}{H(\boldsymbol{q})}\right] = 1 - \varphi_s(\boldsymbol{r}') \otimes \delta(\boldsymbol{r}') = 1 - \varphi_s(\boldsymbol{r}') \tag{6.66}$$

and the light intensity hence becomes $L(\boldsymbol{r}') = A(\boldsymbol{r}') \cdot A^*(\boldsymbol{r}') = 1 - 2\varphi_s(\boldsymbol{r}')$. The originally blurred image [convoluted with $h(\boldsymbol{r})$] is deblurred by the optical filter $1/H(\boldsymbol{q})$. However, this deblurring cannot restore information lost at gaps in the contrast transfer; spatial frequencies are reconstructed only if they are present in the electron micrograph and are larger than the noise.

The filter can be divided into an amplitude and a phase part:

$$\frac{1}{H(\boldsymbol{q})} = \frac{1}{|H(\boldsymbol{q})|}\exp(-\mathrm{i}\phi_q) \; . \tag{6.67}$$

A negative sign of $H(\boldsymbol{q})$ can be included in the phase factor by recalling that $\exp(\mathrm{i}\pi) = \exp(-\mathrm{i}\pi) = -1$.

The use of a light-optical filter with $|H(\boldsymbol{q})| = 1$ and $\exp(-\mathrm{i}\phi_q) = -1$ was proposed to correct those spatial-frequency intervals for which the sign of CTF is negative [6.162]. This method generates the image that would have been produced by a system with $|B(q)|$ as CTF and was applied to TEM micrographs [6.163]. A mask that creates the desired phase shift can be produced by recording a diffractogram in the focal plane of L_2 by a photoresist layer on a glass plate; the resist persists in annular rings for those spatial frequencies for which the sign of $B(q)$ is negative. The filter is immersed in a fluid of suitable refractive index so that the optical phase difference in the resist has the value π.

If, furthermore, an amplitude filter of the form $1/|H(q)| \propto 1/|\sin W(q)|$ is used, CTF in the transfer bands can be equalized in magnitude. However, small gaps around the zeros of $\sin W(q)$ have to be tolerated. This method was also tested for TEM micrographs [6.164–167].

A phase filter can also be created by means of an amplitude grating. The distance between the slits must be so small that the twin images in the first-order diffracted beams are separated from those produced by the transmitted and undiffracted wave with no overlap. In the transfer intervals with a negative sign of $B(q)$, the grating is shifted by half the slit separation. This causes a phase shift π of the twin images, whereas the phase of the transmitted wave remains unchanged. This phase shift is a direct consequence of the translation theorem (3.47) for Fourier transforms. It is also possible to construct a combined amplitude and phase filter from a two-dimensional grating of transparent rectangles that vary in size (amplitude) and position (phase). Such binary filters can be calculated and plotted by a computer [6.168–170].

The noise amplitude in periodic structures can be decreased by introducing in the Fourier plane a mask that contains holes at the diffraction maxima [6.171–173]. Thus, if negatively stained particles show a superposition of structures from the back and front for example, these can be separated by selecting the corresponding diffraction maxima [6.174]. Care will be needed to avoid introducing artificial periodicities by this filtering method [6.175].

Another simpler method for decreasing the noise in periodic images is to produce a suitable multiple exposure of a photographic copy of the image by moving the negative or the copy by multiples of the specimen periodicity [6.176, 177] or by an n-fold rotation of the micrograph by multiples of $2\pi/n$ if the structure shows an n-fold rotational symmetry [6.178]. This method is sometimes known as stroboscopy, because the same effect can be obtained by mounting the micrograph on a turntable that rotates at a frequency f and illuminating it with a source that flashes at a frequency nf. An n-fold rotational symmetry will then be detected by the eye. Artificial structures may be produced, if the correct n is established by varying n in the stroboscopic superposition. Finally, we mention that it is often more useful to superimpose different micrographs, which can be done more accurately digitally (Sect. 6.5.5) because objective criteria for alignment can then be applied.

6.5.5 Digital Image Processing

All of the optical analogue methods described in Sect. 6.5.4 can also be exploited on a digital computer if the image intensity is first stored in a matrix array. This requires sampling and quantization of the intensity by means of a scanning microdensitometer; alternatively, a television pick-up tube or an image intensifier may be used. Modern computers can store $2^n \times 2^n$ (n = 8–10) matrix elements with $2^8 = 256$ different grey levels. For comparison, we note that a photographic emulsion 50 cm^2 in area contains 5×10^6 image elements. It is not possible to describe all of the various digital procedures in detail here. Our aim in this section is to give some idea of what is, in principle, possible.

Digital processing becomes of special interest if two or more micrographs of a series are used to get information about the specimen. The amplitude and phase distribution of the specimen can then be separated (see below), for example, or the three-dimensional structure of a specimen can be reconstructed from a tilt series (Sect. 6.5.6). Near the resolution limit, each micrograph will require a two-dimensional restoration procedure. For this, the methods of Sect. 6.5.4 can be recast in digital form. A two-dimensional fast-fourier transform using the Cooley-Tukey algorithm provides the diffraction pattern of the micrograph, which contains information about the contrast-transfer gaps, the defocus, spherical aberration, paraxial astigmatism and image aberrations of higher order. Filtering in Fourier space can be applied, followed by an inverse Fourier transform, to improve the image. The resulting image amplitude is complex; phase information will not be lost, unlike the case of optical analogue recording. The ultimate aim of a restoration is to acquire knowledge about the specimen amplitude and phase: $\psi_s(\boldsymbol{r}) = a_s(\boldsymbol{r}) \exp[i\varphi_s(\boldsymbol{r})]$, without transfer gaps. Apart from the case of weak-phase, weak-amplitude objects, this problem is nonlinear; it is reviewed in detail in [6.179–181]. Procedures that set out from various sets of initial data have been investigated:

1) Use of the diffraction pattern $\propto |F(q)|^2$ and a bright (or dark) field image $\propto |\psi|^2$ (Gerchberg-Saxton algorithm) [6.182, 183]. The method requires a periodic specimen [6.184] and has been applied to negatively stained catalase [6.185] and to periodic magnetic structures [6.186].
2) Two (or more) micrographs recorded at different values of defocus and containing $|\psi_1|^2$ and $|\psi_2|^2$ [6.187, 188]. The main idea behind this method is that the Fourier transforms of ψ_1 and ψ_2 are $F_1(\boldsymbol{q})\exp[-\mathrm{i}W_1(q)]$ and $F_2(\boldsymbol{q})\exp[-\mathrm{i}W_2(q)]$ respectively. Therefore, F_1 and F_2 are related by $F_1 = F_2\exp[-\mathrm{i}(W_2 - W_1)]$. The difference between the wave aberrations is independent of C_s and other lens aberrations and contains only the defocusing term, in which Δz is the focus difference between the two micrographs. Practical difficulties are the accuracy required aligning the micrographs to about one-half of the desired resolution (see below) and the contribution from inelastic scattering.
3) Bright- and dark-field micrographs taken under identical electron-optical conditions [6.184, 189].
4) Two micrographs with complementary-half-plane diaphragms. The basic idea of this method has already been discussed as single-sideband holography in Sect. 6.5.2.

Methods 1–3 are iterative methods, in which an initial approximation for amplitude and phase is guessed. Considerable thought has been given to the problem of achieving rapid convergence and a unique solution for the phase, especially in the presence of unavoidable noise [6.190].

The first step in any digital computation involving a series of micrographs with the same or different defocus is to align the individual micrographs in orientation and position. A preliminary adjustment can be made by use of characteristic image details. For exact alignment, cross-correlation is needed [6.191]. The cross-correlation of two functions $f_1(\boldsymbol{r})$ and $f_2(\boldsymbol{r})$ is the integral

$$\phi_{12}(r) = \iint f_1(\boldsymbol{r}) f_2(\boldsymbol{r}' + \boldsymbol{r})\, d^2\boldsymbol{r}' \; . \tag{6.68}$$

Setting $f_1 = f_2$ gives the auto-correlation function. This integral will have a maximum at $\boldsymbol{r} = 0$ for two similar, exactly aligned images, because the integrand is then positive-definite over the whole area. The integral will show regularly spaced maxima for periodic structures. If two otherwise similar micrographs are not exactly aligned, the position of the maximum indicates the necessary shift (see example in Fig. 6.34). Two micrographs taken at different values of defocus may give a very broad correlation maximum, which makes determination of the shift vector $\boldsymbol{r}$ more difficult [6.192, 193].

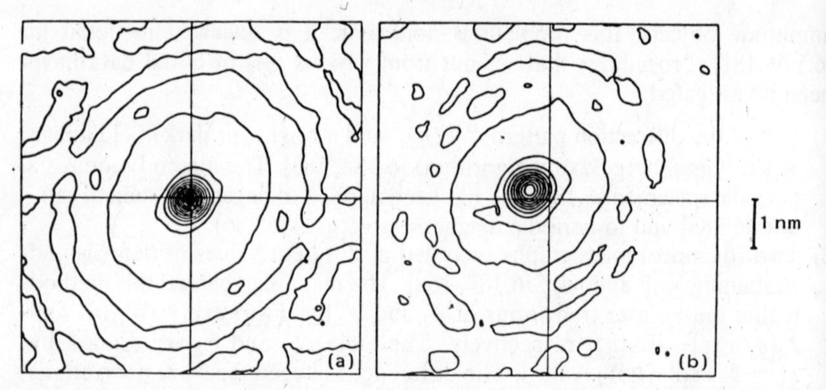

Fig. 6.34 a, b. Example of the cross-correlation of electron micrographs of carbon foils: (**a**) auto-correlation of one micrograph with a correlation peak at $x = 0$, $y = 0$, (**b**) cross-correlation of two successive micrographs, indicating an image shift between the two exposures of $x = 0.35$ mm, $y = 0.17$ mm [6.191]

For determination of the defocus, spherical aberration and astigmatism, it is necessary to calculate the Fourier transforms $F_1(\boldsymbol{q})$ and $F_2(\boldsymbol{q})$ of $f_1(\boldsymbol{r})$ and $f_2(\boldsymbol{r})$, respectively. It is therefore of interest to note that the cross-correlation is the inverse Fourier transform of the *Wiener spectrum* $W_{12}(\boldsymbol{q}) = F_1(\boldsymbol{q}) \cdot F_2^*(\boldsymbol{q})$:

$$\phi_{12}(\boldsymbol{r}) = \mathcal{T}^{-1}[W_{12}(\boldsymbol{q})] = \mathcal{T}^{-1}[F_1(\boldsymbol{q}) \cdot F_2^*(\boldsymbol{q})] \,. \tag{6.69}$$

This method of aligning two micrographs with equal defocus can be used in an image-difference method designed to provide information about radiation damage in the specimen between two exposures or to subtract from a macromolecule image the image of a clean supporting film obtained beforehand [6.192, 194]. However, successive micrographs of a clean carbon film show variations in structure, caused by contamination and radiation damage.

For the detection of n-fold rotational symmetry the following method can be used [6.195]. The image intensity $I(r, \varphi)$ in polar coordinates is expanded in a Fourier series

$$I(r, \varphi) = \sum_{-\infty}^{+\infty} g_n(r) \mathrm{e}^{in\varphi} \tag{6.70}$$

and the strength of an n-fold component can be calculated from

$$\mathrm{P_n} = \int |g_n(\mathrm{r})|^2 r \, dr \,. \tag{6.71}$$

The presence of unique n-fold symmetry will be indicated by pronounced maxima of P_n for one value of n and its multiples.

The Fourier coefficients $F(\theta, \varphi)$ of a two-dimensional periodic specimen with a unit cell characterized by two translation vectors vary in a defocus series, and the Fourier coefficients of a micrograph are proportional to $|F(\theta, \varphi) \sin W(\theta)|$. The value of $|F(\theta, \varphi)|$ and the amplitude component transferred, if any, can be evaluated from the series by the method of least squares. These corrected Fourier coefficients can be used to calculate a periodic image that represents an average over all unit cells and micrographs. There is an optimum increase of the signal-to-noise ratio because noise due to the electron statistics, the grain of the photographic emulsion and the inhomogenities of the specimen, scatter diffusely over the whole Fourier plane (diffraction pattern). This method is applied for catalase crystals, for example [6.196]. This technique can also be employed to sharpen micrographs taken at very low electron exposures to reduce radiation damage [6.197, 198].

The signal-to-noise ratio for aperiodic specimens can be improved by the following scheme [6.167, 199, 200]. If we consider the amplitude distribution in the image to be the specimen function $\psi_s(\boldsymbol{r})$ convoluted with the blurring (point-spread) function $h(\boldsymbol{r})$ (3.74) superimposed on an additive noise distribution $n(\boldsymbol{r})$,

$$a(\boldsymbol{r}) = \psi_s(\boldsymbol{r}) \otimes h(\boldsymbol{r}) + n(\boldsymbol{r}) , \tag{6.72}$$

the Fourier transform becomes

$$A(\boldsymbol{q}) = F(\boldsymbol{q}) \cdot H(\boldsymbol{q}) + N(\boldsymbol{q}) \tag{6.73}$$

with $H(\boldsymbol{q}) = -M(\boldsymbol{q}) \sin W(\boldsymbol{q})$ (3.73) and $F(\boldsymbol{q}) = \mathcal{T}\{\psi_s(\boldsymbol{r})\}$.

Instead of applying only the filter function $H^{-1}(q)$ as in (6.66), the filter is multiplied by a further weighting function $H_W(q)$; after inverse Fourier transformation, we obtain

$$a'(\boldsymbol{r}) = \psi_s(\boldsymbol{r}) \otimes h_w(\boldsymbol{r}) + n(\boldsymbol{r}) \otimes h_w(\boldsymbol{r}) \otimes \bar{h}(\boldsymbol{r}) \tag{6.74}$$

with $h_w = \mathcal{T}^{-1}\{H_w\}$ and $\bar{h} = \mathcal{T}^{-1}\{H^{-1}\}$.

The convolution of $\psi_s(\boldsymbol{r})$ with $h_w(\boldsymbol{r})$ in the first term inevitably decreases the resolution. Resolution will not be lost only if H_w = const and $h_w = \delta(0)$. We therefore conclude that each noise-filtering operation will be a compromise between a loss of resolution (blurring of the image points) and a reduction of the noise amplitude.

The signal-to-noise amplitude can be calculated from the Wiener spectra $W_0 = |F(q)|^2$ and $W_n = |N(q)|^2$ and takes the following values

before filtering | after filtering

$$\frac{\mathrm{S}}{\mathrm{N}} = \frac{\iint W_0(\boldsymbol{q})|H(\boldsymbol{q})|d^2\boldsymbol{q}}{\iint W_n(\boldsymbol{q})d^2\boldsymbol{q}} \qquad \frac{\mathrm{S}}{\mathrm{N}} = \frac{\iint W_0(\boldsymbol{q})|H_w(\boldsymbol{q})|^2 d^2\boldsymbol{q}}{W_n(\boldsymbol{q})|H(\boldsymbol{q})|^{-2}|H_w(\boldsymbol{q})|^2 d^2\boldsymbol{q}} . \tag{6.75}$$

The following weighting functions have been used:

$$H_w(q) = \exp\left[\alpha\left(1 - \frac{1}{|\sin W(q)|}\right)\right] \qquad [6.200]$$

$$H_w(q) = \frac{H^*(q)}{|H(q)|^2 + W_n(q)/W_0(q)} \quad \textit{Wiener's optimum filter } [6.201, 202].$$

Such weighting functions can also be applied in optical analogue reconstruction as the combination $H(q)^{-1} H_w(q)$ [6.167]. A disadvantage of the Wiener filter is that it requires a knowledge of the Wiener spectrum $W_0(q)$ and hence $F_0(q)$; this information is often not readily available.

Noise-reduced images of aperiodic structures similar in appearance but randomly distributed over the micrograph (e.g., macromolecules or virus particles with a site of preferential attachment to the supporting film) can be obtained by averaging over a sufficiently large number of particles after alignment in position and orientation. For this, the computer must be furnished with a motif-detection capability [6.203]. The method becomes reasonably practicable when applied interactively on an image-analysing computer [6.204–208]. The particles are selected by eye and centered by the cross-correlation methods described above. This means that the cross-correlation maxima for different shifts and rotations have to be calculated. If a low-dose exposure is employed to reduce radiation damage, a subsequent high-dose picture can be used for pre-alignment and selection of particles that have the most-satisfactory appearance.

Figure 6.35 shows an application of noise reduction to ribosomes [6.209]. The ribosomes are randomly distributed and can be separated into left- and right-oriented particles (Fig. 6.35 a). Figure 6.35 b shows a series of left-oriented images after alignment: 77 particles were used for averaging. Figure 6.35 e and f show the averages of 38 and 39 arbitrarily selected particles and c) the average of all 77 particles at a resolution of 1/1.4 nm^{-1}. Figure 6.35 d shows the result of further averaging over neighbouring image points, with a resolution of 1/3.2 nm^{-1}.

Van Heel [6.210] presented a method that allows automatic detection of single particles against an extremely noisy background – biological macromolecules in low-dose micrographs, for example. Each image element is replaced by the image variance in its environment.

These operations are considerably faciliated by the use of a high-level computer language specially designed for image processing [6.211]. A number of such languages are in fairly widespread use: *Semper* [6.212], *Spider* [6.213], and *Imagic* [6.214].

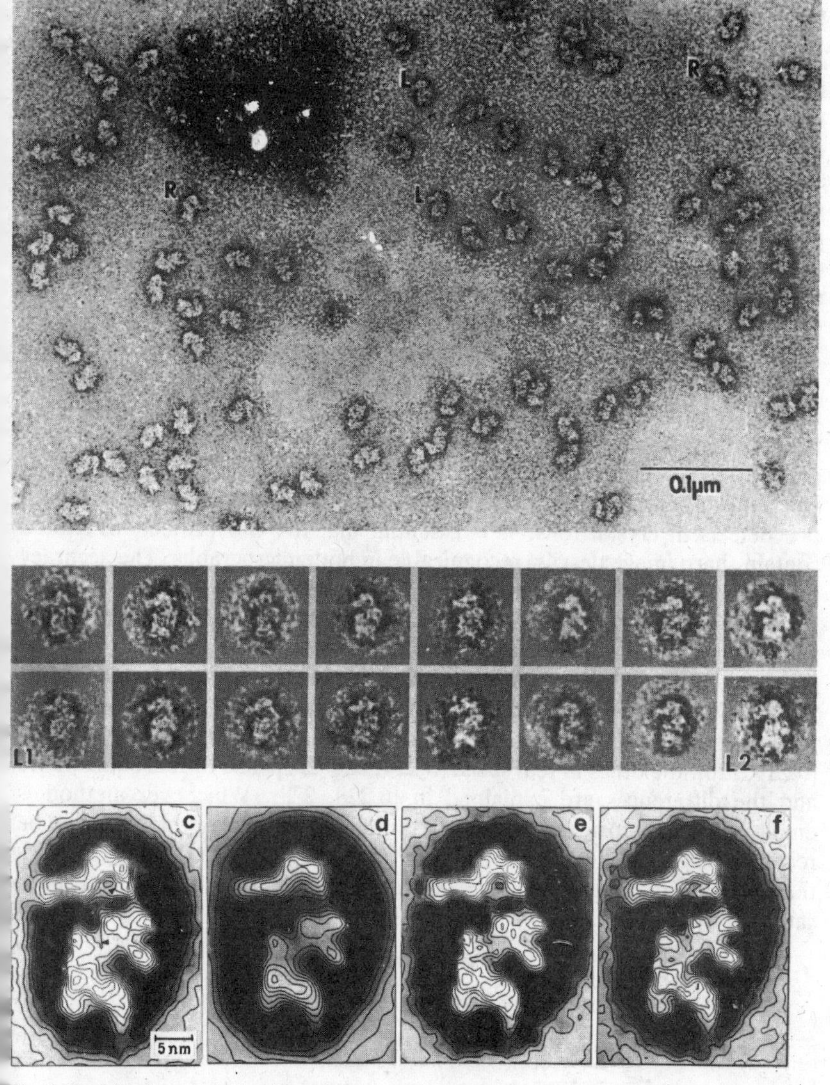

Fig. 6.35. (a) Micrograph showing left-oriented (L) and right-oriented (R) 40 S ribosomal subunits of HeLa cells, (b) gallery of 16 of 77 L particles after alignment. (c) Average obtained from all 77 particles displayed at $1/1.4\ nm^{-1}$ resolution and (d) with $1/3.2\ nm^{-1}$ resolution, (e) and (f) averages from independent sets of 38 and 39 particles [6.209]

6.5.6 Three-Dimensional Reconstruction

After two-dimensional image restoration by means of the methods described in Sects. 6.5.4, 5, a tilt series can be used to obtain a three-dimensional reconstruction; the techniques employed are comparable to those of x-ray computer tomography, since both reconstruct internal structures from a series of projections. This method should not be confused with stereo micrographs, which needs only two tilted micrographs. For that, two specimen points A and B with a height difference $\Delta z = z_B - z_A$ are imaged with different separations Δx_1 and Δx_2 at tilt angles $\pm\gamma$ (Fig. 6.36). In this simple case of parallel projection, the parallax

$$p = (x_{B2} - x_{A2}) - (x_{B1} - x_{A1}) = \Delta x_2 - \Delta x_1 = 2M\,\Delta z \sin\gamma \tag{6.76}$$

is directly proportional to the height difference Δz if the tilt axis passes through the centre of the image area observed. For further details of stereographic reconstruction, see [6.215–218]. The method can be applied to surface replicas, thick biological sections, aggregates of small particles and lattice defects in crystal foils. It is essential that the two projections should contain sharp image details, recognizable in both micrographs. The accuracy for tilt angles $\gamma = \pm 10°$ is of the order of $\Delta z = \pm 3$ nm for $\Delta x = \pm 1$ nm.

Three-dimensional reconstruction does not, on the contrary, need sharp image details. The aim is to reconstruct the specimen density distribution $\varrho(x, y, z)$ from a series of projections. The method is in use mainly for macromolecules and biological structures.

Two types of methods are employed, one operates in Fourier space [6.219, 220], the other in real space [6.214, 221–227]. The formal equivalence and the differences are explained in [6.228, 229]. Whichever method is employed, only information available in the micrographs of the series can be reconstructed. The Fourier method will be discussed in more detail, because the information content and the information gaps can be evaluated more satisfactorily.

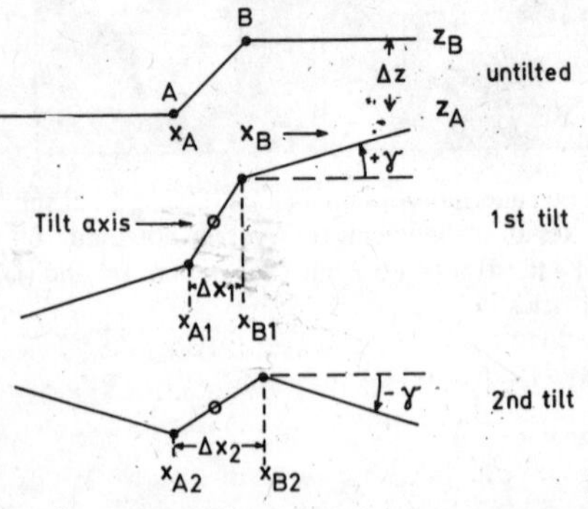

Fig. 6.36. Basis relation for stereometric reconstruction

The specimen is represented by the mass-density distribution $\varrho(x,y,z)$, the three-dimensional Fourier transform of which is

$$F(q_x,q_y,q_z) = \iiint \varrho(x,y,z) \exp[-2\pi i(q_x x + q_y y + q_z z)]\, dx\, dy\, dz\,. \tag{6.77}$$

For specimens with cylindrical symmetry, it is better to use cylindrical polar coordinates. The Fourier transform then becomes a Fourier-Bessel transform [6.230].

In central section through the Fourier space, *e.g.* in the q_x,q_y plane ($q_z = 0$), the transform reduces to

$$F(q_x,q_y,0) = \iint \underbrace{\left[\int \varrho(x,\,y,\,z)\, dz\right]}_{\sigma(x,y)} \exp[-2\pi i(q_x x + q_y y)]\, dx\, dy\,. \tag{6.78}$$

The integral $\sigma(x,y)$ is a projection in the z direction (mass-thickness distribution) of the mass-density distribution ϱ; this produces an image intensity $I(x,y) = I_0 \exp[-\sigma(x,y)/x_k]$, applying (6.6) for scattering contrast. Equation (6.78) tells us that a two-dimensional Fourier transform of $\varrho(x,y)$ yields a central section through Fourier space. If $F(q_x,q_y,q_z)$ is known from a tilt series, equivalent to a bundle of central sections through the Fourier space, $\varrho(x,y,z)$ can be calculated by an inverse Fourier transform. This model reveals one of the difficulties of three-dimensional reconstruction: the maximum tilt angle that can be applied using a specimen goniometer is $\pm 45°$ so that a double cone of the Fourier space with an aperture of 45° remains "vacant".

For many specimens, symmetry relations can be used to reduce the necessary number n of micrographs. Thus for a specimen with helical symmetry a single micrograph ($n = 1$) is sufficient (T4 phage tail); for icosahedral symmetry (e.g., tomato bushy stunt virus) we find $n = 2$ and in the absence of symmetry $n = 30$ (e.g., ribosomes [6.231]). A rule of thumb $n = \pi D/d$ is proposed where D is the diameter and d the resolution [6.219].

For further details and applications of this method which has reached a high degree of refinement, see review articles [6.232–237].

6.6 Lorentz Microscopy

6.6.1 Lorentz Microscopy and Fresnel Diffraction

It was shown in Sects. 2.1.3 and 3.1.5 that the angular deflection ε (2.18b) of an electron beam by a transverse magnetic field can be calculated either by evaluating the Lorentz force on the electron or by introducing the phase shift (3.8) caused by the magnetic vector potential or the enclosed magnetic flux Φ_m. By use of an arbitrary origin ($\boldsymbol{r} = 0$) in the specimen plane, the phase shift caused by the magnetic field can be written (3.8)

$$\varphi_m(\boldsymbol{r}) = -\frac{2\pi e}{h}\phi_m = -\frac{2\pi e}{h}Btx\,, \tag{6.79}$$

where t is the film thickness and x is a coordinate in the object plane normal to $\boldsymbol{B}_s$. Thus for a ferromagnetic film of iron with a thickness t = 50 nm and a spontaneous magnetization B_s = 2.1 T, (2.18b) gives a deflection angle ε = 0.1 mrad for 100 keV electrons. A phase difference $\varphi_m = \pi$ corresponding to a path difference $\lambda/2$ is found for x = 20 nm. A ferromagnetic film is therefore a pure but not necessarily weak-phase object and can be studied with the theory of phase contrast.

As in (3.39), the wave amplitude behind the specimen is modified from $\psi_0 \exp(2\pi ikz)$ to

$$\psi(x) = \psi_0 \exp[i\varphi_m(x)]\exp(2\pi ikz)\,. \tag{6.80}$$

In the focal plane of the objective lens, the amplitude distribution is given by the Fourier transform

$$F(q_x) = \int\limits_{-\infty}^{+\infty} \exp[i\varphi_m(x)]\exp(-2\pi iq_x x)dx\,. \tag{6.81}$$

From the translation theorem for Fourier transforms (3.47), this has the form of a δ-function at $q_x = \varphi_m/2\pi x = eBt/h$. Substituting $h = mv\lambda$ and recalling that $q = \theta/\lambda$, we obtain the same angular deflection $\varepsilon = \theta$ for $t = L$ as in (2.18b).

Because $F(q)$ is concentrated within a very small range of values of q, only defocusing terms of the wave aberration $W(q)$ in (3.67) need to be considered and (3.73) gives

$$\psi_m(x') = \frac{\psi_0}{M}\int F(q)\exp(i\pi\lambda\Delta z q^2)\exp(2\pi iq_x x')\,dq_x\,. \tag{6.82}$$

Substituting for $F(q)$ from (6.81), we find

$$\begin{aligned}\psi_m(x') &= \frac{\psi_0}{M}\int\{\int \exp[i\varphi_m(x)]\exp(-2\pi iq_x x)dx\}\exp(i\pi\lambda\Delta z q_x^2)\exp(2\pi iq_x x')\\ &= \frac{\psi_0}{M}\int(\int \exp\{i\pi[2q_x(x'-x) + \lambda\Delta z q_x^2]\}dq_x)\exp[i\varphi_m(x)]dx\,.\end{aligned} \tag{6.83}$$

Introducing $q'^2 = 2\lambda\Delta z[q + (x'-x)/(\lambda\Delta z)]^2$, we can rewrite (6.83)

$$\psi_m(x') = \frac{\psi_0}{M}\frac{1}{\sqrt{2\lambda\Delta z}}\int\limits_{-\infty}^{+\infty}\underbrace{\left[\int\limits_{-\infty}^{+\infty}\exp(i\pi q'^2/2)dq'\right]}_{1+i}\exp\left\{i\left[\varphi_m(x') - \pi\frac{(x'-x)^2}{\lambda\Delta z}\right.\right. \tag{6.84}$$

The inner integral is one of the Fresnel integrals of (3.37). A further substitution, $u = \sqrt{2/\lambda\Delta z}\,(x' - x)$, gives

$$\psi_m(x') = \frac{\psi_0}{M}\frac{1+i}{2}\int_{-\infty}^{+\infty} \exp(i\varphi_m)\exp(-i\pi u^2/2)\,du\,. \tag{6.85}$$

This is none other than Fresnel diffraction from the phase distribution caused by the magnetization. This can also be seen at $\varphi_m = 0$; the integral contributes a further factor $1 + i$ and we have $|\psi_m|^2 = |\psi_0|^2/M^2$. For a uniform magnetization (B_s = const), φ_m is a linear function of x (6.79). This again results in a uniform intensity distribution in the image but the specimen coordinates are shifted by $\varepsilon M \Delta z$, which can also be deduced from particle optics by considering a plane at a distance Δz behind the specimen (Fig. 6.38). Contrast effects in the image are possible for only non-uniform distributions of $\boldsymbol{B}_s$, such as magnetic domain walls, across which the direction of $\boldsymbol{B}_s$ changes, or magnetization ripple, in which the magnetic field exhibits periodic small-angle deviations from the mean value of $\boldsymbol{B}_s$.

6.6.2 Imaging Modes of Lorentz Microscopy

All modes of Lorentz microscopy require the illumination aperture α_i to be smaller than the deflection angle ε; otherwise, the illumination aperture would become incoherent for small values of ε. Apertures $\alpha_i \simeq 10^{-2}$ mrad can be attained by strongly exciting condenser lens 1 (Sect. 4.2.1). Because the gun brightness is conserved (4.11), the current density at the specimen plane is reduced; hence, very long exposure times are needed for large magnifications. However, in most applications of Lorentz microscopy, the imaging of domain walls for example, magnification 10 000 is sufficient. The holographic method (Sect. 6.5.3) requires the higher brightness of a field-emission gun. Furthermore, the magnetic distribution must not be disturbed by the magnetic field of the objective lens (see method c). For reviews devoted to Lorentz microscopy see [6.238–241].

a) Small-Angle Electron Diffraction

In small-angle electron diffraction (Sect. 9.3.4), a diffraction pattern is recorded with $\alpha_1 \simeq 10^{-2}$ mrad and a large camera length. A ferromagnetic layer with uniaxial anisotropy consists of domains with antiparallel directions of $\boldsymbol{B}_s$ separated by 180° walls. The primary beam splits into two spots with an angular separation $\pm\,\varepsilon$ [6.242, 243]. Four different distributions of $\boldsymbol{B}_s$ parallel to $\langle 100 \rangle$ are possible in a [100] epitaxial iron film, causing splitting into four spots (Fig. 6.37). In polycrystalline films with varying directions of magnetization, the splitting of the primary beam results in a circular or sickle-shaped diffraction pattern. The splitting $2\,\varepsilon$ can be used to determine the specimen

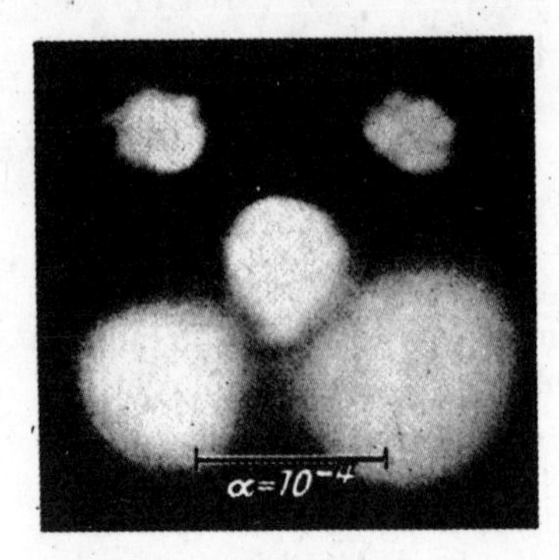

Fig. 6.37. Small-angle electron diffraction pattern of a (100) oriented epitaxially grown iron film on NaCl. The primary beam splits into four beams, corresponding to the ⟨100⟩ directions of spontaneous magnetisation

thickness by means of (2.18) [6.244]. However, the angle of divergence can be decreased by reducing the density and the saturation induction B_s that can occur in thin films.

Periodic domain-wall spacings, 25 μm in a cobalt foil for example, can create diffraction maxima, which can be interpretated quantitatively in terms of diffraction at a phase grating [6.245, 246].

b) Foucault Mode

In the case of a 180° domain wall, the antiparallel magnetisation directions produce two spots in the focal plane of the objective lens; these are separated by a distance $\simeq 2\varepsilon f$. If the objective diaphragm is moved, one of the spots can be suppressed, and the corresponding domain becomes dark in the image [6.247, 248]. In this mode, the domain wall is imaged as a boundary between dark and bright areas (Fig. 6.39c, d). The objective lens then works in focus, so that the specimen is imaged on the final screen without any defocusing. The method is limited to magnifications below 10000 by the gun brightness [6.249].

Some means of ensuring that the objective diaphragm is situated in the focal plane should be provided. Any displacement decreases the area in which the Foucault contrast can be observed. A thin-foil diaphragm (Sect. 4.4.1) is preferable; this also decreases charging of the diaphragm. Introduction of a phase shift in the focal plane also produces domain contrast similar to that seen in the Fresnel mode, but with the specimen in focus [6.250].

The Foucault mode can also be used to study much larger extended stray fields around thin wires or small compact specimens [6.251–257].

In another version of the Foucault mode [6.258] the small-angle deflection caused by the inner potential (refractive index) is exploited. Small crystals act like prisms, resulting in a splitting of the electron diffraction spots and the primary beam, if the illumination aperture is very small. One or more of these deflected beams can be absorbed by the diaphragm and can cause a contrast difference that depends on the inclination of the crystal faces. For more-complex specimens, the contrast is caused by the local gradient of the optical path length; thus a dark contour for a positive gradient and a bright contour for a negative gradient lead to a pseudo-topographic image contrast.

c) Fresnel Mode

The intensity distribution at a distance Δz below or above the specimen is imaged by under- or overfocusing, respectively [6.259, 260]. The principle of this mode will first be explained in terms of the particle model (geometric theory), in which the electron trajectories are deflected by an angle ε proportional to the local value of $\int B\, dz$. Figure 6.38 shows that the electron trajectories either converge (left) or diverge (right) at a distance Δz below the specimen with antiparallel domain walls. The left domain wall will appear as a bright line and the right one as a dark line in the defocused image. Defocusing in the opposite direction reverses the contrast (Fig. 6.39 a, b). The width of the gap or of the overlap $b \simeq 2\varepsilon\Delta z$ of the divergent and convergent wall image, respectively, should be large enough to be detectable at medium magnifications, so that with $b = 0.1$ µm and $\varepsilon = 0.1$ mrad, for example, a defocusing Δz of 0.5 mm is needed.

The geometric theory cannot be used for detailed image analysis; the wave-optical theory has to be used. The most striking effect is the appearence of *biprism fringes* in the convergent image (Figs. 6.38 and 40), in which two coherent waves overlap with a convergence angle 2ε [6.261]. The fringe spacing Δx can be calculated with (3.27) and $\beta = \varepsilon$. Use of (2.18), $mv = h/\lambda$ and $L = t$ gives

$$\Delta x = \frac{h}{2eBt} \,. \tag{6.86}$$

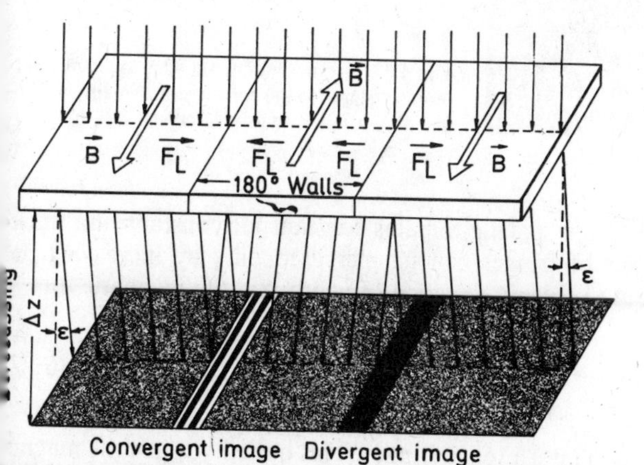

Fig. 6.38. Deflections of electron trajectories in a magnetic film with 180° domain walls that form a convergent (*left*) and divergent (*right*) wall image at a defocus Δz (Fresnel mode)

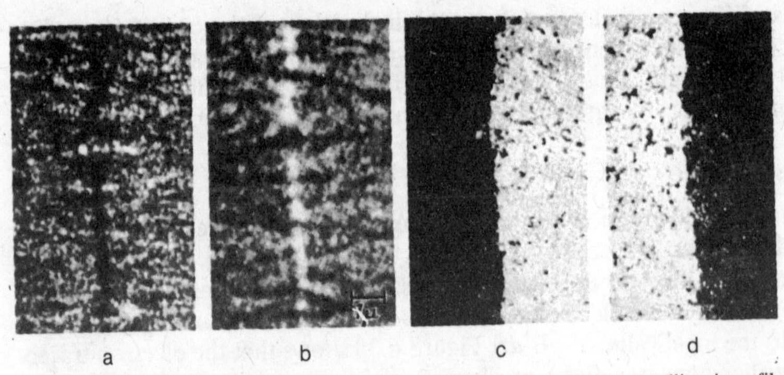

Fig. 6.39 a–d. Micrographs that show a 180° domain wall of a polycrystalline iron film in the Fresnel mode with (**a**) under- and (**b**) overfocussing and in the Foucault mode absorbing the left (**c**) and the right (**d**) deflected beam

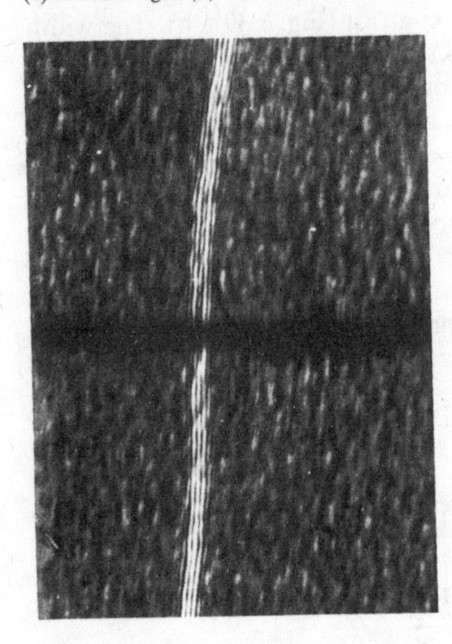

Fig. 6.40. Fresnel mode, showing biprism interference fringes in the convergent domain-wall image of (100) oriented single-crystal iron film

This means that the fringe spacing remains constant for constant film thickness, and the number of fringes can be increased only by increasing the overlap (defocusing Δz). Inside the zone of width Δx, the magnetic flux

$$\phi_{\mathrm{m}} = \Delta x B t = \frac{h}{2e} \tag{6.87}$$

is enclosed between two interference fringes. This quantity is just the magnetic-flux quantum (fluxon, Sect. 3.1.5). This fluxon-criterion [6.262] can be used to estimate the spacing and number of fringes observable.

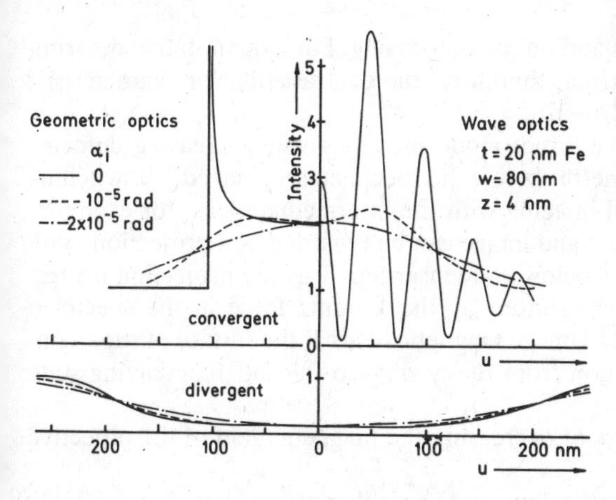

Fig. 6.41. Calculated intensity profiles of convergent and divergent domain-wall images across a 180° wall of width w = 80 nm in a 20 nm Fe foil at a defocus Δz = 4 mm [(α_i) illumination aperture]

Figure 6.41 shows a comparison of the intensity profile across a 180° domain wall of width w = 80 nm in a t = 20 nm thick Fe foil at a defocusing Δz = 4 mm calculated by use of the geometric theory (left) and wave optics (right). The differences for a very coherent beam (zero-illumination aperture α_i) are obvious. An aperture of $\alpha_i = 10^{-2}$ mrad is already sufficient to blur the biprism fringes of the wave-optical theory, and geometric and wave optics result in similar intensity profiles.

The intensity profile of the divergent wall image is also affected by use of the wave-optical theory (Fig. 6.41, bottom) [6.240, 263]. Criteria for which the purely geometric theory is valid are satisfied [6.240, 264]. In the geometric theory, the width of the divergent wall image is enlarged, in the first-order approximation, by the wall width w, whereas in the convergent image, it is decreased by w ($b = 2\varepsilon\Delta z \pm w$). Differences between the widths of the divergent and convergent images can be used to estimate the wall thickness [6.265–268]. This method also requires a knowledge of the influence of elastic and inelastic small-angle scattering on image contrast [6.263]. Thicker films are better studied in a high-voltage electron microscope, in which small-angle scattering has less effect [6.269].

Comparison of convergent-wall images that contain biprism fringes with wave-optical calculations based on models of the magnetisation distribution inside the domain wall can be used to test the model and to distinguish between Néel walls and Bloch walls [6.270–273]. Additionally, the small-angle diffraction pattern can be used for the reconstruction of the magnetization profile in stripe domains [6.186, 274]. Tilt of the specimen by ± 45° allows the B_z component to be determined, as well [6.275].

Besides domain walls, periodic fluctuations in the magnetisation (ripple) can be observed. A ripple structure is not seen in single-crystal films or in electrolytically polished foils but is mostly observed in evaporated, polycrystalline films. The contrast of the ripple structure depends strongly on the length of the periodicities and on the defocusing. For a quantitative determination of the ripple spectrum, therefore, the contrast-transfer characteristic has to be considered [6.276–278].

The sensitivity of the Fresnel mode increases with increasing defocus. When planes a few centimetres below the specimen are imaged, by switching off the objective lens and imaging with the intermediate lens, for example, the relation between object and image can be regarded as a projection, with the demagnified crossover below condenser lens 1 as the projection centre. The projected shadows are shifted by the Lorentz force in the specimen plane. This has been used to image magnetic fields at the surface of superconductors during the transition from the normal to the superconducting state [6.279, 280].

There are various ways of decreasing the magnetic field of the objective lens at the specimen:

1) Switching off the objective lens and using the intermediate lens [6.281],
2) Positioning the specimen some millimetres in front of the objective lens and reducing the lens excitation [6.282, 283].
3) Use of a specially designed objective lens of long focal length and small bore to ensure that B falls off rapidly [6.284].

Coils can be used to produce a magnetic field parallel to the film and to observe the movement of ferromagnetic domain walls (see, e.g., [6.265]). Such a field normal to the electron beam also causes a deflection. Two further coils are therefore inserted above and below the specimen with opposite excitation to compensate the deflection and to hold the beam on axis.

d) Diffraction Contrast

The deflection by the Lorentz force also changes slightly the excitation error of Bragg reflections. This can result in a lateral shift of any bend contours (Sect. 8.1.1) that cross a domain wall in a single-crystal film [6.285].

e) STEM Modes

The theorem of reciprocity (Sect. 4.5.3) tells us that all modes of Lorentz microscopy can also be used in the STEM mode. For the Fresnel mode, it will be necessary to use a very small detector aperture [6.286, 287]. The advantages of STEM are that a direct record of the intensity profiles across domain-wall images is available even if the image intensity is too low for direct viewing. Contamination marks can be printed on the specimen and the deflection by the Lorentz force can be read directly in terms of the change of the spacing of these marks. An additional mode, applicable only in STEM, is the use of two half-plane detectors or a quadrant detector; the signal A-B produces differential contrast similar to that of the Foucault mode [6.288].

f) Reconstruction and Holographic Methods

Because each interferogram or defocused image may also be regarded as a hologram, the phase and the magnetisation distribution can be reconstructed [6.289, 290]. For example, an inversion method can be used to obtain information about the magnetisation in a domain wall from a divergent-wall image [6.291] or the Gerchberg-Saxton algorithm (Sect. 6.5.5) is applied to reconstruct the distribution in stripe domains [6.186]. Another holographic recording and reconstruction method is off-axis holography [6.155] (Sect. 6.5.3).

g) Stroboscopic Mode

The dynamic properties of domain walls in high-frequency magnetic fields (1–30 MHz) can be investigated by stroboscopy. The short strobe pulses of the stroboscopic illumination must be synchronized to the ac magnetic field applied to the specimen but with a variable time shift (phase angle). This is achieved by chopping the electron beam. Two methods are in use. In the first, a negative cut-off voltage is applied to the Wehnelt electrode of the electron gun and periodically overridden by a positive strobe pulse [6.292]. Alternatively, the electron beam is deflected by the static electric field of a parallel-plate condenser and returned on-axis by applying a voltage pulse that has a few nanoseconds duration [6.293, 294]. The technique can be used to investigate the forced and free oscillations of domain walls and Bloch lines, with a time resolution of the order of nanoseconds. The method allows the "mass" and relaxation times of domain walls and Bloch lines to be determined quantitatively [6.294, 295].

A further application of stroboscopy in electron microscopy is measurement of surface potentials in integrated circuits up to frequencies in the gigahertz range by means of the voltage contrast of a scanning electron microscope [6.296, 297] or by cathodoluminescence and electron-beam-induced currents [6.298, 299].

6.6.3 Imaging of Electrostatic Specimen Fields

The Fresnel mode of Lorentz microscopy also can be used for the investigation of electrostatic fields caused by charging of the specimen, by ferroelectric domains or by the electric-field strength in the depletion layer of p-n junctions.

Electrostatic fields are generated by electron bombardment in non-conducting specimens. *Mahl* and *Weitsch* [6.300, 301] detected a fluctuating granulation in the shadow projection (Fresnel mode with very large defocusing) of collodion, formvar and SiO supporting films, which disappeared after a conducting film of metal or carbon was deposited by evaporation. Charging can also be avoided by an additional bombardment with low-energy electrons of a few 100 eV [6.301]. From the deflection of the electron beam, local field strengths of the order of 10^6 V cm^{-1} can be estimated. Fluctuating charging occurs only if the beam also hits the specimen grid [6.302]. Otherwise, a stronger charging of uniform magnitude causes a larger deflection. The fluctuations can be explained in terms of a charge-compensation mechanism due to the secondary electrons produced at the specimen grid. This compensation fluctuates statistically. In an analogous method, *Warrington* [6.303] avoids image distortion when imaging glass foils by introducing an Al-C layer a few millimetres above the specimen.

Small particles, such as MgO, NaCl or polystyrene spheres, on a carbon or metal film become charged relative to the supporting film [6.304, 305]. This charging acts like a lens, it focuses the electron rays some 3–6 cm below the specimen. The charge on NaCl crystals can be estimated to be + 2 V which corresponds to a field strength at the surface of the order of 10^4–10 V/cm. *Jönsson* and *Hoffmann* [6.306] also observed positive charging by secondary-electron emission of insulated layers on a conductive support.

Ferroelectric polarisation is associated with a larger lattice deformation than that by the magnetostriction of ferromagnetics. Ferroelectric domains in ferroelectrics can therefore be distinguished by diffraction contrast and edge fringes on oblique domain boundaries [6.307–311]. *Tanaka* and *Honjo* [6.312] demonstrated by defocusing (Fresnel mode) the action of internal electric field in boundaries with a head-to-head direction of polarisation. However, the deflection angle is smaller than 10^{-2} mrad.

The electric-field strength inside the depletion layer of a p-n junction has been imaged with the Foucault mode by *Titchmarsh* and *Booker* [6.313] and with the Fresnel mode by *Merli* et al. [6.314, 315].

7. Kinematical and Dynamical Theory of Electron Diffraction

The theoretical treatment of electron diffraction at crystals needs the concepts of lattice planes and the reciprocal lattice, as in x-ray diffraction. Kinematical theory leads to the Bragg condition and to a description of the influence of the structure of a unit cell and of the external size of a crystal on the diffracted amplitude in terms of structure and lattice amplitudes, respectively. The observed diffraction pattern is equivalent to the points of intersection of the Ewald sphere of radius $1/\lambda$ with the reciprocal-lattice nodes.

The dynamical theory considers the interaction between the primary and reflected waves. For example, when the Bragg condition is satisfied, the two-beam case results in a complementary oscillation of the primary and reflected intensities with increasing thickness. On taking into account the boundary condition at the surface and the crystal periodicity of the wave-field inside the crystal, the solution of the Schrödinger equation becomes a Bloch-wave field. An example of the effect of inelastic scattering is the difference between the interaction probability for Bloch waves with nodes and antinodes at the nuclei. This results in the effect known as anomalous absorption. The critical-voltage phenomenon is a typical dynamical effect, which can cancel the intensity of Bragg reflections at a voltage that depends sensitively on the structure amplitude.

Inelastic scattering between the Bragg reflections is also influenced by the crystal periodicity and results in Kikuchi lines and bands. Diffraction by amorphous specimens produces diffuse diffraction maxima, which depend on the density distribution of atoms. Polycrystalline specimens can create Debye-Scherrer rings.

7.1 Fundamentals of Crystallography[1]

7.1.1 Bravais Lattice and Lattice Planes

A crystal lattice consists of a regular array of *unit cells*, which are the smallest building blocks of the lattice. This unit cell is a parallelepiped, built up from three non-complanar, fundamental translation vectors $\boldsymbol{a}_1$, $\boldsymbol{a}_2$, $\boldsymbol{a}_3$. The whole crystal lattice can be generated by translation of the unit cell through mul-

1 For further details the reader is referred to [7.1]

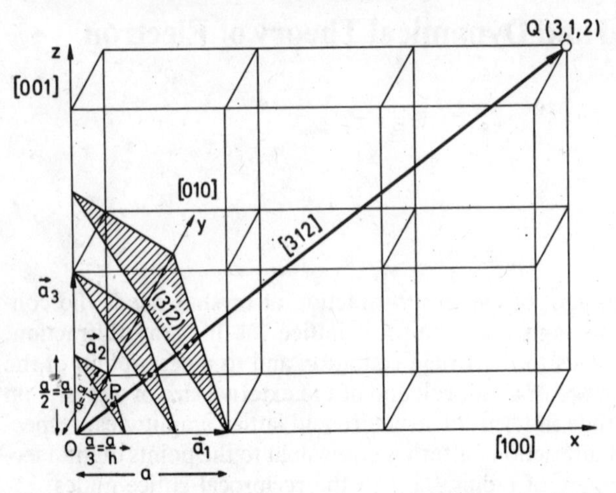

Fig. 7.1. Construction of a crystal by translation of the unit cell with fundamental vectors $\boldsymbol{a}_1$, $\boldsymbol{a}_2$, $\boldsymbol{a}_3$. Example of lattice directions [100], [010], [001] and [312] and lattice planes with Miller indices (312)

tiples of the $\boldsymbol{a}_i$ (Fig. 7.1). The origins of the unit cells, therefore, can be described by a translation vector

$$\boldsymbol{r}_g = m\boldsymbol{a}_1 + n\boldsymbol{a}_2 + o\boldsymbol{a}_3 \quad (m, n, o \text{ integer}) . \tag{7.1}$$

Alternatively, the lattice may be characterized by the values of $|\boldsymbol{a}_i| = a, b, c$ and the angles α, β, γ between the axes (Table 7.1). The unit cell is said to be *primitive* if one single atom in the unit cell is sufficient to describe the positions of all other atoms by $\boldsymbol{r}_g$. The unit cell normally contains more than one $(k = 1, \ldots, n)$ atoms at the positions

$$\boldsymbol{r}_k = u_k\boldsymbol{a}_1 + v_k\boldsymbol{a}_2 + w_k\boldsymbol{a}_3 , \tag{7.2}$$

$\boldsymbol{r}_1 = (0{,}0{,}0)$ and $\boldsymbol{r}_2 = \left(\frac{1}{2}, \frac{1}{2}, \frac{1}{2}\right)$ in a body-centred cubic lattice, for example, (Table 7.1). All other lattice points (open circles) belong to neighbouring unit cells. Table 7.1 lists the unit cells of the most-important crystal structures and the coordinates (u_k, v_k, w_k). The position of an atom in a Bravais translation lattice can be described by the vector sum $\boldsymbol{r}_g + \boldsymbol{r}_k$.

The cubic lattices have the advantage that their structure can be described by a Cartesian coordinate system. However, it is worth mentioning that the face- and body-centred cubic lattices can also be described by a primitive unit cell which is, however, trigonal in shape (Table 7.1).

The direction in crystals is expressed as a vector that connects the origin O to the origin Q of another unit cell; the components of the vector are scaled so that all are integer, as small as possible (e.g., [312], [100] etc. in Fig. 7.1).

Table 7.1. List of the most common crystal types (Bravais translation lattices). (Structure of the unit cell, lattice-plane spacings d_{hkl} and structure factors $|F|^2$)

1. Cubic lattices $a = b = c$; $\alpha = \beta = \gamma = 90°$; $d = \dfrac{a}{\sqrt{h^2 + k^2 + l^2}}$

a) Body-centred cubic lattice (e.g. Cr, Fe, Mo, W)
The unit cell consists of atoms at

$(0,0,0)$ and $\left(\frac{1}{2},\frac{1}{2},\frac{1}{2}\right)$

$|F|^2 = 0$ if $(h + k + l)$ odd

$|F|^2 = 4f_{\mathrm{Cr}}^2$ if $(h + k + l)$ even

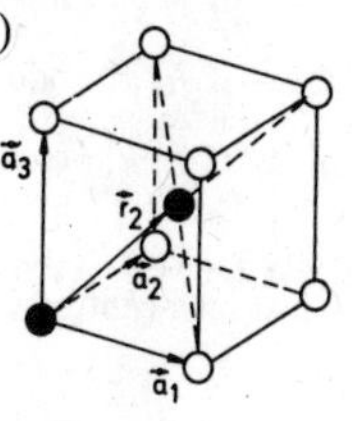

Cubic unit cell

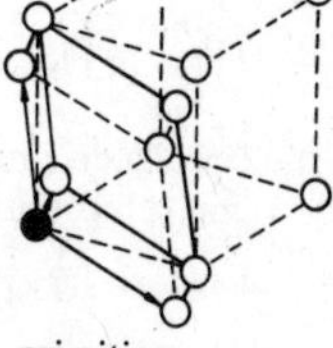

primitive unit cell

b) Face-centred cubic lattice (e.g. Al, Ni, Cu, Ag, Au, Pt)
The unit cell consists of atom at

$(0,0,0)$, $\left(\frac{1}{2},\frac{1}{2},0\right)$, $\left(\frac{1}{2},0,\frac{1}{2}\right)$, $\left(0,\frac{1}{2},\frac{1}{2}\right)$

$|F|^2 = 0$ if h, k, l mixed (even and odd)

$|F|^2 = 16\, f_{\mathrm{Al}}^2$ if h, k, l all even or odd

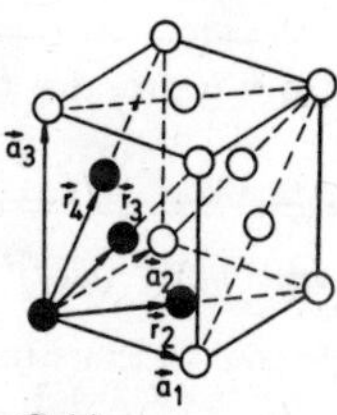

Cubic unit cell

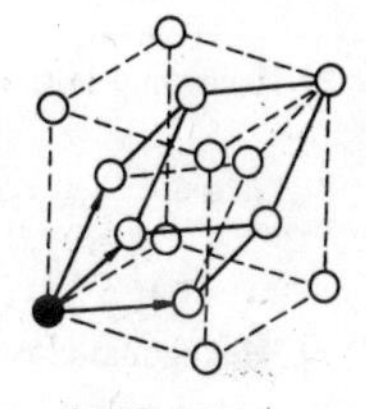

primitive unit cell

c) Diamond structure (e.g. C, Si, Ge)
The unit cell consists of two face-centred lattices shifted by one quarter of the body diagonal: $\left(\frac{1}{4},\frac{1}{4},\frac{1}{4}\right)$

$|F|^2 = 0$ if h, k, l mixed

$|F|^2 = 64\, f_{\mathrm{Ge}}^2$ if h, k, l even and $(h + k + l) = 4n$

$|F|^2 = 32\, f_{\mathrm{Ge}}^2$ if h, k, l odd

$|F|^2 = 0$ if h, k, l even and $(h + k + l) = 4(n + 1/2)$

d) Caesium chloride structure (e.g. CsCl, TlCl)
The unit cell consists of two primitive cubic lattices shifted by half a body diagonal: $\left(\frac{1}{2},\frac{1}{2},\frac{1}{2}\right)$

Cs: $(0,0,0)$; Cl: $\left(\frac{1}{2},\frac{1}{2},\frac{1}{2}\right)$

$F^2 = (f_{\mathrm{Cs}} + f_{\mathrm{Cl}})^2$ if $(h + k + l)$ even

$F^2 = (f_{\mathrm{Cs}} - f_{\mathrm{Cl}})^2$ if $(h + k + l)$ odd

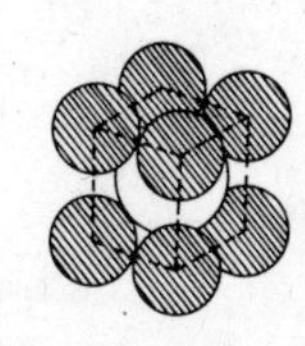

CsCl

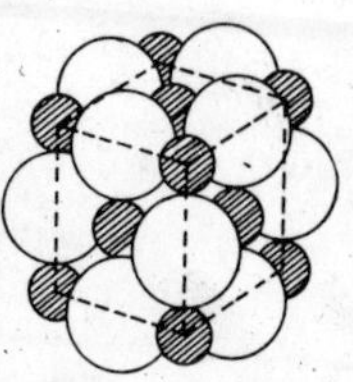

NaCl

Table 7.1 (continued)

e) Sodium chloride structure (e.g., NaCl, LiF, MgO)
Unit cell consists of two face-centred Na and Cl sublattices shifted by half a body diagonal: $\left(\frac{1}{2}, \frac{1}{2}, \frac{1}{2}\right)$

$|F|^2 = 0$ if h, k, l mixed

$|F|^2 = 16\,(f_{Na} - f_{Cl})^2$ if h, k, l odd

$|F|^2 = 16\,(f_{Na} + f_{Cl})^2$ if h, k, l even

f) Zincblende structure (e.g., ZnS, CdS, InSb, GaAs)
Unit cell consists of two face-centred Zn and S sublattices shifted by one quarter of a body diagonal: $\left(\frac{1}{4}, \frac{1}{4}, \frac{1}{4}\right)$

$|F|^2 = 0$ if h, k, l mixed

$|F|^2 = 16\,(f_{Zn}^2 + f_S^2)$ if h, k, l odd

$|F|^2 = 16\,(f_{Zn} + f_S)^2$ if h, k, l even and $(h + k + l) = 4n$

$|F|^2 = 16\,(f_{Zn} - f_S)^2$ if h, k, l even and $(h + k + l) = 4(n + 1/2)$

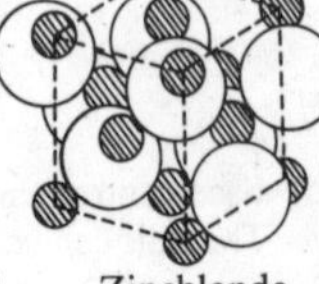
Zincblende

2. Hexagonal lattices, $a = b \neq c,\ \alpha = \beta = 90°,\ \gamma = 120°$

$$d = \frac{a}{\sqrt{\frac{4}{3}(h^2 + k^2 + hk) + (a/c)^2 l^2}}$$

a) Hexagonal close-packed structure (e.g. Mg, Cd, Co, Zn)
The unit cell consists of atoms at (0,0,0), $\left(\frac{1}{3}, \frac{1}{3}, \frac{1}{2}\right)$

$|F|^2 = 0$ if l odd, $(h + 2k) = 3n$

$|F|^2 = 4f^2$ if l even, $(h + 2k) = 3n$

$|F|^2 = 3f^2$ if l odd, $(h + 2k) = 3n + 1$ or $3n + 2$

$|F|^2 = f^2$ if l even, $(h + 2k) = 3n + 1$ or $3n + 2$

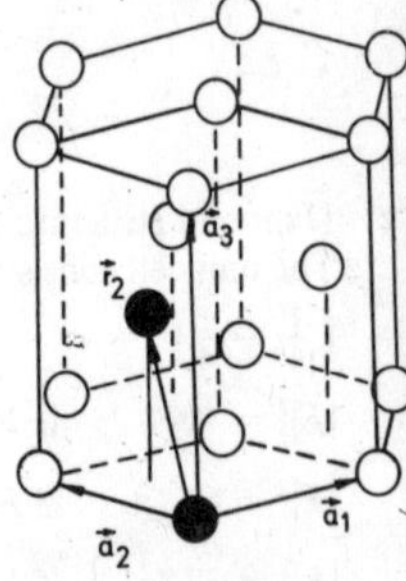

b) Wurtzite structure (e.g., ZnS, ZnO)
Unit cell consists of two hexagonal close-packed Zn and S sublattices shifted by $\left(\frac{1}{3}, \frac{1}{3}, \frac{1}{8}\right)$

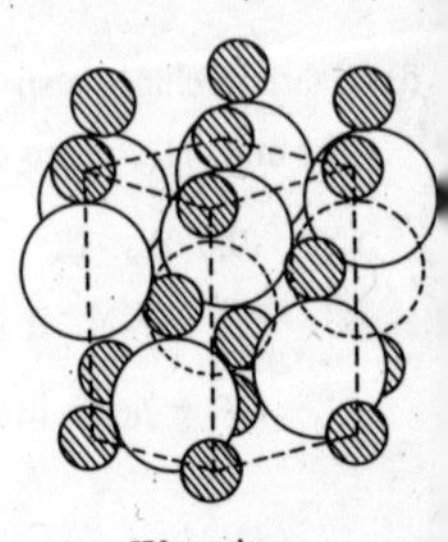
Wurtzite

Table 7.1 (continued)

3. Tetragonal lattices, $a = b \neq c$, $\alpha = \beta = \gamma = 90°$

$$d = \frac{a}{\sqrt{h^2 + k^2 + (a/c)^2 l^2}}$$

4. Orthorhombic lattices, $a \neq b \neq c$, $\alpha = \beta = \gamma = 90°$

$$d = \frac{1}{\sqrt{(h/a)^2 + (k/b)^2 + (l/c)^2}}$$

5. Trigonal lattices, $a = b = c$, $\alpha = \beta = \gamma = 120°$

$$d = a\sqrt{\frac{1 - 3\cos^2\alpha + 2\cos^3\alpha}{B\sin^2\alpha + 2C(\cos^2\alpha - \cos\alpha)}} \; ; \quad \mathrm{B} = h^2 + k^2 + l^2 \quad \mathrm{C} = hk + kl + hl$$

6. Monoclinic lattices, $a \neq b \neq c$, $\alpha = \gamma = 90°, \beta \neq 90°$

$$d = \frac{1}{\sqrt{A/\sin^2\beta + k^2/b^2}} \; ; \quad A = \frac{h^2}{a^2} + \frac{l^2}{c^2} - \frac{2hl}{ac}\cos\beta$$

7. Triclinic lattices, $a \neq b \neq c$; $\alpha \neq \beta \neq \gamma \neq 90°$

$$d = abc\sqrt{\frac{1 - \cos^2\alpha - \cos^2\beta - \cos^2\gamma + 2\cos\alpha\cos\beta\cos\gamma}{q_{11}h^2 + q_{22}k^2 + q_{33}l^2 + q_{12}hk + q_{13}hl + q_{23}kl}}$$

$$q_{11} = b^2c^2\sin^2\alpha \; ; \; q_{22} = a^2c^2\sin^2\beta \; ; \; q_{33} = a^2b^2\sin^2\gamma$$

$$q_{12} = 2abc^2(\cos\alpha\cos\beta - \cos\gamma)$$

$$q_{13} = 2ab^2c(\cos\alpha\cos\gamma - \cos\beta)$$

$$q_{23} = 2a^2bc(\cos\beta\cos\gamma - \cos\alpha)$$

Lattice planes are parallel, equidistant planes through the crystal with the same periodicity as the unit cells. Examples of three equidistant lattice planes are shown in Fig. 7.1, a further plane goes through the origin O. Such a set of lattice planes can be characterized by *Miller indices*. The plane closest to the one that passes through the origin intercepts the fundamental translation vectors $\boldsymbol{a}_i$ at points that may be written: $\boldsymbol{a}_1/h$, $\boldsymbol{a}_2/k$, $\boldsymbol{a}_3/l$ (h, k, l integers); otherwise, the system of parallel planes could not have the same periodicity as the lattice, because this requirement implies that there must be an integral number h, k, l of interceptions of parallel planes that divide the translation vectors $\boldsymbol{a}_i$ of the unit cell into equal parts. The triple (hkl) is the set of Miller indices, which are the reciprocal intercepts in units of $|\boldsymbol{a}_i|$. The intercepts in Fig. 7.1 are $a_1/h = a_1/3$, $a_2/k = a_2/1$, $a_3/l = a_3/2$ and so the Miller indices are (312). Miller indices are always enclosed in parentheses to distinguish them

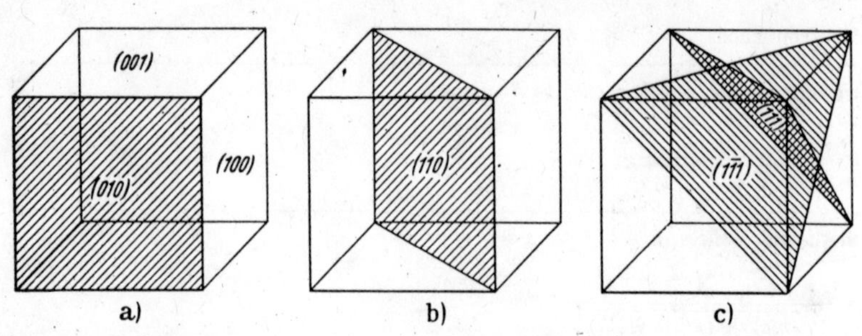

Fig. 7.2 a–c. Example of lattice planes in a cubic crystal: (**a**) cubic planes $\{100\}$, (**b**) dodecahedral planes $\{110\}$ and (**c**) octahedral planes $\{111\}$

from directions, which are always denoted by square brackets. Only in cubic lattices is the $[hkl]$ direction normal to the (hkl) lattice planes.

If a lattice plane intersects one or two axes at infinity, which means that the plane is parallel to one or two of the $\boldsymbol{a}_i$, then the corresponding Miller indices are zero (Fig. 7.2). If the plane cuts one of the axes on the negative side of the origin the corresponding Miller indices are negative. This is indicated by placing a minus sign above the index, e.g. $(1\bar{1}1)$ in Fig. 7.2c, which shows further examples of indices in a cubic lattice.

For hexagonal lattices, four indices $(hkil)$ are often used; these are obtained from the intercepts with the c axis and the three binary axes inclined at 120° to one another. The indices h, k and i satisfy the relation $i = -(h + k)$.

If we wish to refer to a full set of equivalent lattice planes, all six cubic faces of a cubic crystal for example: (100), (010), (001), $(\bar{1}00)$, $0\bar{1}0)$, $(00\bar{1})$, we enclose the Miller indices in braces (curly brackets): $\{100\}$. Thus we might say: The $\{111\}$-planes in a face-centred cubic lattice are close-packed planes. A full set of crystallographic equivalent directions or axes, all directions parallel to the fundamental vectors $\boldsymbol{a}_i$ for example, is denoted by angle brackets, thus: $\langle 100 \rangle$.

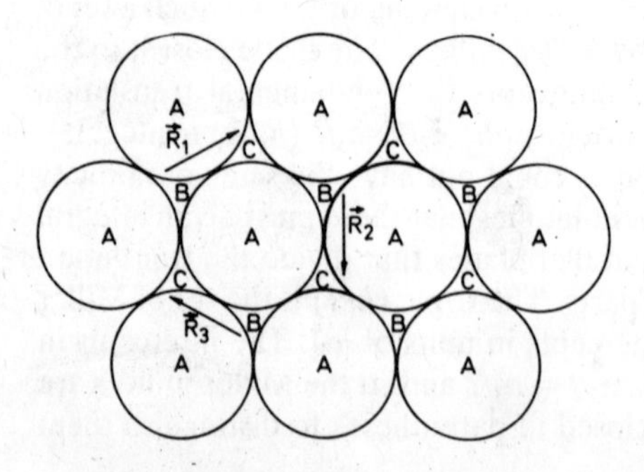

Fig. 7.3. Close-packed atoms in a layer A and positions of the atoms in the neighbouring layer B or C. $\boldsymbol{R}_i$ ($i = 1, 2, 3$) are the displacement vectors of the layers in a close-packed lattice

Close-packed structures such as the face-centred cubic and the hexagonal close-packed structures are of special interest. Figure 7.3 shows that there are two possible sets of positions, B and C, at which a second close-packed plane can be stacked above the plane with atoms at the positions A. The face-centred cubic lattice can be characterized by the sequence ABCABC..., and the {111}-planes are then close-packed, whereas the close-packed hexagonal structure follows the sequence ABAB... and the close-packed planes are now (0001). This corresponds to a ratio $c/a = \sqrt{8/3} = 1.63$. However, the measured value of this ratio for hexagonal crystals is slightly different, as a result of binding forces depending on the crystallographic directions parallel and normal to the close-packed planes.

7.1.2 The Reciprocal Lattice

The reciprocal-lattice concept is important for the understanding and interpretation of electron-diffraction patterns. There are different ways of introducing the reciprocal lattice, which will be shown to be equivalent.

We start with an intuitive, graphical construction. Each point of the reciprocal lattice will be related to a set of lattice planes of the crystal lattice with Miller indices (hkl). This point can be constructed by plotting a vector $\boldsymbol{n}$ normal to the (hkl) planes and of length $1/d_{hkl}$ from the origin O of the reciprocal lattice. This procedure is illustrated in Fig. 7.4 for a two-dimensional projection of a lattice ($\boldsymbol{a}_3$ normal to the plane built up by the vectors $\boldsymbol{a}_1$ and $\boldsymbol{a}_2$). The lattice planes $(hk0) = (320)$ intercept this plane in parallel straight lines d_{hk0} apart. Figure 7.4 shows that all points of a reciprocal lattice can be described by the reciprocal translation vectors $\boldsymbol{a}_1^*$ and $\boldsymbol{a}_2^*$ with $|\boldsymbol{a}_1^*| = 1/d_{100} = 1/a_1$ and $|\boldsymbol{a}_2^*| = 1/d_{010} = 1/a_2$, and that $\boldsymbol{g} = h\boldsymbol{a}_1^* + k\boldsymbol{a}_2^*$ is a reciprocal-lattice vector.

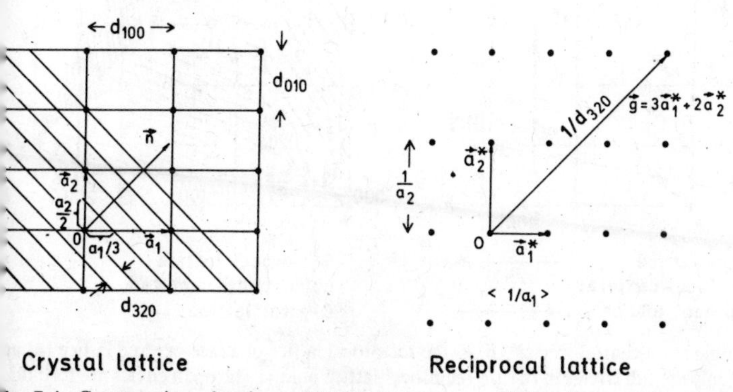

Fig. 7.4. Construction of reciprocal-lattice vectors $\boldsymbol{g}$ parallel to the normal to the crystal-lattice planes with indices (hkl) and of length $1/d_{hkl}$

The next method is a more abstract, mathematical construction. If the $\boldsymbol{a}_i$ are the fundamental translation vectors of a primitive unit cell, the lattice vectors $\boldsymbol{a}_i$ and the translation vectors $\boldsymbol{a}_j^*$ of the reciprocal lattice are related by

$$\boldsymbol{a}_i \cdot \boldsymbol{a}_j^* = \delta_{ij} = \begin{cases} 0 \text{ if } i \neq j \\ 1 \text{ if } i = j \end{cases} \qquad (i, j = 1, 2, 3) \; . \tag{7.3}$$

This system of nine equations has the solution

$$\boldsymbol{a}_1^* = \frac{\boldsymbol{a}_2 \times \boldsymbol{a}_3}{V_e} \; , \tag{7.4}$$

$\boldsymbol{a}_2^*$ and $\boldsymbol{a}_3^*$ being obtained by cyclic exchange of the indices. $V_e = \boldsymbol{a}_1 \cdot (\boldsymbol{a}_2 \times \boldsymbol{a}_3)$ is the volume of the unit cell. This shows that the vector $\boldsymbol{a}_1^*$ is normal to $\boldsymbol{a}$ and $\boldsymbol{a}_3$.

The reciprocal lattice of a primitive cubic lattice with lattice constant a i again a primitive cubic; the lattice constant of the reciprocal unit cell is $1/a$ The reciprocal lattice of a face-centred cubic (fcc) lattice can be deduced b considering the primitive trigonal cell of Table 7.1; we see that any funda mental vector of the primitive trigonal unit cell of the body-centred cubi (bcc) lattice is normal to two fundamental vectors of the primitive trigona unit cell of the fcc lattice. The condition (7.4) for the fcc lattice therefore, i fulfilled by a bcc lattice. Conversely, the reciprocal lattice of a bcc lattice i fcc (Fig. 7.5).

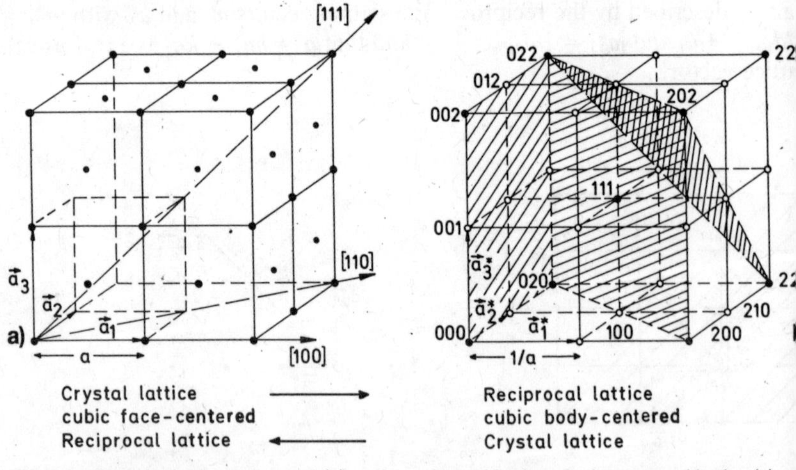

Fig. 7.5. Cubic body-centred crystal (**b**) as the reciprocal lattice of a face-centred lattice (**a**) ar vice versa. Only the full circles in (b) are reciprocal-lattice points. The open circles are forbidd by the extinction rules for the structure amplitude F. The shaded planes are used in Sect. 7.1.3 construct the Laue zones (Fig. 7.7)

In most crystals, a primitive unit cell cannot be found. Not all of the reciprocal-lattice points predicted by (7.4) can, in fact, be observed when the fundamental vectors of a non-primitive unit cell are used. Some points will disappear. In the reciprocal lattice of a fcc structure (Fig. 7.5 b), for example, the reciprocal-lattice points 200, 220 etc. are allowed, but 100, 210 etc. are "forbidden". The reasons for this will become clearer when we meet the zero rules ($F = 0$) of the structure amplitude F in Sect. 7.2.2 or the interpretation of the reciprocal lattice as the three-dimensional Fourier transform of the crystal lattice (see the end of this section).

Let us now consider some important laws, which can be derived from the definitions (7.3) and (7.4) of the reciprocal lattice:

1) The reciprocal-lattice vector $\boldsymbol{g} = h\boldsymbol{a}_1^* + k\boldsymbol{a}_2^* + l\boldsymbol{a}_3^*$ (h, k, l integers) is normal to the (hkl) lattice planes.
Proof: Figure 7.1 shows that two non-parallel vectors in the (hkl) plane can be obtained as differences between the points at which the fundamental lattice vectors intersect this plane:
$\boldsymbol{r}_1 = \boldsymbol{a}_1/h - \boldsymbol{a}_2/k$, $\boldsymbol{r}_2 = \boldsymbol{a}_1/h - \boldsymbol{a}_3/l$. The scalar products of these vectors with $\boldsymbol{g}$ are zero, which means that $\boldsymbol{g} \perp \boldsymbol{r}_1, \boldsymbol{r}_2$; $\boldsymbol{g}$ is, therefore, also normal to all other vectors that lie in the (hkl) plane and can hence be described by a linear combination of $\boldsymbol{r}_1$ and $\boldsymbol{r}_2$.

2) The length of the reciprocal-lattice vector $\boldsymbol{g}$ is equal to the reciprocal-lattice-plane spacing $1/d_{hkl}$.
Proof: Let $\boldsymbol{u}_\mathrm{n}$ be the unit vector normal to the (hkl) plane and hence parallel to $\boldsymbol{g}$, which means that we can write: $\boldsymbol{u}_\mathrm{n} = \boldsymbol{g}/|\boldsymbol{g}|$. From Fig. 7.1 we see that d_{hkl} is equal to the projection of $\boldsymbol{a}_1/h$, $\boldsymbol{a}_2/k$ or $\boldsymbol{a}_3/l$ on the unit vector $\boldsymbol{u}_\mathrm{n}$:

$$d_{hkl} = \boldsymbol{u}_\mathrm{n} \cdot \boldsymbol{a}_1/h = \frac{\boldsymbol{g}}{|\boldsymbol{g}|} \cdot \frac{\boldsymbol{a}_1}{h} = \frac{1}{|\boldsymbol{g}|} . \tag{7.5}$$

This definition of d_{hkl} as $1/|\boldsymbol{g}|$ can be used to calculate the distance d_{hkl} for more-complicated crystal structures (Sect. 7.1.3).

3) If the following system of equations

$$\boldsymbol{a}_i \cdot \boldsymbol{g} = h_i \quad (i = 1, 2, 3\ ;\ h_{1,2,3} = h, k, l) \tag{7.6}$$

is given, the solution will be

$$\boldsymbol{g} = h\boldsymbol{a}_1^* + k\boldsymbol{a}_2^* + l\boldsymbol{a}_3^* . \tag{7.7}$$

Proof: Substitute (7.7) into (7.6) and use (7.3).

The third way to introduce the reciprocal lattice is to define it as the Fourier transform of the crystal lattice. The Fourier integral (3.43) of a three-dimensional crystal lattice with δ-functions at the origins of the unit cells becomes a sum over the discrete lattice points $\boldsymbol{r}_g$ (7.1),

$$F(q) = \sum_{m,n,o} \exp(-2\pi i \boldsymbol{q}\cdot\boldsymbol{r}_g) = \sum_{m,n,o} \exp[-2\pi i \boldsymbol{q}\cdot(m\boldsymbol{a}_1 + n\boldsymbol{a}_2 + o\boldsymbol{a}_3)] , \tag{7.8}$$

where m, n, o, are integers. This sum will be non-zero only if the products $\boldsymbol{q}\cdot\boldsymbol{a}_i$ in the exponent are all integers. If we call these integers h_i, we recover the system of equations (7.6) and have non-vanishing values of $F(q)$ for $\boldsymbol{q} = \boldsymbol{g}$.

7.1.3 Calculation of Lattice-Plane Spacings

The most-general method for calculating the distances between lattice planes, d_{hkl}, is to use (7.5) with $\boldsymbol{g} = h\boldsymbol{a}_1^* + k\boldsymbol{a}_2^* + l\boldsymbol{a}_3^*$

$$\begin{aligned} 1/d_{hkl} = |\boldsymbol{g}|^2 &= h^2|\boldsymbol{a}_1^*|^2 + k^2|\boldsymbol{a}_2^*|^2 + l^2|\boldsymbol{a}_3^*|^2 \\ &+ 2(hk\boldsymbol{a}_1^*\cdot\boldsymbol{a}_2^* + kl\boldsymbol{a}_2^*\cdot\boldsymbol{a}_3^* + lh\boldsymbol{a}_3^*\cdot\boldsymbol{a}_1^*) . \end{aligned} \tag{7.9}$$

For the simple case of the cubic lattice with $a = b = c$ and $\alpha = \beta = \gamma = 90°$, the $\boldsymbol{a}_i^*$ are orthogonal and the last bracket in (7.9) becomes zero. Further, we have $|\boldsymbol{a}_1^*| = |\boldsymbol{a}_2^*| = |\boldsymbol{a}_3^*| = 1/a$ so that

$$d_{hkl} = a/\sqrt{h^2 + k^2 + l^2} . \tag{7.10}$$

The most-complex case is the triclinic lattice with $a \neq b \neq c$ and $\alpha \neq \beta \neq \gamma \neq 90°$; this has the lowest symmetry of all the Bravais translation lattices listed in Table 7.1. The first term of (7.9) becomes

$$h^2|\boldsymbol{a}_1^*|^2 = h^2\frac{|\boldsymbol{a}_2\times\boldsymbol{a}_3|^2}{V_e^2} = h^2\frac{(bc\sin\alpha)^2}{V_e^2} \tag{7.11}$$

and similarly for the second and third terms. The first term in the bracket of (7.9) can be transformed in this manner:

$$hk\boldsymbol{a}_1^*\cdot\boldsymbol{a}_2^* = \frac{hk}{V_e^2}(\boldsymbol{a}_2\times\boldsymbol{a}_3)\cdot(\boldsymbol{a}_3\times\boldsymbol{a}_1) . \tag{7.12}$$

By use of the vector relation $(\boldsymbol{a}\times\boldsymbol{b})\cdot(\boldsymbol{c}\times\boldsymbol{d}) = (\boldsymbol{a}\cdot\boldsymbol{c})(\boldsymbol{b}\cdot\boldsymbol{d}) - (\boldsymbol{a}\cdot\boldsymbol{d})(\boldsymbol{b}\cdot\boldsymbol{c})$, the product in (7.12) becomes

$$\begin{aligned} (\boldsymbol{a}_2\times\boldsymbol{a}_3)\cdot(\boldsymbol{a}_3\times\boldsymbol{a}_1) &= (\boldsymbol{a}_2\cdot\boldsymbol{a}_3)(\boldsymbol{a}_3\cdot\boldsymbol{a}_1) - (\boldsymbol{a}_2\cdot\boldsymbol{a}_1)\,a_3^2 \\ &= abc^2(\cos\alpha\cos\beta - \cos\gamma) . \end{aligned} \tag{7.13}$$

The volume of the unit cell

$$V_e = \boldsymbol{a}_1 \cdot (\boldsymbol{a}_2 \times \boldsymbol{a}_3) = \boldsymbol{a}_3 \cdot (\boldsymbol{a}_1 \times \boldsymbol{a}_2) = \boldsymbol{a}_2 (\boldsymbol{a}_3 \times \boldsymbol{a}_1) \tag{7.14}$$

reduces to

$$V_e = a_{3z} ab \sin\gamma \tag{7.15}$$

if $\boldsymbol{a}_1 \| x$ and $\boldsymbol{a}_2$ lie in the x-y plane of a Cartesian coordinate system. The x component $a_{3x} = c\cos\beta$, and the y component a_{3y} can be obtained from the relation

$$\begin{aligned}\boldsymbol{a}_2 \cdot \boldsymbol{a}_3 &= a_{2x}a_{3x} + a_{2y}a_{3y} + a_{2z}a_{3z} \\ &= bc\cos\gamma\cos\beta + ba_{3y}\sin\gamma + 0 = bc\cos\alpha\ .\end{aligned}$$

Solving for a_{3y}, we find

$$a_{3y} = \frac{c(\cos\alpha - \cos\gamma\cos\beta)}{\sin\gamma}\ . \tag{7.16}$$

Using the relation $c^2 = a_{3x}^2 + a_{3y}^2 + a_{3z}^2$, solving for a_{3z}^2 and substituting for a_{3x} and a_{3y}, we find

$$a_{3z}^2 = \frac{c^2}{\sin^2\gamma}(1 - \cos^2\alpha - \cos^2\beta - \cos^2\gamma + 2\cos\alpha\cos\beta\cos\gamma)\ . \tag{7.17}$$

Substituting (7.17) into (7.15) and (7.15, 12, 11) into (7.9), we obtain the last formula of Table 7.1, from which all the other formulae for systems of higher symmetry can be derived.

7.1.4 Construction of Laue Zones

The product of a translation vector $\boldsymbol{r}_g$ (7.1) of the crystal lattice and a reciprocal-lattice vector $\boldsymbol{g}$ (7.7)

$$\boldsymbol{g} \cdot \boldsymbol{r}_g = mh + nk + ol = N \tag{7.18}$$

is an integer. If $N = 0$, all $\boldsymbol{g}$ for a given value of $\boldsymbol{r}_g$ lie in a plane through the origin of the reciprocal lattice and are normal to the zone axis $\boldsymbol{r}_g$. The system of lattice planes that belongs to these values of $\boldsymbol{g}$ forms a bundle of planes that have the zone axis as a common line of intersection (Fig. 7.6 a). The reciprocal lattice plane that contains the corresponding $\boldsymbol{g}$ is called the zero *Laue zone*. For $N = 1, 2, \ldots$ the first, second and higher Laue zone (FOLZ, and HOLZ), respectively, are obtained, which are parallel to the zero Laue zone (Fig. 7.6b). This means that the Laue zones are parallel sections through the reciprocal lattice.

The construction of Laue zones is very useful for indexing and computation of electron-diffraction patterns. Either triplets of integers *hkl* are sought that fulfil the condition (7.18) and have non-zero structure amplitude (Sect. 7.2.2), or a model of the reciprocal lattice like that of Fig. 7.5 b can be used.

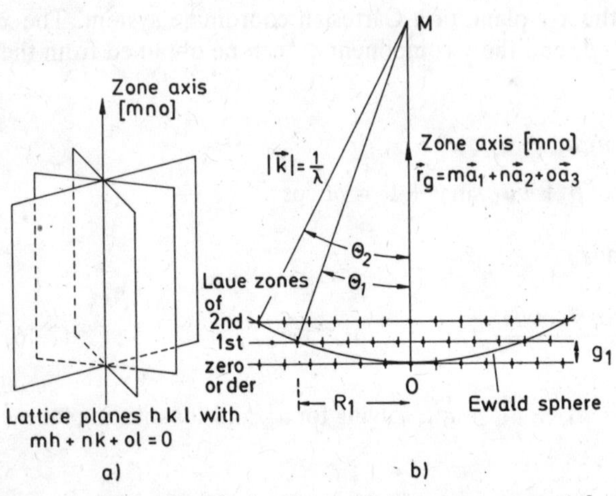

Fig. 7.6. (**a**) Bundle of lattice planes with the common zone axis [*mno*]. (**b**) Position of the zero and high-order Laue zones in the reciprocal lattice. The angles θ_1 and θ_2 at which the Ewald sphere cuts the high-order Laue zones are discussed in Sect. 9.4.6

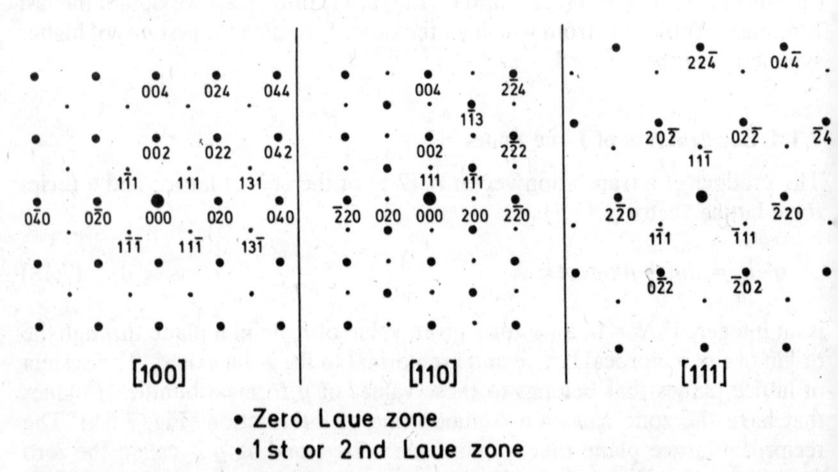

Fig. 7.7. Examples of the construction of zero and first or second Laue zones for the [100], [110] and [111] zone axes of a cubic face-centred lattice

From Fig. 7.5 b, for example, the reciprocal-lattice points for a fcc lattice can be read; these are situated on the Laue zones for the zone axes [*mno*] = [100], [110] and [111]. Figure 7.7 shows indexed zone patterns for the zero, first and second Laue zones.

7.2 Kinematical Theory of Electron Diffraction

7.2.1 Bragg Condition and Ewald Sphere

The Laue conditions $\boldsymbol{q} \cdot \boldsymbol{a}_i = h_i$ (integers), which result from the Fourier transform (7.8) of the crystal lattice, guarantee that the scattered plane waves with wave vector $\boldsymbol{k}$ do indeed overlap and interfere constructively so that their amplitudes sum. With $\boldsymbol{q} = \boldsymbol{k} - \boldsymbol{k}_0$ (3.42) and $\boldsymbol{q} = \boldsymbol{g}$ from (7.8), the forgoing Laue conditions can be solved for $\boldsymbol{q} = \boldsymbol{k} - \boldsymbol{k}_0$ by use of (7.6) and (7.7), which gives

$$\boldsymbol{k} - \boldsymbol{k}_0 = \boldsymbol{g} = h\boldsymbol{a}_1^* + k\boldsymbol{a}_2^* + l\boldsymbol{a}_3^* \ . \tag{7.19}$$

This is the Bragg condition in vector notation. The vector $\boldsymbol{g} = \boldsymbol{k} - \boldsymbol{k}_0$ is normal to the bisector of the angle between $\boldsymbol{k}_0$ and $\boldsymbol{k}$ (Fig. 7.8). On the right-hand side of (7.19), we have the reciprocal-lattice vector $\boldsymbol{g}$, which is normal to the lattice planes (*hkl*). It follows that $\boldsymbol{k} - \boldsymbol{k}_0$ is parallel to this normal. The glancing angles of incidence and of scattering θ_B relative to the lattice planes (Fig. 7.8) must be equal. Although this is strictly an interference phenomenon, the result can be interpreted as a reflection at the lattice planes. It differs from light-optical reflection in that only a fixed glancing angle θ_B is allowed. This Bragg angle θ_B can be calculated by examining the magnitude

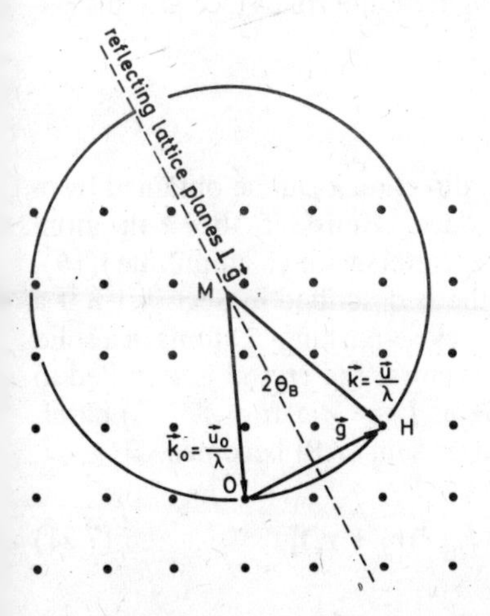

Fig. 7.8. Ewald sphere of radius $k = 1/\lambda$ in a reciprocal lattice. A Bragg reflection is excited if the sphere intersects a reciprocal lattice point, e.g. *H*

of (7.19). Figure 7.8 shows that $|\boldsymbol{k} - \boldsymbol{k}_0| = 2\sin\theta_B/\lambda$ and $|\boldsymbol{g}| = 1/d_{hkl}$ (7.5). Thus (7.19) results in the well-known Bragg condition

$$2\, d_{hkl} \sin\theta_B = \lambda\ . \tag{7.20}$$

The Bragg condition is valid for x-rays and electrons. Typical back-reflections with $2\,\theta_B$ close to 180° can be obtained with x-rays, whereas the same lattice planes typically give forward-reflection for 100 keV electrons since the wavelength is so much smaller. The scattering amplitude for x-rays is approximately isotropic for all scattering angles, and back-reflection can be observed, whereas the scattering amplitude for electrons decreases with increasing angle and the Bragg reflections are limited to a cone with an aperture of the order of a few 10^{-2} rad.

Formula (7.19) can be used to generate a construction first applied by Ewald. A vector $\boldsymbol{k}_0 = \overline{\mathrm{MO}}$ is drawn with one end at the origin O of the reciprocal lattice and with a length $|\boldsymbol{k}_0| = 1/\lambda$ (Figs.7.6 and 8). The other end M (excitation point) of $\boldsymbol{k}_0$ is taken as the centre of a sphere of radius $1/\lambda$. Diffraction will be observed only if this *Ewald sphere* intersects one or more points $\boldsymbol{g}$ of the reciprocal lattice (e.g., H in Fig. 7.8). The direction $\boldsymbol{k} = \overline{\mathrm{MH}}$ will be the direction of the scattered wave and $\boldsymbol{k} - \boldsymbol{k}_0 = \boldsymbol{g}$ is the vector that connects the endpoints of $\boldsymbol{k}$ and $\boldsymbol{k}_0$.

The Ewald sphere in Fig. 7.8 has been drawn with a small radius, as for x-rays. In electron diffraction, the radius of the Ewald sphere, $1/\lambda = 240\ \mathrm{nm}^{-1}$ for 80 keV electrons, is much larger than the distances between the reciprocal-lattice points, e.g. $1/a = 2.8\ \mathrm{nm}^{-1}$ for copper (Fig. 7.6b).

If the incident electron beam is parallel to a zone axis, the diffraction pattern (e.g., Fig. 7.24) contains Bragg reflections near the primary beam from the zero Laue zone; at larger Bragg angles, circles of reflections occur where the Ewald sphere cuts the first and higher Laue zones (see also discussion of HOLZ patterns in Sect. 9.4.6).

7.2.2 Structure and Lattice Amplitude

The amplitude of the scattered wave in the direction $\boldsymbol{k}$ can be obtained from the Fourier transform of the crystal lattice. Consider the k-th atom ($k = 1, \ldots, n$) inside one unit cell; this atom scatters with an amplitude $f_k(\theta)$, which can be calculated by one of the methods described in Sect. 4.1.3 if a screened Coulomb potential modified by the close-packing of atoms in a solid (e.g., the muffin-tin model) is used. Furthermore, the crystal is assumed to be parallelepipedal in shape, with edge lengths $L_i = M_i a_i$ ($i = 1, 2, 3$) parallel to the fundamental vectors $\boldsymbol{a}_i$. The Fourier sum (7.8) becomes

$$F(q) = \sum_{m=1}^{M_1} \sum_{n=1}^{M_2} \sum_{o=1}^{M_3} \sum_{k=1}^{n} f_k \exp[-2\pi \mathrm{i}\,(\boldsymbol{k} - \boldsymbol{k}_0) \cdot (\boldsymbol{r}_g + \boldsymbol{r}_k)]\ . \tag{7.21}$$

The summation over k, which corresponds to the different atoms of the unit cell, can be extracted and (7.21) becomes

$$F(q) = \underbrace{\sum_{k=1}^{n} f_k \exp[-2\pi i(\boldsymbol{k}-\boldsymbol{k}_0)\cdot \boldsymbol{r}_k]}_{F} \cdot \underbrace{\sum_m \sum_n \sum_o \exp[-2\pi i(\boldsymbol{k}-\boldsymbol{k}_0)\cdot \boldsymbol{r}_g]}_{G} . \quad (7.22)$$

The first factor F is called the *structure amplitude* and depends only on the positions of the atoms inside the unit cell. The second factor G is called the *lattice amplitude* and depends on only the external shape of the crystal.

The structure amplitude will be of interest only for the Bragg condition $\boldsymbol{k}-\boldsymbol{k}_0 = \boldsymbol{g}$. It will not be altered by small deviations from the geometry of the Bragg condition, unlike G as our later calculations will show. Substituting for $\boldsymbol{r}_k$ from (7.2), we find

$$F = \sum_{k=1}^{n} f_k \exp(-2\pi i \boldsymbol{g}\cdot \boldsymbol{r}_k) = \sum_{k=1}^{n} f_k \exp[-2\pi i(u_k h + v_k k + w_k l)] . \quad (7.23)$$

The value of F will now be calculated for some typical examples:

a) Body-Centred Cubic Lattice

The coordinates of the two atoms in the unit cell (Table 7.1) are $\boldsymbol{r}_1 = (0,0,0)$ and $\boldsymbol{r}_2 = \left(\frac{1}{2},\frac{1}{2},\frac{1}{2}\right)$. Substitution in (7.23) gives

$$F = f\{1 + \exp[-\pi i(h+k+l)]\}.$$

Using the relation

$$\exp(-i\pi n) = \begin{cases} 1 & \text{if } n \text{ is an even integer} \\ -1 & \text{if } n \text{ is an odd integer} \end{cases}$$

we find

$$F = \begin{cases} 2f & \text{if } h+k+l \text{ is even} \\ 0 & \text{if } h+k+l \text{ is odd.} \end{cases}$$

b) Face-Centred Cubic Lattice

The coordinates of the four atoms of the unit cell are $\boldsymbol{r}_1 = (0,0,0)$, $\boldsymbol{r}_2 = \left(\frac{1}{2},\frac{1}{2},0\right)$, $\boldsymbol{r}_3 = \left(\frac{1}{2},0,\frac{1}{2}\right)$, $\boldsymbol{r}_4 = \left(0,\frac{1}{2},\frac{1}{2}\right)$. Hence

$$F = f\{1 + \exp[-\pi i(h+k)] + \exp[-\pi i(h+l)] + \exp[-\pi i(k+l)]\} .$$

The rules odd + odd = even etc. show that

$$F = \begin{cases} 4f & \text{if the } h,\ k,\ l \text{ are either all even or all odd} \\ 0 & \text{if the } h,\ k,\ l \text{ are mixed (odd and even).} \end{cases}$$

c) NaCl Structure

The unit cell consists of two sodium and chlorine face-centred sublattices that are shifted by one half of the body diagonal $\left(\frac{1}{2}, \frac{1}{2}, \frac{1}{2}\right)$ of the unit cell. This shift can be considered by introducing a common phase factor for the chlorine sublattice

$$\begin{aligned} F = \{&f_{\mathrm{Na}} + f_{\mathrm{Cl}} \exp[-\pi \mathrm{i}(h + k + l)]\} \\ &\{1 + \exp[-\pi \mathrm{i}(h + k)] + \exp[-\pi \mathrm{i}(h + l)] + \exp[-\pi \mathrm{i}(k + l)]\} , \end{aligned}$$

which results in

$$F = \begin{cases} 4(f_{\mathrm{Na}} + f_{\mathrm{Cl}}) & \text{if } h,\ k,\ l \text{ are even} \\ 4(f_{\mathrm{Na}} - f_{\mathrm{Cl}}) & \text{if } h,\ k,\ l \text{ are odd} \\ 0 & \text{if } h,\ k,\ l \text{ are mixed} \end{cases}$$

Similar calculations can be made for other crystal structures (Table 7.1). The three types of cubic lattices show different zero rules, i.e., different sets of (hkl) for which $F = 0$. These exclude some of the $\boldsymbol{g}$ values of the reciprocal lattice that would not be found if a primitive unit cell with only one atom at the origin of each unit cell were used. The two atoms in the body-centred cell and the four atoms in the face-centred cell all scatter either in phase (constructive interference) or in antiphase leading to $F = 0$ (destructive interference).

For more complicated structures (e.g., NaCl), the non-zero reflections can have different structure amplitudes. In the case of KCl, the difference $(f_{\mathrm{K}} - f_{\mathrm{Cl}})$ for h, k, l odd becomes very small. It is zero for x-rays because both the K^+ and the Cl^- ions have the same electron configuration as argon. In electron diffraction, there is a small residual difference between the scattering amplitudes due to the difference in the nuclear charges Ze.

We now consider the triple sum of the lattice amplitude G in (7.22) and we allow small deviations from the exact Bragg condition $\boldsymbol{k} - \boldsymbol{k}_0 = \boldsymbol{g}$. This deviation is described by the *excitation error* $\boldsymbol{s} = (s_x, s_y, s_z)$; this vector in the reciprocal lattice connects the lattice point $\boldsymbol{g}$ to the Ewald sphere in the direction parallel to the incident wave. The magnitude of s and the tilt angle $\Delta\theta$ out of the exact Bragg condition are related by (Fig. 7.9)

$$s = g\Delta\theta = \frac{\Delta\theta}{d_{hkl}} = \frac{2\sin\theta_{\mathrm{B}}}{\lambda}\Delta\theta . \tag{7.24}$$

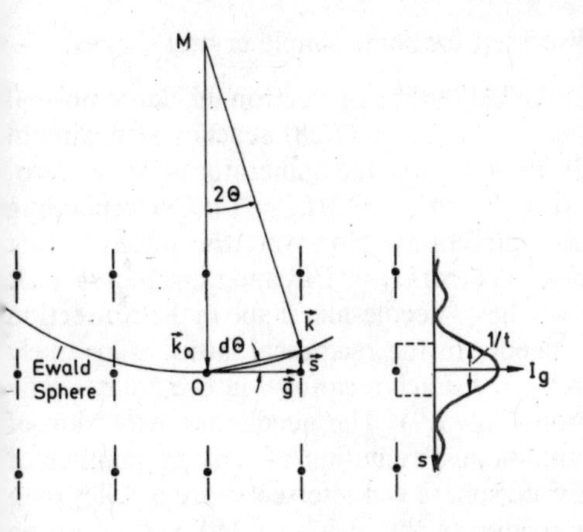

Fig. 7.9. Introduction of the excitation error s and convolution of the reciprocal-lattice points with the needle-shaped square of the lattice amplitude $|G|^2$ for a thin foil of thickness t

Substituting $\boldsymbol{k} - \boldsymbol{k}_0 = \boldsymbol{g} + \boldsymbol{s}$ in (7.22) and recalling that $\boldsymbol{g} \cdot \boldsymbol{a}_i = n$ (integer) and $\exp(-2\pi i n) = 1$, we find the lattice amplitude

$$G = \sum_{m=1}^{M_1} \sum_{n=1}^{M_2} \sum_{o=1}^{M_3} \exp[-2\pi i (\boldsymbol{g} + \boldsymbol{s}) \cdot \boldsymbol{r}_g] = \sum_{m,n,o} \exp(-2\pi i \boldsymbol{s} \cdot \boldsymbol{r}_g) \ . \quad (7.25)$$

The phase $2\pi i \boldsymbol{s} \cdot \boldsymbol{r}_g$ varies very slowly as we move through the crystal from one unit cell to another. The triple sum can therefore be replaced by an integral over the crystal volume (V_e: volume of unit cell),

$$G = \frac{1}{V_e} \int_{-L_1/2}^{+L_1/2} \int_{-L_2/2}^{+L_2/2} \int_{-L_3/2}^{+L_3/2} \exp[-2\pi i (s_x x + s_y y + s_z z)] \, dx \, dy \, dz \ . \quad (7.26)$$

Setting $V_e = a_1 a_2 a_3$ and integrating with respect to x, we find

$$G_x = \frac{1}{a_1} \int_{-L_1/2}^{+L_1/2} \exp(-2\pi i s_x x) \, dx$$

$$= \frac{1}{\pi s_x a_1} \frac{\exp(\pi i s_x L_1) - \exp(-\pi i s_x L_1)}{2i} = \frac{\sin(\pi s_x M_1 a_1)}{\pi s_x a_1} \quad (7.27)$$

and correspondingly for the y and z directions. This is the typical formula, well known in light optics, for the diffraction at a grating with M_1 slits with a spacing a_1. The total diffracted intensity becomes

$$I_g = \lambda^2 |F|^2 |G|^2 = \lambda^2 |F|^2 \frac{\sin^2(\pi s_x M_1 a_1)}{(\pi s_x a_1)^2} \frac{\sin^2(\pi s_y M_2 a_2)}{(\pi s_y a_2)^2} \frac{\sin^2(\pi s_z M_z a_3)}{(\pi s_z a_3)^2} \ . \quad (7.28)$$

The form of $|G|^2$ will now be discussed for some simple crystal shapes:

1) *Thin Crystal Foils (Disc)* with the z direction of electron incidence normal to the surface (Fig. 7.10a). The last factor in (7.28) reaches a maximum value of M_3^2 for $s_z = 0$; it first falls to zero when the numerator becomes zero, which occurs when $\pi s M_3 a_3 = \pi$ or $s_z = 1/M_3 a_3 = 1/L_3 = 1/t$. Corresponding values are found for the x and y directions. However, the intensity first becomes zero at much lower excitation errors $s_x = 1/L_1$ and $s_y = 1/L_2 = 1/D$. The function $|G(s_x, s_y, s_z)|^2$ therefore has a needle-like shape in the z direction (Fig. 7.10a). The length of the needle in the reciprocal lattice is inversely proportional to the foil thickness $L_3 = t$. Each reciprocal-lattice point will be convoluted with this $|G|^2$ function (Fig. 7.9). The needle-like extension of the lattice points provokes simultaneous excitation of a large number of Bragg reflections, because the Ewald sphere can intersect more needles than points (Figs. 7.6b and 9). Such needles in the reciprocal lattice can also be observed for plate-like precipitates in alloys. If the plates are inclined to the foil this can cause elongations of the Bragg spots or the latter may appear to be shifted (Fig. 9.25).

2) *Needle-Like Crystals* with long axis in the z direction (Fig. 7.10b). The first zero of $|G|^2$ is reached for small s_z and larger s_x and s_y. $|G|^2$ is thus a disc normal to the z-axis.

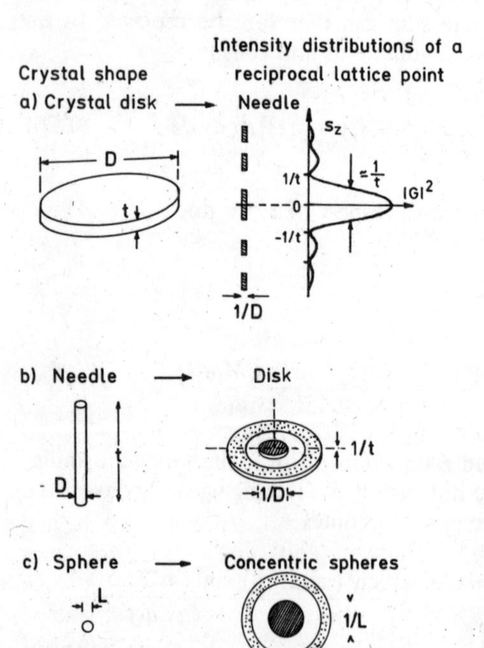

Fig. 7.10. Shape of the square of the lattice amplitude $|G|^2$ with which reciprocal-lattice points have to be convolved for (**a**) a crystal disc (thin foil of thickness t), (**b**) a needle of length t and (**c**) a sphere of diameter d

3) *Small Cubes (Spheres)* with edge (diameter) L. The extension of $|G|^2$ is the same in all directions $s_x = s_y = s_z = 1/L$ (Fig. 7.10c). If the Ewald sphere intersects a reciprocal-lattice point, a broadened diffraction spot is observed. Debye-Scherrer rings will be broadened by

$$\frac{\Delta r}{r} = \frac{1/L}{1/d} = \frac{d}{L}. \tag{7.29}$$

This relation can be used to estimate particle dimensions in the range $L =$ 0C–5 nm from the broadening Δr of the rings.

Finally we can arrive at the concept of convolution of each reciprocal-lattice point with $|G|^2$ by reasoning based on the Fourier transform and in particular, on the convolution theorem (3.51), which transforms a convolution of two functions to a product of their Fourier transforms; likewise a product is transformed to a convolution.

A crystal can be described by the following expression

$[p(\boldsymbol{r}) \otimes f(\boldsymbol{r})] \cdot g(\boldsymbol{r})$, which involves the three functions $p(\boldsymbol{r})$, $f(\boldsymbol{r})$ and $g(\boldsymbol{r})$;
$p(\boldsymbol{r})$ denotes a set of δ-functions at the origins of the unit cells;
$f(\boldsymbol{r})$ is a set of δ-functions located at the k atoms within a single unit cell.
The convolution $[p(\boldsymbol{r}) \otimes f(\boldsymbol{r})]$ represents an infinite lattice, in which each origin is convoluted with the content of a unit cell. Finally, $g(\boldsymbol{r}) = 1$ inside and 0 outside the crystal.

The Fourier transform yields

$$T\{[p(\boldsymbol{r}) \otimes f(\boldsymbol{r})] \cdot g(\boldsymbol{r})\} = [P(\boldsymbol{q}) \cdot F(\boldsymbol{q})] \otimes G(\boldsymbol{q}) \tag{7.30}$$

in which $P(\boldsymbol{q})$ denotes a set of δ-functions at the reciprocal-lattice points; $F(\boldsymbol{q})$ is the structure amplitude. The presence of points at which $F = 0$ reduces the number of reciprocal lattice points, if the unit cell contains more than one atom. $G(\boldsymbol{q})$ is the lattice amplitude. Each reciprocal-lattice point is convoluted with this function.

7.2.3 Column Approximation

In discussions of the contrast of defects in crystalline specimens, it is useful to consider not only the amplitude $F(q)$ in the Fraunhofer-diffraction plane but also the intensity at a point P just below the specimen (Fresnel diffraction).

We assume $r \to \infty$ in the formula for Fresnel diffraction (Sect. 3.2.1), which means that the point source is at infinity and hence that we have a plane incident wave of amplitude ψ_0. There are dz/V_e unit cells per unit area in an element of thickness dz, each of which scatters with a scattering angle $\theta = 2\theta_B$ to calculate the contribution $d\psi_g$ of the layer dz to the diffracted amplitude at the point P (Fig. 7.11a). Strictly $\psi_0/\cos\theta$ should be used instead

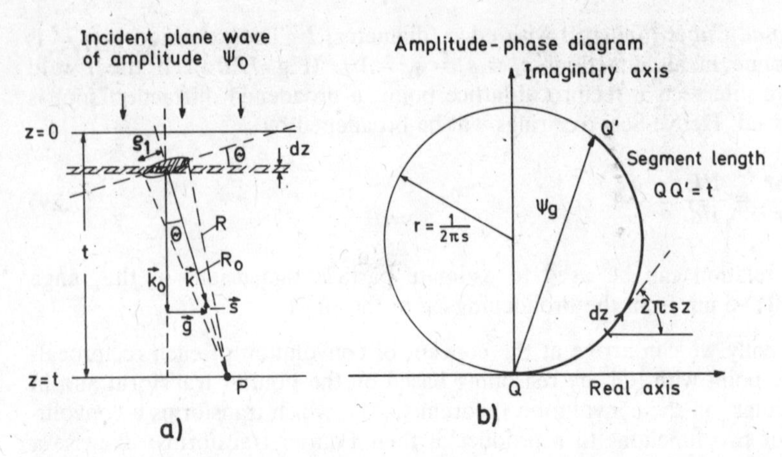

Fig. 7.11. (**a**) Column approximation for calculating the amplitude ψ_g of a diffracted wave at a point P on the bottom of a crystal foil of thickness $z = t$ [$(2\varrho_1)$ diameter of the first Fresnel zone], (**b**) Amplitude-phase diagram for calculating $I_g = \psi_g \psi_g^*$

of ψ_0 to correct for the cross-section of the wavefront. This correction can, however, be neglected because $\cos\theta \simeq 1$ for the small values of θ in question. Equation (3.31) becomes

$$d\psi_g = \psi_0 \frac{dz}{V_e} \int_S F(\theta) \frac{e^{2\pi ikR}}{R} dS = \psi_0 \frac{2\pi dz}{V_e} \int_{R_0}^{R} F(\theta) e^{2\pi ikR} dR$$

$$= i\psi_0 \frac{\lambda F(\theta)}{V_e} e^{2\pi ikR_0} dz = \frac{i\pi}{\xi_g} \psi_0 e^{2\pi ikR_0} dz \tag{7.31}$$

with $dS = 2\pi R\, dR$. We have introduced the *extinction distance* ξ_g (Table 7.2), defined by

$$\xi_g = \frac{\pi V_e}{\lambda F(\theta)} . \tag{7.32}$$

It was shown in Sect. 3.2.1 that the main contribution to the integral in (7.31) comes from the first Fresnel zone of radius $\varrho_1 = \sqrt{\lambda R_0}$ (Fig. 7.11 a). For a distance (foil thickness) $R_0 = 100$ nm and $\lambda = 3.7$ pm (100 keV electrons), we find $\varrho_1 = 0.6$ nm. This means that only a column with a diameter of 1–2 nm is contributing to the amplitude at the point P, and the method is therefore called the *column approximation*.

The amplitude ψ_g of a Bragg reflection is obtained by integrating (7.31) over the thickness. Using $\boldsymbol{k} = \boldsymbol{k}_0 + \boldsymbol{g} + \boldsymbol{s}$, $R_0 = t - z$ and $|\psi_0| = 1$, we find

$$\psi_g = \mathrm{i}\frac{\pi}{\xi_g}\exp(2\pi\mathrm{i}k_0t)\int_0^t \exp[-2\pi\mathrm{i}(\boldsymbol{g}+\boldsymbol{s})\cdot\boldsymbol{z}]\,dz$$

$$= \mathrm{i}\frac{\pi}{\xi_g}\exp(2\pi\mathrm{i}k_0t)\int_0^t \exp(-2\pi\mathrm{i}sz)\,dz\,. \tag{7.33}$$

The integral may be evaluated as for (7.27) and the diffraction intensity at the point P ($I_0 = 1$) becomes

$$I_g = \psi_g\psi_g^* = \frac{\pi^2}{\xi_g^2}\frac{\sin^2(\pi ts)}{(\pi s)^2}\,. \tag{7.34}$$

Substituting for ξ_g from (7.32), we obtain the same dependence on the excitation error s as in (7.28).

The last integral in (7.33) can also be solved graphically by use of an amplitude-phase diagram (Sect. 3.2.1), in which lengths dz are added vectorially with slope $2\pi sz$ in the complex-number plane (Fig. 7.11 b); the result is a circle of radius r. The length of the circular segment QQ′ is t. The circle is closed when the phase factor $\exp(2\pi\mathrm{i}sz)$ reaches unity, which occurs for $sz = 1$ and so, the perimeter $2\pi r$ of the circle, is equal to $1/s$. It follows that $r = 1/2\pi s$. The amplitude ψ_g is proportional to the square of the length of the chord $\overline{QQ'}$. For foil thicknesses $t = n/s$ (n integer), Q and Q′ coincide and the diffraction intensity I_g becomes zero; when the thickness is further increased, $I_g = R$ again increases and subsequently oscillates as shown in Fig. 7.14 b.

7.3 Dynamical Theory of Electron Diffraction

7.3.1 Limitations of the Kinematical Theory

The kinematical theory is valid for only very thin films for which the reflection intensity I_g is small and the decrease of the primary beam intensity I_0 can be neglected. If the Bragg condition is exactly satisfied ($s = 0$), we obtain from (7.34) $I_g = \pi^2t^2/\xi_g^2$ and $I_0 = 1$. The intensity I_g increases as t^2, and the condition $I_g \ll 1$ will be fulfilled only for $t < \xi_g/10$. If $s \neq 0$, the intensity I_g oscillates with increasing t and reaches maximum values of $1/\xi_g^2s^2$ (Fig. 7.14 b). The condition $I_g \ll 1$ can be satisfied with $s \gg 1/\xi_g$. In Sect. 7.3.4, we shall see that, in this case, the kinematical and dynamical theories lead to identical results.

Furthermore, it must not be forgotten that the case in which only one Bragg reflection is excited, which is called the two-beam case (including the primary beam with $\boldsymbol{g} = 0$), is unusual; normally, a larger number $n > 2$ of reflections must be considered (n-beam case). Numerous small reflection intensities I_g can reduce the intensity of the primary beam more strongly than in the two-beam case. The n-beam case therefore restricts the validity of the kinematical theory to even smaller thicknesses. In the Bragg condition ($s = 0$), the dynamical theory predicts an oscillation of the intensities I_0 and I_g. A strong Bragg reflection will excite neighbouring reflections with a larger amplitude than the primary beam. In many practical situations, therefore, the interaction of 30–100 Bragg reflections has to be considered. Furthermore, the intensity does not remain localized in the Bragg reflections. The diffuse electron scattering between the Bragg diffraction spots by inelastic and thermal diffuse scattering causes an absorption (Sect. 7.4.2).

7.3.2 Formulation of the Dynamical Theory as a System of Differential Equations

This formulation of dynamical theory was first used by *Darwin* [7.2, 4] for x-ray diffraction and transferred by *Howie* and *Whelan* [7.3] to electron diffraction.

We first discuss the two-beam case. An incident wave of amplitude ψ_0 and a diffracted wave of amplitude ψ_g fall on a layer of thickness dz inside the crystal foil. After passing through this layer, the amplitude ψ_0 will be changed by $d\psi_0$ and ψ_g by $d\psi_g$. These changes can be calculated from Fresnel-diffraction theory using the column approximation. The contributions of ψ_0 and ψ_g to $d\psi_0$ and $d\psi_g$ can be obtained by use of (7.31–33) with the extinction distances (7.32) $\xi_0 = \pi V_e/\lambda F(0)$ and $\xi_g = \pi V_e/\lambda F(2\,\theta_B)$. The result is a linear system of differential equations (Howie-Whelan equations):

$$\begin{aligned} \frac{d\psi_0}{dz} &= \frac{i\pi}{\xi_0}\psi_0 + \frac{i\pi}{\xi_g}\psi_g e^{2\pi isz} \\ \frac{d\psi_g}{dz} &= \frac{i\pi}{\xi_g}\psi_0 e^{-2\pi isz} + \frac{i\pi}{\xi_0}\psi_g\,. \end{aligned} \tag{7.35}$$

The second term of the first equation results from the scattering of the diffracted wave back into the primary beam; the sign of the excitation error s is the reverse of that for scattering in the opposite direction (first term in the second equation). This system of equations can be extended to the n-beam case by introducing the relative excitation errors s_{g-h} and extinction distances ξ_{g-h}:

$$\frac{d\psi_g}{dz} = \sum_{h=g_1}^{g_n} \frac{i\pi}{\xi_{g-h}}\psi_h \exp(2\pi i s_{g-h} z) \quad \text{for} \quad \boldsymbol{g} = \boldsymbol{g}_1, \ldots, \boldsymbol{g}_n\,; \quad \boldsymbol{g}_1 = 0\,. \tag{7.36}$$

In the final result, we are interested in only the reflection intensities I_g and we can, therefore, use the transformation

$$\psi_0' = \psi_0 \exp(-\mathrm{i}\pi z/\xi_0) \; ; \qquad \psi_g' = \psi_g \exp(2\pi \mathrm{i} s z - \mathrm{i}\pi z/\xi_0) \; . \tag{7.37}$$

These new quantities contain only an additional phase factor, which cancels out when we multiply by the complex conjugate amplitude. Substitution of (7.37) into (7.35) yields the simpler formulae

$$\begin{aligned} \frac{d\psi_0'}{dz} &= \frac{\mathrm{i}\pi}{\xi_g}\psi_g' \\ \frac{d\psi_g'}{dz} &= \frac{i\pi}{\xi_g}\psi_0' + 2\pi \mathrm{i} s \psi_g' \; . \end{aligned} \tag{7.38}$$

The boundary conditions for these differential equations at the entrance surface of the foil ($z = 0$) are $|\psi_0| = 1$ and $|\psi_g| = 0$. We discuss a solution of (7.38) together with the solution of the eigenvalue problem in Sect. 7.3.4.

The system of differential equations for the n-beam case can be solved by the Runge-Kutta or a similar numerical method; alternatively an analogue computer can be used [7.5] with a reduced number of reflections. The multislice method [7.6] uses elements of finite thickness Δz and projects the potential inside the layer Δz onto the lower boundary of the layer. The space between the layers is treated as vacuum. Equations (7.36) become a system of difference equations. The size of Δz for a given permissible error is determined by the ξ_g and the crystal structure.

7.3.3 Formulation of the Dynamical Theory as an Eigenvalue Problem

This formulation was first used by *Bethe* [7.7] for x-ray diffraction. The Schrödinger equation (3.24) is solved with a potential $V(r)$ that is a superposition of all of the atomic potentials (Fig. 3.3) and therefore has the same periodicity as the lattice. This means that $V(r)$ can be expanded as a Fourier sum:

$$V(r) = -\sum_g V_g \exp(2\pi \mathrm{i}\boldsymbol{g}\cdot\boldsymbol{r}) = -\frac{h^2}{2m}\sum_g U_g \exp(2\pi \mathrm{i}\boldsymbol{g}\cdot\boldsymbol{r}) \; . \tag{7.39}$$

A value V_g can be attributed to each point $\boldsymbol{g}$ of the reciprocal lattice. The V_g [eV] and U_g [cm^{-2}] are related to the structure amplitude $F(\theta)$ ($\theta = 2\theta_\mathrm{B}$) of the kinematical theory because, in the Born approximation, $F(\theta)$ is also a Fourier transform of the scattering potential $V(r)$ (Sect. 5.1.3):

$$V_g = \frac{\lambda^2 E}{2\pi V_\mathrm{e}}\frac{E + 2E_0}{E + E_0}F(\theta) = \frac{h^2}{2\pi m V_\mathrm{e}}F(\theta) \; ; \qquad U_g = \frac{F(\theta)}{\pi V_\mathrm{e}} \; . \tag{7.40}$$

The extinction distance ξ_g introduced by (7.32) can be written as follows ... $\xi_{g,100}$ denotes the extinction distance for $E = 100$ keV,

$$\xi_g = \frac{\pi V_e}{\lambda F(\theta)} = \frac{\lambda E}{2V_g} \frac{E + 2E_0}{E + E_0} = \frac{h^2}{2m\lambda V_g} = \frac{1}{\lambda U_g}, \tag{7.41}$$

$$\xi_g = \xi_{g,100} \frac{m_{100}\lambda_{100}}{m\lambda} = \xi_{g,100} \frac{v}{v_{100}}. \tag{7.42}$$

Equation (7.42) allows us to transfer tabulated values of ξ_g for $E = 100$ keV (Table 7.2) to other electron energies. The influence of lattice vibrations (see the Debye-Waller factor $\exp(-2M)$ in Sect. 7.5.3) and the thermal expansion of the lattice causes a slow increase of ξ_g with temperature [7.8, 9].

If (7.39) for $V(\boldsymbol{r})$ is substituted in (3.24), the solutions will also reflect the lattice periodicity. Such solutions of the Schrödinger equation are called *Bloch waves*

$$b^{(j)}(\boldsymbol{k},\boldsymbol{r}) = \sum_g C_g^{(j)} \exp[2\pi i(\boldsymbol{k}_0^{(j)} + \boldsymbol{g}) \cdot \boldsymbol{r}]. \tag{7.43}$$

We have to sum over the n excited points $\boldsymbol{g} = \boldsymbol{g}_1, \ldots, \boldsymbol{g}_n$ of the reciprocal lattice, including the incident direction ($\boldsymbol{g}_1 = 0$). The number $j = 1, \ldots, n$ of different Bloch waves is needed to describe the propagation of electron waves in a crystal and to fulfill the boundary condition at the vacuum-crystal interface. This requires a superposition of n^2 different waves with wave vectors $\boldsymbol{k}_0^{(j)} + \boldsymbol{g}$ and amplitude factors $C_g^{(j)}$.

We substitute (6.39) and (6.43) into (3.24) and introduce the abbreviation

$$K = 1/h[2m_0E(1 + E/2E_0) + 2m_0V_0(1 + E/E_0)]^{1/2} \tag{7.44}$$

for the wave vector inside the crystal, which is obtained from the sum of the kinetic energy and the coefficient $V_0 = eU_i$ (inner potential) (Sect. 3.1.3) of the Fourier expansion (7.39). This gives

$$4\pi^2[K^2 - (\boldsymbol{k}_0^{(j)} + \boldsymbol{g})^2 + \sum_{h \neq 0} U_h \exp(2\pi i\boldsymbol{h} \cdot \boldsymbol{r})] C_g^{(j)} \exp[2\pi i(\boldsymbol{k}_0^{(j)} + \boldsymbol{g}) \cdot \boldsymbol{r}] = 0 \tag{7.45}$$

for all $\boldsymbol{g}$. This system of equations can be satisfied if the coefficients of identical exponential terms simultaneously become zero. This results after collecting up terms containing the factor $\exp[2\pi i(\boldsymbol{k}_0^{(j)} + \boldsymbol{g}) \cdot \boldsymbol{r}]$. We obtain the fundamental equations of dynamical theory

$$[K^2 - (\boldsymbol{k}_0^{(j)} + \boldsymbol{g})^2] C_g^{(j)} + \sum_{h \neq 0} U_h C_{g-h}^{(j)} = 0\,; \quad \boldsymbol{g} = \boldsymbol{g}_1, \ldots, \boldsymbol{g}_n. \tag{7.46}$$

The $\boldsymbol{k}_0^{(j)} + \boldsymbol{g}$ are the wave vectors of the Bloch waves, the magnitudes of which are not identical with K. As in kinematical theory (Fig. 7.8), we obtain the excitation points M_j as the starting points of the vectors $\boldsymbol{k}_0^{(j)}$, which end at the origin O of the reciprocal lattice. For calculation of the position of M_j, we desribe its position by the following vector (Fig. 7.12)

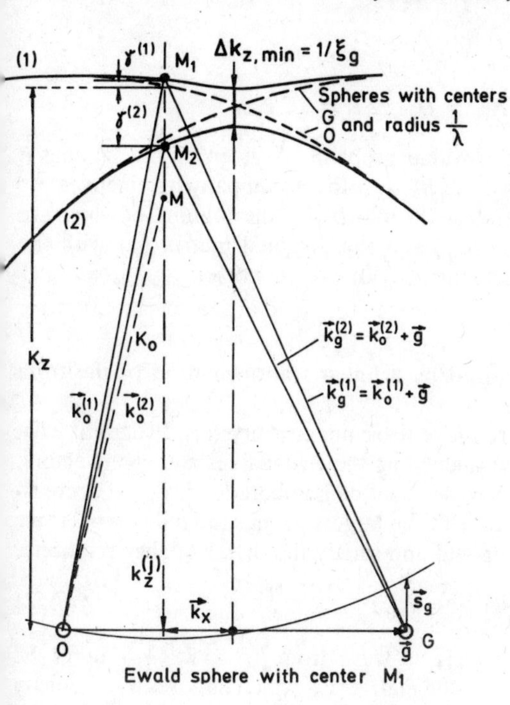

Fig. 7.12. Branches $(j) = (1)$ and (2) of the dispersion surface for the two-beam case with a least distance $\Delta k_{z,\min} = 1/\xi_g$ in the Bragg condition ($k_z = 0$). Construction of the excitation points M_1 and M_2 on the dispersion surface for the tilt parameter k_x and the four wave vectors $\boldsymbol{k}_0^{(j)}$ and $\boldsymbol{k}_g^{(j)} = \boldsymbol{k}_0^{(j)} + \boldsymbol{g}$ $(i = 1, 2)$ to the reciprocal-lattice points O and $\boldsymbol{g}$

$$\boldsymbol{k}^{(j)} = \boldsymbol{k}_z^{(j)} + \boldsymbol{k}_x = (K_z + \gamma^{(j)})\,\boldsymbol{u}_z + k_x\boldsymbol{u}_x\;, \tag{7.47}$$

where $\boldsymbol{u}_x$, $\boldsymbol{u}_z$ are unit vectors. The component k_x depends on the tilt angle $\Delta\theta$ or the excitation error s_g as follows

$$\Delta\theta = k_x/K = s_g/g\;. \tag{7.48}$$

Recalling that $K \gg g$, $K + k_z^{(j)} \simeq 2K$ and introducing the difference $(k_z^{(j)} - K_z) = \gamma^{(j)}$ from (7.47), we find that the first factor of (7.46) becomes

$$[K^2 - (\boldsymbol{k}_0^{(j)} + \boldsymbol{g})^2] = (K + |\boldsymbol{k}_0^{(j)} + \boldsymbol{g}|)(K - |\boldsymbol{k}_0^{(j)} + \boldsymbol{g}|) \simeq 2K(s_g - \gamma^{(j)})\;. \tag{7.49}$$

s_g is negative when the reciprocal-lattice point g is outside the Ewald sphere, as in Fig. 7.12. By use of (7.49), the system of equations (7.46) can be written in matrix form, after dividing by $2K$ [7.10–12]

$$\begin{pmatrix} A_{11}A_{12}\dots A_{1n} \\ A_{21}A_{22}\dots A_{2n} \\ \dots\dots\dots\dots \\ A_{n1}A_{n2}\dots A_{nn} \end{pmatrix} \begin{pmatrix} C_1^{(j)} \\ C_2^{(j)} \\ \dots \\ C_n^{(j)} \end{pmatrix} = \gamma^{(j)} \begin{pmatrix} C_1^{(j)} \\ C_2^{(j)} \\ \dots \\ C_n^{(j)} \end{pmatrix} \quad \text{for} \quad j = 1, \dots, n \tag{7.50}$$

with the matrix elements

$$A_{11} = 0,\ A_{gg} = s_g,\ A_{hg} = A_{gh} = U_{g-h}/2K = 1/2\,\xi_{g-h}\ .$$

This is the equation for an eigenvalue problem. A given matrix $[A]$ has n different eigenvalues $\gamma^{(j)}$ $(j = 1, \ldots, n)$ with the accompanying eigenvectors $C_g^{(j)}$ $(\boldsymbol{g} = \boldsymbol{g}_1, \ldots, \boldsymbol{g}_n)$. If we introduce the matrix $[C]$, the columns of which are the eigenvectors, so that $C_{gj} = C_g^{(j)}$ and the diagonal matrix $\{\gamma\}$ with the eigenvalues $\gamma^{(j)}$ as diagonal elements, (6.50) can be written

$$[A]\,[C] = [C]\,\{\gamma\}\ . \tag{7.51}$$

A matrix $[A]$ is thus diagonalized by a linear transformation of the form $[C^{-1}]\,[A]\,[C]$.

The matrix $[A]$ is symmetric for centrosymmetric crystals. Programs exist for defining the matrix $[A]$ and calculating the eigenvalues and eigenvectors. It is suggested that the Bloch waves should be numbered in order of decreasing $k_z^{(j)}$ [7.13]. The Bloch waves with the largest $\gamma^{(j)}$ has the index $j = 1$, etc.

The eigenvectors are orthogonal and satisfy the orthogonality relations

$$\sum_g C_g^{(i)} C_g^{(j)} = \delta_{ij}\ ; \qquad \sum_j C_g^{(j)} C_h^{(j)} = \delta_{gh}\ . \tag{7.52}$$

Changing the direction of the incident wave from $\boldsymbol{k}_0^{(j)}$ to $\boldsymbol{k}_0^{(j)} - \boldsymbol{h}$ alters the sequence of the column vectors in the matrix $[C]$, which imposes a periodicity condition on the $C_g^{(j)}$

$$C_g(\boldsymbol{k}_0^{(j)}) = C_{g+h}(\boldsymbol{k}_0^{(j)} - \boldsymbol{h})\ . \tag{7.53}$$

The n eigenvalues $\gamma^{(j)}$ correspond to n Bloch waves (7.43) with wave vectors $\boldsymbol{k}_0^{(j)} + \boldsymbol{g}$. Their starting points do not lie on a sphere of radius K around O but at modified points M_j given by (7.47). For different tilts of the specimen – equivalent to varying values of k_x in (7.47) – the points $M_j(k_x)$ lie on the *dispersion surface*. The starting points of the wave vectors $\boldsymbol{k}_0^{(j)} + \boldsymbol{g}$ on this dispersion surface can be obtained by the following construction. The $\boldsymbol{K}_0$ vector parallel to the incident direction determines the point M in Fig. 7.12. Through M a straight line is drawn parallel to the crystal normal. The points of intersection with the n-fold dispersion surface are the excitation points M_j, which lie above one another in the case of normal incidence. This geometrical construction results from the boundary condition that the tangential components of the waves have to be continuous at the crystal boundary. For non-normal incidence, the excitation points M_j are no longer above one another (see e.g., [7.14].

The total wave function, the solution of (7.46), will be a linear combination of the Bloch waves $b^{(j)}(\boldsymbol{k}, \boldsymbol{r})$ (7.43) with the Bloch-wave excitation amplitudes $\varepsilon^{(j)}$, i.e.,

$$\psi_{\text{tot}} = \sum_j \varepsilon^{(j)} b^{(j)}(\boldsymbol{k}, \boldsymbol{r}) = \sum_j \varepsilon^{(j)} \sum_g C_g^{(j)} \exp\left[2\pi \mathrm{i}\,(\boldsymbol{k}_0^{(j)} + \boldsymbol{g}) \cdot \boldsymbol{r}\right]\ . \tag{7.54}$$

The amplitude ψ_g of a particular reflected wave can be obtained by summing over all $j = 1, \ldots, n$ waves from the excitation points M_j to the corresponding reciprocal-lattice points $\boldsymbol{g}$:

$$\psi_g = \sum_j \varepsilon^{(j)} C_g^{(j)} \exp[2\pi i (\boldsymbol{k}_0^{(j)} + \boldsymbol{g}) \cdot \boldsymbol{r}] \quad \text{or}$$
$$\psi_g = \sum_j \varepsilon^{(j)} C_g^{(j)} \exp(2\pi i \gamma^{(j)} z) \tag{7.55}$$

if a constant phase factor is omitted. The excitation amplitudes $\varepsilon^{(j)}$ of the Bloch waves can be obtained from the boundary condition at the entrance of the incident plane wave into the crystal. The phase factors in (7.55) are all equal to unity for $z = 0$ and the plane wave in vacuum and the Bloch-wave field in the crystal must be continuous. This requires

$$\psi_0(0) = \sum_j \varepsilon^{(j)} C_0^{(j)} = 1 ;$$
$$\psi_g(0) = \sum_j \varepsilon^{(j)} C_g^{(j)} = 0 \quad \text{for all } \boldsymbol{g} \neq 0 \tag{7.56}$$

or in a matrix formulation for the two-beam case

$$\begin{pmatrix} C_0^{(1)} & C_0^{(2)} \\ C_g^{(1)} & C_g^{(2)} \end{pmatrix} \begin{pmatrix} \varepsilon^{(1)} \\ \varepsilon^{(2)} \end{pmatrix} = \begin{pmatrix} \psi_0(0) \\ \psi_g(0) \end{pmatrix} = \begin{pmatrix} 1 \\ 0 \end{pmatrix} , \tag{7.57}$$

which can be readily expanded to the n-beam case

$$[C]\underline{\varepsilon} = \underline{\psi(0)} , \tag{7.58}$$

where $\underline{\varepsilon}$ and $\underline{\psi(0)}$ are column vectors of n components.

Comparison with the first of the orthogonality relations (7.52) shows that the boundary conditions (7.56–58) can be satisfied by writing $\varepsilon^{(j)} = C_0^{(j)}$ for normal incidence. In a more-general formulation, the $\varepsilon^{(j)}$ can be calculated from (7.58) by multiplying by the inverse matrix $[C^{-1}]$, which is identical with the transposed matrix $[\tilde{C}]$ because of the orthogonality of the $C_g^{(j)}$:

$$\underline{\varepsilon} = [C^{-1}]\, \underline{\psi(0)} . \tag{7.59}$$

7.3.4 Discussion of the Two-Beam Case

In order to bring out the most-important results of the dynamical theory, we now solve and discuss the two-beam case in detail, though it will be a poor approximation in practice. For high electron energies, the curvature of the

Ewald sphere is so small that a large number of reflections (30–100) are excited simultaneously.

In kinematical theory, the centres M of the various Ewald spheres (Fig. 7.8) lie on a sphere of radius $k = 1/\lambda$ around the origin O of the reciprocal lattice if the direction of the incident wave is varied. When the intensity of the diffracted beam is increased by increasing the thickness and becomes larger than the intensity of the primary beam, the former can be treated as the primary wave and a sphere of radius k can also be drawn around the reciprocal-lattice point $\boldsymbol{g}$ as the geometrical surface that describes all possible values of the excitation points M (Fig. 7.12). As will be shown below, the two spheres do not intersect each other but withdraw from one another in a characteristic manner.

For the two-beam case, the fundamental equations of the dynamical theory (7.46 and 50) are

$$\begin{aligned} -\gamma^{(j)} C_0^{(j)} + \frac{U_g}{2K} C_g^{(j)} &= 0 \\ \frac{U_g}{2K} C_0^{(j)} + (-\gamma^{(j)} + s) C_g^{(j)} &= 0 . \end{aligned} \tag{7.60}$$

Such a homogeneous linear system of equations for the $C_g^{(j)}$ has a non-zero solution if and only if the determinant of the coefficients is zero:

$$\begin{vmatrix} -\gamma^{(j)} & \dfrac{U_g}{2K} \\ \dfrac{U_g}{2K} & (-\gamma^{(j)} + s) \end{vmatrix} = \gamma^{(j)2} - s\gamma^{(j)} - U_g^2/4K^2 = 0 . \tag{7.61}$$

This is a quadratic equation for the eigenvalues $\gamma^{(j)}$. In the n-beam case this characteristic equation is of order n. Before discussing the solution, it will be shown that the Howie-Whelan equations (Sect. 7.3.2) lead to the same characteristic equation. If we substitute for ψ_g' and $d\psi_g'$ from the first equation of (7.38) into the second equation, we obtain

$$\frac{d^2\psi_0'}{dz^2} - 2\pi \mathrm{i} s \frac{d\psi_0'}{dz} + (\pi/\xi_g)^2 \psi_0' = 0 . \tag{7.62}$$

A similar equation is obtained for ψ_g'. If we look for a solution of the form $\psi' = A \exp(2\pi \mathrm{i}\gamma^{(j)} z)$, the same equation as (7.61) is found for the $\gamma^{(j)}$ because $\xi_g = K/U_g$ (7.41). This shows that the two different ways of treating the dynamical theory lead to the same solution.

Solving the quadratic equation (7.61) gives

$$\gamma^{(j)} = \frac{1}{2}\,[s - (-1)^j \sqrt{(U_g/K)^2 + s^2}] = \frac{1}{2}\,[s - (-1)^j \sqrt{1/\xi_g^2 + s^2}]$$
$$= \frac{1}{2\,\xi_g}\,[w - (-1)^j \sqrt{1 + w^2}] \tag{7.63}$$

in which the parameter $w = s\xi_g$ characterizes the tilt out of the Bragg condition ($w = 0$). This solution is plotted in Fig. 7.13a as a function of w and in Fig. 7.12 for an Ewald sphere of a relatively small radius. The two circles around O and G in Fig. 7.12 correspond to the straight lines (asymptotes of the hyperbola) in Fig. 7.13a. The two Ewald spheres (asymptotes) do not intersect but approach most closely for the Bragg condition $w = 0$; their separation is then

$$\Delta k_{z,\,\min} = \gamma^{(1)} - \gamma^{(2)} = U_g/K = 1/\xi_g\,. \tag{7.64}$$

By use of the eigenvalues $\gamma^{(j)}$, the linear system of equations (7.60) can be solved for the $C_g^{(j)}$. For the amplitude $\varepsilon^{(j)} C_g^{(j)} = C_0^{(j)} C_g^{(j)}$ of the four Bloch waves with wave vectors $\boldsymbol{k}_0^{(j)} + \boldsymbol{g}$ for normal incidence, we obtain

$$C_0^{(j)} C_0^{(j)} = \frac{1}{2}\left[1 + (-1)^j \frac{w}{\sqrt{1 + w^2}}\right]; \quad C_0^{(j)} C_g^{(j)} = -\frac{1}{2}\,\frac{(-1)^j}{\sqrt{1 + w^2}}\,. \tag{7.65}$$

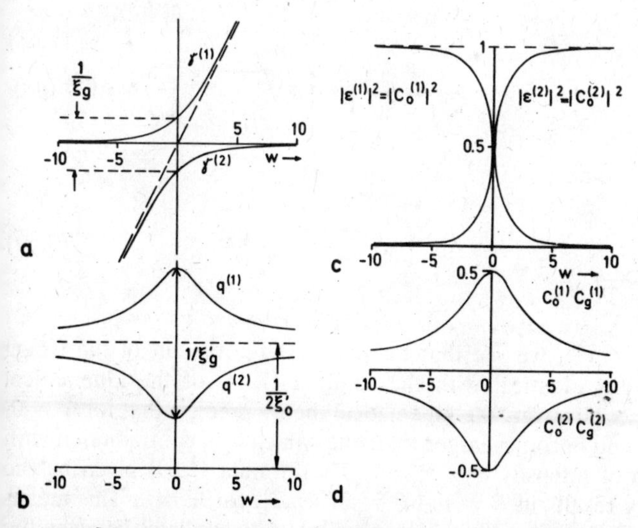

Fig. 7.13a–d. Dependence of the Bloch-wave parameters of the two-beam case on the tilt parameter $w = s\xi_g$ out of the Bragg condition ($w = 0$). (**a**) $\gamma^{(j)}$; $\gamma^{(2)} - \gamma^{(1)}$ is the distance between the two branches of the dispersion surface (Fig. 7.12). (**b**) Absorption parameters $q^{(j)}$ and (**c**), (**d**) wave amplitudes of the four excited Bloch waves

In the Bragg condition $w = 0$, all four waves have the amplitude 1/2 (Fig. 7.13 c, d).

Sometimes the substitution $w = \cot\beta$ is used for the two-beam case. The matrix $[C]$ of the eigenvectors then becomes

$$[C] = \begin{pmatrix} C_0^{(1)} & C_0^{(2)} \\ C_g^{(1)} & C_g^{(2)} \end{pmatrix} = \begin{pmatrix} \sin(\beta/2) & \cos(\beta/2) \\ \cos(\beta/2) & -\sin(\beta/2) \end{pmatrix} \tag{7.66}$$

For calculating the intensity $I_0 = \psi_0\psi_0^*$ of the primary beam which we call the *transmission* T and the intensity of the reflected beam $I_g = \psi_g\psi_g^*$ or *reflection* R, we use (7.55) and substitute the specimen thickness t for the z-component of the vector $\boldsymbol{r}$

$$\begin{aligned} \psi_0(t) &= \sum_{j=1}^{2} C_0^{(j)} C_0^{(j)} \exp(2\pi i k_z^{(j)} t) \\ \psi_g(t) &= \sum_{j=1}^{2} C_0^{(j)} C_g^{(j)} \exp(2\pi i k_z^{(j)} t) \exp(2\pi i g x) \, . \end{aligned} \tag{7.67}$$

Substituting the values given in (7.63, 65) and omitting the common phase factor $\exp(2\pi i K_z t)\exp(\pi i w t/\xi_g)$, we find

$$\begin{aligned} \psi_0(t) &= \cos\left(\pi\sqrt{1+w^2}\,\frac{t}{\xi_g}\right) - \frac{iw}{\sqrt{1+w^2}}\sin\left(\pi\sqrt{1+w^2}\,\frac{t}{\xi_g}\right) \\ \psi_g(t) &= \frac{i}{\sqrt{1+w^2}}\sin\left(\pi\sqrt{1+w^2}\,\frac{t}{\xi_g}\right)\exp(2\pi i g x) \end{aligned} \tag{7.68}$$

The intensities become

$$\underbrace{\psi_g\psi_g^*}_{R} = \underbrace{1-\psi_0\psi_0^*}_{1-T} = \frac{1}{1+w^2}\sin^2\left(\pi\sqrt{1+w^2}\,\frac{t}{\xi_g}\right) \tag{7.69}$$

Recalling that $w = s\xi_g$, we see that for $w \gg 0$ (large tilt out of the Bragg condition) (7.69) is identical with the solution (7.34) of the kinematical theory. Otherwise, however, the kinematical theory predicts that for $w = 0$, R increases as t^2 and becomes larger than one, which is in contradiction with the conservation of intensity $T + R = 1$. The formula (7.69) given by the dynamical theory results in $R = 1 - T = \sin^2(\pi t/\xi_g)$ for $w = 0$. This means that, even in the Bragg position, the electron intensity oscillates between the primary beam and the Bragg reflected beam with increasing film thickness (Fig. 7.14 a) ("pendellösung" of the dynamical theory). We now clearly see the meaning of the extinction distance ξ_g (Table 7.2); it is the periodicity in

Table 7.2. Extinction distances ξ_g [nm] for $E = 100$ keV. (Th: *Thomas* et al. [7.15], H: *Hirsch* et al. [7.16])

Cubic face-centred lattice and NaCl structure								
hkl	111	200	220	311	222	400	331	
Al	56.3	68.5	114.4	147.6	158.6	202.4	235.7	Th
Cu	28.6	32.6	47.3	57.9	61.5	76.4	88.1	Th
Ni	26.8	30.6	44.6	54.7	58.1	72.0	82.9	Th
Ag	24.2	27.2	38.6	47.4	50.4	63.0	73.0	Th
Pt	14.7	16.6	23.2	27.4	28.8	34.3	38.5	H
Au	18.3	20.2	27.8	33.6	35.6	43.5	49.5	Th
Pb	24.0	26.6	35.9	41.8	43.6	50.5	55.5	H
LiF	171.7	64.5	94.2	219.9	121.0	146.3	335.2	H
MgO	272.6	46.1	66.2	1180	85.2	103.3	1075	H
Cubic body-centred lattice								
hkl	110	200	211	220	310	222	400	
Cr	28.8	42.3	55.5	68.6	81.6	94.7	121.9	Th
Fe	28.6	41.2	53.5	65.8	78.0	90.4	116.2	Th
Nb	26.1	38.3	49.9	61.4	72.9	84.6	108.5	Th
Mo	22.9	33.6	43.2	52.7	62.0	72.3	89.7	Th
Ta	20.2	27.5	33.9	40.0	45.9	51.8	63.8	Th
W	18.0	24.5	30.2	35.5	41.0	46.2	55.6	Th
Diamond structure								
hkl	111	220	311	400	331	511 333	400	
C	47.6	66.5	124.5	121.5	197.2	261.3	215.1	H
Si	60.2	75.7	134.9	126.8	204.6	264.5	209.3	H
Ge	43.0	45.2	75.7	65.9	102.8	127.3	100.8	H
Hexagonal lattice								
hkl	$\bar{1}100$	$11\bar{2}0$	$\bar{2}200$	$\bar{1}101$	$\bar{2}201$	0002	$\bar{1}102$	
Mg	150,9	140,5	334.8	100.1	201.8	81.1	231.0	H
Co	46,7	42,9	102.7	30.6	62.0	21.8	70.2	H
Zn	55,3	49,7	118.0	35.1	70.4	26.0	76.2	H
Zr	59,4	49,3	115.1	37.9	69.1	51.7	83.7	H
Cd	51,9	43,8	102.3	32.4	60.8	24.4	68.3	H

depth of this oscillation. There are thicknesses $t = (n + 1/2)\xi_g$ for which the intensity is completely concentrated in the Bragg reflection and others, $t = n\xi_g$, for which the whole intensity returns to the direction of incidence. These oscillations result from the superposition of the two waves with wave vectors $\boldsymbol{k}_0^{(1)} + \boldsymbol{g}$ and $\boldsymbol{k}_0^{(2)} + \boldsymbol{g}$ which are somewhat different in magnitude: $|\boldsymbol{k}_z^{(1)} - \boldsymbol{k}_z^{(2)}| = 1/\xi_g$ (7.64).

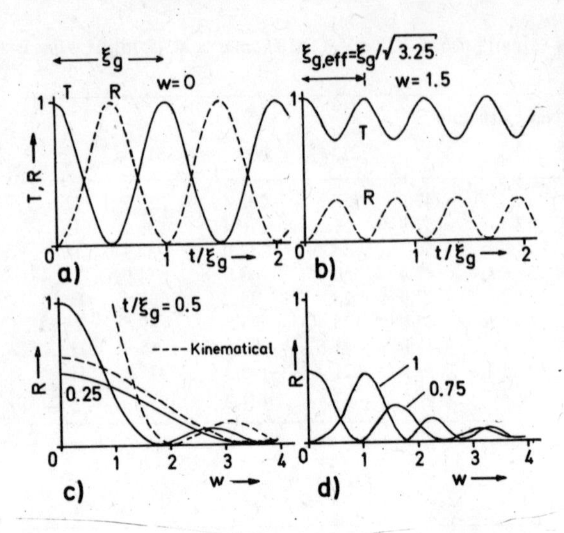

Fig. 7.14 a–d. Dynamical two-beam case without absorption. (**a**) and (**b**) thickness dependence of the transmitted (T) and Bragg-reflected intensity (R) (a) in the Bragg condition $w = 0$ and (b) for a tilt parameter $w = s\xi_g = 1.5$. (**c**) and (**d**) tilt dependence (rocking curve) of R for the different relative thicknesses t/ξ_g (between 0.25 and 1). (---) comparison with the kinematical theory

For $w \neq 0$, the amplitude (7.69) of the oscillation decreases as $(1 + w^2)^{-1}$ and the depth of the oscillations can be described by a reduced, effective extinction distance (Fig. 7.14 b)

$$\xi_{g,\text{eff}} = \xi_g / \sqrt{1 + w^2} \,. \tag{7.70}$$

The dependence of T and R on the tilt angle $\Delta\theta$ of the specimen or the excitation error s or tilt parameter w for a fixed thickness t is called a *rocking curve* (Fig. 7.14 c, d). In the absence of absorption, the condition $T + R = 1$ is everywhere satisfied, and T and R are, as can be seen from (7.69), symmetric in w. (This will cease to be the case for T when we consider absorption in the next section.) Figure 7.14 c, d show R for $t/\xi_g = 0.25$–1. We observe that $R = 0$ for $w = 0$ and $t/\xi_g = 1$ (Fig. 7.14 a). If the specimen is tilted ($w \neq 0$), R increases again (Fig. 7.14 d), reaching a maximum at $w \simeq 1$. The distances Δw between the minima ($R = 0$) of the rocking curve become narrower with increasing t. Figure 7.14 c also contains the results of the kinematical theory (dashed lines). Deviations from the kinematical theory are observed for larger values of t/ξ_g, especially when w is small.

The relation $E = \hbar^2 k^2/2\,m$ between energy and momentum $p = \hbar k$ can be used to reveal an analogy between the dispersion surface as a function of k_x and the Fermi surface of low-energy conduction electrons. If there is no interaction between the electrons and the lattice (no excitation of a low-order reflection), the dispersion surface degenerates to a sphere around the origin of the reciprocal lattice (Fermi surface of free electrons). In the theory of conduction electrons, the Fermi surface also splits into energy bands with forbidden gaps if there exists an interaction with the lattice potential, and dE/dk becomes zero at the boundary of the Brillouin zone for which the

Bragg condition is satisfied. The same behaviour can be seen in Fig. 7.12. The splitting $\Delta k_{z,\min}$ of the energy gap is directly proportional to V_g and, therefore, to the interaction with the crystal lattice.

7.4 Dynamical Theory Considering Absorption

7.4.1 Inelastic-Scattering Processes in Crystals

If the electron energy falls from the initial value E_n to the final value E_m during a scattering process with an energy loss $\Delta E = E_n - E_m$, the dispersion surfaces for these two energies are different (Fig. 7.15). The surfaces have the same shape because $\Delta E \ll E_n$ but are shifted by $\Delta k_z = (\boldsymbol{k}_n - \boldsymbol{k}_m)_z = \Delta E/hv$ [7.28]. The excitation point P corresponds to the excitation of a Bloch wave of type $j = 2$. The transitions 1–4 are called:

1. Elastic
3. Inelastic

} interband transition

2. Elastic
4. Inelastic

} intraband transition

The symmetry (type) of Bloch waves is changed in an interband transition and is preserved in an intraband transition. The vector $\overline{\mathrm{PQ}}$ corresponds to q' in Fig. 5.11 and according to (5.51)

$$q'^2 = \overline{\mathrm{PQ}}^2 = \overline{\mathrm{RQ}}^2 + \overline{\mathrm{PR}}^2 = K^2(\theta^2 + \theta_{\mathrm{E}}^2) . \tag{7.71}$$

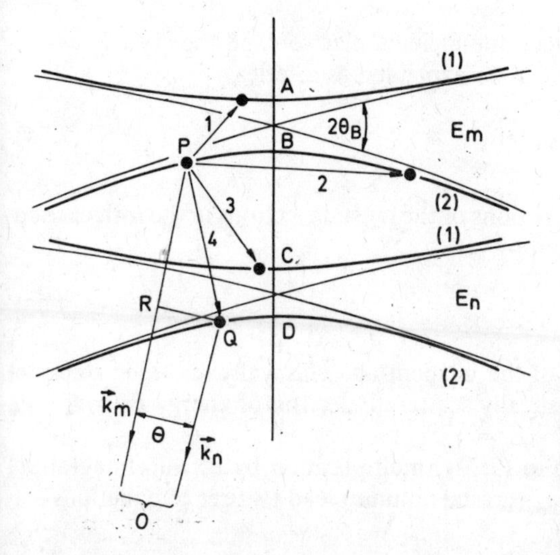

Fig. 7.15. Shift of the dispersion surface caused by electron excitation $m \rightarrow n$ with energies E_m and E_n, respectively, and interband (1,3) and intraband (2,4) transitions

The shift Δk_z of the dispersion surfaces for E_n and E_m depends on the scattering process. For thermal diffuse scattering (electron-phonon scattering), the difference can be neglected. Sometimes thermal diffuse scattering is attributed to inelastic scattering but it can also be attributed to elastic scattering because it becomes similar to Rutherford scattering for large scattering angles. (Estimates of the energy loss ΔE for small-angle scattering ($\Delta E \leq 10^{-2}$ eV) are to be found in Table 5.1.) For the Al 15 eV plasmon loss, the shift becomes $\Delta k_z = \Delta E/hv = 2 \times 10^{-2}\ \mathrm{nm}^{-1}$ at $E = 100$ keV, whereas the distance $\overline{\mathrm{AB}} = \overline{\mathrm{CD}}$ between the branches of the dispersion surface is $1/\xi_g = 1/56.3 = 1.8 \times 10^{-2}\ \mathrm{nm}^{-1}$: the distance and the shift are thus of the same order of magnitude.

Inelastic scattering by a single atom (Sect. 5.2.2) was treated by a model in which the incident wave is plane and the scattered wave is also plane far from the scattering atom. In a crystal, the primary-wave function $\psi_m\ (m = 0)$ as well as the scattered wave function ψ_n are Bloch waves, which are solutions of the Schrödinger equation (3.25) for the complete system of incident electron (coordinate $\boldsymbol{r}$) and atomic electrons (coordinates $\boldsymbol{r}_j$) and nuclei ($\boldsymbol{r}_k$) [7.28, 29].

$$\left[-\frac{\hbar^2}{2m}\nabla^2 + H_\mathrm{c} + H'\right]\Psi = E\Psi\,. \tag{7.72}$$

The first term of the Hamiltonian represents the propagation of free electrons, the term H_c the interaction of the bound electrons and ions and

$$H' = \frac{1}{4\pi\varepsilon_0}\left(\sum_j \frac{e^2}{|\boldsymbol{r}-\boldsymbol{r}_j|} - \sum_k \frac{e^2 Z_k}{|\boldsymbol{r}-\boldsymbol{r}_k|}\right) \tag{7.73}$$

the interaction energy between the incident electron and the crystal.

The total wave function can be expanded as a series

$$\Psi(\boldsymbol{r},\boldsymbol{r}_j,\boldsymbol{r}_k) = \sum_n a_n(\boldsymbol{r}_j,\boldsymbol{r}_k)\,\psi_n(\boldsymbol{r})\,, \tag{7.74}$$

where the a_n are the wave functions of the crystal electrons in the n-th excited state of energy ε_n determined by

$$H_\mathrm{c} a_n = \varepsilon_n a_n \tag{7.75}$$

$\psi_0(\boldsymbol{r})$ is the wave function of the incident and elastically scattered electron and $\psi_n(\boldsymbol{r})$ that of the inelastically scattered electron of energy $E_n = E - \varepsilon_n$ for an energy loss $\Delta E = E - E_n = \varepsilon_n$.

Substitution of (7.74) into (7.72), multiplication by a_n^* and integration over the coordinates $\boldsymbol{r}_j$ and $\boldsymbol{r}_k$ (crystal volume) lead to a set of equations for the ψ_n

$$\left[-\frac{\hbar^2}{2m}\nabla^2 - E_n + H_{nn}\right]\psi_n = -\sum_{m\neq n} H_{nm}\psi_m\,; \quad n = 0, 1, \ldots \tag{7.76}$$

with the matrix elements

$$H_{nm}(\boldsymbol{r}) = \int_V a_n^*(\boldsymbol{r}_j, \boldsymbol{r}_k)\, H'(\boldsymbol{r}, \boldsymbol{r}_j, \boldsymbol{r}_k)\, a_m(\boldsymbol{r}_j, \boldsymbol{r}_k)\, d^3\boldsymbol{r}_j\, d^3\boldsymbol{r}_k = \langle a_n|H'|a_m\rangle\,. \tag{7.77}$$

The diagonal elements

$$H_{nn}(\boldsymbol{r}) = H_{00}(\boldsymbol{r}) = -\sum_g V_g\, \mathrm{e}^{-2\pi i \boldsymbol{g}\cdot\boldsymbol{r}} \tag{7.78}$$

represent the usual potential $V(\boldsymbol{r})$ in (7.39). The off-diagonal elements H_{nm} that appear on the right-hand side of (7.76) characterize the probability of an inelastic transition from (a_m, ψ_m) to (a_n, ψ_n), caused by the Coulomb interaction H' and are small compared to H_{nn}. It can be seen that the normal case of elastic electron scattering with the Bloch-wave solution is obtained if all of the off-diagonal elements H_{nm} are zero and no inelastic scattering occurs.

The H_{nm} have the same periodicity as the lattice and can be expanded in a Fourier series

$$H_{nm}(\boldsymbol{r}) = \exp(-2\pi i \boldsymbol{q}_{nm}\cdot\boldsymbol{r}) \sum_g H_g^{nm} \exp(2\pi i \boldsymbol{g}\cdot\boldsymbol{r})\,, \tag{7.79}$$

where q_{nm} is the wave vector of the crystal excitation created in the transition $m \rightarrow n$.

If all of the $H_{nm}(\boldsymbol{r})$ with $n \neq m$ are small compared to $H_{00}(\boldsymbol{r})$ and the amplitudes ψ_n of the inelastically scattered waves are small compared to ψ_0 the set of equations (7.76) can be written [7.29]

$$\left[-\frac{\hbar^2}{2m}\nabla^2 - E_0 + H_{00}\right]\psi_0 = -\sum_{n\neq 0} \mathrm{H}_{0n}\psi_n \tag{7.80 a}$$

$$\left[-\frac{\hbar^2}{2m}\nabla^2 - E_n + H_{nn}\right]\psi_n = -H_{n0}\psi_0\,. \tag{7.80 b}$$

Yoshioka [7.29] also omitted H_{nn} in (7.80 b) and solved the system with the aid of the Green's function for scattered spherical waves [7.30]. However, in a crystal, both the incident and scattered waves can propagate only as Bloch waves. The solution with a Green's function constructed of Bloch waves is discussed in [7.31]. *Howie* [7.28] considered a long-range interaction potential H' for the excitation of plasmons. Single-electron excitation is discussed in [7.32] and electron-phonon scattering in [7.33, 34].

Solutions ψ_n of (7.80) in the form of a series of Bloch waves $b^{(i)}$ (7.43) can be tried. These Bloch waves are solutions of $[-(\hbar^2/2m)\nabla^2 - E_n + H_{nn}]\psi_n = 0$ but have z-dependent amplitudes $\varepsilon_n^{(i)}(z)$

$$\begin{aligned}\psi_n(\boldsymbol{r}) &= \sum_i \varepsilon_n^{(i)}(z)\, b^{(i)}(\boldsymbol{k}_n^{(i)}, \boldsymbol{r}) \\ &= \sum_i \varepsilon_n^{(i)}(z) \sum_g C_g^{(i)}(\boldsymbol{k}_n^{(i)}) \exp[2\pi i(\boldsymbol{k}_n^{(i)} + \boldsymbol{g})\cdot \boldsymbol{r}] \,.\end{aligned} \tag{7.81}$$

Substituting (7.81) into (7.80), neglecting the small terms $d^2\psi_n/dz^2$, multiplying both sides by $b^{(j)*}$ and integrating over the x, y plane containing the reciprocal-lattice vectors $\boldsymbol{g}$, we obtain the following relation between the Bloch-wave amplitudes $\varepsilon^{(j)}$ of the incident and $\varepsilon^{(i')}$ of the scattered wave (the latter indicated by a dash)

$$\frac{d\varepsilon_n^{(i')}}{dz} = \sum_{m\neq n} \sum_j c_{mn}^{i'j}\, \varepsilon_m^{(j)} \,. \tag{7.82}$$

The matrix elements

$$c_{mn}^{i'j} = -\frac{im}{h^2[k_n^{(i')}]_z} \exp[2\pi i(k_m^{(j)} - k_n^{(i')} - q_{nm}^{i'j})z] \sum_{h,g} C_g^{(i')} H_{g-h}^{mn} C_h^{(j)} \tag{7.83}$$

describe the transition probabilities between branches j and i' of the dispersion surfaces ($i' = j$: intraband and $i' \neq j$; interband transition).

It can be seen from (7.83) that the scattering in a crystal depends on both eigenvector components $C_g^{(i')}$ and $C_h^{(j)}$. This is none other than a reciprocity theorem [7.35], which means that if a primary wave travels in the reverse direction along the path of the scattered wave, it will be scattered with the same probability in the former primary direction. In a crystal, the interaction is inevitably a scattering from one Bloch-wave field into another because incident and scattered waves have to exhibit the lattice periodicity. The scattered intensity, therefore, depends not only on the scattering angle θ, as for a single atom or for amorphous material, but also on the excitation probabilities of the Bloch waves in the incident and scattered directions. This observation will be used in the discussion of the intensities of Kikuchi lines and bands (Sect. 7.5.4).

Plasmon scattering is not concentrated at the nucleus but within a relatively large volume of 1–10 nm diameter. This process is therefore limited to small scattering angles, and in (7.83) only one term H_0^{mn} need be considered. The sum over the matrix elements in (7.83) gives $\sum_g C_g^{(i')} H_0^{mn} C_g^{(j)} = H_0^{nm}\delta_{i'j}$, which is non-zero only for $i' = j$. This means that plasmon scattering causes only intraband scattering. Image contrast by Bragg reflection is consequently preserved [7.28]. The same is true for the ionisation processes in inner shells, so long as $H_0^{mn} \gg H_g^{mn}$ [7.32].

Interband scattering can be observed for large scattering angles, which is equivalent to a narrower localization of the scattering process near the nucleus [7.32]. On the other hand, electron-phonon scattering (thermal diffuse scattering) is predominately interband scattering for small scattering angles and intraband scattering for large scattering angles between the Bragg reflections [7.34]; it, therefore, contributes mainly to the absorption parameters of the Bloch-wave field.

7.4.2 Absorption of the Bloch-Wave Field

The transition from the initial state $m = 0$ to any $n \neq m$ and i' in (7.82) results in an exponential decrease of $\varepsilon_0^{(j)}$, with the value of $\varepsilon_0^{(j)}(0) = C_0^{(j)}$ at the entrance surface of the crystal ($z = 0$) being determined by the boundary conditions discussed at the end of Sect. 7.3.3.

This exponential decrease can also be incorporated in the Schrödinger equation (3.24) and in the fundamental equations of dynamical theory (7.46) by introducing an additive imaginary lattice potential V_g': $V_g \to V_g + \mathrm{i}V_g'$ or by replacing U_g by $U_g + \mathrm{i}U_g'$ [7.29]. The V_g' values can be converted to U_g' as in (7.39). Returning to (7.41), $V_g + \mathrm{i}V_g'$ obliges us to replace $1/\xi_g$ by $1/\xi_g + \mathrm{i}/\xi_g'$, where ξ_0' is the mean absorption distance and the ξ_g' are anomalous absorption distances. Values of the imaginary Fourier coefficients are listed in Table 7.3. For relative values of ξ_g/ξ_g', see also [7.36, 37].

The V_g are assumed to be independent of electron energy, because they are defined as Fourier coefficients of the lattice potential $V(r)$. The V_g' are proportional to v^{-1}. In view of (7.42), this implies that $\xi_g \propto v$ and ξ_0', $\xi_g' \propto v^2$, which can be confirmed experimentally [7.19, 38, 39].

Replacing $1/\xi_g$ by $1/\xi_g + \mathrm{i}/\xi_g'$ in (7.35), we obtain the form of the Howie-Whelan equations in which these absorption effects are considered. In the formulation of the dynamical theory as an eigenvalue problem (Sect. 7.3.3), the matrix $[A]$ in (7.50) now contains the components

$$A_{11} = \mathrm{i}U_0'/2K; \quad A_{gg} = s_g + \mathrm{i}U_0'/2K; \quad A_{gh} = (U_{g-h} + \mathrm{i}U_{g-h}')/2K$$

with the result that the eigenvalues become complex: instead of $\gamma^{(j)}$, we write $\gamma^{(j)} + \mathrm{i}q^{(j)}$. The characteristic equation (7.61) for the complex eigenvalues becomes more complicated. Assuming that $\xi_0', \xi_g' \gg \xi_g$, approximately the same values are obtained for the real part $\gamma^{(j)}$ (7.63), and for the two-beam case the imaginary absorption parameters become

$$q^{(j)} = \frac{1}{2}\left[\frac{1}{\xi_0'} - \frac{(-1)^j}{\xi_g'\sqrt{1+w^2}}\right], \tag{7.84}$$

which are plotted in Fig. 7.13 b.

Table 7.3. Imaginary Fourier coefficients V_g' [eV] for different substances at $E = 100$ keV and room temperature (and in a few cases at 150 K). The absorption distances ξ_0' is equal to $340/V_0'$ nm and $\mu_0 = V_0'/54$ nm^{-1}

	V_0'	$\boldsymbol{g} = hkl$ and V_g'				Ref.
Al	–	111	220	311		
theor.	0.85	0.18	0.14	0.13		[7.17]
	0.58	0.16				[7.18]
exp.	0.37	0.23	–	–		[7.19]
	0.6	–	0.11	0.13		[7.20]
	0.54	0.17				[7.18]
Si		220	311	422		
theor.	0.70	0.11	0.07	0.08		[7.17]
exp.	0.68	0.11	0.08	0.08		[7.21]
	0.62	0.14	0.08	–		[7.20]
Cu		111	200	220	311	
theor.	3.48	0.83	0.79	0.68	0.63	[7.17]
exp.	1.48	0.81	0.92	–	–	[7.22]
						[7.23]
	1.35	–	–	0.49	0.45	[7.20]
Ge		220	400	422		
theor.	1.56	0.54	0.48	0.43		[7.17]
exp.	1.25	0.52	–	0.36		[7.21]
	1.35	–	0.32	–		
Au		220	331	440		
theor.	7.57	2.8	2.3	1.87		[7.17]
exp.	2.64	2.0	–	1.5		[7.24]
	–	–	1.62	–		[7.20]
150 K:	6.71		1.81	–		[7.17]
	2.5		1.12			[7.25]
MgO		200				
theor.	1.8	0.16				[7.17]
exp.	1.5	0.13				[7.26]
NaCl		220	420			
theor.	1.63	0.20	0.14			[7.17]
exp.	–	0.21	0.15			[7.27]
PbTe (150 K)		422				
theor.	4.7	0.98				[7.17]
exp.	1.8	0.67				[7.25]

The $q^{(j)}$ can be calculated for the n-beam case. If in the additional elements of the matrix $[A]$, the U_g' will be smaller than 0.1 U_g, then the familiar first-order perturbation method of quantum mechanics can be applied, giving

$$q^{(j)} = \left\langle b^{(j)}(\boldsymbol{k},\boldsymbol{r}) \left| \frac{U'}{2K} \right| b^{(j)}(\boldsymbol{k},\boldsymbol{r}) \right\rangle = \frac{1}{2K} \sum_g \sum_h U'_{g-h} C_h^{(j)} C_g^{(j)} , \tag{7.85}$$

in which the $C_g^{(j)}$ are the components of the eigenvectors of the unperturbed matrix, without the complex components, corresponding to the situation in which absorption is disregarded. The Bloch-wave formula (7.43) has to be modified to

$$b^{(j)}(\boldsymbol{k},\boldsymbol{r}) = \exp(-2\pi q^{(j)}z) \sum_g C_g^{(j)} \exp[2\pi i(\boldsymbol{k}_0^{(j)} + \boldsymbol{g}) \cdot \boldsymbol{r}] . \tag{7.86}$$

The first factor with $q^{(j)}$ in the real exponent describes an exponential decrease of the Bloch-wave amplitude with increasing depth z below the surface. From the first term of (7.84), we know that all Bloch-wave amplitudes decrease as $\exp(-\pi z/\xi_0')$. Differences of the $q^{(j)}$ due to the second term of (7.84) can be understood from the following Bloch-wave model. Let us combine the four possible waves of the two-beam case (Fig. 7.12) but not as we did in (7.55), when we calculated the amplitudes of the primary and Bragg-reflected beams. Now, the waves with wave vectors $\boldsymbol{k}_0^{(1)}$ and $\boldsymbol{k}_g^{(1)}$ form "Bloch wave 1" and $\boldsymbol{k}_0^{(2)}$ and $\boldsymbol{k}_g^{(2)}$ "Bloch wave 2" where $\boldsymbol{k}_g^{(1,2)} = \boldsymbol{k}_0^{(1,2)} + \boldsymbol{g}$. This superposition of two inclined waves propagates in the direction of the angle bisector, which is parallel to the reflecting lattice planes hkl. The superposition shows interference fringes in the x direction, perpendicular to the direction of propagation, with a periodicity equal to the lattice-plane spacing d_{hkl}. From (7.65), we see that all of the $C_0^{(j)} C_h^{(j)}$ ($j = 1, 2$; $\boldsymbol{h} = 0, \boldsymbol{g}$) take the values $\pm 1/2$ in the Bragg condition ($w = 0$). The $C_h^{(j)}$ are symmetric for $j = 1$ (equal signs) and antisymmetric for $j = 2$ (opposite signs). Substitution in (7.43) and (7.86) results in

$$\begin{aligned} |b^{(1)}| &\propto \cos(\pi \boldsymbol{g} \cdot \boldsymbol{r}) = \cos(\pi x/d_{hkl}) \\ |b^{(2)}| &\propto \sin(\pi \boldsymbol{g} \cdot \boldsymbol{r}) = \sin(\pi x/d_{hkl}) \end{aligned} \tag{7.87}$$

because the $q^{(j)}$ are equal for the same branch (j) of the dispersion surface.

The probability density $|b^{(j)}|^2$ of the Bloch waves 1 and 2, is therefore proportional to $\cos^2(\pi x/d_{hkl})$ and $\sin^2(\pi x/d_{hkl})$, respectively. This results in minima (nodes) at the lattice plane for the antisymmetric wave ($j = 2$) and in maxima (antinodes) for the symmetric wave ($j = 1$) (Fig. 7.16). This is important for the absorption of these Bloch waves. In particular, thermal diffuse scattering is caused by deviations from the ideal lattice structure due to thermal vibrations of the lattice. Because the amplitude of lattice vibrations is small, the symmetric Bloch wave 1 with maxima at the nuclei will be scattered more strongly than the antisymmetric wave, the values of the absorption parameters $q^{(1)}$ will be larger and the Bloch-wave amplitude will decrease more rapidly with increasing z; the antisymmetric Bloch wave 2 with nodes at the nuclei interacts less strongly ($q^{(2)} < q^{(1)}$, see also Fig. 7.13b). A consequence of thermal diffuse scattering is that ξ_0' and ξ_g' depend on temperature.

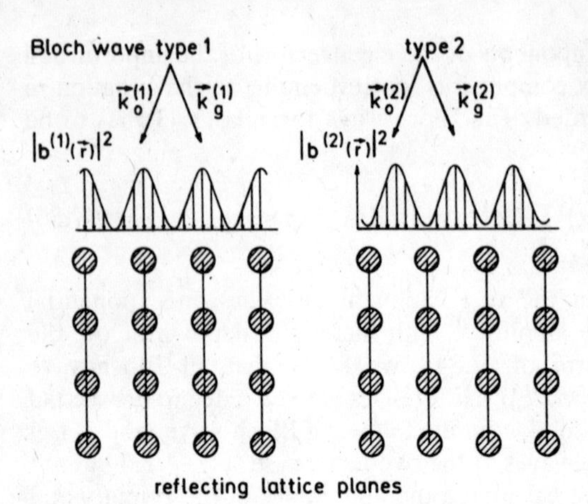

Fig. 7.16. Squared amplitudes (interaction probabilities) $\psi\psi^*$ of Bloch waves type 1 and type 2 in the two-beam case (independent Bloch-wave model) with antinodes and nodes at the nuclei and lattice planes, respectively

There are tilt angles out of the Bragg position for which the antisymmetric Bloch waves show a large transmission (low $q^{(2)}$) and a large excitation amplitude $\varepsilon^{(2)}$ (anomalous transmission). As shown in Fig. 7.13, this is the case for positive excitation parameters w for which the squared amplitude $|\varepsilon^{(2)}|^2 = |C_0^{(2)}|^2$ is large and $q^{(2)} < q^{(1)}$.

The following analytical formulae for T and R can be derived for the two-beam case [7.40]; using the abbreviations

$$\mu_g = 2\pi U_g' K = 2\pi/\xi_g' \quad \text{and} \quad \mu_0 = 2\pi/\xi_0' ,$$

we find

$$T = \frac{e^{-\mu_0 z}}{2(1+w^2)}\left[(1+2w^2)\cosh\frac{\mu_g z}{\sqrt{1+w^2}}\right.$$

$$\left. + 2w\sqrt{1+w^2}\sinh\frac{\mu_g z}{\sqrt{1+w^2}} + \cos\left(2\pi\frac{\sqrt{1+w^2}}{\xi_g}z\right)\right] \tag{7.88 a}$$

$$R = \frac{e^{-\mu_0 z}}{2(1+w^2)}\left[\cosh\frac{\mu_g z}{\sqrt{1+w^2}} - \cos\left(2\pi\frac{\sqrt{1+w^2}}{\xi_g}z\right)\right] . \tag{7.88 b}$$

Certain characteristic differences are found when the above expressions are compared with the two-beam case without absorption, (7.69). The rocking curve of the transmission is not symmetric about the Bragg position (Fig. 7.18). This asymmetry is a consequence of the second term in the square bracket of (7.88 a). The reflected intensity R still remains symmetric [there are no terms in odd powers of w in (7.88 b)]. The relation $T + R = 1$ is no longer valid. The amplitude of the pendellösung fringes decreases with increasing thickness (Fig. 7.17). Only a broad transmission band (anomalous transmission) with extremely weak "pendellösung" fringes is observed for very thick specimens (Fig. 7.18 b).

In the case of anomalous transmission, only the antisymmetric Bloch wave 2 with low $q^{(2)}$ remains after passage through a thick specimen layer. This observation might seem to imply that the Bloch waves can be treated as approximately independent. However, this independent Bloch-wave model cannot satisfy the boundary condition at the entrance plane of the crystal and experiment shows that there are situations in which the whole Bloch-wave

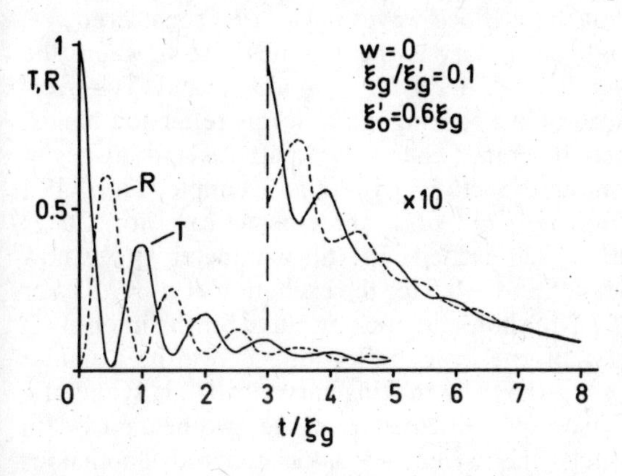

Fig. 7.17. Thickness dependence of the transmitted (T) and Bragg-reflected (R) intensities in the Bragg condition ($w = 0$) with absorption

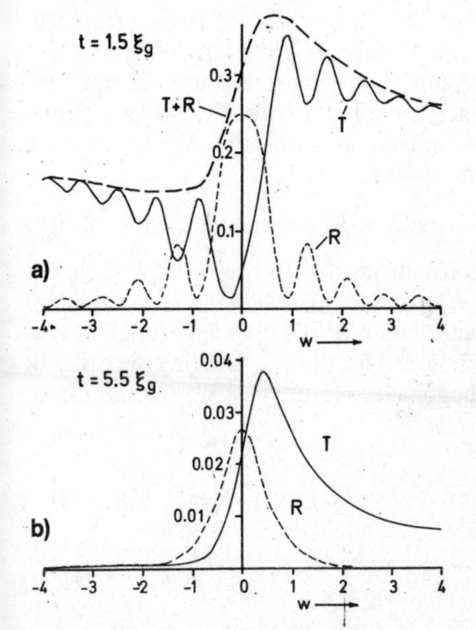

Fig. 7.18. Tilt dependence (rocking curve) of the transmitted (T) and Bragg-reflected (R) intensities for the dynamical two-beam case with absorption for foil thicknesses (**a**) $t = 1.5\,\xi_g$ and (**b**) $t = 5.5\,\xi_g$

field (dependent Bloch-wave model) has to be considered – in study of the probability of large-angle backscattering, for example [7.41].

7.4.3 Dynamical *n*-Beam Theory

The n-beam case must normally be treated numerically, by solving the eigenvalue problem (7.50) for a large number of reciprocal lattice points $\boldsymbol{g}$ near the Ewald sphere or by applying the Howie-Whelan equations in the multislice method. The dispersion surface splits into n branches, and $j = 1, \ldots, n$ absorption parameters $q^{(j)}$ of the n Bloch waves have to be considered.

A special case of the n-beam theory is the systematic row, where the reflections $-n\boldsymbol{g}, \ldots, -2\boldsymbol{g}, -\boldsymbol{g}, 0, +\boldsymbol{g}, +2\boldsymbol{g}, \ldots, +n\boldsymbol{g}$ are excited. The overlap of $-\boldsymbol{g}$ and $+\boldsymbol{g}$ reflections in the rocking curve forms reflection bands, which can be seen in Kikuchi diagrams, channelling patterns and images of bent crystal foils (bend contours, Sect. 8.1.1). As an example, Fig. 7.19a shows the absorption parameters $q^{(j)}$ of a three-beam case for Cu at $E = 100$ keV with $\boldsymbol{g} = \bar{2}\bar{2}0$, 0, 220 excited. The tilt parameter k_x/g is now zero for the symmetric incidence and $+0.5$ for the excitation of $\boldsymbol{g} = 220$. For a thickness $t = 40$ nm, Fig. 7.20a shows the rocking curve for the intensity I_0 of the primary beam and I_{220} of one Bragg reflection. $I_{\bar{2}\bar{2}0}$ will be a similar curve with its centre at $k_x/g = -0.5$. The rocking curve for I_{220} is symmetric about the Bragg position (see also discussion of the two-beam case in Sect. 7.4.2) whereas I_0 exhibits the asymmetry associated with anomalous absorption. This can be seen from Fig. 7.19. For $k_x/g < 0.5$, the Bloch-wave intensity $|C_0^{(1)}|^2$ with a large value of $q^{(1)}$ is larger than $|C_0^{(2)}|^2$, which results in stronger absorption. For $k_x/g > 0.5$, the intensity $|C_0^{(2)}|^2$ is larger for the lower parameter $q^{(2)}$, leading to a greater value of the transmission. (The intensity $|C_0^{(3)}|^2$ is so low in this special case that it is not included in Fig. 7.19b.) Tilting to negative values of k_x/g results in anomalous absorption for $k_x/g > -0.5$ and anomalous transmission for $k_x/g < -0.5$.

In the sum of intensities $\sum_g I_g = I_{-g} + I_0 + I_g$, the "pendellösung" fringes cancel. However, the sum is not constant, as it is in the dynamical theory without absorption, but varies with k_x/g and a defect band of reduced intensity is created (Fig. 7.20a). This cancellation of the "pendellösung" fringes can be shown analytically by using (7.55, 86) for the ψ_g, changing the order of summation and employing the orthogonality relations (7.52) for the eigenvector components:

$$\sum_g I_g = \sum_g \psi_g \psi_g^* = \sum_g \left| \sum_j C_0^{(j)} C_g^{(j)} \exp(-2\pi q^{(j)} z) \exp[2\pi i (\boldsymbol{k}^{(j)} + \boldsymbol{g}) \cdot \boldsymbol{r}] \right|^2$$

$$= \sum_g \sum_{i,j} C_0^{(i)} C_0^{(j)} C_g^{(i)} C_g^{(j)} \exp[-2\pi (q^{(i)} + q^{(j)}) z] \exp[2\pi i (\boldsymbol{k}^{(i)} - \boldsymbol{k}^{(j)}) \cdot \boldsymbol{r}]$$

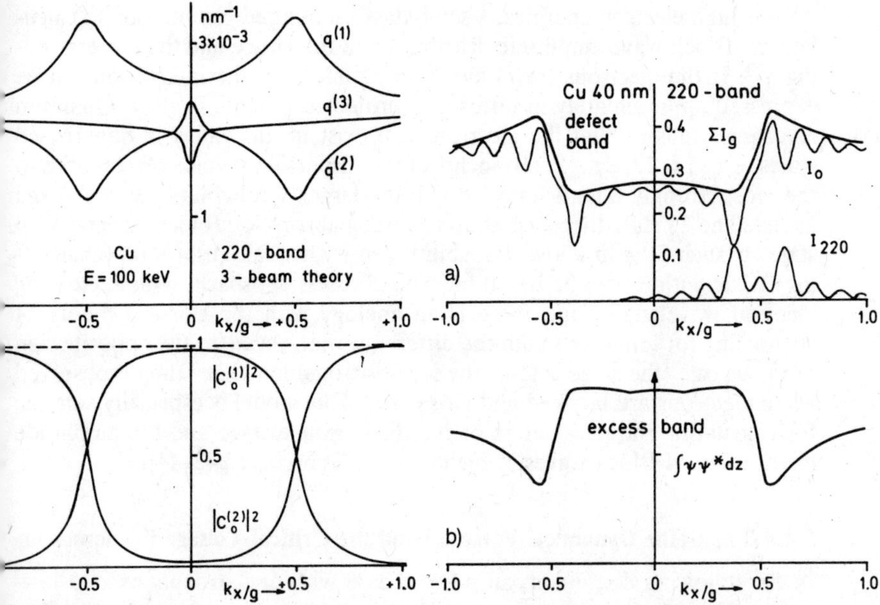

Fig. 7.19. Dependence of (**a**) absorption parameters $q^{(j)}$ and (**b**) the squared amplitudes $|C_0^{(j)}|^2$ of the Bloch waves on the tilting parameters k_x/g for a 220 band of Cu at $E = 100$ keV (Three-beam case with excitation of $\bar{2}\bar{2}0$, 0, 220)

Fig. 7.20. Dependence of (**a**) the primary-beam intensity I_0 and Bragg-reflected intensity I_g ($\boldsymbol{g} = 220$) (rocking curve) and $\sum_g I_g$ (defect band) and (**b**) the large-angle scattering probability $\int \psi\psi^*\,dz$ (excess band) as a function of the tilt parameter k_x/g for the three-beam case of Cu at $E = 100$ keV

$$= \sum_{i,j} \Big[\underbrace{\sum_g C_g^{(i)} C_g^{(j)}}_{\delta_{ij}} \Big] C_0^{(i)} C_0^{(j)} \exp[-2\pi(q^{(i)} + q^{(j)})z] \exp[2\pi i(\boldsymbol{k}_0^{(i)} - \boldsymbol{k}_0^{(j)}) \cdot \boldsymbol{r}] ,$$

and hence

$$\sum_g I_g = \sum_i |C_0^{(i)}|^2 \exp(-4\pi q^{(i)} z) . \tag{7.89}$$

Inside the defect band, only the Bloch-wave intensity $|C_0^{(1)}|^2$ with a large value of $q^{(1)}$ is excited and the intensity therefore decreases.

Scattering effects that are strongly localized at the nuclei (e.g., scattering through large angles and backscattering or excitation of an inner shell with an energy-loss maximum or x-ray emission) are proportional to the probability density $\psi\psi^*$ and to $\int \psi\psi^*\,dz$ in a finite thickness t. Inside the defect band, the intensity $|C_0^{(1)}|^2$ of the Bloch wave with antinodes at the nuclei is large, which causes increased large-angle scattering or an increased probability of inner-shell ionisation, resulting in an excess band (Fig. 7.20 b).

For high electron energies, a semi-classical method can be used to calculate the Bloch-wave amplitude distribution in the lattice and the corresponding $q^{(j)}$. If the electrons travel along an atomic row, the interaction can be expressed approximately in terms of a projected potential valley. Quantum mechanics tells us that the electrons can exist in this valley in quantizised states, e.g. $1s, 2s, 2p, \ldots$. Branch 1 of the dispersion surface corresponds to the most strongly bound state $1s$ with the largest probability density at the nuclei. The $1s$ state therefore shows a strong absorption (blocking) and a $2p$ state (branch 2) with a low probability density a weak absorption (channeling). Calculations can be based on semi-classical approach, with decreasing electron wavelength, and there is an analogy with the classical theory of channeling for ion beams with the difference that, owing to the opposite sign of the charge (the same is also true for positrons) ion channeling is observed when electrons are blocked and vice versa. This model is especially suitable for calculating the parameters of the dispersion surface and the amplitude distribution of Bloch waves at high electron energies [7.42–44].

7.4.4 The Bethe Dynamical Potential and the Critical-Voltage Phenomenon

In the following discussion, an n-beam case with one strongly excited low-order reflection ($\boldsymbol{g}_1 = 0$ and $\boldsymbol{g}_2 = \boldsymbol{g}$) and a number of weakly excited reflections $\boldsymbol{g}_n = \boldsymbol{h}_n$ $(n \geq 2)$ is considered (Bethe's approximation [7.7]). In the fundamental equations (7.46), the first equations are the same and in those for $\boldsymbol{g}_n$ $(n \geq 2)$ only terms with the largest amplitude factors $C_0^{(j)}$ and $C_g^{(j)}$ need be considered in a first-order approximation:

$$[K^2 - (\boldsymbol{k}_0^{(j)} + \boldsymbol{g})^2]\, C_g^{(j)} + \sum_{h \neq 0} U_h C_{g-h}^{(j)} = 0 \quad \textit{for } \boldsymbol{g}_1 = 0 \quad \text{and } \boldsymbol{g}_2 = \boldsymbol{g}$$

$$[K^2 - (\boldsymbol{k}_0^{(j)} + \boldsymbol{h})^2]\, C_h^{(j)} + U_h C_0^{(j)} + U_{g-h} C_g^{(j)} = 0 \quad \text{for } \boldsymbol{h} = \boldsymbol{g}_3, \ldots, \boldsymbol{g}_n \,. \tag{7.90}$$

The last equations are not coupled and can be solved for the $C_h^{(j)}$; these can then be substituted in the first equations for $\boldsymbol{g}_1 = 0$ and $\boldsymbol{g}_2 = g$. This yields a two-beam case analogous to (7.60) but with a corrected potential coefficient $U_{g,\text{dyn}}$ or *dynamic potential*

$$V_{g,\text{dyn}} = V_g - \frac{2m_0}{h^2}(1 + E/E_0) \sum_{h \neq 0} \frac{V_g V_{g-h}}{K^2 - (\boldsymbol{k} + \boldsymbol{h})^2} = V_g - V_{g,\text{corr}} \,. \tag{7.91}$$

Depending on the sign of the denominator in (7.91), $V_{g,\text{dyn}}$ may be larger or smaller than V_g. The latter will be decreased when the reciprocal-lattice point is inside the Ewald sphere. From the relation $\xi_g = (\lambda U_g)^{-1}$ (7.41), the extinction distances will also be changed by this dynamical interaction.

The value of $V_{g,\text{dyn}}$ in (7.91) decreases with increasing electron energy and can vanish for a certain critical voltage V_c or energy $E_c = eV_c$ if the excitation error is positive (reciprocal-lattice point inside the Ewald sphere). At this voltage, $\xi_g \to \infty$, which means that the corresponding reflection will be not excited. This effect was first observed in electron-diffraction patterns [7.45, 46]. The minimum distance $\Delta k_{z,\min}$ between the branches of the dispersion surface falls to zero and the branches intersect.

The use of (7.91) alone can explain the existence of a critical voltage but is not sufficient for an accurate calculation. With increasing energy, first the dynamical potential $V_{2g,\mathrm{dyn}}$ of a systematic row $(0, \pm g, \pm 2g, \ldots)$ vanishes because the two terms in (7.91) cancel for the second-order reflection (e.g., 400 in a 200 row) when the $2g$ reflection is fully excited. We therefore assume a three-beam case 0, g, $2g$ [7.47] with $s_{2g} = 0$ and $s_g = K - \sqrt{K^2 - g^2} \simeq g^2/2K$. The eigenvalue equation (7.50) can be rewritten

$$\begin{pmatrix} -2K\gamma & U_g & U_{2g} \\ U_g & (g^2 - 2K\gamma) & U_g \\ U_{2g} & U_g & -2K\gamma \end{pmatrix} \begin{pmatrix} C_0 \\ C_g \\ C_{2g} \end{pmatrix} = 0 . \tag{7.92}$$

The solution can be simplified by considering the symmetry of the matrix, which leads us to distinguish the two cases

a) $C_0 = C_{2g}$, $C_g \neq 0$ and
b) $C_0 = -C_{2g}$, $C_g = 0$.

Substituting in (7.92), we obtain the reduced system of equations:

$$\text{a)}\quad \begin{aligned} (U_{2g} - 2K\gamma)\, C_0 + \quad\quad\; U_g\, C_g &= 0 \\ 2\,U_g\, C_0 + (g^2 - 2K\gamma)\, C_g &= 0 \end{aligned}$$

$$\text{giving}\quad K\gamma^{(1,2)} = \frac{1}{4}\left[U_{2g} + g^2 \pm \sqrt{8\,U_g^2 + (U_{2g} - g^2)^2}\right] .$$

$$\text{b)}\quad -2K\Delta k_z C_0 - U_{2g}\, C_0 = 0 \quad \text{giving} \quad K\gamma^{(3)} = -\frac{1}{2}\, U_{2g} .$$

The values of γ for case a) are obtained by setting the determinant equal to zero. The difference

$$\gamma^{(2)} - \gamma^{(3)} = \frac{1}{4K}\left[3\,U_{2g} + g^2 - \sqrt{8\,U_g^2 + (U_{2g} - g^2)^2}\right]$$

vanishes when the quantity in the square brackets is zero. This yields the condition

$$U_g^2 = U_{2g}^2 + U_{2g}\, g^2 . \tag{7.93}$$

When U_g is replaced by use of (7.39), the critical energy $E = E_c$ appears in the factor $m = m_0\,(1 + E/E_0)$ and solving for E results in

$$E_c = eV_c = \left[\frac{h^2 g^2 V_{2g}}{2m_0 (V_g^2 - V_{2g}^2)} - 1\right] E_0 \,. \tag{7.94}$$

The critical voltage can also be obtained by calculating the Bloch-wave amplitudes $C_0^{(j)}$, because the symmetry of Bloch waves 2 and 3 changes for the second-order reflection when the electron energy exceeds the critical energy [7.15]. We find

	$V < V_c$	$V > V_c$
$j = 2$	symmetric	antisymmetric
$j = 3$	antisymmetric	symmetric

and

$V \lessgtr V_c$ when $|C_0^{(2)}| \gtrless |C_0^{(3)}|$ — for the second-order reflection at $k_x = 0$ (symmetrical incidence of the electron beam)

$V \lessgtr V_c$ when $|C_0^{(3)}| \gtrless |C_0^{(4)}|$ — for the third-order reflection at $k_x = 0.5\,g$ (Bragg condition for the first order).

In Table 7.4, some experimental values of the critical voltage are listed. Most of the values V_c are greater than 100 keV and hence in the HVEM range. Vanishing of the reflection $2g$ in the Bragg condition can be observed with Kikuchi lines in electron-diffraction patterns [7.48], with bend contours in electron micrographs or in convergent-beam diffraction patterns [7.49, 50]. Figure 7.21 shows the rocking curve near the 222 Bragg condition for a 300 nm Cu foil. The critical voltage V_c can be determined with an accuracy of a few kilovolts.

The following applications of the critical-voltage effect are of interest:

1) Accurate measurement of V_c and study of the excitation of other reflections predicted by the dynamical theory allow us to measure the coefficients V_g of the lattice potential and to calculate the structure amplitude $F(\theta)$ at the corresponding scattering angle $\theta = 2\,\theta_B$ by using (7.40) [7.47, 50–56]. The differences between $F(\theta)$ and the calculated values of $f(\theta)$ for free atoms can be used to obtain information about variations in the electron-density distribution caused by the packing of the atoms in a solid (e.g., for Ge and Si [7.57]).

2) The critical voltage depends on the composition of an alloy. It decreases from $V_c = 590$ kV for pure Ni to $V_c \simeq 450$ kV for a Ni-10 mol% Au alloy, for example [7.58]; this can be used for local measurements of concentration.

Table 7.4 Examples of experimental critical voltages V_c [kV] at room temperature [7.15]

	Cubic face-centred metals			
hkl	111	200	220	331
Al	425, 430	895, 918	–	–
Co	276	555	1745	2686
Ni	295, 298	588, 587	1794	2730
Cu	310, 325	600, 605	1750	2700
Ag	55	225	919	1498
Au	(<0)	108	726	1266
Cu_3Au	175	425	–	–

	Cubic body-centred metals		
hkl	110	200	211
V	230, 238	–	–
Cr	259, 265	1238	–
Fe	305	1249	–
Nb	35	749	1595
Mo	35	789	1729
Ta	–	651	–
W	–	660	>1100

	Diamond cubic crystals		
hkl	111	220	400
Si	1113	>1150	–
Ge	925	1028	>1100
GaP	1026	1098	>1100

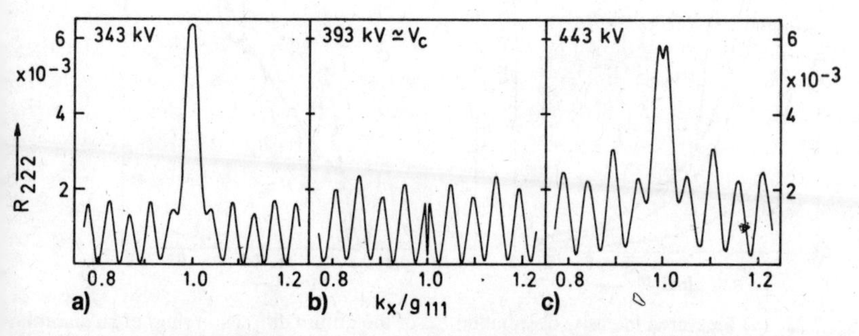

Fig. 7.21. Calculated rocking curve near the 222 Bragg condition (**a**) 50 kV below the critical voltage V_c, (**b**) at $V_c = 393$ kV with the critical-voltage effect at $k_x/g_{111} = 1$ and (**c**) 50 kV beyond the critical voltage of copper ($t = 300$ nm) [7.49]

3) Measurements of V_c at different temperatures give $F(\theta)\exp(-M)$ (Sect. 7.5.3) and can be used to obtain an accurate value of the Debye temperature θ_D [7.47] and its dependence on orientation in non-cubic crystals (e.g., Zn, Cd [7.59]) or alloys (e.g., Cu-Al, Cu-Au [7.60]).

4) The study of V_c can shed light on ordering in solids. Thus an increase of V_c from 166 kV to 175 kV is observed during the transition from a disordered to an ordered state in $AuCu_3$ [7.15]. Ordering state of short-range in Fe-Cr and Au-Ni alloys act like frozen-in lattice vibrations and make a temperature-independent contribution to the Debye temperature. By measurement of the temperature dependence of V_c the contributions from distortion of the long-range and from thermal vibrations can be separated if it is assumed that the corresponding $\langle u^2 \rangle$ can be added [7.58, 61].

These few examples show how the critical-voltage effect offers interesting possibilities for the quantitative analysis of metals and alloys.

7.5 Intensity Distributions in Diffraction Patterns

7.5.1 Diffraction at Amorphous Specimens

The diffraction patterns of amorphous films – carbon supporting films, polymers, silicon and aluminium oxide, glass and ceramics – consist of diffuse rings (Fig. 7.22 a). Each amorphous structure contains a nearest-neighbour ordering, which can be described by a radial density function $\varrho(r)$. The

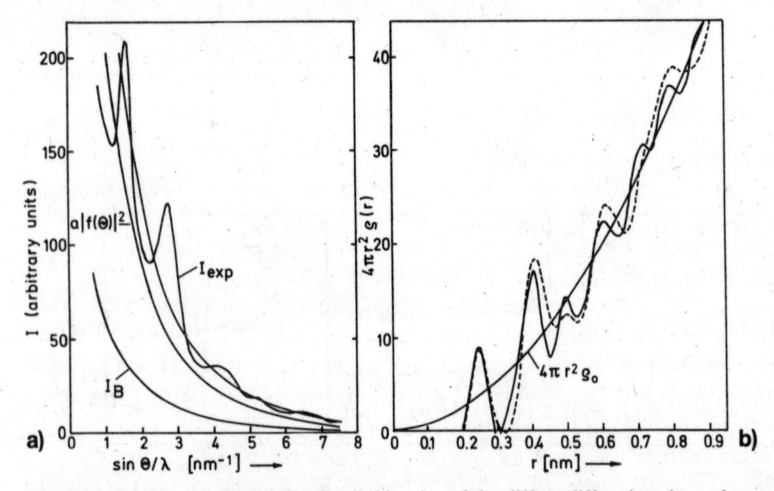

Fig. 7.22. (**a**) Measured intensity distribution I_{exp} of the diffuse diffraction rings of an amorphous Ge film ($E = 60$ keV) and fits of the backgrounds $a|f(\theta)|^2$ and I_B. (**b**) Calculation of the radial density distribution $4\pi r^2 \varrho(r)$ oscillating about the mean value $4\pi r^2 \varrho_0$. (---) values obtained by x-ray diffraction for comparison [7.62]

probability of finding the centres of neighbouring atoms inside a spherical shell between r and $r + dr$ is $4\pi r^2 \varrho(r)\, dr$ (Fig. 7.22 b). This probability distribution can be deduced from the diffraction pattern. The following method is due to *Leonhardt* et al. [7.62].

The observed intensity distribution I_{exp} in the diffraction pattern oscillates about the intensity distribution $a|f(\theta)|^2$ that would be seen if all of the atoms scattered independently, without interference:

$$I_{exp} = a|f(\theta)|^2 \left[1 + \int_0^\infty 4\pi r^2 \varrho(r) \frac{\sin(sr)}{sr} ds \right] + I_B \tag{7.95}$$

with the abbreviation $s = 4\pi \sin\theta/\lambda$; I_B is a continuous background caused by incoherent scattering. After producing a normalized function

$$i(s) = \frac{I_{exp} - [a|f(\theta)|^2 + I_B]}{a|f(\theta)|^2} \tag{7.96}$$

from the experimentally observed distribution I_{exp}, the transformation (7.95) can be reversed, giving

$$4\pi r^2 \varrho(r) = 4\pi r^2 \varrho_0 + \frac{2r}{\pi} \int_0^\infty i(s)\, s \sin(sr)\, ds\,. \tag{7.97}$$

Figure 7.22 a shows an example of the measured intensity distribution I_{exp} and the fitted curves $a|f(\theta)|^2$ and I_B which are used to obtain the radial density distribution in Fig. 7.22 b.

7.5.2 Intensity of Debye-Scherrer Rings

Polycrystalline specimens with random crystal orientations produce diffraction spots distributed randomly in azimuth. If the irradiated area is large and/or the crystal size is small, the high density of diffraction spots forms a continuous Debye-Scherrer ring for each allowed set of *hkl* values with non-vanishing structure amplitude F (Fig. 9.24).

The intensity of these Debye-Scherrer rings is obtained by averaging over all crystal orientations, that is, by integrating over s in (7.28) or (7.69). The kinematical theory gives the following expression for the integrated intensity I_{hkl} [A] of the total ring with Miller indices *hkl* [7.63, 64]:

$$I_{hkl} = j_s \frac{2\pi^2 m^2 e^2}{h^4} KNV_e p_{hkl} |V_g|^2 \lambda^2 d_{hkl} \tag{7.98}$$

with j_s being the current density in the specimen plane, K the number of crystals with in average N unit cells, V_e the volume of the unit cell, p_{hkl} the multiplicity of the hkl planes, e.g. $p_{100} = 6$ due to the six possible cubic planes, $p_{110} = 12$, $p_{111} = 8$ etc., and $V_g = V_g \exp(-M)$, the lattice potential with the Debye-Waller factor (Sect. 7.5.3).

This implies that the ring intensity depends on only the total number KN of unit cells in the electron beam but not on the shape and dimension of the crystals. The intensity ratios of various rings are independent of wavelength and crystal dimensions, and depend on p_{hkl}, V_g and d_{hkl}.

If dynamical two-beam theory is employed, the following differences relative to kinematical theory are found [7.65]:

$$\frac{I_{\text{dyn}}}{I_{\text{kin}}} = \frac{1}{A_{hkl}} \int_0^{A_{hkl}} \mathrm{J}_0(2x)\, dx \tag{7.99}$$

where J_0 is the Bessel function and $A_{hkl} = 2\pi e m_0 t V_g / h^2$.

The dynamical theory predicts the same results as the kinematical theory ($I_{\text{dyn}}/I_{\text{kin}} \simeq 1$) if A_{hkl} is small either because the crystal diameter t is small or the wavelength λ is short. The intensity ratios of different rings do not remain independent because they also depend on A_{hkl}. The curves in Fig. 7.23 show the decrease of $I_{\text{dyn}}/I_{\text{kin}}$ with increasing A_{hkl} and measurements of the ratios for evaporated Al films with small and large crystals [7.64]. The ratio $I_{\text{dyn}}/I_{\text{kin}}$

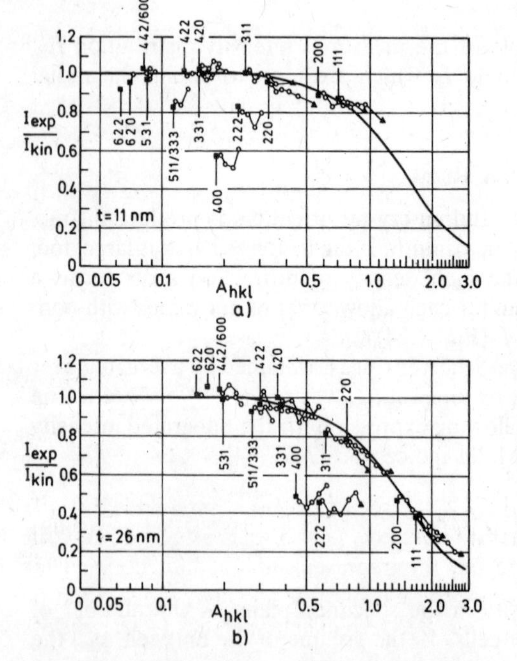

Fig. 7.23. Comparison of the diffracted intensity I_{exp} of Debye-Scherrer rings of evaporated Al films of thicknesses (**a**) $t = 11$ nm and (**b**) $t = 26$ nm with the value I_{kin} calculated by the kinematical theory. The decrease of $I_{\text{exp}}/I_{\text{kin}}$ for large A_{hkl}, and high-order systematic reflections (e.g. 222, 400) indicates deviations from the kinematical theory. (—) calculations of the intensity ratio with the dynamical theory (7.99)

is close to unity for crystals of the order 10 nm (Fig. 7.23 a). The intensities of Bragg reflections with low indices are the first to decrease for larger crystals (Fig. 7.23 b), whereas the kinematical theory still holds for high indices.

Reflections of higher order (e.g., 400 or 222) do not lie on the curves because reflections of low order are also excited. The two-beam theory is no longer valid and many-beam theory has to be used.

These laws governing Debye-Scherrer ring intensities assume that the orientation of the crystals is random with no preferential orientation (texture). A texture causes strong changes of the ring intensities and also an azimuthal variation of ring intensity for oblique electron incidence (Sect. 9.4.2).

7.5.3 Influence of Thermal Diffuse Scattering

Thermal vibrations of the atoms (nuclei) cause a distortion of the lattice periodicity and produce the following effects:

1) Decrease of the effective potential coefficients V_g by the Debye-Waller factor with increasing temperature, thus influencing the extinction distances ξ_g and the critical voltage V_c.
2) Increase of the absorption parameters $q^{(j)}$ of the dynamical theory and decrease of the absorption distances ξ_0' and ξ_g' with increasing temperature.
3) Thermal diffuse scattering in the background between and near the Bragg spots.

These interactions can be treated as electron-phonon scattering because the lattice vibrations are quantized (phonons of momentum $p = hk$ and energy $E = h\omega$).

The influence of lattice vibrations on the potential coefficients V_g can be understood from the following simplified model. If the atoms are shifted through a distance $\boldsymbol{u}(\boldsymbol{r})$ from their equilibrium positions $\boldsymbol{r}$, the potential is changed in first order only by a translation $V(\boldsymbol{r}) \rightarrow V(\boldsymbol{r} + \boldsymbol{u}(\boldsymbol{r}))$. Using the Fourier expansion (7.39) of the lattice potential, each Fourier coefficient V_g contains the following factor ($\langle\,\rangle$: mean value):

$$\begin{aligned}\langle e^{2\pi i \boldsymbol{u}\cdot\boldsymbol{g}}\rangle &= \langle 1 + 2i\boldsymbol{u}\cdot\boldsymbol{g} - 2\pi^2(\boldsymbol{u}\cdot\boldsymbol{g})^2 + \dots\rangle \\ &= 1 - 2\pi^2\langle(\boldsymbol{u}\cdot\boldsymbol{g})^2\rangle + \dots \simeq e^{-2\pi^2\langle u^2\rangle g^2} = e^{-M_g}\,. \end{aligned} \tag{7.100}$$

The mean value $\langle u\rangle$ in the Taylor series becomes zero and only the quadratic term in $\langle(\boldsymbol{u}\cdot\boldsymbol{g})^2\rangle$ has a non-zero value. This is usually again written as an exponential function but this representation cannot be used for large vibration amplitudes because these cannot be adequately described by the simple translation model. The *Debye-Waller factor* $\exp(-M_g)$ results formally in a reduction of the Fourier coefficient V_g to $V_g \exp(-M_g)$ and the reflection amplitudes of the kinematical theory will be decreased by the same factor; the diffraction intensities are thus attenuated by $\exp(-2M_g)$. An increase of ξ_g is also expected due to (7.32) and is observed experimentally [7.9].

The mean-square value $\langle u^2 \rangle$ of the lattice vibration amplitude depends on the phonon spectrum of the crystal. The Debye model for a monatomic cubic crystal gives

$$\langle u^2 \rangle = \frac{3h^2}{4\pi^2 Mk\theta_D}\left(\frac{1}{4} + \frac{T^2}{\theta_D^2}\int_0^{\theta_D/T}\frac{x\,dx}{e^x - 1}\right), \tag{7.101}$$

where M is the atomic mass and θ_D the Debye temperature. The term in brackets is tabulated (*International Tables for X-Ray Crystallography*, Vol. II). The term 1/4 results from the zero-point vibrations, which are present even at $T = 0$ as the quantum-mechanical treatment of a harmonic oscillator shows. The quantity $\langle u^2 \rangle$ will also depend on $\boldsymbol{g}$ for non-cubic crystals and when different types of atoms are present in the unit cell.

The absorption parameters $q^{(j)}$ or imaginary parts V_g' of the lattice potential (Sect. 7.4.2) contain a large contribution from thermal diffuse scattering and depend strongly on temperature (see, for example, calculations [7.17, 66, 67] and experiments [7.22]).

Scattering between the Bragg spots suspend the destructive interference in ideal crystals. The background caused by thermal diffuse scattering between the spots increases with increasing temperature as does the background near strongly excited Bragg spots by virtue of transverse and longitudinal phonons of long wavelength [7.67–71].

The influence of lattice vibrations is small at large scattering angles. An exception is the contrast of the excess Kikuchi bands in electron back-scattering patterns (EBSP, Sect. 9.3.4), which decreases with increasing temperature. Temperature-independent large-angle scattering is found as a result of summing the scattering processes at individual nuclei, which means that the scattering process is concentrated near a nucleus.

The background of electron-diffraction patterns of polycrystalline films depends more strongly on temperature than the Debye-Waller factor predicts [7.72, 73]. The influence of thermal diffuse scattering on the background of diffraction patterns of polycrystalline films has been investigated for Al [7.74], Ag [7.75] and Au [7.76]. The intensities of the primary and Bragg-reflected beams also depend more strongly on temperature than would be expected from the influence of the Debye-Waller factor (7.100) [7.77–80]. These effects can be explained by the dependence of the absorption parameters $q^{(j)}$ on temperature.

7.5.4 Kikuchi Lines and Bands

The background between the Bragg-diffraction spots of a diffraction pattern contains a structure that can be characterized as excess and defect Kikuchi lines and bands (Fig. 7.24).

Excess and defect Kikuchi lines are formed by the following mechanism. Inelastically scattered electrons can be Bragg reflected at lattice planes with reciprocal lattice vector $\boldsymbol{g}$ if the Bragg angle is $\pm\theta_B$ or if the direction of incidence $\boldsymbol{k}_0'$ lies on one of the *Kossel cones*, which have an aperture $90° - \theta_B$ and the $\boldsymbol{g}$ direction (normal to the lattice planes) as axis (Fig. 7.25). The Bragg-reflected beam $\boldsymbol{k}_g'$ also lies on the opposite cone in the plane defined by $\boldsymbol{g}$ and $\boldsymbol{k}_0'$ and results in a bright excess Kikuchi line along the hyperbola in which the Kossel cone intersects the plane of observation. The lines are approximately straight owing to the low value of θ_B. The Bragg reflection decreases the intensity of the incident direction $\boldsymbol{k}_0'$ and the intersection of the

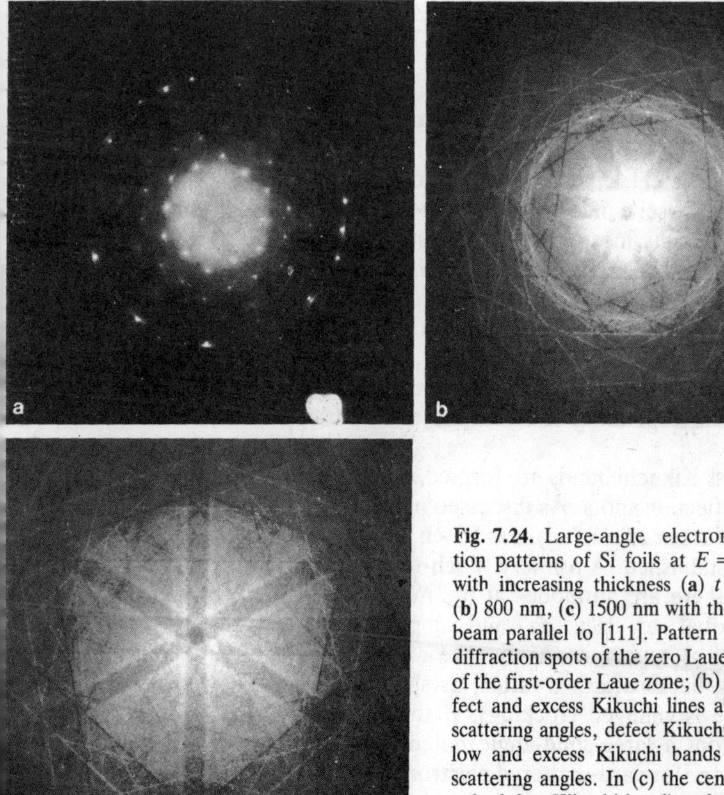

Fig. 7.24. Large-angle electron diffraction patterns of Si foils at $E = 100$ keV with increasing thickness (**a**) $t = 80$ nm, (**b**) 800 nm, (**c**) 1500 nm with the electron beam parallel to [111]. Pattern (a) shows diffraction spots of the zero Laue zone and of the first-order Laue zone; (b) shows defect and excess Kikuchi lines at medium scattering angles, defect Kikuchi bands at low and excess Kikuchi bands at larger scattering angles. In (c) the centre shows only defect Kikuchi bands and the region of defect and excess Kikuchi lines is shifted towards larger angles

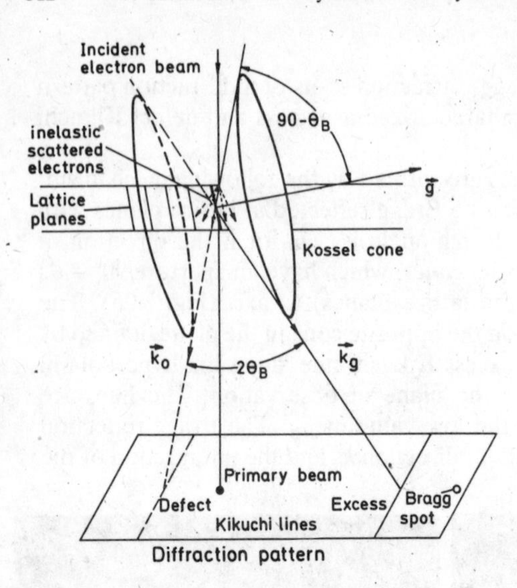

Fig. 7.25. Generation of excess and defect Kikuchi lines by Bragg scattering of diffusely scattered electrons regarded as the intersection of the Kossel cones of angular aperture $90° - \theta_B$ with the plane of observation

incident Kossel cone with the plane of observation results in a dark defect Kikuchi line. This mechanism, therefore, generates a set of corresponding excess and defect Kikuchi lines separated by an angular distance $2\theta_B$. The system of Kossel cones behaves as though fixed to the crystal, which means that the Kikuchi lines move if the crystal is tilted, whereas the position of the Bragg-reflection spot is fixed in the plane of observation (angle $2\theta_B$ with the primary beam), and the Bragg spots are visible only for a limited tilting range around the Bragg position of the primary beam. In the Bragg position (excitation error $s = 0$), one Kikuchi line coincides with the diffraction spot and the other with the primary beam. The displacement between the Bragg spot and the corresponding Kikuchi line can be used to measure the excitation error.

Excess Kikuchi bands are formed in each scattering process, between the Bragg reflection spots. As discussed in Sect. 7.4.1, the inelastically scattered electrons are scattered again as Bloch waves, so that a theorem of reciprocity can be established. A primary Bloch-wave field has a larger scattering probability if there are antinodes at the nuclei. The scattered intensity becomes proportional to $\int \psi\psi^* dz$ and depends on the tilting parameter k_x (Fig. 7.20b), which means that the whole intensity of large-angle scattering including backscattering varies, as shown in Fig. 7.20b if the direction of incidence is changed (rocking). If the rocking pattern forms a raster, the signal from a large-solid-angle detector designed to collect backscattered electrons or forward-scattered electrons with scattering angles ≥ 5–$10°$ generates a channeling pattern, with excess-band intensity distributions as in Fig. 7.20b. When we observe a stationary electron-diffraction pattern, the direction of the primary beam is fixed. The scattering of electrons into larger angles depends on the Bloch-wave intensity at the nuclei but the scattered Bloch-wave field also shows antinodes at the nuclei. The scattered intensity depends on the observation angle and the Bloch-wave intensity that would appear if the electron beam struck the crystal opposite the direction of observation. The whole angular distribution of scattered electrons is therefore not uniform but modulated by a system of excess Kikuchi bands. This can be seen if we observe electron back-scattering patterns (EBSP) recorded on a screen near the specimen in a SEM (Sect. 9.3.4).

The excess Kikuchi bands can also show a contrast reversal that apparently converts them into defect bands (Fig. 7.24 b, c) by the following mechanism [7.16, 81]. A set of defect and excess Kikuchi lines can be observed so long as more electrons hit the lattice planes from one side than from the other (see centre of Fig. 7.24 b). With increasing foil thickness, this will be the case for large scattering angles, for which the scattered intensity distribution decreases with increasing scattering angle (Fig. 7.24 c). For small scattering angles, however, the angular distribution of scattered electrons is so diffuse and uniform that equal numbers of electrons hit the lattice planes from both Kossel cones, and cancellation of the pattern of defect and excess lines is expected. However, the resulting intensity distribution will be a defect Kikuchi band due to the influence of anomalous absorption. For one direction of observation near the Kossel cone, the intensity will be the sum of the intensities of the incident and Bragg-reflected beams, that is, $T + R$ for the two-beam and $\sum_g I_g$ for the n-beam case. The dependence of this sum on the tilt angle has already been discussed on Fig. 7.20 a and results in a defect band.

These are the basic mechanisms – in a somewhat simplified presentation – whereby excess and defect Kikuchi lines and bands are formed; their appearance in diffraction patterns depends on the distribution of the diffusely scattered electrons. With increased foil thickness (Fig. 7.24) the central region of defect Kikuchi bands expands to larger scattering angles [7.82–84] and excess and defect Kikuchi lines are observed in only that annular region for which the scattered intensities decrease strongly and different intensities hit the two sides of the lattice planes. This is also the region in which the central excess bands show a contrast reversal, which converts them to defect Kikuchi bands, produced by the mechanism discussed above. More details about the quantitative treatment of these diffraction phenomena are to be found in [7.35, 81, 85–89].

8. Diffraction Contrast and Crystal-Structure Imaging

A crystal can be imaged with the primary beam (bright field) or with a Bragg reflection (dark field). The local intensity depends on the thickness, resulting in thickness contours, and on the tilt of the lattice planes, resulting in bend contours, which can be described by the dynamical theory of electron diffraction. In certain cases, the intensity of a Bragg reflection depends so sensitively on specimen thickness that atomic surface steps can be observed.

If the objective aperture is increased so that the primary beam and one or more Bragg reflections contribute to the image intensity, the lattice planes or the crystal structure can be imaged. Reliable interpretation of these crystal-lattice images is possible only if combined with computer simulations.

The most important application of diffraction contrast is the imaging of lattice defects such as dislocations, stacking faults, phase boundaries, precipitates and defect clusters. The contrast depends on the Bragg reflection used and its excitation error, the type of the fault and the depth inside the foil. This allows us to determine quantitatively the Burgers vector of a dislocation or the displacement vector of a boundary. The resolution, of the order of ten nanometres when using a strongly excited Bragg reflection, can be reduced to the order of one nanometre by the weak-beam technique, which allows us to measure the width of dissociated dislocations, for example. At high resolution, the interpretation of defect images again needs the help of computer simulations.

8.1 Diffraction Contrast of Crystals Free of Defects

8.1.1 Edge and Bend Contours

The diffraction angle $\theta_g = 2\,\theta_B$ between primary and reflected beams is normally larger than the objective aperture α_o; for Cu, for example, the 111 lattice spacing d is 0.208 nm and so at $E = 100$ keV ($\lambda = 3.7$ pm), we find $\theta_g \simeq \lambda/d = 1.8$ mrad. This means that the Bragg-diffracted beams are halted by the objective diaphragm in the bright-field mode and do not contribute to the image intensity (Fig. 4.20 a). The image intensity thus becomes equal to that of the primary beam (transmission T), which depends on the excitation error s_g or tilt parameter $w = s\xi_g$ of the Bragg reflection and on the specimen thickness t. Furthermore, contributions to the image can also come from electrons that have been scattered elastically or inelastically into the diffuse background of the diffraction pattern and pass through the objective diaphragm. Like the scattering contrast of amorphous specimens, this contribution depends on the objective aperture α_o [8.1, 2]. Crystals show a higher transmission in regions without strong Bragg reflections than do amorphous films of equal mass-thickness because the diffuse scattering between the Bragg reflections is reduced by destructive interference [8.3, 4].

We now use the results of the dynamical theory of electron diffraction (Figs. 7.17 and 18) to discuss diffraction contrast. The "pendellösung" of the dynamical theory (Fig. 7.17) can be seen as edge contours (Fig. 8.1) in the image of specimen edges of electropolished metals or of small cubic crystals (MgO), for example. A high intensity will be observed for thicknesses (Fig. 8.2 a) at which the Bragg-reflected intensity is scattered back to the primary beam; maximum transmission T thus occurs at depths equal to integral multiples of the extinction distance $\xi_{g,\mathrm{eff}}$, (7.70) and Fig. 7.14. The spacing of the edge contours is greatest in the Bragg position ($w = 0$) and decreases with increasing positive or negative tilt, owing to the decrease of the effective extinction distance $\xi_{g,\mathrm{eff}}$. This dependence on the tilt can be seen in Fig. 8.1 on a bent-crystal edge. With the aid of a tilting stage or specimen goniometer, the specimen can be brought into the Bragg position, whereupon the largest spacing is observed. The extinction distances can be measured when the edge profile is known or the local thickness can be determined by use of tabulated values of ξ_g (Table 7.3). However, the extinction distance for $w = 0$ can also be influenced by the excitation of other Bragg reflections (see dynamical Bethe potentials in Sect. 7.4.4). A quantitative measurement of the intensity of edge contours can be used to determine the absorption distances ξ_0' and ξ_g' [8.1]. A large number of contours can be observed in high-voltage electron microscopy [8.5, 6] because ξ_g increases as v, and ξ_0', ξ_g' are proportional to v^2 (Sect. 7.4.2).

Edge contours can also be observed at crystal boundaries that are oblique to the foil surface (Fig. 8.2 b). When the orientation of the second crystal is such that it does not show strong Bragg reflections, the diffraction contrast of the first crystal at the local depth of the boundary will not be changed much by the second crystal and will suffer only the exponential absorption.

Bent-crystal foils of constant thickness show extended and curved bend contours (Figs. 8.2 c and 8.3) due to the local variation of the excitation errors. Lines of equal intensity represent lines of equal inclination of the lattice planes to the electron beam. In some cases, it will be possible to reconstruct the three-dimensional curvature of a crystal (e.g., for lens-shaped cavities in PbI_2 crystals [8.7]). When the bend contours are nearly straight, the image intensity across the contour is a direct image of the rocking curve (Figs. 7.18 and 20 a). A bending radius R of the crystal corresponds to a tilt $\Delta\theta = x/R$ at a distance x from the exact Bragg position. The width of the bend contour is therefore inversely proportional to R, and the contours move when the specimen is tilted in a goniometer. Bending and tilting can also be

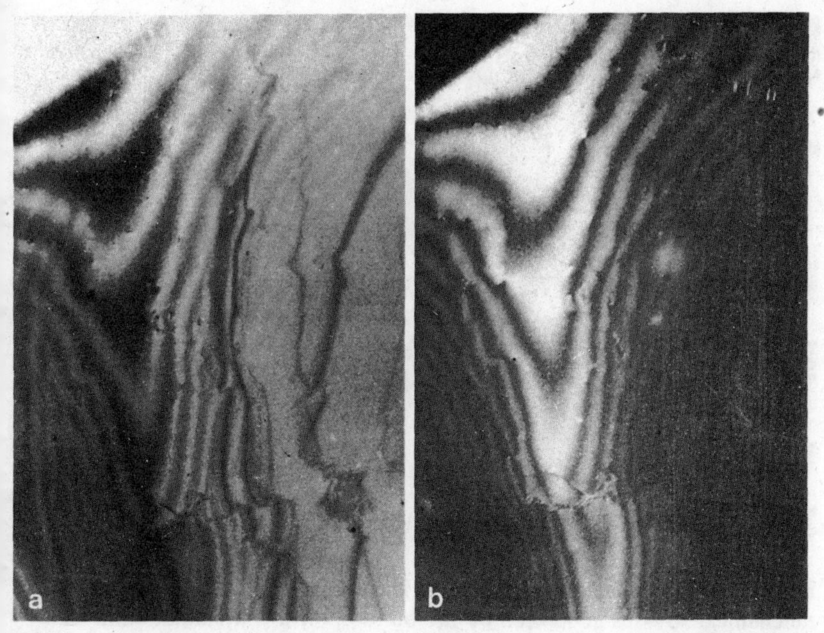

Fig. 8.1. (**a**) Bright- and (**b**) dark-field micrograph of edge contours at a bent Al foil of increasing thickness (top → bottom) with a maximum extinction distance ξ_g at the Bragg position and smaller $\xi_{g,\mathrm{eff}}$ for positive and negative tilt

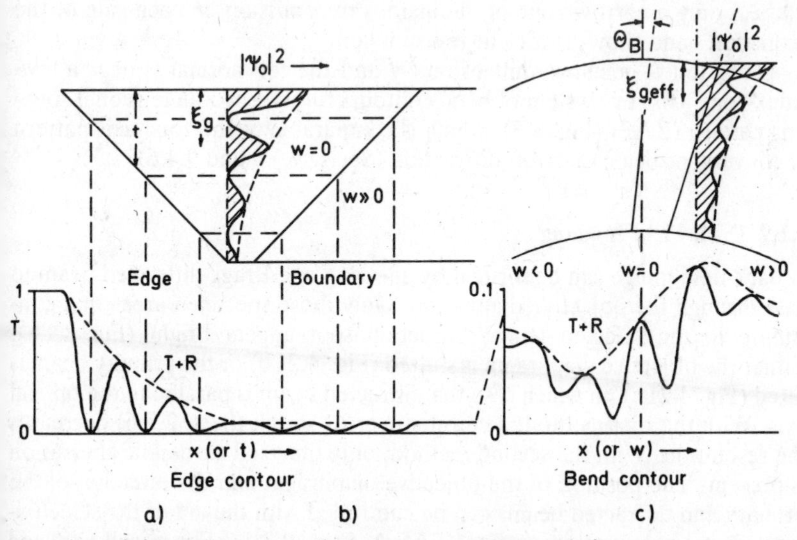

Fig. 8.2 a–c. Oscillations of $T = |\psi_0|^2$ in a crystal foil and generation of edge contours at (**a**) a crystal wedge and (**b**) a boundary. (**c**) Formation of bend contours in a bent foil

Fig. 8.3. Bend contours in a NIMONIC 75 thin foil, forming a [100] zone-axis pattern (Courtesy by P. Tambuyser, Joint Res. Centre, Petten, Netherlands)

induced by specimen heating caused by intense electron irradiation or by mechanical stresses generated by the build-up of a contamination layer.

Broad dark bands are normally superpositions of low-index *hkl* and $\bar{h}\bar{k}\bar{l}$ reflections, for which the regions of diffraction contrast overlap (Figs. 7.20a and 8.3). Subsidiary maxima of the rocking curve and high-order reflections can be observed for medium foil thicknesses. For large specimen thicknesses ($t \gg \xi_0'$) only a narrow zone of anomalous transmission on each side of the extinction band shows sufficient transmission.

If the foil is bent two-dimensionally and the foil normal is near a low-index-zone axis, the system of bend contours forms a two-dimensional zone-axis pattern (ZAP) (Fig. 8.3), which is comparable to the zone-axis pattern of convergent-beam electron diffraction (Sects. 9.3.4 and 9.4.6).

8.1.2 Dark-Field Imaging

A dark-field image can be formed by allowing one Bragg diffracted beam to pass through the objective diaphragm. Only those specimen areas that contribute to the selected Bragg-diffraction spot appear bright (Fig. 8.1b). Either the objective diaphragm is shifted (Fig. 4.21b) or the primary beam is tilted (Fig. 4.21c), in which case the diffracted beam is parallel to the optical axis. With the first method, a chromatic-error streak (Sect. 2.3.6) decreases the resolution. With the second method, only the axial chromatic aberration is present. The position of the objective diaphragm and the direction of the primary and diffracted beams can be controlled with the aid of the selected-area electron-diffraction pattern (SAED, Sect. 9.3.1). For a well-adjusted microscope, the transition from the bright-field to the dark-field mode can be effected simply by switching on the currents in the deflection coils used for beam tilting. The theorem of reciprocity indicates that the intensities of the primary and diffracted beams should not be changed if the directions of the two beams are interchanged. A method that does not involve changing the diffraction condition has been described by *Shirota* et al. [8.8]: here, the chromatic-error streak in the shifted-diaphragm method is compensated by the chromatic aberration of the intermediate lens up to diffraction angles $\theta_i = 2.5°$, by deflecting the beam in the first intermediate image plane, and dark-field micrographs with a resolution of 3 nm for $\theta_g = 2°$ and 1.7 nm for 1° are obtainable.

Decreasing the focal length of the intermediate lens in order to pass from SAED, in which the area selected is determined by the shadow of the objective diaphragm, to the image-forming mode offers the possibility of determining the Miller indices of an edge or bend contour. The excitation error s_g can be measured by observing the relative position of Kikuchi lines and Bragg-reflection spots (Sect. 9.4.4). Such information about the crystal orientation is needed to determine the Burgers vector of dislocations and other defects (Sect. 8.5.4).

Crystal phases with different crystal structure or orientation and hence with different diffraction spots can be separated in the dark-field mode by selecting the corresponding diffraction spots in the SAED pattern. If necessary, the diameter of the objective diaphragm can be decreased to 5 μm, but this will not be favourable for high resolution owing to the contamination and charging of such a small diaphragm. Figure 8.4 shows an example in which microtwins in Ni-layers, electrolytically deposited on a copper single crystal are identified [8.9]. Double reflection at the matrix and the microtwins produces typical twin reflections such as *a* and *b* in Fig. 8.4c, f. Dark-field micrographs obtained with these twin reflections indicate the different orientations (Fig. 8.4a, b) of the twin lamellae which are superposed in the bright-field micrograph (*e*). Figure 8.4d shows the contribution of the 220 Bragg reflection to the dark extinction contour across the imaged area.

The dark-field images of different phases or crystal orientations with different Bragg angles are shifted parallel to the $\boldsymbol{g}$ vector and proportional to $C_s\theta_g^3M - \Delta z\theta_gM$ when defocusing. When two micrographs are examined with a stereo viewer, the specimen structures appear to have different heights. This method can be useful for separating different phases or orientations [8.10, 11].

Dark-field imaging of lattice defects with weak beams is the most effective way for obtaining high-resolution micrographs of lattice defects (Sect. 8.5.3).

8.1.3 The STEM Mode and Multi-Beam Imaging

As shown in Sect. 4.5.3, the ray diagram of the STEM mode is the reciprocal of that of the TEM mode (Fig. 4.26). The illumination aperture α_i of the

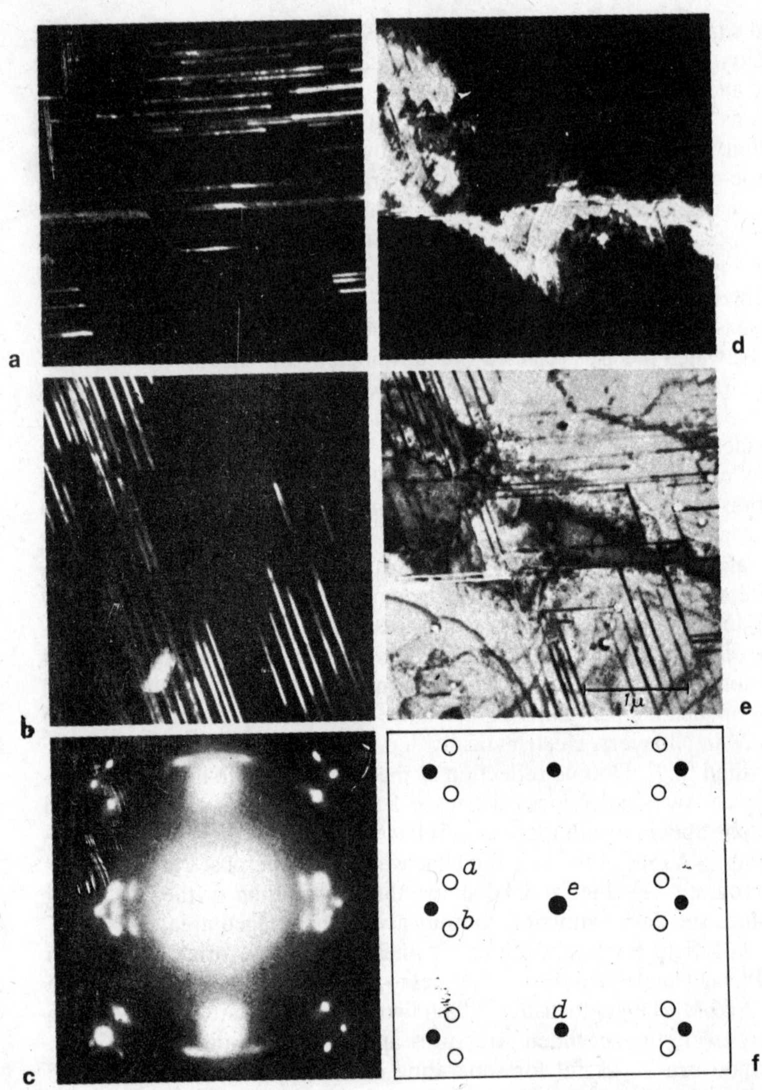

Fig. 8.4a–f. Example of dark-field micrographs obtained by selecting diffraction spots in th diffraction pattern (**c**) and (**f**) of an electrolytically deposited nickel film on a copper substrate (**a**) and (**b**) selection of two sets of microtwins, (**e**) bright-field and (**d**) dark-field micrograp with the 200 Bragg reflection

TEM mode corresponds to the detector aperture α_d of the STEM mode. Fo high-resolution STEM, the electron-probe aperture α_p must be large, of th same order as the objective aperture α_o of a TEM (Sect. 4.2.2). The sam contrast would be seen in the TEM and STEM modes if a small detecto

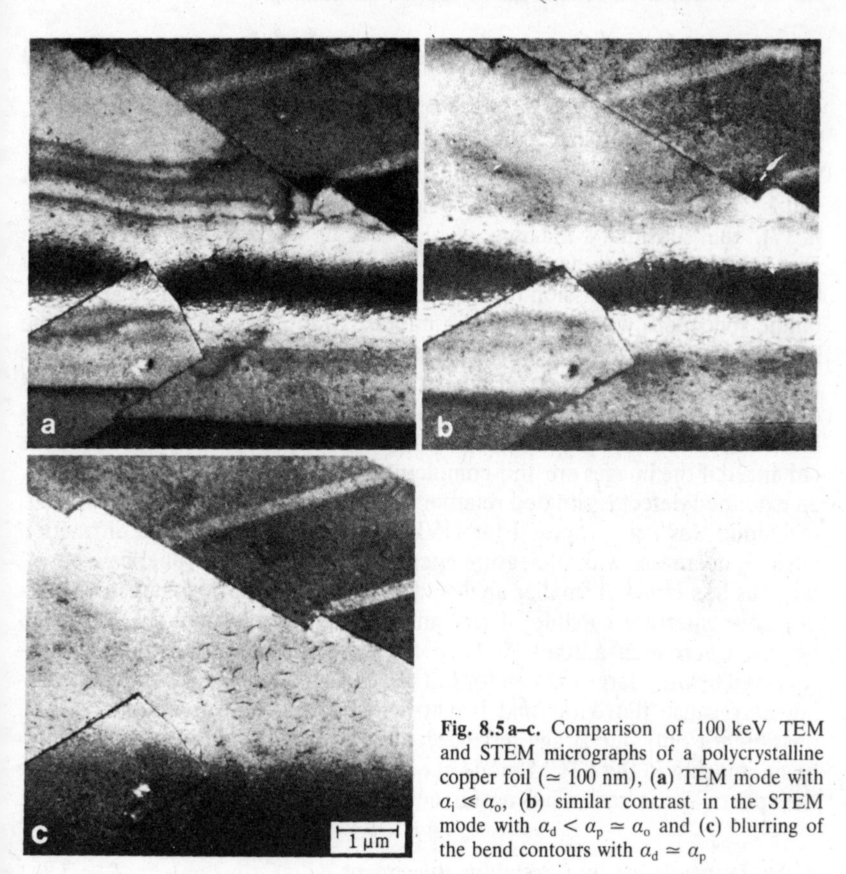

Fig. 8.5 a–c. Comparison of 100 keV TEM and STEM micrographs of a polycrystalline copper foil (≃ 100 nm), (**a**) TEM mode with $\alpha_i \ll \alpha_o$, (**b**) similar contrast in the STEM mode with $\alpha_d < \alpha_p \simeq \alpha_o$ and (**c**) blurring of the bend contours with $\alpha_d \simeq \alpha_p$

aperture α_d were used in STEM (Fig. 8.5 a, b). However, a large fraction of the incident cone of aperture α_p would then not pass through the detector aperture α_d even in the absence of a specimen. It is therefore better to work with $\alpha_d \simeq \alpha_p$ (see also discussion of the contrast of amourphous specimens in the STEM mode, Sect. 6.1.4). By collecting all of the incident electrons in this way the angle of incidence (excitation error) to the lattice planes varies widely and the image contrast is a superposition (convolution) of images with a broad spectrum of excitation errors s_g. This superposition is incoherent when interference between the different rays of the incident electron probe does not occur. The contrast of the weak edge contours at thicker parts of the edge is thus reduced owing to the variation of $\xi_{g,\mathrm{eff}}$ for different values of s_g. Bend contours are also blurred and high-order reflections disappear (Fig. 8.5 c) [8.12–16]. The diffraction contrast of a polycrystalline specimen becomes more uniform inside a single crystal and only differences of anomalous transmission can be seen for different crystal orientations; the contrast of lattice defects (dislocations) is, however, preserved though with reduced contrast (Figs. 8.5 and 8.23).

If the detector aperture α_d is comparable to or larger than the Bragg-diffraction angle $\theta_g = 2\theta_B$, the primary beam and one or more Bragg-diffracted beams can pass through the detector aperture and the image contrast will be a superposition of these beams. In this *multi-beam imaging* mode [8.17], the subsidiary maxima of the rocking curve vanish (see $\sum I_g$ in Fig. 7.20 a) and the image intensity is influenced by only the dependence of the anomalous transmission on the excitation error. A STEM image of crystalline specimens will normally be a mixture of multi-beam imaging and the blurring caused by having a broad spectrum of excitation errors [8.15]. The multi-beam imaging can be of advantage for the imaging of lattice defects (Sects. 8.4.2 and 8.5.2). If the intensities in the bright- and dark-field images are complementary, the defect contrast will be cancelled, whereas it is enhanced if the images are anti-complementary. This allows us to decide how an extended defect is situated relative to the top and bottom of the foil. This technique was first proposed for HVEM [8.18]. In HVEM, the diffraction angle θ_g decreases with increasing energy and, because the spherical aberration has less effect at smaller angles, a sharp image can be obtained with an objective aperture capable of transmitting the primary and the diffracted beams, whereas in a 100 keV-TEM the spherical aberration for a Bragg-reflected beam is large enough to shift the dark-field image. Though defocusing may cause the dark- and bright-field images to overlap, this will be possible for only one Bragg reflection and not for all simultaneously. Multi-beam imaging in the STEM mode can also be employed at 100 keV because the spherical aberration has no further influence behind the specimen.

8.1.4 Transmission by Crystalline Specimens

Crystals with a layer structure can be prepared as thin foils by cleavage and some materials can be prepared as epitaxially grown films on a single-crystal substrate. The most-widely used methods for thinning metals, alloys, minerals and other materials are chemical etching, electrolytic polishing and ion-beam etching. It is necessary to prepare specimens with thicknesses in the range 0.1–0.5 μm for imaging crystal defects at $E = 100$ keV and less than 10 nm for high-resolution studies of crystal structure (Sect. 8.2.2). These three thinning methods yield sufficiently thin areas only in wedge-shaped foils near holes. For many applications – observation of dislocation structure and in-situ experiments, for example – the question arises whether such thin foils are representative of the bulk material. Dislocations can be rearranged and/or migrate to the foil surface, and in-situ precipitation and electron irradiation experiments, for example, are influenced by the surface, which plays the role of a sink for mobile point defects.

The useful specimen thickness is limited by the decrease of the Bragg-diffraction intensities and by the chromatic aberration associated with energy losses. Absorption of the Bloch-wave field and the probability of energy-loss decrease with increasing electron energy. Larger thicknesses can be used in HVEM. A very important consequence of this increase of the useful thickness is that a very much larger specimen area can be examined with high voltage than with 100 keV, with which the transmission is mostly limited to very small zones near holes.

The many-beam dynamical calculations show that the anomalous transmission depends strongly on the excitation condition and varies from element to element [8.19, 20]. Thus the transmission curves for a 111 row of aluminium (Fig. 8.6) reach a maximum at 1 MeV and at 5 MeV at tilts on the positive side of the 111 Bragg position whereas at 3 MeV, the maximum occurs at the symmetric excitation ($k_x/g = 0$). (The "pendellösung" oscillations drawn for $E = 1$ MeV are smoothed out at 3 and 5 MeV.) These effects are caused by differences of the Bloch-wave channeling [8.21–23]. Figure 8.7 indicates the foil thicknesses of Al, Fe and Au for which $T = 0.001$ of the primary beam at the beam tilts of maximum transmission. This shows that optimum transmission can be obtained for Al at $E \simeq 3$ MeV, for Fe at $\simeq 1.5$ MeV and for gold at 10 MeV though the increase is modest beyond 3 MeV. It is therefore useful to be able to vary the accelerating voltage of a HVEM and to match the voltage to the specimen orientation or to tilt the specimen into an optimum orientation.

The definition of the maximum useful thickness depends on the criterion adopted and the type of contrast. The $T = 0.001$ criterion of Fig. 8.7 is arbitrary but agrees rather well with experiment. In practice, the transmission is also influenced by the number of elastically and inelastically scattered electrons in the cone of half angle α_o (objective aperture).

In the Kikuchi patterns of MoS_2 lamellae, *Uyeda* and *Nonoyama* [8.24] observed that the Bragg spots disappear when $t \geq 2.7\,\beta^2$ ($\beta = v/c$, t measured in μm) and that the anomalous transmission near low-order Bragg

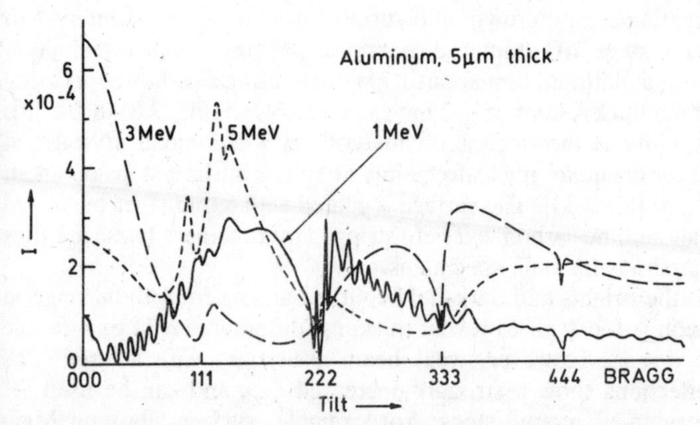

Fig. 8.6. Bright-field rocking curves for a 5 μm Al foil at 1 MeV, 3 MeV and 5 MeV [8.19]

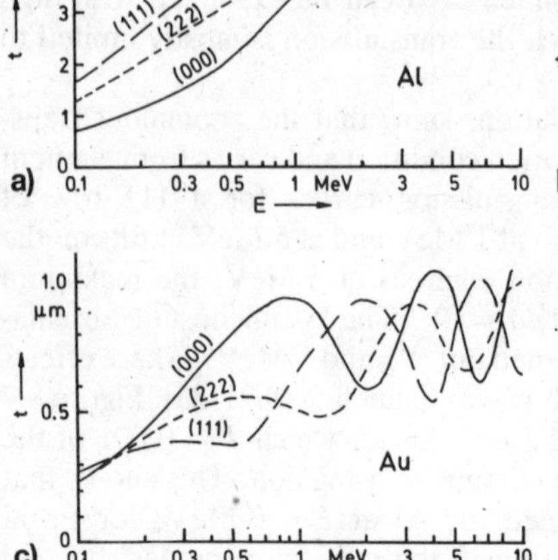

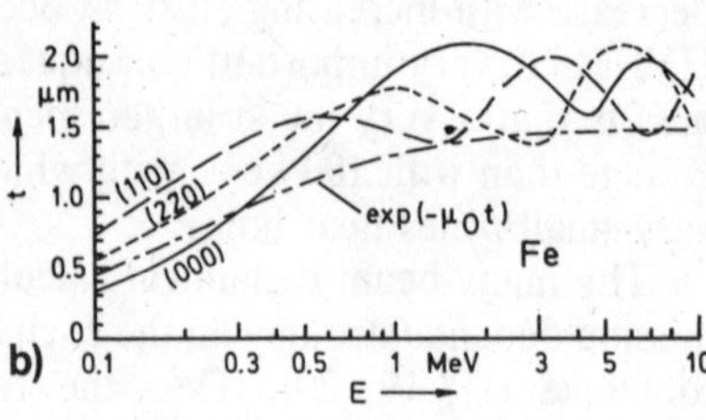

Fig. 8.7 a–c. Specimen thickness t that corresponds to a transmission $T = 1 \times 10^{-3}$ for three different orientations [symmetric (000) position and on the positive side of the indicated (hkl) positions at the maximum of the rocking curve] for (**a**) Cu, (**b**) Fe and (**c**) Au [8.19]

positions vanishes for $t \gtrsim 5.7\,\beta^2$; dislocations are visible up to these thicknesses. *Thomas* [8.25] used the visibility of stacking faults in Si and Fe as criterion and found that the transmission increases faster than β^2 ($t = 9$ μm for Si and $t = 2$ μm for Fe at $E = 1$ MeV). *Thomas* and *Lacaze* [8.26] found a further appreciable increase of the transmission in Si for energies above 1 MeV but only a slight increase for Fe. Further results for Al, Cu, Fe, Mo, W and Au have been reported by *Fujita* et al. [8.27, 28].

8.1.5 Imaging of Atomic Surface Steps

When investigating crystal growth and surface structures, it is of interest to resolve surface steps of atomic dimensions. By the replica technique (shadowing with a platinum film about 1 nm thick, using a carbon supporting film about 10 nm thick), steps of 1–2 nm can be resolved (Fig. 8.8) [8.29, 30]. Another possibility is the decoration method, in which small crystals, of silver or gold for example, nucleate mainly at atomic surface steps on alkali halides (Fig. 8.9) [8.31–33]. The surface is coated with a thin layer of carbon and the crystals and the carbon layer are stripped off together. These are thus preparation methods for bulk specimens.

Normally, the bright- and dark-field contrast arising from the change in the transmission is too faint to reveal thickness differences d of one atomic step because $d \ll \xi_g$. However, weak-beam excitations with a large s_g or forbidden reflections show a strongly decreased $\xi_{g,\mathrm{eff}}$ and can be used for dark-field imaging of atomic steps. For example, surface steps on MgO become visible by imaging with a weakly excited 200 and a strongly excited

Fig. 8.8. Carbon-palladium replica of a deformed copper single-crystal (8% stretched). The mean step height is 1.8 nm. The carbon film is reinforced with a collodion backing and shadowed with a palladium film (S. Mader)

600 reflection or 400 and 200, respectively [8.34]. With so-called forbidden reflections, steps on Au and Si foils can be detected by contrast differences [8.35–38] though it will be shown below that this case can also be interpreted as a weak-beam excitation.

The principle of the method will now be discussed by use of the example of an [111] oriented Au film. Figure 8.10a shows the unit cell with the ABC packing sequence (Fig. 7.3) in a hexagonal notation. The structure amplitude (7.23) becomes

$$F = f\{1 + \exp[2\pi i(h + k + l)/3] + \exp[4\pi i(h + k + l)/3]\} \tag{8.1}$$

$$= 0 \quad \text{if} \quad h + k + l = 3n + 1 \, .$$

This is the condition for a *forbidden reflection* in hexagonal notation. These reflections are situated between the primary beam and 220 Bragg spots (cubic notation). However, F is zero only if the number of close-packed layers N is a multiple of three: $N = 3$ m; this means that there must be an integral number of complete unit cells in hexagonal notation. With one layer more or less ($N = 3$ m $\pm$ 1), F does not vanish because $F = 0$ is a consequence of des-

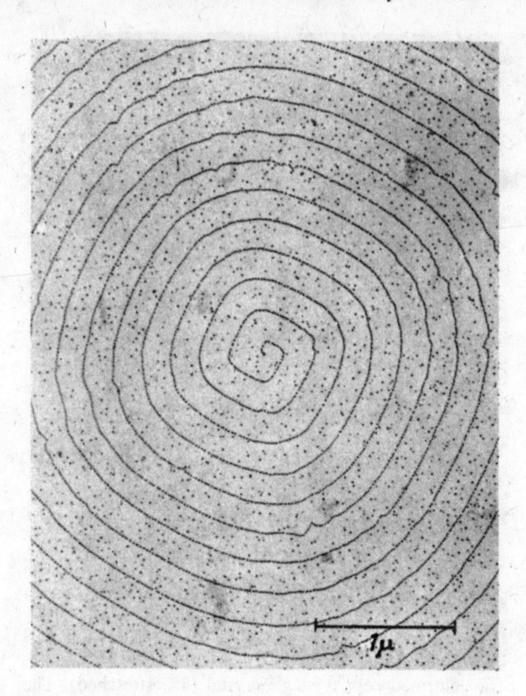

Fig. 8.9. Gold decoration of surface steps (0.28 nm) on NaCl generated by a screw dislocation after sublimation for 6 h at 350 °C (H. Bethge)

tructive interference between the scattered waves for all of the atoms of a unit cell. These additional or missing layers therefore lead to non-zero values of F and are imaged bright in a dark-field micrograph if this weak Bragg spot is selected.

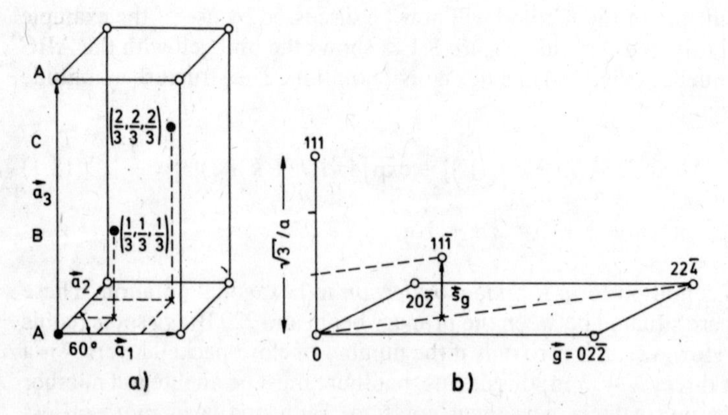

Fig. 8.10. (**a**) "Hexagonal" unit cell of a fcc lattice, (**b**) Reciprocal lattice in cubic notation showing that the excitation error s_g is $(\sqrt{3}/a)/3$ for the $11\bar{1}$ reflection

In another equivalent explanation, this is interpreted as a weak-beam effect with a large excitation error s_g. In the reciprocal lattice shown in Fig. 8.10 b, the first Laue zone contains a $1\bar{1}1$ reflection, which is allowed (*hkl* odd in cubic notation) but normally not excited owing to the large excitation error s_g, which denotes the distance from the Ewald sphere in the zero Laue zone if the electron beam is parallel to the [111] zone axis. This means that s_g is equal to the distance between the first and zero Laue zones, $s_g = (\sqrt{3}/a)/3$. Substituting for s_g in (7.34) or (7.69) gives the same result because $w = s_g \xi_g \gg 1$ and

$$I_g \propto \sin^2(\pi t s_g) = \sin^2(\pi \sqrt{3}\, t/3\, a) \tag{8.2}$$

so that I_g becomes zero if $t = 3\ \mathrm{m} \times (a\sqrt{3}/3)$. The last factor in the bracket is, however, just the distance between the close-packed layers, namely, one third of the space diagonal $a\sqrt{3}$ in the cubic unit cell, which is the same result as that given by (8.1). For a more exact formulation, the weak-beam excitation in the presence of strong excitations has to be treated by dynamical n-beam calculation*.

8.2 Crystal-Structure Imaging

8.2.1 Lattice-Plane Fringes

The crystal lattice can be imaged and resolved, provided that the information about the lattice structure – that is the primary beam and one or more Bragg reflections – can pass through the objective diaphragm and that the contrast is not destroyed by insufficient spatial and temporal coherence. In the two-beam case, only the primary beam, one Bragg reflection and the diffuse background of the diffraction pattern contribute to the image. Primary beam and Bragg reflection are inclined at an angle $\theta_g = 2\,\theta_B$ ($2\,d \sin\theta_B = \lambda$). The plane waves interfere in the image plane with smaller angular separation θ_g/M (M: magnification), between the primary and reflected waves. The distance between the maxima of the resulting two-beam interference fringes is Md. *Menter* [8.39] was the first to succeed in resolving lattice planes of copper-phthalocyanine ($d = 1.2$ nm). Today it is possible to resolve lattice fringes with $d \simeq 0.1$ nm. Figure 8.11 a shows lattice-plane fringes from an evaporated *Au* film with a 20° crystal boundary ($d_{200} = 0.204$ nm). Diffraction contrast from the dense dislocations at the grain boundary is superposed on the lattice-fringe image.

For more-extensive calculation and discussion of the contrast of lattice fringes in the two-beam case [8.41, 42], we use (7.68) for the amplitudes ψ_0 and ψ_g. These equations can be rewritten:

* Surface steps can also be imaged in the reflexion mode [8.138]

$$\psi_0 = |\psi_0| e^{-i\phi} \quad \text{with } |\psi_0|^2 = T \text{ and } \tan\phi = \frac{w}{\sqrt{1+w^2}} \tan\left[\frac{\pi\sqrt{1+w^2}}{\xi_g} t\right] \tag{8.3 a}$$

$$\psi_g = i|\psi_g| e^{2\pi i g x} \quad \text{with } |\psi_g|^2 = R\,. \tag{8.3 b}$$

The superposition of the two waves in the image planes results in the image amplitude

$$\psi = \psi_0 + \psi_g\, e^{-iW(\theta_g)} \tag{8.4}$$

including the wave aberration $W(\theta_g)$ (3.66) in the diffracted beam. With $|\boldsymbol{g}| = 1/d$, the image intensity becomes

$$I(x) = \psi\psi^* = |\psi_0|^2 + |\psi_g|^2 - 2|\psi_0|\,|\psi_g| \sin\left[\frac{2\pi x}{d} + \phi - W(\theta_g)\right]. \tag{8.5}$$

This means that the image contrast depends on the thickness t and the tilt parameter w of the crystal foil. The highest-contrast can be observed for thicknesses $t = \xi_g/4, 3\xi_g/4, 5\xi_g/4$ etc., where $|\psi_0| = |\psi_g|$ in the Bragg position ($w = 0$); for intermediate thicknesses with $|\psi_0| = 0$ and $|\psi_g| = 0$, the contrast vanishes. It also decreases with increasing tilt parameter w, because the maximum possible value of $I_g = |\psi_g|^2$ decreases.

The phase contribution ϕ in (8.5) can vary with x in edge contours (varying t) and/or bend contours (varying w) and causes a shift of the lattice-fringe positions. The variation of the observed lattice-fringe spacing d' can be calculated [8.43] by writing

$$\frac{2\pi}{d'} = \frac{d}{dx}\left(\frac{2\pi x}{d} + \phi\right) = \frac{2\pi}{d} + \frac{d\phi}{dx}\,. \tag{8.6}$$

In edge and bend contours $d\phi/dx$ is normally so small that this correction can be neglected. This effect must, however, be taken into account if an accurate value of d is being determined by counting a large number of fringes (see below).

With the primary beam on axis ($W = 0$) and the diffracted beam at an angle $\theta_g = 2\theta_B \simeq \lambda/d$ to the axis, the wave-aberration term $W(\theta_g)$ in (8.5) causes a shift of the fringe position by one lattice-plane spacing if $W(\theta_g) = 2\pi$. This defocusing shift does not affect the visibility (resolution and contrast) of the fringes. However, the contast of closely spaced lattice planes or large spatial frequencies is influenced by the attenuation characterized by the contrast-transfer envelope $K_c(q)$ in (6.46), caused by the chromatic aberration and by the energy spread ($\Delta E \simeq 1$ eV) of the electron beam. The resolution can therefore be increased by using tilted illumination with the primary beam tilted at an angle θ_B to the axis [8.44]. The wave aberration $W(\theta_g)$ will be the same for both waves and the contrast-transfer envelope $K_c(q)$ decreases the fringe contrast to the same value for twice the spatial frequency or half the lattice-plane spacing. In this way, lattice-plane spacings down to 0.1 nm can be resolved.

The variation of the observed lattice-plane spacing d' can be used to measure the spacing d directly in a micrograph. The fringe separation may vary not only by virtue of the effect discussed in (8.6) but also because the composition of the alloy or mineral varies. If this is the case, it will be of great interest to determine the local variations of d by the lattice-fringe spacing method. In a homogeneous material, the lattice spacing d can be determined more accurately by electron diffraction and the fringe distance can be used for calibration of the magnification. As discussed in Sect. 4.4.3, this method is accurate to within $\pm 2\%$ and the limited reproductibility of the magnification contributes a further error of $\pm 1\%$.

If the lattice-fringe spacing can be determined with an accuracy δ, the final absolute error Δd can be reduced by measuring the spacing of a larger number N of fringes

$$d \pm \Delta d = \frac{Nd \pm \delta}{N} \quad \rightarrow \quad \frac{\Delta d}{d} = \frac{1}{N}\frac{\delta}{d} \,. \tag{8.7}$$

The measurement of δ/d will not be more accurate than 5%. The variation of d in an alloy can be characterized by the slope $\gamma = (\Delta d/\Delta c)/d$ in a $d(c)$ plot, where c is the concentration, giving $\Delta d/d = \gamma \Delta c$. The number of fringe spacings that must be measured to detect a variation Δc of concentration can then be obtained from (8.7)

$$N = \frac{\delta}{d}\frac{1}{\gamma \Delta c} \tag{8.8}$$

and is inversely proportional to Δc. Numerical examples are

$$\gamma = 0.1\,, \qquad \Delta c = 10\% \qquad N = 5$$

$$\left.\begin{array}{ll} \gamma = 0.1 & \Delta c = 1\% \\ \gamma = 0.01 & \Delta c = 10\% \end{array}\right\} N = 50$$

The following experimental studies will illustrate the applicability of this method. As a first example, we mention the spinodal alloys, which exhibit modulations in composition with a wavelength of the order of 10 nm. An Au-Ni alloy for which the value of γ is relatively large, $\gamma = 0.15$, was investigated by *Sinclair* et al. [8.45]. For the investigation of Cu-Ni-Cr [8.46] with $\gamma = 0.001$, it is necessary to average over a large number N of fringes. In this case, the averaged value was not determined directly from the micrograph but by optical diffraction.

Another application is the imaging of lattice planes in and near crystal defects, which complements the information given by the contrast effects discussed in Sect. 8.4.5. The latter are caused by lattice displacements, which affect the amplitude of the primary and diffracted beams. Translational antiphase boundaries, for example, cause a fringe shift [8.47]; Guinier-Preston zones of Al-Cu consisting of single atomic planes enriched in Cu in the Al lattice can be imaged [8.48]. Compositional differences of precipitation and grain-boundary segregation can shed light on the generation of these defects [8.49]. A combination of energy-dispersive x-ray microanalysis and lattice-fringe studies is reported for a ceramic Mg-Si-Al-O-N alloy [8.50].

8.2.2 Crystal-Structure Imaging

The imaging of lattice planes only requires a coherent superposition of the primary beam and a reflected beam (or a systematic row) in the image plane. If several non-systematic reflections are used, a cross grating of lattice planes can be resolved (Fig. 8.11 b); in the case of gold a resolution $d = 0.204$ can be obtained if, as first proposed by *Komoda* [8.51], four reflections (000, 200, 020, and 220) enter the objective lens with the optic axis in the centre of the square defined by the four diffraction spots.

The crystal structure could, in principle, be determined by electron diffraction by applying an inverse Fourier transform to the measured Bragg

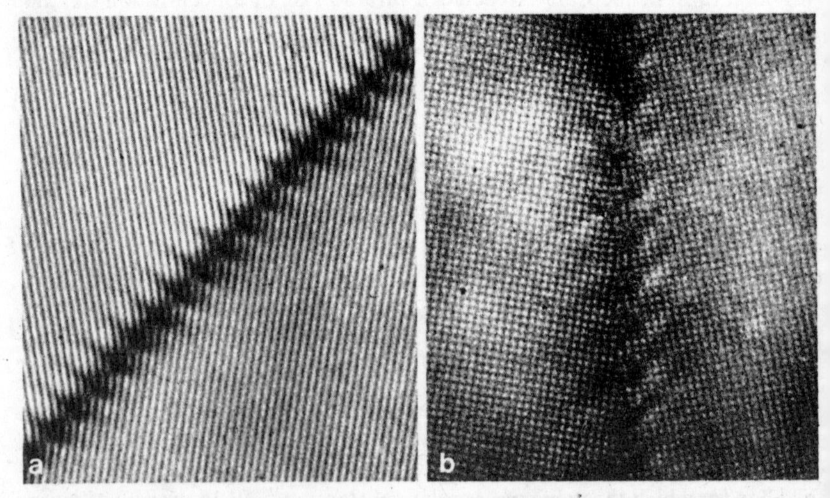

Fig. 8.11 a, b. High-resolution image of 20° [001] tilt boundaries in epitaxially grown Au films on NaCl bicrystals. (**a**) 0.204 nm lattice fringes in common [110] orientation, (**b**) crossed lattice image at a boundary in [010] orientation [8.40]

intensities, but the phase information is lost. It is an advantage of crystal-structure imaging that the inverse Fourier transform – from the diffraction pattern to the image – takes place directly inside the microscope without any loss of phase. Imaging of the crystal structure therefore becomes a special case of phase-contrast image formation. The main limitations of crystal-structure imaging with a large number of reflections are

1) the additional phase shifts $W(\theta_g)$, which are different for the various reflections,
2) the departures from kinematical theory when thick specimens are used, and
3) the attenuation of the contrast-transfer function by spatial and temporal incoherence (Sect. 6.4.2).

For thin crystals, a phase-grating theory can be applied. By use of (3.20) for the refractive index, the wave function behind the specimen is found to be

$$\psi_s(x,y) = \psi_0 \exp\left[-\mathrm{i}\,\frac{2\pi}{\lambda E}\,\frac{E+E_0}{E+2E_0}\int_0^t V(x,y,z)dz\right]. \tag{8.9}$$

In the weak-phase-grating theory, appropriate for very thin crystals, the exponential is

$$\psi_s(x,y) = \psi_0 \left[1-\mathrm{i}\,\frac{2\pi}{\lambda E}\,\frac{E+E_0}{E+2E_0}\int_0^t V(x,y,z)dz + \dots\right] \tag{8.10}$$

and the phase-contrast theory of Sect. 6.2.4 can be applied, except that the integration over the Fourier plane now becomes a summation over the limited number of diffraction spots.

In the dynamical theory, the Bloch-wave formulation can be used to calculate the intensity in the image plane:

$$\begin{aligned} I(r) = \sum_{g,h}\sum_{i,j} C_0^{(i)} C_0^{(j)} C_g^{(i)} C_h^{(j)} \exp[-2\pi(q^{(i)}+q^{(j)})t]\,\cdot \\ \exp\{\mathrm{i}[2\pi(\gamma^{(i)}-\gamma^{(j)})t + 2\pi(\boldsymbol{g}-\boldsymbol{h})\cdot\boldsymbol{r} - W(\theta_g,\Delta z) + W(\theta_h,\Delta z)]\}\,. \end{aligned} \tag{8.11}$$

The phase shift $W(\theta_g)$ contains the spherical-aberration constant C_s, and it is impossible to image a large number of hkl Bragg reflections around an $[mno]$ zone axis with $mh + kl + ol = 0$ with the same phase shifts, so that the superposition of the reflection amplitudes in the image becomes incorrect. These difficulties do not arise for specimens with very large unit cells $\geqq 1$ nm with a large number of reflections at low θ_g. The practical implementation of this technique and some typical applications are summarized in [8.52–54]. A good demonstration that the phase-grating approximation corresponds to a projection of the crystal structure is the fact that tunnels in the 3 × 3 and 3 × 4 block structures of ternary oxides of Nb-W or Ti and other materials are seen as bright spots (Fig. 8.12). The channels that contain Nb atoms are imaged as white dots, the regions where the octahedra have common edges, or shear planes, exhibit grey contrast and the region near the

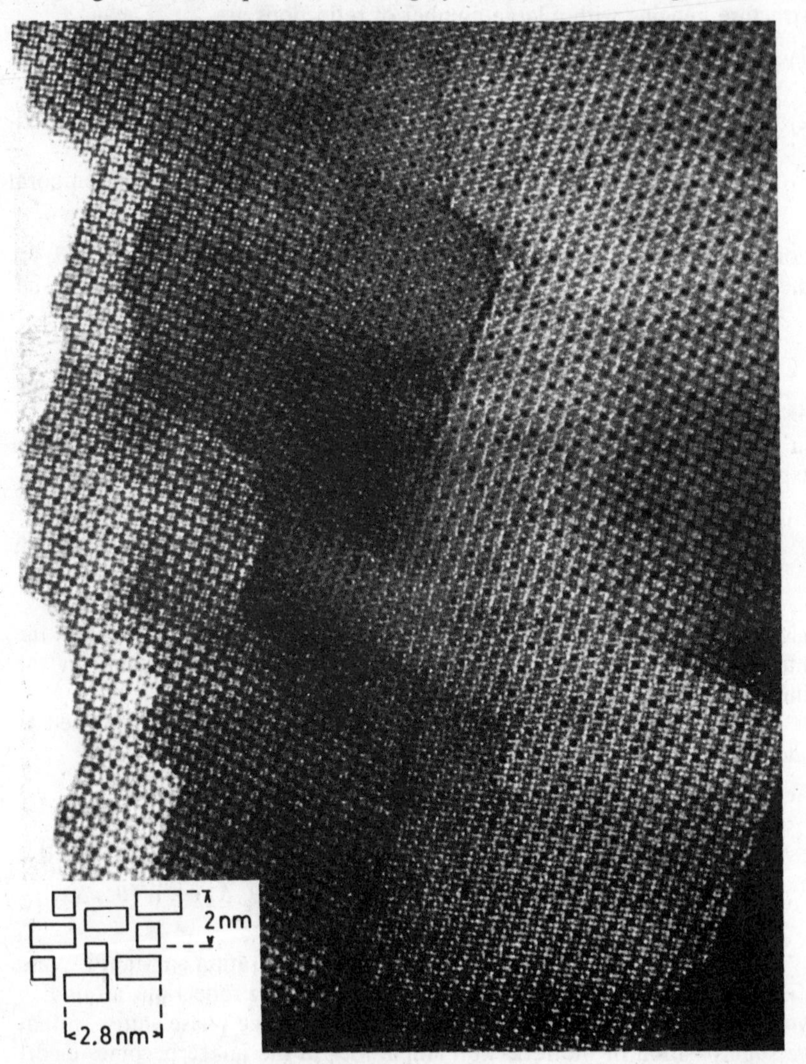

Fig. 8.12. Crystal-structure imaging of $Nb_{22}O_{54}$ containing octahedra of NbO_6, which form 3 × 3 and 3 × 4 blocks (see inset for dimension), and its change with increasing specimen thickness (from left to right) and defocusing (from top to bottom). The crystal structure is imaged correctly at the top left. A recurrence of the same structure can be seen at the bottom right [8.55]

tetrahedrally coordinated metal atom, surrounded by eight metal atoms in the nearest octahedra, is black.

For high-resolution imaging, the value of the defocusing Δz that gives a broad main transfer band of the contrast-transfer function (CTF) (Sect. 6.4.1) will be most satisfactory, just as for amorphous specimens. Once again, the limiting factor will be the decrease of the CTF at large q caused by the finite illumination aperture and energy spread (spatial and temporal partial coherence, respectively). A resolution of 0.35 – 0.4 nm is possible with $C_c = 1.8$ mm and $\alpha_i \simeq 10$ mrad at $E = 100$ keV, for example. *Bursill* and *Wilson* [8.56] succeeded in resolving 0.3 nm in hollandite by reducing α_i to 0.2 mrad and keeping the energy spread of the electron source narrow. Though C_s is larger for HVEM lenses, the phase shift $W(\theta_g)$ decreases as $\theta_g^4/\lambda \sim \lambda^3$ with increasing energy, and the attraction of HVEM for crystal-structure imaging is clear. At $E = 1$ MeV, for example, an objective lens with $C_s = 4.2$ mm has a main transfer interval ranging from $q = 1/0.43$ to $1/0.16\ \mathrm{nm}^{-1}$ [8.57].

Experiments and calculations for very thin specimens confirm that, for thicknesses less than 5 nm, the crystal can be regarded as a weak-phase specimen, which is equivalent to saying that kinematical theory is valid. The maximum allowable thickness increases with increasing size of the unit cell. Model calculations are necessary for each particular structure, to check whether the weak-phase approximation is indeed valid. The interaction between the electrons and the crystal potential is weakened as the electron energy is increased, so that thicker crystals can be observed in the weak-phase approximation.

For thick specimens, the Bragg-diffraction intensities vary with thickness and excitation errors (dynamical theory), and the intensity distribution in the image of the lattice shows drastic changes, which can result in imaging artifacts and even contrast reversal of the main structure. If the thickness is still further increased, to $\simeq 50$ nm, the structure that is observed for thicknesses below 5 nm may reappear [8.52, 58]. In Fig. 8.12, this can be observed on the wedge-shaped crystal at the bottom, right-hand side. A many-beam computer simulation, considering phase shifts and dynamical theory, is therefore a necessary complement to any experimental high-resolution work on crystal-structure imaging. Computation methods using the multi-slice method have been developed [8.59, 60].

8.2.3 Moiré Fringes

When two crystal foils overlap and are rotated a few degrees relative to another or when their lattice constants are different, a pattern of interference fringes is observed. These are known as Moiré fringes and were first observed by *Mitsuishi* et al. [8.61] on graphite lamellae. The generation of these patterns will be discussed for the example of two crystal foils of different lattice constant and parallel orientation. Such specimens can be prepared by epitaxy (e.g., palladium with $d_{220} = 0.137$ nm on a single-crystal gold film with $d_{220} = 0.144$ nm, Fig. 8.13 [8.62]). However, the two crystal foils need not to be in direct contact. We expect to see a double electron-diffraction pattern with Bragg reflections at different distances OP and OQ from the origin O (primary beam) (Fig. 8.14 a). The diffracted beam in the first layer can be further diffracted in the second. This results in a shifted diffraction pattern, with the direction P as the new primary beam (Fig. 8.14 b). For all strong Bragg reflections, this produces the diffraction pattern of Fig. 8.14 c.

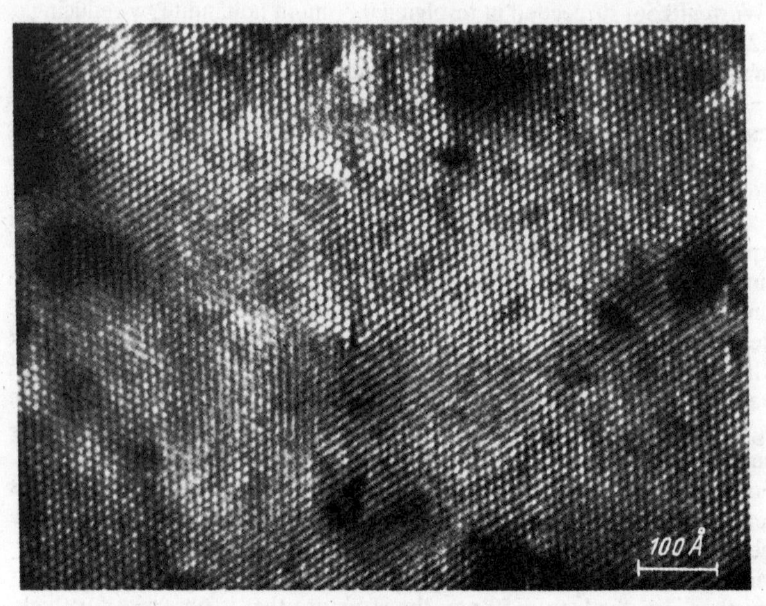

Fig. 8.13. Moiré-effect in separately prepared and superposed single crystal films of palladium and gold; parallel and rotation Moiré fringes are seen [Courtesy by G. A. Basset et al.]

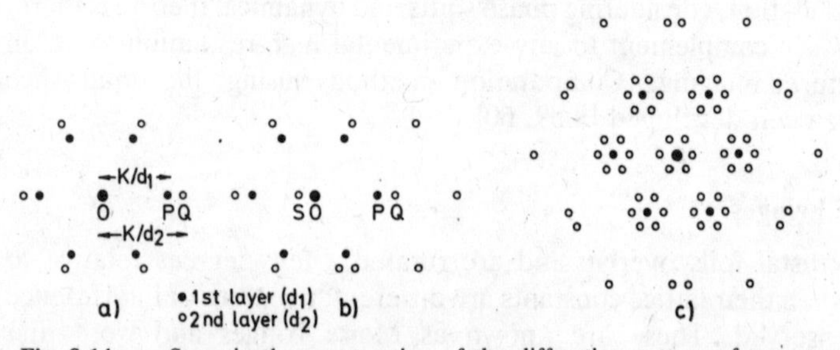

Fig. 8.14 a–c. Steps in the construction of the diffraction pattern of superposed crystals with different lattice-plane spacings d_1 and d_2. (**a**) Double diffraction pattern produced by the primary beam O only, (**b**) the beam P in the first crystal is diffracted in the second and produces a spot S near O. (**c**) All six diffraction spots equivalent to P act as a primary beam in the second crystal

The diffraction spots around the primary beam O can pass through the normal objective diaphragm and contribute to the image contrast. These diffraction spots can formally be attributed to a material with a larger lattice constant d_M. If K denotes a diffraction constant proportional to the camera length, the distance OS becomes

$$\mathrm{OS} = \mathrm{PS} - \mathrm{OP} = \mathrm{OQ} - \mathrm{OP} = \frac{K}{d_2} - \frac{K}{d_1} = \frac{K}{d_M} \tag{8.12}$$

and the Moiré fringes have the *apparent lattice constant*

$$d_M = \frac{d_1 d_2}{d_1 - d_2} \tag{8.13}$$

($d_M = 2.9$ nm for the example of the Pd-Au double layer).

When the two crystals are rotated by a small angle α with the foil normal as axis, the Moiré-fringe spacing becomes

$$d_M = \frac{d_1 d_2}{(d_1^2 + d_2^2 - 2\,d_1 d_2 \cos\alpha)^{1/2}} \,. \tag{8.14}$$

A two-dimensional Moiré pattern is formed when several Bragg reflections show double reflection. *Rang* [8.63] discussed further possibilities of a rotation Moiré with the rotation axis parallel to the foil. This effect can be observed in lens-shaped cavities of crystal lamellae.

For an exact discussion of the image contrast, it is necessary to use the dynamical theory of electron diffraction [8.42, 64].

The generation of Moiré fringes can also be understood in terms of a more intuitive model, in which the imaging of lattice planes is regarded as a projection. Figures 8.15 a and b contain two parallel lattices one of which (a) has a lattice constant d_1 while the other (b) has a lattice constant d_2 and an additional edge dislocation. The superposition of transparent foils results in Fig. 8.15 c. Dark and bright areas of width d_M alternate when the effect of the dislocation is neglected. The narrow lines with the lattice-plane spacing will be resolved only in high resolution; normally, only the fringes with the larger period d_M are resolved. In light optics, the same impression is obtained by looking at Fig. 8.15 from a larger distance. Figure 8.15 d–f represent the optical analogue of a rotation Moiré of two lattices with equal lattice constants ($d_1 = d_2$). In this case, the dislocation is also imaged but with the wrong orientation. Care is therefore necessary when interpretating the Moiré patterns of lattice defects. An idea of the various possible images can easily be obtained by shifting and rotating superposed transparent layers, as in Fig. 8.15 [8.65].

Recent work shows that the Moiré effect can be successfully used for the investigation of epitaxy [8.66, 67]. Moiré fringes also can be observed between a matrix and precipitates (Sect. 8.6.1).

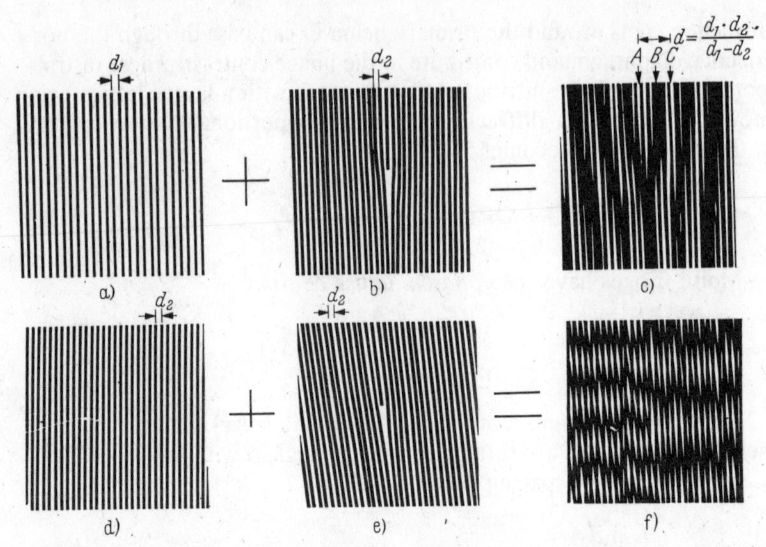

Fig. 8.15 a–f. Optical analogue that demonstrates the imaging of dislocations with the Moiré effect. Lattices with lattice-plane spacings d_1 (**a**) d_2 (**b**) superpose to form (**c**). Two lattices (**d**) and (**e**) of equal d_2 superpose to give a rotation Moiré (**f**)

8.3 Calculation of Diffraction Contrast of Lattice Defects

8.3.1 Kinematical Theory and the Howie-Whelan Equations

The most simple theory for the investigation of the image contrast of dislocations, stacking faults and other defects uses the kinematical column approximation (Sect. 7.2.3) [8.68, 69]. A unit cell at a depth z near a lattice defect is assumed to be displaced by a vector $\boldsymbol{R}(z)$ relative to an ideal lattice without a defect (Fig. 8.16). Equation (7.33) for the reflected amplitude becomes

$$\begin{aligned}\psi_g &= \frac{\mathrm{i}\pi}{\xi_g}\int_0^t \exp\{-2\pi\mathrm{i}(\boldsymbol{g}+\boldsymbol{s})\cdot[z+\boldsymbol{R}(z)]\}\,dz \\ &= \frac{\mathrm{i}\pi}{\xi_g}\int_0^t \exp\{-2\pi\mathrm{i}[sz+\boldsymbol{g}\cdot\boldsymbol{R}(z)]\}\,dz\ ,\end{aligned} \tag{8.15}$$

in which we have written $\exp(-2\pi\mathrm{i}\boldsymbol{g}\cdot z) = 1$ because the product of a reciprocal-lattice vector $\boldsymbol{g}$ and a translation vector $\boldsymbol{r}$ of a crystal lattice is always an integer; $\boldsymbol{s}\cdot\boldsymbol{R}$ has been neglected, because it is the product of two quantities that are small relative to g and z. The integral of $\exp(-2\pi\mathrm{i}sz)$ was analysed graphically in Sect. 7.2.3 by the amplitude-phase diagram (APD) and resulted in a circle of radius $r = (2\pi s)^{-1}$. The phase and hence the radius of curvature of the APD is changed by the additive term $-2\pi\boldsymbol{g}\cdot\boldsymbol{R}(z)$ in the

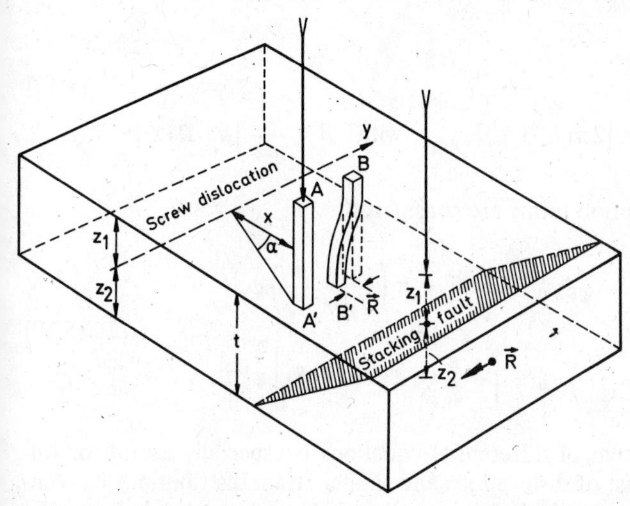

Fig. 8.16. A crystal foil of thickness t contains a screw dislocation parallel to the foil surface at a depth z_1 ($z_1 + z_2 = t$). The column AA′ of the ideal lattice is deformed to BB′ by a displacement vector $\boldsymbol{R}(z)$. A stacking fault displaces the lower part of the lattice by a constant displacement vector $\boldsymbol{R}$, relative to the upper part

exponent of (8.15). Examples of such modified APDs are shown in Fig. 8.17 for a stacking fault and in Fig. 8.24 for a dislocation. Equation (8.15) results in the important rule that the scalar product $\boldsymbol{g} \cdot \boldsymbol{R}$ has to be non-zero for the imaging of lattice defects. Bragg reflections $\boldsymbol{g}$ for which the contrast of the lattice defect disappears because $\boldsymbol{g} \cdot \boldsymbol{R} = 0$ can be used to determine the direction and magnitude of the displacement $\boldsymbol{R}$.

A local lattice distortion can be included in the dynamical theory by adding a term $2\pi \boldsymbol{g} \cdot \boldsymbol{R}(z)$ to the phase $\phi = 2\pi s z$ in (7.35) and (7.36) becomes

$$\frac{d\psi_g}{dz} = \sum_{h=g_1}^{g_n} \frac{\mathrm{i}\pi}{\xi_{g-h}} \psi_h \exp[2\pi \mathrm{i} s_{h-g} z + 2\pi \mathrm{i} (\boldsymbol{h}-\boldsymbol{g}) \cdot \boldsymbol{R}(z)] \,. \tag{8.16}$$

By use of the transformations

$$\begin{aligned} \psi_0' &= \psi_0 \exp(-\mathrm{i}\pi z/\xi_0) \,; \\ \psi_g' &= \psi_g \exp[2\pi \mathrm{i} s z - \mathrm{i}\pi z/\xi_0 + 2\pi \mathrm{i} \boldsymbol{g} \cdot \boldsymbol{R}(z)] \end{aligned} \tag{8.17}$$

analogous to those employed in the two-beam case (7.37), which do not affect the intensities, because $|\psi_0'|^2 = |\psi_0|^2$ and $|\psi_g'|^2 = |\psi_g|^2$, (8.16) becomes

$$\frac{d\psi_0'}{dz} = \frac{\mathrm{i}\pi}{\xi_g}\psi_g'$$
$$\frac{d\psi_g'}{dz} = \frac{\mathrm{i}\pi}{\xi_g}\psi_0' + [2\pi\mathrm{i}(s+\beta)]\psi_g' \qquad \text{with } \beta = \frac{d}{dz}[\boldsymbol{g}\cdot\boldsymbol{R}(z)] \tag{8.18}$$

or when the absorption terms are considered

$$\frac{d\psi_0'}{dz} = -\frac{\pi}{\xi_0'}\psi_0' + \pi\left(\frac{\mathrm{i}}{\xi_g} - \frac{1}{\xi_g'}\right)\psi_g'$$
$$\frac{d\psi_g'}{dz} = \pi\left(\frac{\mathrm{i}}{\xi_g} - \frac{1}{\xi_g'}\right)\psi_0' + \left[-\frac{\pi}{\xi_0'} + 2\pi\mathrm{i}(s+\beta)\right]\psi_g' \,. \tag{8.19}$$

This linear system of differential equations is especially useful for calculating the contrast of dislocations and similar strain distributions but cannot be used for stacking faults because the constant phase shift between the two regions above and below such a fault is lost with the transform (8.17).

Equation (8.19) shows that a strain field modifies the excitation error s to $s + \beta$ and alters the deviation from the Bragg condition because the lattice planes are locally bent more into or out of the Bragg position (see also Fig. 8.26). The column approximation can be used for strongly excited Bragg reflections. For the weak-beam method in which the excitation errors are large and resolution better than 2–5 nm is required, a direct solution of the Schrödinger equation has to be used (Sect. 8.5.3).

8.3.2 Matrix-Multiplication Method

The amplitude ψ_g of a Bragg reflection at a depth z in a single-crystal foil can be calculated by use of (7.55) when the eigenvalues $\gamma^{(j)}$, the components $C_g^{(j)}$ of the eigenvectors, and the Bloch-wave excitation amplitudes $\varepsilon^{(j)}$ are known. As for (7.51 and 57), this equation can be written in matrix form. Using (7.58), we have

$$\underline{\psi}(z) = [C]\,\{\exp(2\pi\mathrm{i}\gamma z)\}\,\underline{\varepsilon} = [C]\,\{\exp(2\pi\mathrm{i}\gamma z)\}\,[C^{-1}]\,\underline{\psi}(0)$$
$$= [S]\,\underline{\psi}(0)\,; \qquad \underline{\psi}(0) = \begin{pmatrix}1\\0\end{pmatrix}. \tag{8.20}$$

The matrix $[S] = [C]\,\{\exp(2\pi\mathrm{i}\gamma z)\}\,[C^{-1}]$ is called the scattering matrix. The anomalous absorption of Bloch waves can be introduced by replacing $\gamma^{(j)}$ by $\gamma^{(j)} + \mathrm{i}q^{(j)}$. This method can be used for the calculation of lattice-defect contrast by dividing the foil into distinct layers; the kth layer is between z_{k-1}

and z_k. The amplitude at the exit surface can be calculated from the amplitude at the entrance surface, by use of (8.20),

$$\underline{\psi}(z_k) = [S_k]\,\underline{\psi}(z_{k-1}) \quad \text{and} \quad \underline{\psi}(t) = [S_n][S_{n-1}] \dots [S_1]\,\underline{\psi}(0)\,. \tag{8.21}$$

Lattice defects are considered by replacing U_g by $U_g \exp(2\pi i \boldsymbol{g} \cdot \boldsymbol{R})$ in the off-diagonal elements of the matrix $[A]$ (7.50) or by using a fault matrix $[F] = \{\exp(2\pi i \boldsymbol{g} \cdot \boldsymbol{R})\}$ and replacing $[A]$ by $[F^{-1}][A][F]$. This modifies $[C]$ to $[F^{-1}][C]$ and S to $[S_k] = [F^{-1}][S][F]$. This method is of interest for cases where the foil contains large undisturbed regions or volumes of constant displacement vector $\boldsymbol{R}$ [8.70, 71]. Thus planar faults can be described by shifting the lower part of the foil, of thickness t_2, relative to the upper part, of thickness t_1, by a constant displacement vector $\boldsymbol{R}$. For the two-beam case, the fault matrix takes the form

$$[F] = \begin{pmatrix} 1 & 0 \\ 0 & \exp(2\pi i \boldsymbol{g} \cdot \boldsymbol{R}) \end{pmatrix}; \quad [F^{-1}] = \begin{pmatrix} 1 & 0 \\ 0 & \exp(-2\pi i \boldsymbol{g} \cdot \boldsymbol{R}) \end{pmatrix}. \tag{8.22}$$

The amplitudes behind the foil are

$$\underline{\psi}(t) = [S_2][S_1]\,\underline{\psi}(0) = [F^{-1}][S(t_2)][F][S(t_1)]\,\underline{\psi}(0)\,. \tag{8.23}$$

The matrix formulation can also be used to obtain the following symmetry rule for lattice-defect contrast [8.72]. Friedel's law for centrosymmetric crystals implies that the matrices $[A]$ and $[S]$ are symmetric. If the displacement in a column at depth z is $R(z)$ in one foil, and if the $R(z)$ of a second foil is obtained by inversion with respect to a point at the centre of the column, equal contrast in bright-field results when $R(z) = -R(t-z)$. Symmetry rules for dark-field images are discussed in [8.73].

8.3.3 Bloch-Wave Method

For some applications, in which a better understanding of the alterations of the Bloch-wave field caused by lattice defects may be helpful, it is useful to rewrite the Howie-Whelan equations (8.18) for the reflection amplitudes ψ_g as equations for the Bloch-wave excitation amplitudes $\varepsilon^{(j)}$ (7.54) [8.74]. The ψ_g and $\varepsilon^{(j)}$ are related by (7.55) or $\underline{\psi}(0) = [C]\,\underline{\varepsilon}$ at the entrance surface and $\underline{\psi}(t_1) = [F^{-1}][C]\{\exp(2\pi i \gamma z)\}\,\underline{\varepsilon}$ directly below a planar fault at a depth t_1. Substitution in (8.23) gives the Bloch-wave amplitude $\varepsilon'^{(j)}$ at the depth t_1:

$$\underline{\varepsilon}' = \{\exp(-2\pi i \gamma t_1)\}[C^{-1}][F][C]\{\exp(2\pi i \gamma t_1)\}\,\underline{\varepsilon}(0)\,. \tag{8.24}$$

The $\varepsilon^{(j)}$ and $\varepsilon'^{(j)}$ are constant inside an undisturbed lattice volume.

Each Bloch wave has to be multiplied by the exponential absorption term $\exp(-2\pi q^{(j)} z)$ of (7.86). The fault matrix $[F]$ is equal to the unit matrix $[E]$ in a fault-free crystal and in this case, $\varepsilon^{(j)} = \varepsilon'^{(j)}$; there is thus no interband scattering (Sect. 7.4.1), and the Bloch waves propagate independently. When defects are present, the fault matrix $[F] \neq [E]$ and, in the two-beam case, the excitation amplitude $\varepsilon'^{(1)}$ becomes a linear combination of $\varepsilon^{(1)}$ and $\varepsilon^{(2)}$ so that interband scattering does ocur. In Sect. 8.4.2, this reasoning will be used to discuss stacking-fault contrast.

If the crystal is not a planar fault but varies continuously in depth, the following alteration of the Bloch-wave amplitude across a depth element Δz is obtained

$$\begin{aligned} \underline{\varepsilon}' &= \underline{\varepsilon} + \frac{d}{dz}\,\underline{\varepsilon} \\ &= \{\exp(-2\pi\mathrm{i}\gamma z)\}[C^{-1}]\left([E] + \frac{d}{dz}[F]\,\Delta z\right)\{\exp(2\pi\mathrm{i}\gamma z)\}\,\underline{\varepsilon}\,. \end{aligned} \tag{8.25}$$

If the terms of $[F]$ can be expanded as $\exp(\mathrm{i}\alpha_g) = 1 + \mathrm{i}\alpha_g + \ldots$, with $\alpha_g = 2\pi\boldsymbol{g}\cdot\boldsymbol{R}$ and β_g small, (8.18) gives

$$\frac{d}{dz}\,\underline{\varepsilon} = \{\exp(-2\pi\mathrm{i}\gamma z)\}[C^{-1}]\{2\pi\mathrm{i}\beta_g\}[C]\{\exp(2\pi\mathrm{i}\gamma z)\}\,\underline{\varepsilon}\,. \tag{8.26}$$

This equation for the Bloch waves is equivalent to the Howie-Whelan equations (8.18). Multiplying the matrices and using the abbreviation $\Delta\gamma_{ij} = \gamma^{(i)} - \gamma^{(j)}$, we find

$$\frac{d}{dz}\,\varepsilon^{(j)} = \sum_i \varepsilon^{(i)}(z)\exp(2\pi\mathrm{i}\Delta\gamma_{ij}z)\sum_g C_g^{(j)}C_g^{(i)}\,2\pi\mathrm{i}\,\beta_g \quad (j = 1, \ldots, n)\,. \tag{8.27}$$

For weak interband scattering by a lattice defect, the initial value $\varepsilon^{(j)}(0) = C_0^{(j)}$ can be used at the entrance surface and integration over a column in the z direction gives

$$\varepsilon'^{(j)}(t) = \varepsilon^{(j)}(0) + 2\pi\mathrm{i}\sum_i\sum_g C_0^{(i)}C_g^{(j)}C_g^{(i)}\int_0^t \beta_g\exp(2\mathrm{i}\Delta\gamma_{ij}z)\,dz\,. \tag{8.28}$$

If, for small defects, β_g decreases strongly inside the foil, the limits of integration in (8.28) can be extended to $\pm\infty$, and the integral becomes a Fourier transform [8.75–77].

If $\Delta\gamma_{ij}$ is large or, in other words, the extinction distance $\xi_{ij} = 1/\Delta\gamma_{ij}$ responsible for the transition $i \to j$ is small, then only those columns for which β_g changes appreciably inside one extinction distance will contribute. This means that only the core of the defect will be imaged. This is the situation in weak-beam imaging (Sect. 8.5.3).

8.4 Planar Lattice Faults

8.4.1 Kinematical Theory of Stacking-Fault Contrast

The structure of a stacking fault will be illustrated by two examples, the face-centred cubic (fcc) lattice and the close-packed hexagonal lattice. A close-packed plane can be positioned on sites B or C of a layer A (Fig. 7.3). The fcc lattice with the {111} planes as close-packed planes can be described by the layer sequence ABCABC... and the hexagonal lattice by the sequence ABAB... The lattice contains a stacking fault if one part of the crystal is shifted relative to the other by the displacement vector $\boldsymbol{R}_i$ $(i = 1, 2, 3)$ of Fig. 7.3. The displacement vector $\boldsymbol{R}_1$, for example, transfers an A layer into a B layer and the new sequence is ABCABC|BCABC... The line (|) indicates the position of the stacking fault. This fault can also be generated by removing an A layer from the crystal. It is called an intrinsic stacking fault. An extrinsic stacking fault arises when an additional layer is introduced, a B layer in the sequence ABCABC|BABC, or when intrinsic faults occur in two neighbouring planes.

In the kinematical theory of image contrast, the integral in (8.15) can be solved by the amplitude-phase diagram (APD) (Fig. 7.11), which gives a simple graphical solution of the contrast effects to be expected [8.69]. In Fig. 8.16, an inclined stacking fault crosses the foil, and the part below the fault is displaced by a constant vector $\boldsymbol{R} = a\,(u,v,w)$. In (8.15), this lower part contributes with an additional phase shift

$$\alpha = 2\pi \boldsymbol{g} \cdot \boldsymbol{R} = 2\pi a\,(hu + kv + lw) \tag{8.29}$$

because in a fcc lattice $\boldsymbol{R} = \langle 112 \rangle\, a/6$ and the hkl are all even or all odd (to satisfy the extinction rules of the structure amplitude F, see Table 7.1). This

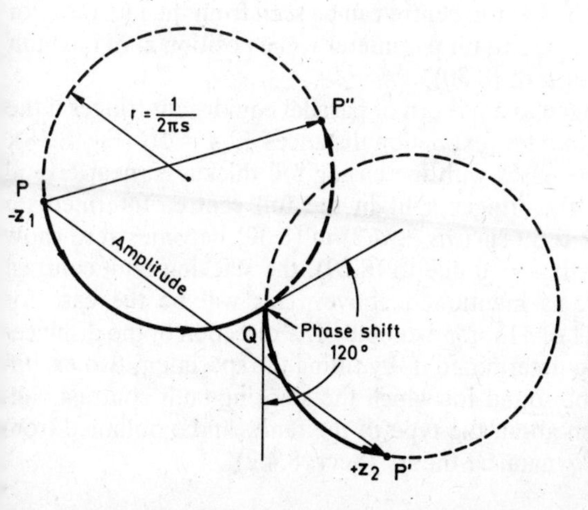

Fig. 8.17. Influence of a stacking fault at Q at a depth z_1 on the amplitude-phase diagram (phase shift by $-2\pi/3$ equivalent to a kink of 120°)

gives $\alpha = 2\pi n/3$ with $n = 0, \pm 1, \pm 2, \ldots$ The phase shift α is normally limited to the interval $-\pi \le \alpha \le \pi$ so that only the values $\alpha = 0, \pm 2\pi/3$ need be distinguished when considering the influence on image contrast. The stacking fault becomes invisible for $\alpha = 0$ because there will be no difference between (8.15) and the equation for a perfect crystal (7.33). In some crystal structures, hexagonal AlN or rutile for example, and in antiphase boundaries of ordered alloys (Sect. 8.4.3), stacking faults with $\alpha = \pi$ also have to be considered [8.78, 79].

A non-zero value of α has to be added to all phases below the stacking fault. This means that at the point Q of an APD corresponding to an $\alpha = 2\pi/3$ fault, a kink of 120° has to be introduced for a 220 reflection, for example (Fig. 8.17). The intensity of the Bragg-reflected beam is proportional to $\overline{\mathrm{PP}'}^2$, and will be greater than that of the undisturbed crystal, which is proportional to $\overline{\mathrm{PP}''}^2$, for the example shown in Fig. 8.17. The stacking fault will then be brighter in the dark field. The total curve length in the APD is equal to the foil thickness t and, therefore, the 120° kink moves for an inclined fault in such a way that $t = t_1 + t_2$ and t_1 is the depth of the fault in the column considered. The kinematical theory therefore predicts that the contrast will be symmetric about the foil centre.

If $I_g = \psi_g \psi_g^*$ is calculated from (8.15), the integral can be split into two parts, with the limits 0, z_1 and z_1, t. By introducing the distance of the fault from the foil centre, $z' = z_1 - t/2$, the following formula can be obtained [8.80]

$$I_g = \frac{1}{(\xi_g s)^2} [\sin^2(\pi t s + \alpha/2) + \sin^2(\alpha/2) - 2\sin(\alpha/2)\sin(\pi t s + \alpha/2)\cos(2\pi s z')] \tag{8.30}$$

and the symmetry relative to the foil centre can be seen from the fact that, for a fixed value of foil thickness t and tilt parameter s, the position z' of the fault appears only in the last term of (8.30).

A stacking fault is imaged as a pattern of parallel equidistant fringes if the foil thickness is larger than a few extinction distances $\xi_{g,\mathrm{eff}} = 1/s$ (Fig. 8.18). The number of fringes increases with increasing foil thickness, near crystal edges for example, and the fringes split in the foil centre. Intermediate thicknesses for which the term $\sin(\pi t s + \alpha/2)$ in (8.30) becomes zero show vanishing fringe contrast. If $\alpha = 0$ due to (8.29), the stacking-fault contrast vanishes for all s and t, as mentioned above. This will be the case for $\boldsymbol{R} = [1\bar{2}1]\,a/6$ and $\boldsymbol{g} = 31\bar{1}$ or $1\bar{1}3$, for example. The direction of the displacement vector $\boldsymbol{R}$ can thus be determined if, by tilting the specimen, two excitations of different $\boldsymbol{g}$ can be found for which the stacking-fault contrast vanishes. Further information about the type of the fault can be obtained from the intensity laws of the dynamical theory (Sect. 8.4.2).

Fig. 8.18. Stacking faults in a metal foil (Courtesy by P. B. Hirsch)

The number of fringes also increases with increasing tilt parameter s, owing to the decrease of $\xi_{g,\mathrm{eff}}$, and the contrast difference between the dark and bright fringes is less.

For two overlapping stacking faults, the phase $\alpha = 2\pi/3$ of one fault is doubled in the region where the projected images of the faults overlap. The total phase $\alpha = 4\pi/3$ is equivalent to $\alpha = -2\pi/3$ and the contrast is correspondingly modified; dark and bright fringes are interchanged (Fig. 8.18). The total α becomes zero for three overlapping faults and the contrast vanishes in the overlap region for all $\boldsymbol{g}$. The fault contrast always vanishes for two overlapping faults with $\alpha = \pi$.

8.4.2 Dynamical Theory of Stacking-Fault Contrast

The dynamical theory has to be used near a Bragg position [8.71, 80]. The contrast in the two-beam case can easily be calculated by the matrix method of Sect. 8.3.2 using (8.23). The matrices $[F]$ and $[F^{-1}]$ are defined in (8.22) and $[S]$ is given by (8.20) with the $[C]$ matrix of (7.66). The abbreviation

$\Delta k = \gamma^{(1)} - \gamma^{(2)} = 1/\xi_{g,\mathrm{eff}}$ is used, and a common phase factor $\exp(\pi \mathrm{i} \Delta k t)$, unimportant for the calculation of intensity, is omitted:

$$
\begin{aligned}
\psi_0(t) &= \cos(\pi k t) - \mathrm{i}\cos\beta\,\sin(\pi\Delta k t) \\
&\quad + \frac{1}{2}\sin^2\beta\,(\mathrm{e}^{\mathrm{i}\alpha}-1)\,[\cos(\pi\Delta k t) - \cos(2\pi\Delta k z')] \\
\psi_g(t) &= \mathrm{i}\sin\beta\,\sin(\pi\Delta k t) + \frac{1}{2}\sin\beta\,(1-\mathrm{e}^{-\mathrm{i}\alpha})\,. \\
&\quad \{\cos\beta\,[\cos(\pi\Delta k t) - \cos(2\pi\Delta k z')] - \mathrm{i}\sin(\pi\Delta k t) \\
&\quad + \mathrm{i}\sin(2\pi\Delta k z')\}\,.
\end{aligned}
\tag{8.31}
$$

β becomes equal to $\pi/2$ in the exact Bragg condition $w = 0$. This results in the following values of the bright-field intensity $I_0 = \psi_0\psi_0^*$

$$
\begin{aligned}
&\text{for}\quad \alpha = \frac{3\pi}{2}:\ I_0 = \frac{3}{4}\cos^2\left(\frac{2\pi z'}{\psi_g}\right) + \frac{1}{4}\cos^2\left(\frac{\pi t}{\xi_g}\right) \\
&\text{for}\quad \alpha = \pi \quad :\ I_0 = \cos^2\left(\frac{2\pi z'}{\xi_g}\right)
\end{aligned}
\tag{8.32}
$$

The image intensity remains symmetric about the foil centre because z' appears only in cosine terms, and because absorption has been neglected, we have $I_0 + I_g = 1$. However, the cosine term containing z' is quadratic in (8.32) so that the number of fringes is double that found for large w in the kinematical theory [linear cosine term in (8.30)]. The fringe spacing now corresponds to alterations of $\xi_g/2$ in the stacking-fault depth. The transition from the pure kinematical theory ($w \gg 1$) to the dynamical theory in Bragg position ($w = 0$) proceeds by the formation of subsidiary fringes. The amplitude of these fringes increases with decreasing w, resulting in equal amplitudes and twice the number of fringes in the Bragg position.

The influence of anomalous absorption has been calculated [8.81] and the results are confirmed in experiments with larger specimen thicknesses. The following effects can be observed for an $\alpha = 2\pi/3$ stacking fault:

1) The dark-field fringe pattern becomes asymmetric but the bright-field pattern remains symmetric about the foil centre (Fig. 8.19). At the electron entrance side (top), the dark- and bright-field fringes are anticomplementary, that is, maxima and minima appear at the same places for the same depths of the stacking fault whereas at the exit side (bottom), the bright- and dark-field fringes are complementary; maxima occur in bright field, whereas minima are seen in dark field and vice versa (Fig. 8.20 a, b). This contrast effect can be used to decide how the stacking fault is inclined in the foil. In

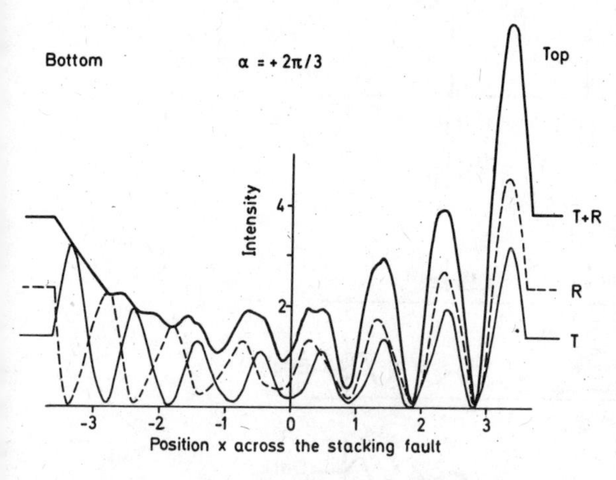

Fig. 8.19. Intensity of the primary beam T and the Bragg reflection R (bright- and dark-field intensities, respectively) across an inclined stacking fault. Note the complementarity of bright- and dark-field intensities at the bottom and anti-complementarity at the top. We recall that with multi-beam imaging ($T + R$), the contrast cancels when bright- and dark-field images are complementary

multi-beam imaging (Sect. 8.1.3), the fringes persist at the top of the foil but are cancelled at the bottom ($T + R$ in Figs. 8.19 and 8.20 c).

2) The fringe contrast decreases in the central region of the foil in the bright- and dark-field mode and completely vanishes in thick foils. Although,

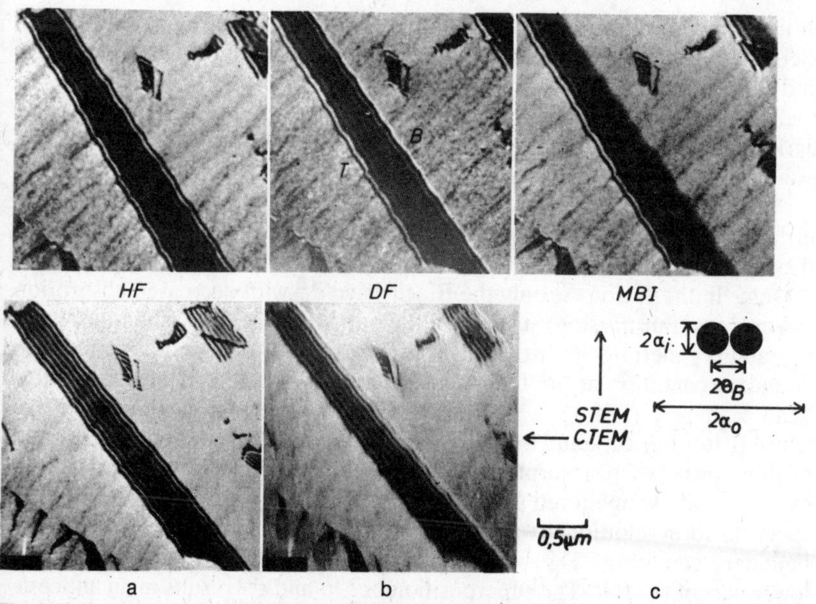

Fig. 8.20. Imaging of a stacking fault in a Cu-7wt.% Al alloy in the STEM and TEM modes (top and bottom, respectively) with the indicated values of the electron-probe aperture $\alpha_p = \alpha_i$, the Bragg angle θ_B and the detector aperture $\alpha_d = \alpha_0$. **(a)** Bright field, **(b)** dark field obtained by selecting only the primary and reflected beams, respectively, and **(c)** $T + R$ multi-beam image with vanishing fringe contrast at the bottom due to the complementarity of bright- and dark-field intensities

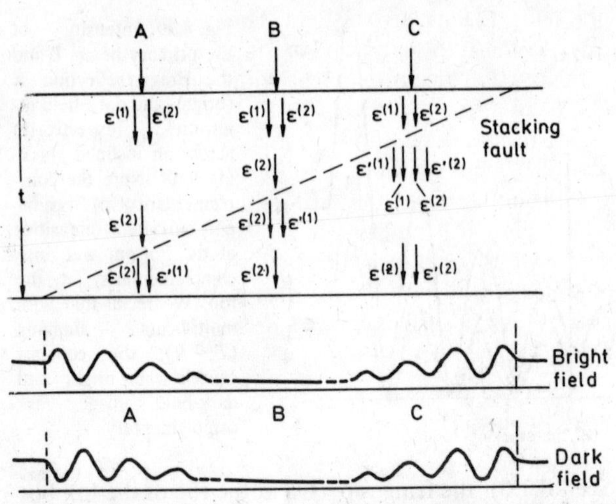

Fig. 8.21. Explanation of the complementarity of dark and bright field at the bottom (*A*) and the anti-complementarity at the top (*C*) of a stacking fault and the vanishing stacking-fault contrast in the centre (*B*) of a thick foil due to the differences of absorption of the Bloch waves

in the Bragg position, twice as many fringes are seen in the central part for medium thicknesses, only the normal number of fringes is observed at the top and the bottom.

3) The first fringe in bright field is bright for $\alpha = +2\pi/3$ (Fig. 8.20 a) and dark for $\alpha = -2\pi/3$. If the direction of $\boldsymbol{g}$ is known, this can be used to establish the stacking-fault type [8.81–83].

Observations 1 and 2 can be understood in terms of the dynamical theory including absorption from the following model (Fig. 8.21). In the top part of the foil, first the Bloch waves with the excitation amplitudes $\varepsilon^{(1)}$ and $\varepsilon^{(2)}$ are present. In the region A, only the Bloch wave $\varepsilon^{(2)}$ with decreased absorption (anomalous transmission) survives at the bottom of the foil. When the stacking fault is penetrated a further division into $\varepsilon^{(2)}$ and $\varepsilon'^{(1)}$ occurs to satisfy the boundary condition at the fault plane; this can also be interpreted as interband scattering (Sect. 8.3.3). These Bloch waves create a complementary fringe pattern in dark and bright field, analogous to the edge contours found in thin layers when absorption is neglected. In region C, the Bloch waves exist with nearly unaltered amplitudes $\varepsilon^{(1)}$ and $\varepsilon^{(2)}$ at the boundary. They are split and form additional waves with amplitudes $\varepsilon'^{(1)}$ and $\varepsilon'^{(2)}$ to satisfy the boundary condition. The waves $\varepsilon'^{(1)}$ and $\varepsilon^{(1)}$ are strongly absorbed in the lower part of the foil. The superposition of $\varepsilon'^{(2)}$ and $\varepsilon^{(2)}$ results in an anticomplementary fringe pattern in bright and dark field. Correspondingly, only $\varepsilon^{(2)}$ remains at bottom in the central region B and the fringe contrast vanishes.

The following differences can be observed for π stacking faults. There is no difference of contrast for $\pm\pi$. Bright- and dark-field images are always symmetric about the foil centre and anticomplementary. The central fringe is invariably bright in bright field and dark in dark field. The fringes are parallel to the central line and not to the intersection of the fault with the surface. Only two new fringes per extinction distance appear at the surfaces when the thickness is increased.

In contrast to Fig. 8.21, these effects can be explained by the fact that $\alpha = \pi$ corresponds to a displacement vector $\boldsymbol{R}$ of half a lattice-plane spacing. Bloch waves of type 1 with antinodes at the lattice planes become Bloch waves of type 2 with nodes below the fault plane. This means that the fault causes complete interband scattering $\varepsilon^{(1)} \rightarrow \varepsilon^{(2)}$ and $\varepsilon^{(2)} \rightarrow \varepsilon^{(1)}$ and there is no splitting into twice the number of Bloch waves, as there was for a $2\pi/3$ fault.

Stacking faults limited by partial dislocations or dissociated dislocations are discussed in Sect. 8.5.4.

8.4.3 Antiphase and Other Boundaries

Stacking-fault-like contrast with $\alpha = 0, \pm\pi$ or $\pm 2\pi/3$ can also be observed at antiphase boundaries in ordered alloys [8.84–86]. Fig. 8.22 shows an example of the geometry of an antiphase boundary in an AuCu alloy. The imaging of periodic antiphase boundaries (AuCu II phase) by the extra diffraction spots will be discussed in Sect. 9.4.5 (Figs. 9.26 and 9.27). Here we consider only the contrast of single boundaries such as those present in the AuCu I phase, generated by the primary beam and a single Bragg reflection.

The contrast is influenced by the displacement vector $\boldsymbol{R}$, the Bragg reflection $\boldsymbol{g}$ of the ordered structure and its structure amplitude $F(\theta_g)$. The examples of Table 8.1 show that antiphase boundaries can be imaged with the superlattice reflections but not with the fundamental reflections, which are also present in the disordered state with a random distribution of the atoms on the lattice sites. This is a possible way of distinguishing antiphase boundaries from stacking faults.

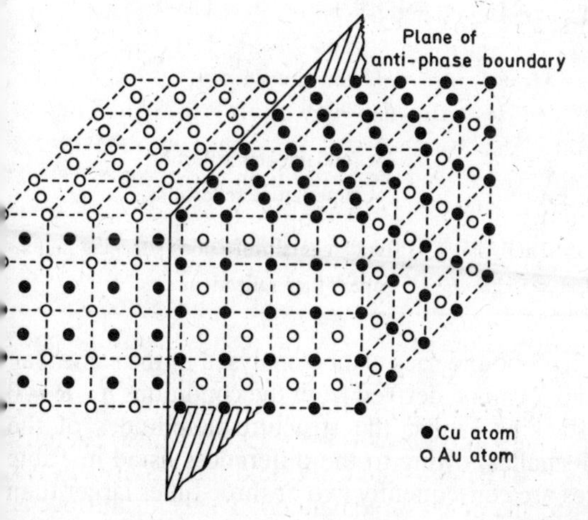

Fig. 8.22. Atomic positions at an anti-phase boundary in the fcc lattice of an AuCu alloy

Table 8.1. Examples of observable phase shifts $\alpha = 2\pi \boldsymbol{g} \cdot \boldsymbol{R}$ at antiphase boundaries in different crystal systems

1) Au-Cu I phase (tetragonal distorted fcc lattice)

$$\text{Cu: } \boldsymbol{r}_1 = (0,0,0)\,, \quad \boldsymbol{r}_2 = \left(\frac{1}{2},\frac{1}{2},0\right); \qquad \text{Au: } \boldsymbol{r}_3 = \left(\frac{1}{2},0,\frac{1}{2}\right), \quad \boldsymbol{r}_4 = \left(0,\frac{1}{2},\frac{1}{2}\right)$$

Fundamental reflection: $F = 2(f_{Au} + f_{Cu})$ *hkl* mixed (odd and even)
Superlattice reflections: $F = 2(f_{Au} - f_{Cu})$ *hkl* = even, even, odd or odd, odd, even

$$\boldsymbol{R} = \frac{a}{2}(011)\,, \quad \frac{a}{2}(01\bar{1})\,, \quad \frac{a}{2}(101)\,, \quad \frac{a}{2}(10\bar{1})$$

$$\alpha = \begin{cases} 0 \text{ for the fundamental reflections} \\ 0, \pm\pi \text{ for } hkl \text{ mixed (superlattice reflections)} \end{cases}$$

2) B 2-structure (CsCl-type, FeAl, β-CuZn)

$$A\text{: } \boldsymbol{r}_1 = (0,0,0)\,, \quad B\text{: } \boldsymbol{r}_2 = \left(\frac{1}{2},\frac{1}{2},\frac{1}{2}\right)$$

Fundamental reflections: $F = f_A + f_B$ $h + k + l$ even
Superlattice reflections: $F = f_A - f_B$ $h + k + l$ odd

$$\boldsymbol{R} = \frac{a}{2}\langle 111 \rangle$$

$$\alpha = \pi(h + k + l) = \begin{cases} 0 & \text{Fundamental reflections} \\ 0, \pm\pi & \text{Superlattice reflections} \end{cases}$$

3) L 1_2-structure (Cu_3Au, Ni_3Al, Ni_3Mn)

$$\text{B: } \boldsymbol{r}_1 = (0,0,0) \quad \text{A: } \boldsymbol{r}_2 = \left(\frac{1}{2},\frac{1}{2},0\right), \quad \boldsymbol{r}_3 = \left(\frac{1}{2},0,\frac{1}{2}\right), \quad \boldsymbol{r}_4 = \left(0,\frac{1}{2},\frac{1}{2}\right)$$

Fundamenttal reflections $F = 3f_A + f_B$ *hkl* odd or even
Superlattice reflections $F = f_A - f_B$ *hkl* mixed

$$\boldsymbol{R} = \frac{a}{2}\langle 110 \rangle\,, \quad \alpha = \begin{cases} 0 & \text{for fundamental reflections} \\ 0, \pm\pi & \text{for superlattice reflections} \end{cases}$$

$$\boldsymbol{R} = \frac{a}{6}\langle 112 \rangle \quad = \begin{cases} 0, \pm 2\pi/3 & \text{fundamental reflections (stacking faults)} \\ 0, \pm\pi/3, \pm 2\pi/3 & \text{superlattice reflections} \end{cases}$$

The contrast of π- and $2\pi/3$-boundaries can be analysed in the same way as that of stacking faults, and $\boldsymbol{R}$ can be derived from the conditions $\boldsymbol{g} \cdot \boldsymbol{R} = 0$ of non-visibility. Unlike stacking faults, the structure amplitudes of the superlattice reflections are smaller, owing to the differences listed in Table 8.1. The extinction distances are consequently two or three times larger than those of the fundamental reflections, which results in a reduced number of fringes.

Stacking-fault-like α-fringes can also be observed in non-centrosymmetric crystals at the boundary between enantiomorphic phases [8.87–90]. Domain contrast is also observed for some orientations.

So-called δ-fringes are observed when the boundary cannot be described by a displacement vector $\boldsymbol{R}$ but instead, either the two phases have different lattice constants or there is a tilt of the lattice with $\Delta \boldsymbol{g} = \boldsymbol{g}_1 - \boldsymbol{g}_2 \neq 0$. We then have $\delta = s_1\xi_{g1} - s_2\xi_{g2}$ [8.91–94]. This contrast can also be interpreted as Moiré fringes (Sect. 8.2.3); such effects are observed at ferroelectric domain boundaries [8.95] and antiferromagnetic domain boundaries in NiO [8.96]. δ-fringes can also be observed at precipitates when the active Bragg reflections of matrix and precipitate are almost the same [8.97]. If $\Delta \boldsymbol{g}$ is larger and a single reflection is strongly excited, only thickness (edge) contours (Sect. 8.1.1) are observed.

The following characteristic differences between δ-fringes and α-fringes can be used to distinguish the two types. The spacing of δ-fringes can be different at the top and bottom if $\xi_{g1} \neq \xi_{g2}$. The dark-field image becomes symmetric for $\xi_{g1} = \xi_{g2}$. The fringes are parallel to the surface so that, with increasing thickness, new fringes are generated in the centre. The contrast of the outer fringes is dark or bright, depending on the sign of δ. Further differences can occur in the magnitude of the contrast modulation and in the background at the top and bottom of the foil.

Twin boundaries are also of interest because they are frequently encountered. Whereas single non-overlapping boundaries show edge contours in one part of the crystal for large differences of s_1 and s_2, many-beam excitation can cause complicated fringe patterns [8.98]. In thin twin lamellae, which show no separation of the top and bottom boundary in the projected image, the crystal parts above and below the lamellae are displaced by a vector $\boldsymbol{R}$, which depends on the number of lattice planes in the lamellae; $\boldsymbol{R}$ can also become zero, therefore, whereupon the contrast vanishes for all Bragg reflections of the matrix.

Larger cavities in crystals also cause strong diffraction contrast [8.99]; the resulting contrast may be brighter or darker than that of the matrix, depending on the excitation of the matrix reflection and on the depth in the foil.

Examples of further applications are the imaging of boundaries in minerals [8.100] and of martensitic transformation [8.101].

8.5 Dislocations

8.5.1 Kinematical Theory of Dislocation Contrast

Dislocations parallel to the foil surface can be imaged as dark lines in bright field and bright lines in dark field and inclined dislocations with a dotted-line

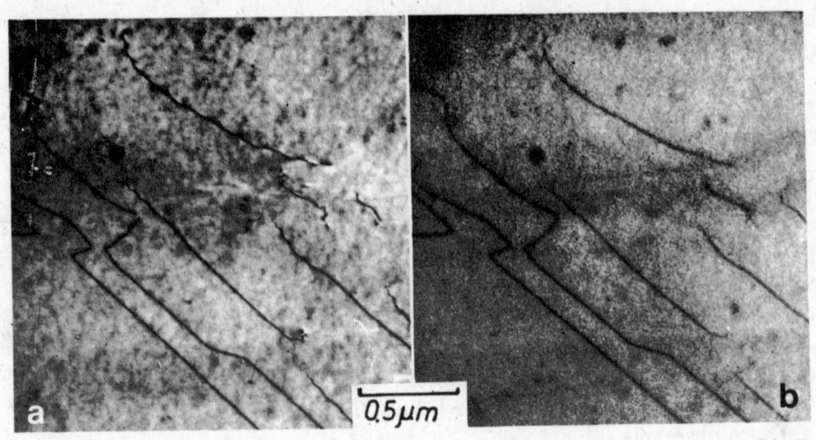

Fig. 8.23 a, b. Dislocations in an Al foil imaged in (**a**) the 100 keV TEM-mode and (**b**) the STEM mode. Dislocations parallel to the surface show uniform contrast, inclined dislocations show alternating contrast in the TEM mode and blurring of the alternating contrast in the STEM mode

or zig-zag contrast if $t \gg \xi_g$ (Fig. 8.23). Most of the contrast effects observable with large excitation errors can be described on the basis of kinematical theory, whereas the dynamical theory (Sect. 8.5.2) is necessary near the Bragg position and for weak-beam excitations.

The local displacement vector $\boldsymbol{R}(z)$ must be known before the contrast can be calculated. A simple analytical formula for R can be established for a screw dislocation (Fig. 8.16) with a Burgers vector $\boldsymbol{b}$ parallel to the unit vector $\boldsymbol{u}$ along the dislocation line. The column AA′ of a perfect crystal is bent to BB′ by a screw dislocation in the y direction. The unit cells are displaced in the y direction parallel to the Burgers vector $\boldsymbol{b}$ of the dislocation. A circle around the dislocation line on a lattice plane does not close, its end being shifted by $\boldsymbol{b}$ relative to the origin. Assuming that the isotropic theory of elasticity is applicable, the displacement vector of a screw dislocation becomes

$$\boldsymbol{R} = \boldsymbol{b}\,\frac{\alpha}{2\pi} = \frac{\boldsymbol{b}}{2\pi}\arctan(z/x) \tag{8.33}$$

This displacement of a general dislocation, for which $\boldsymbol{b}$ is not parallel to $\boldsymbol{u}$ and the glide plane is parallel to the foil surface, becomes

$$\boldsymbol{R} = \frac{1}{2\pi}\left(\boldsymbol{b}\alpha + \frac{1}{4(1-\nu)}\{\boldsymbol{b}_\mathrm{e} + \boldsymbol{b}\times\boldsymbol{u}[2(1-2\nu)\ln|r| + \cos(2\alpha)]\}\right), \tag{8.34}$$

where ν is Poisson's ratio and $\boldsymbol{b}_e$ is the edge component of the Burgers vector; isotropic eleasticity theory is again assumed to be valid and relaxation of elastic strain due to the free foil surfaces is neglected.

When (8.33) is substituted into (8.15), the scalar product $\boldsymbol{g} \cdot \boldsymbol{b}$ is an integer n, because it is the product of a reciprocal-lattice vector and a crystal vector; $\boldsymbol{g} \cdot \boldsymbol{b} = n = 1$ for $\boldsymbol{g} = 111$ and $\boldsymbol{b} = [110]a/2$ in a fcc lattice, for example. This yields

$$\psi_g = \frac{\mathrm{i}\pi}{\xi_g} \int\limits_{-z_1}^{-z_2} \exp[-\mathrm{i}(2\pi sz + n \arctan z/x)]\, dz = \frac{\mathrm{i}\pi}{\xi_g} \int\limits_{-z_1}^{-z_2} \mathrm{e}^{\mathrm{i}\varphi} dz\,. \qquad (8.35)$$

The origin of the z-coordinate is placed at the depth of the dislocation line ($z_1 + z_2 = t$). The phase φ, the slope of the element of arclength dz in an APD, becomes smaller or larger than the phase $2\pi sz$ of a perfect crystal, depending on the sign of s and x. If $sx < 0$, the curvature at $z = 0$ becomes smaller (Fig. 8.24 a) than the curvature $1/r = 2\pi s$ for the perfect crystal. If $sx > 0$, the curvature becomes greater (Fig. 8.24 b). The value of arctan (z/x) saturates to a constant value $\pi/2$ for large z. The APD again tends asymptotically to a circle with the same radius $r = 1/2\pi s$, for large z. The scattered amplitude ψ_g is proportional to the length of the line $\overline{\mathrm{PP'}}$. The arc length between P and P′ is equal to $z_1 + z_2 = t$. If such APDs are plotted for all values of $2\pi sx$, the intensity distribution $I(x) = \psi_g \psi_g^* = (\pi/\xi_g)^2 \overline{\mathrm{PP'}}^2$ normal to the dislocation line can be obtained for the dark-field mode and different values of $\boldsymbol{g} \cdot \boldsymbol{b} = n$ (Fig. 8.25). In this figure, only the distance AB of the centres of the asymmetric circles are plotted, which cancels oscillations caused by the particular depth z_1 of the dislocations and gives an average kinematical image.

Analogous calculations can be made for edge dislocations, or mixed dislocations with the glide plane parallel to the foil surface [8.102]. Apart from the width of the intensity maximum, no other important differences occur.

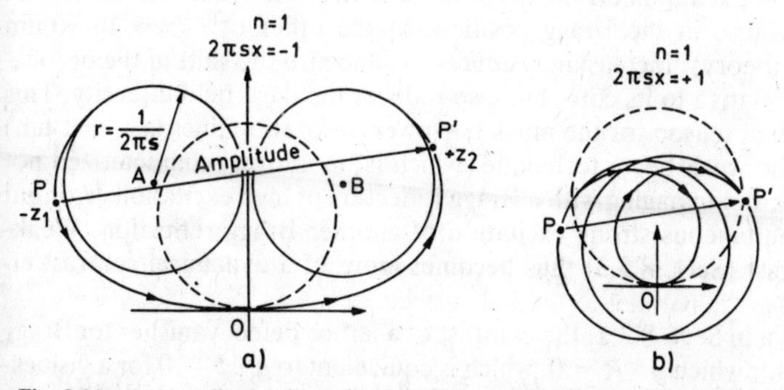

Fig. 8.24 a, b. Amplitude-phase diagram for a column through the crystal close to a screw dislocation with $\boldsymbol{g} \cdot \boldsymbol{b} = n = 1$ for **(a)** $2\pi sx = -1$ and **(b)** $2\pi sx = +1$ [8.69]

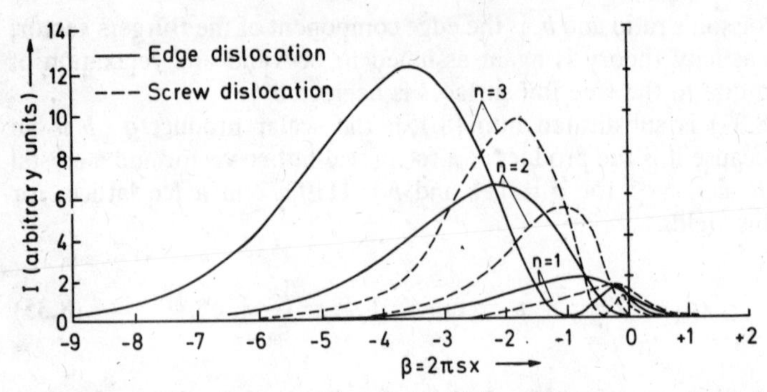

Fig. 8.25. Intensity profiles proportional to $\overline{AB}^2$ in Fig. 8.24 across the dark-field images of edge and screw dislocations with different values of $\boldsymbol{g} \cdot \boldsymbol{b} = n$. The centre of the dislocation is at $x = 0$ [8.69]

This simple kinematical theory predicts the following characteristics concerning the position and the width of the dislocation image. The image of the dislocation is not at the core ($x = 0$) but the maximum in the dark-field mode is shifted to one side, by a distance of the same order as the half width. Figure 8.26 demonstrates schematically that the maximum will be on the side of the dislocation on which the lattice planes are bent nearer to the exact Bragg position. The position of the dislocation image therefore changes to the opposite side when a dislocation crosses the bend contour of the Bragg reflection (shaded in Fig. 8.27 a), which changes the sign of s. For $\boldsymbol{g} \cdot \boldsymbol{b} = n = 2$, the dynamical theory indicates that a double image can arise in the Bragg position. Dislocation loops change their size for different signs of s (in-line and out-line contrasts, Fig. 8.27 b).

The half width $x_{0.5}$ of the dislocation image is of the order of $1/2\pi s$ but depends also on the type of dislocation. The half width of an edge dislocation is approximately twice that of a screw dislocation for equal values of s (Fig. 8.25). For low excitation errors, the width will be of the order of a few tens of nanometres and, in the Bragg position, of the order of $\xi_g/\pi \simeq$ 10–20 nm (dynamical theory). Increasing s reduces $x_{0.5}$ and also the shift of the dislocation image relative to its core, but also reduces the dark-field intensity. This is the principal reason for the much narrower dislocation lines ($x_{0.5} \simeq 1$ nm) seen with the weak-beam technique, which is, moreover, characterized not only by dark-field imaging with a Bragg reflection of high excitation error but also by simultaneous strong excitation of another Bragg reflection. Weak-beam contrast (Sect. 8.5.3) thus becomes more of a dynamical contrast effect.

As shown in Sect. 8.3.1, the contrast of a lattice defect vanishes for Bragg reflections for which $\boldsymbol{g} \cdot \boldsymbol{R} = 0$, which is equivalent to $\boldsymbol{g} \cdot \boldsymbol{b} = 0$ for a dislocation. Figure 8.28 shows schematically that, in the presence of an edge dislocation, the lattice planes are bent for $\boldsymbol{g}_1$ but not for $\boldsymbol{g}_2$ and $\boldsymbol{g}_3$. This $\boldsymbol{g} \cdot \boldsymbol{b} = 0$

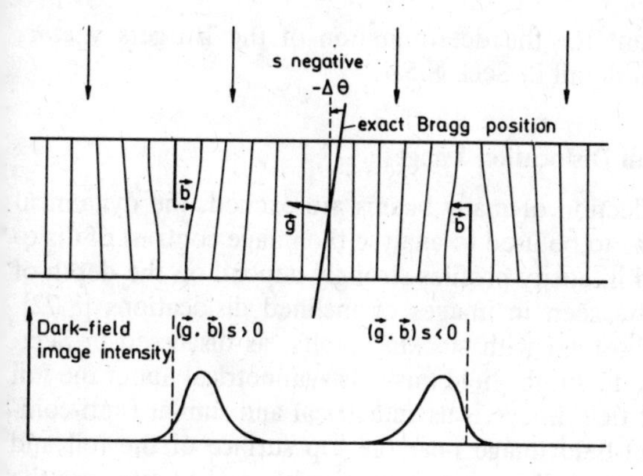

Fig. 8.26. Diagram showing on which side of its core the dislocation image is situated

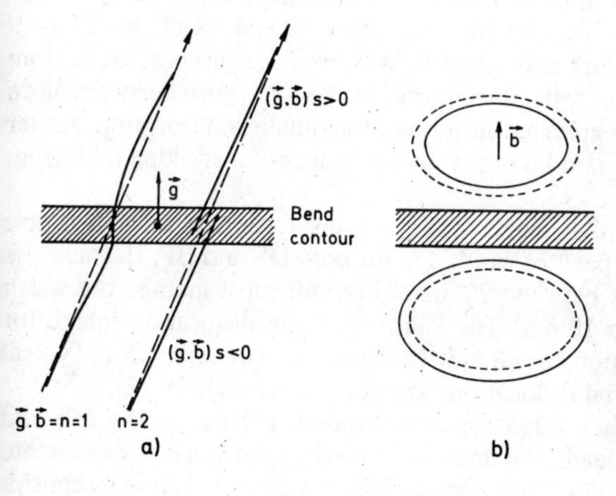

Fig. 8.27. (**a**) Behaviour of a dislocation image crossing a bend contour and (**b**) of a dislocation-loop image for positive and negative values of $(\boldsymbol{g} \cdot \boldsymbol{b})s$ as seen by an observer looking from below the foil

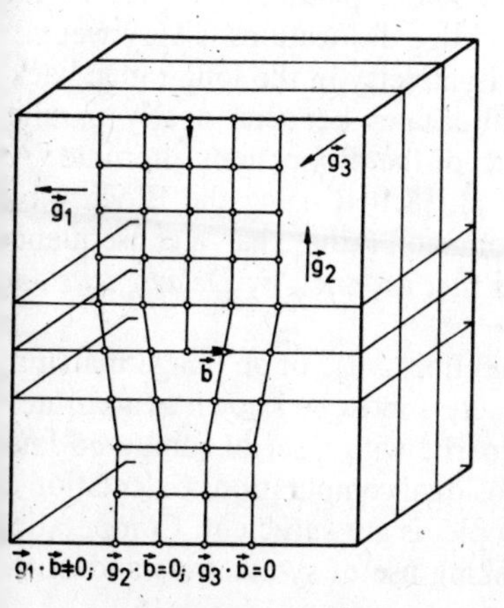

Fig. 8.28. Demonstration of the $\boldsymbol{g} \cdot \boldsymbol{b} = 0$ rule for an edge dislocation. Only the lattice planes that belong to $\boldsymbol{g}_1$ are strongly bent, so that $\boldsymbol{g}_1 \cdot \boldsymbol{b} \neq 0$ whereas $\boldsymbol{g}_2 \cdot \boldsymbol{b} = \boldsymbol{g}_3 \cdot \boldsymbol{b} = 0$

rule is therefore important for the determination of the Burgers vector, which will be discussed in detail in Sect. 8.5.4.

8.5.2 Dynamical Effects in Dislocation Images

If a strong low-order reflection or many beams are excited, the dynamical theory with absorption has to be used to analyse the image contrast of dislocations. The contrast and intensity profiles strongly depend on the depth of the dislocation, as can be seen in images of inclined dislocations [8.72]. Similar effects can be observed with stacking faults, as discussed in Sect. 8.4.2 and Fig. 8.20. Thus the bright-field image is symmetrical about the foil centre, whereas the dark-field image is asymmetrical and similar (anti-complementary) to the bright-field image near the top surface of the foil and complementary at the bottom. This asymmetry can be used to attribute the ends of an inclined dislocation to the top and bottom surfaces. In the multi-beam image (MBI) that can be formed with a STEM (Sect. 8.1.3) or a HVEM, the sum of the dark and bright-field intensities cancels at the bottom surface [8.15]. For thicker foils, the contrast in the foil centre becomes more uniform, which can be explained in terms of anomalous absorption, rather like the vanishing of the middle region of the fringes of a stacking fault (Fig. 8.21).

For $\boldsymbol{g} \cdot \boldsymbol{b} = n = 1$, the image of a screw dislocation at the foil centre consists of a sharp dark peak of width $\xi_g/5$ for both DF and BF. Because the extinction distance ξ_g is between 20 and 50 nm for most metals, the width ranges from about 4 to 10 nm. The splitting of the dislocation image for $n = 2$ in the Bragg position has already been discussed in Sect. 8.5.1. Typical contrast effects for partial dislocations are also discussed in [8.72].

Though $\boldsymbol{g} \cdot \boldsymbol{b}$ vanishes for a screw dislocation normal to the foil and parallel to the electron beam, the dislocation can be seen as a black and white spot near the Bragg position. This contrast effect can be explained by considering surface relaxation and changes in the lattice parameter caused by the foil surface [8.103]. The same contrast for edge dislocations is very weak.

Dark dislocation lines often show an asymmetry in the long-range background intensity. Thus if this contrast alternates between nearly parallel dislocations, the signs of the $\boldsymbol{g} \cdot \boldsymbol{b}$ products of the dislocations alternate. As shown by the calculations of *Wilkens* et al. [8.104] using the Bloch wave method, this is an intrinsic-contrast phenomenon rather than a consequence of stress relaxation at the foil surfaces, as first assumed by *Delavignette* and *Amelinckx* [8.105].

Owing to the complexity of dynamical effects, use of an image-matching method is indispensable; the image parameters must be known as accuratel as possible and the micrographs are compared with a set of computed faul images. Inclined defects need a two-dimensional computation. Calculation c line profiles across a dislocation, for example, is not sufficient. Computatio time can be considerably reduced by making use of symmetries and of th fact that the Howie-Whelan equations (8.18) are linear. This means that there can be only two independent solutions of these equations and that once these are known, all other solutions (e.g., for different depths of the defect) can be obtained by forming a linear combination with the initial condition $\psi_0 = 1$ and $\psi_g = 0$ at the top surface. Such computation methods are described in [8.106–109]. The usual way of obtaining half-tone images is to use a line printer, generating the grey scale by overprinting different symbols. A more satisfactory result is obtained if a film writer is available; in it, a light-emitting diode "writes" the image directly onto film.

8.5.3 Weak-Beam Imaging

The Bragg contrast of dislocations using $\boldsymbol{g}$ vectors with low values of *hkl* and a small excitation error s_g creates broad images (Fig. 8.29 a) with a half-width $x_{0.5} \simeq 5$–20 nm. Dense dislocation networks or weakly dissociated dislocations consisting of two partial dislocations on either side of a stacking fault cannot be resolved. Because the product of s and x appears in the abscissa $(2\pi sx)$ of Fig. 8.25, the width of dislocation images falls to 1.5–2 nm if large excitation errors are employed: $s_g \geq 0.2\ \mathrm{nm}^{-1}$ (Fig. 8.29 b).

The basic principle of this weak-beam imaging technique [8.110–112] is demonstrated in Fig. 8.30 for a two-beam case. In Fig. 8.30 a, the lattice planes are in the Bragg position ($s = 0$), giving the usual depth oscillation of

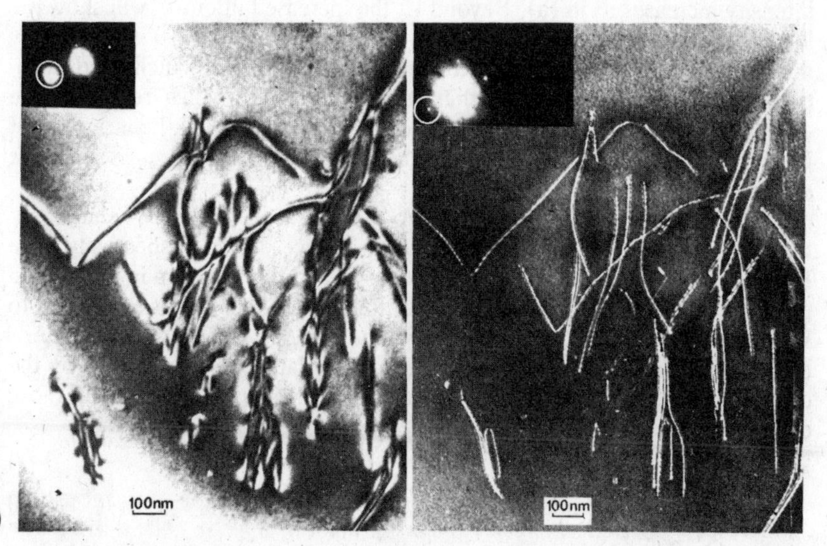

Fig. 8.29. (**a**) Dislocations in heavily deformed silicon imaged with a strong $2\bar{2}0$ diffracted beam. (**b**) Weak-beam $2\bar{2}0$ dark-field image of the same area showing the increase of the resolution of dislocation detail. The insets show the diffraction conditions used to form the images [8.113]

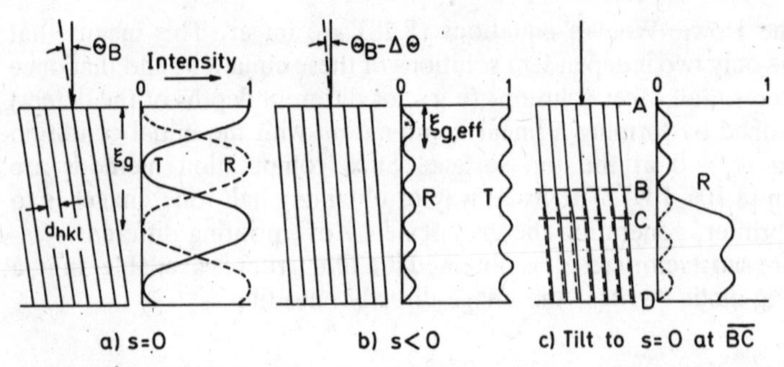

Fig. 8.30 a–c. Schematic explanation of weak-beam contrast: (**a**) Oscillation of the transition T and reflected beam intensity R in the Bragg position ($s = 0$), (**b**) reduction of $\xi_{g,\text{eff}}$ and of the oscillation amplitudes of T and R for $s < 0$, (**c**) reflected intensity in the presence of a lattice-plane tilt to $s \simeq 0$ between B and C

I_0 and I_g with a repeat distance ξ_g. In Fig. 8.30 b a large excitation error s causes a decrease of the amplitude of the oscillation and the periodicity drops to $\xi_{g,\text{eff}} \simeq 1/s \ll \xi_g$ [see also Fig. 7.14 b and Eq. (7.69)]. The strain field on one side of a dislocation is idealized in Fig. 8.30 c by regions (columns) AB and CD without distortion, and in the region BC the lattice planes are tilted into the Bragg position. The corresponding intensity I_g initially follows the low-intensity oscillation of periodicity $\xi_{g,\text{eff}} = 1/s$ but below the depth B, the intensity increases as in (a). Beyond C, the increased intensity will show the low-intensity oscillations of (b) again. A region BC with $s \simeq 0$ is found only near the core of a dislocation. The model also explains why inclined dislocations show profile oscillations that correspond to ξ_g, which results in the alternating contrast seen in Fig. 8.29 a, whereas in the weak-beam image of Fig. 8.29 b, the repeat distance is much shorter because $\xi_{g,\text{eff}}$ is so much smaller.

This principle can also be applied to the two-beam Howie-Whelan equations (7.38) and (8.18) without and with a lattice defect, respectively. The Bragg position in (7.38) corresponds to $s = 0$ and the last term in the second equation becomes zero. The same condition for (8.18), which is equivalent to $s = 0$ in the region BC of Fig. 8.30, will be $s + \beta = 0$. The tilt parameter s can be compensated by an opposite tilt of the lattice planes caused by the displacement field $\boldsymbol{R}(z)$ of the defect. The image peak occurs for those columns in which

$$s + \boldsymbol{g} \cdot \frac{d\boldsymbol{R}}{dz} = 0 \tag{8.36}$$

at a turning point of $\boldsymbol{g} \cdot d\boldsymbol{R}/dz$, i.e. where $\boldsymbol{g} \cdot d^2\boldsymbol{R}/dz^2 = 0$. When the displacement fields $\boldsymbol{R}(z)$ of screw and edge dislocations (8.33, 34) are substituted, the image maximum is found to be at

$$x_W = -\frac{\boldsymbol{g}\cdot\boldsymbol{b}}{2\pi s}\left(1+\frac{\varepsilon}{2(1-\nu)}\right) \quad \text{with } \varepsilon = \begin{cases}1 & \text{edge}\\ 0 & \text{screw}\end{cases} \text{ dislocation}, \tag{8.37}$$

whereas for $\boldsymbol{g}\cdot\boldsymbol{b} = 2$, the average kinematical image in Fig. 8.25 (distance between the centres of the initial and final circles in the APD of Fig. 8.24) results in a maximum at

$$x_K = -\frac{a}{2\pi s} \quad \text{with } a = \begin{cases}2.1 & \text{edge}\\ 1.0 & \text{screw}\end{cases} \text{ dislocation}. \tag{8.38}$$

Values of x_W and x_K are indicated in Fig. 8.31, which shows a six-beam calculation of the image profile of an undissociated edge dislocation in copper at various depths. Dynamical effects influence the image profiles, especially if another reciprocal lattice vector $\boldsymbol{g}$ is strongly excited: a double-image line may appear, the lines may change their position for different depths of inclined dislocations, or the contrast may disappear even if $\boldsymbol{g}\cdot\boldsymbol{b} \neq 0$. Such effects have to be considered for quantitative measurement of the size of dissociated dislocations or dislocation dipoles [8.113], and it will also be necessary to check the results against computed images, based on the assumed model and the excitation parameters.

The column approximation is of limited validity for the high-resolution conditions of weak-beam imaging and can predict incorrect image profiles. A direct method of solving the Schrödinger equation proposed by *Howie* and

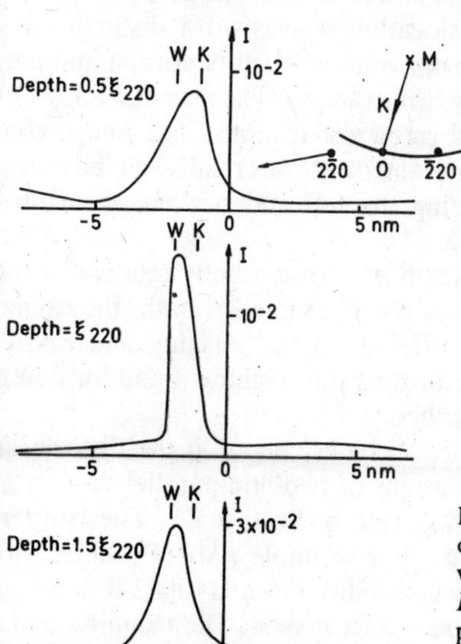

Fig. 8.31. Computed weak-beam-image profiles of an undissociated edge dislocation in copper at various depth ($\boldsymbol{g}\cdot\boldsymbol{b} = 2$, $\boldsymbol{g} = 2\bar{2}0$, $t = 2\xi_g$ $E = 100$ keV, isotropic elasticity). (W, K) image positions predicted by (8.37) and (8.38), respectively [8.112]

Basinski [8.114] assumes that the wave function $\psi_s(r)$ and the V_g vary very slowly over distances equal to the dimensions of a unit cell. Thus, the maxima of weak-beam dislocation profile can shift to the opposite side with increasing foil thickness [8.115].

The strong excitation of another $\boldsymbol{g}$ vector together with the positions of Kikuchi lines can be used to reach a value $s_g \geq 0.2\ \text{nm}^{-1}$ necessary for the weak-beam diffraction condition. In copper, for example, a $\bar{2}\bar{2}0$ or 600 reflection has to be strongly excited to get $s_g \simeq 0.2\ \text{nm}^{-1}$ for $\boldsymbol{g} = 220$ [$g(\bar{g})$ and $g(3g)$ condition, respectively]. Conditions for other materials can be established using the Ewald-sphere construction.

Whereas the bright-field mode is normally used when imaging with low excitation error, it is necessary to use the dark-field mode for the weak-beam technique, which has the advantage that the dislocation image is narrow and well contrasted but the disadvantage that a long exposure time, of 10–30 s, is needed compared to 1–2 s for a bright-field exposure.

The weak-beam technique has been applied to many problems, the quantitative measurement of stacking-fault energy by the separation of the Shockley partial dislocations, for example, in Ag,Cu [8.116]; in Au [8.117]; in Ni [8.118]; in stainless steel [8.119]. Dislocations in Ge and Si are generally found dissociated and the dependence of the dissociation width on orientation is in good agreement with the anisotropic theory of elasticity [8.120].

8.5.4 Determination of the Burgers Vector

A knowledge of the direction and magnitude of the Burgers vector $\boldsymbol{b}$ is important for the interpretation of dislocation images – for distinguishing between edge and screw dislocations, with Burgers vectors normal and parallel to the dislocation line, respectively, for example. The sign of $\boldsymbol{b}$ allows us to distinguish right-hand and left-hand screw dislocations. The image contrast depends on s, $\boldsymbol{g}$ and $\boldsymbol{b}$ and their signs and directions relative to the image must be known. Methods of recognizing the top and bottom of inclined dislocations are discussed in Sect. 8.5.2.

The first step is to align the diffraction pattern correctly relative to the image, because the angle of rotation φ is not the same for both, the excitations of the intermediate lens being different in the imaging and SAED modes. In Sect. 9.4.4, methods for performing this alignment and for determining the crystal orientation are described.

The direction of the Burgers vector can be calculated if the dislocation contrast disappears for the Bragg excitations of two-non-parallel vectors $\boldsymbol{g}_1$ and $\boldsymbol{g}_2$ due to the $\boldsymbol{g} \cdot \boldsymbol{b} = 0$ cancellation rule of Sect. 8.5.1. The Burgers vector will then be parallel to $\boldsymbol{g}_1 \times \boldsymbol{g}_2$. For example $h\bar{h}l$ reflections with $h \neq 0$, l are needed to distinguish the invisibility of an $[110]a/2$-dislocation from that of other $\langle 110 \rangle a/2$-dislocations in a fcc crystal. The magnitude of $\boldsymbol{b}$ can be obtained from the behaviour of the dislocation image when crossing the corresponding bend contour (e.g., $\boldsymbol{g} \cdot \boldsymbol{b} = 1$ and 2 in Fig. 8.27 a). The sign of the Burgers vector is obtained from the sign of s and the position (side) of the dislocation image relative to the dislocation centre. The direction can be evaluated geometrically by use of Fig. 8.26. The position of the image line can be established by tilting the specimen to an opposite s and changing the dislocation image to the opposite side [8.121]. The sign of s can be determined by observing the position of the Kikuchi line relative to the correlated Bragg-diffraction spot because the system of Kikuchi lines (Kossel cones) remains fixed relative to the crystal foil during tilt, whereas the Bragg-diffraction spots do not change in position but only in intensity. The excitation error s is defined as positive (see also Sect. 7.3.3) if the reciprocal lattice point is inside the Ewald sphere and the Kikuchi line outside the diffraction spot. Kikuchi line and spot coincide for $s = 0$. The shift Δx of the Kikuchi line is directly proportional to the tilt $\Delta\theta$ of the Bragg position: $\Delta x = L\Delta\theta$ (L: diffraction camera length).

Partial dislocations which enclose a stacking fault can result from a dissociation of a dislocation, e.g.

$$\frac{a}{2}[\bar{1}01] \rightarrow \frac{a}{6}[\bar{1}\bar{1}2] + \frac{a}{6}[\bar{2}11] . \tag{8.39}$$

They become invisible if $\boldsymbol{g} \cdot \boldsymbol{b} = \pm 1/3$ [8.70, 72]. An example of a dissociated dislocation network is shown in Fig. 8.32, in Fig. 8.32a $\boldsymbol{g} \cdot \boldsymbol{b} = \pm 1/3$ for all partial dislocations and only the stacking-fault contrast (dark) can be seen. On tilting the specimen and imaging with the $\boldsymbol{g}$ excitations (indicated in the diffraction patterns of Fig. 8.32b–d), the stacking-fault contrast and one of the three partial dislocations vanish. (The size or width of the stacking fault area can be used to determine the stacking-fault energy [8.122]. With large stacking-fault energy, the width is of the order of a few nanometres, which can only be resolved by the weak-beam technique.)

In practice, difficulties can arise in finding a second $\boldsymbol{g}$ for which the contrast disappears. Complexities in the form of residual contrasts may appear if the edge component of the Burgers vector is large and $\boldsymbol{g} \cdot (\boldsymbol{b} \times \boldsymbol{u})$ should also be zero if the displacement $\boldsymbol{R}(z)$ can be described by (8.34). Furthermore, elastic anisotropy and displacement relaxation at the foil surface have to be considered. Ambiguties can be associated with the determination of $\boldsymbol{b}$ from a disappearance condition with large $\boldsymbol{g}$, such as $\boldsymbol{g} \geq 311$ in copper [8.123]. Any detailed investigation of Burgers vectors must, therefore, be accompanied by computation of defect micrographs. Asymmetries in the image of inclined dislocations related to the sign of $\boldsymbol{g} \cdot \boldsymbol{b}$ or $\boldsymbol{g} \cdot (\boldsymbol{b} \times \boldsymbol{u})$ can then be exploited [8.124].

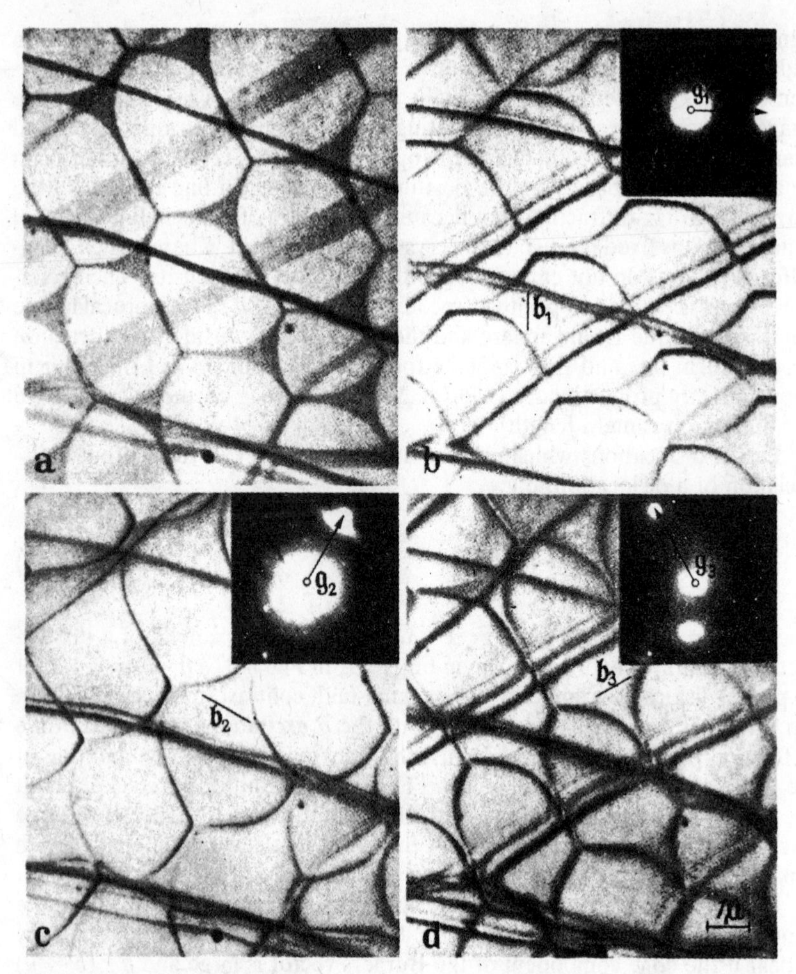

Fig. 8.32 a–d. Hexagonal network of stacking faults in graphite bounded by partial dislocations, imaged with different Bragg reflections excited (see insets). (**a**) Stacking-fault contrast only, (**b**), (**c**) and (**d**) vanishing contrast for one of the three types of partial dislocations with different Burgers vectors [8.122]

8.6 Lattice Defects of Small Dimensions

8.6.1 Coherent and Incoherent Precipitates

Three types of precipitates can be distinguished, which depend on the fit between the lattice of the precipitate and the matrix: coherent, partially coherent and incoherent precipitates (Fig. 8.33). The resulting types of con-

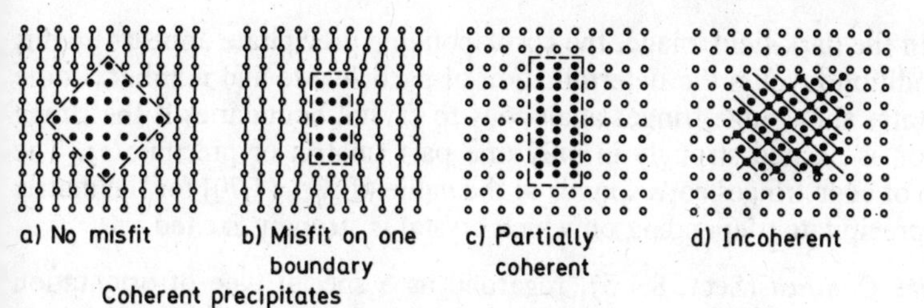

Fig. 8.33. Schematical view of lattice distortions near coherent, partially coherent and incoherent precipitates

trast can be separated into the following effects; these normally appear superposed but one of these often dominates.

1) *Scattering Contrast.* For materials of large density and high atomic number, many electrons are scattered through large angles and absorbed by the objective diaphragm. A precipitate of larger density therefore appears darker in bright field. On this scattering contrast, however, stronger dynamical contrast is superimposed.

2) *Structure-Factor Contrast* caused by differences between the extinction distances ξ_g of the matrix and ξ_g^* of the precipitate. This contrast dominates in coherent precipitates without misfit. Coherent precipitates with misfit create a lattice strain field around the particle; this causes strain contrast, see point 6 below. For $w = s\xi_g = 0$ (Bragg position), the structure-factor contrast acts like a formal change of the specimen thickness (t: foil thickness, Δt^*: thickness of the precipitate):

$$t_{\text{eff}} = t + \xi_g \Delta t^* \left(\frac{1}{\xi_g^*} - \frac{1}{\xi_g} \right) = t + \Delta t \, . \tag{8.40}$$

Differentiating $I_g = R = \sin^2(\pi t/\xi_g)$ given by equation (7.69) and muliplying by Δt, we obtain the following intensity variation [8.125]

$$\Delta I_g = - \Delta I_0 = \frac{dI_g}{dt} \Delta t = \pi \Delta t^* \left(\frac{1}{\xi_g^*} - \frac{1}{\xi_g} \right) \sin \left(\frac{2\pi t}{\xi_g} \right) , \tag{8.41}$$

provided that $\Delta t^* \ll \xi_g$. Maximum visibility with uniform bright contrast occurs where $t/\xi_g = 1/4, 5/4, \ldots$ and with dark contrast where $t/\xi_g = 3/4, 7/4, \ldots$ if $\xi_g^* > \xi_g$. This structure-factor contrast will clearly be most effective in thin areas of the foil, because the Bloch-wave absorption (not considered in (8.41)) decreases the effective contrast produced by small changes of thickness.

3) *Orientation Contrast* can be observed for partially coherent and incoherent precipitates if they show a strong and the matrix a weak reflection, or vice

versa. In the dark-field image, the corresponding precipitate appears bright; this condition favours the determination of particle size and number. Large precipitates show edge fringes analogous to crystal boundaries if the Bragg reflection is excited strongly in only one part (matrix or precipitate). The spacing of edge fringes corresponds to the value of $\xi_{g,\text{eff}}$ (7.70) for the matrix or the precipitate, depending on which crystal is strongly excited.

4) *Moiré Contrast* (Sect. 8.1.5), regarded as a special type of orientation contrast, will be seen if double reflection in the matrix and the precipitate results in extra diffraction spots near the primary beam, producing Moiré fringes when superposed in the image.

5) *Displacement-Fringe Contrast* is observed if, for example, a plate-like precipitate causes a normal displacement vector in a partially coherent precipitate:

$$\boldsymbol{R}_\text{n} = \delta \Delta t\, \boldsymbol{u}_\text{n} - n\, \boldsymbol{b}_\text{n} \, . \tag{8.42}$$

This displaces the matrix-lattice planes in opposite directions on either side of the precipitate, where $\delta = 2(a_1 - a_2/(a_1 + a_2)$ is the misfit parameter and n is the number of dislocations in the peripheral interface with a Burgers-vector component $\boldsymbol{b}_\text{n}$ parallel to the normal. A phase shift $\alpha = 2\pi \boldsymbol{g} \cdot \boldsymbol{R}_\text{n}$ is generated as for the stacking-fault contrast, but $\boldsymbol{R}_\text{n}$ is not necessarily a lattice vector. Otherwise, many characteristics of stacking-fault contrast can be transferred to this case. Fringes are observed when the plates are inclined to the foil surface. This contrast becomes most marked if the matrix is excited strongly and the precipitate weakly.

6) *Matrix Strain-Field Contrast.* Coherent precipitates often create large strains in the matrix: Guinier-Preston zones II in Al-Cu, for example, or Co precipitates in Cu-Co alloys. Besides scattering, structure factor and displacement-fringe contrast, a long-range strain contrast can be observed [8.126]. The displacement vector of the strain field of a precipitate of radius r_0 is

$$\boldsymbol{R} = \varepsilon \boldsymbol{r} \quad \text{for} \quad r < r_0 \qquad \text{and} \quad \boldsymbol{R} = \frac{\varepsilon r_0^3}{r^3} \boldsymbol{r} \quad \text{for} \quad r > r_0 \, , \tag{8.43}$$

with $\varepsilon = 3K\delta\,[3K + 2E(1+\nu)]^{-1}$ where K denotes the bulk modulus of the precipitate, and E and ν are Young's modulus and Poisson's ratio for the matrix. Small cubic or tetrahedral precipitates also provoke such a symmetric strain field over a larger distance. Substitution of the radial displacement field $\boldsymbol{R}$ in (8.15) shows that a line exists along which $\boldsymbol{g} \cdot \boldsymbol{R} = 0$ and there is, hence, no contrast (Fig. 8.34), which results in a butterfly or coffee-bean contrast. The no-contrast line is normal to $\boldsymbol{g}$ and changes direction if other $\boldsymbol{g}$ are used by tilting the specimen. Plate-like precipitates create an anisotropic

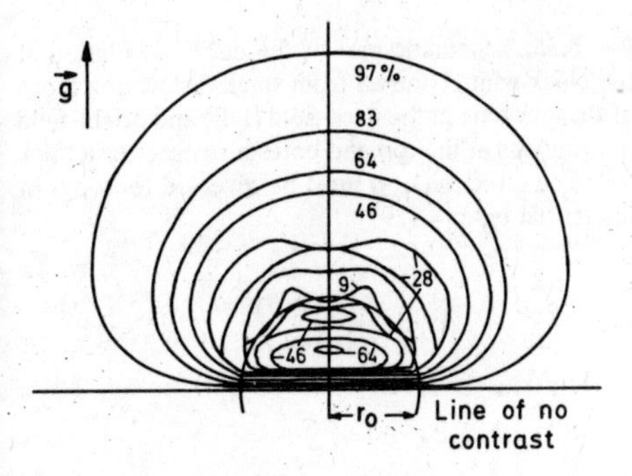

Fig. 8.34. Calculated bright-field intensity contours in the upper half-plane around a spherically, symmetrically strained precipitate of radius $r_0 = 0.25\,\xi_g$ in the centre of a foil of thickness $t = 5\,\xi_g$ and a distortion $\varepsilon g \xi_g = 10.2$. The figures indicate the intensity as percentage of the background [8.126]. The figure has to be mirrored at the line of no contrast, resulting in a "butterfly" contrast

strain field and the no-loss line does not change much if $\boldsymbol{g}$ is varied. The width of equal contrast is proportional to $\varepsilon g r_0/\xi_g$ and can be used to calculate the misfit of a precipitate [8.125, 126].

8.6.2 Defect Clusters

Frenkel defects (pairs of vacancies and interstitials) can be produced by bombardment with high-energy radiation. Single vacancies or interstitials cannot be resolved. The elastic strain of the lattice is too small and the phase contrast is insufficient. The defect clusters must be 1–2 nm in size for the strain field to give observable diffraction contrast effects that can be used for an analysis of the type of cluster. Large clusters result in dislocation loops, stacking-fault tetrahedra or cavities.

The symmetry of the strain field, the orientation of the loops, their Burgers vector $\boldsymbol{b}$, the distinction between interstitial and vacancy clusters and their depth- and size-distributions are of interest.

Kinematical image conditions result in black spots [8.127]. In the Bragg position, alternating black-white contrast is observed if the faults are near the top or bottom regions of the foil [8.128]. In the centre of the foil, Bloch-wave absorption (see stacking fault contrast, Sect. 8.3.2) leads to black points, only [8.129].

As an example, typical contrast effects of small edge-dislocation loops of the vacancy type will now be described. A vector $\boldsymbol{l}$ from the dark to the bright spot of a bright-field image is introduced and the following contrast effects result [8.130–132].

1) $\boldsymbol{l}$ is parallel to the projection of $\boldsymbol{b}$ onto the foil plane and with only few exceptions, is independent of the direction of the excited $\boldsymbol{g}$. (For strain fields with spherical symmetry, $\boldsymbol{l}$ is always parallel or antiparallel to $\boldsymbol{g}$, see also Sect. 8.6.1).

2) The sign of $\boldsymbol{g} \cdot \boldsymbol{l}$ depends on the depth of the loop in the foil and differs for loops of vacancy and interstitial type. The diagram of Fig. 8.35 shows the

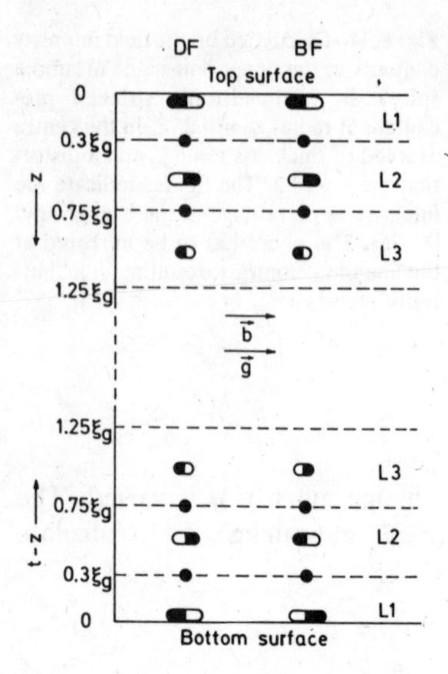

Fig. 8.35. Schematic plot of the depth oscillation of the black-white contrast from small dislocation loops of vacancy type in the dark-field (DF) and bright-field (BF) modes at the top and bottom surfaces of a thick foil. The direction of $\boldsymbol{g}$ must be reversed for loops of interstitial-type [8.129]

typical contrast behaviour. The zone L_1, for example, has a length $\xi_g/3 \simeq$ 5–10 nm. As for stacking faults and dislocations, dark- and bright-field micrographs are complementary only at the bottom surface of the foil. It is therefore necessary to know the sign of $\boldsymbol{g} \cdot \boldsymbol{l}$ and the depth before clusters of vacancies and interstitials can be distinguished. The depth can be measured with a high accuracy of ± 2 nm by recording a stereo pair [8.130–132].

3) Loops with $\boldsymbol{g} \cdot \boldsymbol{b} = 0$ can exhibit weak residual contrast. The $\boldsymbol{g} \cdot \boldsymbol{b} = 0$ rule for dislocations has to be used with care for these defects too. Computer simulations is essential for a detailed analysis [8.133, 134].

Disordered zones in displacement cascades in irradiated ordered alloys can be successfully investigated in dark-field using superlattice reflections, which reveal the disordered regions as dark spots [8.135, 136].

Small cavities in irradiated material (diameter $\simeq$ 1–2 nm) show a dependence on depth like that discussed as structure-factor contrast in Sect. 8.6.1 but can be best observed with slight defocusing, because a phase-contrast contribution has to be considered [8.137].

9. Analytical Electron Microscopy

X-ray spectrometers can be coupled to a transmission electron microscope to record x-ray quanta emitted from the specimen. With an energy-dispersive spectrometer, quantitative analysis is possible for elements with atomic numbers above ten.

Electron spectrometers incorporated in the column or placed below the final screen can be used to record electron-energy-loss spectra, which contain information about the electronic structure and the elemental composition of the specimen. Elements for which $Z \geq 4$ can be detected and the technique is more efficient than x-ray analysis because the spectrometer can collect a large fraction of the inelastically scattered electrons, which are concentrated within small scattering angles. An electron-energy filter allows us to work in the energy-selecting imaging mode, which can be used for elemental mapping, for example.

Electron-diffraction methods are employed for the identification of substances by measuring the lattice-plane spacings and for the determination of crystal orientations in polycrystalline films (texture) or single-crystal foils. The determination of structure is possible but difficult due to strong dynamical effects. Extra spots and streaks caused by antiphase structures or plate-like precipitates, for example, may also be observed when imaging a selected area.

The selected-area diffraction technique, in which an area of the order of one micrometre is selected by a diaphragm in the first intermediate image, is a standard method. The introduction of additional scan and rocking coils into an instrument capable of producing an electron probe of the order of a few nanometres, renders micro-area diffraction techniques feasible. In particular, the convergent-beam technique and the observation of high-order Laue zone lines inside the primary spot provide further information about the crystal symmetry.

With a scanning attachment, the cathodoluminescence and the electron-beam-induced current (EBIC) of semiconductor devices can be recorded and imaged. This allows us to transfer methods that have proved successfully in scanning electron microscopy to transmission microscopes, where they complement the techniques permitting elemental and crystal analysis and the high-resolution imaging of lattice defects.

9.1 X-Ray Microanalysis in a TEM

9.1.1 Wavelength-Dispersive Spectrometry

A wavelength-dispersive spectrometer (WDS) makes use of the Bragg reflection of x-rays by a single crystal ($2d \sin\theta_B = n\lambda$). Better separation of narrow characteristic lines and a larger solid angle of collection of $\Delta\Omega \simeq 10^{-3}$ sr can be obtained by focusing. The electron-irradiated spot on the specimen acts as an entrance slit, while the analysing crystal and the exit slit are mounted on a Rowland circle of radius R (Fig. 9.1). The lattice planes of the crystal are bent so that their radius is $2R$ and the surface of the crystal is ground to a radius R. Behind the exit slit, the x-ray quanta are recorded by a proportional counter. The detection efficiency of the Bragg reflection and of the proportional counter is about 10–30%. The number of electron-ion pairs generated and the resulting current pulse created by charge multiplication in the high electric field near the central wire of the counter are proportional to the quantum energy $E_x = h\nu$. Bragg reflections of higher order ($n > 1$) can be eliminated by pulse-height discrimination, because such reflections are caused by quanta of lower wavelength or higher quantum energy. For the recording of a spectrum or for altering the focusing condition to another wavelength, the exit slit, the proportional counter and the analysing crystal can be moved by means of a pivot mechanism. X-ray microanalysers are equipped with two or three spectrometers, which permit simultaneous

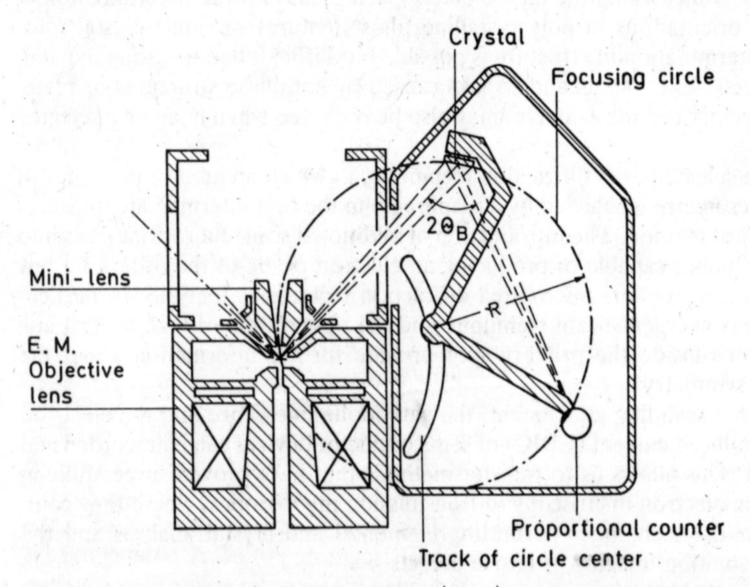

Fig. 9.1. Incorporation of a wavelength-dispersive x-ray spectrometer with large take-off angle in a transmission electron microscope

recording of different wavelengths. Crystals with different lattice spacings have to be used so that the whole wavelength range can be analysed. Special proportional counters with very thin mylar windows are available for analysing the weak $K\alpha$ radiation of low-atomic-number elements; the lower limit is beryllium.

Exact adjustment of the specimen height is necessary to satisfy the focusing condition. Small variations of height or a shift of the electron beam can cause large variations of the x-ray intensity recorded. Wavelength-dispersive spectrometers, therefore, are used mainly in x-ray microanalysers; they can also be mounted in TEM or SEM, though energy-dispersive spectrometers are more commonly used with these instruments.

Figure 9.1 shows a cross-section through the objective lens and one of the two crystal spectrometers used in the AEI/Kratos EMMA instruments [9.1, 2]. By using a third condenser lens (a LePoole mini-lens [9.3]), the electron-probe diameter can be decreased to 0.1 μm with probe currents of 2–5 nA at 100 keV [9.4]. The x-ray take-off angle is 45°. Wavelength-dispersive spectrometers of lower resolution have been described in [9.5, 6].

9.1.2 Energy-Dispersive Spectrometer

X-rays are now absorbed in the detector and create high-energy photoelectrons, which again produce charge carriers by either ionising gas atoms in a proportional counter or creating electron-hole pairs in semiconductors. The proportionality of the number of charge carriers to the x-ray quantum energy $E_x = h\nu$ and the possibility of analysing the pulse height of the collected charge form the basis of the energy-dispersive spectrometer (EDS) [9.7]. If the x-ray quantum is absorbed by ionisation of an inner shell, any residual x-ray energy contributes to the kinetic energy of the photoelectron. Any remaining energy is either transferred to an Auger electron or to an x-ray quantum of energy E'_x, characteristic of the detector atoms (x-ray fluorescence); this is Ar $K\alpha$ in a proportional counter or Si $K\alpha$ in a Si(Li) detector. If this x-ray quantum is again absorbed in the detector, the whole quantum energy E_x serves to create charge carriers. If this fluorescence x-ray quantum can leave the detector, its energy E'_x does not contribute to the number of charge carriers created. This can cause a weak escape peak of energy $E_x - E'_x$ in a recorded pulse-height spectrum.

A gas-flow proportional counter, filled with a mixture of 90% argon and 10% methane, is one of the possible types of energy-dispersive detector. The advantages are: the internal amplification (10^3–10^4) of the number of electron-ion pairs created, by repeated secondary ionization in the high electric field near the central wire, and the short dead time, $\tau \simeq 10^{-5}$ s, which is the minimum time between two recognizably separate pulses. A disadvantage is the large average energy $\overline{E}_i = 27$ eV needed for ionization. The poor statistics of the relatively small number $N = E_x/\overline{E}_i$ of electron-ion pairs generated causes the FWHM (full width at half maximum) of a characteristic x-ray peak to be large, typically of the order of 1 keV, so that $K\alpha$ lines can be resolved only for elements with atomic numbers separated by three or more units. This is sufficient for the discrimination of x-ray quanta when a proportional counter is used as a detector in a WDS, but not for elemental analysis.

The best resolution is obtained with EDS when a lithium-drifted silicon detector is used (Fig. 9.2). This consists of a reverse-biased p-i-n junction (p-type, intrinsic and n-type), 2–5 mm in diameter. The intrinsic zone generated by the diffusion of Li atoms is 3–5 mm wide. Electron-hole pairs created by x-ray absorption in this active volume can be separated by applying a reverse bias of the order of 1 kV. The charge pulse collected is proportional to the quantum energy E_x and is converted to a voltage pulse by means of a field-effect transistor (FET), which acts as a charge-sensitive amplifier. The Si(Li) crystal and the FET are permanently cooled by liquid nitrogen, the first to avoid diffusion of Li atoms and the second to reduce noise. Further reduction of noise and an improvement of the preamplifier time constant is provided by charge accumulation. This means that the output signal of the preamplifier consists of voltage steps proportional to E_x. When the voltage reaches a preset level, a light-emitting diode is switched on and the leakage current induced in the FET by the emitted light returns the output voltage to zero (pulsed optical feed-back).

The main amplifier not only amplifies the output of the pre-amplifier but also shapes the pulses. The processing-time constant is of the order of 5–10 μs and is longer than the rise-time of a voltage step in the pre-amplifier. A pulse-pileup rejector cancels the pulses when the time between two voltage steps is equal to or smaller than the processing-time constant of the main amplifier. A live-time corrector prolongs the preset counting time by one processing-time constant per pulse-pileup rejection.

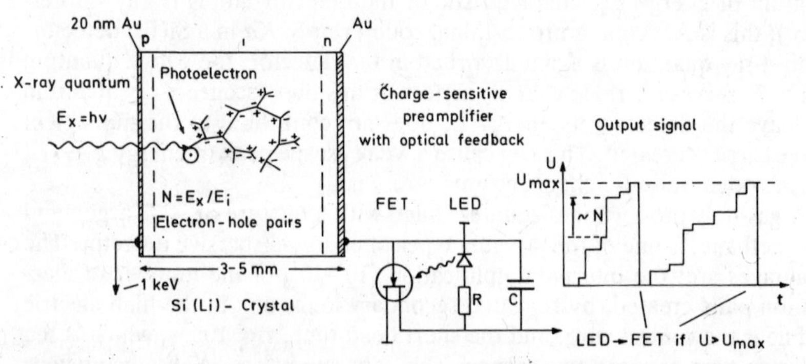

Fig. 9.2. Creation of electron-hole pairs in a Si(Li) energy-dispersive x-ray detector coupled to a charge-sensitive preamplifier with an optical feed-back circuit switching the output signal to zero by means of a ligth-emitting diode (LED), which irradiates the field-effect transistor (FET) if $U \geq U_{max}$

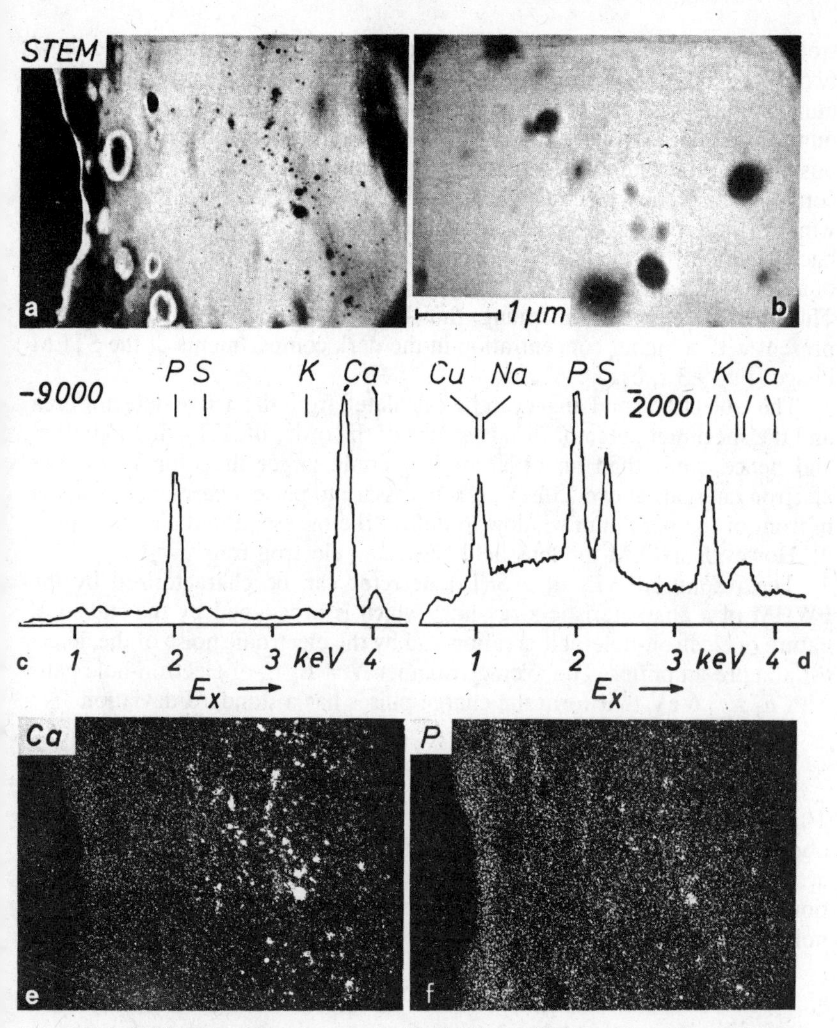

Fig. 9.3. (**a**) STEM dark-field micrograph of a freeze-dried section (about 5 μm thick) of a partially calcified aorta, (**b**) the same at larger magnification. The x-ray spectrum (400 s recording time) (**c**) of the dark mineralized compartments shows a greater Ca and P concentration than the spectrum (**d**) of the cell matrix (The figures 2000 and 9000 indicate the number of counts per channel). (**e**) and (**f**) x-ray elemental maps obtained with the Ca $K\alpha$ and P $K\alpha$ lines respectively (by courtesy of E.R. Krefting)

The shaped pulses from the main amplifier are collected, sorted, stored and displayed by a multichannel analyser (MCA). Pulse amplitudes from 0 to 10 V are linearly converted to numbers between 0 and 512 or 1024 by analogue-to-digital conversion. These numbers then serve as adresses for the storage locations in the 512 or 1024 channel memory. Figure 9.3 c and d show two recorded x-ray spectra for $E_x = 1$–4 keV. Most EDS are equipped with a minicomputer that can typically identify the characteristic peaks, count the number of quanta in different preset energy windows, subtract the continuous background radiation, separate overlapping lines and, in a final stage, a complete correction program (Sect. 9.1.3). The pulses from a preset energy window that contain the counts of a characteristic line and the continuous background below the line can be used to generate bright spots on the CRT of a STEM attachment during a simultaneous scan of the specimen and CRT. This provides elemental mapping; thus Fig. 9.3 d, e show that Ca and P are present with a higher concentration in the dark compartments of the STEM image (Fig. 9.3 a, b).

The collection solid angle can be calculated from the active detector area and the specimen-detector distance. It is of the order of $\Delta\Omega = 0.1$ to 0.01 sr and hence more than an order of magnitude larger than for WDS. For electron energies above 25 keV, it is necessary to place a permanent magnet in front of the beryllium window to deflect the backscattered electrons [9.8, 9]. However in TEM the lens field acts as an electron trap.

The resolution ΔE_x of a Si(Li) detector can be characterized by the FWHM of a characteristic x-ray line, which is broadened by the statistical nature of electron-hole-pair creation and by the electronic noise of the detector and pre-amplifier. The average number $N = E_x/\bar{E}_i$ of electron-hole pairs with $\bar{E}_i = 3.6$ eV that form the charge pulses has a standard deviation

$$\sigma_x = \sqrt{FN} = \sqrt{FE_x/\bar{E}_i} \,. \tag{9.1}$$

The standard deviation would be $\sigma_x = \sqrt{N}$ if the number of created pairs obeyed Poisson statistics. The Fano factor ($F = 0.05$ for silicon) takes into account departures from the Poisson distribution due to the gradual deceleration of the photoelectrons. Denoting the standard deviation of the electronic noise by σ_n, the FWHM becomes

$$\Delta E_x = 2.35\sqrt{\sigma_x^2 + \sigma_n^2} \,. \tag{9.2}$$

Values of $\Delta E_x = 100$–150 eV for $E_x = 5.9$ keV (Mn $K\alpha$) are typical.

The resolution of EDS is thus more than ten times worse than that of WDS and not all overlapping lines can be separated, especially those of the L series of one element and of the K series of another. Furthermore, an energy window of the order of the FWHM is necessary to ensure that all the characteristic quanta of a given line are counted. For EDS, a relatively large number of background counts are recorded from the continuous x-ray spectrum.

The detection efficiency $\eta(E_x)$ of a Si(Li) detector is nearly 100% in the range $E_x = 3$–15 keV (Fig. 9.4). The decrease at low E_x is caused by the absorption of x-rays in the thin Be window that separates the high vacuum

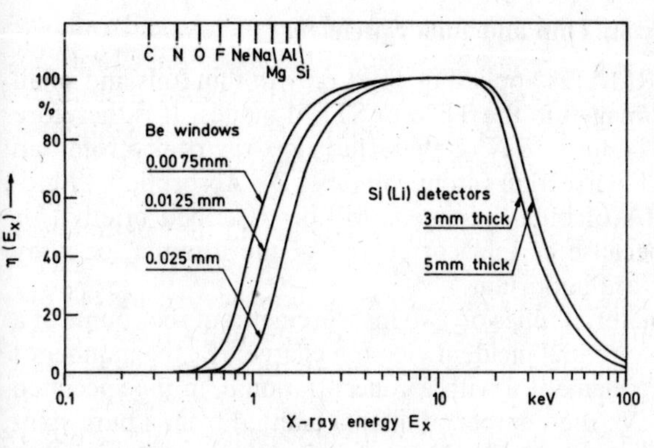

Fig. 9.4. Detection efficiency $\eta(E_x)$ of a Si(Li) energy-dispersive spectrometer for various thicknesses of the Be window and the detector

from the Si crystal and cryostat. For this reason, $E_x = 1$ keV is the lower limit and $K\alpha$ quanta of elements down to Na ($Z = 11$) can be analysed. Windowless detectors can analyse quantum energies as low as a few hundred eV; maxima produced by $K\alpha$ quanta from C, N, O can be recorded, but the lines overlap since their FWHM is too large [9.10, 11]. The decrease of efficiency at high quantum energies is caused by the increasing probability of penetrating the crystal without photoionization. In nuclear physics, therefore, Ge(Li) crystals that have a stronger absorption are in use for analysing x-ray and γ-quanta up to 100 keV.

Unlike WDS, for which the irradiated point has to be adjusted on the Rowland circle, EDS does not need any mechanical adjustment, and can hence be used when scanning a larger and/or rough specimen. Energy-dispersive spectroscopy, is therefore commonly used in SEM. In TEM an EDS has the advantage of occupying little space. It is merely necessary to design the cryostat tube with the Be window and the Si(Li) crystal in such a way that it collects the x-ray quanta from the transparent specimen in the pole-piece gap of the objective lens with as large a solid angle and take-off angle as possible (Fig. 4.24). A further advantage of EDS is that all lines are recorded simultaneously. For these reasons, many x-ray microanalysers that work with WDS are also equipped with an EDS to provide a rapid assessment of the elements present in the specimen. Its collection angle being smaller, WDS normally needs a larger electron-probe current, of the order of 10^{-8}–10^{-7} A, whereas EDS can work with lower currents. However, thanks to the preselection of the radiation by the analysing crystal, WDS can work with a higher count rate for a particular characteristic line. This implies lower statistical errors and more accurate determination of low elemental concentrations. EDS can work only at rates of a few thousand counts per second.

9.1.3 X-Ray Emission from Thin and Bulk Specimens

X-ray microanalysis (XRMA) is applied in TEM only to thin foils and small particles, which can be imaged in the TEM or STEM modes. It is therefore sufficient to discuss the count rate N of characteristic x-ray peaks from thin films. The so-called ZAF correction (atomic number **Z** – **A**bsorption – **F**luorescence) for the XRMA of bulk specimens will be described briefly (for details see [9.12–14]) because certain corrections of the formula for x-ray emission from thin foils are important.

The XRMA of bulk material relies on two measurements of the counts of a characteristic x-ray line for equal incident electron charge: N_a, the number of quanta emitted from the element a with a concentration c_a in the specimen under investigation and N_s, the number of quanta emitted from a pure standard of the element a. The ratio $k = N_a/N_s$ is only a first-order approximation for the concentration c_a.

The number of x-ray quanta dN_a generated per incident electron along an element of path length $dx = \varrho ds$ of the electron trajectory can be calculated using the ionisation cross-section σ_a, the x-ray fluorescence yield ω_a (Sect. 5.4.2) and the ratio p_a of the observed line intensity to the intensity of all lines of the series, $p_{K\alpha} = N_{K\alpha}/(N_{K\alpha} + N_{K\beta})$, for example. With the notation $n_a = \varrho c_a N_A/A_a$ the number of atoms of element a per unit volume (ϱ is the density and A_a is the atomic weight of a), dN_a becomes

$$dN_a = \omega_a p_a \sigma_a n_a ds = \omega_a p_a \frac{c_a N_A}{A_a} \frac{\sigma_a}{dE/dx} dE . \tag{9.3}$$

The stopping power

$$S = \left| \frac{dE}{dx} \right| = \frac{N_A e^4}{8\pi\varepsilon_0^4 E} \sum_{i=a,b,\ldots} c_i \frac{Z_i}{A_i} \ln (E/E_i) \tag{9.4}$$

is a modified form of the Bethe formula (5.79) with $i = a, b, \ldots$ elements in the specimen, where E_i denotes the mean ionisation energy. X-ray quanta can be generated along the whole trajectory, so long as the electron energy E' is greater than the ionisation energy (E_I) of the $I = K$, L or M shell. Integrating (9.3) then gives

$$N_a = \omega_a p_a c_a \frac{N_A}{A_a} R \int_{E_I}^{E} \frac{\sigma_a}{S} dE' , \tag{9.5}$$

where R is the *backscattering correction factor:*

$$R = \frac{\text{total number of quanta actually generated in the specimen}}{\text{total number of quanta generated if there were no backscattering}}$$

This quantity depends on the energy distribution of backscattered or transmitted electrons.

The ratio $k = N_a/N_s$ becomes

$$k = c_a R_a \int_{E_I}^{E} \frac{\sigma_a}{S_a} dE' / R_s \int_{E_I}^{E} \frac{\sigma_a}{S_s} dE' = c_a \frac{1}{k_Z} . \tag{9.6}$$

The values of R for the specimen and the standard will in general be different. Therefore, k_Z also depends on the differences between the stopping powers S.

For thin films of mass-thickness $x = \varrho t$, (9.5) simplifies to

$$N_a = \omega_a p_a \sigma_a c_a \frac{N_A}{A_a} x . \tag{9.7}$$

The atomic-number correction k_Z is not very different from unity and can be neglected if the following conditions are satisfied:

1) The ionisation cross-section σ_a does not increase because of a decrease of the mean electron energy in the specimen by energy losses;
2) The path lengths are not increased by multiple scattering;
3) Differences in the backscattering and transmission do not influence the ionisation probability.

The absorption correction is necessary because of the depth distribution $n(z)$ of x-ray emission and the absorption of x-rays inside the specimen. If the x-ray detector collects x-rays emitted with a take-off angle ψ relative to the foil surface, x-ray quanta emitted at a depth z below the surface have to penetrate an effective length $z \operatorname{cosec} \psi$ and only a fraction $\exp[-(\mu/\varrho)\, z \operatorname{cosec} \psi] = \exp(-\chi z)$ survives. The fraction of x-ray quanta leaving the specimen is

$$f(\chi) = \frac{\int_0^\infty n(z) \exp(-\chi z)\, dz}{\int_0^\infty n(z)\, dz} \quad \text{with} \quad \chi = (\mu/\varrho) \operatorname{cosec} \psi , \tag{9.8}$$

and (9.6) becomes

$$k = c_a \frac{1}{k_Z} \frac{f(\chi_a)}{f(\chi_s)} = c_a \frac{1}{k_Z k_A} . \tag{9.9}$$

Assuming that x-ray quanta are generated uniformly in a foil of mass-thickness $x = \varrho t$, (9.8) becomes

$$f(\chi) = \frac{1}{\chi x}(1 - e^{-\chi x}) \simeq 1 - \frac{1}{2}\chi x + \ldots \tag{9.10}$$

For thin specimens, therefore, the absorption correction can be neglected only if $\chi x \ll 1$. The correction is essential for thick specimens and low quantum energies, especially if $E_{x,a}$ is just above $E_{I,b}$ for a matrix of atoms b, that is, just beyond a jump of μ/ϱ in Fig. 5.38 [9.15–18]. When the x-ray quanta are detected through the pole-piece gap at an angle of 90° to the electron beam, the specimen has to be tilted to increase ψ. Care must be taken with specimens that are severely bent.

If characteristic quanta of an element b are generated with an energy $E_{x,b} \geq E_{I,a}$, these quanta and also the fraction of the continuum that satisfies this condition can be absorbed by photo-ionisation of atoms of element a, thus causing an increase of N_a by x-ray fluorescence. Equation (9.9) becomes

$$k = \frac{N_a}{N_s} = c_a \frac{1}{k_Z k_A} \frac{1 + r_a}{1 + r_s} = c_a \frac{1}{k_Z k_A k_F} . \tag{9.11}$$

where the ratio $r = N_f/N_d$ denotes the ratio of the quanta generated by fluorescence (N_f) and directly by the electron beam (N_d). The contribution N_f from fluorescence is generated in a much larger volume than N_d and can normally be neglected in thin films [9.19]; it can however be significant in FeCr alloys for example, where Cr $K\alpha$ fluorescence is excited by Fe $K\alpha$ [9.20].

Electron waves in single crystals are Bloch waves with nodes or antinodes at the nuclear sites depending on the crystal orientation (Sect. 7.4.2). Any x-ray emission is therefore anisotropic (depending on the tilt angle) because the ionisation of an inner shell is localised near the nuclei. The number of x-ray quanta emitted can vary by as much as 50% when the specimen is tilted through a few degrees near a zone axis [9.21–23]. If two atoms a and b occupy different lattice sites, the Bloch-wave amplitude can be different for a and b. Quantitative results on single-crystal foils, therefore, are reliable only if the result is insensitive to a small specimen tilt.

The number of quanta emitted, N_a in (9.5) and (9.7), has to be multiplied by the number of incident electron $n_0 = I\tau/e$ and the collection efficiency $\Delta\Omega/4\pi$ determined by the finite solid angle $\Delta\Omega$ of detection and the efficiency $\eta(E_x)$ of the detector (Fig. 9.4) to establish the count rate; (9.7) becomes

$$N_a = \omega_a p_a \sigma_a c_a \frac{N_A}{A_a} \frac{\Delta\Omega}{4\pi} \eta(E_x) x \frac{I\tau}{e} . \tag{9.12}$$

An important problem for XRMA of thin films in TEM is the contribution from continuous and characteristic quanta generated not in the irradiated area but anywhere in the whole specimen and specimen cartridge by electrons scattered at diaphragms above and below the specimen and from x-ray fluorescence due to x-ray quanta generated in the column. Additional diaphragms have to be inserted at suitable levels to absorb scattered electrons and x-rays whereas the objective diaphragm should be removed during XRMA. The number of unwanted quanta can be further reduced by constructing the specimen holder from light elements such as Be, Al or high-strength graphite [9.24].

9.1.4 Standardless Methods for Thin Specimens

The procedure involving the use of bulk pure-element standards for the XRMA of bulk specimens, as described in Sect. 9.1.3, can, in principle, be adapted for the XRMA of thin films [9.25, 26]. However, in many applications it is the ratio c_a/c_b of the concentrations that is of interest, so that only the ratio N_a/N_b of the counts of the peaks of elements a and b is needed. Equations (9.7 and 10) give [9.15, 16, 27, 28]

$$\frac{N_a}{N_b} = \frac{\omega_a p_a \sigma_a \eta(E_a) A_a}{\omega_b p_b \sigma_b \eta(E_b) A_b} \frac{c_a}{c_b} \frac{1 - \frac{1}{2}\chi_a x}{1 - \frac{1}{2}\chi_b x} = k_{ab}\left[1 - \frac{1}{2}(\chi_a - \chi_b) x\right] \frac{c_a}{c_b} . \tag{9.13}$$

This ratio method is independent of the local mass-thickness x of the specimen if the absorption correction can be neglected. In their work on an Al-Zn-Mg-Cu alloy using the ratio method *Thomson* et al. [9.29] found a dependence of the N_{Cu}/N_{Al} ratio on the foil thickness, which could be interpretated by assuming that a Cu-rich surface layer about 15 nm thick was present. The local thickness of a specimen can be measured by allowing contamination spots to form on the top and bottom of the foil. Tilting the specimen separates the two spots, and the thickness can be calculated from the tilt angle and the separation [9.16].

The correction factor $k_{ab} = k_a/k_b$ in (9.13) cannot be calculated very accurately (Fig. 9.5) [9.18], because accurate values of the ionisation cross-sections σ and fluorescent yields ω are not known. The quantity k_a is commonly determined experimentally by measuring the count rates of pure-element films reduced to equal thickness and incident electron charge $q = I\tau$. Further $k_{a,Si}$ values have been reported by *Schreiber* and *Wims* [9.30] for the *K*, *L* and *M* lines.

The diameter of the smallest possible area analysed is limited by the relation $I_p \propto d_p^{8/3}$ (4.21) between the probe current and diameter and by the spatial broadening of the electron probe by multiple scattering (Sect. 5.3.3). Resolutions of 10–50 nm can be achieved in STEM mode of TEM; this is sufficient for measuring segregation and composition profiles at grain boundaries, for example. [9.31, 32]. When the counting time is long, there is a

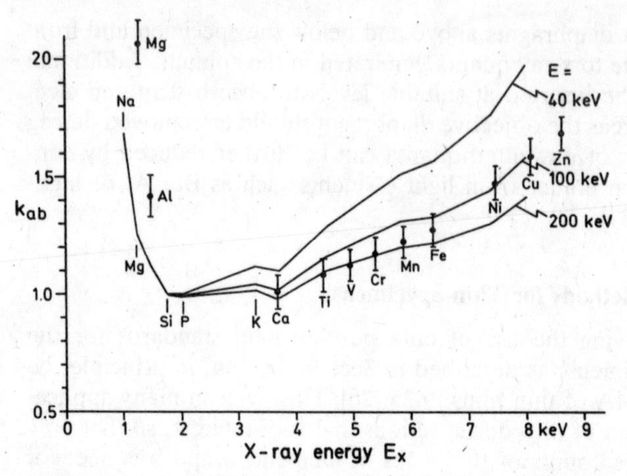

Fig. 9.5. Ratio k_{ab} relative to Si as a function of x-ray quantum energy E_x of the $K\alpha$-lines, measured at 100 keV and calculated for 40, 100 and 200 keV [9.18]

danger of specimen- or electron-probe drift and this can limit the number of spots in a line scan [9.33].

The Cliff-Lorimer ratio method cannot be applied to organic specimens because the characteristic peak of the carbon matrix has such a low energy that the quanta are absorbed in the Be window of EDS; even with WDS, is it not practical to compare the counts of an element with those of C $K\alpha$ since the x-ray fluorescence yield φ_K is poor and the absorption inside the specimen is strong. The Hall method [9.34–36] suggested that the counts N_B of the continuous background should be used to provide a signal proportional to the mass inside the irradiated volume and to eliminate the local mass-thickness, which can show large variations in biological specimens, especially in freeze-dried preparations.

For this method the relative weight fraction of element a becomes

$$c_a = k_H \frac{N_a}{N_B} \tag{9.14}$$

The constant k_H can be measured by using a standard of similar composition and having a known concentration $c_{a,s}$:

$$c_a = \frac{N_a/N_b}{N_{a,s}/N_{b,s}} c_{a,s} . \tag{9.15}$$

The thicknesses of the specimen and the standard need not be known. The only requirements are that both should be "thin" and that the analytical conditions should be the same for specimen and standard. The energy window of the continuum should be near the characteristic peak to achieve similar absorption conditions. Known concentrations of elements can be embedded in resins as salts or organo-metallic compounds for comparison with thin embedded sections, see e.g. [9.37, 38]. Similar standards can be used for freeze-dried or frozen hydrated specimens. However, a severe problem is the radiation damage suffered by biological specimens, especially the loss of mass of the organic material (Sect. 10.2.2) [9.36, 39]; XRMA needs a high current density and most of the mass is lost in a few seconds. Due to the non-uniform composition, the mass loss can vary locally, thereby perturbing the standardization based on the Hall method.

9.1.5 Counting Statistics and Sensitivity

The sensitivity of XRMA is limited by the counting statistics. If a quantity that is subject to statistical variations is measured n times, to get a set of values N_i $(i = 1, \dots, n)$ with mean value $\overline{N}$, the standard deviation σ is given by

$$\sigma^2 = \frac{1}{n-1} \sum_{i=1}^{n} (N_i - \overline{N})^2 \tag{9.16}$$

for Poisson statistics and large n,

$$\sigma^2 = \overline{N} \tag{9.17}$$

There is a probability of 68.3% that a measured value N_i will lie in the confidence interval $\overline{N} \pm \sigma$ and a probability of 95% that it lies in $\overline{N} \pm 2\sigma$.

The analytical sensitivity and the minimum mass fraction will now be estimated as examples.

The analytical sensitivity is the smallest difference $\Delta c = c_1 - c_2$ of concentrations that can be detected. $\overline{N}_1$ and $\overline{N}_2$ denote the expected mean values of counts for the two concentration. We assume that $\overline{N}_1 \simeq \overline{N}_2 = \overline{N} \gg \overline{N}_B$, where $\overline{N}_B$ corresponds to the background. The two values $\overline{N}_1$ and $\overline{N}_2$ are significantly different at the 95% confidence level if

$$\overline{N}_1 - \overline{N}_2 \geq 2\sqrt{\sigma_1^2 + \sigma_2^2} \simeq 2\sqrt{2\overline{N}} \tag{9.18}$$

and the analytical sensitivity becomes [9.40]

$$\frac{\Delta c}{c} = \frac{\overline{N}_1 - \overline{N}_2}{\overline{N}} \simeq 3\overline{N}^{-1/2} \tag{9.19}$$

Thus a sensitivity $\Delta c/c$ of 1% requires $\overline{N} \geq 10^5$ counts. The mean value $\overline{N}$ is proportional to the product of electron-probe current and counting time, but this is limited by radiation damage and contamination: the counting time cannot exceed about 10 min.

If the concentration of an element is as low as about 1 wt.% $\overline{N}$ is no longer much larger than $\overline{N}_B$ if EDS is being used and it ceases to be clear whether an element is present in the sample. The question then arises, what minimum-detectable mass fraction (MMF) will be detectable within a certain confidence interval. Equation (9.18) becomes

$$\overline{N} - \overline{N}_B \geq 2\sqrt{\sigma^2 + \sigma_B^2} \simeq 2\sqrt{\overline{N} + \overline{N}_B} \,. \tag{9.20}$$

When a standard with concentration c_s of the element being investigated is used, the MMF becomes, with $\overline{N} \simeq \overline{N}_B$

$$\text{MMF} = \frac{\overline{N} - \overline{N}_B}{\overline{N}_s - \overline{N}_{B,s}} c_s \simeq \frac{c_s}{\overline{N}_s - \overline{N}_{B,s}} 3\sqrt{\overline{N}_B} \,. \tag{9.21}$$

Estimates based on (9.21) and especially comparisons of EDS, WDS and energy-loss spectroscopy (ELS), have the disadvantage that certain assumptions have been made that are not true in all practical cases – assumptions about the magnitude $\overline{N}_B$ of the background as an example. For example, a calculated value of 3–5% for MMF is found for elements in a 100 nm thick Si foil irradiated at $E = 100$ keV with $j = 20$ A cm^{-2}, a spot size of 10 nm diameter and a counting time of 100 s, whereas the minimum-detectable mass is of the order of 0.5–1 × 10^{-19} g [9.41].

Typical values for WDS in an x-ray microanalyser with electron probe diameter of 1 μm are $\simeq$ 100 ppm for MMF and $\simeq 10^{-15}$ g for the minimum-detectable mass. The MMF is thus more favourable in an x-ray microanalyser but the minimum-detectable mass is greater.

9.2 Energy-Loss Spectroscopy and Energy-Selecting Microscopy

9.2.1 Electron Spectrometers and Filters

Electron spectrometers of high resolution are needed to resolve the relatively low energy losses between $\Delta E = 0$–2000 eV. The energy spread $\Delta E \simeq 1$–2 eV of a thermionic electron gun is normally adequate to record a spectrum of energy losses larger than a few eV. Energy spectrometers for elemental analysis by inner-shell ionisation need a resolution of only 5–20 eV, though in some cases fine structures can be seen in the spectrum if the resolution is better. A higher resolution needs a field-emission gun ($\Delta E \simeq 0.2$ eV) or the electron beam has to be monochromatized (e.g., by a Wien filter). The most-important types of electron spectrometers and filters will now be described (see also reviews [9.42–44]).

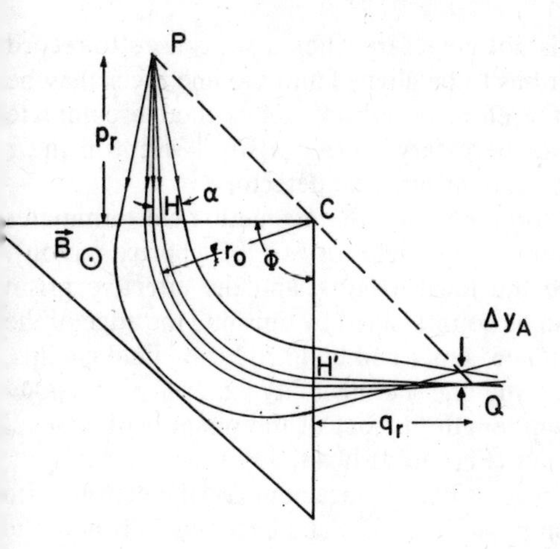

Fig. 9.6. Radial focusing property of a 90° electron prism with second-order aberration Δy_A for large α

Electron-Prism Spectrometer. Electron prisms consist of transverse agnetic or electric fields. In a transverse magnetic field, the radius of the ajectories is proportional to the momentum (2.13). In a radial electric field e radius is proportional to the electron energy. A spectrometer whould ave a high resolution $\Delta E/E$ for two energies E and $E - \Delta E$ (ΔE is the nergy loss) and a large angle of acceptance, which means a large entrance perture α. The two aims can be reconciled only by designing the sector field give additional focusing and by correcting the second-order aberrations. A oint source P will then be imaged by the spectrometer as a sharp image oint Q, at which a slit can be placed in front of an electron detector (Fig. 6). Details of the electron-optical theory of sector fields are to be found in .44–46].

The central beam in a magnetic sector field (Fig. 9.6) is bent into the form f a circle of radius $r_0 = mv/eB$ with centre C. If the incident and exit direc- ons are normal to the edges of the sector field with sector angle ϕ, focusing ccurs for small α (paraxial rays), and the points P, C and Q are collinear Barber's rule). The distances p_r = PH and q_r = H'Q (focal lengths) are ven by $p_r = q_r = r_0/\tan(\phi/2)$ for a symmetric prism. There is no focusing of e momentum components in the z direction parallel to the magnetic field.

Tilting the magnet edge ($\varepsilon \neq 0$) (Fig. 9.7 a) has the same effect (in first rder) as adding a quadrupole lens with focal length $f = \pm r_0 \cot \varepsilon$ for compo- ents of the momenta in the radial (+) and axial (−) directions, respectively .47]. The focal lengths are, therefore,

$$\frac{1}{p_r} = \frac{1}{q_r} = \frac{1}{r_0}\left(\tan\frac{\phi}{2} - \tan\varepsilon\right); \qquad \frac{1}{p_z} = \frac{1}{q_z} = \frac{1}{r_0}\tan\varepsilon , \tag{9.22}$$

o-called *double-stigmatic focusing* can be obtained if $p_r = q_r = p_z = q_z = 2\,r_0$ nd $\tan(\phi/2) = 2\tan\varepsilon$. For $\phi = 90°$ gives $\tan\varepsilon = 0.5$ or $\varepsilon = 26.5°$ (Fig. 9.7 a).

Focusing in the z direction is not necessary when a slit is used to record the spectrum. However, the slit has to be aligned and the line focus may be curved. Double focusing will therefore be advantageous though complete focusing in the z direction is not necessary; indeed a small width in the z direction can be desirable to avoid damaging the detector.

It has been assumed in the foregoing that the magnetic field terminates abruptly at the edges (sharp-cutoff-fringe-field, or SCOFF approximation). The real fringe fields influence the focal lengths, and the effective prism angle ϕ becomes larger. This can be counteracted by finishing the edge of the magnetic pole-piece plates with a 45° taper and by introducing field clamps, which are constructed from the same high-permeability material as the pole-pieces and placed at half the gap length in front of the prism, with a small hole for the incident and exit rays (Fig. 9.7 a) [9.48, 49].

The *dispersion* $\Delta y/\Delta E$ is defined as the displacement Δy of electrons with energy $E - \Delta E$ in the dispersion plane (Fig. 9.7) and becomes (p is here the electron momentum)

$$\frac{\Delta y}{\Delta p} = \frac{4r_0}{p}\,; \qquad \frac{\Delta y}{\Delta E} \doteq \frac{2r_0}{E}\,\frac{1 + E/E_0}{1 + E/2E_0} \tag{9.23}$$

for a symmetric prism with $\phi = 90°$. We find $\Delta y/\Delta E = 1$ μm/eV for $E = 100$ keV and $r_0 = 5$ cm.

On increasing α, a second-order angular aberration $\Delta y_{\mathrm{A}} = B\alpha^2$ becomes apparent (Fig. 9.6). This aberration can be corrected in radial direction by curving the edges of the magnet (Fig. 9.7 b) [9.46, 48, 50, 51]. The total width Δs of the zone occupied by the zero-loss electrons in the dispersion plane is determined by the size of the image of the entrance slit or diaphragm, which is blurred, owing to the energy width of the electron gun and also by the second-order aberrations. This width and the dispersion (9.23) limit the *resolution* $\Delta E_{\mathrm{r}} = \Delta s/(\Delta y/\Delta E)$. With a thermionic electron gun at 100 keV, a resolution of 1–5 eV is obtainable.

A magnetic prism spectrometer is normally situated below the viewing screen of TEM. The lens system can be used to adapt optimally the different operating modes of TEM to the spectrometer [9.52, 53]. Thus, any corrections needed to focus the beam on the exit slit can be made by means of a pre-spectrometer lens or a quadrupole lens placed between spectrometer and exit slit. In the TEM-imaging mode, the objective aperture α_0 is reduced to $\alpha \doteq \alpha_0/M$ by the magnification M. This allows us to work with $\alpha_0 \simeq 10$ mrad which is important in connection with inner-shell ionisations (Sect. 9.2.4).

The resolution of a prism spectrometer can be increased by decelerating the electrons in a retarding field to an energy of the order of 1 keV. Magnetic [9.54–56] and electrostatic prisms [9.57] are used in this way. Electrostatic prisms consist of radial electric fields between concentric cylindrical or spherical electrodes. An electrostatic prism spectrometer without retardation is described in [9.58].

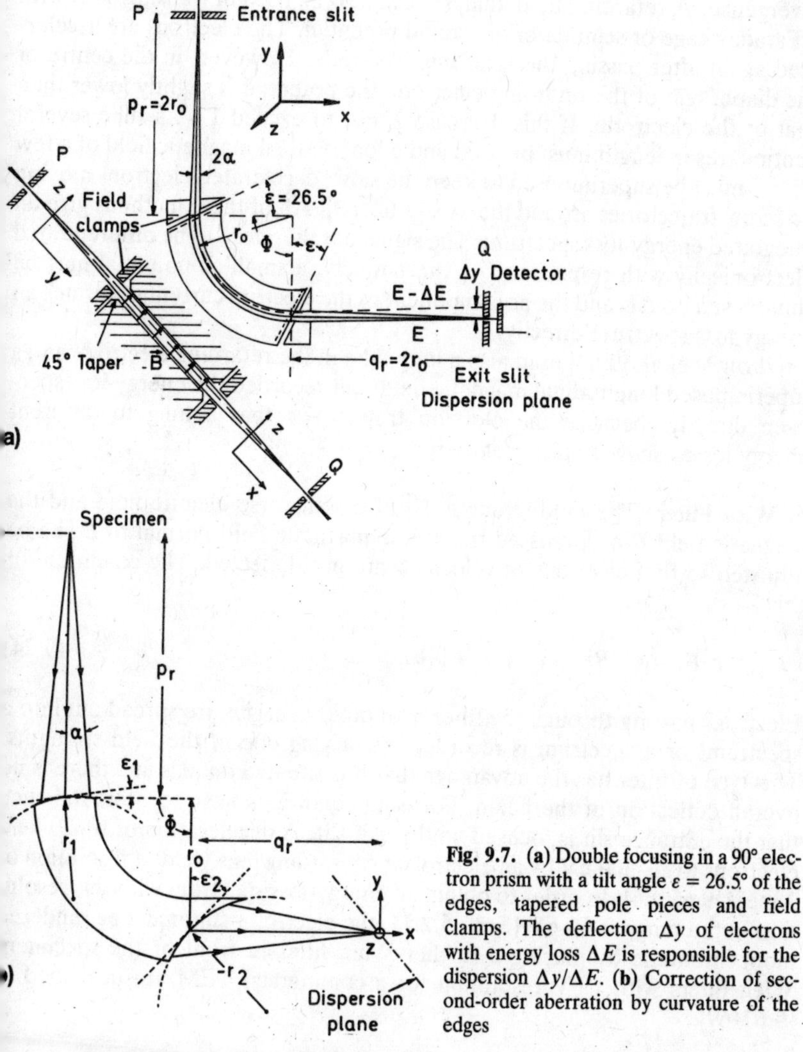

Fig. 9.7. (**a**) Double focusing in a 90° electron prism with a tilt angle $\varepsilon = 26.5°$ of the edges, tapered pole pieces and field clamps. The deflection Δy of electrons with energy loss ΔE is responsible for the dispersion $\Delta y/\Delta E$. (**b**) Correction of second-order aberration by curvature of the edges

) **Retarding-Field Spectrometers.** A retarding field analyser consists of a iaphragm held at a potential $-U + \Delta U$ (where $-$U is the cathode potenal) [9.59–61]. Only electrons with energy losses $\Delta E < e\Delta U$ can pass the otential barrier created by the retarding electrode. Different methods can e used for signal detection. A Faraday cage may be mounted behind the iaphragm and held at a high voltage; alternatively a Faraday cage with two oles is mounted in front of the retarding electrode and transmits the primary eam while collecting the reflected electrons by virtue of their larger angular divergence. A retarding field analyser can also be used in transmission, with a Faraday cage or scintillator at ground potential. The electrons are accelerated again after passing the retarding electrode. However, in the centre of the diaphragm of the retarding electrode the potential is slightly lower than that of the electrode. If this decrease is not to exceed 1 V, a tube several centimetres in length must be used and a longitudinal magnetic field of a few 10^{-2} T must be superimposed to keep the slow, decelerated electrons moving on screw trajectories around the axis [9.62, 63]. All of these methods give an integrated energy-loss spectrum. The signal has therefore to be differentiated electronically with respect to U. Alternatively, a small ac potential may be superposed on ΔU and the ac component of the electron current generates an energy loss spectrum directly.

Browne et al. [9.64] used also a long tube in the retarding electrode and a superimposed longitudinal magnetic field, but recorded the energy-loss spectrum directly, because the electron trajectories that belong to different energy losses show a spiral dispersion.

c) Wien Filter. The field strength $|\boldsymbol{E}|$ of a transverse electric field and the magnetic field $\boldsymbol{B}$ of a crossed transverse magnetic field normal to $\boldsymbol{E}$ can be adjusted so that electrons of velocity v are not deflected. The condition for this is

$$F = e\,|\boldsymbol{E}| = ev\,|\boldsymbol{B}| \quad \rightarrow \quad v = |\boldsymbol{E}|/|\boldsymbol{B}| \,. \tag{9.24}$$

Electrons passing through the filter with other energies are spread out into a spectrum, or a spectrum is recorded by varying one of the field strengths This type of filter has the advantage that it is situated on axis and there is no overall deflection of the beam. Focusing conditions have to be found such that the entrance slit is focused on the exit slit. A deceleration of 10–20 keV electrons to 20–300 eV by an electrostatic retarding lens yields a resolution o 2 meV [9.65, 66]. In order to obtain an energy-loss spectrum with this resolu tion in the range 0–10 eV (Sect. 5.2.1), the electron beam must be rendered monochromatic by placing a further Wien filter in front of the specimen Wien filters with 1 eV resolution for a commercial TEM are described i [9.67–69].

d) Use of Electrostatic and Magnetic Lenses for Electron Spectrometry Figure 9.8 shows the principle of a Möllenstedt electrostatic energy analyse [9.70–74]. If electrons enter an electrostatic lens at an off-axis distance r, th trajectories depend in a complicated manner on that distance (Fig. 9.8 b The approximately linear deflection region between points 2 and 3 is th interesting range of r for energy analysis. If the electron beam is limited by circular diaphragm in the case of a lens with rotational symmetry or by a sl with a cylindrical lens, the deflection depends only on the distance r but als and very sensitively, on electron energy, so that the image of the circul

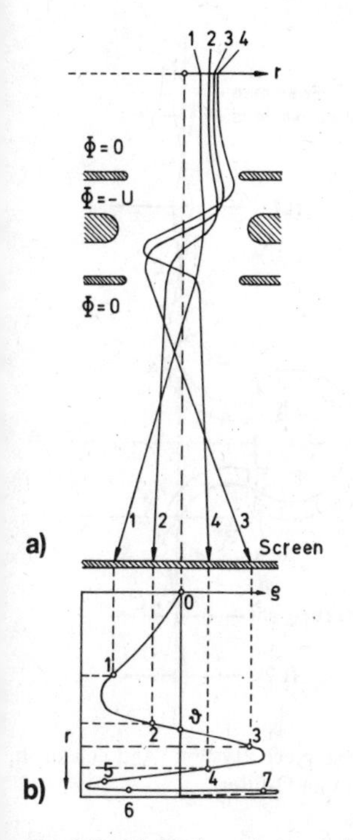

Fig. 9.8. (**a**) Trajectories and deflection ϱ of electron rays in an electrostatic lens (Möllenstadt analyser) for different radial distances r, (**b**) dependence of s on r [9.70]

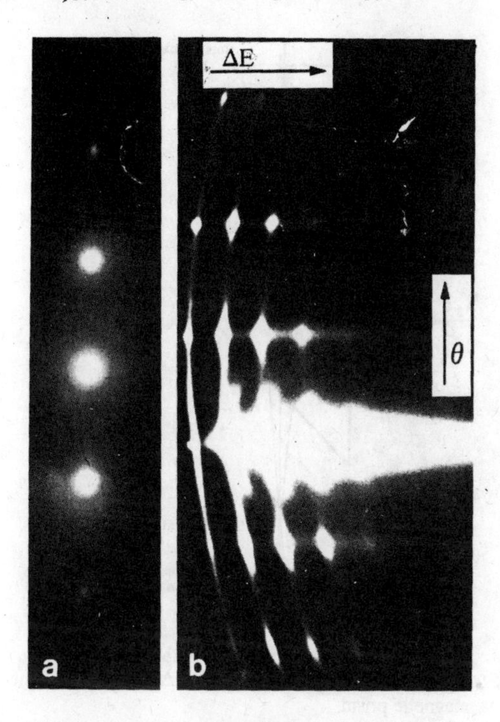

Fig. 9.9. Energy-loss spectrum (**b**) of 100 keV electrons with the Bragg-reflection row (**a**) across the entrance slit of a Möllenstedt energy analyser. The dispersion of the plasmon losses can be seen as a stronger curvature of the multiple plasmon losses near the primary beam [9.42]

diaphragm or slit is spread out into an energy-loss spectrum. Figure 9.9 b shows the energy-loss spectrum of an Al foil with a row of Bragg reflections (Fig. 9.9 a) imaged on the entrance slit [9.42].

In principle, a magnetic lens can also be used for this purpose [9.75]. However, the smaller chromatic aberration of a magnetic lens reduces the resolution by one order of magnitude. If the selector diaphragm in front of the intermediate lens of TEM is shifted off axis and the caustic ring is magnified by means of the projector lens, it is possible to achieve a resolution of 1–2 eV [9.76, 77]; however, a specially designed intermediate lens is needed for this.

A magnetic cylindrical lens capable of providing a resolution of 1 eV has been proposed by *Ichinokawa* [9.78]. The Möllenstedt electrostatic filter lens is limited to energies of 50–100 keV by problems of electrical breakdown, whereas the Ichinokawa filter can also be used in HVEM [9.79, 80].

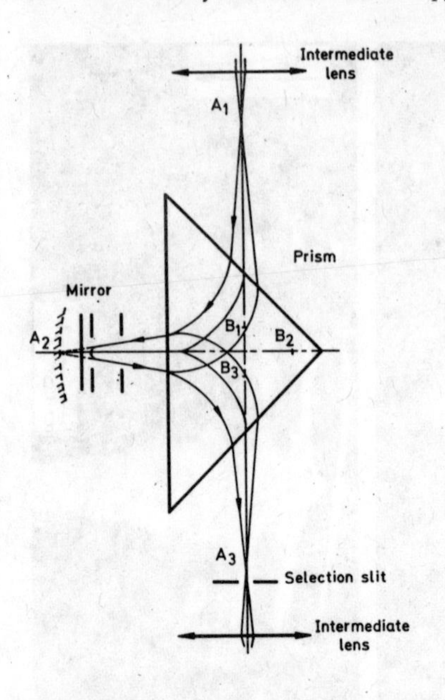

Fig. 9.10. Castaing-Henry filter with a retarding electrode at a potential $-(U + \Delta U)$ and a magnetic prism

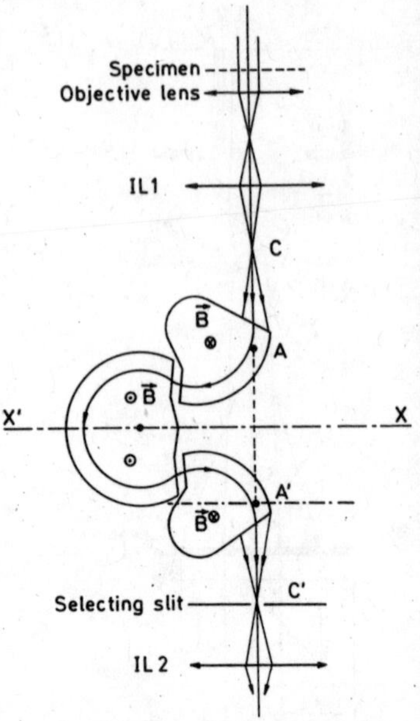

Fig. 9.11. Pole-piece system and electron trajectories in an Ω-filter

e) Castaing-Henry and Ω-Filter. *Castaing* and *Henry* [9.81, 82] combined a retarding-field electrode (electron mirror) and a magnetic prism, as shown in Fig. 9.10. A subsequent lens can be used to image either the dispersion plane A_3 that contains the energy-loss spectrum or the virtual intermediate image at B_3. An energy-filtered image 1 μm in diameter or a filtered diffraction pattern can be recorded by placing an energy-selecting slit across the energy spectrum with a resolution of 1 eV.

An analogous magnetic filter can be designed by use of four 90° prisms [9.83, 84]. This *Ω* filter (Fig. 9.11) can be introduced between two lenses of a TEM and can also be used in HVEM [9.85, 86].

The Castaing-Henry and the *Ω* filter share with the Wien filter the advantage of conserving the optic axis of the microscope: there is no net deflection of the main beam.

9.2.2 The Recording and Analysis of Energy-Loss Spectra

An energy-loss spectrum can be recorded on a photographic emulsion, which has the advantage of providing a simultaneous record of all energy losses; this is counterbalanced by the inconvenience of the delayed readout, caused by the time needed for photographic development and densitometry. The dispersion of the spectrometer has to be at least of the order of 20 μm/eV, because the resolution of photographic emulsions is limited by electron diffusion (Sect. 4.6.2). A further problem is the limited dynamic range, that is the limited range of linearity between density and incident charge density. Thus K-shell losses can be recorded only by over-exposure of the plasmon region.

For magnetic-prism spectrometers, a scintillator-photomultiplier combination or a semiconductor detector is normally used for recording. The spectrum is scanned over a slit of variable width in the dispersion plane, by use of scanning coils behind the spectrometer. This system is nearly ideal in that it has low noise and a high recording speed. A scintillator can be used either in the single-electron counting mode or in the analogue-signal mode. The P-46 or P-47 scintillators (cerium-doped yttrium aluminium garnet) have a time constant of 75 ns and pulse rates up to 10^7 are possible. The scintillation pulses of 100 keV electrons are large enough for it to be possible to separate single pulses from the noisy background by means of a discriminator. A disadvantage is that only one energy window is recorded at a time. This mplies that a much-longer recording time ($\simeq 100\times$) is needed when a detecor system with simultaneous recording of all energy losses is employed. This nay be unacceptable for beam-sensitive organic materials. The high-intensity ow-loss part of the spectrum (plasmon losses) can be scanned in a shorter ime or with a reduced beam current. An ideal detector would consist of a inear array of photon or electron detectors giving truly parallel detection, ike a photographic emulsion but with the advantage of electronic readout. A inear photodiode array might be used, for example [9.87, 88]. A disadvanage is the large dynamic range of the spectrum (10^4–10^7), and bombardment f a photodiode array can cause an irreversible increase in the diode dark urrent. A scintillator-coupled SIT vidicon is sensitive enough to detect two igh-energy electrons with a spatial resolution of 100 μm and an energy esolution of 5 eV [9.89].

The advantages of energy-loss spectroscopy will become fully apparent nly when such simultaneous recording of the spectrum becomes possible. 'he gain due to the larger probability of detecting an inner shell ionisation by LS is lost if the whole spectrum cannot be recorded simultaneously because ne charge density needed for recording a spectrum has to be multiplied by ne number of energy windows covered, whereas x-ray microanalysis by nergy-dispersive spectrometry possesses to the full the advantage of simulneous recording.

The spectrum can be recorded directly on a TV screen or a pen recorder. igital storage in a multi-channel analyser allows the signal-averaging mode ith a repeated scan to be used instead of a slow scan, which has the advange of averaging over a gradual drift. Storage in a mini-computer, as for ray microanalysis, permits us to apply all of the procedures required for uantitative analysis [9.90, 91]: edge identification, background subtraction, rrections for multiple scattering and for the aperture and energy windows ed (Sect. 9.2.4), deconvolution of the multiple-loss spectrum [9.92, 93], calculation of optical constants using the Kramers-Kronig relation [9.93] extraction of the density function from the extended fine structure (EXAFS of inner-shell ionisations (Sect. 5.2.4).

9.2.3 Information from Low-Energy Losses

It was shown in Sect. 5.2.3 that the energy-loss spectrum in the 0–50 e region can be described by the dielectric theory. One application of ELS i determination of optical constants in the ultraviolet [9.94]. The loss spectr of different substances show characteristic differences. However, the spectr are not so specific that they can be used for elemental analysis. Furthermore the spectra contain multiple losses and surface-plasmon losses, which depen on the foil thickness. Nevertheless ELS can be used in some cases as a additional analytical tool for SiC [9.95], for example, or glass [9.96] c organic molecules [9.97, 98].

The displacement of sharp plasmon losses can be used to give local me surements of the electron concentration N because the plasmon energ $\Delta E = \hbar\omega_p$ is proportional to $N^{1/2}$, (5.65). Figure 9.12 shows, as an exampl the shift of the plasmon energy in an Al-Mg alloy [9.99] for which the diffe ent phases can be identified by their energy losses. The plasmon loss Δ depends linearly on concentration in the α, γ and δ phases, which allows t local concentrations to be measured with a spatial resolution of the order 10 nm. Thus, the variation of Mg concentration was measured near larg angle boundaries after quenching of an Al-7at% Mg alloy [9.100], and t variation of Cu in the Al-rich phase near $CuAl_2$ precipitates in an eutectic A $CuAl_2$ alloy [9.101]. The plasmon losses of Al-Zn alloys have also be investigated [9.102].

The position of a plasmon loss can be determined with an accuracy 0.1 eV ($\simeq 1/40$ of the FWHM) [9.103]. However, the background of t

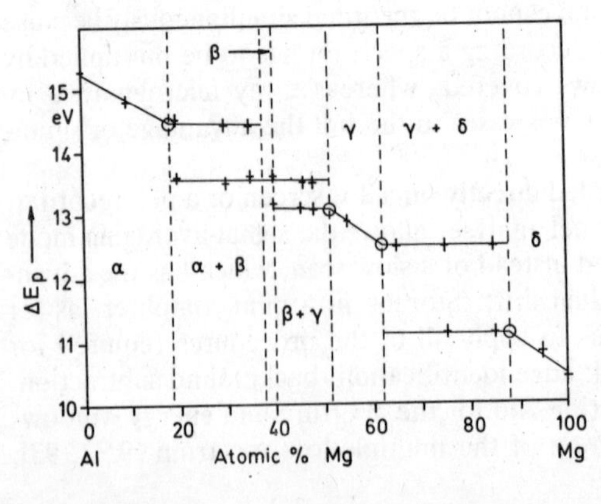

Fig. 9.12. Variation of the plas loss energy ΔE_p with compositio the Al-Mg system [9.99]

energy-loss spectrum caused by contamination can produce spurious shifts of the same order [9.104], and strains in inhomogeneous alloys can cause shifts relative to calibrations made with a homogeneous alloy [9.105]. Shifts of ± 0.1 eV have been observed at distances ± 20 nm from a dislocation [9.106].

Even when the high-energy losses are used for elemental analysis, it is important to record the low-loss region so that the foil thickness can be determined and because the inner-shell losses are convoluted with this loss spectrum. The thickness may be obtained either from the exponential decrease of the zero loss, which is proportional to $\exp[-N_A \varrho t \sigma(\alpha)/A]$, or from the ratio of the intensities of the first plasmon and zero loss, which is proportional to t/Λ_p (5.69). If the amplitude of the first plasmon loss is smaller than that of the zero loss, then $t \leq \Lambda_p$. For quantitative analysis of inner-shell ionisations, the thickness should not be greater than the mean free path Λ_p of the plasmon losses.

Another piece of information that can be extracted from the low-loss region is the ratio of the inelastic to the elastic cross-sections, which depends on the atomic number [Table 5.3 and Eq. (5.58)]. This ratio can also be used to investigate radiation damage in organic substances [9.107].

9.2.4 Elemental Analysis by Inner-Shell Ionisations

The basic laws of inner-shell ionisation are summarized in Sect. 5.2.4. Energy-loss spectroscopy (ELS) can be used for low-Z elemental analysis in the range $0 < \Delta E < 2$ keV [9.97, 108, 109]. Compared to x-ray microanalysis, ELS has following advantages:

1) Each inner-shell ionisation results in an electron-energy loss whereas the number of emitted x-ray quanta is obtained by multiplying by the fluorescence yield ω, which is very small for low Z; for example, $\omega_K = 2.3 \times 10^{-3}$ for C ($E_K = 280$ eV) and $\omega_K = 2.2 \times 10^{-2}$ for Na ($E_K = 1.1$ keV).

2) The angular distribution of the inelastically scattered electrons is concentrated within a cone of semi-angle $\theta_E = \Delta E/2E$ (Sect. 5.2.4) and a large fraction of the electrons that contain information about an inner-shell ionisation can be collected and analysed by an electron spectrometer. The characteristic x-rays are, on the contrary, emitted isotropically and the solid angle of collection $\Delta\Omega_c$ is small. Only a tiny fraction, $\Delta\Omega_c/4\pi \simeq 10^{-4} - 10^{-5}$ for WDS and $10^{-2} - 10^{-3}$ for EDS, can be detected.

3) The low-energy x-ray quanta $E_x < 1$ keV are absorbed in the Be-window of a Si(Li) detector unless a window-less system is employed. In principle, it is possible to analyse elements down to Be with WDS and a proportional counter with an ultrathin window but the values of ω and $\Delta\Omega$ are exceedingly small.

For ELS, the formula (9.12) for x-ray microanalysis has to be modified to

$$N_a = \sigma_a c_a \frac{N_A}{A_a} \eta(\alpha, \Delta E_w) x n_0 = n_a \sigma_a \eta(\alpha, \Delta E_w) n_0 , \tag{9.25}$$

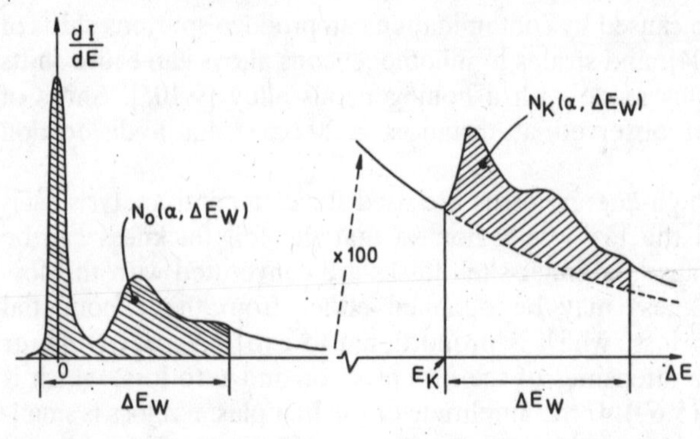

Fig. 9.13. Quantitative microanalysis by electron-energy-loss spectroscopy with the signals N_0 $(\alpha, \Delta E_w)$ and N_K $(\alpha, \Delta E_w)$; α denotes the collection aperture of the spectrometer and ΔE_w the width of the energy window

where n_a is the number of atoms of element a per unit area, η is the collection efficiency, which depends on α, the collection angle of the spectrometer, and on ΔE_w, the width of the energy window.

For quantitative EDS, the numbers of electrons $N_0(\alpha, \Delta E_w)$ and $N_K(\alpha, \Delta E_w)$ in the ranges $0 \leq \Delta E \leq \Delta E_w$ and $E_K \leq \Delta E \leq E_K + \Delta E_w$, respectively, are measured (Fig. 9.13). Setting $N_0(\alpha, \Delta E_w) \simeq n_0$ will cause little error because most of the electrons are concentrated in the low-energy window and any decrease caused by multiple scattering and by the use of a limited collecting angle acts on both N_0 and N_K.

The background just preceding the excitation edge $\Delta E = E_K$ has to be extrapolated for $\Delta E \geq E_K$ so that N_K is the number of electrons with K losses only. Because the background decreases as ΔE^{-n} (n = 3.5–4.5, depending on the magnitude of α), double-logarithmic plot gives a straight line (Fig. 5.10), which can easily be extrapolated beyond $\Delta E \geq E_K$ [9.110, 111]. Though it is possible to estimate the total number of energy losses and to measure the total cross-section σ_K by using large values of ΔE_w and α (see references in Sect. 5.2.4), only a small window, of the order of ΔE_w = 50 eV, is needed for quantitative analysis; this contains a fraction

$$\eta_w = 1 - \left(1 + \frac{\Delta E_w}{E_K}\right)^{1-n} \simeq (n-1)\frac{\Delta E_w}{E_K} + \dots \qquad \text{for} \quad E_w \ll E_K \tag{9.26}$$

of all K-loss electrons [9.112].

The angular distribution of the K-loss electrons is concentrated within a cone of semiangle $\theta_E = E_K/2E$; a spectrometer with a collecting angle α collects the fraction

$$\eta_a = \frac{\ln(1 + \alpha^2/\theta_E^2)}{\ln(2/\theta_E)} \tag{9.27}$$

for apertures in the range $\theta_E \ll \alpha < \theta_f = (E_K/E)^{1/2}$ (see Sect. 5.2.4 for the meaning of θ_f).

The efficiency in (9.25) becomes

$$\eta(\alpha, \Delta E_w) = \eta_w \eta_a . \tag{9.28}$$

A fraction of 0.1–0.5 can readily be collected under optimum conditions. This is orders of magnitude more than the fraction of x-rays collected.

In these calculations, multiple elastic and inelastic scattering in the foil has not been considered. Increasing the foil thickness broadens the angular distribution because the mean-free-path lengths Λ_{el} and Λ_p for elastic scattering and plasmon losses, respectively, are orders of magnitude smaller than Λ_K for K-shell ionisation. If these scattering processes occur before and/or after a K-shell ionisation, therefore, the number of electrons with K-shell losses inside the collection angle α decreases; this is partly counterbalanced by the initial increase proportional to the foil thickness (9.25). A maximum of K-loss electrons inside an aperture α is observed for $t \simeq \Lambda_p$ because for low-Z elements $\Lambda_p < \Lambda_{el}$ [9.112, 113]. This means that ELS is firmly confined to thin films, and ELS of thicker films is possible only in a HVEM, the thickness being proportional to the increase of Λ_p with energy (Fig. 5.15). A further advantage of ELS in HVEM is that θ_E decreases with increasing E [9.114]. Observation of the ratio of the zero and plasmon-loss intensities is important to ensure that the limits of the useful thickness range have not been exceeded. Multiple inelastic scattering also results in a convolution of the excitation edge with the loss spectrum in the low-loss range (plasmon losses) and the number of K-loss electrons inside the window ΔE_w falls.

As in x-ray microanalysis (Sect. 9.1.4), a ratio method can be applied [9.115] if the relative abundance n_a/n_b of two elements is required, and (9.25) gives

$$\frac{n_a}{n_b} = \frac{N_a}{N_b} \frac{\sigma_b \eta_b(\alpha, \Delta E_w)}{\sigma_a \eta_a(\alpha, \Delta E_w)} \tag{9.29}$$

Just as x-ray emission depends on crystal orientation (Sect. 9.1.3), so too does the number of K-loss electrons depend on the orientation of the incident electron beam relative to the lattice planes, being proportional to the square of the Bloch-wave amplitude at the nuclei [9.116, 117].

The background below the excitation edge increases with α, ΔE_w and t, and the signal of the K-shell excitation decreases with decreasing concentration c_a and saturates when ΔE_w and α are increased sufficiently (Fig. 9.14) [9.110, 113]. The conditions for maximum signal-to-background ratio do not coincide with those for optimum signal-to-noise ratio, as shown by an analysis of oxygen and boron in B_2O_3, for example [9.118]. A small collection aperture α is desirable for a large signal-to-background ratio (Fig. 9.15 a) because the K-loss electrons are concentrated at low scattering angles whereas the background is spread over larger angles. The use of a low value of α, however, decreases the number of electrons detected and increases the

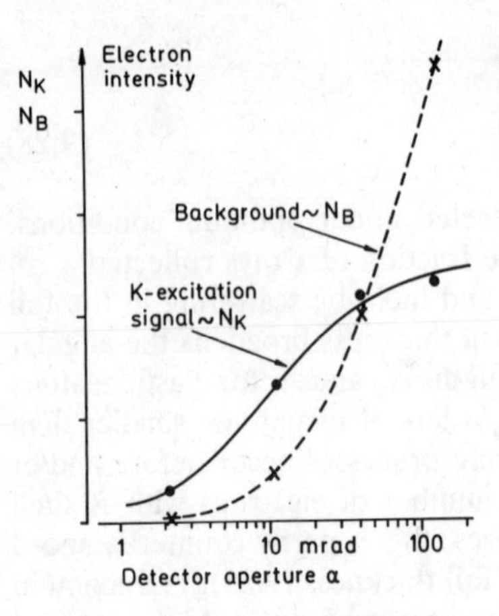

Fig. 9.14. K-loss signal and background in carbon (E = 80 keV) as a function of detector aperture [9.110]

statistical noise. The signal-to-noise ratio reaches a maximum for α = 10–20 mrad (Fig. 9.15 b); this value of the spectrometer collection aperture can just be used without too strong a decrease of the energy resolution by aberrations.

The smallest-detectable mass has been estimated to be between 10^{-22} and 10^{-20} g and the lowest concentration to be 10^{-4}–10^{-5} [9.112, 119]. Thus, a signal from the M-ionisation loss of Fe can be recorded from a single ferritin

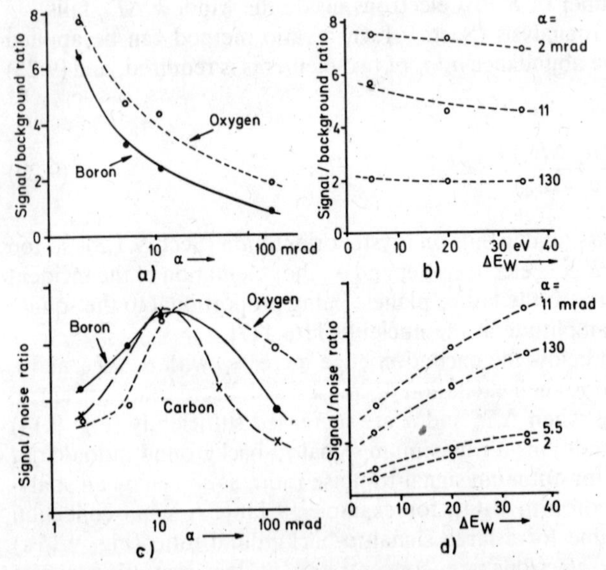

Fig. 9.15 a–d. Signal-to-background ratios and signal-to-noise ratios of the energy-loss spectra of oxygen and boron in B_2O_3 as a function of detector aperture α and width ΔE_w of the energy window [9.118]

molecule with an electron probe 50 nm in diameter; it indicates the presence of about 5×10^{-19} g Fe, corresponding to some 5000 iron atoms.

9.2.5 Energy-Selecting Microscopy

Whereas in energy-loss spectroscopy (ELS) the loss spectrum is recorded with a stationary electron spot or beam, the aim of energy-selecting microscopy (ESM) is to take an energy-filtered micrograph by selecting only the zero-loss electrons or those, that correspond to a plasmon loss or to an inner-shell ionisation by means of a fixed energy window. The following methods can be employed:

1) the retarding field of an electrostatic lens [9.120] can be used as a high-pass energy filter. However, the central electrode must have a small bore (0.1–0.3 mm), to keep the potential drop between the electrode and the axis small.
2) A filter lens with a Möllenstedt analyser and a selecting slit in the dispersion plane [9.121], a Castaing-Henry mirror-prism system or an Ω-filter may be employed.
3) Scanning transmission electron microscope may be equipped with an energy-loss spectrometer (Fig. 4.25).
4) An image or diffraction pattern may be scanned across the entrance diaphragm of an electron spectrometer. For an image, this method has the disadvantage that a very large irradiation charge density $q = j\tau$ is required, which makes it unusable for radiation-sensitive specimens. However, for robust specimens, it can be used to compare a directly energy-filtered image and a conventional one. When a diffraction pattern is being filtered, this technique is known as the Grigson mode (Sect. 9.3.4) and the deflection system consists of Grigson coils.
5) An image [9.122, 123] or a diffraction pattern (Fig. 9.9) can be located on the entrance slit of a spectrometer (y direction); recording the spectrum in the x direction is an intermediate method between ELS and ESM.

Energy-selecting microscopy is typically used for the following applications:

a) Increase of Contrast and Resolution. The chromatic aberration and the energy losses inside the specimen limit the resolution of organic specimens because the inelastic cross-section is three times larger than the elastic one (Sect. 5.2.2). For membranes, for example, both the contrast and the resolution are reduced. As shown for $E = 60$ keV [9.124], energy filtering with the Castaing-Henry system results in a remarkable increase of contrast and resolution for thicknesses $t \simeq 50$–100 nm. For thick specimens studied in the STEM mode, gain of contrast given by energy filtering is discussed in [9.125]. There is no loss of resolution due to chromatic aberration because there is no imaging lens behind the specimen, but the electron-probe broadening caused by multiple scattering limits the resolution (top-bottom effect, Sect. 5.3.3).

Only the elastically scattered electrons convey high-resolution information, as shown with a STEM [9.126]. The elastic signal can resolve single atoms, whereas the inelastic signal gives an image of much-lower resolution because the strongly forward scattering cannot transmit information about high spatial frequencies. It is not obvious that an energy filter offers any advantage in the high-resolution TEM of thin specimens, because the phase contrast of the elastically scattered electrons already dominates over the more diffuse, inelastic part of the image.

b) Material Discrimination in the Low-Loss Region. By forming images with the different plasmon losses, in turn, different phases can be separated and identified. Thus, evaporated Al films have been investigated [9.127]. With an energy window at $\Delta E = 22$ eV, bright areas between the crystals showed the presence of aluminium oxide. Images formed with the first plasmon loss $\Delta E = 15$ eV of Al showed bright areas in the thinner parts of the crystals, whereas with the second plasmon loss $\Delta E = 30$ eV, they occured in the thicker parts. Images of polystyrene spheres likewise show bright areas, which move towards the thicker central region of the sphere with increasing ΔE [9.124]. In an Al-7.6% Zn-2.6%Mg alloy studied by ESM, the η precipitates appeared dark against the matrix with $\Delta E = 14.6 \pm 1$ eV, but when their plasmon loss $\Delta E = 22.5 \pm 2.5$ eV was selected, they appeared bright against a darker matrix [9.128]

c) Elemental Mapping with Inner-Shell Ionisations. Elemental mapping is in widespread use in x-ray microanalysis. The electron beam is scanned over a specimen area in SEM or STEM, and the characteristic x-ray quanta of a particular element are used to modulate the intensity of a TV tube scanned in synchronism (Fig. 9.3 e, f). A disadvantage is that the image is very noisy if x-rays are used and its resolution is poor because a large probe current is necessary. Elemental mapping with ELS can be achieved with a much better signal-to-noise ratio and resolution. The latter is limited only by the non-localisation of inelastic scattering. An energy-selecting window can be set directly beyond an edge excitation ($\Delta E \geq E_I$) and a second one may be set just below the edge if desired ($\Delta E < E_I$). The contrast can be increased and the background reduced by subtracting the latter from the former [9.129]. Another way of distinguishing the contribution of an edge excitation from the background is to scan the energy window by means of deflection coils placed between spectrometer and detector slit (e.g., at 2 kHz); large 2 kHz amplitudes are produced in the presence of an edge excitation, but far from the edge the amplitude is low because the background varies much more slowly. The sensitivity can be increased by filtering the signal with a phase-sensitive (lock-in) amplifier [9.130].

This technique has been applied to a variety of specimens. Concentrations of Be with $\Delta E_K = 115 \pm 5$ eV have been identified in an Al-Be composite material [9.86]. In biology, phosphorus can be demonstrated in the virus membrane bilayer with a resolution of 0.3–0.5 nm and a sensitivity of 2 $\times 10^{-21}$ g ($\simeq$ 150 P atoms), by using the P $L_{2,3}$ edge (ΔE = 150 $\pm$ 7 eV) 9.131].

l) Preservation of Diffraction Contrast After Inelastic Scattering. An impor-ant problem in the interpretation of the diffraction contrast of crystalline pecimens and their lattice defects is the contribution of inelastically scat-ered electrons; this depends on whether the scattering process is an ntraband scattering (preservation) or interband scattering inside the ranches of the dispersion surface (non-preservation of diffraction contrast) Sect. 7.4.1). Experiments without energy filtering showed that the contrast s preserved for low-angle scattering between the Bragg reflections but not or electrons scattered through large angles [9.132, 133]. With energy filter-ng, the preservation of diffraction contrast (edge and bend contours and tacking-fault fringes) after inelastic scattering was demonstrated [9.121, 34–136].

Conversely, little or no diffraction contrast was observed with thermal liffuse scattering, which is mainly interband at low scattering angles [9.123, 35].

.3 Electron-Diffraction Modes

.3.1 Selected-Area Electron Diffraction (SAED)

'he cone of diffracted electrons with an aperture of the order of a few 0 mrad can pass through the small pole-piece bores of the final lenses only if he back focal plane of the objective lens that contains the first diffraction attern is focused on the screen. Figure 4.20 shows the ray diagram of this echnique [9.137–139]. A selector diaphragm of diameter d situated in the ntermediate-image plane (magnification $M \simeq$ 20–50) in front of the inter-ediate or diffraction lens selects an area of the specimen of diameter d/M. 'his zone can be selected in the normal bright-field mode (Fig. 4.20 a), in hich the primary beam passes through the objective diaphragm. When the ntermediate-lens excitation is decreased, its focal length is increased and the iffraction pattern in the focal plane of the objective lens can be focused on he final image screen after the objective diaphragm is removed (Fig. 4.20 b). he excitations of the later projection lenses are unchanged. Those lenses agnify either the second intermediate image or the diffraction pattern ehind the intermediate lens. Figure 9.16 shows an example of SAED from μm diameter areas of Al and Al_2Cu in a diamond-cut section of an Al-Cu utectic alloy.

The diameter of the area selected cannot be decreased below 1 μm owing the spherical aberration of the objective lens. The intermediate images of

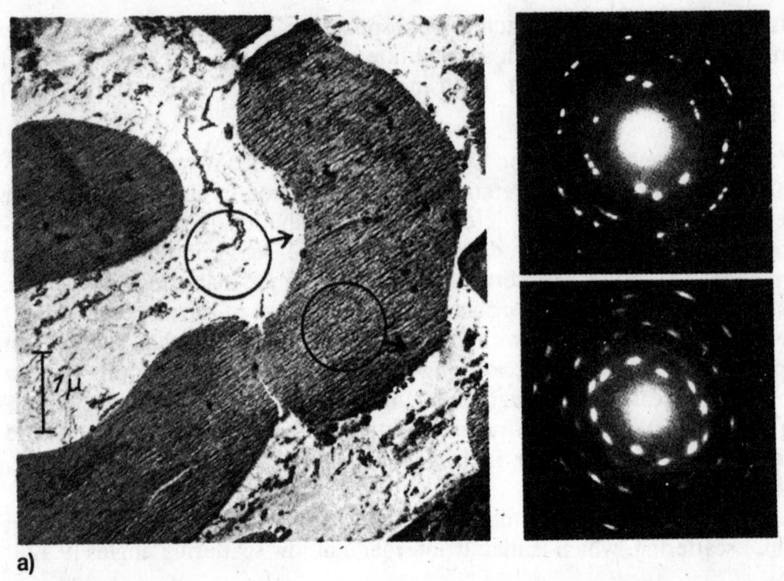

Fig. 9.16. Example of selected-area electron diffraction (SAED) from a thin section of an A Cu eutectic cut with a microtome, (**b**) and (**c**) contain the SAED pattern of the circles indicat in (**a**)

the Bragg reflections (dark-field images) are shifted relative to the brigh field image [9.140, 141] by a distance

$$\Delta s = (C_s\theta_g^3 - \Delta z\theta_g)M \ , \qquad (9.3$$

as can be seen from (3.64), which depends on the defocusing Δz and t spherical-aberration constant C_s ($\theta_g = 2\,\theta_B$ and θ_B is the Bragg angle). It is, course, possible to compensate the shift caused by sperical aberration by suitable choice of the defocus Δz, but only for one Bragg reflection, not f the whole diffraction pattern, simultaneously. The consequence is that Bra reflections of high order with large θ_g do not come from the area that w selected in the bright-field mode. Thus for $2\,\theta_B = 50$ mrad and $C_s = 2$ m the shift Δs is 0.25 μm. This limits SAED to areas greater than 0.5–1 μm diameter. The diffraction angle $\theta_g = 2\,\theta_B$ decreases linearly with λ as t electron energy is increased. Even though the spherical-aberration consta C_s is larger in HVEM, the first term in (9.30) is much smaller than fo 100 keV TEM, and areas with diameters ≥ 0.1 μm can be selected withc running into this problem. A further selection error results when the positi of the intermediate image is shifted by a stray magnetic excitation of t objective lens caused by changing the intermediate lens from the imaging the diffraction mode.

Diffraction patterns from smaller areas can only be obtained by using the rocking-beam technique (Sect. 9.3.2) or by producing a small electron probe (Sect. 9.3.3).

The resolution $d/\Delta d$ of an SAED pattern can be defined in terms of the smallest lattice-spacing difference Δd that can be resolved and can be estimated from the ratio $r/\Delta r$. Here, r denotes the distance from the spot to the pattern centre, $r = 2\theta_B f = \lambda f/d$ (f: focal length of the objective lens) and Δr is the diameter of the spot, which is equal to the diameter $2\alpha_i f$ of the primary beam (α_i: illumination aperture):

$$\frac{d}{\Delta d} = \frac{r}{\Delta r} = \frac{\lambda}{2\alpha_i d} . \tag{9.31}$$

Thus for $\lambda = 3.7$ pm, $d = 0.1$ nm and $\alpha_i = 0.1$ mrad, we find $d/\Delta d = 200$. The resolution can be increased only by decreasing α_i but this reduces the pattern intensity.

The spherical aberration of the objective lens can cause barrel and spiral distortion of the SAED pattern (Fig. 2.17) [9.142, 143] but this is, however, smaller than 1% for magnetic lenses; an elliptic distortion can arise due to astigmatism of the intermediate lens. The most severe distortion is caused by the projector lens, which can generate pincushion or barrel distortion, depending on the lens excitation. For the accurate determination of lattice spacings d, the diffraction (camera) length L must be calibrated by use of TlCl, for example, as a diffraction standard (Sect. 9.4.1). The spacing d can be calculated from (9.33) by use of the measured distance r of the diffraction spot or Debye-Scherrer ring from the centre. However, the value of the correction factor is not found to be 3/8, in practice; it can be much larger [9.144], because the pincushion or barrel distortion of the projector lens causes the same kind of change in the spot distance r as the difference between $\tan 2\theta_B$ and $\sin\theta_B$, which is the reason for the factor 3/8 in (9.33).

9.3.2 Electron Diffraction Using a Rocking Beam

A rocking beam, with varying angles of incidence γ in the specimen plane, can be generated by means of scan coils situated between the specimen and the final condenser lens. The following two diffraction techniques can be employed:

1) In the first technique, described by *Fujimoto* et al. [9.145], the specimen area (0.2–4 μm) that contributes to the electron-diffraction pattern (EDP) is defined by a selector diaphragm at the first intermediate image, as in the SAED technique (Sect. 9.3.1). The lenses below the objective produce a magnified image of the first EDP, formed in the focal plane of the objective lens. When the incident beam is rocked, the primary beam and all of the diffraction spots shift, so that the diffraction pattern is scanned bodily across

the final screen plane. A small fixed detector diaphragm selects the direction in the EDP that just passes the microscope on-axis. The detector signal records the electron intensity and then is fed to the intensity modulation of a cathode-ray tube, which is scanned in synchronism with the rocking, using deflection currents in the x and y directions proportional to those through the rocking coils. The theorem of reciprocity tells us that the intensity of the on-axis beam is the same as in the stationary SAED technique [9.146].

The angular resolution of the recorded EDP can be varied by altering either the diameter of the detector diaphragm or the magnification (diffraction or camera length) of EDP in the final screen plane, but it can never be smaller than the illumination aperture of the rocking beam.

This technique has the advantage that EDP is recorded in the scanning mode and the spot intensities can therefore be displayed directly, by the *Y*-modulation technique, for example [9.147, 148]; furthermore, the selection error of SAED is avoided because all the beams recorded pass through the microscope on axis. The diameter of the selected area is, however, limited by the diameter of the selector diaphragm, which cannot be smaller than 5 μm, owing to charging and contamination.

2) In a second technique, described by *van Oostrum* et al. [9.149], a highly magnified image ($M \geq 100\,000\times$) is formed on the final screen (detector plane). The diameter d of the detector diaphragm selects a small area of the image, which, back-projected into the specimen plane, can be much smaller than the selected area in SAED (e.g., 3–10 nm for $M = 100\,000$ and $d = 0.3$–1 mm as in [9.150–152]).

The angular resolution is provided by the diaphragm in the focal plane of the objective lens, which again cannot be smaller than 5 μm. If the primary beam passes through this diaphragm, a bright-field image appears in the detector plane. The primary beam is intercepted by the diaphragm if the beam is tilted. A diffraction spot passes through the diaphragm if the tilt angle $\gamma = 2\,\theta_B$, and can then generate a dark-field image in the detector plane. This means that during the rocking, a bright-field image is seen followed by a series of dark-field images that correspond to different diffraction spots. The dark-field images are not shifted relative to the bright-field images as they are with the SAED technique because all the beams recorded are on axis.

A further advantage of the van Oostrum technique is that the transition from the image to EDP can be made simply by switching on the rocking coils. This avoids any possible error in SAED due to stray-field excitation of the objective lens caused by switching over the intermediate lens from imaging to diffraction.

In conclusion, we see that the Fujimoto technique can give a better angular resolution but the area selected is limited by the diameter of the selector diaphragm, whereas the van Oostrum technique can select smaller areas but the angular resolution is limited by the size of the objective diaphragm.

9.3.3 Electron Diffraction Using an Electron Probe

The selection error of SAED can also be avoided by using small electron probes. It was shown in Sect. 4.2.3 that electron-probe diameters ≤ 0.1–$1\ \mu m$ can be obtained only by introducing an additional condenser lens; this may be a mini-lens placed in front of the objective lens or in the prefield of a highly excited objective lens [9.153, 154]. Probe diameters of 2–5 nm can be produced with a thermionic electron gun in a 100 keV TEM, whereas diameters less than 1 nm require a field-emission gun. The smallest possible probe diameter will be obtained with a large probe aperture α_p, of the order of a few 10 mrad. Decreasing this aperture increases the spot diameter. It is, therefore, impossible to obtain EDPs with sharp diffraction spots from areas as small as 2–10 nm because apertures α_p less than 1 mrad would be needed [9.155, 156].

Figures 9.17 a, c and 18 a–c show how the EDP changes from a spot pattern to a convergent beam and a Kossel pattern, as the electron probe aperture α_p is increased. A spot pattern requires $\alpha_p \ll \theta_B$ (Figs. 9.17 a and 9.18 a) so that the primary beam and the Bragg-reflection spots are sharp. The system of Kikuchi lines and bands will not be affected by the illumination aperture because they are normally generated by the cone of electrons diffusely scattered inside the specimen, and the necessary angular divergence is aided by any initial divergence (convergence) of the incident beam.

Increasing the aperture ($\alpha_p < \theta_B$, Figs. 9.17 b and 18 b) increases the spot diameter. The circular spots of the primary beam and the Bragg-diffraction

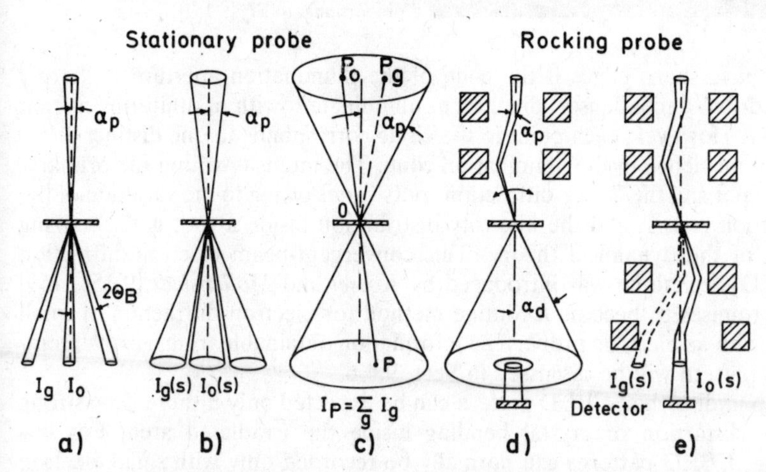

Fig. 9.17 a–e. Electron-diffraction techniques that use a small electron probe. Increasing the probe aperture of a stationary probe produces (**a**) a Bragg-diffraction spot-pattern, (**b**) a convergent beam electron-diffraction (CBED) pattern and (**c**) a Kossel pattern in which the intensity at a point P is proportional to ΣI_g. (**d**) A rocking probe of small aperture α_p produces a spot-, CBED- or Kossel pattern, depending on the detector perture α_d. (**e**) Double-rocking technique

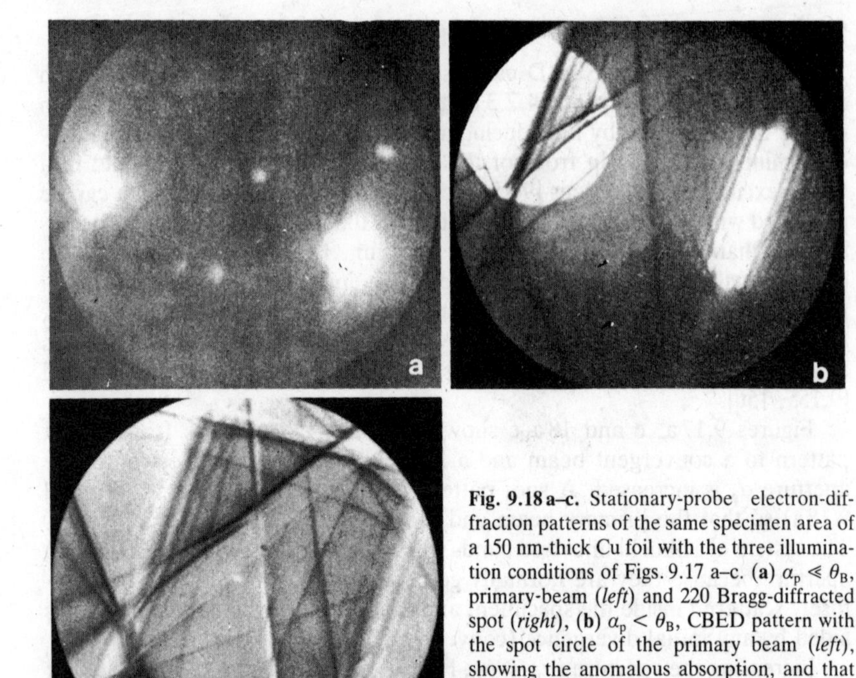

Fig. 9.18 a–c. Stationary-probe electron-diffraction patterns of the same specimen area of a 150 nm-thick Cu foil with the three illumination conditions of Figs. 9.17 a–c. (**a**) $\alpha_p \ll \theta_B$, primary-beam (*left*) and 220 Bragg-diffracted spot (*right*), (**b**) $\alpha_p < \theta_B$, CBED pattern with the spot circle of the primary beam (*left*), showing the anomalous absorption, and that of the Bragg reflection (*right*), showing the symmetric pendellösung fringes. (**c**) $\alpha_p > \theta_B$, Kossel pattern with an intensity distribution proportional to ΣI_g [9.157]

spots have sharp edges if the cone of the illumination aperture is sharply limited by a condenser diaphragm illuminated with a uniform current density. However, each point in the circle corresponds to one distinct direction of incidence in the illumination cone. The intensity within the primary-beam spot and the Bragg-diffraction spots varies owing to the variation of the excitation errors, and the intensity distribution inside a spot is the rocking curve of the dynamical theory. This convergent-beam electron-diffraction (CBED) technique was introduced by *Kossel* and *Möllenstedt* [9.158, 159] and promises to become a routine method for electron diffraction of small areas with an electron probe. The information obtainable from a convergent-beam pattern will be discussed in Sect. 9.4.6.

An undisturbed CBED pattern can be expected only if there is no strong lattice distortion or crystal bending inside the irradiated area. For this reason, CBED patterns can normally be recorded only with small electron probes. A further limitation is the rapid growth of a contamination needle on the irradiated area, which can be reduced either by specimen heating [9.160] or specimen cooling [9.161]; see also the discussion of contamination in Sect. 10.4.2.

A further increase of probe aperture, $\alpha_p > \theta_B$ (Fig. 9.18 c and 19 a), increases the overlap of the spots [9.162, 163]. The intensity at a point P of the Kossel pattern does not consist of only the contribution from the primary-beam direction P_0. In addition, the directions P_g contribute with the corresponding Bragg-diffraction intensities. The intensity $I_P = \sum_g I_g$, including $\boldsymbol{g} = 0$, is a multi-beam rocking curve; according to Fig. 7.20 a, there is cancellation of the "pendellösung" fringes, but an intensity distribution in the form of defect Kossel bands persists, caused by the dependence of anomalous transmission on the excitation error, see also (7.89). Residual "pendellösung"-fringe contrast in Fig. 9.18 c can result from buckling of the Cu foil inside the irradiated area.

Fig. 9.19 a–d. Electron diffraction pattern near the 111 pole of a Si foil: (**a**) stationary-probe Kossel pattern recorded with the transmitted electrons, (**b**) rocking-probe diffraction pattern with the transmitted electrons, (**c**) with the transmitted electrons scattered through large angles ($\theta > 10°$) and (**d**) with the backscattered electrons ($\theta > 90°$) [9.157]

The following nomenclature is proposed [9.164] to distinguish between Kikuchi and Kossel bands. In EDPs with Kikuchi bands, the angular divergence is caused by scattering in directions between the diffraction spots, whereas Kossel bands are produced by the convergence of the external-probe aperture. Kossel patterns without diffraction spots can be obtained for all film thicknesses, whereas Kikuchi bands appear only in thicker specimens.

These results show that information about the crystal structure does not necessarily require the use of small electron-probe apertures α_p and that small electron probes with large apertures are more suitable for CBED and Kossel-pattern studies. A conventional spot pattern does not always provide the most information about crystal structure and symmetry.

Electron-probe diffraction patterns can be displayed on a CRT by post-specimen deflection of the diffraction pattern across a detector diaphragm or by rocking the electron probe by means of scan coils placed in front of the specimen (Fig. 9.17 d). However, it is not easy to avoid shifting the electron probe during rocking. The spherical aberration of the final probe-forming lens causes an unavoidable shift and the electron probe moves along a caustic figure. This shift can be compensated by adding a contribution to the deflection-coil current proportional to θ^3 [9.165], and the shift can be kept less than 0.1 μm in TEM. The rocking causes EDP to move across the detector plane, like the rocking beam in the Fujimoto technique (Sect. 9.3.2). If a small probe aperture $\alpha_p \ll \theta_B$ is used, a spot pattern will be recorded by a small detector diaphragm ($\alpha_d \ll \theta_B$) as the theorem of reciprocity shows; with $\alpha_d < \theta_B$, a convergent beam pattern is obtained and with $\alpha_d > \theta_B$, a Kossel pattern. However, in order to avoid the shift caused by spherical aberration, it is better to use a stationary probe and a post-specimen deflection. The advantage of rocking the electron probe is that types of information not obtainable from a conventional EDP, become accessible. Figure 9.19 a shows a stationary electron-probe Kossel pattern of the 111 pole of a Si foil. Using a rocking probe and a large detector aperture, a similar pattern can be recorded with the cone of electrons scattered through small angles (Fig. 9.19 b). These diagrams contain defect Kikuchi bands due to anomalous electron transmission. Suppose now that an annular detector is placed below the specimen [9.166]; electrons scattered through large angles, normally not used in TEM, will be collected and a contrast reversal is observed (Fig. 9.19 c). The diagram now contains excess Kikuchi bands generated by direct scattering of electrons out of the Bloch-wave field into larger angles. The pattern is the same as that recorded with backscattered electrons (Fig. 9.19 d) like in SEM, where it is known as an electron-channelling pattern (ECP). The only difference is that the noise is larger in the BSE pattern (Fig. 9.19 d) than in the pattern recorded with forward-scattered electrons (Fig. 9.19 c) because many fewer electrons are backscattered than forward scattered through large angles; we recall that the Rutherford cross section varies with angle as $\sin^{-4}(\theta/2)$.

If two semiconductor detectors are used, one annular in shape, the other occupying the central region, the first will record a Kossel pattern and the second a spot pattern [9.167, 168]. The two signals can be added or subtracted to obtain a Kossel pattern on the CRT with superposed bright or dark diffraction spots. This faciliates accurate determination of orientation because the relative position of the Kossel bands relative to the spots and the position of the central beam in the diagram can be established with high accuracy.

Another interesting variant is the double-rocking method of *Möllenstedt* and *Meyer* [9.169, 170] (Fig. 9.17 e). A second post-specimen scan-coil system is arranged in such a way that the primary beam falls on the detector at all rocking angles. A convergent-beam pattern can be obtained for the primary beam over a very much larger angular range undisturbed by the overlap of other Bragg reflections. In a similar fashion, a convergent-beam pattern can be obtained from a Bragg reflection if the detector diaphragm is shifted and only the diffracted intensity I_g recorded (see also zone axis patterns in Sect. 9.3.4).

9.3.4 Further Diffraction Modes in TEM

a) Small-Angle Electron Diffraction. Small-angle x-ray diffraction is successfully used for the investigation of periodicities and particles of the order of 10–100 nm. *Mahl* and *Weitsch* [9.171] used this method with electrons. Diffraction at spacings of $d = 100$ nm requires a primary beam with an angular divergence (illumination aperture) smaller than the diffraction angle: $\alpha_i < 2\theta_B = \lambda/d = 0.03$ mrad for $E = 100$ keV. With a double-condenser system, $\alpha_i = 0.01$ mrad can be achieved (Sect. 4.2.1) and the neighbourhood of the primary beam can be magnified by the projector lens (thereby increasing of the camera length to several meters). An extensive account of small-angle electron diffraction has been given in [9.172].

A typical application of this method is the study of evaporated films with isolated crystals, showing diffuse rings with diameters inversely proportional to the mean value of the crystal separation [9.173]. Further applications are periodicities in collagen (Fig. 9.20), conglomerates of latex spheres and virus particles [9.174, 175], catalase [9.176] and high polymers [9.177]. Organic specimens have to be coated with a metal conductive layer to prevent charging, which would perturb the primary beam. The periodicity can also be resolved in a micrograph. The same information is obtainable by laser diffraction on the developed film or plate (Sect. 6.4.6). Small-angle electron diffraction is also important for Lorentz microscopy (Sect. 6.6.2) because the primary beam is split by the Lorentz force inside the magnetic domains.

b) Scanning Electron Diffraction. For the quantitative interpretation of EDPs it can be useful, to record the intensities directly by scanning the diffraction pattern across a detector diaphragm (Grigson mode). The detec-

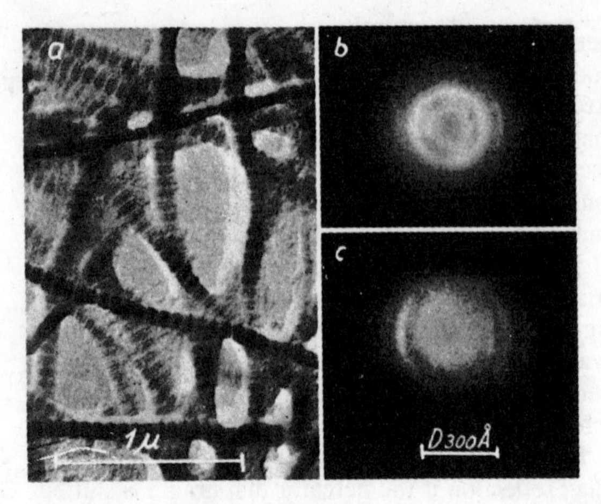

Fig. 9.20. Small-angle electron-diffraction patterns (**b** and **c**) of a shadow-cast collagen specimen (**a**) showing the periodicities along the collagen fibres [9.170]

tor may be a Faraday cage, a semiconductor detector or a scintillator-photomultiplier combination [9.45, 178, 179]. This method also can be used for energy filtering of EDP [9.180] by means of a retarding-field filter [9.178, 181] or a magnetic electron prism [9.45]. This method is of special interest for ultra-high vacuum experiments [9.182]. The EDP can also be recorded digitally and by means of one of the two beam-rocking methods described in Sect. 9.3.2 [9.183].

c) Reflection High-Energy Electron Diffraction (RHEED). Bulk material can be investigated by allowing the electron beam to fall obliquely on the specimen at a glancing angle $\theta < \theta_B$ (Fig. 9.21). The electrons penetrate only a few atomic layers into the material [9.184]. Because the interaction volume is

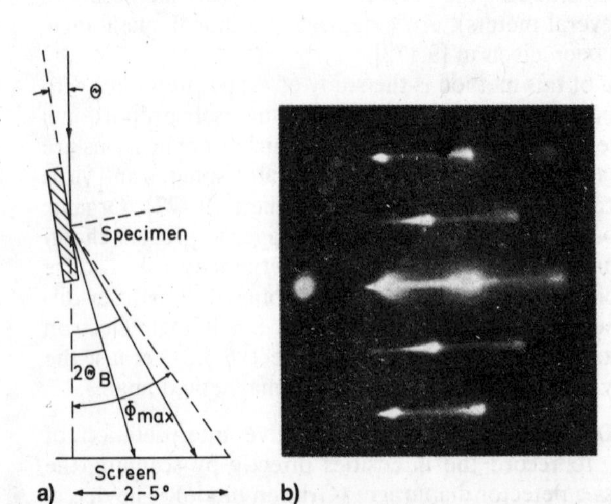

Fig. 9.21. (**a**) Schematic ray diagram for recording a reflexion high-energy electron diffraction (RHEED) diagram, (**b**) example of a RHEED pattern of a NaF film evaporated on a (100) NaF cleavage plane at 270 °C (H. Raether)

thin, the reciprocal-lattice points will be drawn out to needles normal to e surface, and the Bragg spots will be elongated. Plane surfaces also show ikuchi lines and bands. The influence of refraction, which shifts the Bragg pots to smaller Bragg angles, has to be considered. The method is as sensive to surface layers as low-energy electron diffraction (LEED). The surface an be cleaned by heating or by electron and/or ion bombardment .185–187]. Charging effects can cause problems with insulating materials. hey can be avoided by irradiating the specimen with 200–1000 eV electrons om a separate source, which increases secondary-electron production and voids the build-up of a large negative charge, or by heating (increase of lectrical conductivity).

) **Electron-Back-Scattering Pattern and Electron-Channelling Pattern.** ncreasing the angle θ between beam and specimen (Fig. 9.21 a) to 5°–30° and ncreasing ϕ_{max} to 20°–30° yields electron-back-scattering pattern (EBSP) .188], which contain excess Kikuchi bands (Fig. 9.22). Increasing θ still urther causes contrast reversal of defect bands for small take-off angles ϕ .189, 190] analogous to the extension of defect bands in the central area of transmission EDP with increasing thickness (Sect. 7.5.4, Fig. 7.24).

Electron-back-scattering patterns can be obtained in TEM by positioning e specimen about 10 cm above the photographic plate and deflecting the lectron beam onto the specimen by means of deflection coils situated below e projector lens [9.190]. Recording of an electron-channeling pattern ECP) [9.191, 192] by rocking the electron beam and recording the back-cattered or forward-scattered electrons (Sect. 9.3.3, Fig. 9.19 c, d) is related the method of EBSP by the theorem of reciprocity [9.190, 193].

) **Zone-Axis Pattern (ZAP).** In convergent-beam electron diffraction CBED) (Sect. 9.3.3) each point corresponds to another direction of electron ncidence inside the convergent electron probe. If a crystal foil is bent two-imensionally and the electron beam hits the foil nearly parallel to a low-ndexed zone axis, then each point of the foil in a bright-field image corre-ponds also to another direction of incidence of the parallel electron beam, elative to the lattice planes (Fig. 7.3). If the two main radii of curvature of he foil have the same sign and equal order of magnitude (dome- or cup-haped), the bend contours (Sect. 8.1.1) which form a zone-axis pattern ZAP) show the same intensity distribution as a CBED pattern. A similarity xists also to the double-rocking technique (Sect. 9.3.3 and Fig. 9.17 e) ecause the tilt of the foil can be larger than the Bragg angles, which would esult in overlap of the reflection circles in CBED.

Contrary to CBED which produces a two-dimensional rocking curve in ourier space (diffraction plane), the ZAP is a real-space method. Whereas a BED pattern is formed by an electron probe of a few nanometres in diame-er, a ZAP extends over a few micrometres.

ZAPs represent a lot of information about the crystal symmetry [9.194–196] nd can be used to determine the crystal class (see also discussion of the high-

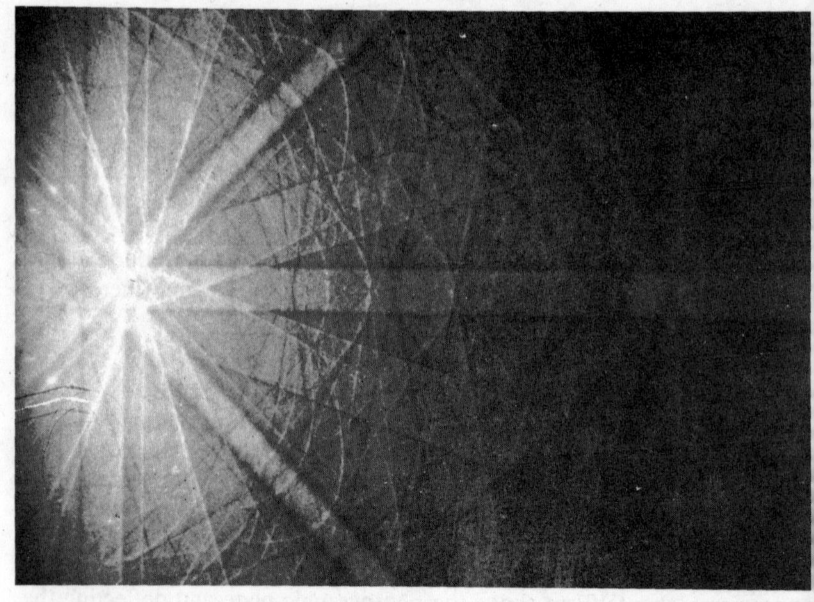

Fig. 9.22. Electron-backscattering pattern (EBSP) with excess Kikuchi bands of a germaniun single crystal with an angle $\theta = 10°$ (Fig. 9.21 a)

order Laue-zone pattern (HOLZ) in Sect. 9.4.6). If the incident electron energy is varied, the intensity distribution near the zone axis changes and gives information about the Bloch-wave channeling and critical voltage [9.196–198].

9.4 Applications of Electron Diffraction

9.4.1 Lattice-Plane Spacings

Calculation of the lattice-plane spacing d_{hkl} from the Bragg condition (7.20) requires a knowledge of the electron wavelength (3.1) and the Bragg angle θ_B. Since θ_B is small, the sine that occurs in the Bragg condition can be replaced by the tangent in a first-order approximation

$$\frac{\lambda}{d} = 2\sin\theta_B \simeq \tan(2\theta_B) = \frac{r}{L}; \qquad d \simeq \frac{\lambda L}{r}, \tag{9.32}$$

where r is the distance of the diffraction spot from the primary beam or the radius of a Debye-Scherrer ring, and L is the diffraction (camera) length. For higher accuracy, a further term of the series expansion can be included,

$$d = \frac{\lambda L}{r}\left[1 + \frac{3}{8}\left(\frac{r}{L}\right)^2 + \dots\right]. \tag{9.33}$$

However, this formula is valid only if the diffraction pattern is magnified without any barrel or pincushion distortion. The pincushion distortion of the projector lens (Sect. 2.3.4) contributes a further term in r^2 so that a larger value than 3/8 may be found for the constant in (9.33) during calibration [9.144].

The diffraction (or camera) length L cannot be measured directly in the SAED mode. The product λL must therefore be determined by calibration with a substance of known lattice constant. Only substances with the following properties should be used for calibration:

1) many sharp rings with known *hkl*;
2) chemically stable and no change of lattice parameters under electron irradiation;
3) correspondance with x-ray lattice constant;
4) easy preparation.

Table 9.1 contains some suitable calibration substances and their lattice parameters. The value of TlCl has been compared with x-ray data in high-precision experiments ($\Delta d/d = \pm 3 \times 10^{-5}$) [9.199].

When single-crystal standards are used, the reciprocal-lattice points have a needle-like shape parallel to the normal of the foil (Fig. 7.9 and 10a). If the crystal normal is inclined to the electron beam, the intersections of these needles with the Ewald sphere can alter the position of Bragg-diffraction spots [9.200].

Table 9.1. Calibration standards for electron diffraction

Standard	Lattice type	Lattice constant [nm]	Ref.
LiF	NaCl	a = 0.4020	[9.204]
TlCl	CsCl	0.3841	[9.199, 205]
MgO	NaCl	0.4202	[9.205]
ZnO	Wurtzite	a = 0.3243	[9.206]
		c = 0.5194	
Au	fcc	a = 0.4	
Si	Diamond	0.54307	[9.203]

Tilt coils for the electron beam can be used for the calibration of diffraction patterns, because the deflection angle θ is proportional to the coil current. The diffraction pattern is shifted by $L\theta$ [9.201, 202]. Double exposures with known and previously calibrated θ allow the distortion of the EDP to be determined. Alternatively, the lattice spacing can be measured directly in the microscope with an accuracy of 0.1% by recording the x and y coil currents needed to bring the diffracted beam on axis [9.203].

9.4.2 Texture Diagrams

For many polycristalline specimens, the distribution of crystal orientations is not random; instead, one lattice plane may lie preferentially parallel to the specimen plane. In this plane, however, the crystals are rotated randomly around a common axis $F = [mno]$, the fibre axis of the fibre texture.

The existence of this fibre axis means that the reciprocal-lattice points are distributed around concentric circles centred on the fibre axis (Fig. 9.23 a). (They would lie on a sphere for a totally random distribution.)

If the electron beam is parallel to the fibre axis, the Ewald sphere intersects the circles and a Debye-Scherrer ring pattern is observed, which does not show all possible hkl but only those that fulfil the condition $(hkl)[mno] = hm + kn + lo = 0$.

The limitation of the number of observable rings can simulate extinction rules $|F|^2 = 0$. As a consequence, wrong conclusions may be drawn about the crystal structure. In a weak fibre texture, the fibre axis will be not more than a preferential direction; all possible rings now appear but with the wrong intensity ratios. The diffraction pattern of an evaporated Au film (Fig. 9.24 a), for example, shows a 220 ring that is more intense than would be expected from a random distribution.

A fibre texture can be recognized clearly if the direction of electron incidence E is tilted through an angle β relative to the fibre axis F; in practice, the specimen normal is tilted relative to the electron beam in an object

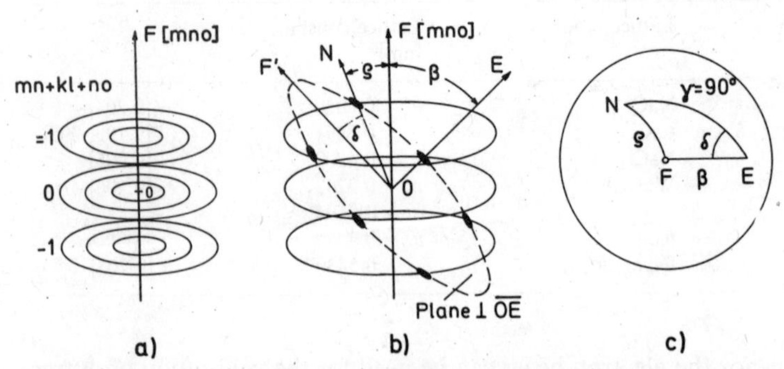

Fig. 9.23. (**a**) Distribution of the $\boldsymbol{g}$ vectors of the crystallites in reciprocal space for a foil with a fibre texture with $[mno]$ as a fibre axis. (**b**) Intersection of the Ewald sphere (---) with these rings resulting in sickle-shaped segments of the Debye-Scherrer rings. (**c**) Relations between the angles [(E) electron-beam direction, (N) intersection of Ewald sphere and reciprocal lattice, (F) fibre axis]

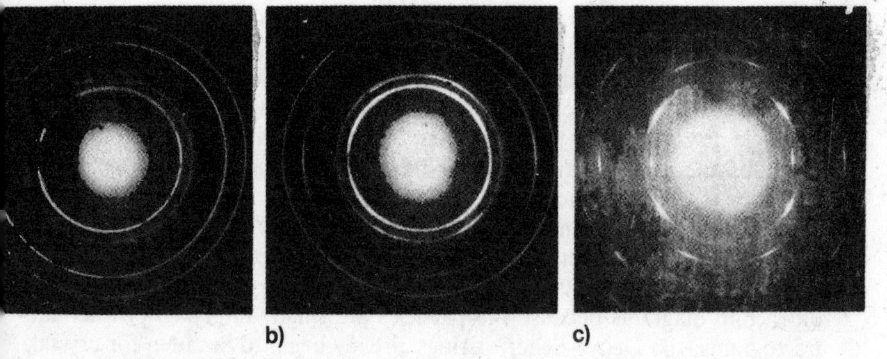

Fig. 9.24. (**a**) Electron-diffraction pattern at normal incidence of an evaporated Au film with a weak (111) fibre texture, (**b**) the same specimen tilted 45° to the electron beam showing a weak dependence of the ring intensities on the azimuth. (**c**) Stronger fibre texture of an evaporated Zn film tilted at 45° to the electron beam

goniometer. The intersections of the Ewald sphere with the concentric circles then become sickle-shaped (Fig. 9.23 b), as can be seen for the weak fibre texture of a gold film in Fig. 9.24 b and for the stronger texture of a zinc film in Fig. 9.24 c. The sickles become narrower for a larger tilt angle $\beta = 45$–$60°$. The texture should be most clearly detectable with $\beta = 90°$, but this is possible only in a reflexion high-energy electron-diffraction (RHEED) experiment (Sect. 9.3.4).

Quantitative results about the fibre axis can be obtained by measuring the azimuths δ of the centres of the sickle-shaped ring segments [9.207, 208]. The intersection of the plane that contains the directions E and F with the Ewald sphere defines a direction F′ in the diffraction pattern that gives the origin of δ (Fig. 9.23 b). The angle ϱ between (hkl) and $[mno]$ can be calculated by evaluating their scalar product. The hkl are obtained from the ring diameter. For the spherical triangle FEN in Fig. 9.23 c, with $\gamma = 90°$, we have

$$\cos\varrho = \cos\beta\cos\gamma + \sin\beta\sin\gamma\cos\delta = \sin\beta\cos\delta\,. \tag{9.34}$$

For known values of the tilt angle β and azimuth δ, the value of $\cos\varrho$ can be calculated and compared with theoretical values for different possible fibre axes. A procedure for non-cubic crystals is described in [9.209, 210].

Measurement of the azimuthal distribution of the ring intensities can be used for the quantitative characterisation of a texture [9.179].

9.4.3 Crystal Structure

The method described for measuring the lattice-plane spacings d_{hkl} can be used for direct comparison of different substances. If the chemical composition is known or has been established by x-ray microanalysis or electron-energy-loss spectroscopy, the crystal structure can be identified by comparing the spacings with tabulated x-ray values of the d_{hkl} (A.S.T.M. Index). However, a problem is that the d_{hkl} cannot be measured with an accuracy better than about $\pm 1\%$ and the results may fail to coincide with the tabulated data. In single-crystal EDPs, the symmetry of the spot diagram, the extinction rules and the angles between the diffraction spots can be used for further identification of the structure.

The complete determination of crystal structure by Fourier synthesis using electron-diffraction patterns is superior to x-ray analysis if the material under investigation exists in only small quantities or produces diffuse x-ray diffraction diagrams because the particles are small. In crystal-powder diffractograms, the Debye-Scherrer rings already begin to broaden for crystals smaller than 100 nm. This broadening can be detected in EDP only when the crystals are smaller than 5 nm.

However, Fourier synthesis requires exact values of the reflection intensities and the following difficulties are encountered in electron diffraction:

1) The transition from the kinematical to the dynamical theory is significant for thicknesses or dimensions as small as 5 nm;
2) Forbidden and weak reflections can be excited by multiple diffraction.
3) In small particles or small irradiated areas, only a few reflections, for which the reciprocal-lattice points are near the Ewald sphere, are excited;
4) Foils of large area are often bent and exhibit Bragg-spot intensities that depend randomly on the fraction of the irradiated area that contributes to the Bragg reflection.

Pinsker [9.208 a] and *Vainshtein* [9.208 b] described methods for obtaining quantitative values of the structure factors $|F(\theta)|^2$ from polycrystalline ring and texture patterns for subsequent Fourier synthesis of the atomic positions in the unit cell [9.209]. These methods assume that the kinematical theory is completely valid and that there is no influence on the intensity of high orders of a systematic row (see e.g. Fig. 7.23). In consequence, results have to be interpretated with care [9.211]. A correction using the Blackman formula (7.99) has been applied to Ni_3C, for example, [9.212], see also [9.213].

A detailed description of the problem and difficulties of determining crystal structure by electron diffraction was given by *Cowley* [9.214, 215] who also described a method for measuring the integral intensity of Bragg spots. Each diffraction spot is expanded to a square by a pair of deflection coils, which allows the integrated intensity to be determined more easily by photometry. *Fujime* [9.216] oscillated the specimen around an axis in the specimen plane to average over the reflection range of a reciprocal-lattice point.

Thin crystal lamellae are often parallel to the specimen plane. The diameter of Laue zones (Sect. 9.4.6) can be used to get information about the lattice-plane spacings normal to this plane. An eucentric goniometer that provides tilt angles up to $\pm 45°$ and adjustment of the tilt axis with an accuracy of 0.1–1 μm permits us to explore the three-dimensional structure of the reciprocal lattice [9.217–219]. It will be useful to rotate the crystal around the specimen normal, so that a systematic row is parallel to the tilt axis. In most cases, the unit cell can be reconstructed with only a few tilts, for which the electron beam is again parallel to a zone axis. If necessary, a further tilt around another row can be performed [9.220–222].

For complete structure analysis, not only a goniometer but also the possibilities of convergent-beam diffraction should be used, to measure the lattice-potential coefficients V_g and the crystal thickness from the spacing of the "pendellösung" fringes (Sect. 9.4.6). An electron micrograph contains information about the crystal size and lattice defects, that can be included in the analysis of EDP. For thin crystals ($t \leq 5$ nm), it may be possible to resolve the projected lattice structure directly (Sect. 8.2).

X-ray diffraction is influenced by the electron-density distribution of the atomic shell and electron diffraction by the screened Coulomb potential of the nuclei. The scattering amplitudes of single atoms are proportional to Z for x-rays (5.38) and to $Z^{1/3}$ for electrons (5.39). The difference between light and heavy atoms is smaller for electrons than for x-rays. Light atoms can be localized better in the presence of heavy atoms [9.223, 224]. Here, however, neutron diffraction is much superior because the scattering amplitudes are of the same order for all nuclei.

The different scattering mechanisms for x-rays and electrons can be observed with KCl, for example. Both ions have the electron configuration of argon and the scattering amplitudes for x-rays are equal; the structure amplitude $4(f_K - f_{Cl})$ therefore vanishes for h,k,l odd (Sect. 7.2.2). In electron diffraction, the nuclear charges are different, so that, for odd values of $h,k,l,$ the amplitude is weak but not zero, whereas the amplitude for h,k,l even, $4(f_K + f_{Cl})$, is strong [9.225].

9.4.4 Crystal Orientation

A knowledge of the exact orientation of crystals is important for the investigation of lattice defects and the relative orientations of different phases, or between the matrix and precipitates.

Many diffraction spots are observed if the foil is very thin, so that the reflection range of the reciprocal-lattice points is enlarged, or if the foil is bent. Both effects limit the accuracy of orientation determination. For this, the diffraction spots R_1, R_2, ... at the distances r_n from the central beam O are used to calculate d_n, by taking the product λL from a calibration pattern (Sect. 9.4.1). When the lattice structure and lattice constant are known, the corresponding reciprocal-lattice points $\boldsymbol{g}_n = (h_n, k_n, l_n)$ can be determined. For cubic crystals the ratio method can be applied by using

$$\frac{r_1^2}{r_2^2} = \frac{h_1^2 + k_1^2 + l_1^2}{h_2^2 + k_2^2 + l_2^2} . \tag{9.35}$$

From a table of values of this ratio for all possible combinations of $|g_1|$, $|g_2|$, the indices of both reflections can be identified without knowing λL. To be sure that spots have been indexed with the correct signs, agreement between the observed angle α between OR_1 and OR_2 and the theoretical values,

$$\cos\alpha_{12} = \frac{g_1 \cdot g_2}{|g_1|\,|g_2|} = \frac{h_1h_2 + k_1k_2 + l_1l_2}{(h_1^2 + k_1^2 + l_1^2)^{1/2}(h_2^2 + k_2^2 + l_2^2)^{1/2}} \tag{9.36}$$

must be checked. There may be small differences between α and α_{12} [9.226]. The direction of electron incidence (normal to the foil) is parallel to the vector product of two reciprocal-lattice vectors

$$\boldsymbol{n} \parallel g_1 \times g_2 = (k_1l_2 - k_2l_1\,, \quad l_1h_2 - l_2h_1\,, \quad h_1k_2 - h_2k_1) . \tag{9.37}$$

The needle-like extension of the reciprocal-lattice points (Sect. 7.2.2) widens the tilt range, which can be $\pm 10°$ for low-order reflections [9.227]. The orientation therefore becomes more accurate if a large number of reflections is used, especially high-order reflections at the periphery of EDP. An accuracy of $\pm 3°$ can then be achieved [9.228, 229]. Further improvements can be made by considering the intensities of the diffraction spots [9.227].

The orientation determination becomes unique and more accurate if three reflections g_1, g_2, g_3 are employed [9.230, 231]. A circle is drawn through the three spots to establish the correct numbering of the sequence. If the central beam O is inside the circle the reflections are numbered anticlockwise, otherwise they are numbered clockwise. With this convention, the determinant

$$g_1 \cdot (g_2 \times g_3) = \frac{1}{V} \begin{vmatrix} h_1 & k_1 & l_1 \\ h_2 & k_2 & l_2 \\ h_3 & k_3 & l_3 \end{vmatrix} \tag{9.38}$$

should be positive. If it is not, the signs of the *hkl* must be reversed so that for all combinations of two g_i (9.36) is also obeyed. The direction of the normal antiparallel to the electron beam is then given by the mean value

$$\boldsymbol{n} \parallel |g_1|^2(g_2 \times g_3) + |g_2|^2(g_3 \times g_1) + |g_3|^2(g_1 \times g_2) . \tag{9.39}$$

However, the orientation is not unique if all of the reflections happen to belong to a single zone with odd symmetry [9.226, 232, 233]. After rotation of the crystal through 180°, the same diffraction pattern is obtained. The orientation can be established uniquely from a second EDP obtained after tilting the specimen. A goniometer should also be used if EDP does not contain two or three diffraction spots convenient for calculation. With these precautions, an accuracy of $\pm 0.1\%$ can be attained.

A high accuracy is also obtained if the Kikuchi lines are used [9.226, 231, 234, 235]. A reflection is in the exact Bragg position if the excess Kikuchi line goes through the reflection spot. If the distance from the line to the reflection g_n is a_n, the tilt out of the Bragg position or the excitation error s_g is

$$\Delta\theta = \frac{\lambda}{d}\frac{a_n}{r_n}; \qquad s_g = g\Delta\theta = \frac{\lambda}{d^2}\frac{a_n}{r_n}. \tag{9.40}$$

Equation (9.39) implies exact Bragg positions. If the three terms are multiplied by $\alpha_n = (r_n + 2a_n)r_n$, an accuracy of $\pm 0.1°$ is obtained [9.231]. For determination of relative orientations of two crystals with the Kikuchi pattern, see [9.236].

The relative orientations between a matrix and coherent or partially coherent precipitates can be determined by means of a transfer matrix, which relates the coordinate systems of the two phases [9.237–239].

The orientation can be checked by use of specimen details that are visible in the micrograph, provided that precipitates, stacking faults or dislocations are recognisable in different planes and show traces with measurable relative angles. The traces of structures in octahedral planes can be used, for example, to analyse the accuracy and uniqueness of the orientation determination [9.240, 241].

The diffraction of a specimen area is rotated relative to the image of the same area because of image rotation by magnetic lenses. This rotation angle can be calibrated by use of MoO_3 crystal lamellae. These are prepared as smoke particles by heating a Mo wire in air by a high ac current and collecting the vapour on a glass slide. After floating the crystal film on water and collecting it on a formvar-coated grid, crystal lamellae parallel to the specimen plane are obtained. The long edge of these crystals is parallel to [100]. The rotation angle can be read from a double exposure of an image and the SAED pattern [9.242].

If the orientation of the diffraction pattern relative to the specimen is of interest, the angle of rotation between specimen and micrograph also has to be measured [9.243]. Kikuchi lines can be used also for this purpose [9.244]. The system of Kossel cones is fixed to the crystal so that the Kikuchi lines move if the specimen is tilted about an axis of known orientation. This direction can be compared with the direction of shift of the Kikuchi lines.

9.4.5 Examples of Extra Spots and Streaks

Electron-diffraction patterns contain not only the Bragg diffraction spots that are expected from the structure of the unit cell of a perfect crystal but also additional spots and streaks. Not every effect can be discussed here, because they vary from one specimen to another. A few typical examples will be described, which should provide some guidance in the discussion of particular diffraction patterns.

a) Forbidden Reflections. In dynamical theory, the intensity of a beam diffracted at *hkl* lattice planes can be equal in magnitude to the primary beam. A second Bragg reflection at lattice planes $h'k'l'$ can therefore produce reflections with indices $h\text{-}h'$, $k\text{-}k'$, $l\text{-}l'$. Spots that are forbidden by the extinction rules for the structure amplitude F (Sect. 7.2.2) may therefore be seen: an $00l$ spot (l odd) in FeS_2 [9.245], a 222 spot in Ge by double excitation at the allowed $(\bar{1}11)$ and (311) lattice planes [9.246, 247] or a 00.1 spot in hexagonal cobalt coming from (h0.1) and ($\bar{h}$0.0) [9.216]. This explanation of forbidden reflections is restricted to particular crystal orientations with relatively low values of s_g (accidental interaction). In other situations, more Bragg reflections can be excited simultaneously in dynamical theory, resulting in more-intense forbidden diffraction spots or Debye-Scherrer rings (systematic interaction).

b) Twins and Oriented Precipitates. Twinning results from a mirror reflection of the crystal structure about special lattice planes. For example, face-centred cubic crystals show a twin formation with a mirror reflection about the {111}-planes, and Ag films evaporated on [100] cleavage planes of NaCl or Ni films electrolytically deposited on copper show an epitaxy with frequent twin lamellae. The reciprocal lattice of these twin lamellae can be obtained from the reciprocal lattice of the matrix by mirror reflection about the {111}-planes and additional reciprocal-lattice points occur on one-third of the neighbouring points in the ⟨111⟩ directions. Extra spots are seen (Fig. 8.4), which are strongest if the foil is tilted through about 16° out of the [100] orientation [9.248]. Precipitates with a fixed orientation relative to the matrix can cause similar effects. In order to identify the origin of extra spots, the specimen must be imaged in the dark-field mode, selecting these extra spots only; the parts of the image that contribute to the spot will then appear bright (Sect. 8.1.2).

c) Stacking Faults and Planar Precipitates. The finite extension of a crystal-plate results in a needle-shaped extension of the reciprocal-lattice points, normal to the plate (Sect. 7.2.2). If the electron beam is parallel to this normal, the diffraction pattern is not changed and reflections will appear only over a larger tilt range. However, small shifts in the position of diffraction spots can result from the intersection of the Ewald sphere with the needles (Fig. 7.9). The needles of the reciprocal-lattice points create diffuse streaks in the diffraction pattern if the angle between the normal to the crystal plates and the electron beam is near to 90°. The streaks can extend from one Bragg spot to another. The existence of streaks indicates that the specimen contains planar faults such as stacking faults of high density, precipitate lamellae or

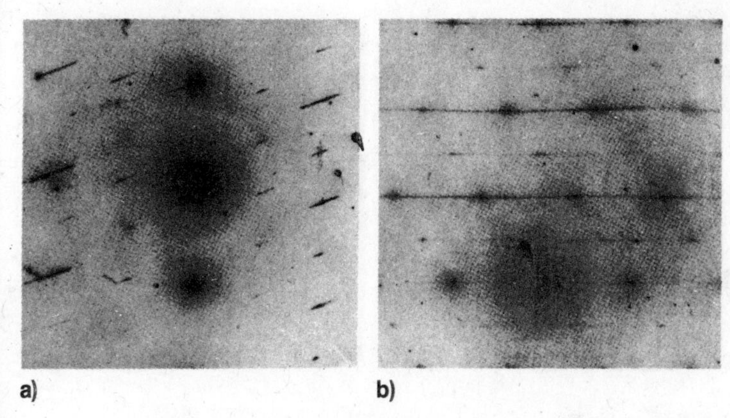

Fig. 9.25 a, b. Selected-area electron-diffraction pattern of plate-like Fe_4N precipitates with stacking faults in the {111} planes in two different orientations: (**a**) {111} oblique to the electron beam and (**b**) one of the planes parallel to the electron beam [9.249]

Guinier-Preston zones. Figure 9.25 shows an example of {111} stacking faults in Fe_4N particles (fcc, a = 0.378 nm) extracted from a Fe-0.1wt.% N alloy heat treated to 370 °C. In Fig. 9.25 a, the {111} planes are inclined more parallel to the foil and in b) their normals are inclined at angle of nearly 90° to the electron beam [9.249].

d) Ordered Alloys with a Super-Lattice Structure. The most important effects will be discussed for the example of Cu-Au alloys. The alloy $AuCu_3$ is not ordered at temperatures $T \geq 388$ °C, where it consists of a solid solution with a random distribution of Au and Cu atoms at the sites of a fcc lattice. The diffraction pattern contains the reflections allowed by extinction rules for this lattice type (Table 7.1). Below this transition temperature, the alloy acquires an ordered structure with the Au atoms at the corners and Cu atoms at the face centres of the cubic unit cell. In this situation, $F = f_{Au} + 3F_{Cu}$ for all even and all odd hkl and $F = f_{Au} - f_{Cu}$ for mixed hkl; for the latter, $F = 0$ in the non-ordered structure. Additional Bragg spots therefore appear below $T = 388$ °C and this can be used for calibration of the specimen temperature, for example (Sect. 10.1.1).

In an AuCu alloy (50 : 50 wt.%), the transition from the random phase to the ordered CuAu II phase, in which alternate (002) planes consist wholly of Cu and wholly of Au atoms, respectively, occurs at 410 °C. Every 2 nm, the Cu and Au atoms change places, and the consequence is a domain structure with anti-domain boundaries (Fig. 8.22). The resulting unit cell is elongated with a spacing of 4 nm and the Bragg-diffraction spots are split in the ⟨100⟩ directions by this increased spacing (Fig. 9.26). The anti-domain structure can be imaged in the dark-field mode (Fig. 9.27) in which one Bragg reflected beam and the surrounding superlattice reflections contribute to the image

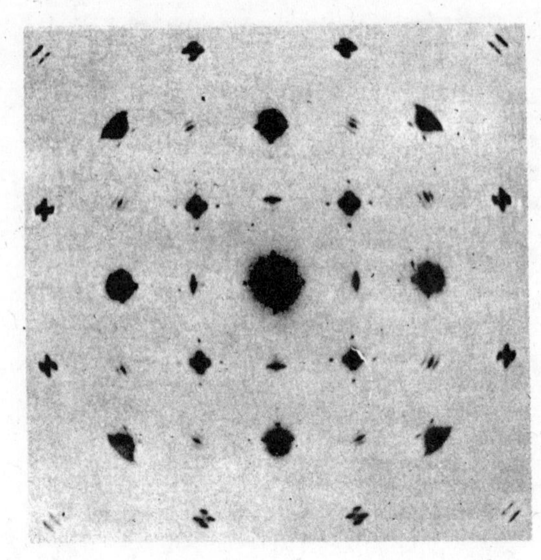

Fig. 9.26. Electron diffraction pattern of an ordered CuAu II film [9.250]

Fig. 9.27. Dark-field imaging of antidomain boundaries 2 nm apart in a CuAu II film as an interference effect between the diffracted beam and its satellites (see Fig. 9.26) [9.251]

intensity. Similar superlattice reflections can be observed in the EDPs of other alloys (see, for example, [9.252]).

Below 380 °C, the phase CuAu I is stable; this consists of a face-centred tetragonal lattice ($c/a = 0.92$) again with alternating (002) planes of Au and Cu atoms but without the closely spaced domain boundaries of the CuAu I phase. These domain boundaries are planar faults and can be identified by their fringe pattern in electron micrographs (Sect. 8.4).

9.4.6 Convergent Beam and High-Order Laue Zone (HOLZ) Diffraction Patterns

A line scan across the circles of a convergent-beam diffraction pattern (Sect. 9.3.3 and Fig. 9.18 b) is nothing else but a rocking curve (Sect. 7.3.4, Figs. 7.18, 20), which contains information about the local specimen thickness t, and the extinction and absorption distances ξ_g and ξ_g'. The thickness can be obtained from the positions of the subsidiary minima of the "pendellösung" fringes. The two-beam rocking curve (7.69) has minima if the argument of the sine term is an integral multiple n of π. The corresponding excitation errors s_n are then given by

$$\text{a)}\ s_n^2 = \frac{n^2}{t^2} - \frac{1}{\xi_g^2} \qquad \text{or} \qquad \text{b)}\ \left(\frac{s_n}{n}\right)^2 = \frac{1}{t^2} - \frac{1}{n^2\xi_g^2} \tag{9.41}$$

Plotting s_n^2 against n^2 [9.253] gives us t from the slope and ξ_g from the intercept with the coordinate; alternatively plotting $(s_n/n)^2$ against $1/n^2$ gives t as the intercept with the ordinate [9.254]. In both cases, the correct starting number n_1 of the first minimum has to be known. For a foil thickness between $m\xi_g$ and $(m+1)\,\xi_g$, the appropriate value is $n_1 = m + 1$. If the wrong value of n_1 is used, the plot does not give a straight line.

However, a two-beam case can never be perfectly realised, and many-beam calculations are necessary. Using rough values of t and ξ_g given by the two-beam method, many-beam best fits must be computed, allowing t, ξ_g and ξ_g' to vary.

A necessary condition for CBED is that the circles do not overlap, which restricts the range of rocking. A larger rocking angle can be employed without overlap by the double-rocking technique (Sect. 9.3.3) or by ZAP patterns (Sect. 9.3.4).

The CBED pattern may also contain a diffuse background with Kikuchi bands. This background has to be subtracted before seeking a best fit with many-beam calculations. A photometric record of the background very near the selected trace through the CBED pattern can be obtained by placing a thin wire across the circular diaphragm in the condenser lens, which casts a shadow across the CBED pattern and allows the diffuse background inside the shadow to be measured [9.255].

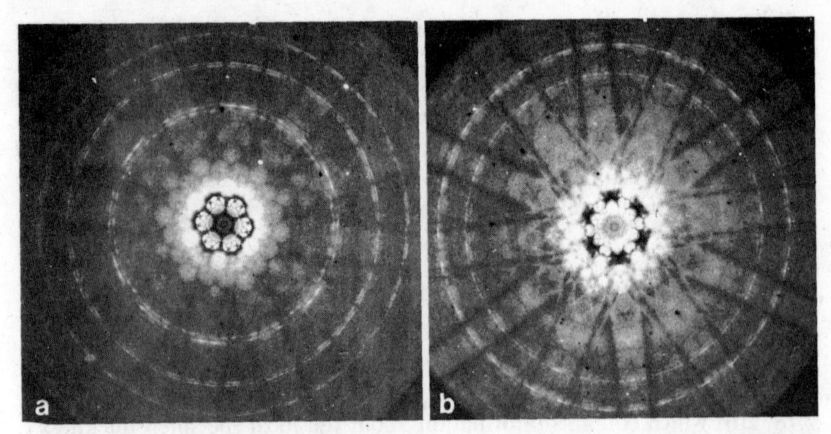

Fig. 9.28. High-order Laue zone (HOLZ) diffraction patterns of 2 H polytypes of (**a**) MoS_2 and (**b**) $MoSe_2$ with structure factors $(f_{Mo} - \sqrt{2} f_C)$, f_{Mo} and $(f_{Mo} + \sqrt{2} f_C)$ (C: chalcogen) for the first to third Laue zones respectively, showing that the first-order Laue zone (FOLZ) of $MoSe_2$ is practically invisible owing to the very small contribution from $f_{Mo} - \sqrt{2} f_{Se}$ [9.258]

A high-order Laue zone (HOLZ) diffraction pattern is obtained when the electron beam is incident on the specimen parallel to a low-index zone axis. The Ewald sphere intersects the needles of the zero-zone reciprocal-lattice points, producing the convergent-beam circles of the Bragg reflections around the primary beam; at larger angular distances, $\sin \theta_n = \lambda R_n$, the next higher Laue zones of order n are intersected in the reciprocal lattice with radii R_n (Figs. 7.6 b, 9.28). These radii can be evaluated from Fig. 7.6 b using the result given by elementary geometry:

$$R_n^2 = g_n \left(\frac{2}{\lambda} - g_n \right) ; \qquad g_n = \frac{n}{a \sqrt{m^2 + n^2 + o^2}} , \tag{9.42}$$

where $2/\lambda$ is the diameter of the Ewald sphere, and g_n denotes the distance between the n-th Laue zone and the zero Laue zone for the zone axis $[mno]$. Unfortunately, the pole-piece diameter of most TEMs restricts the observable diameter of diffraction patterns to about one third of that shown in Fig. 9.28. Quantitative measurement of R_n gives information about the third dimension of the reciprocal lattice, especially for materials with layer structures such as $ZrSe_2$, $TaSe_2$, NbS_2, TaS_2 or MoS_2 [9.256, 257]. For different crystal structures, the HOLZ rings appear with different relative intensities (Fig. 9.28). Laue zones may disappear completely if the structure amplitude F is zero for them but these forbidden reflections can reappear with weak intensity by many-beam systematic interactions. HOLZ reflections correspond to low values of V_g and hence to large values of ξ_g, and their intensities seem to be directly related to the structure factor $|F|^2$ for these beams even

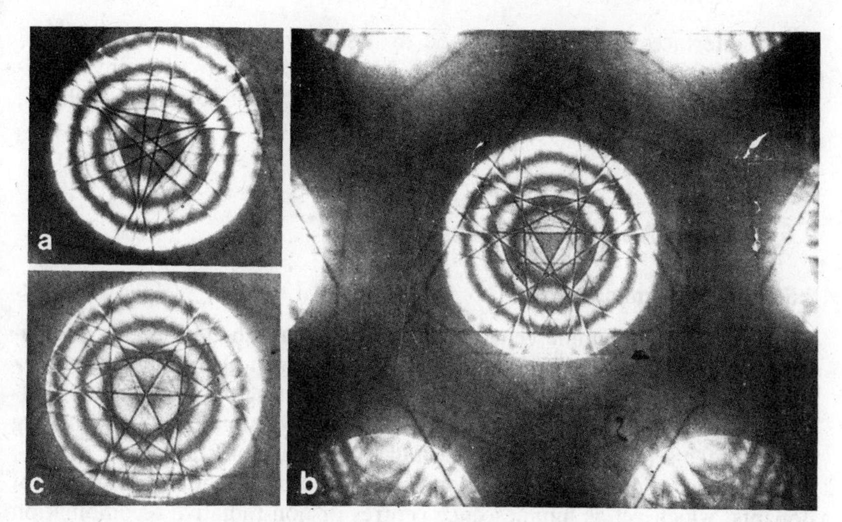

Fig. 9.29. HOLZ lines in the primary circle of a CBED pattern (neighboured six 220 reflections in b) from (111) Si at three different electron energies (**a**) $E = 96.5$ keV (**b**) 100.5 keV, (**c**) 103.5 keV. The bright rings are pendellösung fringes that depend on film thickness and result from the interaction with the Bragg reflections from the zero Laue zone [9.262]

for thicker specimens. However, in crystal thicknesses greater than a few tens of nanometres, dynamical interaction effects can occur and these furnish information about the crystal potential and the dispersion surface [9.258–261]. The high-order reflections decrease more strongly with increasing temperature due to the Debye-Waller factor (Sect. 7.5.3) so that specimen cooling increases the intensity of HOLZ rings and of HOLZ lines in the CBED of the primary beam (see below).

In a CBED pattern, the HOLZ reflections become bright HOLZ lines, and each bright line in the outer HOLZ rings appears inside the primary-beam circle as a dark line, because there is a relationship between excess and defect Kikuchi lines and the HOLZ lines. However, Kikuchi lines are hard to observe for thin specimens and become clearer as the thickness is increased, whereas HOLZ lines are narrower and are most readily visible at thicknesses for which the Kikuchi lines are still weak. The pattern of overlapping lines (Fig. 9.29) depends very sensitively on the electron energy E [9.262] and/or the lattice constant a. The lattice dimension can be evaluated with high precision once the beam voltage has been calibrated with a lattice of known dimensions (e.g., Si); alternatively, a relative change Δa of the lattice constant due to a change of composition or to electron-beam heating [9.263] can be obtained from a shift of the HOLZ lines when the pattern of crossing lines is compared with a set of computer maps. The accuracy is $\Delta a/a = \Delta E/2E = 2 \times 10^{-4}$ at $E = 100$ keV. Thus local concentrations of Al in Cu-Al alloys can be measured with an accuracy of 1 at% [9.257] or strains at planar interfaces can be determined in a similar fashion to the chemical changes [9.264]. The symmetry of HOLZ lines in the centre of a convergent-beam pattern can be used for determination of the crystal space group [9.265, 266].

9.5 Further Analytical Modes

9.5.1 Cathodoluminescence

The use of cathodoluminescence (CL) is a well-established technique in scanning electron microscopy (SEM). The CL signal can also be recorded in a TEM equipped with a STEM attachment by collecting the light quanta emitted. An advantage of this mode is the possibility of simultaneously imaging lattice defects in the STEM mode and examining their influence on CL; alternatively, additional information about variations in the concentration of dopants, which act as luminescence centres or non-radiative recombination centres, can be obtained. A disadvantage is that the CL intensity is attenuated in a foil thickness much smaller than the electron range and, in addition, the film surface acts as a dead layer. The method is restricted to those semiconductors with a high-luminescence yield. Otherwise resolutions of the order of a few tens of nanometres are obtainable thanks to the reduction of electron diffusion.

The low intensity requires an efficient light-collection system with a large solid angle and a lateral selection of the irradiated area only to shield the

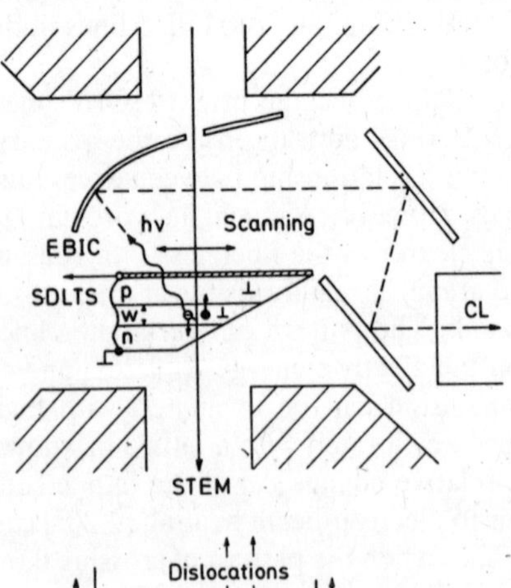

Fig. 9.30. Diagram showing a light-collection system for cathodoluminescence (CL) and the simultaneous recording of an electron-beam-induced current (EBIC) and a STEM signal from a polished wedge-shaped p-n junction

signal from CL contributed by diffusely scattered electrons. An obstacle for the collection of the light quanta is the small dimension of the pole-piece gap. A tapered silver tube or an elliptical mirror (Fig. 9.30) is used to transmit the light to a quartz light pipe and a photomultiplier [9.267]. X-rays can cause CL in the quartz light pipe and this signal must be eliminated by placing additional lead-shielded mirrors between the collection system and the light pipe [9.268].

In diamond, for example, almost all of the luminescence is emitted from dislocations as a result of localized electron states near these defects [9.267]. The CL depends on the crystal orientation and shows bend contours that are similar to channelling effects in energy-loss spectroscopy and x-ray emission; this has been shown for ZnS single-crystal foils [9.269]. In a $Ga_{1-x}Al_xAs$ laser structure, CL can be applied to analyse radiative and non-radiative centres, to record the luminescence spectra from the different parts of the structure, which may be separated by a wedge-shaped etching of the structure, and to compare the CL signal with that obtained in EBIC experiments [9.268].

9.5.2 Electron-Beam-Induced Current (EBIC)

The EBIC mode is widely used in SEM [9.274]. A strong local electric field in a p-n junction (Fig. 9.30) or a Schottky barrier must be present to separate the electron-hole pairs generated by the electron beam. A current can be recorded at zero bias or with a reverse bias, which increases the field strength and the width w of the depletion layer. The EBIC signal consists not only of electron-hole pairs separated in the depletion layer but also of minority carriers, which reach the layer by diffusion. The EBIC signal may decrease at lattice defects, such as dislocations or stacking faults, which act as recombination centres. It is thus of interest to image the lattice defects in the TEM or STEM mode. Combination of the SEM/EBIC and the TEM modes (in different instruments) [9.271] has the advantage that the EBIC mode can be applied first to the bulk-semiconductor device, after which a TEM investigation of the same thin area can give information about the faults. Another possibility is to observe the thin sample in STEM or TEM with a scanning attachment in the STEM and EBIC modes simultaneously [9.268, 272]. Because the active area and depletion layers are of the order of a few micrometres thick, HVEM offers a better penetration of thick regions. Unlike SEM/EBIC experiments, in which the electron-hole pairs are generated in the whole volume of the electron cloud a few micrometres in diameter, the generation in STEM/EBIC is concentrated in the volume irradiated by the primary electron probe, which is only slightly broadened by multiple scattering. This can result in a better resolution though the latter is ultimately limited by the diffusion of the minority carriers.

Another interesting mode is scanning deep-level transient spectroscopy (SDLTS) [9.268, 273], which can provide a profile of the defect concentration in the direction normal to the junction. The STEM electron probe is switched on and off so that the deep levels are filled by the injected carriers when the beam is on. They can be detected by observing the thermally stimulated current transient created when the levels empty during the off time of the beam. The depth of the levels (activation energy for emission) can be determined from the temperature dependence of the transient-time constant, which can be measured by opening two sampling-rate windows at times t_1 and t_2 after the electron-beam chopping pulse. Methods for chopping electron beams are described in Sect. 6.6.2.

10. Specimen Damage by Electron Irradiation

Most of the energy dissipated in energy losses is converted into heat. The rise in specimen temperature can be decreased by keeping the illuminated area small.

The electron excitation of organic molecules causes bond rupture and loss of mass and crystallinity. The damage is proportional to the charge density in C cm^{-2} at the specimen. This limits the high-resolution study of biological specimens.

Inorganic crystals can be damaged by the formation of point defects, such as colour centres in alkali halides, and of defect cluster. High-energy electrons can transfer momentum to the nuclei, which results in displacement onto an interstitial lattice site when the enery transferred exceeds a threshold value of the order of 20–50 eV.

Hydrocarbon molecules condensed from the vacuum of the microscope or deposited on the specimen during preparation and storage can form a contamination layer by radiation damage and cross-linking. Especially when only a small specimen area is irradiated, the hydrocarbon molecules can diffuse on the specimen and are cracked and fixed by the electron beam.

10.1 Specimen Heating

10.1.1 Methods of Measuring Specimen Temperature

The specimen temperature can be calculated only for simple geometries, a circular hole covered with a homogeneous foil, for example (Sect. 10.1.3). In practice, therefore, methods of measuring the specimen temperature during electron irradiation are necessary. These can also be used for calibration of specimen heating and cooling devices. (Table 10.1 summarizes the various methods in use. The continuous methods (I) can be employed in the whole temperature range indicated, while the other methods (II) use a fixed transition temperature as reference.

The first group contains methods based on evaporated thermocouples. However, such methods can only be applied to special geometries due to the difficulty of ensuring proper contact between the evaporated films. Furthermore, the electromotive force of a thermocouple depends on the thickness and structure of the evaporated layers [10.17, 18]. Another possibility is to

Table 10.1. Methods of measuring specimen temperature

Specimen	Temperature	Observed effect	Observing mode	Ref.
I. Continuous methods				
1. Cu-constantan	< 600 °C	Thin film thermocouple	Thermopower	[10.1]
2. Ag-Pd		Thin film thermocouple	Thermopower	[10.2]
3. Au-Ni		Thin film thermocouple	Thermopower	[10.3]
4. Pb	< 300 °C	Thermal expansion	Electron diffraction	[10.4]
5. Ni	< 700 °C	Thermal expansion	Electron diffraction	
6. Al and others		Critical voltage	Electron diffraction	[10.5]
II. Methods using transition temperatures				
a) Reversible indicators				
7. $AuCu_3$	388 °C	Superstructure	Electron diffraction	[10.6]
8. Ag_2S	179 °C	Rhombohedral → cubic	Electron diffraction	[10.7]
9. Ice	–70 °C	Cubic-ice	Electron diffraction	[10.8]
10. Thin In layer	156 °C	Melting	Dark-field mode	[10.9, 10]
11. Thin Pb layer	327 °C	Melting	Dark-field mode	
12. Thin Ge layer	958 °C	Melting	Dark-field mode	
13. Paraffin	70 °C	Melting before damage	Electron diffraction	[10.11]
b) Irreversible indicators				
14. Pb layer	327 °C	Flowing out of the melting zone	Bright-field mode	[10.12]
15. Al_2O_3 layer	600 °C	Amorphous → crystalline	Bright-field mode	
16. Sn layer	232 °C	Melting	Bright-field mode	[10.13]
17. Fe layer	920 °C	bcc → fcc	Bright-field mode	
18. Fe_3C	720 °C	Dissolving in Fe matrix	Bright-field mode	
19. PbO needles	890 °C	Melting	Bright-field mode	[10.14]
20. NaCl crystals	700 °C	Sublimation	Bright-field mode	[10.15]
21. NaCl and other alkali halide crystals		Melting of larger crystals	Bright-field mode	[10.16]

use the thermal expansion of a crystal lattice since the Bragg angle in Debye-Scherrer ring patterns or single-crystal diffraction diagrams decreases with increasing temperature. (Numerical example: for an evaporated Pb film, temperature rise of $\Delta T = 100\,°C$ results in $\Delta d/d = 1.8 \times 10^{-3}$). The temperature of a small specimen area can be determined by selected-area electron diffraction [10.4]. The variation of critical voltage (Sect. 7.4.4) with temperature provides another method for small areas [10.5].

The indicator specimens (II) can be divided into those that are modified reversibly and those for which the modification is irreversible. The first kind have the advantage that the transition temperature can be crossed repeatedly. This makes it easier to determine the current density necessary to bring the specimen to the fixed transition temperature. If the excitation of the condenser lens is not changed, the temperature rise due to electron bombardment is proportional to the incident electron current, which can be changed by means of the Wehnelt bias. This enables us to estimate the temperature for other values of the electron current. However, large variations of specimen temperature ensue when the diameter of the irradiated area is changed (Sect. 10.1.3).

A very simple and straightforward method is biased on the behaviour of evaporated indium layers with an island structure in the vicinity of the melting point. The melting or solidification of the small crystals is indicated by the vanishing or reappearance of the Bragg reflections in dark-field conditions. However, the presence of an evaporated film can alter the rate of heat generation and the heat conductivity of the specimen. The method can be used for local measurement of temperature without greatly affecting the generation of heat and the heat conductivity if a 10 μm-diameter film spot is evaporated through an optically aligned 10 μm diaphragm [10.9, 10].

The specimen temperature in a liquid-helium-cooled stage can be estimated from the condensation of gases (Xe, Kr, O_2, A, N_2, Ne) in the range 10–70 K [10.19]. The gas-sublimation temperatures depend, however, on the gas pressure which is not known precisely because of the cryopumping effect of the cold shield. This temperature increases from 58 K to 68 K for xenon and from 8.5 K to 10 K for neon in the pressure range 10^{-4}–10^{-2} Pa, for example [10.20].

Another method to measure the rise of temperature uses the known temperature dependence of the climb rate of Frank dislocation loops in material with high stacking-fault energy [10.21]. An average local rise of 6 °C was determined if irradiating an Al-1.5 wt.% Mg alloy under normal bright-field operation.

10.1.2 Generation of Heat by Electron Irradiation

A knowledge of the mean contribution $\Delta Q/\Delta x$ of one electron per unit mass thickness ($\Delta x = \varrho t$) to specimen heating is required for a theoretical discussion and calculation of the specimen temperature.

If it is assumed that only plasmon energy losses ΔE with a mean-free-path length Λ contribute to heat generation, $\Delta Q/\Delta x$ can be calculated by use of (5.69):

$$\frac{\Delta Q}{\Delta x}=\frac{\sum_{n=1}^{\infty} n\Delta E P_n(t)}{\varrho t}=\frac{\mathrm{e}^{-t/\Lambda}\Delta E}{\varrho\Lambda}\sum_{n=1}\frac{(t/\Lambda)^{n-1}}{(n-1)!}=\frac{\Delta E}{\varrho\Lambda}\,. \tag{10.1}$$

Thus for Al with a plasmon loss at ΔE = 15.3 eV and a mean-free-path Λ = 70 nm for E = 60 keV, we find $\Delta Q/\Delta x$ = 0.83 eV μg^{-1}cm^2. This value is much smaller than that found experimentally (Fig. 10.1). This means that Bethe losses, which appear in Fig. 5.16 as a continuous background to the energy-loss spectrum, also contribute strongly to specimen heating.

It therefore seems more reasonable to use the Bethe continuous slowing-down approximation (5.79) to estimate the rate of heat generation. This model was applied to the problem of specimen heating [10.22] and yields for non-relativistic energies

$$\frac{\Delta Q}{\Delta x}=\left|\frac{dE}{dx}\right|=7.8\times 10^4\frac{Z}{A}\frac{1}{E}\ln\frac{E}{IZ} \tag{10.2}$$

with E [eV], $\Delta Q/\Delta x$ [eV μg^{-1}cm^{-2}], and $J = IZ \simeq 13.5\,Z$ denotes the mean ionisation energy in eV units. Figure 10.1 contains values calculated from this formula for E = 60 keV. The dependence on the atomic number Z is confirmed by experiment [10.23, 24]. In these experiments, the energy loss converted into heat was measured directly by irradiating a large transparent thermocouple on a supporting film of 1 cm diameter. The energy loss was

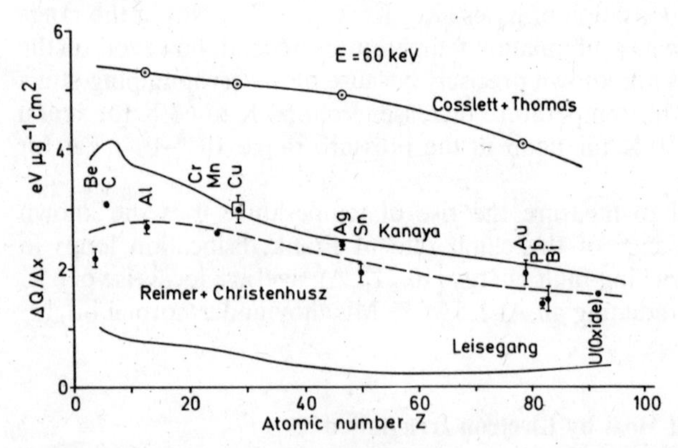

Fig. 10.1. Dependence of energy dissipated per unit mass-thickness $\Delta Q/\Delta x$ [eV/μg cm^{-2}] on the atomic number Z for 60 keV electrons. Full curve calculated with the Bethe formula (10.2) [10.23, 24]. The calculations of *Leisegang* [10.25] and the measurements of *Cosslett* and *Thomas* [10.26] at E = 10–20 keV and transferred to 60 keV under- or overestimate the value of $\Delta Q/\Delta x$

calibrated by measuring the heat generated by an electric current passing through the irradiated film and producing the same temperature. Values calculated by *Leisegang* [10.25] underestimate $\Delta Q/\Delta x$; the experimental values of *Cosslett* and *Thomas* [10.16] obtained at E = 10–20 keV and transferred to 60 eV overestimate $\Delta Q/\Delta x$ since multiple scattering is frequent at low energies. For relativistic energies $E > 60$ keV $\Delta Q/\Delta x$ becomes proportional β^{-2} due to the Bethe formula (5.79).

The heat generated is proportional to the mass-thickness of electron-transparent films. There is a stronger increase for greater thicknesses due to multiple scattering and to the increase of the inelastic scattering probability with decreasing electron energy [10.24, 27]. When bulk specimens such as specimen-grid bars or diaphragms are irradiated, the generation of heat per unit time P saturates for thicknesses of the order of the electron range at a value

$$P = fP_0 = fIU\,, \tag{10.3}$$

where $P_0 = IU$ denotes the total beam power. For bulk copper, for example, $f = 66\%$. A fraction of the power P_0 is lost to the backscattered electrons.

10.1.3 Calculation of Specimen Temperature

The specimen temperature becomes stationary when the heat generated is equal to the heat dissipated by radiation and thermal conduction. This problem of thermal conduction can be solved only for simple geometries such as a circular hole covered with a uniform foil [10.22, 28, 29] or for rod-shaped specimens (needles) [10.14].

The power dissipated by radiation can be estimated using the Stefan-Boltzmann law

$$P_{\mathrm{rad}} = SA\sigma(T^4 - T_0^4)\,, \tag{10.4}$$

where S denotes the specimen surface area (both surfaces must be counted for foils), A is the absorptive power of the black-body radiation which is the same as the emissivity (Kirchhoff's law), σ is the Stefan-Boltzmann constant, T is the temperature of the specimen and T_0 that of the environment. The absorptive power A is equal 1 for only a black body and is of the order of 0.01–0.05 for bulk metals. For thin transparent films, A is still smaller: $A = 9 \times 10^{-4}$ for a 10 nm collodion film, for example [10.30]. The influence of radiation loss can therefore be neglected for thin-foil specimens and has to be considered if the heat dissipation by thermal conduction is reduced by the presence of a large self-supporting area and/or if the temperature is high, because the radiation loss increases with the fourth power of T in (10.4). This can be seen from the following example [10.10]. A SiO foil was over a 400 μm-diameter hole, and indicator spots of Ge (T_{m} = 958 °C) and In

(156 °C) were deposited near the centre. The current densities necessary for melting were in the ratio 9.3 : 1. A ratio 6.9 : 1 would be expected for pure heat conduction and of 82 : 1 for pure radiation loss. The increase of the rate of dissipation of heat in Ge relative to the value for pure heat conduction is just detectable, whereas the same experiment with a 200 μm diaphragm gave the ratio expected for pure heat conduction. The irradiation of small particles on a supporting film of poor heat conductivity produces another extreme case. The temperature then increases to a value at which the radiation loss becomes dominant, with the result that the current density required to melt the particles is proportional to T_s^4 [10.16].

The heat transfer by conduction through an area S caused by a gradient ∇T of the specimen temperature is described by

$$P_C = -\lambda S \nabla T\,, \tag{10.5}$$

where λ is the heat conductivity. Two extreme radiation conditions for a foil over a diaphragm of radius R will now be discussed: uniform illumination and highly localized (small-area) illumination.

a) Uniform Illumination. The whole foil is irradiated with a uniform current density j [A cm^{-2}] by means of a strongly defocused condenser-lens system. The thermal power generated inside a circle of radius r and area πr^2 with the foil centre at $r = 0$ has to be transferred by heat conduction through an area $S = 2\pi r t$ (t: foil thickness). This results in the equilibrium relation:

Power dissipated = power transferred by heat conduction

$$\pi r^2 \frac{j}{e}\frac{\Delta Q}{\Delta x}\varrho t = -\lambda 2\pi r t\frac{dT}{dr} \rightarrow \frac{dT}{dr} = -\frac{j\varrho}{2e\lambda}\frac{\Delta Q}{\Delta x}r \tag{10.6}$$

Integrating (10.6) and using the condition $T = T_0$ at $r = R$, we find

$$T(r,\tau) = T_0 + \frac{j\varrho}{4e\lambda}\frac{\Delta Q}{\Delta x}(R^2 - r^2)(1 - e^{-\tau/\tau_0})\,. \tag{10.7}$$

The last term in the square brackets arises from the equation of heat transfer considering the dependence on time τ with the time constant

$$\tau_0(r) = \frac{\varrho c(R^2 - r^2)}{4\lambda}\,, \tag{10.8}$$

c being the specific heat of the foil. Table 10.2 contains some calculated time constants $\tau_0(0)$ for the foil centre ($r = 0$). A value of 99.3% of the stationary temperature is reached after a time of $5\,\tau_0$. This means that the temperature of an electron microscope specimen rises so rapidly that the stationary value

Table 10.2. Heat conductivity λ, density ϱ, specific heat c and time constant $\tau_0(0)$ at $r = 0$ and for $R = 50$ μm

Substance	λ [J/K cm s]	ϱ [g/cm³]	c [J/g K]	τ_0 [ms]
Carbon film	1.5×10^{-2} [10.31]	2.0	–	–
Formvar film	2.4×10^{-3} [10.31]	1.2	2.0	7.5
Glass (SiO)	10^{-2}	2.2	0.8	1.1
Metal (Cu)	4	8.9	0.36	5×10^{-3}

Time constant of an irradiated grid bar (Cu, 2 mm long): $\tau_0 = 4$ ms

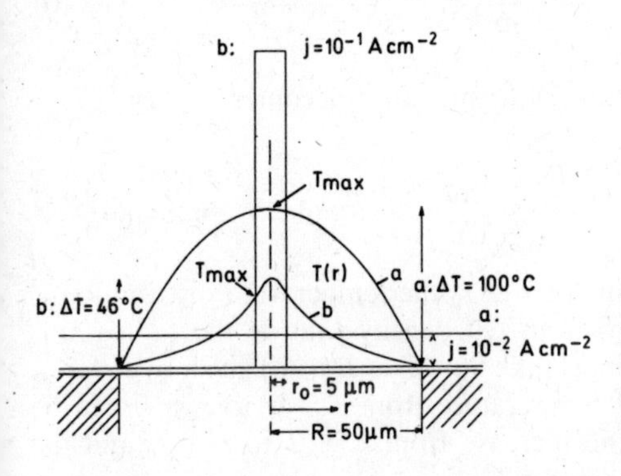

Fig. 10.2. Temperature distribution in a formvar film irradiated with 60 keV electrons for (curve **a**) uniform current density j and (curve **b**) small-area illumination

Table 10.3. Rise of specimen temperature ΔT in the centre of a circular diaphragm ($R = 50$ μm) covered with a supporting film and irradiated with 100 keV electrons

Substance	Uniform illumination $R = 50$ μm, $j = 10^{-2}$ A cm^{-2}	Small-area illumination $r_0 = 0.5$ μm, $j = 1$ A cm^{-2}
Formvar	62 °C	6 °C
Glass (SiO)	27 °C	2.5 °C
Metal (Cu)	0.3 °C	0.03 °C

will be attained immediately after the irradiation conditions are changed. However, the heat conductivity of organic specimens can be altered by radiation damage and normally increases with increasing irradiation as a result of cross-linking; this causes a decrease of temperature with increasing irradiation time and constant illumination conditions. This damage process requires a much higher electron dose $j\tau$ than does the mass loss of organic films for example (Sect. 10.2.2). The heat conductivity of pure carbon and SiO films is not affected by irradiation [10.10].

The stationary temperature distribution $T(r, \infty)$ of (10.7) has a parabolic form (Fig. 10.2, curve a) with the maximum temperature

$$T_{max} = T_0 + \frac{j\varrho}{4e\lambda}\frac{\Delta Q}{\Delta x}R^2 \qquad (10.9)$$

at the centre ($r = 0$) of the diaphragm.

b) Small-Area Illumination. Equation (10.5) has to be modified when irradiating a smaller specimen area by using a more strongly focused condenser lens. If only a small area of radius r_0 is irradiated with a current density j ($r_0 \ll R$), the heat-transfer equation becomes, for $r \geq r_0$

$$\pi r_0^2 \frac{j}{e}\frac{\Delta Q}{\Delta x}\varrho t = -\lambda 2\pi r t\frac{dT}{dr} \rightarrow \frac{dT}{dr} = -\frac{j\varrho}{2e\lambda}\frac{\Delta Q}{\Delta x}\frac{r_0^2}{r}. \qquad (10.10)$$

With $T = T_0$ for $r = R$, the stationary temperature becomes

$$T(r,\infty) = T_0 + \frac{j\varrho}{2e\lambda}\frac{\Delta Q}{\Delta x}r_0^2 \ln\frac{R}{r} \quad \text{for} \quad r \geq r_0. \qquad (10.11)$$

Inside the irradiated area, that is for $r \leq r_0$ the temperature distribution is again parabolic, provided that the current density j inside the radius r_0 is uniform. At $r = r_0$, the solution must take the value $T(r_0,\infty)$ given by (10.11). However, the small increase of temperature from $r = r_0$ to $r = 0$ can be neglected in comparison with the increase from $r = R$ to $r = r_0$ and the temperature in the foil centre becomes

$$T_{max} = T_0 + \frac{j\varrho}{2e\lambda}\frac{\Delta Q}{\Delta x}r_0^2 \ln\frac{R}{r_0}. \qquad (10.12)$$

This is a much smaller temperature rise $\Delta T = T_{max} - T_0$ than that predicted by (10.9) even for larger current densities (thus for a ten-fold increase of current density in Fig. 10.2 ΔT shows only half the value than for uniform illumination; see also the numerical example of Table 10.3.

For more-exact calculations, the current-density distribution within the electron beam, typical Gaussian, has to be considered [10.10, 13]. If the electron beam hits the diaphragm or the specimen-grid bars, the temperature T_0 in (10.9) can be increased sharply, owing to the greater generation of heat (10.3) in bulk material. In practice, therefore, irradiation of bulk parts of the specimen support should be avoided and the area of the foil or supporting film irradiated should be kept as small as possible to limit specimen heating. The estimated values in the last column of Table 10.3 show that under these conditions, the rise of temperatur due to electron irradiation can be kept low.

Whereas an increase of specimen temperature of a few degrees K has no significant effect with the specimen at room temperature (300 K), this can become a large relative increase at liquid-helium temperature (4 K). The radiation damage depends very sensitively on temperature (Sect. 10.2.3) and most of the discrepancy in experimental results at low temperature may be attributed to a temperature rise [10.32]. The heat conductivities of most substances decrease by one or more magnitudes when the temperature is decreased from 300 to 4 K [10.33]. It is very important that there is a good heat conductivity of the supporting film (carbon), good thermal contact between film and grid and support and good cryo-shielding of parts of the microscope at high temperature. The special aspect of frozen-hydrated biological specimens in a cold stage is discussed in [10.34, 35].

The temperature increase caused by a moving electron probe is of interest for scanning transmission electron microsocpy [10.36, 37].

10.2 Radiation Damage of Organic Specimens

10.2.1 Elementary Damage Processes in Organic Specimens

Radiation damage in organic material is caused by all kinds of ionizing radiation. The damage depends on the energy dissipated per unit volume, which is proportional to the number of incident electron $n = j\tau/e$ per unit area, where τ is the irradiation time in seconds. The incident charge density $q = j\tau = en$ [C cm^{-2}] can thus be used to compare different irradiation conditions. The quantity q is called *electron dose* although in radiation chemistry, dose is defined as energy dissipated per unit volume [J cm^{-3}].

A dose of 1 C cm^{-2} corresponds to a number of electrons $n = 6 \times 10^{18}$ cm^{-2} = 6×10^4 nm^{-2} and at $E = 60$ keV to a transferred energy density $n\varrho|\Delta E/\Delta x| = 1.8 \times 10^{25}$ eV cm^{-3} = 1.8×10^4 eV nm^{-3} $\simeq 10^6$ J cm^{-3}, if we insert the values $\varrho = 1$ g cm^{-3} and $|\Delta E/\Delta x| = 3 \times 10^6$ eV g^{-1}cm^2 that correspond to carbon (Fig. 10.1). This last value is given by the Bethe formula (5.79) and (10.2). Most of this energy density is consumed in ionisation processes. From (5.79), we see that this contribution decreases as v^{-2} with increasing electron energy E (see also Sect. 10.2.3).

Current densities of the order of $j = 10^{-2}$ A cm^{-2} are necessary at magnifications $M \simeq 10\,000$, which corresponds to an energy density of 200 eV nm^{-3} in 1 s. High-resolution micrographs require an electron dose of $q = 0.5$ C cm^{-2} to expose a photographic emulsion with a density $S = 1$, which means an energy density of 10^4 eV nm^{-3}! Table 10.4 contains a scale of physical and biological damage effects (Reviews: [10.38–48]). The energy dissipated in an electron microscope specimen after a brief irradiation, therefore corresponds to conditions that, outside a microscope, occur only near the centre of nuclear explosions! Results from radiation chemistry, where the energy densities are much lower, in the order of 1–10 J cm^{-3} [10.38, 49–54], are relevant only to the very early stage of radiation damage.

The primary damage process is inelastic scattering, which causes molecular excitation or ionisation or collective molecular excitations (similar to plasmon excitations of a free-electron gas). The energy dissipated is either

Table 10.4. Scale of radiation-damage processes for 100 keV electrons

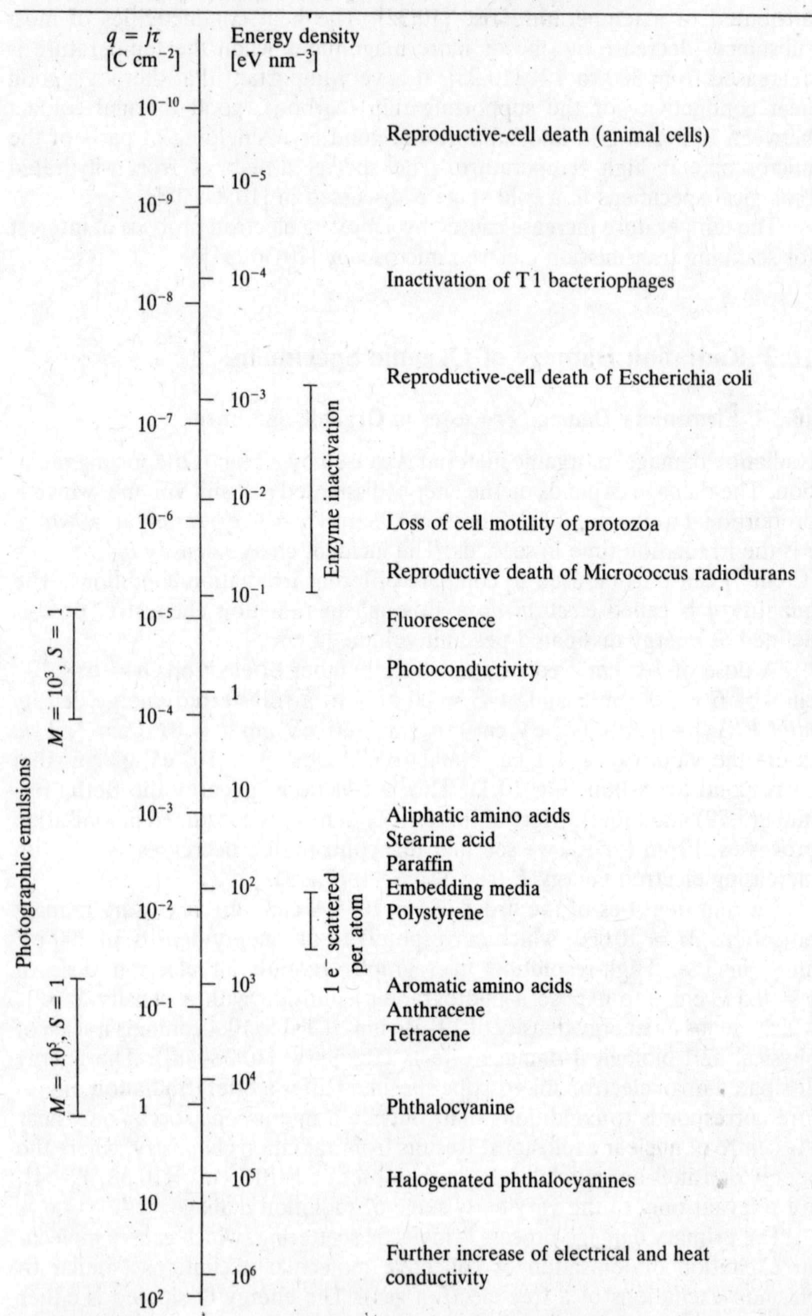

converted to molecular vibrations (heat) or causes bond scission, for example, a loss of hydrogen and production of radicals:

$$MH^* \rightarrow M^{\cdot} + H^{\cdot}$$

(the bonding energy of H to C atoms is in the order of 4 eV) or a bond break in a carbon chain:

$$R\text{–}CH_2\text{–}CH_2\text{–}CH_2^*\text{–}R \rightarrow R\text{–}CH_3 + CH_2\text{=}CH\text{–}R \, .$$

C–H bonds break more frequently in aliphatic chains than in aromatic compounds, owing to the spread of energy dissipation by the π-electron system of benzene rings [10.55]. Bond scission also leads to the production of low-molecular-weight molecules and radicals. These primary processes are unaffected by temperature and cannot be avoided by specimen cooling.

A quantum-mechanical calculation has been applied to radicals produced by cleaving hydrogen from the DNA bases adenine, guanine, cytosin and thymin [10.56]. These results can be used to discuss the change of the fine structure of the carbon *K* edge in electron-energy-loss spectra [10.45].

The loss of hydrogen atoms only would not affect strongly the molecular and crystal architecture, but the scission of carbon chains and of side groups and the secondary processes cause great damage. Secondary processes are, for example, the diffusion of hydrogen atoms to other molecules and the formation of addition radicals at unsaturated bonds or of a hydrogen molecule:

$$R\text{–}H + H^{\cdot} \rightarrow R\text{–}H_2^{\cdot} \quad \text{or} \quad R\text{–}H + H^{\cdot} \rightarrow R^{\cdot} + H_2 \, .$$

Other typical secondary processes are cross-linking of molecular chains, such as

$$\begin{matrix} R\text{–}CH_2\text{=}CH\text{–}R \\ \\ R\text{–}CH_2\text{–}CH_2^*\text{–}R \end{matrix} \rightarrow \begin{matrix} R\text{–}CH_2\text{–}CH\text{–}R \\ | \\ R\text{–}CH_2\text{–}CH\text{–}R \end{matrix} + H \, ,$$

the reactions with radicals and the thermal diffusion and evaporation of fragmented atoms and molecules of low molecular weight, such as H_2, CH_4, CO_2, NH_3. The generation of H_2 by electron irradition was demonstrated by mass spectroscopy of released H_2 from specimens irradiated at low temperature (10 K) and warmed to room temperature [10.57]. These processes, which involve loss of mass can be diminished by specimen cooling. In some cases, the probability that the scission products recombine or cross-link before leaving the specimen can be increased. This cage or frozen-in effect is the most important process whereby the occurrence of secondary processes can be reduced at low temperatures (Sect. 10.2.3).

Table 10.5. Principal bonds broken by radiation damage in pure organic compounds and some G values [10.50, 58]

Type of compound	Main site of attack	G_{-M}	G_{H_2}	G_{CO_2}	G_{CO}	G_{H_2O}	G_{CH_4}
Hydrocarbons							
Saturated	C–H, C–C	6 – 9	3.8 –5.6				0.2–0.7
Unsaturated	C–H, C–C, polymerization or cross-linking	11 –10	0.8 –1.2				0.13
Aromatic	C–H, Side chain C–C	0.2– 1	0.01–0.18				
Alcohols	H–COH, C–COH	3 – 6	3.5 –4.5		0.04–0.23	0.3–0.9	
Ethers	C–H, C–OR	7	2.0 –3.6		0.06–0.13		
Aldehydes and ketones	C–H, C–C=O	7	0.8 –1.2		0.6 –1.6		0.1–2.6
Esters	C–H, O=C–OR	4	0.5 –0.9	0.3–1.6	0.15–1.6		0.4–2.0
Carboxylic acids	C–H, C–COOH	5	0.5 –2.3	0.5–4.0	0.1 –0.5	0.1–2.2	0.5–1.4
					G_{NH_3}		
Amino acids							
Leucine	C–H, C–NH_2, C–COOH	14	0.5	2.8	5.1		
Valine	C–H, C–NH_2, C–COOH	8	0.2	0.6	4.1		

Typical consequences of secondary processes are collapse of crystal structure and of molecular architecture due to mass loss. The final stage is a cross-linked, carbon-rich, polymerized cinder. The continuing increase of the thermal and electrical conductivities at very high doses > 1 C cm^{-2} indicates that the carbonization of the material proceeds and that the rearrangement of atoms and molecules does not stop [10.42].

The mean number of specific reactions (e.g., bond rupture, cross-linking or disappearance of original molecules) that are produced by an energy loss of 100 eV is known in radiation chemistry as the G value. Typical results from radiation chemistry are shown in Table 10.5; G_{-M} corresponds to the disappearance of molecules, G_{H_2} to the appearance of H_2, etc. These values show that G_{-M} decreases in the following order: unsaturated and saturated hydrocarbons, ethers, aldehydes, carboxylic acids, aromatic compounds. The very low value G_{H_2} for aromatics is of special interest, as is the higher value for saturated hydrocarbons, which is nearly of the same order as the G_{-M} value. Thus for a high-resolution micrograph, $G = 1$ with 10^4 eV nm^{-3} provokes 10^2 nm^{-3} damage processes! This estimate gives an idea of the demands that electron microscopists make on their organic specimens.

10.2.2 Quantitative Methods of Measuring Damage Effects

Our knowledge of the individual damage processes is very poor. Even for a particular molecule, the scale of primary and secondary processes is very broad and complex, and it is impossible to describe the whole damage process in detail. For practical electron microscopy, damage processes that can be observed in the final image or in a diffraction pattern and that can be used directly for quantitative measurement of damage are of special interest.

a) Loss of Mass. The transmission T of an amorphous specimen layer decreases as $\exp(-x/x_k)$ (6.6) up to mass-thicknesses $x = \varrho t =$ 30–50 μg cm^{-2} at $E = 100$ keV. The quantity $K = \log_{10}(1/T)$ is proportional to x (Fig. 6.1) and can be used to measure the loss of mass during irradiation [10.59–63]. For high-polymers, collodion, formvar, methacrylate, epon and other embedding media, the mass-thickness shows an approximately exponential decrease down to a residual mass-thickness (Fig. 10.3), which for many substances corresponds roughly to the carbon content. This confirms that many non-carbon atoms can leave the specimen after bond scission. A fraction of the carbon atoms can also leave the specimen as volatile fragments and some of the non-carbon atoms will be bound more strongly. It is necessary to measure the transmission with low current densities, of the order of $j = 10^{-4}$ A cm^{-2}, because the dose that corresponds to the terminal mass loss is of the order of $q_{max} = j\tau = 10^{-2}$ C cm^{-2} for polymers. The percentage of mass lost by polymethacrylate films in Fig. 10.3 shows a systematic decrease with increasing film thickness. This is by no means the rule, however, and protective, evaporated carbon films do not decrease the mass loss for all

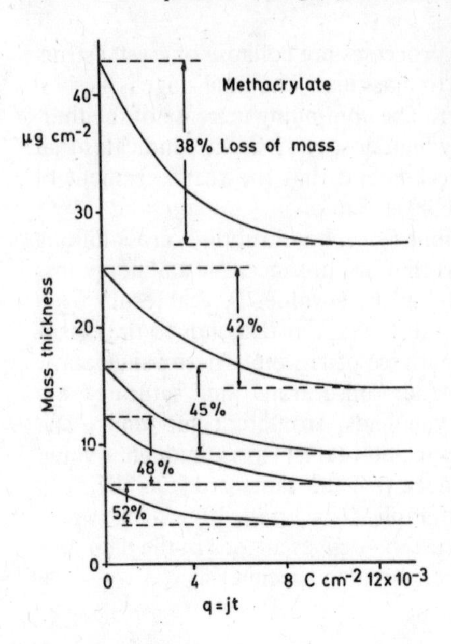

Fig. 10.3. Mass thicknesses of polymethacrylate foils (methyl : butyl = 20 : 80) of different initial mass-thicknesses as a function of electron dose $q = j\tau$, obtained from measurements of electron transmission (E = 60 keV)

substances. The mass loss and other damage effects are proportional to the dose so long as the irradiation conditions not cause appreciable specimen heating. Some substances, such as methacrylate, lose more mass when the temperature is increased by a higher current density [10.64].

In practice, it is prudent to preirradiate and stabilize a biological section with a low current density to avoid an increase of mass loss by specimen heating. For high resolution, the tendency is to irradiate the specimen with a dose as low as possible (Sect. 10.2.4).

Determination of the mass loss by measurement of the transmission is not applicable to organic single or polycrystalline films because they show diffraction contrast, which depends on the accidental excitation error and film thickness. Normally a crystalline film shows a greater averaged transmission T than an amorphous film of equal mass-thickness [6.1], and the transmission T may decrease in the early stage of irradiation [10.65].

Another method of determining the mass loss is by directly weighing the specimen before and after irradiation, but for such experiments a reasonably large mass (10–20 μm films about 1 cm^2 in area) is necessary [10.38, 61, 66]. The change of optical density nt (n: refractive index, t: specimen thickness) can also be measured by interferometric methods or interference colours [10.67, 68].

b) Fading of Electron-Diffraction Intensities. Damage to single molecules also causes distortion of the crystal lattice, which results in a decrease of the electron-diffraction intensity of Debye-Scherrer rings or single-crystal spots

[10.69, 70] and in some cases in a shift and broadening of the reflections as well [10.71].

Most evaporated films of organic compounds are crystalline. This method is, therefore widely used to investigate the influence of chemical structure on resistance to radiation damage, and the dependence on electron energy and specimen temperature (Sect. 10.2.3). The dose necessary for complete disordering of the lattice is also of interest, for estimation of irradiation conditions in which diffraction contrast can be observed or lattice planes in organic crystals resolved.

Figure 10.4 shows examples of decrease of intensity of Debye-Scherrer rings with increased charge density. The lattice of aliphatic compounds is destroyed at low $q_{max} \simeq 10^{-3}$–10^{-2} C cm^{-2} when irradiated with 100 keV electrons. The extrapolated terminal dose $q_{max} \simeq 10^{-1}$ C cm^{-2} increases considerably if the compounds contain benzene rings (aromatic compounds). Crystals of phthalocyanine molecules and its metal derivatives, which were used by *Menter* [8.39] to resolve the lattice-plane fringes of the order of 1.2 nm for the first time, have a very large value of q_{max} = 1–2 C cm^{-2} for the 1.2 nm lattice planes. However, a larger lattice-plane spacing is less affected by lattice distortions than a smaller one. If we assume that the damage can be described by frozen-in lattice vibrations, although this is rather a crude model, the first decrease of intensity can be described by a Debye-Waller factor (7.100) [10.72]

$$I_g = I_{g0} \exp(-4\pi^2 \langle u^2 \rangle g^2) \,. \tag{10.13}$$

This equation can be used to estimate an average displacement $\langle u^2 \rangle$. When the dose q required to obtain the same value of $\langle u^2 \rangle = (0.05\ \text{nm})^2$ is calculated, the values for phthalocyanine are found to be no larger than for other aromatic compounds. Only the halogen-substituted phthalocyanines

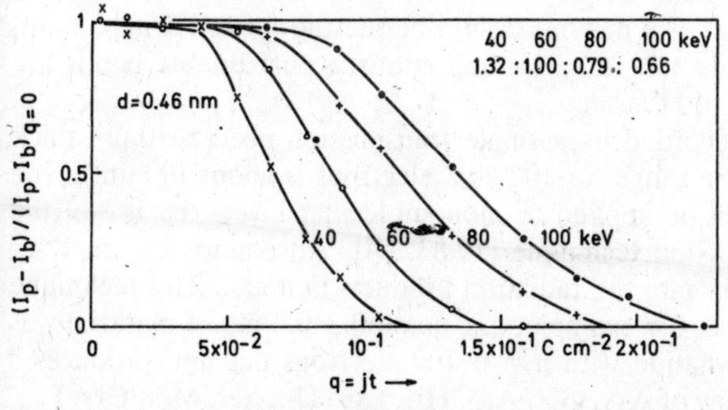

Fig. 10.4. Decrease of the electron diffraction intensity of the d = 0.46 nm lattice planes in tetracene films with increasing electron dose $q = j\tau$ for different electron energies E = 40–100 keV [(I_p) peak intensity of the Debye-Scherrer ring, (I_b) background intensity of the diffraction pattern]

(e.g., $CuCl_{16}C_{32}N_8$) show an exceptional, larger resistance with $q_{max} = 15$–$40\ C\ cm^{-2}$ (see also [10.73]). They can be used for high-resolution study of the crystal lattice [10.74, 75] and the damage effects can be observed directly in the image at molecular dimension [10.76].

The decrease of the diffraction intensities with increasing dose has to be interpretated with care because the shape of the fading curve can depend strongly on the film structure, the texture in polycrystalline films for example. So-called *latence doses* (range of q in which the fading shows very little decrease, Fig. 10.3 for example) can be artifacts caused by the specimen structure and by the transition between the conditions in which the dynamical and the kinematical theories of electron diffraction are applicable. However, the extrapolated terminal doses q_{max} (Table 10.6) are in agreement with the results of other methods. Care should be used when interpreting the shape of the fading curve.

c) Spectroscopic Methods. Changes of the structure of molecules can be detected by their influence on the photon-absorption and electron energy-loss spectra. The absorption spectra of evaporated dye films can be measured inside one irradiated mesh of a supporting grid ($0.1 \times 0.1\ mm^2$). The terminal dose for the damage, as indicated by the absorption maxima (Fig. 10.5) coincides with the value given by electron diffraction (Table 10.6) [10.38, 81]. Photoconductivity and cathodoluminescence are very sensitive to radiation damage: the photoconductivity of phthalocyanine, for example, is lost for $q_{max} \simeq 5 \times 10^{-5}\ C\ cm^{-2}$ [10.38].

Infrared absorption spectra need a larger specimen area and the specimen has to be irradiated in the final image plane. The absorption maxima in the far infrared caused by vibrations and rotations of the molecules disappear first, at relatively low dose q (Fig. 10.6) [10.38, 61, 66]. This indicates that cross-linking and scission of molecules occurs. Absorption maxima in the near infrared can be attributed to typical groups like $-CH_3$, $-CH_2-$, $=CH_2-$, $=C-$, $-COOH$, $OH-$. Figure 10.6 shows the decrease of the maxima and the arrival of a new maximum beside the $-COOR$ maximum, which can be attributed to $\rangle C{=}CH_3$ and confirms that double bonds are generated by the loss of H atoms.

Infrared spectra recorded by a single transmission need 5–10 μm thick specimens (the electron range for 100 keV electrons is about 100 μm). The infrared technique can be applied to monomolecular layers, by use of the multiple-internal reflection technique [10.83, 84]. An amino acid analyser was also used to investigate the radiation products in detail. This technique was applied to catalase, for example, and some 0.2–0.4 mg of material was required [10.85]. Irradiation with $n = 0$–100 electrons per nm^2 produces a more or less rapid decay of Asx, Glx, Arg, His, Lys, Thr, Ser, Met, Cys, Pro, Tyr (e.g. −40% for Lys); the amount of Val, Leu, Ile, Phe remains nearly constant whereas the amount of Gly, Als, Abu increases (e.g., +25% for Ala). This is an indication of possible transformations of amino acids, con-

Table 10.6. The electron dose q needed at 300 K for complete destruction of organic substances, measured by different methods (ML: mass loss, C: contrast, ED: electron diffraction, LA: light absorption, ELS: electron-energy-loss spectroscopy)

Substance	q [C cm^{-2}]	E [keV]	Method	Refs.
Amino acids				
Glycine	1.5×10^{-3}	60	ED	[10.69]
l-Valine	1.5×10^{-3}	80	ED	[10.77]
Leucine	$1.5–2\times10^{-3}$	60	ED	[10.69]
Aliphatic hydrocarbons				
Stearic acid	$2–3\times10^{-3}$	60	ED	[10.69]
Paraffin	$3–5\times10^{-3}$	60	ED, C	[10.69]
	1×10^{-2}	100	ED	[10.65]
Polymethacrylate	$0.8–1\times10^{-2}$	60	C	[10.59]
Polyoxymethylene	$7–8\times10^{-3}$	80/100	ED	[10.70, 78]
Polyethylene	$0.7–1\times10^{-2}$	100	ED	[10.70, 78, 79]
Nylon 6	1.2×10^{-2}	100	ED	[10.70]
Polyvinylformal	1×10^{-2}	75	ML	[10.61]
Polyamide	$1.5–2\times10^{-2}$	75	ML	[10.61]
Polyester	2×10^{-2}	75	ML	[10.61]
Fluorinated ethylene polymer	$0.5–1\times10^{-2}$	75	ML	[10.61]
Tetrafluorinated ethylene polymer	$1–1.5\times10^{-2}$	75	ML	[10.61]
Gelatin	1×10^{-2}	75	ML	[10.61]
Bases of nucleic acids				
Adenosine	1×10^{-2}	80	ED	[10.77]
Adenine	5×10^{-1}	100	ED	[10.72]
Cytosine	3×10^{-1}	20	ELS	[10.80]
	4×10^{-2}	20	ED	[10.80]
Guanine	6×10^{-1}	20	ELS	[10.80]
Uracil	$1–2\times10^{-1}$	100	ED	[10.72]
Aromatic compounds and dyes				
Anthracene	$6–8\times10^{-2}$	60	ED	[10.69]
Tetracene (naphthacene)	2×10^{-1}	100	ED	[10.65, 69]
Pentacene + Tetracene	$3–5\times10^{-2}$	60	LA	[10.81]
Coronene	5×10^{-2}	100	ED	[10.47]
Indigo	1.5×10^{-2}	60	ED	[10.69]
	5×10^{-1}	60	LA	[10.81]
Cloranile	10^{-1}	100	ED	[10.72]
Bromanile	$3–5\times10^{-1}$	100	ED	[10.72]
Hexabromobenzene	5–7	100	ED	[10.72]
Phthalocyanine	1×10^{-1}	100	ED (d = 0.6 nm)	[10.69]
Cu-Phthalocyanine	4×10^{-1}	100	ED (d = 0.6 nm)	[10.72]
	1–2	60	ED (d = 1.2 nm)	[10.69]
	2–3	60	LA	[10.81]
$CuCl_{16}$-phthalocyanine	25–35	100	ED	[10.82]

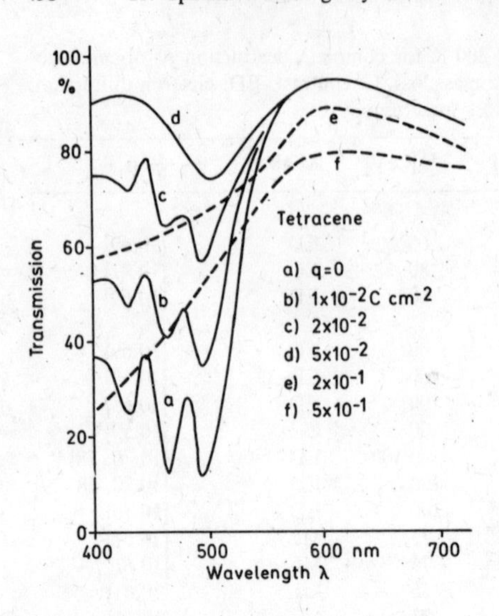

Fig. 10.5. Light-absorption spectra of a 560 nm-thick tetracene film for various electron doses q (E = 60 keV) [10.81]

firmed by the appearance of Abu (α-amino butyric acid) which does not occur in the native, unirradiated catalase. These experiments show that transformations are important in irradiated homo- and heteropolypeptides and that secondary and tertiary chemical reactions can be observed that are not found when single amino acids are irradiated.

The application of electron-energy-loss spectroscopy to the study of radiation damage has the advantage of providing direct elemental analysis of C, N or O atoms, among others, and the decrease of their contribution to the

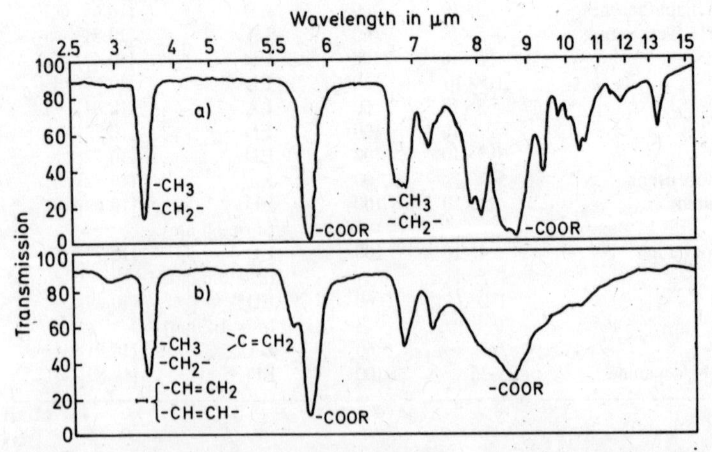

Fig. 10.6. Infrared absorption spectra of a polymethacrylate foil (14 µm thick) (**a**) not irradiated an (**b**) irradiated with a dose of $q = 4 \times 10^{-3}$ C cm^{-2} at E = 60 keV [10.38]

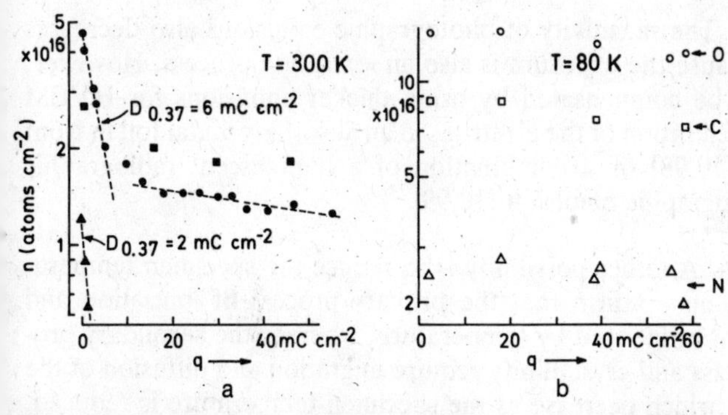

Fig. 10.7. Amounts of carbon (■,□), nitrogen (▲,△) and oxygen (●,○) per unit area of a thin collodion film, depeneing on the incident electron dose q at (**a**) $T = 300$ K and (**b**) $T = 80$ K (liquid-nitrogen cooling), measured by electron-energy-loss spectroscopy at $E = 80$ keV [10.87]

energy-loss spectrum can be followed during electron irradiation [10.86–88] (Fig. 10.7). Energy- or wavelength-dispersive x-ray spectroscopy can also be used to investigate the loss of heavy atoms, such as S and P [10.89], whereas the decrease of the background intensity indicates the loss of mass [9.36, 39; 10.90].

Investigation of irradiated organic films with a laser micro-probe mass analyser (LAMMA) [10.91] shows that the $(M + H)^+$-peak in the positive-ion spectrum decreases with increasing dose of electron pre-irradiation. The terminal doses agree with those found by electron-diffraction fading. Some smaller molecular fragments first increase, which indicates scission products. A disadvantage of this method is that the mass spectrum already contains molecular fragments in the unirradiated state. Other sensitive methods are electron-spin resonance [10.54] and nuclear magnetic resonance [10.92].

A further method for investigation of radiation damage is inelastic-tunneling spectroscopy [10.93]. This technique reveals the vibrational modes of organic compounds that are included in the insulating layer of a metal-insulator-metal (MIM) tunneling diode. The COH functional group in β-D fructose is disrupted and the C=C bond increases, for example.

10.2.3 Methods of Reducing the Radiation Damage Rate

a) High-Voltage Electron Microscopy. One way of reducing the damage rate is to use high-voltage electron microscopy (HVEM). The reduction in energy dissipation, $|dE/ds| \propto v^{-2}$, which results from the variations of the inelastic cross-section (Sect. 5.2.2) and the Bethe loss formula (6.79) with energy has been confirmed by experiments [10.70, 79, 82, 94, 95] that use the electron-diffraction method. A gain of two in the end-point electron dose $q = j\tau$ that causes complete fading of the electron-diffraction pattern can be obtained if E is increased from 100 to 200 keV, but when E is increased to 1 MeV, the gain is only three. The sensitivity of photographic emulsions also decreases by this factor, because the exposure is also an ionisation process. However, this decrease can be compensated by using thicker emulsions for HVEM [10.96, 97], by deceleration of the electrons in an absorbing metal foil in front of the emulsion [10.98] or a combination of a luminescent radiographic screen and a photographic emulsion [10.99].

b) Cryo-Protection. Another possibility is to reduce the specimen temperature. All experiments confirm that the primary process of ionisation and bond rupture is not influenced by temperature, whereas the secondary processes of loss of mass and crystallinity require migration and diffusion of the reaction products, which decrease as the specimen temperature is reduced. However, the degree of improvement varies from one substance to another. Discrepancies of the reported gains on cooling to liquid-helium temperature (4 K) can be attributed to uncertainties of exact specimen temperature; particularly when the specimens were held near a threshold temperature for diffusion and reactions of fragmented atoms and radicals, which seems to be of the order of 5–20 K.

Orth and *Fischer* [10.71] reported an increase of the damage rate for polyethylene when the temperature was increased from 300 K to 370 K. *Grubb* and *Groves* [10.78] observed a decrease of the damage rate by a factor 0.3 or a three-fold gain due to cryoprotection when polyethylene was cooled to 18 K. There was, however, no significant change of the damage rate of polyoxymethylene; polyethylene predominantly cross links, polyoxymethylene is damaged by bond rupture. *Glaeser* [10.77] found no gain for l-valine and a gain of 1.5 for adenosine when it was cooled to 10–20 K, whereas *Salih* and *Cosslett* [10.100] found that the end-point dose for l-valine increases between 300 K and 20 K by a factor 1.8 and for coronene and anthracene by a factor 3 to 4. Gains of five and three were observed for the purple membrane and catalase, respectively, when they were cooled to 150 K [10.46].

Siegel [10.65] reported experiments at 4, 70 and 300 K on paraffin, and tetracene with gains of two and four, respectively, at 4 K. If, however, the temperature of a specimen initially irradiated at 4 K is raised to 300 K without further irradiation, the fading of the electron-diffraction intensity is the same as would have been observed if the specimen were irradiated with the same electron dose at 300 K. This confirms that the mobility of fragments is strongly decreased at 4 K. Radiation-induced recombination processes cannot be excluded, but these do not necessarily recreate the original structure. Measurements of the number of C, O and N atoms by energy-loss spectroscopy (Fig. 10.7) also show that mass loss is strongly reduced at liquid-nitrogen temperature [10.87, 88].

Mass loss has also been measured by quite a different method [10.101] ^{14}C-labelled T 4 phages and E.coli bacteria were irradiated at 4 K and 300 K. The residual ^{14}C content was measured by the autoradiographic method (deposition of a photographic emulsion on the specimen grid and development of the grains exposed by the β emission of ^{14}C). Exposures up to 1 C cm^{-2} show no significant loss when irradiated at 4 K, whereas the loss at 300 K is of the order of 30%.

Experiments with the specimen inside a superconducting lens of the shielding type or cryo-lens show that much larger gains of cryoprotection can be obtained if the temperature rise caused by electron irradiation is limited by the use of carbon supporting films with good thermal contact and radiation shielding. A gain of 10 was found for details smaller than 0.6 nm in crystals of the crotoxin complex embedded in glucose [10.102] and a gain of 30 was obtained for paraffin and phenylalanine and of 70 for the 0.47 nm lattice spacing of l-valine. For the 0.075 nm spacing, however, the gain fell to 4.3 [10.103]. A gain up to 330 was reported for adenosine-5-monophosphate (AMP) [10.104]. These results demonstrate that cryoprotection at 4 K and the use of the HVEM can be useful for investigating biological materials, with greatly reduced secondary damage. These findings need further confirmation. *Dietrich* et al. [10.105] demonstrated that etching of carbon foils by knock-on collisions (Sect. 10.3.2) can be observed at 4 K.

c) Hydrated Organic Specimens. Biological material is normally observed in the dehydrated state. For ultramicrotomy, the specimens are dried, usually in a series of baths of increasing alcohol concentration, after which that intermediate, dehydrating fluid is replaced by a resin, which can be polymerised. Material not prepared in this way loses water in the vacuum of the microscope. There is, however, an ever-increasing interest in observing biological material in the native state. One way of achieving this is to use an environmental cell that encloses a partial pressure of water [10.106]; alternatively, the specimen may be frozen and cryosections cut by use of a cryo-ultramicrotome [10.107–109].

It is known from the radiation chemistry of aqueous systems that the G value for the formation of H and OH radicals and of H_2O_2 is high [$G(-H_2O) \simeq 4.5$ for the liquid state], and that these products cause strong secondary reactions with the biological material. However, this G value decreases to 3.4 in ice at 263 K, to 1.0 at 195 K and to 0.5 at 73 K; this can be attributed to an increase of the molecular recombination of the water-molecule fragments. Model experiments in different states have been done with catalase crystals [10.46]. The outer diffraction spots that correspond to higher resolution faded away first. The electron dose required for complete fading was 300 C cm^{-2} for frozen catalase at $\simeq$ 150 K, which is ten times greater than the value for wet catalase at 300 K. No significant difference was observed for glucose-dried (embedded) catalase at 150 K. This confirms that, at low temperatures, the effect of radiolysis in the presence of water can be neglected, and the damage rate shows the same magnitude in the wet and dried states. After complete fading of the diffraction pattern, voids and bubbles are formed in frozen-hydrated catalase, which do not appear in thin, dried samples nor inside pure ice.

d) Elemental Substitution. The replacement of the H atoms by Cl or Br in benzene or phthalocyanine, for example, increases the end-point dose of electron-diffraction intensity by more than one order of magnitude [10.69, 72, 82] (Table 10.6). End-point doses considerably larger than 1 C cm^{-2} allow crystal-structure imaging of these substances with a high resolution [10.74, 75]. This can be explained by the reduced mobility of halogen atoms *(cage effect)* and the ability of recombination. Halogens are reagents for carbon double bonds, for example.

e) Conductive Coatings and Ultrahigh Vacuum. Coating with a thin evaporated layer of carbon can result in a reduction of the mass loss, especially in substances with a high fraction of molecular scission products such as polymethacrylate [10.59]. Gold-sandwiched coronene crystals were found to have increased radiation resistance, by a factor of five [10.110]. This finding needs further confirmation.

A great improvement of the radiation resistance of indanthrene olive *T* was reported by *Hartman* et al. [10.111] when it was irradiated in ultra-high vacuum. The authors attributed the entire damage to a water-gas reaction process. This finding was not confirmed by *Salih* and *Cosslett* [10.100]. Adsorbed gases can, indeed, react with organic specimens and cause an etching at low specimen temperatures (Sect. 10.4.2); the damage rate due to this process is, however, much less than that caused by ionisation processes inside the specimen.

10.2.4 Radiation Damage and High Resolution

High resolution is possible only by elastic scattering. As shown in Table 10.4, a mean electron dose of the order of 10^{-2} C cm^{-2} is necessary at E = 100 keV for one elastic scattering process per atom. The fraction of scattered electrons that can be used to provide image contrast depends on the operating mode used (e.g., bright or dark field TEM or STEM). Exposure of photographic emulsion to a density $S = 1$ at $M = 100\,000$ needs a dose of 0.1–1 C cm^{-2}.

In order to estimate the minimum dose $q_{\min}$, we consider that $n_0 = j\tau/e = q/e$ electrons are incident per unit area. A fraction f contributes to the image background (e.g., $f = 1 - \varepsilon$, $\varepsilon \ll 1$ for the bright-field or $f = 10^{-2}$–10^{-3} for the dark-field TEM mode). The number $N = nd^2 = fn_0d^2$ forms the image element of area d^2. We assume that the image contrast C is caused by a difference Δn in the number of electrons:

$$C = \frac{\Delta n}{n} = \frac{\Delta n}{fn_0} \tag{10.14}$$

and hence

$$\Delta N = \Delta n d^2 = fn_0 d^2 C \ . \tag{10.15}$$

The shot noise of the background signal (Poisson statistics) is $N^{1/2}$. For the signal to be significant, the signal-to-noise ratio $\varkappa$ must be larger than 3–5, known as the Rose equation [10.112].

$$\frac{\text{Signal}}{\text{Noise}} = \frac{\Delta N}{N^{1/2}} = Cd\,(fn_0)^{1/2} = Cd\,(fq/e)^{1/2} > \varkappa\,. \tag{10.16}$$

Solving for q, we find

$$q_{\min} = \frac{e\varkappa^2}{fd^2C^2} \tag{10.16a}$$

as the minimum dose with which a specimen detail of area d^2 can be detected. Numerical example: Assuming bright field ($f \simeq 1$), $\varkappa = 5$, $d = 1$ nm, $C = 10\%$, we obtain $q_{\min} = 4 \times 10^{-2}$ C cm^{-2}. With hollow-cone illumination, the contrast will fall to $C \simeq 3\%$ (see CTF in Fig. 6.23, 24) and if we wish to resolve $d = 0.5$ nm, we find $q_{\min} = 1.6$ C cm^{-2}.

These examples demonstrate that the square of d and C in the denominator of (10.17) can have a considerable influence on $q_{\min}$ and that the minimum dose is of the same order of magnitude as the dose required for the exposure of a photographic emulsion.

These doses are so high that severe damage to biological specimens is inevitable, especially because taking a micrograph consists of three steps: searching, focusing and recording, and the first two needs a larger dose than the third. Minimum-exposure techniques have therefore been developed, in which the grid is scanned at low magnifications with a strongly reduced electron dose; the microscope is focused on a different specimen region, after which the beam is switched to the area of interest by means of deflection coils and a shutter is opened for exposure [10.113–116].

Another way of decreasing the electron dose involves the low-exposure averaging technique, in which the noise is reduced by averaging over a large number of identical structures. This technique is therefore particularly suitable for periodic specimens [10.117, 118]. If R denotes the number of repeated unit cells, the Rose equation (10.16) becomes $Cd\,(fRq/e)^{1/2} > \varkappa$ and the minimum dose $q_{\min}$ is then reduced by a factor $1/R$. The fog level of the emulsions becomes a serious limitation for low exposure, and nuclear track emulsions have been found better than the emulsions used normally at 100 keV [10.118, 119]. With the aid of cross-correlation methods, the technique can also be used for non-periodic specimens (see also Sect. 6.5.5 and Fig. 6.35) [10.120].

10.3 Radiation Damage of Inorganic Specimens

10.3.1 Damage by Electron Excitation

Electron excitations in metals and in most covalent semiconductors are reversible and cause no damage. Only electron-nucleus collisions (knock-on processes, Sect. 10.3.2) can cause atomic displacements.

In ionic crystals (e.g., alkali halides), the most important excitations are:

1) inner shell ionisation,
2) plasmon losses as collective oscillations of the valence electrons,
3) ionisation of valence electrons and
4) creation of locally bound electron-hole pairs (excitons).

The probability of inner shell ionisations (1) is low and the energy is lost in x-ray and Auger-electron emission. Plasmons (2) are the most-probable excitations but are much too delocalized to cause any localized transfer of energy; they may, however, decay into more localized single-electron excitations of exciton character. Mobile electrons (3) excited into higher states of the conduction band recombine with the less-mobile holes (of large effective mass) via intermediate exciton states. These secondary and the primary excited excitons (4) are responsible for radiolysis. Figure 10.8 shows as an example a possible radiolytic sequence in NaCl [10.121]. A localized hole behaves like a chlorine atom and, in some exciton states with energy around 7 eV, this neutralized anion is tightly localized in a Cl_2^- bond between neighbouring anions while the excited electron stays in hydrogen-like orbitals of large diameter near the surrounding Na^+ cations (Fig. 10.8 a). The Cl_2^- moves by hole tunnelling and interstitial propagation (Fig. 10.8 b, c), which results in a Frenkel pair that consists of a crowdian interstitial (H centre) and an anion vacancy with a trapped electron (F centre).

Similar primary processes also cause radiolysis in other alkali halides and alkaline-earth fluorides (CaF_2, MgF_2). In MgO, however, radiolysis cannot be observed, because the displacement energy for the ions is greater than the available energy of the excitons.

Radiolysis is strongly dependent on temperature if the transition to a Frenkel pair requires an activation energy of the order of 0.1 eV in KI, NaCl and NaBr. This causes a strong decrease of radiolysis at low temperatures, which allows observation of defects in specimens cooled below 50 K [10.122, 123].

Other processes consist of either recombination of the Frenkel pairs or defect accumulation of interstitial halogen atoms, resulting in interstitial dislocation loops even at 50 K. The kinetics of these processes depend on the mobility of the defects (temperature) and on their concentration. As shown in Sect. 10.2.1, the energy dissipated by electron bombardment in a TEM is very much greater than in other irradiation experiments with x-rays or UV

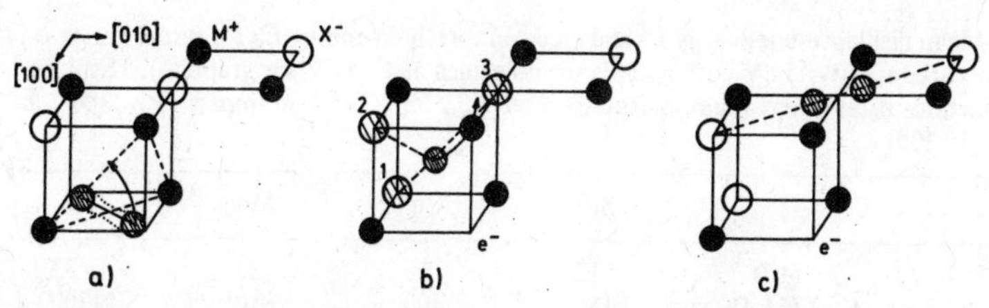

Fig. 10.8 a–c. Radiolysis sequence in alkali halides having the NaCl structure: (**a**) formation of a Cl_2^- bond, (**b**) intermediate state, leading to the final state (**c**), consisting of an interstitial Cl_2^- cathion (H centre) and an anion vacancy with a trapped electron (F centre)

quanta. Results obtained with these radiations are therefore not directly comparable with those observed during electron irradiation in a TEM.

Radiolysis and secondary processes are not proportional to only the dose $q = j\tau$, unlike damage in organic material. Some secondary processes may not be observed at low current densities but only at greater j, with which the rate of production of Frenkel pairs is greater. At high temperatures, the enhanced mobility of anion vacancies and halogen molecules leads to the formation of colloidal metal inclusions and halogen bubbles, whereas the dislocation loops grow and form a dense dislocation network [10.124, 125]. Thus, in CaF_2, voids about 10 nm in diameter condense into a superlattice [10.126–128].

Irradiation of quartz causes a radiolytic transformation into an amorphous state (vitrification). The mechanism is not understood in detail but rupture of Si–O bonds causes rotations of $[SiO_4]$ tetrahedra in the silicate structure [10.129].

Some decomposition products can be identified by electron diffraction, and degradation by electron-beam heating may not lead to the same result as thermal decomposition, owing to the additional action of excitation processes. The following examples illustrate this:

$LiCl \rightarrow Li + Li_2O + Li_2CO_3$ [10.130]
$KMnO_4 \rightarrow MnO_2 \rightarrow MnO$ [10.131]
$AgCl \rightarrow Ag$; $AgNO_3 \rightarrow Ag$; $Cu_2O \rightarrow Cu$; $PbCO_3 \rightarrow Pb$
but no decomposition of Ag_2SO_4, $AuCl_2$, PbO, PbCl [10.132]
$CaSO_4 \cdot 2H_2O \rightarrow CaSO_4 \rightarrow CaO + CaS$ [10.133]
$Mn(OH)_2 \rightarrow \delta\text{-}MnOOH \rightarrow Mn_3O_4 \rightarrow MnO$ [10.134]
$AgN \rightarrow Ag$ [10.135]

10.3.2 Radiation Damage by Knock-On Collisions

Besides the damage produced by ionisation (Sects. 10.2 and 10.3.1) and specimen heating (Sect. 10.1), radiation damage by knock-on collisions has to be considered in HVEM, for reviews see [10.125, 136–139].

Table 10.7. Mean displacement energy E_d and electron threshold energy E_{th} for direct knock-on of atoms (for carbon, E_d = 5 eV corresponds to molecules and 10 eV for graphite). Displacement cross-sections σ_d and maximum energy transfer E_{max} for a head-on impact (θ = 180°) at E = 1 MeV [10.139]

Element	C	Si	Cu	Mo	Au
E_d [eV]	5 (10)	13	19	33	33
E_{th} [keV]	27.2 (54.4)	145	400	810	1300
σ_d [10^{-24} cm^2]	89 (43.6)	69	59	7.2	–
E_{max} [eV]	366	155	68	45	22

It was shown in Sect. 5.1.2 that during an elastic collision between a beam electron and a nucleus, the energy transferred, ΔE, can become greater than the mean (polycrystalline value) of the displacement energy E_d (Table 10.7). As a result of such a displacement, a Frenkel pair that consists of a vacancy and an interstial atom is produced. Similarly, an atom can be pushed into a neighbouring vacancy or an interstitial can be moved to another interstitial site. Neighbouring knocked-on atoms can transfer momentum to a mobile (thermally activated atom). These processes result in radiation-induced or -enhanced diffusion. The displacement energies are greater than for similar processes caused by thermal activation because an atom is pushed to an interstitial position across the saddle point of neighbouring atoms so quickly that the lattice cannot relax by lattice vibrations.

With increasing electron energy E, the energy transfer ΔE first becomes greater than E_d for a scattering angle $\theta = 180°$ at the threshold energy E_{th}. With increasing atomic mass A, the values of E_{th} become greater (Table 10.7 and Fig. 5.3) but the cross-sections σ_d for displacement increase more rapidly for greater A at high energies. The threshold energy also depends on the direction of the knock-on momentum and is least in the close-packed directions $\langle 110 \rangle$, $\langle 100 \rangle$ and $\langle 11\bar{2}0 \rangle$ for face-centred, body-centred and hexagonal close-packed metals, respectively. Thus, for Cu, E_d is 20 eV near $\langle 110 \rangle$ and increases to 45 eV near $\langle 111 \rangle$ but can be as low as 10 eV for a narrow angular region about 10° away from $\langle 110 \rangle$ [10.140–142].

The concentration of displacements $c_d = n_d/n$ (n: number of atoms per unit volume) is proportional to the current density j_a at the nuclei and the irradiation time τ:

$$c_d = \sigma_d j_a \tau / e \ . \tag{10.17}$$

Irradiation of copper for 1 min with a 5 μm spot and a current of 0.2 μA gives $c_d = 0.25\%$ at E = 600 keV and $c_d = 1.25\%$ at E = 1 MeV. However, unlike the ionisation damage, the cross-sections σ_d are so small that damage can be avoided even when working at high resolution. Alternatively, the damage can be exacerbated by increasing the current density and the irradiation time, and a HVEM becomes a powerful tool for investigating radiation-damage effects in-situ because the production of Frenkel pairs can be three to four orders of magnitude greater than with electron accelerators or in a nuclear reactor. Very high electron energies will be useful for studying damage in materials of high atomic number (e.g., 2.5 MeV for Au [10.143]).

The dependence of c_d on the current density j_a at the nuclei enhances the production rate up to a factor four if the Bloch-wave intensity is greatest at the nuclei. Hence, there is a sensitive dependence of c_d on crystal orientation, with a maximum near the Bragg position, where c_d becomes proportional to the intensity of excess Kikuchi bands (Fig. 7.20 b) [10.144, 145].

Defect clusters are observable only by TEM as secondary-damage products. The diffusion of vacancies and interstitials depends strongly on temperature. Whereas interstitials are highly mobile during irradiation at all temperatures, vacancies need higher temperatures (room temperature in Cu for example) to become mobile. A decrease of specimen temperature to 4 K can be used to stabilize a defect structure obtained at high temperature, or the accumulation of defects generated at 4 K can be observed as the specimen temperature is increased.

If the thermally activated diffusion of point defects does not lead to a recombination of Frenkel pairs, then defect clusters, interstitial-dislocation loops, stacking-fault tetrahedra or voids may be formed. Surfaces, dislocations and grain boundaries are sinks for point defects, which decrease the defect concentration over distances of 100–150 nm.

In alloys, atomic displacements can cause radiation-enhanced diffusion; in Al-Cu or Al-Zn alloys, for example, radiation-enhanced precipitation may result. Disordering can be observed in ordered alloys (e.g. Ni_3Mn), but ordering may occur simultaneously as a result of radiation-enhanced diffusion (e.g., Fe-Ni, Au_4Mn).

It should be mentioned that the formation of defect clusters can also be observed during prolonged irradiation in a 100 keV TEM. This damage may be attributable to negative ions accelerated between cathode and anode [10.146].

10.4 Contamination

10.4.1 Origin and Sources of Contamination

Radiation damage of adsorbed hydocarbon molecules on the specimen surface causes a carbon-rich, polymerized film to form; this grows on electron-irradiated areas of the specimen by cross-linking. In competition with this contamination, reactions with activated, adsorbed H_2O, O_2, N_2 molecules cause etching of carboneous material. Depending on specimen preparation, partial pressures, specimen temperature and irradiation conditions, growth of either sign (positive for contamination and negative for etching) may predominate. It is not easy to work at equilibrium (zero growth) (Review: [10.147]).

Various sources of hydrocarbon molecules are

1) Adsorbed layers on the specimen, introduced during preparation or by atmospheric deposition.
2) Vacuum oils from the rotary and diffusion pumps.
3) Grease and rubber O-rings and adsorbed layers (e.g. finger marks) on the microscope walls.

These contributions to contamination can be kept small by taking the following precautions:

Even pure specimens become contaminated by hydrocarbon molecules if exposed to the air for a period of a day. This type of contamination cannot be reduced by the liquid-nitrogen-cooled anticontamination blade (or cold finger) inside the microscope. Yet it is the most important source of contamination, especially when small electron probes < 0.1 μm are used. Contamination of this kind will be introduced with the specimen even under ultra-high-vacuum conditions. Washing the specimen cartridge and the specimen in methyl alcohol is the simplest way to eliminate this contaminant [10.148].

The partial pressure of vacuum-oil molecules can be reduced by using oil of low vapour pressure, a good baffle between the diffusion pump and the column, and by switching over from the rotary to the diffusion pump at a pressure of about 10 Pa to avoid back streaming of the rotary-pump oil. The best solution is to use an oil-free turbomolecular pump.

Grease must be avoided or be used very sparingly. Vacuum leaks can never be cured by heavy greasing but only by carefully polishing the sealing surfaces. Viton rings should be used in preference to rubber. All surfaces should be washed with methyl alcohol, which evaporates completely in air. Finger marks should be avoided by using gloves.

The etching process is affected by the composition of the residual gas and by the partial pressure of water that comes from photographic materials. Emulsions should be pre-evacuated in the presence of a water-absorbing material (e.g. P_2O_5) and raised to atmospheric pressure for only a very short time [10.149]. Emulsions are a very uncertain source of water vapour. The ideal solution is to expose the emulsion outside the microscope column with a fibre-optic plate between the transparent fluorescent screen and the emulsion (Sect. 4.6.2).

10.4.2 Methods for Decreasing Contamination

The precautions mentioned in the preceding section are essential if the contamination rate is to be effectively reduced. Further improvement can be obtained by the following methods:

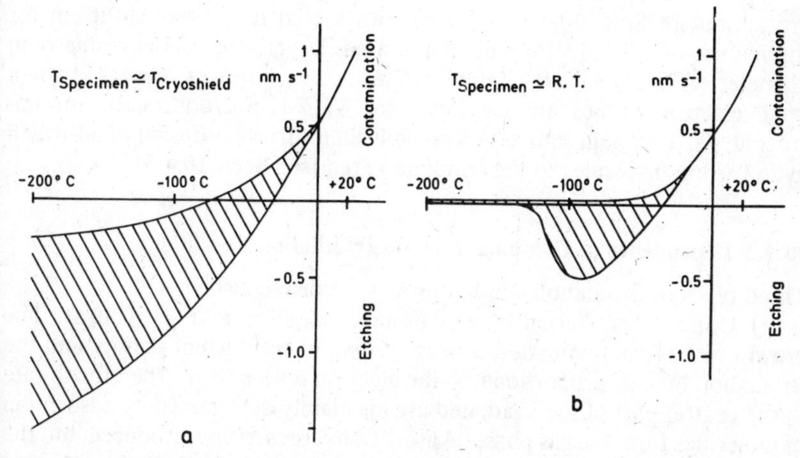

Fig. 10.9. Rate of carbon-atom deposition by contamination (positive sign) and removal by etching (negative sign) as a function of temperature T for (**a**) specimen cooling with the specimen and the cryoshield at the same temperature and (**b**) an anticontamination blade at the temperature T and the specimen at room temperature [10.152, 153]

1) Specimen Heating to 200–300 °C increases the desorption of hydrocarbon molecules and decreases the contamination [10.150]. When small electron probes are employed, it is, however necessary to heat the specimen initially to 400–500 °C.

2) Specimen Cooling [10.151]. The specimen cartridge and additional components act as a cryoshield and decrease the adsorption of hydrocarbon molecules by decreasing their partial pressures near the specimen. At low temperatures, the contamination changes over to etching of carbon or organic material (Fig. 10.9 a); this is attributed to a radiation-induced chemical reaction with adsorbed residual gas molecules (H_2O, N_2, O_2, CO, H_2), resulting in volatile compounds with carbon. The observed spread of experimental results in Fig. 10.9 is a consequence of the variable composition of the residual gas. When metal or inorganic specimens are cooled below −100 °C, there is no contamination. If the cryoshield is not sufficient, an ice film may grow on the specimen, but this will vanish if a small electron probe is being employed.

3) Anticontamination Blades *(cold fingers)* cooled with liquid nitrogen surround the specimen and act as cryoshields but the specimen is still at room temperature. Cooling first causes a decrease of the partial pressure of hydrocarbons and etching of the specimen predominates (Fig. 10.9 b) [10.152, 153]. Further cooling also reduces the partial pressure of the residual gas molecules that are responsible for etching. At low temperatures of the cold finger either contamination or etching continues at a very low rate. The precautions of Sect. 10.4.1, together with a cold finger, are sufficient for routine use of a TEM with illuminated areas larger than one micrometre in diameter. A much higher contamination rate is, however, observed when small electron probes are employed for STEM, microdiffraction or microanalysis. The main source of trouble is then surface diffusion of adsorbed hydrocarbon molecules to the irradiated area (see Sect. 10.4.3).

10.4.3 Dependence of Contamination on Irradiation Conditions

Three types of irradiation conditions may be distinguished:

1) Uniform irradiation of a reasonably large area (Fig. 10.10 a). The growth rate of contamination is proportional to the current density and the irradiation time or proportional to the electron dose $q = j\tau$. The growth rate in the central part of the irradiated area is mainly determined by adsorption of molecules from the gas phase. Adsorbed hydrocarbons introduced into the microscope with the specimen are damaged in the first stage of irradiation and are fixed by cross-linking.

2) Irradiation of a small area with uniform current density (Fig. 10.10 b). If the diameter of the area illuminated is reduced to a few micrometres by imaging the condenser diaphragm on the specimen, an annular contamination spot can be observed with more contamination at the periphery of the irradiated area than at the centre. This phenomenon can be explained by surface diffusion of adsorbed hydrocarbons, which are cross-linked when they are struck by the electron beam and have little opportunity to diffuse to the centre of the illuminated area. The contamination rate is higher than that found with uniform illumination of a larger area because the whole foil is a

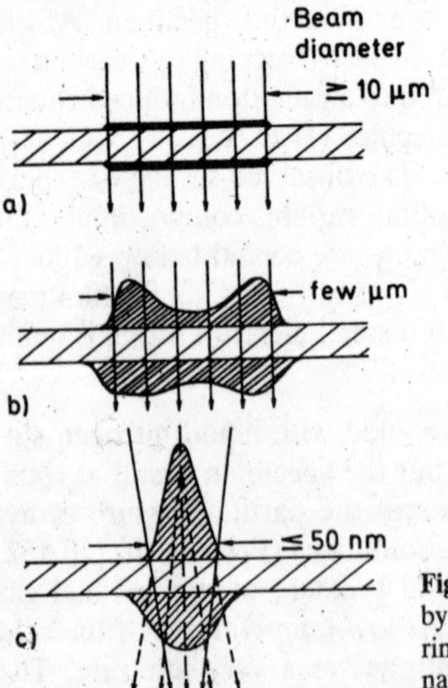

Fig. 10.10. Build-up of (**a**) a uniform contamination layer by uniform irradiation of a larger area, (**b**) a contamination ring, by uniform irradiation of small area and (**c**) a contamination needle, by irradiation with a small electron probe

source, which supplies molecules by diffusion. There is no reason to assume that the electric fields caused by secondary-electron emission have any influence, as claimed by *Fourie* [10.154]. The model calculations of *Müller* [10.155] are fully adequate to explain the generation of contamination rings (see end of this section).

3) Irradiation with an electron probe (Fig. 10.10c). The diameter of the annular zone of higher contamination rate (described above) decreases if the electron-probe diameter is less than 0.1 μm and a needle-shaped contamination spot is formed. This spot can be broadened at the bottom by multiple scattering of electrons at the top of the contamination needle. The contamination cone on the bottom of the foil is also normally broadened by multiple scattering [10.156]. It is very difficult to combat this type of contamination. An excellent vacuum, washing the specimen with methyl alcohol and specimen heating or cooling are necessary precautions. A cold finger, which helps to decrease the contamination rate under irradiation condition 1, is normally not sufficient for this illumination mode. It has been found that the growth of contamination needles increases, the longer the specimen stays in the microscope vacuum. If a larger area (grid mesh) is subsequently irradiated, the contamination rate again decreases, which indicates that the adsorbed molecules from the vacuum are fixed by irradiation and cannot diffuse further to the irradiated spot.

The higher contamination rate with small electron probes can be exploited for micro-writing [10.155, 157], or the local foil thickness can be determined by tilting the irradiated specimen through about 45° to separate the top and bottom cones (needles) of contamination [9.16].

All of these processes and different illumination conditions can be described by a single differential equation [10.155]. The number k of molecules that are cross-linked to the surface per unit time and per unit area is proportional to the density n of the adsorbed molecules and to the current denstiy j

$$k = n\sigma \frac{j}{e} . \tag{10.18}$$

The number ν of molecules incident on unit area in unit time can be calculated from the partial pressure p:

$$\nu = \frac{p}{(2\pi mkT)^{1/2}} . \tag{10.19}$$

These molecules are adsorbed and desorbed again with a time constant τ_0. The equilibrium concentration will be

$$n_\infty = \nu\tau_0 . \tag{10.20}$$

Four mechanisms can change the concentration n:

1) Adsorption of molecules $(\partial n/\partial\tau)_1 = \nu$
2) Desorption $(\partial n/\partial\tau)_2 = -\,n/\tau_0$
3) Diffusion $(\partial n/\partial\tau)_3 = \lambda\,\nabla n$
4) Contamination $(\partial n/\partial\tau)_4 = -\,(j/e)\sigma n$

This may be expressed by the partial differential equation

$$\frac{\partial n}{\partial\tau} = \nu - \frac{n}{\tau_0} + \lambda\,\nabla n - \frac{j}{e}\,\sigma n\,. \tag{10.21}$$

Applying the equilibrium condition $\partial n/\partial\tau = 0$ and imposing rotational symmetry, we obtain

$$\nu - \frac{n}{\tau_0} + \lambda\left(\frac{d^2n}{dr^2} + \frac{1}{r}\frac{dn}{dr}\right) - \frac{j}{e}\sigma n = 0\,. \tag{10.22}$$

By use of the abbreviations $\varrho = (\tau_0\lambda)^{1/2}$, $\alpha = \sigma/e\lambda$, $\nu/\lambda = n_\infty/\varrho^2$, this equation can be written as

$$\frac{d^2n}{dr^2} + \frac{1}{r}\frac{dn}{dr} - \frac{1}{\varrho^2}n\,(r) - \alpha j\,(r)\;n\,(r) + \frac{n_\infty}{\varrho^2} = 0 \tag{10.23}$$

If uniform irradiation j_0 is assumed in the area $r \leq R$, the equation can be solved in this inner area,

$$n_1 = n_\infty\left[1 - C_1\mathrm{K}_0\left(\frac{r}{\varrho}\right)\right] \tag{10.24}$$

and in the outer, non-irradiated area $r \geq R$,

$$n_2 = n_\infty\left[\frac{\varrho_0^2}{\varrho^2} + C_2\mathrm{I}_0\left(\frac{r}{\varrho_0}\right)\right] \quad \text{with} \quad \varrho_0 = \varrho\,(1 + \alpha^2\varrho^2 j_0)^{-1/2} \tag{10.25}$$

The constants C_1 and C_2 can be determined from the boundary conditions at $r = R$: $n_1(R) = n_2(R)$ and $dn_1(R)/dr = dn_2(R)/dr$. I_0 and K_0 are the modified Bessel (Hankel) functions. An example of the resulting distributions $n\,(r)$ and the thickness $t_c = \sigma n\,(r)\,j\tau/e$ of the contamination layer is plotted in Fig. 10.11. It shows how the irradiated area acts as a sink for hydrocarbons. The contamination rate is proportional to $n\,(r)$ and the contamination ring discussed earlier is predicted. Increasing the current density j by a factor two (a → b) and (c → d), the density $n\,(r)$ decreases to approximately half the value observed at low j. This results in a growth of contamination thickness t_c

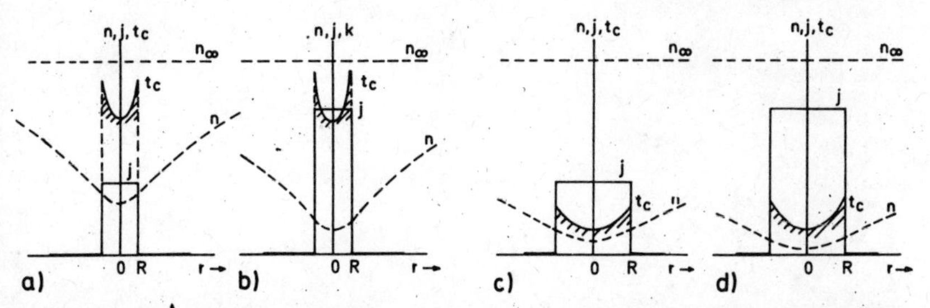

Fig. 10.11 a–d. Equilibrium of the concentration $n(r)$ of adsorbed organic molecules, the irradiated area $r \leq R$ acts as a sink for mobile molecules, so that a contamination layer of thickness $t_0 \propto n(r)\, j\tau$ is formed (n_∞ is the equilibirum concentration at large distance). The irradiated area shows in (**c**) and (**d**) twice the diameter $2R$ as in (**a**) and (**b**) and the current density j is in (**b**) and (**d**) a factor of two larger than in (**a**) and (**c**) [10.155]

which will be proportional to only the irradiation time τ $[n(r)j = \text{const}]$ and not to the dose $q = j\tau$. Constant j and increasing radius R of the irradiated area (a → c) and (b → d) results in a decrease of $n(r)$, which is approximately proportional to R^{-2}. The proportionality of $n(r)$ to $(jR^2)^{-1} = I_p^{-1}$ results in a constant total mass of contamination per unit time. This demonstrates a saturation effect, caused by the delay of the diffusion of hydrocarbon molecules from the unirradiated part of the foil.

Scanning a larger area in the STEM mode will result in an uniform contamination layer for fast scans when $n \simeq n_\infty$. For slow scans, the discussion of the contamination rate is more complicated because the decrease of n is asymmetrical and is left behind the moving electron probe.

References

Chapter 1

Emission, Reflection and Mirror Electron Microscopy

1.1 H. Düker: Emissions-Elektronenmikroskope. Acta Phys. Austriaca *18*, 232 (1964)
1.2 G. Möllenstedt, F. Lenz: Electron emission microscopy. Adv. Electron. Electron Phys. *18*, 251 (1963)
1.3 L. Wegmann: "Photoemissions-Elektronenmikroskopie," in *Handbuch der zerstörungsfreien Materialprüfung*, ed. by E.A.W. Müller (Oldenbourg, München 1969) R.31,1; Mikroskopie *26*, 99 (1970); Optik *26*, 99 (1970)
1.4 L. Wegmann: The photoemission electron microscope, its technique and applications. J. Micr. *96*, 1 (1972)
1.5 O.H. Griffith, G.H. Lesch, G.F. Rempfer, G.B. Birrel, C.A. Stafford, P.C. Jost, T.B. Mariott: Photoelectron microscopy: a new approach to mapping organic and biological surfaces. Proc. Nat. Acad. Sci. USA *69*, 551 (1972)
1.6 S. Grund, W. Engel, P. Teufel: Photoelektronen-Emissionsmikroskop und Immunofluoreszenz. J. Ultrastruct. Res. *50*, 284 (1975)
1.7 E. Ruska, H.O. Müller: Über Fortschritte bei der Abbildung elektronenbestrahlter Oberflächen. Z. Phys. *116*, 366 (1940)
1.8 C. Fert: Observation directe des surfaces en microscopie électronique par reflexion. Optik *13*, 378 (1956)
1.9 V.E. Cosslett, D. Jones: A reflexion electron microscope. J. Sci. Instrum. *32*, 86 (1955)
1.10 C. Fert: "Observation directe de la surface d'un échantillon massif en microscopie électronique," in [Ref.1.57, Vol.1] p.277
1.11 W. Stenzel: "Eine Fünfelektroden-Filterlinse für Reflexions-Elektronenmikroskope," in *Electron Microscopy 1968*, Vol.1, ed. by D.S. Bocciarelli (Tipografia Poliglotta Vaticana, Rome 1968) p.115
1.12 A.B. Bok: "Mirror Electron Microscopy: Theory and Applications," in [Ref.1.32], p.655
1.13 A.B. Bok, J.B. Le Poole, J. Roos, H. DeLang, H. Bethge, J. Heydenreich, M.E. Barnett: "Mirror Electron Microscopy," in *Advances in Optical and Electron Microscopy*, Vol.4, ed. by R. Barer, V.E. Cosslett (Academic, New York 1971) p.161
1.14 G.V. Spivak, V.P. Ivannikov, A.E. Luk'yanov, E.I. Rau: Development of scanning mirror electron microscopy for quantitative evaluation of electric microfields. J. Microsc. Spectrosc. Electron *3*, 89 (1978)
1.15 J. Witzani, E.M. Hörl: Scanning electron mirror microscopy. Scanning *4*, 53 (1980)

Scanning Electron Microscopy and X-Ray Microanalysis

1.16 C.W. Oatley, W.C. Nixon, R.F.W. Pease: Scanning electron microscopy. Adv. Electron. Electron Phys. *21*, 181 (1965)
1.17 C.W. Oatley: *Scanning Electron Microscopy I. The Instrument* (University Press, Cambridge 1972)

1.18 D.B. Holt, M.D. Muir, P.R. Grant, I.M. Boswarva (eds.): *Quantitative Scanning Electron Microscopy* (Academic, London 1974)
1.19 O.C. Wells: *Scanning Electron Microscopy* (McGraw-Hill, New York 1974)
1.20 J.I. Goldstein, H. Yakowitz: *Practical Scanning Electron Microscopy* (Plenum, New York 1975)
1.21 L. Reimer, G. Pfefferkorn: *Raster-Elektronenmikroskopie*, 2nd ed. (Springer, Berlin, Heidelberg, New York 1977)
1.22 F. Maurice, L. Meny, R. Tixier (eds.): *Microanalyse et Microscopie Electronique à Balayage* (Les Editions de Physique, Orsay 1978) [English transl.: *Microanalysis and Scanning Electron Microscopy*, 1979]
1.23 J.I. Goldstein, D.E. Newbury, P. Echlin, D.C. Joy, C. Fiori, E. Lifshin: *Scanning Electron Microscopy and X-Ray Microanaylsis* (Plenum, New York 1981)
1.24 S.J.B. Reed: *Electron Microprobe Analysis* (University Press, Cambridge 1975)
1.25 K.F.J. Heinrich: *Electron Beam X-Ray Microanalysis* (Van Nostrand, New York 1981)
1.26 P. Lechene, R.R. Warner (eds.): *Microbeam Analysis in Biology* (Academic, New York 1979)
1.27 M.A. Hayat (ed.): *X-Ray Microanalysis in Biology* (McMillan, London 1981)
1.28 T.E. Hutchinson, A.B. Somlyo: *Microprobe Analysis of Biological Systems* (Academic, New York 1981)
1.29 O. Johari (ed.): *Scanning Electron Microscopy 1968-1977* (IIT Research Institute, Chicago); 1978 and following years (SEM Inc., AMF O'Hare)
1.30 G. Pfefferkorn (ed.): *Beiträge zur elektronenmikroskopischen Direktabbildung von Oberflächen*, Vol.1 (1968) and following years (Remy, Münster)
1.31 W.C. Nixon (ed.): *Scanning Electron Microscopy: Systems and Applications* (The Institute of Physics, London 1973)

Transmission Electron Microscopy

(for electron optics see [2.4-7])

1.32 S. Amelinckx, R. Gevers, G. Remaut, J. Van Landuyt: *Modern Diffraction and Imaging Techniques in Material Science* (North-Holland, Amsterdam 1970) 2nd ed. in 2 Vols., published in 1978
1.33 W. Baumeister, W. Vogell (eds.): *Electron Microscopy at Molecular Dimensions* (Springer, Berlin, Heidelberg, New York 1980)
1.34 H. Bethge, J. Heydenreich (eds.): *Elektronenmikroskopie in der Festkörperphysik* (Springer, Berlin, Heidelberg, New York 1982)
1.35 O. Brümmer, J. Heydenreich, K.H. Krebs, H.G. Schneider (eds.): *Handbuch Festkörperanalyse mit Elektronen, Ionen und Röntgenstrahlen* (Vieweg, Braunschweig 1980)
1.36 J.M. Cowley: *Diffraction Physics*, 2nd ed. (North-Holland, Amsterdam 1981)
1.37 I. Dietrich: *Superconducting Electron-Optical Devices* (Plenum, New York 1976)
1.38 J.R. Fryer: *The Chemical Applications of Transmission Electron Microscopy* (Academic, London 1979)
1.39 P. Goodman (ed.): *Fifty Years of Electron Diffraction* (Reidel, Dordrecht 1981)
1.40 P. Grivet: *Electron Optics II: Instruments* (translated by P.W. Hawkes) (Pergamon, Oxford 1972)
1.41 P.J. Grundy, G.A. Jones: *Electron Microscopy in the Study of Materials* (Edwards Arnold, London 1976)
1.42 P.W. Hawkes (ed.): *Image Processing and Computer-Aided Design in Electron Optics* (Academic, London 1973)
1.43 P.W. Hawkes (ed.): *Computer Processing of Electron Microscope Images*, Topics Curr. Phys., Vol.13 (Springer, Berlin, Heidelberg, New York 1980)
1.44 P.W. Hawkes: *Electron Optics and Electron Microscopy* (Taylor & Francis, London 1972)
1.45 R.D. Heidenreich: *Fundamentals of Transmission Electron Microscopy* (Wiley, New York 1964)
1.46 M. von Heimendahl: *Einführung in die Elektronenmikroskopie* (Vieweg, Braunschweig 1970)
1.47 P.B. Hirsch, A. Howie, R.B. Nicholson, D.W. Pashley, M.J. Whelan: *Electron Microscopy of Thin Crystals* (Butterworths, London 1965)

1.48 F. Hornbogen: *Durchstrahlungselektronenmikroskopie fester Stoffe* (Chemie, Weinheim 1971)
1.49 W. Hoppe, R. Mason (eds.): Unconventional electron microscopy for molecular structure determination. Adv. Struct. Res. Diffr. Methods *7* (1979)
1.50 J.J. Hren, J.I. Goldstein, D. Joy: *Introduction to Analytical Electron Microscopy* (Plenum, New York 1979)
1.51 H.E. Huxley, A. Klug: New developments in electron microscopy. Philos. Trans. Roy. Soc. B*261*, 1-230 (1971)
1.52 B. Jouffrey (ed.): *Methodes et Techniques Nouvelles d'Observation en Métallurgie Physique* (Societe Francaise du Microscopie Electronique, Paris 1972)
1.53 D.H. Kay: *Techniques for Electron Microscopy*, 2nd ed. (Blackwell, Oxford 1965)
1.54 J.K. Koehler (ed.): *Advanced Techniques in Biological Electron Microscopy* (Springer, Berlin, Heidelberg, New York 1973); *Specific Ultrastructural Problems* (Springer, Berlin, Heidelberg, New York 1978)
1.55 R.H. Lange, J. Blödorn: *Das Elektronenmikroskop, TEM + REM* (Thieme, Stuttgart 1981)
1.56 M.H. Loretto, R.E. Smellman: *Defect Analysis in Electron Microscopy* (Chapman and Hall, London 1975)
1.57 C. Magnan (ed.): *Traité de microscopie electronique*, Vols.1,2 (Herman, Paris 1961)
1.58 G.A. Meek: *Practical Electron Microscopy for Biologists*, 2nd ed. (Wiley, London 1976)
1.59 D.L. Misell: *Image Analysis, Enhancement and Interpretation* (North-Holland, Amsterdam 1970)
1.60 D.C. Pease: *Historical Techniques for Electron Microscopy*, 2nd ed. (Academic, New York 1964)
1.61 J. Picht, J. Heydenreich: *Einführung in die Elektronenmikroskopie* (VEB Verlag Technik, Berlin 1966)
1.62 L. Reimer: *Elektronenmikroskopische Untersuchungs- und Präparationsmethoden*, 2nd ed. (Springer, Berlin, Heidelberg, New York 1967)
1.63 E. Ruska: Die frühe Entwicklung der Elektronenlinsen und der Elektronenmikroskopie. Acta Historica Leopoldina *12* (Barth, Leipzig 1979); The early development of electron lenses and electron microscopy. Microsc. Acta Suppl. *5* (1980)
1.64 W.O. Saxton: *Computer Techniques for Image Processing in Electron Microscopy* (Academic, New York 1978)
1.65 G. Schimmel: *Elektronenmikroskopische Methodik* (Springer, Berlin, Heidelberg, New York 1969)
1.66 B.M. Siegel (ed.): *Modern Developments in Electron Microscopy* (Academic, New York 1964)
1.67 B.M. Siegel, D.R. Beaman (eds.): *Physical Aspects of Electron Microscopy and Microbeam Analysis* (Wiley, New York 1975)
1.68 F.S. Sjöstrand: *Electron Microscopy of Cells and Tissues I. Instrumentation and Techniques* (Academic, New York 1967)
1.69 J.C.H. Spence: *Experimental High-Resolution Electron Microscopy* (University Press, Oxford 1981)
1.70 G. Thomas: *Transmission Electron Microscopy of Metals* (Wiley, New York 1962)
1.71 G. Thomas, J. Washburn: *Electron Microscopy and Strength of Crystals* (Interscience, New York 1963)
1.72 G. Thomas, M.J. Goringe: *Transmission Electron Microscopy of Metals* (Wiley, New York 1979)
1.73 J.N. Turner (ed.): *Methods in Cell Biology*, Vol.22, Three-Dimensional Ultrastructure in Biology (Academic, New York 1981)
1.74 U. Valdrè: *Electron Microscopy in Material Science* (Academic, New York 1971)
1.75 U. Valdrè, E. Ruedl (eds.): *Electron Microscopy in Materials Science*, Part I-IV (Commission of the European Communities, Brussels 1975)
1.76 V.E. Cosslett, R. Barer (eds.): *Advances in Optical and Electron Microscopy*, Vol.1 (Academic, London 1966) series continued
1.77 A.M. Glauert (ed.): *Practical Methods in Electron Microscopy* (North-Holland, Amsterdam 1972) series continued
1.78 J.D. Griffiths (ed.): *Electron Microscopy in Biology*, Vol.1 (Wiley, New York 1981) series continued

1.79 A.M. Hayat: *Principles and Techniques of Electron Microscopy* (Van Nostrand, New York 1970)

Conferences

1.80 *Proc. Annual Meeting of EMSA* (Electron Microscopy Society of America) (Claitor's Publ. Div., Baton Rouge, LO 1967 and following years)
1.81 Proc. EMAG (Electron Microscope Analysis Group, UK); *Electron Microscopy and Analysis* (Institute of Physics, London 1971); *Developments in Electron Microscopy and Analysis 1975* (Academic, London 1976); Meetings of 1977, 1979 and 1981 (Institute of Physics, London)
1.82 *Electron Microscopy*: Proc. Stockholm Conference 1956, ed. by F.J. Ströstrand J. Rhodin (Almqvist and Wiksells, Stockholm 1957)
1.83 *Proc. Europ. Reg. Conf. on Electron Microscopy*, Delft, Vols.1,2, ed. by A.L. Houwink, B.J. Spit (Nederlandse Vereniging voor Electronenmicroscopie, Delft 1960)
1.84 *Electron Microscopy 1964*, Proc. 3rd Europ. Reg. Conf., Vols.A,B, ed. by M. Titlbach (Czechoslovak Acad. Sci., Prague 1964)
1.85 *Electron Microscopy 1968*, Vols.1,2, ed. by D.S. Bocciarelli (Tipografia Poliglotta Vaticana, Rome 1968)
1.86 *Electron Microscopy 1972*, Proc. 5th Europ. Congr. Electron Microscopy, Manchester (Institute of Physics, London 1972)
1.87 *Electron Microscopy 1976*, Proc. 6th Europ. Congr. Electron Microscopy, Jerusalem, Vol.1, ed. by D.G. Brandon, Vol.2, ed. by Y. Ben Shaul (Tal International, Jerusalem 1976)
1.88 *Electron Microscopy 1980*, Proc. 7th Europ. Congr. Electron Microscopy, The Hague, Vols.1-4, ed. by P. Brederoo et al. (Seventh European Congress on Electron Microscopy Foundation, Leiden 1980)

International Congresses on Electron Microscopy

1.89 *Comptes Rendus du Premier Congrès International de Microscopie Electronique* Paris 1950, Revue d'Optique Theorique et Instrumentale, Paris (1953)
1.90 *Proc. 3rd Intern. Conf. Electron Microscopy*, London 1954, ed. by R. Ross (Royal Microscopical Soc., London 1956)
1.91 *Vierter Internationaler Kongreß für Elektronenmikroskopie*, Berlin 1958, Vols.1,2, ed. by W. Bargmann et al. (Springer, Berlin, Göttingen, Heidelber 1960)
1.92 *Electron Microscopy 1962*, 5th Intern. Congr. Electron Microscopy, Philadelp 1962, Vols.1,2, ed. by S.S. Breese (Academic, New York 1962)
1.93 *Electron Microscopy 1966*, 6th Intern. Congr. Electron Microscopy, Vols.1,2, ed. by R. Uyeda (Maruzen, Tokyo 1966)
1.94 *Microscopie Electronique 1970*, Grenoble, Vols.1,2,3, ed. by P. Favard (Société Francaise de Microscopie Electronique, Paris 1970)
1.95 *Electron Microscopy 1974*, Canberra, Vols.1,2, ed. by J.V. Sanders, D.J. Goodchild (Australian Acad. Sci., Canberra 1974)
1.96 *Electron Microscopy 1978*, Toronto, Vols.1,2,3, ed. by J.M. Sturgess (Microscopical Soc. Canada, Toronto 1978)

Scanning Transmission Electron Microscopy

1.97 R.E. Burge, M.T. Browne, S. Lackovic, J.F.L. Ward: "STEM Imaging at High Resolution: the Influence of Detector Geometry," in *Scanning Electron Microscopy 1979/I*, ed. by O. Johari (SEM Inc., AMF O'Hare 1979) p.127
1.98 A.V. Crewe: The current state of high resolution scanning electron microsco Q. Rev. Biophys. *3*, 137 (1970)
1.99 A.V. Crewe: Scanning transmission electron microscopy. J. Micr. *100*, 247 (1

High Voltage Electron Microscopy

(for HVEM conferences see [4.47-51])

1.100 V.E. Cosslett: Current developments in high voltage electron microscopy J. Micr. *100*, 233 (1974)

1.101 V.E. Cosslett: "Recent Progress in High Voltage Electron Microscopy," in [Ref.1.32], p.341

1.102 G. Dupouy: "Electron Microscopy at Very High Voltages," in *Advances in Optical and Electron Microscopy*, Vol.2, ed. by R. Barer, V.E. Cosslett (Academic, New York 1966)

1.103 G. Dupouy: Advantages of megavolt electron microscopy in biological research. Ultramicroscopy *2*, 199 (1977)

1.104 R.M. Fisher, T. Imura: New applications and extensions of the unique advantages of HVEM for physical and materials research. Ultramicroscopy *3*, 3 (1978)

1.105 M.V. King, D.F. Parsons, J.N. Turner, B.B. Chang, A.J. Ratkowski: Progress in applying the high voltage electron microscopy to biomedical research. Cell Biophys. *2*, 1 (1980)

1.106 J.B. LePoole, A.B. Bok, P.J. Rus: "A Compact 1 MV-Electron Microscope," in *Microscopie Electronique 1970*, Vol.1, ed. by P. Favard (Societe Francaise Microscopie Electronique, Paris 1970) p.113

1.107 P. van Zuylen, L.A. Fontijn: "Preliminary Results with the TPD 1000 kV Electron Microscope," in *High Voltage Electron Microscopy*, ed. by P.R. Swann, C.J. Humphreys, M.J. Goringe (Academic, London 1974) p.114

Chapter 2

2.1 W. Raith: "Untersuchungen zur Spin-Polarisation von Elektronenstrahlen," in *Electron Microscopy 1962*, 5th Intern. Congr. Electron Microscopy, Vol.1, ed. by S.S. Breese (Academic, New York 1962) p.AA-6

2.2 K. Tradowsky: "Messungen an polarisierten Elektronenstrahlen mit Elektron-Elektron-(Möller)-Streuung," in *Electron Microscopy 1962*, 5th Intern. Congr. Electron Microscopy, Vol.1, ed. by S.S. Breese (Academic, New York 1962) p.AA-5

2.3 J. Kessler: *Polarized Electrons* (Springer, Berlin, Heidelberg, New York 1976)

2.4 W. Glaser: *Grundlagen der Elektronenoptik* (Springer, Wien 1952)

2.5 A. Septier: *Focusing of Charged Particles* (Academic, New York 1967)

2.6 P. Grivet: *Electron Optics*, Part 1: Optics, Part 2: Instruments, 2nd ed., (translated by P.W. Hawkes) (Pergamon, Oxford 1972)

2.7 P.W. Hawkes (ed.): *Properties of Magnetic Electron Lenses*, Topics Appl. Phys., Vol.18 (Springer, Berlin, Heidelberg, New York 1982)

2.8 W. Glaser: Strenge Berechnung magnetischer Linsen der Feldform $H = H_0/[1 + (z/a)^2]$. Z. Phys. *117*, 285 (1941)

2.9 J. Dosse: Strenge Berechnung magnetischer Linsen mit unsymmetrischer Feldform nach $H = H_0/[1 + (z/a)^2]$. Z. Phys. *117*, 316 (1941)

2.10 W.D. Riecke: "Ein Kondensorsystem für eine starke Objektivlinse," in *Electron Microscopy 1962*, 5th Intern. Congr. Electron Microscopy, Vol.1, ed. by S.S. Breese (Academic, New York 1962) p.KK-5

2.11 E. Ruska: Über die Auflösungsgrenzen des Durchstrahlungs-Elektronenmikroskops. Optik *22*, 319 (1965)

2.12 S. Suzuki, K. Akashi, H. Tochigi: "Objective Lens Properties of Very High Excitation," in *26th Annual Meeting of EMSA* (Claitor's Publ. Div., Baton Rouge, LO 1968) p.320

2.13 W.D. Riecke: "Objective Lens Design for TEM. A Review of the Present State of the Art," in *Electron Microscopy 1972* (The Institute of Physics, London 1972) p.98

2.14 W. Kamminga: Properties of magnetic objective lenses with highly saturated pole pieces. Optik *45*, 39 and *46*, 226 (1976)

2.15 A. Septier: "Superconducting Lenses," in *Electron Microscopy 1972* (The Institute of Physics, London 1972) p.104

2.16 P. Bonjour: "Superconducting Lenses: Present Trends and Design," in *Electron Microscopy 1976*, Vol.1, ed. by D.G. Brandon (Tal International, Jerusalem 1976) p.73
2.17 I. Dietrich: "Superconducting Lenses," in *Electron Microscopy 1978*, Vol.3, ed. by J.M. Sturgess (Microscopical Soc. Canada, Toronto 1978) p.173
2.18 W.D. Riecke: "Practical Lens Design", in [Ref.2.7, p.164]
2.19 T. Mulvey, C.D. Newman: "Versatile Miniature Electron Lenses," in *Electron Microscopy 1972* (The Institute of Physics, London 1972) p.116
2.20 T. Mulvey, M.J. Wallington: Electron lenses. Rep. Prog. Phys. *36*, 347 (1973)
2.21 T. Mulvey: "Imaging System for Conventional Electron Microscopes," in *Electron Microscopy 1974*, Vol.1, ed. by J.V. Sanders, D.J. Goodchild (Australian Acad. Sci., Canberra 1974) p.16
2.22 T. Mulvey: "Unconventional Lens Design," in [Ref.2.7, p.359]
2.23 V.E. Cosslett: Probe size and probe current in the scanning transmission electron microscope. Optik *36*, 85 (1972)
2.24 W. Kunath, W.D. Riecke: Zur Bestimmung der Öffnungsfehlerkoeffizienten magnetischer Objektivlinsen. Optik *23*, 322 (1966)
2.25 L. Albert: Zur Phasenschiebung starker Elektronenlinsen bei endlicher Vergrößerung. Optik *24*, 18 (1966)
2.26 W. Kunath, W.D. Riecke, E. Ruska: "Spherical Aberration of Saturated Strong Objective Lenses," in *Electron Microscopy 1966*, Vol.1, ed. by R. Uyeda (Maruzen, Tokyo 1966) p.139
2.27 T. Yanaka, M. Watanabe: "Aberration Coefficients of Extremely Asymmetrical Objective Lenses," in *Electron Microscopy 1966*, Vol.1, ed. by R. Uyeda (Maruzen, Tokyo 1966) p.141
2.28 C.E. Hall: Method of measuring spherical aberration of an electron microscope objective. J. Appl. Phys. *20*, 631 (1949)
2.29 K. Heinemann: In-situ measurement of objective lens data of a high resolution electron microscope. Optik *34*, 113 (1971)
2.30 G. Liebmann: Measured properties of strong "unipotential" electron lenses. Proc. Phys. Soc. B*62*, 213 (1949)
2.31 K.J. Hanszen: Vergleichende Betrachtungen über den Öffnungsfehler symmetrische und asymmetrischer Elektronen-Einzellinsen auf Grund von Vermessungen der Austrittsstrahltangenten. Z. Naturforsch. A*13*, 409 (1958)
2.32 S. Leisegang: Zum Astigmatismus von Elektronenlinsen. Optik *10*, 5 (1953)
2.33 W. Glaser, H. Grümm: Die Kaustikfläche von Elektronenlinsen. Optik *7*, 96 (1950)
2.34 D. Kynaston, T. Mulvey: The correction of distortion in the electron microscope. Br. J. Appl. Phys. *14*, 199 (1963)
2.35 J. Dosse: Über optische Kenngrößen starker Elektronenlinsen. Z. Phys. *117*, 722 (1941)
2.36 V.E. Cosslett: Energy loss and chromatic aberration in electron microscopy. Z. Angew. Phys. *27*, 138 (1969)
2.37 L. Reimer, P. Gentsch: Superposition of chromatic error and beam broadening in TEM of thick carbon and organic specimens. Ultramicroscopy *1*, 1 (1975)
2.38 M. Fotino: "Evaluation of Factors Affecting the Resolution in Thick Biological Specimens in High-Voltage TEM," in *Electron Microscopy 1976*, Vol.1, ed. by D.G. Brandon (Tal International, Jerusalem 1976) p.277
2.39 S. Katagiri: Experimental investigation of chromatic aberration in the electron microscope. Rev. Sci. Instrum. *26*, 870 (1955)
2.40 O. Scherzer: Sphärische und chromatische Korrektur von Elektronen-Linsen. Optik *2*, 114 (1947)
2.41 A. Septier: Lentille quadrupolaire magnéto-électrique corrigée de l'aberration chromatique. Aberration d'ouverture de ce type de lentilles. C.R. Acad. Sci. Paris *256*, 2325 (1963)
2.42 H. Rose: Über den sphärischen und den chromatischen Fehler unrunder Elektronenlinsen. Optik *25*, 587 (1967)
2.43 H. Rose: Elektronenoptische Aplanate. Optik *34*, 285 (1971)
2.44 H. Koops, G. Kuck, O. Scherzer: Erprobung eines elektronenoptischen Achromators. Optik *48*, 225 (1977)
2.45 H. Koops: "Aberration Correction in Electron Microscopy," in *Electron Microscopy 1978*, Vol.3, ed. by J.M. Sturgess (Microscopical Soc. Canada, Toronto 1978) p.185
2.46 O. Rang: Der elektrostatische Stigmator, ein Korrektiv für astigmatische Elektronenlinsen. Optik *5*, 518 (1949)

Chapter 3

3.1 C. Jönsson, H. Hoffmann, G. Möllenstedt: Messung des mittleren inneren Potentials von Be im Elektronen-Interferometer. Phys. Kondens. Mater. *3*, 193 (1965)
3.2 M. Keller: Ein Biprisma-Interferometer für Elektronenwellen und seine Anwendung. Z. Phys. *164*, 274 (1961)
3.3 R. Buhl: Interferenzmikroskopie mit Elektronenwellen. Z. Phys. *155*, 395 (1959)
3.4 H. Hoffmann, C. Jönsson: Elektroneninterferometrische Bestimmung der mittleren inneren Potentiale von Al, Cu und Ge unter Verwendung eines neuen Präparationsverfahrens. Z. Phys. *182*, 360 (1965)
3.5 K.H. Gaukler, R. Schwarzer: Verbessertes Verfahren zur Bestimmung des mittleren inneren Potentials aus Reflexions-Kikuchi-Diagrammen. Optik *33*, 215 (1971)
3.6 Z.G. Pinsker: *Electron Diffraction* (Butterworths, London 1953)
3.7 K. Molière, H. Niehrs: Interferenzbrechung von Elektronenstrahlen. I. Zur Theorie der Elektroneninterferenzen an parallelepipedischen Kristallen. Z. Phys. *137*, 445 (1954)
3.8 H.J. Altenheim, K. Molière: Interferenzbrechung von Elektronenstrahlen. II. Die Feinstruktur der Interferenzen von Magnesiumoxid-Kristallen. Z. Phys. *139*, 103 (1954)
3.9 O. Rang: Ferninterferenzen von Elektronenwellen. Z. Phys. *136*, 465 (1953)
3.10 G. Möllenstedt: Elektronenmikroskopische Sichtbarmachung von Hohlstellen in Einkristall-Lamellen. Optik *10*, 72 (1953)
3.11 L. Marton, J.A. Simpson, J.A. Suddeth: An electron interferometer. Rev. Sci. Instrum. *25*, 1099 (1954)
3.12 G. Möllenstedt, H. Düker: Fresnelscher Interferenzversuch mit einem Biprisma für Elektronenwellen. Naturwissenschaften *42*, 41 (1955)
3.13 G. Möllenstedt, H. Düker: Beobachtungen und Messungen an Biprisma-Interferenzen mit Elektronenwellen. Z. Phys. *145*, 377 (1956)
3.14 H. Düker: Lichtstarke Interferenzen mit einem Biprisma für Elektronenwellen. Z. Naturforsch. A*10*, 256 (1955)
3.15 G. Möllenstedt, M. Keller: Elektroneninterferometrische Messungen des inneren Potentials. Z. Phys. *148*, 34 (1957)
3.16 G. Möllenstedt, G. Wohland: "Direct Interferometric Measurement of the Coherence Length of an Electron Wave Packet Using A Wien Filter," in *Electron Microscopy 1980*, Vol.1, ed. by P. Brederoo, G. Boom (Seventh European Congr. on Electron Microscopy Foundation, Leiden 1980) p.28
3.17 P.W. Hawkes: "Coherence in Electron Optics," in *Advances in Optical and Electron Microscopy*, Vol.7, ed. by R. Barer, V.E. Cosslett (Academic, London 1978) p.101
3.18 H.A. Ferwerda: "Coherence of Illumination in Electron Microscopy," in *Imaging Processes and Coherence in Physics*, Lecture Notes in Physics, Vol.112, ed. by M. Schlenker, M. Fink, J.-P. Goedgebuer, C. Malgrange, J.-C. Vienot, R.H. Wade (Springer, Berlin, Heidelberg, New York 1980) p.85
3.19 G. Möllenstedt, R. Buhl: Ein Elektronen-Interferenz-Mikroskop. Phys. Bl. *13*, 357 (1957)
3.20 J. Faget, C. Fert: Microscopie interférentielle et mesure de la différence de phase ontroduite par une lame en optic electronic. C. R. Acad. Sci. Paris *244*, 2368 (1957)
3.21 T. Hibi, S. Takahashi: Electron interference microscope. J. Electromicr. Jpn. *12*, 129 (1963)
3.22 P.N.T. Unwin: "Interference Microscopy with a Six Lens Electron Microscope," in *Microscopie Electronique 1970*, Vol.1, ed. by P. Favard (Société Francaise de Microscopie Electronique, Paris 1970) p.65
3.23 G. Möllenstedt, H. Lichte: "Doppler Shift of Electron Waves," in *Electron Microscopy 1978*, Vol.1, ed. by J.M. Sturgess (Microscopical Soc. Canada, Toronto 1978) p.178
3.24 H. Lichte: Ein Elektronen-Auflicht-Interferenzmikroskop zur Präzisionsmessung von Unebenheiten und Potentialunterschieden auf Oberflächen. Optik *57*, 35 (1980)
3.25 R.G. Chambers: Shift of an electron interference pattern by enclosed magnetic flux. Phys. Rev. Lett. *5*, 3 (1960)

3.26 H.A. Fowler, L. Marton, J.A. Simpson, J.A. Suddeth: Electron interferometer studies of iron whiskers. J. Appl. Phys. *32*, 1153 (1961)
3.27 H. Boersch, H. Hamisch, K. Grohmann, D. Wohlleben: Experimenteller Nachweis der Phasenschiebung von Elektronenwellen durch das magnetische Vektorpotenti. Z. Phys. *165*, 79 (1961)
3.28 W. Bayh: Messung der kontinuierlichen Phasenschiebung von Elektronenwellen im kraftfeldfreien Raum durch das magnetische Vektorpotential einer Wolframwendel. Z. Phys. *169*, 492 (1962)
3.29 G. Schaal, C. Jönsson, E.F. Krimmel: Weitgetrennte kohärente Elektronen-Wellenzüge und Messung des Magnetflusses $\phi_0 = h/e$. Optik *24*, 529 (1967)
3.30 H. Boersch, B. Lischke: "Electron Interferometric Measurements of Quantized Magnetic Flux Trapped in Superconducting Tubes," in *Microscopie Electronique 1970*, Vol.1, ed. by P. Favard (Société Francaise de Microscopie Electronique Paris 1970) p.69
3.31 B. Lischke: Bestimmung des Fluxoidquants in supraleitenden Hohlzylindern. Z. Phys. *237*, 469 (1970) and *239*, 360 (1970)
3.32 H. Boersch: Fresnelsche Beugung im Elektronenmikroskop. Phys. Z. *44*, 202 (19
3.33 K.J. Hanszen, B. Morgenstern: Fresnelsche Beugungssäume im elektronenmikroskopischen Bild einer Objektkante und Kontrastübertragungsdaten des Mikroskopobjektivs. Optik *24*, 442 (1967)
3.34 J. Hillier, E.G. Ramberg: The magnetic electron microscope objective: contou phenomena and the attainment of high resolving power. J. Appl. Phys. *18*, 48 (1947)
3.35 L. Reimer, H. Rüberg: Der Einfluß von Objektparametern auf die Intensitätsverteilung in elektronenoptischen Fresnelsäumen an Kanten. Optik *40*, 29 (1974)
3.36 J.N. Turner, H.M. Johnson, R. Slon, P.F. Parsons: Electron edge diffraction from transparent carbon films. Optik *41*, 54 (1974)
3.37 A.R. Wilson, L.A. Bursill, A.E.C. Spargo: Fresnel diffraction effects on hig resolution (< 3 Å) images: Effect of spherical aberration on the Fresnel fringe. Optik *52*, 313 (1979)
3.38 P.W. Hawkes: Note on the sign of the wave aberration in electron optics. Optik *46*, 357 (1976) and *48*, 253 (1977)
3.39 M.E.C. MacLachlan: The sign of the wave aberration in electron optics. Optik *47*, 363 (1977)
3.40 O. Scherzer: The theoretical resolution limit of the electron microscope. J. Appl. Phys. *20*, 20 (1949)
3.41 K.J. Hanszen: Generalisierte Angaben über die Phasenkontrast- und Amplituden kontrast-Übertragungsfunktion für elektronenmikroskopische Objektive. Z. Angew. Phys. *20*, 427 (1966)
3.42 K.J. Hanszen: "The Optical Transfer Theory of the Electron Microscope: Fundamental Principles and Applications," in *Advances in Optical and Electro Microscopy*, Vol.4, ed. by R. Barer, V.E. Cosslett (Academic, London 1971) p.

Chapter 4

4.1 M.E. Haine, P.A. Einstein, P.H. Borcherds: Resistance bias characteristics of the electron microscope gun. Br. J. Appl. Phys. *9*, 482 (1958)
4.2 L.W. Swanson, L.C. Crouser: Total-energy distribution of field-emitted electrons and single-plane work functions for tungsten. Phys. Rev. *163*, 622 (1967)
4.3 H. Boersch: Experimentelle Bestimmung der Energieverteilung in thermisch ausgelösten Elektronenstrahlen. Z. Phys. *139*, 115 (1954)
4.4 K.H. Loeffler: Energy-spread generation in electron-optical instruments. Z. Angew. Phys. *27*, 145 (1969)
4.5 R.W. Ditchfield, M.J. Whelan: Energy broadening of the electron beam in the electron microscope. Optik *48*, 163 (1977)
4.6 H. Rose, R. Spehr: On the theory of the Boersch effect. Optik *57*, 339 (1980)
4.7 K.H. Gaukler, R. Speidel, F. Vorster: Energieverteilungen von Elektronen aus einer Feldemissionskathode. Optik *42*, 391 (1975)
4.8 D.B. Langmuir: Theoretical limitations of cathode-ray tubes. Proc. IRE *25*, 977 (1937)

4.9 J. Dosse: Theoretische und experimentelle Untersuchungen über Elektronenstrahler. Z. Phys. *115*, 530 (1940)
4.10 W. Glaser: *Grundlagen der Elektronenoptik* (Springer, Wien 1952)
4.11 J.A. Swift, A.C. Brown: SEM electron source: pointed tungsten filaments with long life and high brightness. Scanning *2*, 42 (1979)
4.12 A.N. Broers: Electron gun using long-life LaB_6 cathode. J. Appl. Phys. *38*, 1991 (1967); Some experimental and estimated characteristics of the LaB_6 rod cathode electron gun. J. Phys. E*2*, 273 (1969)
4.13 H. Ahmed: "The Use of LaB_6 and Composite Boride Cathodes in Electron Optical Instruments," in *Electron Micrsocopy and Analysis*, ed. by W.C. Nixon (The Institute of Physics, London 1971) p.30
4.14 R. Vogt: Richtstrahlwert und Energieverteilung der Elektronen aus einem Elektronenstrahlerzeuger mit LaB_6-Kathode. Optik *36*, 262 (1972)
4.15 S.D. Ferris, D.C. Joy, H.J. Leamy, C.K. Crawford: "A Directly Heated LaB_6 Electron Source," in *Scanning Electron Microscopy 1975*, ed. by O. Johari (IIT Research Institute, Chicago 1975) p.11
4.16 S. Nakagawa, T. Yanaka: "A Highly Stable Electron Probe Obtained with LaB_6 Cathode Electron Gun," in *Scanning Electron Microscopy 1975*, ed. by O. Johari (IIT Research Institute, Chicago 1975) p.19
4.17 C.K. Crawford: "Mounting Methods and Operating Characteristics for LaB_6 Cathodes," in *Scanning Electron Microscopy 1979/I*, ed. by O. Johari (SEM Inc., AMF O'Hare 1979) p.19
4.18 P.H. Schmidt, D.C. Joy, L.D. Longinotti, H.J. Leamy, S.D. Ferris, Z. Fisk: Anisotropy of thermionic electron emission values of LaB_6 single-crystal emitter cathodes. Appl. Phys. Lett. *29*, 400 (1976)
4.19 M.E. Haine, P.A. Einstein: Characteristics of the hot cathode electron microscope gun. Br. J. Appl. Phys. *3*, 40 (1952)
4.20 L.H. Veneklasen, B.M. Siegel: Oxygen-processed field emission source. J. Appl. Phys. *43*, 1600 (1972)
4.21 J.W. Butler: "Digital Computer Techniques in Electron Microscopy," in *Electron Microscopy 1966*, Vol.1, ed. by R. Uyeda (Maruzen, Tokyo 1966) p.191
4.22 A.V. Crewe, D.N. Eggenberger, J. Wall, L.M. Welter: Electron gun using a field emission source. Rev. Sci. Instrum. *39*, 576 (1968)
4.23 E. Munro: "Design of Electrostatic Lenses for Field-Emission Electron Guns," in *Electron Microscopy 1972* (The Institute of Physics, London 1972) p.22
4.24 D. Kern, D. Kurz, R. Speidel: Elektronenoptische Eigenschaften eines Strahlerzeugungssystemes mit Feldemissionskathode. Optik *52*, 61 (1978)
4.25 G.H.N. Riddle: Electrostatic einzel lenses with reduced spherical aberration for use in field-emission gun. J. Vac. Sci. Technol. *15*, 857 (1978)
4.26 J. Orloff, L.W. Swanson: An asymmetric electrostatic lens for field-emission microprobe applications. J. Appl. Phys. *50*, 2494 (1979)
4.27 F.H. Plomp, L. Veneklasen, B.M. Siegel: "Development of a Field Emission Electron Source for an Electron Microscope," in *Electron Microscopy 1968*, Vol.1, ed. by D.S. Bocciarelli (Tipografia Poliglotta Vaticana, Rome 1968) p.141
4.28 L.H. Veneklasen, B.M. Siegel: "A Field Emission Illuminating System for Transmission," in *Electron Microscopy 1970*, Vol.2, ed. by P. Favard (Société Francaise de Microscopie Electronique, Paris 1970) p.87
4.29 T. Someya, T. Goto, Y. Marada, M. Watanabe: "Development of Field Emission Electron Gun for High Resolution 100 kV Electron Microscope," in *Electron Microscopy 1972* (The Institute of Physics, London 1972) p.20
4.30 W. Engel, W. Kunath, S. Krause: "Properties of Three Electrode Accelerating Lenses for Field Emission Electron Guns," in *Electron Microscopy 1974*, Vol.1, ed. by J.V. Sanders, D.J. Goodchild (Australian Acad. Sci., Canberra 1974) p.118
4.31 J.R.A. Cleaver: Field emission electron gun system incorporating single-pole magnetic lenses. Optik *52*, 293 (1979)
4.32 M. Troyon: "A Magnetic Field Emission Electron Probe Forming System," in *Electron Microscopy 1980*, Vol.1, ed. by P. Brederoo, G. Boom (Seventh European Congr. on Electron Microscopy Foundation, Leiden 1980) p.56
4.33 M.E. Haine: The electron optical system of the electron microscope. J. Sci. Instrum. *24*, 61 (1947)

4.34 W.D. Riecke: Zur Zentrierung des magnetischen Elektronenmikroskops. Optik *24* 397 (1966)
4.35 W.D. Riecke: "Instrument Operation for Microscopy and Microdiffraction," in *Electron Microscopy in Materials Science*, Part 1, ed. by U. Valdrè, E. Ruedl (Commission European Communities, Brussels 1976) p.19
4.36 V.E. Cosslett: Probe size and probe current in the STEM. Optik *36*, 85 (1972)
4.37 V.E. Cosslett, M.E. Haine: "The Tungsten Point Cathode as an Electron Source in *Proc. 3rd Intern. Conf. Electron Microscopy*, ed. by R. Ross (Royal Microscopical Soc., London 1956) p.639
4.38 L.H. Veneklasen: Some general considerations concerning the optics of the field emission illumination system. Optik *36*, 410 (1972)
4.39 J.R.A. Cleaver, K.C.A. Smith: "Two-Lens Probe Forming Systems Employing Field Emission Guns," in *Scanning Electron Microscopy 1973*, ed. by O. Johari (IIT Research Inst., Chicago 1973) p.49
4.40 M. Müller, Th. Koller: Preparation of aluminium oxide films for high resolution electron microscopy. Optik *35*, 287 (1972)
4.41 D. Dorignac, M.E.C. MacLachlan, B. Jouffrey: Low-noise boron supports for hig resolution electron microscopy. Ultramicroscopy *4*, 85 (1979)
4.42 S. Iijima: Thin graphite supporting films for high resolution electron microscopy. Micron *8*, 41 (1977)
4.43 W. Baumeister, M.H. Hahn: Suppression of lattice periods in vermiculite singl crystal specimen supports for high resolution electron microscopy. J. Micr. *101*, 111 (1974)
4.44 U. Valdrè, M.J. Goringe: *Electron Microscopy in Material Science* (Academic, New York 1971) p. 207
4.45 U. Valdrè: "General Considerations on Specimen Stages, in *Electron Microscopy 1972* (The Institute of Physics, London 1972) p.317
4.46 J.A. Venables: "In-Situ Experiments in Electron Microscopes," in *Electron Microscopy 1972* (The Institute of Physics, London 1972) p.344
4.47 P.R. Swann (ed.): *Proceedings Symposium on High Voltage Electron Microscope 1972*, published in J. Micr. *97*, Parts 1 and 2 (1973)
4.48 P.R. Swann, C.J. Humphreys, M.J. Goringe (eds.): *High Voltage Electron Microscopy* (Academic, London 1974)
4.49 B. Jouffrey, P. Favard (eds.): *Microscopie Electronique à Haute Tension* (Société Francaise de Microscopie Electronique, Paris 1976)
4.50 T. Imura, H. Hashimoto (eds.): *High Voltage Electron Microscopy* (Japanese Society of Electron Microscopy, Kyoto 1977)
4.51 P. Brederoo, J. van Landuyt (eds.): *Electron Microscopy 1980*, Vol.4: High Voltage (Seventh European Congr. on Electron Microscopy Foundation, Leiden 1980)
4.52 H.G. Heide: Principle of a TEM specimen device to meet highest requirements: specimen temperature 5-300 K, cryo transfer, condensation protection, specime tilt, stage stability for highest resolution. Ultramicroscopy *6*, 115 (1981)
4.53 J.E. Eades: A helium-cooled specimen stage for electron microscopy. J. Phys. E*15*, 184 (1982)
4.54 D.F. Parsons, V.R. Matricardi, J. Subjeck, I. Uydess, G. Wray: High-voltage electron microscopy of whet whole cancer and normal cells: Visualization of cytoplasmic structure and surface projections. Biochim. Biophys. Acta *290*, 110 (1972)
4.55 J. Stabenow: Herstellung dünnwandiger Objektivaperturblenden für die Elektronenmikroskopie. Naturwissenschaften *54*, 163 (1967)
4.56 J. Kala, J. Podbrdský: Thin foil apertures with very small openings for electron microscopy. J. Phys. E*4*, 609 (1971)
4.57 E. Schabtach: A method for the fabrication of thin foil apertures for electro microscopy. J. Micr. *101*, 121 (1974)
4.58 C.F. Oster, D.C. Skillman: "Determination and Control of Electron Microscopic Magnification," in *Electron Microscopy 1962*, 5th Intern. Congr. Electron Microscopy, Vol.1, ed. by S.S. Breese (Academic, New York 1962) p.EE-3
4.59 G.F. Bahr, E. Zeitler: The determination of magnification in the electron microscope. Lab. Invest. *14*, 880 (1965)
4.60 P.F. Elbers, J. Pieters: "Accurate Magnification Determination in the Siemens Elmiskop I," in *Electron Microscopy 1964*, Proc. 3rd Europ. Reg. Conf., Vol.A, ed. by M. Titlbach (Czechoslovak Acad. Sci., Prague 1964) p.123

4.61 W.C.T. Dowell; Die Bestimmung der Vergrößerung des Elektronenmikroskops mittels Elektroneninterferenz. Optik *21*, 26 (1964)
4.62 R. Luftig: An accurate measurement of the catalase crystal period and its use as an internal marker for electron microscopy. J. Ultrastruct. Res. *20*, 91 (1967)
4.63 N.G. Wrigley: The lattice spacing of crystalline catalase as an internal standard of length in electron microscopy. J. Ultrastruct. Res. *24*, 454 (1968)
4.64 J. Porstendörfer, J. Heyder: Elektronenmikroskopische Untersuchungen an Latex-Teilchen. Optik *35*, 73 (1972)
4.65 J.B. LePoole, P. Stam: "An Objective Method for Focusing," in *Proc. 3rd Intern. Congr. Electron Microscopy London 1954*, ed. by R. Ross (Royal Microscopical Soc., London 1956) p.666
4.66 H. Koike, K. Ueno, M. Suzuki: "Scanning Device Combined with Conventional Electron Microscope," in *Proc. 29th Ann. Meeting of EMSA* (Claytor's Publ. Div., Baton Rouge LO 1971) p.28
4.67 L. Reimer, P. Hagemann: "The Use of Transmitted and Backscattered Electrons in the Scanning Mode of a TEM," in *Developments in Electron Microscopy and Analysis*, ed. by D.L. Misell (The Institute of Physics, London 1977) p.135
4.68 A.V. Crewe, J. Wall, L.M. Welter: A high-resolution STEM. J. Appl. Phys. *39*, 5861 (1968)
4.69 A.V. Crewe, J. Wall: Contrast in a high-resolution STEM. Optik *30*, 461 (1970)
4.70 A.V. Crewe, M. Isaacson, D. Johnson: A high-resolution electron spectrometer for use in transmission scanning electron microscopy. Rev. Sci. Instrum. *42*, 411 (1971)
4.71 A.V. Crewe: "Production of Electron Probes Using a Field Emission Source," in *Progress in Optics*, Vol.11, ed. by E. Wolf (North-Holland, Amsterdam 1973) p.225
4.72 J.M. Cowley: Image contrast in a transmission scanning electron microscope. Appl. Phys. Lett. *15*, 58 (1969)
4.73 E. Zeitler, M.G.R. Thomson: Scanning transmission electron microscopy. Optik *31*, 258 and 359 (1970)
4.74 C. Colliex, A.J. Craven, C.J. Wilson: Fresnel fringes in STEM. Ultramicroscopy *2*, 327 (1977)
4.75 D.C. Joy, D.M. Maher, A.G. Cullis: The nature of defocus fringes in STEM images. J. Micr. *108*, 185 (1976)
4.76 R. Broser-Warminsky, E. Ruska: "Hochauflösende Leuchtschirme für die Elektronenmikroskopie," in *Vierter International Kongreß für Elektronenmikroskopie*, Vol.1, ed. by W. Bargmann et al. (Springer, Berlin, Göttingen, Heidelberg 1960) p.104
4.77 V.E. Cosslett, G.L. Jones, R.A. Camps: "Image Viewing and Recording in High Voltage Electron Microscopy," in [Ref.4.48] p.147
4.78 H.G. Heide: Zur Vorevakuierung von Photomaterial für Elektronenmikroskopie. Z. Angew. Phys. *19*, 348 (1965)
4.79 E. Guetter, M. Menzel: "An External Photographic System for Electron Microscopes," in *Electron Microscopy 1978*, Vol.1, ed. by S.M. Sturgess (Microscopical Soc. Canada, Toronto 1978) p.92
4.80 H. Frieser, E. Klein: Die Eigenschaften photographischer Schichten bei Elektronenbestrahlung. Z. Angew. Phys. *10*, 337 (1958)
4.81 H. Frieser, E. Klein, E. Zeitler: Das Verhalten photographischer Schichten bei Elektronenbestrahlung. Z. Angew. Phys. *11*, 190 (1959)
4.82 R.C. Valentine: "The Response of Photographic Emulsions to Electrons," in *Advances in Optical and Electron Microscopy*, Vol.1, ed. by R. Barer, V.E. Cosslett (Academic, London 1966) p.180
4.83 R.E. Burge, D.F. Garrard: The resolution of photographic emulsions for electrons in the energy range 7-60 keV. J. Phys. E*1*, 715 (1968)
4.84 R.E. Burge, D.F. Garrard, M.T. Browne: The response of photographic emulsions to electrons in the energy range 7-60 keV. J. Phys. E*1*, 707 (1968)
4.85 G.C. Farnell, R.B. Flint: The response of photographic materials to electrons with particular reference to electron micrography. J. Micr. *97*, 271 (1973)
4.86 W. Lippert: Erfahrungen mit der photographischen Methode bei der Massendickenbestimmung im Elektronenmikroskop. Optik *29*, 372 (1969)

4.87 G.L. Jones, V.E. Cosslett: "Sensitivity and Resolution of Photographic Emulsions to Electrons (60-700 keV)," in *Microscopie Electronique 1970*, Vol.1, ed. by P. Favard (Société Francaise de Microscopie Electronique, Paris 1970) p.349
4.88 M. Fotino: "Improved Response of Photographic Emulsions for Electron Micrographs at Higher Voltages," in *Electron Microscopy 1974*, Vol.1, ed. by J.V. Sanders, D.J. Goodchild (Australian Acad. Sci., Canberra 1974) p.104
4.89 P.H. Broerse, P. Kramer, W. Kühl, H.F. Premsela: "Electron Microscopy at Extremely Low Current Densities in the Specimen with a New Light Intensifier," in *Electron Microscopy 1968*, Vol.1, ed. by D.S. Bocciarelli (Tipografia Poliglotta Vaticana, Rome 1968) p.217
4.90 K.H. Hermann, D. Krahl, V. Rindfleisch: Use of TV image intensifiers in electron microscopy. Siemens Forsch. Entwicklungsber. *1*, 67 (1972)
4.91 D.G. Brandon, D. Shechtman, D.N. Seidman: "Preliminary Results with a Channel Plate Image Intensifier in the Electron Microscope," in *Microscopie Electronique 1970*, Vol.1, ed. by P. Favard (Société Francaise de Microscopie Electronique, Paris 1970) p.343
4.92 C.A. English, J.A. Venables: "The Use of Channel Plates as Image Intensifiers in Electron Microscopy," in *Electron Microscopy and Analysis* (The Institute of Physics, London 1971) p.40
4.93 E.L. Thomas, S. Danyluck: A channelplate image intensifier for the electron microscope. J. Phys. E*4*, 843 (1971)
4.94 D.A. Gedcke, J.B. Ayers, P.B. deNee: "A Solid State Backscattered Electron Detector Capable of Operating at T.V. Scan Rates," in *Scanning Electron Microscopy 1978/I*, ed. by O. Johari (SEM Inc., AMF O'Hare 1978) p.581
4.95 M. Kikuchi, S. Takashima: "Multi-Purpose Backscattered Electron Detector," in *Electron Microscopy 1978*, Vol.1, ed. by J.M. Sturgess (Microscopical Soc. Canada, Toronto 1978) p.82
4.96 J. Pawley: "Performance of SEM Scintillation Materials," in *Scanning Electron Microscopy 1974*, ed. by O. Johari (IIT Research Institute, Chicago 1974) p.28
4.97 W. Baumann, A. Niemitz, L. Reimer, B. Volbert: Preparation of P-47 scintillators for STEM. J. Micr. *122*, 181 (1981)

Chapter 5

5.1 W.J. Byatt: Analytical representation of Hartree potentials and electron scattering. Phys. Rev. *104*, 1298 (1956)
5.2 T. Tietz: Über den Mottschen Polarisationseffekt bei der Streuung mittelschneller Elektronen. Nuovo Cimento *36*, 1365 (1965)
5.3 H.L. Cox, R.A. Bonham: Elastic electron scattering amplitudes for neutral atoms calculated using the partial wave method at 10, 40, 70 and 100 kV for Z = 1 to Z = 54. J. Chem. Phys. *47*, 2599 (1967)
5.4 H. Raith: Komplexe Atomstreuamplituden für die elastische Elektronenstreuung an Festkörperatomen. Acta Cryst. A*24*, 85 (1968)
5.5 L. Reimer, K.H. Sommer: Messungen und Berechnungen zum elektronenmikroskopischen Streukontrast für 17 bis 1200 keV-Elektronen. Z. Naturforsch. A*23*, 1569 (1968)
5.6 G. Molière: Theorie der Streuung schneller geladener Teilchen. Z. Naturforsch A*2*, 133 (1947)
5.7 R.J. Glauber: In *Lectures in Theoretical Physics*, ed. by W.E. Brittin, G. Dunham (Interscience, New York 1959) p.315
5.8 E. Zeitler, H. Olsen: Screening effects in elastic electron scattering. Phys. Rev. A*136*, 1546 (1964); Complex scattering amplitudes in elastic electron scattering. Phys. Rev. A*162*, 1439 (1967)
5.9 J. Haase: Berechnung der komplexen Streufaktoren für schnelle Elektronen unter Verwendung von Hartree-Fock-Atompotentialen. Z. Naturforsch. A*23*, 1000 (1968)
5.10 F. Lenz: Zur Streuung mittelschneller Elektronen in kleinste Winkel. Z. Naturforsch. A*9*, 185 (1954)
5.11 F. Arnal, J.L. Balladore, G. Soum, P. Verdier: Calculations of the cross sections of electron interaction with matter. Ultramicroscopy *2*, 305 (1977)

5.12 R.E. Burge, G.H. Smith: A new calculation of electron scattering cross sections and a theoretical discussion of image contrast in the electron microscope. Proc. Phys. Soc. *79*, 673 (1962)
5.13 P.A. Doyle, P.S. Turner: Relativistic Hartree-Fock x-ray and electron scattering factors. Acta Cryst. A*24*, 390 (1968)
5.14 J.A. Ibers, J.A. Hoerni: Atomic scattering amplitudes for electron diffraction. Acta Cryst. *7*, 405 (1954)
5.15 J.A. Ibers: Atomic scattering amplitudes for electrons. Acta Cryst. *11*, 178 (1958)
5.16 J.A. Ibers, B.K. Vainshtein: "Scattering Amplitudes for Electrons," in *International Tables for X-Ray Crystallography*, Vol.3, ed. by K. Lonsdale (Kynoch, Birmingham 1962)
5.17 J. Geiger: "Zur Streuung von Elektronen am Einzelatom," in *Electron Microscopy 1962*, 5th Intern. Congr. Electron Microscopy, Vol.1, ed. by S.S. Breese (Academic, New York 1962) p.AA-12
5.18 N.F. Mott: The polarisation of electrons by double scattering. Proc. Roy. Soc. A*135*, 429 (1932)
5.19 W.A. McKinley, H. Feshbach: The Coulomb scattering of relativistic electrons by nuclei. Phys. Rev. *74*, 1759 (1948)
5.20 J.A. Doggett, L.V. Spencer: Elastic scattering of electrons and positrons by point nuclei. Phys. Rev. *103*, 1597 (1956)
5.21 N. Sherman: Coulomb scattering of relativistic electrons by point nuclei. Phys. Rev. *103*, 1601 (1956)
5.22 M.E. Riley, J. Crawford, C.J. MacCallum, F. Biggs: Theoretical electron-atom elastic scattering cross sections. At. Data Nucl. Data Tables *15*, 443 (1975)
5.23 S.R. Lin: Elastic electron scattering by screened nuclei. Phys. Rev. A*133*, 965 (1964)
5.24 W. Bühring: Computational improvements in phase shift calculations of elastic electron scattering. Z. Phys. *187*, 180 (1965); Elastic scattering by mercury atoms. Z. Phys. *212*, 61 (1968)
5.25 J. Kessler, N. Weichert: The influence of screening on Mott scattering by mercury atoms. Z. Phys. *212*, 48 (1968)
5.26 L. Reimer, E.R. Krefting: "The Effect of Scattering Models on the Results of Monte Carlo Calculations," in *Use of Monte Carlo Calculations in Electron Probe Microanalysis and Scanning Electron Microscopy*, ed. by K.F.J. Heinrich, D.E. Newbury, H. Jakowitz, NBS Special Publ. 460 (US Government Printing Office, Washington 1976) p.45
5.27 J. Kessler: *Polarized Electrons* (Springer, Berlin, Heidelberg, New York 1976)
5.28 J.P. Langmore, J. Wall, M. Isaacson: The collection of scattered electrons in dark field electron microscopy. I. Elastic scattering. Optik *38*, 335 (1973); II. Inelastic scattering. Optik *39*, 359 (1974)
5.29 J. Geiger, K. Wittmaack: Wirkungsquerschnitte für die Anregung von Molekülschwingungen durch schnelle Elektronen. Z. Phys. *187*, 433 (1965)
5.30 H. Boersch, J. Geiger, A. Bohg: Wechselwirkung von Elektronen mit Gitterschwingungen in NH_4Cl and NH_4Br. Z. Phys. *227*, 141 (1969)
5.31 B. Schröder, J. Geiger: Electron-spectrometric study of amorphous Ge and Si in the two-phonon region. Phys. Rev. Lett. *28*, 301 (1972)
5.32 H. Raether: *Solid State Excitations by Electrons*, Springer Tracts Mod. Phys., Vol.38 (Springer, Berlin, Heidelberg, New York 1965) p.84
5.33 R.D. Leapman, V.E. Cosslett: "Energy Loss Spectrometry of Inner Shell Excitations," in *Electron Microscopy 1976*, Vol.1, ed. by D.G. Brandon (Tal International, Jerusalem 1976) p.431
5.34 D.B. Wittry, R.P. Ferrier, V.E. Cosslett: "Microanalysis in the TEM by Selected Area Electron Spectrometry," in *Fifth Intern. Congr. on X-Ray Optics and Microanalysis*, ed. by G. Möllenstedt, K.H. Gaukler (Springer, Berlin, Heidelberg, New York 1969) p.293
5.35 R.P.T. Hills, R.P. Ferrier: "Selected Area Electron Spectrometry," in *Electron Microscopy 1972* (The Institute of Physics, London 1972) p.206
5.36 C. Colliex, B. Jouffrey: Diffusion inélastique des electrons dans un solide par excitation de niveaux atomiques profonds. Philos. Mag. *25*, 491 (1972)
5.37 M. Isaacson: Interaction of 25 keV electrons with the nucleic acid bases adenine, thymine and uracil. J. Chem. Phys. *56*, 1803 and 1813 (1972)

5.38 H. Koppe: Der Streuquerschnitt von Atomen für unelastische Streuung schnelle Elektronen. Z. Phys. *124*, 658 (1948)
5.39 W. Brünger, W. Menz: Wirkungsquerschnitte für elastische und unelastische Elektronenstreuung an amorphen C- und Ge-Schichten. Z. Phys. *184*, 271 (1965)
5.40 W. Lippert: Über das Verhältnis des unelastischen zum elastischen Gesamtstreuquerschnitt für Elektronen bei den Arbeitsbedingungen der Elektronenmikroskopie. Naturwissenschaften *50*, 219 (1963)
5.41 H.G. Badde, H. Kappert, L. Reimer: Wellenoptische Theorie des Ripple-Kontras in der Lorentzmikroskopie. Z. Angew. Phys. *30*, 83 (1970)
5.42 R.E. Burge, D.L. Misell, J.W. Smart: The small-angle scattering of electrons in thin films of evaporated carbon. J. Phys. C*3*, 1661 (1970)
5.43 M. Isaacson, J. Langmore, J. Wall, A.V. Crewe: "Inelastic Scattering in Electron Microscopy," in *31st Annual Meeting of EMSA* (Claitor's Publ. Div., Baton Rouge, LO 1973) p.254
5.44 R.F. Egerton: Measurement of inelastic/elastic scattering ratio for fast electrons and its use in the study of radiation damage. Phys. Status Solidi A*37*, 663 (1976)
5.45 R.F. Egerton, J.G. Philip, M.J. Whelan: "Applications of Energy Analysis in a Transmission Electron Microscope," in *Developments in Electron Microscopy and Analysis*, ed. by J.A. Venables (Academic, London 1976) p.137
5.46 O. Klemperer, J.P.G. Shepherd: Characteristic energy losses of electrons in solids. Adv. Phys. *12*, 355 (1963)
5.47 H. Raether: *Excitation of Plasmons and Interband Transitions by Electrons*, Springer Tracts Mod. Phys., Vol.88 (Springer, Berlin, Heidelberg, New York 1980)
5.48 J. Geiger: *Elektronen und Festkörper* (Vieweg, Braunschweig 1968)
5.49 J. Daniels, C. von Festenberg, H. Raether, K. Zeppenfeld: *Optical Constants of Solids by Electron Spectroscopy*, Springer Tracts Mod. Phys., Vol.54 (Springer, Berlin, Heidelberg, New York 1970)
5.50 H. Boersch, J. Geiger, H. Hellwig, H. Michel: Energieverluste von Elektronen in Metallen in den verschiedenen Aggregatzuständen. Messungen an Al und Hg. Z. Phys. *169*, 252 (1962)
5.51 E. Petri, A. Otte. Direct nonvertical interband and intraband transitions in Al. Phys. Rev. Lett. *34*, 1283 (1975)
5.52 J.J. Ritsko, N.O. Lipari, P.C. Gibbons, S.E. Schnatterly, J.R. Fields, R. Devaty: Observation of electric monopole transitions in tetra-cyanoquinodimethane. Phys. Rev. Lett. *36*, 210 (1976)
5.53 C.H. Chen, J. Silcox: Direct nonvertical interband transitions of large wave vectors in aluminum. Phys. Rev. B*16*, 4246 (1977)
5.54 D. Pines: Collective energy losses in solids. Rev. Mod. Phys. *28*, 184 (1956) *Elementary Excitations in Solids* (Benjamin, New York 1962)
5.55 D. Bohm, D. Pines: A collective description of electron interactions: III. Coulomb interactions in a degenerate electron gas. Phys. Rev. *92*, 609 (1953)
5.56 G. Meyer: Über die Abhängigkeit der charakteristischen Energieverluste von Temperatur und Streuwinkel. Z. Phys. *148*, 61 (1957)
5.57 L.B. Leder, L. Marton: Temperature dependence of the characteristic energy loss of electrons in Al. Phys. Rev. *112*, 341 (1958)
5.58 H. Watanabe: Experimental evidence for the collective nature of the characteristic energy loss of electrons in solids. J. Phys. Soc. Jpn. *11*, 112 (1956)
5.59 C. Kunz: Die Winkelverteilung der charakteristischen Energieverluste von Elektronen, gemessen am 15 eV-Al-Verlust und am 17 eV-Si-Verlust. Phys. Status Solidi *1*, 441 (1961)
5.60 C. Kunz: Über die Winkelabhängigkeit der charakteristischen Energieverluste an Al, Si, Ag. Z. Phys. *167*, 53 (1962)
5.61 C. Kunz: Measurement of characteristic electron energy loss in alkali metals Phys. Lett. *15*, 312 (1965)
5.62 H. Boersch, H. Miessner, W. Raith: Untersuchungen zur Winkelabhängigkeit des 14,7 eV-Energieverlustes von Elektronen in Al. Z. Phys. *168*, 404 (1962)
5.63 J. Geiger: Winkelverteilung der Energieverluste mittelschneller Elektronen in Antimon. Z. Naturforsch. A*17*, 696 (1962)

5.64 P. Schmüser: Anregung von Volumen- und Oberflächen-Plasmaschwingungen in Al und Mg durch mittelschnelle Elektronen. Z. Phys. *180*, 105 (1964)
5.65 T. Kloos: Plasmaschwingungen in Al, Mg, Li, Na und K angeregt durch schnelle Elektronen. Z. Phys. *265*, 225 (1973)
5.66 M. Creuzburg, H. Raether: On the behavior of nuclear energy losses of electrons in alkali halides. Solid State Commun. *2*, 175 (1964)
5.67 K. Zeppenfeld: Anisotropie der Plasmaschwingungen in Graphit. Z. Phys. *211*, 391 (1968)
5.68 C.H. Chen, J. Silcox: Detection of optical surface guided modes in thin graphite films by high-energy electron scattering. Phys. Rev. Lett. *35*, 390 (1975)
5.69 M. Urner-Wille, H. Raether: Anisotropy of the 15 eV plasmon dispersion in Al. Phys. Lett. A*58*, 265 (1976)
5.70 J. Sevely, J.Ph. Perez, B. Jouffrey: "Energy Losses of Electrons Through Al and C Films from 300 keV up to 1200 keV," in *High Voltage Electron Microscopy*, ed. by P.R. Swann, C.J. Humphreys, M.J. Goringe (Academic, New York 1974) p.32
5.71 R.A. Ferrell: Characteristic energy loss of electrons passing through metal foils. Phys. Rev. *107*, 450 (1957)
5.72 L. Marton, J.A. Simpson, H.A. Fowler, N. Swanson: Plural scattering of 20 keV electrons in Al. Phys. Rev. *126*, 182 (1962)
5.73 M. Creuzburg, H. Dimigen: Energieanalyse im Elektroneninterfernzbild von Si-Einkristallen. Z. Phys. *174*, 24 (1963)
5.74 R.E. Burge, D.L. Misell: Electron energy loss spectra of evaporated carbon films. Philos. Mag. *18*, 251 (1968)
5.75 R.E. Burge, D.L. Misell: Convolution effects in electron energy-loss spectra recorded by electron transmission. J. Phys. C*2*, 1397 (1969)
5.76 R.H. Ritchie: Plasma losses by fast electrons in thin films. Phys. Rev. *106*, 874 (1957)
5.77 H. Boersch, J. Geiger, A. Imbusch, N. Niedrig: High resolution investigation of the energy losses of 30 keV electrons in Al foils of various thicknesses. Phys. Lett. *22*, 146 (1966)
5.78 R.B. Pettit, J. Silcox, R. Vincent: Measurement of surface-plasmon dispersion in oxidized Al films. Phys. Rev. B*11*, 3116 (1975)
5.79 T. Kloos: Zur Dispersion der Oberflächenplasmaverluste an reinen und oxydierten Al-Oberflächen. Z. Phys. *208*, 77 (1968)
5.80 E.A. Stern, R.A. Ferrell: Surface plasma oscillations of a degenerate electron gas. Phys. Rev. *120*, 130 (1960)
5.81 M. Creuzburg: Unsymmetrie in der Intensitätsverteilung charakteristischer Energieverluste. Z. Naturforsch. A*18*, 101 (1963); Über die Winkelabhängigkeit und ihre Unsymmetrie von Energieverlusten an Si und Ge. Z. Phys. *174*, 511 (1963)
5.82 C. von Festenberg, E. Kröger: Retardation effects for the electron energy loss probability in GaP and Si. Phys. Lett. A*26*, 339 (1968)
5.83 C.H. Chen, J. Silcox, R. Vincent: Electron energy losses in silicon: bulk and surface plasmons and Čerenkov radiation. Phys. Rev. B*12*, 64 (1975)
5.84 R.A. Ferrell: Predicted radiation of plasma oscillations in metal films. Phys. Rev. *111*, 1214 (1958)
5.85 R.W. Brown, P. Wessel, E.P. Trounson: Plasmon reradiation from Ag films. Phys. Rev. Lett. *5*, 472 (1960)
5.86 U. Bürker, W. Steinmann: Strahlung von Oberflächenplasmonen in Al. Z. Phys. *224*, 179 (1969)
5.87 A.J. Braundmeier, H.W. Williams, E.T. Arakawa, R.H. Ritchie: Radiative decay of surface plasmons from Al. Phys. Rev. B*5*, 2754 (1972)
5.88 G. Sauerbrey, E. Woeckel, P. Dobberstein: Radiation decay of electron-induced surface plasmons in rough surfaces. Phys. Status Solidi B*60*, 665 (1973)
5.89 E.T. Arakawa, N.O. Davis, R.D. Birkhoff: Temperature and thickness dependence of transition radiation from thin Ag foils. Phys. Rev. A*135*, 224 (1964)
5.90 H. Boersch, P. Dobberstein, D. Fritzsche, G. Sauerbrey: Transition radiation, Bremsstrahlung und Plasmastrahlung. Z. Phys. *187*, 97 (1965)
5.91 C. Colliex, V.E. Cosslett, R.D. Leapman, P. Trebbia: Contribution of electron energy loss spectroscopy to the development of analytical electron microscopy. Ultramicroscopy *1*, 301 (1976)

5.92 U. Fano, J.W. Cooper: Spectral distribution of atomic oscillator strengths. Rev. Mod. Phys. *40*, 441 (1968)
5.93 R.D. Leapman, V.E. Cosslett: Extended fine structure above the x-ray edge in electron energy loss spectra. J. Phys. D*9*, L29 (1976)
5.94 B.M. Kincaid, A.E. Meixner, P.M. Platzman: Carbon K edge in graphite measured using electron-energy-loss-spectroscopy. Phys. Rev. Lett. *40*, 1296 (1978)
5.95 D.E. Sayers, E.A. Stern, F.W. Lytle: New technique for investigating non-crystalline structures: Fourier analysis of the extended x-ray absorption fine structure. Phys. Rev. Lett. *27*, 1204 (1971)
5.96 R.F. Egerton: Inelastic scattering of 80 keV electrons in amorphous carbon. Philos. Mag. *31*, 199 (1975)
5.97 M. Gryzinski: Classical theory of atomic collisions. Phys. Rev. *138*, A305, 322 and 336 (1965)
5.98 M. Inokuti: Inelastic collisions of fast charged particles with atoms and molecules - the Bethe theory revised. Rev. Mod. Phys. *43*, 297 (1971)
5.99 R.D. Leapman, V.E. Cosslett: Electron energy loss spectrometry: mean free paths for some characteristic x-ray excitations. Philos. Mag. *33*, 1 (1976)
5.100 Y. Kihn, J. Sevely, B. Jouffrey: Excitation des niveaux atomiques K du carbone, du magnésium et de l'aluminium par des électrons de 60 keV. Philos. Mag. *33*, 733 (1976)
5.101 D.L. Misell, R.E. Burge: Convolution, deconvolution and small-angle plural electron scattering. J. Phys. C*2*, 61 (1969)
5.102 R.A. Crick, D.L. Misell: A theoretical consideration of some defects in electron optical images. A formulation of the problem for the incoherent case. J. Phys. D*4*, 1 (1971)
5.103 K.T. Considine, K.C.A. Smith, V.E. Cosslett: "Measurement of Large Energy Losses at High Voltages," in *Microscopie Electronique 1970*, Vol.2, ed. by P. Favard (Société Francaise de Microscopie Electronique, Paris 1970) p.131
5.104 L. Landau: On the energy loss of fast electrons by ionization. J. Phys. USSR *8*, 201 (1944)
5.105 O. Blunck, S. Leisegang: Zum Energieverlust schneller Elektronen in dünnen Schichten. Z. Phys. *128*, 500 (1950)
5.106 H.D. Maccabee, D.G. Papworth: Correction to Landau's energy loss formula. Phys. Lett. A*30*, 241 (1969)
5.107 Y. Kamiya: "Contrast Effects of Inelastic Scatterings on Images at High Accelerating Voltages," in *Electron Microscopy 1966*, Vol.1, ed. by R. Uyeda (Maruzen, Kyoto 1966) p.95
5.108 L. Reimer, K. Brockmann, U. Rhein: Energy losses of 20-40 keV electrons in 150-650 μg cm^{-2} metal films. J. Phys. D*11*, 2151 (1978)
5.109 H. Bethe: Zur Theorie des Durchganges schneller Korpuskularstrahlen durch Materie. Ann. Phys. *5*, 325 (1930)
5.110 V.E. Cosslett, R.N. Thomas: Multiple scattering of 5-30 keV electrons in evaporated metal films. II. Range-energy relations. Br. J. Appl. Phys. *15*, 1283 (1964)
5.111 L. Reimer, P. Gentsch: Superposition of chromatic error and beam broadening in TEM of thick carbon and organic specimens. Ultramicroscopy *1*, 1 (1975)
5.112 P. Gentsch, H. Gilde, L. Reimer: Measurement of the top bottom effect in scanning transmission electron microscopy of thick amorphous specimens. J. Micr. *100*, 81 (1974)
5.113 H. Koike, K. Ueno, M. Suzuki: "Scanning Device Combined with Conventional Electron Microscope," in *Proc. 29th Ann. Meeting of EMSA* (Claytor's Publ. Div., Baton Rouge, LO 1971) p.28
5.114 M. Fotino: "Evaluation of Factors Affecting the Resolution in Thick Biological Specimens in High-Voltage TEM," in *Electron Microscopy 1976*, Vol.1, ed. by D.G. Brandon (Tal International, Jerusalem 1976) p.277
5.115 K. Jost, J. Kessler: Die Ortsverteilung mittelschneller Elektronen bei Mehrfachstreuung. Z. Phys. *176*, 126 (1963)
5.116 T. Groves: Thick specimens in the CEM and STEM. Resolution and image formation. Ultramicroscopy *1*, 15 and 170 (1975)
5.117 H. Rose: The influence of plural scattering on the limit of resolution in electron microscopy. Ultramicroscopy *1*, 167 (1975)
5.118 H. Rose: "The Influence of Plural Scattering on the Contrast and Resolution in Electron Microscopy," in *Electron Microscopy 1976*, Vol.1, ed. by

D.G. Brandon (Tal International, Jerusalem 1976) p.254
5.119 L. Reimer, H. Gilde, K.H. Sommer: Die Verbreiterung eines Elektronenstrahles (17-1200 keV) durch Mehrfachstreuung. Optik *30*, 590 (1970)
5.120 W. Bothe: Die Streuung von Elektronen in schrägen Folien. Sitzungsber. Heidelb. Akad. Wiss., 7. Abhandlung, 307 (1951)
5.121 H.W. Thümmel: *Durchgang von Elektronen- und Betastrahlung durch Materieschichten* (Akademie, Berlin 1974)
5.122 J.I. Goldstein, J.L. Costley, G.W. Lorimer, S.J.B. Reed: "Quantitative X-Ray Analysis in the Electron Microscope," in *Scanning Electron Microscopy 1977*, Vol.1, ed. by O. Johari (IIT Research Institute, Chicago 1977) p.315
5.123 T. Just, H. Niedrig, H. Yersin: Schichtdickenbestimmung mittels Elektronen-Rückstreuung. Z. Angew. Phys. *25*, 89 (1968)
5.124 H. Niedrig, P. Sieber: Rückstreuung mittelschneller Elektronen an dünnen Schichten. Z. Angew. Phys. *31*, 27 (1971)
5.125 F.J. Hohn, H. Niedrig: Elektronenrückstreuung an dünnen Metall- und Isolatorschichten. Optik *35*, 290 (1972)
5.126 H.G. Badde, H. Drescher, E.R. Krefting, L. Reimer, H. Seidel, W. Bühring: "Use of Matt Scattering Cross Sections for Calculating Backscattering of 10-100 keV Electrons," in *Electron Microscopy and Analysis*, ed. by W.C. Nixon (The Institute of Physics, London 1971) p.74
5.127 H. Seiler: Einige aktuelle Probleme der Sekundärelektronenemission. Z. Angew. Phys. *22*, 249 (1967)
5.128 H. Drescher, L. Reimer, H. Seidel: Rückstreukoeffizient und Sekundärelektronen-Ausbeute von 10-100 keV-Elektronen und Beziehungen zur Raster-Elektronenmikroskopie. Z. Angew. Phys. *29*, 331 (1970)
5.129 L. Reimer, H. Drescher: Secondary electron emission of 10-100 keV electrons from transparent films of Al and Au. J. Phys. D*10*, 805 (1977)
5.130 P. Kirkpatrick, L. Wiedmann: Theoretical continuous x-ray energy and polarization. Phys. Rev. *67*, 321 (1945)
5.131 N.F. Mott, H.S.W. Massey: *The Theory of Atomic Collisions*, 3rd ed. (University Press, Oxford 1965)
5.132 C.R. Worthington, S.G. Tomlin: The intensity of emission of characteristic x-radiation. Proc. Phys. Soc. A*69*, 401 (1956)
5.133 J.W. Motz, R.C. Placious: K-ionization cross sections for relativistic electrons. Phys. Rev. A*136*, 662 (1964)
5.134 C.J. Powell: Cross sections for ionization of inner-shell electrons by electrons. Rev. Mod. Phys. *48*, 33 (1976)
5.135 C.J. Powell: "Evaluation of Formulas for Inner-Shell Ionization Cross Sections, in *Use of Monte Carlo Calculations in Electron Probe Microanalysis and Scanning Electron Microscopy*, ed. by K.F.J. Heinrich, D.E. Newbury, H. Jakowitz, NBS Special Publ.460 (US Government Printing Office, Washington 1976) p.97
5.136 J.A. Bearden: X-ray wavelengths. Rev. Mod. Phys. *39*, 78 (1967); J.A. Bearden, A.F. Burr: Reevaluation of x-ray atomic energy levels. Rev. Mod. Phys. *39*, 125 (1967)
5.137 N.A. Dyson: *X-Rays in Atomic and Nuclear Physics* (Longman, London 1973)
5.138 R.W. Fink, R.C. Jopson, H. Mark, C.D. Swift: Atomic fluorescence yields. Rev. Mod. Phys. *38*, 513 (1966)
5.139 W. Bambynek, B. Crasemann, R.W. Fink, H.U. Freund, H. Mark, C.D. Swift, R.E. Price, R.V. Rao: X-ray fluorescence yields, Auger and Coster-Kronig transition probabilities. Rev. Mod. Phys. *44*, 716 (1972)
5.140 H.U. Freund: Recent experimental values for K-shell x-ray fluorescence yields. X-Ray Spectrom. *4*, 90 (1975)
5.141 T.K. Kelly: Mass absorption coefficients and their relevance in electron probe microanalysis. Trans. Inst. Min. Metall. B*75*, 59 (1966)
5.142 K.F.J. Heinrich: "X-Ray Absorption Uncertainty", in *The Electron Microprobe*, ed. by T.D. McKinley, K.F.J. Heinrich, D.B. Wittry (Wiley, New York 1966) p.296
5.143 A. Sommerfeld: Über die Beugung und Bremsung der Elektronen. Ann. Phys. *11*, 257 (1931)

Chapter 6

6.1 L. Reimer: Deutung der Kontrastunterschiede von amorphen und kristallinen Objekten in der Elektronenmikroskopie. Z. Angew. Phys. *22*, 287 (1967)
6.2 F. Lenz: Zur Streuung mittelschneller Elektronen in kleinste Winkel. Z. Naturforsch. A*9*, 185 (1954)
6.3 C.E. Hall: Scattering phenomena in electron microscope image formation. J. Appl. Phys. *22*, 655 (1951)
6.4 W. Lippert. Experimentelle Studien über den Kontrast im Elektronenmikroskop. Optik *11*, 412 (1954)
6.5 W. Lippert: Über die "elektronenmikroskopische Durchlässigkeit" dünner Schichten. Optik *13*, 506 (1956)
6.6 L. Reimer: Zur Elektronenabsorption dünner Metallaufdampfschichten im Elektronenmikroskop. Z. Angew. Phys. *9*, 34 (1957)
6.7 L. Reimer: Messung der Abhängigkeit des elektronenmikroskopischen Bildkontrastes von Ordnungszahl, Strahlspannung und Aperturblende. Z. Angew. Phys. *13*, 432 (1961)
6.8 L. Reimer, K.H. Sommer: Messungen und Berechnungen zum elektronenmikroskopischen Streukontrast für 17-1200 keV Elektronen. Z. Naturforsch. A*23*, 1569 (1968)
6.9 E. Zeitler, G.F. Bahr: Contributions to the quantitative interpretation of electron microscope pictures. Exp. Cell Res. *12*, 44 (1957)
6.10 W. Lippert: Bemerkungen zur elektronenmikroskopischen Dickenmessung von Kohleschichten. Z. Naturforsch. B*17*, 335 (1962)
6.11 W. Schwertfeger: Zur Kleinwinkelstreuung von mittelschnellen Elektronen beim Durchgang durch amorphe Festkörperschichten. Thesis, Tübingen (1974)
6.12 W. Lippert: Zur Brauchbarkeit der Bornschen Näherung bei der Berechnung der Elektronenstreuung für den Bereich der Elektronenmikroskopie. Naturwissenschaften *49*, 534 (1962)
6.13 W. Lippert, W. Friese: "Zur Darstellbarkeit des Kontrastes mit Hilfe der Lenzschen Theorie," in *Electron Microscopy 1962*, 5th Intern. Congr. Electron Microscopy, ed. by S.S. Breese (Academic, New York 1962) p.AA-1
6.14 V.E. Cosslett: "High Voltage Electron Microscopy: Increase in Penetration with Voltage," in *Electron Microscopy 1968*, Vol.1, ed. by D.S. Bocciarelli (Tipografia Poliglotta Vaticana, Rome 1968) p.59
6.15 G. Dupouy, F. Perrier, P. Verdier: Amélioration du contraste des images d'objets amorphes minces en microscopie électronique. J. Microscopie *5*, 655 (1966)
6.16 R.F. Whiting, F.P. Ottensmeyer: Heavy atoms in model compounds and nucleic acids imaged by dark field TEM. J. Mol. Biol. *67*, 173 (1972)
6.17 C.E. Hall: Dark-field electron microscopy. I. Studies of crystalline substances in dark-field. J. Appl. Phys. *19*, 198 (1948)
6.18 J. Dubochet, M. Ducommun, M. Zollinger, E. Kellenberger: A new preparation method for dark-field electron microscopy of biomacromolecules. J. Ultrastruct. Res. *35*, 147 (1971)
6.19 G.J. Brakenhoff, N. Nanninga, J. Pieters: Relative mass determination from dark-field electron micrographs with an application to ribosomes. J. Ultrastruct. Res. *41*, 238 (1972)
6.20 W. Krakow, L.A. Howland: A method for producing hollow cone illumination electronically in the conventional transmission microscope. Ultramicroscopy *2*, 53 (1976)
6.21 E. Zeitler, M.G.R. Thomson: Scanning transmission electron microscopy. Optik *31*, 258 and 359 (1970)
6.22 L. Reimer, P. Gentsch, P. Hagemann: Anwendung eines Rasterzusatzes zu einem TEM. I. Grundlagen und Abbildung amorpher Objekte. Optik *43*, 431 (1975)
6.23 C.E. Hall: Electron densitometry of stained virus particles. J. Biophys. Biochem. Cytol. *1*, 1 (1955)
6.24 E. Krüger-Thiemer: Ein Verfahren für elektronenmikroskopische Massendickemessungen an nichtkristallinen Objekten. Z. Wiss. Mikr. *62*, 444 (1955)
6.25 N.R. Silvester, R.E. Burge: A quantitative estimation of the uptake of two new electron stains by the cytoplasmic membrane of ram sperm. J. Biophys. B chem. Cytol. *6*, 179 (1959)

6.26 L. Reimer, P. Hagemann: Recording of mass thickness in STEM. Ultramicroscopy *2*, 297 (1977)
6.27 M.K. Lamvik: Electronmicroscopic mass determination using photographic isodensity techniques. Ultramicroscopy *1*, 187 (1976)
6.28 E. Zeitler, G.F. Bahr: A photometric procedure for weight determination of submicroscopic particles. J. Appl. Phys. *33*, 847 (1962)
6.29 G.F. Bahr, E. Zeitler: The determination of the dry mass in populations of isolated particles. Lab. Invest. *14*, 955 (1965)
6.30 F.S. Sjöstrand: The importance of high resolution electron microscopy in tissue cell ultrastructure research. Sci. Tools *2*, 25 (1955)
6.31 B. von Borries, F. Lenz: "Über die Entstehung des Kontrastes im elektronenmikroskopischen Bild", in *Electron Microscopy*, Proc. Stockholm Conference 1956, ed. by F.J. Sjöstrand, J. Rhodin (Almqvist and Wiksells, Stockholm 1957) p.60
6.32 F. Thon: Elektronenmikroskopische Untersuchungen an dünnen Kohlefolien. Z. Naturforsch. A*20*, 154 (1965)
6.33 F. Thon: Zur Defokussierungsabhängigkeit des Phasenkontrastes bei der elektronenmikroskopischen Abbildung. Z. Naturforsch. A*21*, 476 (1966)
6.34 F. Lenz, W. Scheffels: Das Zusammenwirken von Phasen- und Amplitudenkontrast in der elektronenmikroskopischen Abbildung. Z. Naturforsch. A*13*, 226 (1958)
6.35 A. Howie, O.L. Krivanek, M.L. Rudee: Interpretation of electron micrographs and diffraction patterns of amorphous materials. Philos. Mag. *27*, 235 (1973)
6.36 G.J. Brakenhoff: On the sub-nanometre structure visible in high-resolution dark-field electron microscopy. J. Micr. *100*, 283 (1974)
6.37 A. Oberlin, M. Oberlin, M. Maubois: Study of thin amorphous and crystalline carbon films by electron microscopy. Philos. Mag. *32*, 833 (1975)
6.38 L. Reimer, H. Gilde: "Scattering Theory and Image Formation in the Electron Microscope," in *Image Processing and Computer-Aided Design in Electron Optics*, ed. by P.W. Hawkes (Academic, London 1973) p.138
6.39 L. Albert, R. Schneider, H. Fischer: Elektronenmikroskopische Sichtbarmachung von <10 Å großen Fremdstoffeinschlüssen in elektrolytisch abgeschiedenen Nickelschichten mittels Phasenkontrast durch Defokussierung. Z. Naturforsch. A*19*, 1120 (1964)
6.40 M. Rühle, M. Wilkens: "Defocusing Contrast of Cavities," in *Electron Microscopy 1972* (The Institute of Physics, London 1972) p.416
6.41 L. Reimer, H. Gilde: Electron optical phase contrast of small gold particles. Optik *41*, 524 (1975)
6.42 O. Scherzer: The theoretical resolution limit of the electron microscope. J. Appl. Phys. *20*, 20 (1949)
6.43 M.E. Haine: Contrast arising from elastic and inelastic scattering in the electron microscope. J. Sci. Instrum. *34*, 9 (1957)
6.44 R.D. Heidenreich, R.W. Hamming: Numerical evaluation of electron microscopical image phase contrast. Bell Syst. Tech. J. *44*, 207 (1965)
6.45 C.B. Eisenhandler, B.M. Siegel: Imaging of single atoms with the electron microscope by phase contrast. J. Appl. Phys. *37*, 1613 (1966)
6.46 R. Langer, W. Hoppe: Die Erhöhung von Auflösung und Kontrast im Elektronenmikroskop mit Zonenkorrekturplatten. Optik *24*, 470 (1966); *25*, 413 and 507 (1967)
6.47 L. Reimer: Elektronenoptischer Phasenkontrast. Z. Naturforsch. A*24*, 377 (1969)
6.48 H. Niehrs: Optimale Abbildungsbedingungen und Bildintensitätsverlauf bei einer Elektronenmikroskopie von Atomen. Optik *30*, 273 (1969); *31*, 51 (1970)
6.49 D.L. Misell: Image formation in the electron microscope. J. Phys. A*4*, 782 and 798 (1971)
6.50 D.L. Misell: Image resolution and image contrast in the electron microscope. J. Phys. A*6*, 62, 205 and 218 (1973)
6.51 T. Kobayashi, L. Reimer: Computation of electron microscopical images of single organic molecules. Optik *43*, 237 (1975)
6.52 W. Chiu, R.M. Glaeser: Single atom image contrast: conventional dark-field and bright-field electron microscopy. J. Micr. *103*, 33 (1975)
6.53 A. Pitt: "Dark Field Image Calculation," in *Electron Microscopy and Analysis 1979*, ed. by T. Mulvey (The Institute of Physics, London 1980) p.269

6.54 H. Hoch: Dunkelfeldabbildung von schwachen Phasenobjekten im Elektronenmikroskop. Optik *47*, 65 (1977)
6.55 W. Krakow: Computer experiments for tilted beam dark-field imaging. Ultramicroscopy *1*, 203 (1976)
6.56 H. Hashimoto, A. Kumao, K. Hino, H. Yotsumoto, A. Ono: Images of Th atoms in TEM. Jpn. J. Appl. Phys. *10*, 1115 (1971)
6.57 R.M. Henkelman, F.P. Ottensmeyer: Visualization of single heavy atoms by dark field electron microscopy. Proc. Nat. Acad. Sci. USA *68*, 3000 (1971)
6.58 F.P. Ottensmeyer, E.E. Schmidt, T. Jack, J. Powell: Molecular architecture: the optical treatment of dark field electron micrographs of atoms. J. Ultrastruct. Res. *40*, 546 (1972)
6.59 F. Thon, D. Willasch: Imaging of heavy atoms in dark field electron microscopy using hollow cone illumination. Optik *36*, 55 (1972)
6.60 K.J. Hanszen: Problems of image interpretation in electron microscopy with linear and nonlinear transfer. Z. Angew. Phys. *27*, 125 (1969)
6.61 K.J. Hanszen: "The Relevance of Dark Field Illumination in Conventional and Scanning TEM," PTB-Bericht A Ph-7 (Physikalisch-Technische Bundesanstalt, Braunschweig 1974)
6.62 D.L. Misell: Image resolution in high voltage electron microscopy. J. Phys. D*6*, 1409 (1973)
6.63 H. Formanek, M. Müller, M.H. Hahn, T. Koller: Visualization of single heavy atoms with the electron microscope. Naturwissenschaften *58*, 339 (1971)
6.64 J.R. Parsons, H.M. Johnson, C.W. Hoelke, R.R. Hosbons: Imaging of uranium atoms with the electron microscope by phase contrast. Philos. Mag. *27*, 1359 (1973)
6.65 W. Baumeister, M.H. Hahn: Electron microscopy of monomolecular layers of thorium atoms. Nature *241*, 445 (1973)
6.66 S. Iijima: Observation of single and clusters of atoms in bright field electron microscopy. Optik *48*, 193 (1977)
6.67 E.B. Prestridge, D.J.C. Yates: Imaging the rhodium atom with a conventional high resolution electron microscope. Nature *234*, 345 (1971)
6.68 D. Dorignac, B. Jouffrey: "Atomic Resolution at 3 MV," in *Microscopie Electronique à Haute Tension*, ed. by B. Jouffrey, P. Favard (Société Francaise de Microscopie Electronique, Paris 1976) p.143
6.69 D. Dorignac, B. Jouffrey: "Iron Single Atom Images," in *Electron Microscopy 1980*, Vol.1, ed. by P. Brederoo, G. Boom (Seventh European Congr. on Electron Microscopy Foundation, Leiden 1980) p.112
6.70 M. Retsky: Observed single atom elastic cross sections in a scanning electron microscope. Optik *41*, 127 (1974)
6.71 M. Isaacson, J.P. Langmore, H. Rose: Determination of the non-localization of the inelastic scattering of electrons by electron microscopy. Optik *41*, 92 (1974)
6.72 A.V. Crewe, J.P. Langmore, M.S. Isaacson: "Resolution and Contrast in the STEM," in *Physical Aspects of Electron Microscopy and Microbeam Analysis*, ed. by B. Siegel, D.R. Beaman (Wiley, New York 1975) p.47
6.73 M. Isaacson, M. Utlaut, D. Kopf: "Analog Computer Processing of STEM Images," in *Computer Processing of Electron Microscope Images*, Topics Curr. Phys., Vol.13, ed. by P.W. Hawkes (Springer, Berlin, Heidelberg, New York 1980) p.257
6.74 A.V. Crewe, J. Langmore, M. Isaacson, M. Retsky: "Understanding Single Atoms in STEM," in *Electron Microscopy 1974*, Vol.1, ed. by J.V. Sanders, D.J. Goodchild (Australian Acad. Sci., Canberra 1974) p.260
6.75 M.S. Isaacson, J. Langmore, N.W. Parker, D. Kopf, M. Utlaut: The study of the adsorption and diffusion of heavy atoms on light element substrates by means of the atomic resolution STEM. Ultramicroscopy *1*, 359 (1976)
6.76 J.S. Wall, J.F. Hainfeld, J.W. Bittner: Preliminary measurements of uranium atom motion on carbon films at low temperatures. Ultramicroscopy *3*, 81 (1978)
6.77 K.J. Hanszen, B. Morgenstern, K.J. Rosenbruch: Aussagen der optischen Übertragungstheorie über Auflösung und Kontrast im elektronenmikroskopischen Bild. Z. Angew. Phys. *16*, 477 (1964)
6.78 K.J. Hanszen, B. Morgenstern: Die Phasenkontrast- und Amplitudenkontrast-Übertragung des elektronenmikroskopischen Objektivs. Z. Angew. Phys. *19*, 215 (1965)

6.79 K.J. Hanszen: Generalisierte Angaben über die Phasenkontrast- und Amplitudenkontrast-Übertragungsfunktionen für elektronenmikroskopische Objektive. Z. Angew. Phys. *20*, 427 (1966)

6.80 K.J. Hanszen: "The Optical Transfer Theory of the Electron Microscope: Fundamental Principles and Applications", in *Advances in Optical and Electron Microscopy*, Vol.4, ed. by R. Barer, V.E. Cosslett (Academic, London 1971) p.1

6.81 K.J. Hanszen: "Contrast Transfer and Image Processing," in *Image Processing and Computer-Aided Design in Electron Optics*, ed. by P.W. Hawkes (Academic, London 1973) p.16

6.82 P.W. Hawkes: "Coherence in Electron Optics," in *Advances in Optical and Electron Microscopy*, Vol.7, ed. by R. Barer, V.E. Cosslett (Academic, London 1978) p.101

6.83 P.W. Hawkes: Electron image processing: a survey. Computer Graphics and Image Processing *8*, 406 (1978); *18*, 58 (1982)

6.84 K.J. Hanszen, L. Trepte: Der Einfluß von Strom- und Spannungsschwankungen sowie der Energiebreite der Strahlelektronen auf Kontrastübertragung und Auflösung des Elektronenmikroskopes. Optik *32*, 519 (1971)

6.85 K.J. Hanszen, L. Trepte: Die Kontrastübertragung im Elektronenmikroskop bei partiell kohärenter Beleuchtung. Optik *33*, 166 and 182 (1971)

6.86 J. Frank: The envelope of electron microscopic transfer functions for partially coherent illumination. Optik *38*, 519 (1973)

6.87 R.H. Wade, J. Frank: Electron microscope transfer functions for partially coherent axial illumination and chromatic defocus spread. Optik *49*, 81 (1977)

6.88 W.O. Saxton: Spatial coherence in axial high resolution conventional electron microscopy. Optik *49*, 51 (1977)

6.89 H. Yoshida, A. Ohshita, H. Tomita: Determination of spatial and temporal coherence functions from a single astigmatic image. Jpn. J. Appl. Phys. *20*, 2427 (1981)

6.90 W. Hoppe, D. Köstler, D. Typke, N. Hunsmann: Kontrastübertragung für die Hellfeld-Bildrekonstruktion mit gekippter Beleuchtung in der Elektronenmikroskopie. Optik *42*, 43 (1975)

6.91 K.H. Downing: Note on transfer functions in electron microscopy with tilted illumination. Optik *43*, 199 (1975)

6.92 S.C. McFarlane: The imaging of amorphous specimens in a tilted-beam electron microscope. J. Phys. C*8*, 2819 (1975)

6.93 R.H. Wade. Concerning tilted beam electron microscope transfer functions. Optik *45*, 87 (1976)

6.94 P.W. Hawkes: Electron microscope transfer functions in closed form with tilted illumination. Optik *55*, 207 (1980)

6.95 W. Krakow: "Calculation and Observation of Atomic Structure for Tilted Beam Dark-Field Microscopy," in *Developments in Electron Microscopy and Analysis*, ed. by J.A. Venables (Academic, London 1976) p.261

6.96 W. Hoppe: Towards three-dimensional electron microscopy at atomic resolution. Naturwissenschaften *61*, 239 (1974)

6.97 W.K. Jenkins, R.H. Wade: "Contrast Transfer in the Electron Microscope for Tilted and Conical Bright Field Illumination," in *Developments in Electron Microscopy and Analysis 1977*, ed. by D.L. Misell (The Institute of Physics, London 1977) p.115

6.98 W. Kunath: Signal-to-noise enhancement by superposition of bright-field images obtained under different illumination tilts. Ultramicroscopy *4*, 3 (1979)

6.99 W. Kunath, K. Weiss: "Hollow Cone Illumination in Bright Field Imaging of Ferritin," in *Electron Microscopy 1980*, Vol.1, ed. by P. Brederoo, G. Boom (Seventh European Congr. on Electron Microscopy Foundation, Leiden 1980) p.114

6.100 O. Scherzer: Zur Theorie der Abbildung einzelner Atome in dicken Objekten. Optik *38*, 387 (1973)

6.101 W.O. Saxton, W.K. Jenkins, L.A. Freeman, D.J. Smith: TEM observations using bright field hollow cone illumination. Optik *49*, 505 (1978)

6.102 H. Rose: Nonstandard imaging methods in electron microscopy. Ultramicroscopy *2*, 251 (1977)

6.103 J. Fertig, H. Rose: On the theory of image formation in the electron microscope. Optik *54*, 165 (1979)

6.104 H. Rose: Phase contrast in STEM. Optik *39*, 416 (1974)
6.105 N.H. Dekkers, H. deLang: Differential phase contrast in a STEM. Optik *41*, 452 (1974)
6.106 W.C. Stewart: On differential phase contrast with an extended illumination source. J. Opt. Soc. Am. *66*, 813 (1976)
6.107 H. Rose: Image formation by inelastically scattered electrons in electron microscopy. Optik *45*, 139 (1976)
6.108 P.W. Hawkes: Half-plane apertures in TEM, split detectors in STEM and ptychography. J. Opt. Paris *9*, 235 (1978)
6.109 G.R. Morrison, J.N. Chapman: "STEM Imaging with a Quadrant Detector," in *Electron Microscopy 1981*, ed. by M.J. Goringe (The Institute of Physics, London 1981) p.329
6.110 W. Hoppe: Ein neuer Weg zur Erhöhung des Auflösungsvermögens des Elektronenmikroskops. Naturwissenschaften *48*, 736 (1961)
6.111 F. Lenz: Zonenplatten zur Öffnungsfehlerkorrektur und zur Kontrasterhöhung. Z. Phys. *172*, 498 (1963)
6.112 F. Thon, B.M. Siegel: "Zonal Filtering in Optical Reconstruction of High Resolution Phase Contrast Images," in *Microscopie Electronique 1970*, Vol.1, ed. by P. Favard (Société Francaise de Microscopie Electronique, Paris 1970) p.13
6.113 H. Tochigi, H. Nakatsuka, A. Fukami, K. Kanaya: "The Improvement of the Image Contrast by Using the Phase Plate in the TEM," in *Microscopie Electronique 1970*, Vol.1, ed. by P. Favard (Société Francaise de Microscopie Electronique, Paris 1970) p.73
6.114 H.M. Johnson, D.F. Parsons: "In-Focus Phase Contrast Electron Microscopy," in *Microscopie Electronique 1970*, Vol.1, ed by. P. Favard (Société Francaise de Microscopie Electronique, Paris 1970) p.71
6.115 P.N.T. Unwin: An electrostatic phase plate for the electron microscope. Ber. Bunsenges. Phys. Chem. *74*, 1137 (1970)
6.116 W. Krakow, B.M. Siegel: Phase contrast in electron microscope images with an electrostatic phase plate. Optik *42*, 245 (1975)
6.117 G. Möllenstedt, R. Speidel, W. Hoppe, R. Langer, K.-H. Katerbau, F. Thon: "Electron Microscopical Imaging Using Zonal Correction Plates," in *Electron Microscopy 1968*, Vol.1, ed. by D.S. Bocciarelli (Tipografia Poliglotta Vaticana, Rome 1968) p.125
6.118 F. Thon, D. Willasch: "Hochauflösungs-Elektronenmikroskopie mit Spezialaperturblenden und Phasenplatten," in *Microscopie Electronique 1970*, Vol.1, ed. by P. Favard (Société Francaise de Microscopie Electronique, Paris 1970) p.3
6.119 K.-H. Müller: Phasenplatten für Elektronenmikroskope. Optik *45*, 73 (1976)
6.120 H.G. Badde, L. Reimer: Der Einfluß einer streuenden Phasenplatte auf das elektronenmikroskopische Bild. Z. Naturforsch. A*25*, 760 (1970)
6.121 D. Willasch: High resolution electron microscopy with profiled phase plates. Optik *44*, 17 (1975)
6.122 L. Reimer, H.G. Badde, E. Drewes, H. Gilde, H. Kappert, H.J. Höhling, D.B. von Bassewitz, A. Rössner: Laserbeugung an elektronenmikroskopischen Aufnahmen. Forschungsber. Landes Nordrhein Westfalen Nr.2314 (1973)
6.123 J.E. Berger, D. Harker: Optical diffractometer for production of Fourier transforms of electron micrographs. Rev. Sci. Instrum. *38*, 292 (1967)
6.124 O.L. Krivanek: A method for determining the coefficient of spherical aberration from a single electron micrograph. Optik *45*, 97 (1976)
6.125 W. Krakow, K.H. Downing, B.M. Siegel: The use of tilted specimens to obtain the contrast transfer characteristics of an electron microscope imaging system. Optik *40*, 1 (1974)
6.126 L. Reimer, H.G. Heine, R.A. Ajeian: Optimalbedingungen für den Beugungsnachweis von Defokussierungsstrukturen in elektronenmikroskopischen Aufnahmen. Z. Naturforsch. A*24*, 1846 (1969)
6.127 L. Reimer, H. Kappert: Bestimmung der Domänenwanddicke aus defokussierten elektronenoptischen Aufnahmen von ferromagnetischen Schichten. Z. Angew. Phys. *26*, 58 (1969)
6.128 J. Frank: Nachweis von Objektbewegungen im lichtoptischen Diffraktogramm von elektronenmikroskopischen Aufnahmen. Optik *30*, 171 (1969)

6.129 J. Frank: Observation of the relative phases of electron microscopic phase constrast zones with the aid of the optical diffractometer. Optik *35*, 608 (1972)
6.130 L. Reimer, B. Volbert, P. Bracker: Quality control of SEM micrographs by laser diffractometry. Scanning *1*, 233 (1978)
6.131 K.H. Herrmann, D. Krahl: 'Real-time'-Elektronenbildwandlung mit Thermoplastschichten. Optik *45*, 231 (1976)
6.132 P. Bonhomme, A. Beorchia, B. Meunier, F. Dumont, D. Rossier: Incoherent reading light tests of a Pockels-effect imaging device used in an 'in-line' optical processor of microscopical electron images. Optik *45*, 159 (1976)
6.133 A. Beorchia, P. Bonhomme, N. Bonnet: Modulation transfer function and detective quantum efficiency of Electrotitus. Optik *55*, 11 (1980)
6.134 D. Gabor: Microscopy by reconstructed wave-fronts. Proc. Roy. Soc. A*197*, 454 (1949); Proc. Phys. Soc. B*64*, 449 (1950)
6.135 A. Tonomura, A. Fukuhara, H. Watanabe, T. Komoda: Optical reconstruction of image from Fraunhofer electron-hologram. Jpn. J. Appl. Phys. *7*, 295 (1968)
6.136 J. Munch: Experimental electron holography. Optik *43*, 79 (1975)
6.137 K.J. Hanszen, G. Ade, R. Lauer: Genauere Angaben über sphärische Längsaberration, Verzeichnung in der Pupillenebene und über die Wellenaberration von Elektronenlinsen. Optik *35*, 567 (1972)
6.138 K.J. Hanszen: "Neuere theoretische Erkenntnisse und praktische Erfahrungen über die holographische Rekonstruktion elektronenmikroskopischer Aufnahmen," PTB-Bericht A Ph-4 (Physikalisch-Technische Bundesanstalt, Braunschweig 1973)
6.139 G. Ade: Erweiterung der Kontrastübertragungstheorie auf nicht-isoplanatische Abbildungen. Optik *50*, 143 (1978)
6.140 K.J. Hanszen: Holographische Rekonstruktionsverfahren in der Elektronenmikroskopie und ihre kontrastübertragungstheoretische Deutung. Optik *32*, 74 (1970)
6.141 A. Lohmann: Optische Einseitenbandübertragung angewandt auf das Gabor-Mikroskop. Opt. Acta *3*, 97 (1956)
6.142 K.J. Hanszen: Einseitenband-Holographie. Z. Naturforsch. A*24*, 1849 (1969)
6.143 W. Hoppe, R. Langer, F. Thon: Verfahren zur Rekonstruktion komplexer Bildfunktionen in der Elektronenmikroskopie. Optik *30*, 538 (1970)
6.144 W. Hoppe: Zur 'Abbildung' komplexer Bildfunktionen in der Elektronenmikroskopie. Z. Naturforsch. A*26*, 1155 (1971)
6.145 F. Thon: "Hochauflösende elektronenmikroskopische Abbildung amorpher Objekte mittels Zweistrahlinterferenzen," in *Electron Microscopy 1968*, Vol.1, ed. by D.S. Bocciarelli (Tipografia Poliglotta Vaticana, Rome 1968) p.127
6.146 K.H. Downing: "Compensation of Lens Aberrations by Single-Sideband Holography," in *Proc. 30th Ann. EMSA Meeting* (Claitor's Publ. Div., Baton Rouge LO 1972) p.562
6.147 P. Sieber: "High Resolution Electron Microscopy with Heated Apertures and Reconstruction of Single-Sideband Micrographs," in *Electron Microscopy 1974*, Vol.1, ed. by J.V. Sanders, D.J. Goodchild (Australian Acad. Sci., Canberra 1974) p.274
6.148 K.H. Downing, B.M. Siegel: Discrimination of heavy and light components in electron microscopy using single-sideband holographic techniques. Optik *42*, 155 (1975)
6.149 E.N. Leith, J. Upatnieks: Reconstructed wavefronts and communication theory. J. Opt. Soc. Am. *52*, 1123 (1962)
6.150 G. Möllenstedt, H. Wahl: Elektronenholographie und Rekonstruktion mit Laserlicht. Naturwissenschaften *55*, 340 (1968)
6.151 H. Tomita, T. Matsuda, T. Komoda: Electron microholography by two-beam method. Jpn. J. Appl. Phys. *9*, 719 (1970)
6.152 H. Tomita, T. Matsuda, T. Komoda: Off-axis electron micro-holography. Jpn. J. Appl. Phys. *11*, 143 (1972)
6.153 A. Tonomura, J. Endo, T. Matsuda: An application of electron holography to interference microscopy. Optik *53*, 143 (1979)
6.154 J. Endo, T. Matsuda, A. Tonomura: Interference electron microscopy by means of holography. Jpn. J. Appl. Phys. *18*, 2291 (1979)
6.155 A. Tonomura, T. Matsuda, J. Endo, T. Arii, K. Mihama: Direct Observation of fine structure of magnetic domain walls by electron holography. Phys. Rev. Lett. *44*, 1430 (1980)

6.156 K.J. Hanszen: "Experience and Results Obtained in Electron Microscopical Holography by Using a Reference Beam in the Light Optical Reconstruction Step," in *Electron Microscopy 1980*, Vol.1, ed. by P. Brederoo, G. Boom (Seventh European Congr. on Electron Microscopy Foundation, Leiden 1980) p.136
6.157 K.J. Hanszen, R. Lauer, G. Ade: "Discussions of the Possibilities and Limitations of In-Line and Off-Axis Holography in Electron Microscopy," PTB-Bericht A Ph-15 (Physikalisch-Technische Bundesanstalt, Braunschweig 1980)
6.158 K.J. Hanszen, R. Lauer: "Holographic Phase Determination of Strong Objects," in *Electron Microscopy 1980*, Vol.1, ed. by P. Brederoo, G. Boom (Seventh European Congr. on Electron Microscopy Foundation, Leiden 1980) p.140
6.159 K.J. Hanszen: Holography in electron microscopy. Adv. Electron. Electron Phys. *59*, 1 (1982)
6.160 K.J. Hanszen: "Lichtoptische Anordnungen mit Laser-Lichtquellen als Hilfsmittel für die Elektronenmikroskopie," in *Electron Microscopy 1968*, Vol.1, ed. by D.S. Bocciarelli (Tipografia Poliglotta Vaticana, Rome 1968) p.153
6.161 J. Rogers: "The Design and Use of an Optical Model of the Electron Microscope," in *Proc. of the ICO-11 Conference* (Madrid, 1978)
6.162 A. Maréchal, P. Croce: Un filtre de fréquences spatiales pour l'amélioration du contraste des images optiques. C. R. Acad. Sci. Paris *237*, 607 (1953)
6.163 M.H. Hahn: Eine optische Ortsfrequenzfilter- und Korrelationsanlage für elektronenmikroskopische Aufnahmen. Optik *35*, 326 (1972)
6.164 G.W. Stroke, M. Halioua: Attainment of diffraction-limited imaging in high-resolution electron microscopy by 'a posteriori' holographic image sharpening. Optik *35*, 50 (1972)
6.165 G.W. Stroke, M. Halioua: Image deblurring by holographic deconvolution with partially-coherent low-contrast objects and application to electron microscopy. Optik *35*, 489 (1972)
6.166 G.W. Stroke, M. Halioua: Image improvement in high-resolution electron microscopy with coherent illumination (low-contrast objects) using holographic image-deblurring deconvolution. Optik *37*, 192 and 249 (1973)
6.167 G.W. Stroke, M. Halioua, F. Thon, D. Willasch: Image improvement in high-resolution electron microscopy using holographic image deconvolution. Optik *41*, 319 (1974)
6.168 A.W. Lohmann, D.P. Paris: Computer generated spatial filters for coherent optical data processing. Appl. Opt. *7*, 651 (1968)
6.169 A.W. Lohmann, D.P. Paris: Binary Fraunhofer holograms, generated by computer Appl. Opt. *6*, 1739 (1967)
6.170 R.E. Burge, R.F. Scott: Binary filters for high resolution electron microscopy. Optik *43*, 53 (1975); *44*, 159 (1976)
6.171 S. Boseck, H. Hager: Beseitigung des spatialen Rauschens in elektronenmikroskopischen Aufnahmen durch lichtoptische Filterung. Optik *28*, 602 (1968)
6.172 S. Boseck, R. Lange: Ausschöpfung des Informationsgehaltes von elektronenmikroskopischen Aufnahmen biologischer Objekte mit Hilfe des Abbeschen Beugungsapparates, gezeigt am Beispiel kristallartiger Strukturen. Z. Wiss. Mikr. *70*, 66 (1970)
6.173 J.B. Bancroft, G.J. Hills, R. Markham: A study of the self-assembly process in a small spherical virus. Virology *31*, 354 (1967)
6.174 A Klug, D.J. deRosier: Optical filtering of electron micrographs: reconstruction of one-sided images. Nature *212*, 29 (1966)
6.175 C.A. Taylor, J.K. Ranniko: Problems in the use of selective optical spatial filtering to obtain enhanced information from electron micrographs. J. Micr. *100*, 307 (1974)
6.176 R. Markham, J.H. Hitchborn, G.J. Hills, S. Frey: The anatomy of tobacco mosaic virus. Virology *22*, 342 (1964)
6.177 R.C. Warren, R.M. Hicks: A simple method of linear integration for resolving structures in periodic lattices. J. Ultrastruct. Res. *36*, 861 (1971)
6.178 R. Markham, S. Frey, G.J. Hills: Methods for the enhancement of image detail and accentuation of structure in electron microscopy. Virology *20*, 88 (1963)

6.179 D.L. Misell: "The Phase Problem in Electron Microscopy," in *Advances in Optical and Electron Microscopy*, Vol.7, ed. by R. Barer, V.E. Cosslett (Academic, London 1978) p.185
6.180 W.O. Saxton: Computer techniques for image processing in electron microscopy. Adv. Electron. Electron Phys. Suppl. *10*, 289 (1978)
6.181 W.O. Saxton: "Recovery of Specimen Information for Strongly Scattering Objects," in *Computer Processing of Electron Microscope Images*, Topics Curr. Phys., Vol.13, ed. by P.W. Hawkes (Springer, Berlin, Heidelberg, New York 1980) p.35
6.182 R.W. Gerchberg, W.O. Saxton: Phase determination from image and diffraction plane pictures in the electron microscope. Optik *34*, 275 (1971)
6.183 R.W. Gerchberg, W.O. Saxton: A practical algorithm for the determination of phase from image and diffraction plane pictures. Optik *35*, 237 (1972)
6.184 J. Frank: A remark on phase determination in electron microscopy. Optik *38*, 582 (1973)
6.185 R.W. Gerchberg: Holography without fringes in the electron microscope. Nature *240*, 404 (1972)
6.186 J.N. Chapman: The application of iterative techniques to the investigation of strong phase objects in the electron microscope. Philos. Mag. *32*, 527 and 541 (1975)
6.187 D.L. Misell: An examination of an iterative method for the solution of the phase problem in optics and electron optics. J. Phys. D*6*, 2200 and 2217 (1973)
6.188 P. Schiske: Phase determination from a focal series and the corresponding diffraction pattern in electron microscopy for strongly scattering objects. J. Phys. D*8*, 1372 (1975)
6.189 D.A. Ansley: Determining the phase of line objects by measuring their intensity in dark field and bright field illumination. Opt. Commun. *8*, 140 (1973)
6.190 P. van Toorn, A.M.J. Huiser, H.A. Ferwerda: Proposals for solving the phase retrieval problem for semi-weak objects from noisy electron micrographs. Optik *51*, 309 (1978)
6.191 R. Langer, J. Frank, A. Feltynowski, W. Hoppe: Anwendung des Bilddifferenzverfahrens auf die Untersuchung von Strukturänderungen dünner Kohlefolien bei Elektronenbestrahlung. Ber. Bunsenges. Phys. Chem. *74*, 1120 (1970)
6.192 J. Frank: "Two-Dimensional Correlation Functions in Electron Microscope Image Analysis," in *Electron Microscopy 1972* (The Institute of Physics, London 1972) p.622
6.193 L.S. Al-Ali: "Translational Alignment of Differently Defocused Micrographs Using Cross-Correlation," in *Developments in Electron Microscopy and Analysis*, ed. by J.A. Venables (Academic, London 1976) p.225
6.194 W. Hoppe, R. Langer, J. Frank, A. Feltynowski: Bilddifferenzverfahren in der Elektronenmikroskopie. Naturwissenschaften *56*, 267 (1969)
6.195 R.A. Crowther, L.A. Amos: Harmonic analysis of electron microscope images with rotational symmetry. J. Mol. Biol. *60*, 123 (1971)
6.196 H.P. Erickson, A. Klug: Measurement and compensation of defocusing and aberrations by Fourier processing of electron micrographs. Philos. Trans. B*261*, 105 (1971)
6.197 A.M. Kuo, R.M. Glaeser: Development of methodology for low exposure, high resolution electron microscopy of biological specimens. Ultramicroscopy *1*, 53 (1975)
6.198 P.N.T. Unwin, R. Henderson: Molecular structure determination by electron microscopy of unstained crystalline specimens. J. Mol. Biol. *94*, 425 (1975)
6.199 J.L. Harris: Image evaluation and restoration. J. Opt. Soc. Am. *56*, 569 (1966)
6.200 J. Frank, P. Bußler, R. Langer, W. Hoppe: Einige Erfahrungen mit der rechnerischen Analyse und Synthese von elektronenmikroskopischen Bildern hoher Auflösung. Ber. Bunsenges. Phys. Chem. *74*, 1105 (1970)
6.201 T.A. Welton: "Computational Correction of Aberrations in Electron Microscopy," in *Proc. 29th Annual Meeting of EMSA* (Claitor's Publ. Div., Baton Rouge, LO 1971) p.94
6.202 T.A. Welton: A computational critique of an algorithm for image enhancement in bright field electron microscope. Adv. Electron. Electron Phys. *48*, 37 (1978)

6.203 W.O. Saxton, J. Frank: Motif detection in quantum noise-limited electron micrographs by cross-correlation. Ultramicroscopy *2*, 219 (1977)
6.204 J. Frank: Averaging of low exposure electron micrographs of nonperiodic objects. Ultramicroscopy *1*, 159 (1979)
6.205 J. Frank: Optimal use of image information using signal detection and averaging techniques. Ann. NY Acad. Sci. *306*, 112 (1978)
6.206 J. Frank: "Reconstruction of Non-Periodic Objects Using Correlation Methods," in *Electron Microscopy 1978*, Vol.3, ed. by J.M. Sturgess (Microscopical Soc. Canada, Toronto 1978) p.87
6.207 J. Frank: "The Role of Correlation Techniques in Computer Image Processing," in *Computer Processing of Electron Microscope Images*, Topics Curr. Phys., Vol.13, ed. by P.W. Hawkes (Springer, Berlin, Heidelberg, New York 1980) p.187
6.208 J. Frank, W. Goldfarb, D. Eisenberg, T.S. Baker: Reconstruction of glutamine synthetase using computer averaging. Ultramicroscopy *3*, 283 (1978)
6.209 J. Frank, A. Verschoor, M. Boublik: Computer averaging of electron micrographs of 40S ribosomal subunits. Science *214*, 1353 (1981)
6.210 M. van Heel: Detection of objects in quantum-noise-limited images. Ultramicroscopy *7*, 331 (1982)
6.211 P.R. Smith: An integrated set of computer programs for processing electron micrographs of biological structures. Ultramicroscopy *3*, 153 (1978)
6.212 W.O. Saxton, T.J. Pitt, M. Horner: Digital image processing: the SEMPER system. Ultramicroscopy *4*, 343 (1979)
6.213 J. Frank, B. Shimkin, H. Dowse: SPIDER - a modular software system for electron image processing. Ultramicroscopy *6*, 343 (1981)
6.214 M. van Heel, W. Keegstra: IMAGIC: a fast, flexible and friendly image analysis software system. Ultramicroscopy *7*, 113 (1981)
6.215 J.G. Helmcke: Theorie und Praxis der elektronenmikroskopischen Stereoaufnahme. Optik *11*, 201 (1954); *12*, 253 (1955)
6.216 J.G. Helmcke, H.J. Orthmann: Fehler bei der Tiefenbestimmung elektronenmikroskopischer Stereoaufnahmen. Optik *11*, 562 (1954)
6.217 R.I. Garrod, J.F. Nankivell: Sources of error in electron stereomicroscopy. Br. J. Appl. Phys. *9*, 214 (1958)
6.218 R.I. Garrod, J.F. Nankivell: Some remarks on the accuracy obtainable in electron stereomicroscopy. Optik *16*, 27 (1959)
6.219 R.A. Crowther, D.J. deRosier, A. Klug: The reconstruction of a three-dimensional structure from projections and its application to electron microscopy. Proc. Roy. Soc. A*317*, 319 (1970)
6.220 G.N. Ramachandran, A.V. Lakshminarayanan: Three-dimensional reconstruction from radiograph and electron micrographs. Proc. Nat. Acad. Sci. USA *68*, 2236 (1971)
6.221 R.A. Crowther, A. Klug: ART and Science or conditions for three-dimensional structure from projections and its application to electron microscopy. J. Theor. Biol. *32*, 199 (1971)
6.222 R. Gordon, R. Bender, G.T. Herman: Algebraic reconstruction techniques (ART for three-dimensional electron microscopy and x-ray photography. J. Theor. Biol. *29*, 471 (1970)
6.223 B.K. Vainshtein: Finding the structure of objects from projections. Sov. Phys. Cryst. *15*, 781 (1971)
6.224 P. Gilbert: Iterative methods for the three-dimensional reconstruction of an object from projections. J. Theor. Biol. *36*, 105 (1972)
6.225 P.F.C. Gilbert: The reconstruction of a three-dimensional structure from projections and its application to electron microscopy. II. Direct methods. Proc. Roy. Soc. B*182*, 89 (1972)
6.226 E. Zeitler: The reconstruction of objects from their projections. Optik *39*, 396 (1974)
6.227 W. Hoppe, H.J. Schramm, M. Sturm, N. Hunsmann, J. Gaßmann: Three-dimensiona electron microscopy of individual biological objects. Z. Naturforsch. A*31*, 645, 1370 and 1380 (1976)
6.228 J.A. Lake: "Reconstruction of Three-Dimensional Structures from Electron Micrographs: The Equivalence of Two Methods," in *Proc. 29th Annual Meeting of EMSA* (Claitor's Publ. Div., Baton Rouge, LO 1971) p.90

6.229 M. Zwick, E. Zeitler: Image reconstruction from projections. Optik *38*, 550 (1973)
6.230 A. Klug, F.H.C. Crick, H.W. Wyckoff: Diffraction by helical structures. Acta Cryst. *11*, 199 (1958)
6.231 D.J. deRosier, A. Klug: Reconstruction of three-dimensional structures from electron micrographs. Nature *217*, 130 (1968)
6.232 B.K. Vainshtein: "Electron Microscopical Analysis of the Three-Dimensional Structure of Biological Macromolecules," in *Advances in Optical and Electron Microscopy*, Vol.7, ed. by R. Barer, V.E. Cosslett (Academic, London 1978) p.281
6.233 N.A. Kiselev: "Reconstruction of the Structure of Encymes from Their Images," in *Electron Microscopy 1978*, Vol.3, ed. by J.M. Sturgess (Microscopical Soc. of Canada, Toronto 1978) p.94
6.234 D.L. Misell: "Image Analysis, Enhancement and Interpretation," *Practical Methods in Electron Microscopy*, Vol.7, ed. by A.M. Glauert (North-Holland, Amsterdam 1978)
6.235 W.O. Saxton: "Digital Processing of Electron Images - a Survey of Motivations and Methods," in *Electron Microscopy 1980*, Vol.1, ed. by P. Brederoo, G. Boom (Seventh Intern. Congr. on Electron Microscopy Foundation, Leiden 1980) p.486
6.236 J.E. Mellema: "Computer Reconstruction of Regular Biological Objects," in *Computer Processing of Electron Microscope Images*, Topics Curr. Phys., Vol.13, ed. by P.W. Hawkes (Springer, Berlin, Heidelberg, New York 1980) p.89
6.237 W. Hoppe, R. Hegerl: "Three-Dimensional Structure Determination by Electron Microscopy (Nonperiodic Specimens)", in *Computer Processing of Electron Microscope Images*, Topics Curr. Phys., Vol.13, ed. by P.W. Hawkes (Springer, Berlin, Heidelberg, New York 1980) p.127
6.238 E. Feldtkeller: Übersicht über das Magnetisierungsverhalten in dünnen Schichten. Z. Angew. Phys. *17*, 121 (1964)
6.239 P.J. Grundy, R.S. Tebble: Lorentz electron microscopy. Adv. Phys. *17*, 153 (1968)
6.240 R.H. Wade: "Lorentz Microscopy or Electron Phase Microscopy of Magnetic Objects," in *Advances in Optical and Electron Microscopy*, Vol.5, ed. by R. Barer, V.E. Cosslett (Academic, London 1973) p.239
6.241 J.P. Jacubovics: "Lorentz Microscopy and Application (TEM and SEM)", in *Electron Microscopy in Materials Science*, Part IV, ed. by U. Valdrè, E. Ruedl (Commission of the European Communities, Brussels 1976) p.1303
6.242 H. Boersch, W. Raith, H. Weber: Die magnetische Ablenkung von Elektronenstrahlen in dünnen Fe-Schichten. Z. Phys. *161*, 1 (1961)
6.243 K. Schaffernicht: Messung der Magnetisierungsverteilungen in dünnen Fe-Schichten durch die Ablenkung von Elektronen. Z. Angew. Phys. *15*, 275 (1963)
6.244 D.H. Warrington, J.M. Rodgers, R.S. Tebble: The use of ferromagnetic domain structure to determine the thickness of iron foils in TEM. Philos. Mag. *7*, 1783 (1962)
6.245 R.H. Wade: Electron diffraction from a magnetic phase grating. Phys. Status Solidi *19*, 847 (1967)
6.246 M.J. Goringe, J.P. Jakubovics: Electron diffraction from periodic magnetic fields. Philos. Mag. *15*, 393 (1967)
6.247 H. Boersch, H. Raith: Elektronenmikroskopische Abbildung Weißscher Bezirke in dünnen ferromagnetischen Schichten. Naturwissenschaften *46*, 574 (1959)
6.248 H.W. Fuller, M.E. Hale: Domains in thin magnetic films observed by electron microscopy. J. Appl. Phys. *31*, 1699 (1960)
6.249 J. Podbrdsky: High resolution in-focus Lorentz electron microscopy. J. Micr. *101*, 231 (1974)
6.250 M.J. Bowman, V.H. Meyer: Magnetic phase contrast from thin ferromagnetic films in the TEM. J. Phys. E*3*, 927 (1970)
6.251 L. Marton: Electron optical observation of magnetic fields. J. Appl. Phys. *19*, 863 (1948)
6.252 L. Marton, S.H. Lachenbruch: Electron optical mapping of electromagnetic fields. J. Appl. Phys. *20*, 1171 (1949)
6.253 L. Marton, J.A. Simpson, S.H. Lachenbruch: Electron optical shadow method of magnetic field mapping. J. Res. NBS *52*, 97 (1954)

6.254 M. von Ardenne: Zur Sichtbarmachung von Störungen oder Inhomogenitäten magnetischer und elektrischer Felder mit der elektronenoptischen Scheidenmethode. Phys. Z. *45*, 312 (1945)
6.255 W. Rollwagen, Ch. Schwink: Die Empfindlichkeit einfacher elektronenoptischer Schlierenanordnungen. Optik *10*, 525 (1953)
6.256 Ch. Schwink: Über neue quantitative Verfahren der elektronenoptischen Schattenmethode. Optik *12*, 481 (1955)
6.257 Ch. Schwink, O. Schärpf: Electron-optic investigation of the magnetic stray field above Bloch walls in cylindric Ni crystals. Phys. Status Solidi *30*, 637 (1968)
6.258 A.G. Cullis, D.M. Maher: High-resolution topographical imaging by direct transmission electron microscopy. Philos. Mag. *30*, 447 (1974)
6.259 M.E. Hale, H.W. Fuller, H. Rubinstein: Magnetic domain observations by electron microscopy. J. Appl. Phys. *30*, 789 (1959)
6.260 H.W. Fuller, M.E. Hale: Determination of magnetization distribution in thin films using electron microscopy. J. Appl. Phys. *31*, 238 (1960)
6.261 H. Boersch, H. Hamisch, D. Wohlleben, K. Grohmann: Antiparallele Weißsche Bereiche als Biprisma für Elektroneninterferenzen. Z. Phys. *159*, 397 (1960); *167*, 72 (1962)
6.262 D. Wohlleben: Diffraction effects in Lorentz microscopy. J. Appl. Phys. *38*, 3341 (1967)
6.263 L. Reimer, H. Kappert: Elektronen-Kleinwinkelstreuung und Bildkontrast in defokussierten Aufnahmen magnetischer Bereichsgrenzen. Z. Angew. Phys. *27*, 165 (1969)
6.264 J.P. Guigay, R.H. Wade: Mainly on the Fresnel mode in Lorentz microscopy. Phys. Status Solidi *29*, 799 (1968)
6.265 E. Fuchs: Magnetische Strukturen in dünnen ferromagnetischen Schichten, untersucht mit dem Elektronenmikroskop. Z. Angew. Phys. *14*, 203 (1962)
6.266 R.H. Wade: The determination of domain wall thickness in ferromagnetic films by electron microscopy. Proc. Phys. Soc. *79*, 1237 (1962)
6.267 R.H. Wade: Investigation of the geometrical-optical theory of magnetic structure imaging in the electron microscope. J. Appl. Phys. *37*, 366 (1966)
6.268 T. Suzuki, A. Hubert: Determination of ferromagnetic domain wall widths by means of high voltage Lorentz microscopy. Phys. Status Solidi *38*, K5 (1970)
6.269 T. Suzuki, M. Wilkens: Lorentz-electron microscopy of ferromagnetic specimens at high voltages. Phys. Status Solidi A*3*, 43 (1970)
6.270 D.S. Hothersall: The investigation of domain walls in thin sections of iron by the electron interference method. Philos. Mag. *20*, 89 (1969)
6.271 D.C. Hothersall: Electron images of domain walls in Co foils. Philos. Mag. *24*, 241 (1971)
6.272 D.C. Hothersall: Electron images of two-dimensional domain walls. Phys. Status Solidi B*51*, 529 (1972)
6.273 P. Schwellinger: The analysis of magnetic domain wall structures in the transition region of Néel and Bloch walls by Lorentz microscopy. Phys. Status Solidi A*36*, 335 (1976)
6.274 J.N. Chapman, R.P. Ferrier, N. Toms: Strong stripe domains. Philos. Mag. *28* 561 and 581 (1973)
6.275 C.G. Harrison, K.D. Leaver: A second domain wall parameter measurable by Lorentz microscopy. Phys. Status Solidi A*12*, 413 (1972)
6.276 R. Ajeian, H. Kappert, L. Reimer: Fraunhofer-Beugung an Lorentz-mikroskopischen Aufnahmen des Magnetisierungs-Ripple. Z. Angew. Phys. *30*, 80 (1970)
6.277 H.G. Badde, H. Kappert, L. Reimer: Wellenoptische Theorie des Ripple-Kontrastes in der Lorentzmikroskopie. Z. Angew. Phys. *30*, 83 (1970)
6.278 T. Suzuki: Investigations into ripple wavelength in evaporated thin films by Lorentz microscopy. Phys. Status Solidi *37*, 101 (1970)
6.279 M. Blackman, A.E. Curzon, A.T. Pawlowicz: Use of an electron beam for detecting superconducting domains of lead in its intermediate state. Nature *200*, 157 (1963)
6.280 G. Pozzi, U. Valdrè: Study of electron shadow patterns of the intermediate state of superconducting lead. Philos. Mag. *23*, 745 (1971)
6.281 E. Fuchs: Abbildung Weißscher Bezirke in dünnen ferromagnetischen Schichten mit dem elektromagnetischen Elektronenmikroskop. Naturwissenschaften *47*, 39 (1960)

6.282 L. Reimer: Die Struktur der magnetischen Bereichsgrenzen in grobkristallinen Eisenschichten. Z. Angew. Phys. *18*, 373 (1965)
6.283 W. Pitsch: Elektronenmikroskopische Beobachtung magnetischer Elementarbereiche in gealterten Eisen-Stickstoff-Legierungen. Arch. Eisenhüttenwes. *36*, 737 (1965)
6.284 W. Liesk: Magnetische Strukturen in dünnen Schichten, beobachtet im Elektronenmikroskop. Z. Angew. Phys. *14*, 200 (1962)
6.285 J.P. Jacubovics: The effect of magnetic domain structure on Bragg reflection in TEM. Philos. Mag. *10*, 277 (1964)
6.286 J.N. Chapman, E.H. Darlington: The application of STEM to the study of thin ferromagnetic films. J. Phys. E*7*, 181 (1974)
6.287 J.N. Chapman, E.M. Waddell, P.E. Batson, R.P. Ferrier: The Fresnel-mode of Lorentz microscopy using a STEM. Ultramicroscopy *4*, 283 (1979)
6.288 J.N. Chapman, P.E. Batson, E.M. Waddell, R.P. Ferrier: The direct determination of magnetic domain wall profiles by differential phase contrast electron microscopy. Ultramicroscopy *3*, 203 (1978)
6.289 A. Olivei: Holography and interferometry in electron Lorentz microscopy. Optik *30*, 27 (1969)
6.290 A. Olivei: Magnetic inhomogeneties and holographic methods in electron Lorentz microscopy. Optik *33*, 93 (1971)
6.291 M.S. Cohen, K.J. Harte: Domain wall profiles in magnetic films. J. Appl. Phys. *40*, 3597 (1969)
6.292 V.I. Petrov, G.V. Spivak, O.P. Pavluchenko: "Transmission Electron Microscope Observation of Domain Pattern of Speedily Remagnetized Thin Ferromagnetic Films," in *Electron Microscopy 1966*, Vol.1, ed. by R. Uyeda (Maruzen, Tokyo 1966) p.615
6.293 V.I. Petrov, G.V. Spivak: On a stroboscopic Lorentz microscope. Z. Angew. Phys. *27*, 188 (1969)
6.294 O. Bostanjoglo, Th. Rosin: Resonance oscillations of magnetic domain walls and Bloch lines observed by stroboscopic electron microscopy. Phys. Status Solidi A*57*, 561 (1980)
6.295 O. Bostanjoglo, Th. Rosin: Resonance oscillations of Bloch lines in permalloy films. Phys. Status Solidi A*66*, K5 (1981)
6.296 G.S. Plows, W.C. Nixon: Stroboscopic scanning electron microscopy. J. Phys. E*1*, 595 (1968)
6.297 E. Menzel, E. Kubalek: "Electron Beam Chopping System in the SEM," in *Scanning Electron Microscopy 1979/I*, ed. by O. Johari (IIT Research Inst., Chicago 1979) p.305
6.298 G.V. Saparin, G.V. Spivak: "Application of Stroboscopic Cathodoluminescence Microscopy," in *Scanning Electron Microscopy 1979/I*, ed. by O. Johari (SEM Inc., AMF O'Hare 1979) p.305
6.299 G.V. Spivak, G.V. Saparin, L.F. Komolova: "The Physical Fundamentals of the Resolution Enhancement in the SEM for CL and EBIC Modes," in *Scanning Electron Microscopy 1977/I*, ed. by O. Johari (IIT Research Inst., Chicago 1977) p.191
6.300 H. Mahl, W. Weitsch: Nachweis von fluktuierenden Ladungen in isolierenden Filmen bei Elektronenbestrahlung. Optik *17*, 107 (1960)
6.301 H. Mahl, W. Weitsch: Versuche zur Beseitigung von Aufladungen auf Durchstrahlungsobjekten durch zusätzliche Bestrahlung mit langsamen Elektronen. Z. Naturforsch. A*17*, 146 (1962)
6.302 G.H. Curtis, R.P. Ferrier: The electric charging of electron microscopical specimens. J. Phys. D*2*, 1035 (1969)
6.303 D.H. Warrington: A simple charge neutralizer for the electron microscope. J. Sci. Instrum. *43*, 77 (1966)
6.304 L. Reimer: Aufladung kleiner Teilchen im Elektronenmikroskop. Z. Naturforsch. A*20*, 151 (1965)
6.305 V. Drahoš, J. Komrska, M. Lenc: "Shadow Images of Charged Spherical Particles," in *Electron Microscopy 1968*, Vol.1, ed. by D.S. Bocciarelli (Tipografia Poliglotta Vaticana, Rome 1968) p.157
6.306 C. Jönsson, H. Hoffmann: Der Einfluß von Aufladungen auf die Stromdichteverteilung im Elektronenschattenbild dünner Folien. Optik *21*, 432 (1964)
6.307 H. Pfisterer, E. Fuchs, W. Liesk: Elektronenmikroskopische Abbildung ferroelektrischer Domänen in dünnen $BaTiO_3$-Einkristallschichten. Naturwissenschaften *49*, 178 (1962)

6.308 H. Blank, S. Amelinckx: Direct observation of ferroelectric domains in $BaTiO_3$ by means of the electron microscope. Appl. Phys. Lett. *2*, 140 (1963)
6.309 E. Fuchs, W. Liesk: Elektronenmikroskopische Beobachtung von Domänenkonfigurationen und von Umpolarisationsvorgängen in dünnen $BaTiO_3$-Einkristallen. J. Phys. Chem. Solidi *25*, 845 (1964)
6.310 R. Ayroles, J. Torres, J. Aubree, C. Roucau, M. Tanaka: Electron-microscope observation of structure domains in the ferroelastic phase of lead phosphate. $Pb_3(PO_4)_2$. Appl. Phys. Lett. *34*, 4 (1979)
6.311 C. Manolikas, S. Amelinckx: Phase transitions in ferroelastic lead orthovanadate as observed by means of electron microscopy and electron diffraction. Phys. Status Solidi A*60*, 607 (1980)
6.312 M. Tanaka, G. Honjo: Electron optical studies of $BaTiO_3$ single crystal films. J. Phys. Soc. Jpn. *19*, 954 (1964)
6.313 J.M. Titchmarsh, G.R. Booker: "The Imaging of Electric Field Regions Associated with p-n Junctions," in *Electron Microscopy 1972* (The Institute of Physics, London 1972) p.540
6.314 P.G. Merli, G.F. Missiroli, G. Pozzi: TEM observations of p-n junctions. Phys. Status Solidi A*30*, 699 (1975)
6.315 C. Capiluppi, P.G. Merli, G. Pozzi, I. Vecchi: Out-of-focus observations of p-n junctions by high-voltage electron microscopy. Phys. Status Solidi A*35*, 165 (1976)

Chapter 7

7.1 B.K. Vainshtein: *Modern Crystallography I*, Springer Ser. Solid-State Sci., Vol.15 (Springer, Berlin, Heidelberg, New York 1981)
7.2 C.G. Darwin: The theory of x-ray reflexion. Philos. Mag. *27*, 315 and 675 (1914)
7.3 A. Howie, M.J. Whelan: Diffraction contrast of electron microscopic images of crystal lattice defects. Proc. Roy. Soc. A*263*, 217 (1961); A*267*, 206 (1962)
7.4 Z.G. Pinsker: *Dynamical Scattering of X-Rays in Crystals*, Springer Ser. Solid-State Sci., Vol.3 (Springer, Berlin, Heidelberg, New York 1978)
7.5 A.W.S. Johnson: The analog computation of dynamic electron diffraction intensities. Acta Cryst. A*24*, 534 (1968)
7.6 J.M. Cowley, A.F. Moodie: The scattering of electrons by atoms and crystals. I. A new theoretical approach. Acta Cryst. *10*, 609 (1957)
D.F. Lynch: Out-of-zone effects in dynamical electron diffraction intensities from Au. Acta Cryst. A*27*, 399 (1971)
P. Goodman, A.F. Moodie: Numerical evaluation of n-beam wave functions in electron scattering by the multi-slice method. Acta Cryst. A*30*, 280 (1974)
7.7 H. Bethe: Theorie der Beugung von Elektronen an Kristallen. Ann. Phys. *87*, 55 (1928)
7.8 G. Thomas, E. Levine: Increase of extinction distance with temperature in Si. Phys. Status Solidi *11*, 81 (1965)
7.9 A. Howie, U. Valdrè: Temperature dependence of the extinction distance in electron diffraction. Philos. Mag. *15*, 777 (1967)
7.10 L. Sturkey: The use of electron-diffraction intensities in structure determination. Acta Cryst. *10*, 858 (1957)
7.11 H. Niehrs: Die Formulierung der Elektronenbeugung mittels einer Streumatrix und ihre praktische Verwendbarkeit. Z. Naturforsch. A*14*, 504 (1959)
7.12 F. Fujimoto: Dynamical theory of electron diffraction in Laue-case. J. Phys. Soc. Jpn. *14*, 1558 (1959); *15*, 859 and 1022 (1960)
7.13 C.J. Humphreys, R.M. Fisher: Bloch wave notation in many-beam electron diffraction theory. Acta Cryst. A*27*, 42 (1971)
7.14 J.P. Spencer, C.J. Humphreys: "Electron Diffraction from Tilted Specimens and Its Application to SEM," in *Electron Microscopy and Analysis*, ed. by W.C. Nixon (The Institute of Physics, London 1971) p.310
7.15 L.E. Thomas, C.G. Shirley, J.S. Lally, R.M. Fisher: "The Critical Voltage Effect and Its Applications," in *High Voltage Electron Microscopy* (Academic, London 1974) p.38

7.16 P.B. Hirsch, A. Howie, R.B. Nicholson, D.W. Pashley, M.J. Whelan: *Electron Microscopy of Thin Crystals* (Butterworths, London 1965)
7.17 G. Radi: Complex lattice potentials in electron diffraction calculated for a number of crystals. Acta Cryst. A*26*, 41 (1970)
7.18 P.A. Doyle: Absorption coefficients for Al 111 systematics: theory and comparison with experiment. Acta Cryst. A*26*, 133 (1970)
7.19 H. Hashimoto: Energy dependence of extinction distance and transmission power for electron waves in crystals. J. Appl. Phys. *35*, 277 (1964)
7.20 L. Reimer, M. Wächter: "Complex Fourier Coefficients of the Crystal Lattice Potential," in *Electron Microscopy 1980*, Vol.3, ed. by P. Brederoo, G. Boom (Seventh European Congr. Electron Microscopy Foundation, Leiden 1980) p.192
7.21 G. Meyer-Ehmsen: Untersuchungen zur normalen und anomalen Absorption von Elektronen in Si- und Ge-Einkristallen bei verschiedenen Temperaturen. Z. Phys. *218*, 352 (1969)
7.22 M.J. Goringe: Temperature dependence of the absorption of fast electrons in Cu. Philos. Mag. *14*, 93 (1966)
7.23 M.J. Goringe, M.J. Whelan: "The Absorption of Fast Electrons in Crystals," in *Electron Microscopy 1966*, Vol.1, ed. by R. Uyeda (Maruzen, Tokyo 1966) p.49
7.24 D. Renard, P. Croce, M. Gandais, M. Sauvin: Etude expérimentale de l'absorption des électrons dans l'or. Phys. Status Solidi B*47*, 411 (1971)
7.25 H.G. Badde, L. Reimer: "Measurement of Complex Structure Potentials in Au and PbTe by Convergent Electron Diffraction," in *Electron Microscopy 1972* (The Institute of Physics, London 1972) p.440
7.26 P. Goodman, G. Lehmpfuhl: Electron diffraction study of MgO h00-systematic interactions. Acta Cryst. *22*, 14 (1967)
7.27 K.G. Gaukler, K. Graff: Struktur- und Absorptionspotentiale von KCl und NaCl aus Beugungsaufnahmen in konvergentem Elektronenbündel. Z. Phys. *232*, 190 (1970)
7.28 A. Howie: Inelastic scattering of electrons by crystals. Proc. Roy. Soc. A*271*, 268 (1973)
7.29 H. Yoshioka: Effect of inelastic waves on electron diffraction. J. Phys. Soc. Jpn. *12*, 618 (1957)
7.30 E.N. Economou: *Green's Functions in Quantum Physics*, 2nd ed., Springer Ser. Solid-State Sci., Vol.7 (Springer, Berlin, Heidelberg, New York, Tokyo 1983)
7.31 G. Radi: Unelastische Streuung in der dynamischen Theorie der Elektronenbeugung. Z. Phys. *212*, 146 (1968)
R. Serneels, D. Haentjens, R. Gevers: Extension of the Yoshioka theory of inelastic electron scattering in crystals. Philos. Mag. A*42*, 1 (1980)
7.32 C.J. Humphreys, M.J. Whelan: Inelastic scattering of fast electrons by crystals. Philos. Mag. *20*, 165 (1969)
7.33 C.R. Hall, P.B. Hirsch: Effect of thermal diffuse scattering on propagation of high energy electrons through crystals. Proc. Roy. Soc. A*286*, 158 (1965)
7.34 P. Rez, C.J. Humphreys, M.J. Whelan: The distribution of intensity in electron diffraction patterns due to phonon scattering. Philos. Mag. *35*, 81 (1977)
7.35 Y. Kainuma: The theory of Kikuchi patterns. Acta Cryst. *8*, 247 (1955)
7.36 R.G. Blake, A. Jostsons, P.M. Kelly, J.G. Napier: The determination of extinction distances and anomalous absorption coefficients by STEM. Philos. Mag. A*37*, 1 (1978)
7.37 J.W. Steeds: Many-beam diffraction effects in gold and measurement of absorption parameters by fitting computer graphs. Phys. Status Solidi *38*, 203 (1970)
7.38 A. Mazel, R. Ayroles: Étude de la distance d' extinction et du coefficient d'absorption des electrons dans des échantillons d'aluminium pour des tensions comprises 50 et 1200 kilovolts. J. Microscopie *7*, 793 (1968)
7.39 G. Dupouy, F. Perrier, R. Uyeda, R. Ayroles, A. Mazel: Mesure du coefficient d'absorption des électrons accélérés sons des tensions comprises entre 100 et 1200 kV. J. Microscopie *4*, 429 (1965)
7.40 H. Hashimoto, A. Howie, M.J. Whelan: Anomalous electron absorption effects in metal foils. Proc. Roy. Soc. A*269*, 80 (1962)
7.41 P. Hagemann, L. Reimer: An experimental proof of the dependent Bloch wave model by large angle electron scattering from thin crystals. Philos. Mag. *40*, 367 (1979)

7.42 M.V. Berry: Diffraction in crystals at high energies. J. Phys. C*4*, 697 (1971)
7.43 M.V. Berry, K.E. Mount: Semiclassical approximations in wave mechanics. Rep. Prog. Phys. *35*, 315 (1972)
7.44 K. Kambe, G. Lehmpfuhl, F. Fujimoto: Interpretation of electron channelling by the dynamical theory of electron diffraction. Z. Naturforsch. A*29*, 1034 (1974)
7.45 F. Nagata, A. Fukuhara: 222 electron reflection from Al and systematic interaction. Jpn. J. Appl. Phys. *6*, 1233 (1967)
7.46 R. Uyeda: Dynamical effects in high voltage electron diffraction. Acta Cryst. A*24*, 175 (1968)
7.47 J.S. Lally, C.J. Humphreys, A.J.F. Metherell, R.M. Fisher: The critical voltage effect in high voltage electron microscopy. Philos. Mag. *25*, 321 (1972)
7.48 L.E. Thomas: Kikuchi patterns in HVEM. Philos. Mag. A*26*, 1447 (1972)
7.49 A.F. Moodie, J.R. Sellar, D. Imeson, C.J. Humphreys: "Convergent Beam Diffraction in the High Voltage Electron Microscope," in *High Voltage Electron Microscopy 1977*, ed. by T. Imura, H. Hashimoto (Japanese Soc. Electron Microscopy, Kyoto 1977) p.191
7.50 J.R. Sellar, D. Imeson, C.J. Humphreys: "Experimental and Theoretical Study of the Convergent-Beam Critical Voltage Effect in High Voltage Electron Diffraction," in *Electron Microscopy 1980*, Vol.1, ed. by P. Brederoo, G. Boom (Seventh European Congr. Electron Microscopy Foundation, Leiden 1980) p.120
7.51 T. Arii, R. Uyeda: Vanishing voltages of the second order reflections in electron diffraction. Jpn. J. Appl. Phys. *8*, 621 (1969)
7.52 T. Arii, R. Uyeda, O. Terasaki, D. Watanabe: Accurate determination of atomic scattering factors of fcc and hcp metals by high voltage electron diffraction. Acta Cryst. A*29*, 295 (1973)
7.53 A. Fukuhara, A. Yanagisawa: Vanishing of 222 Kikuchi line from Ag crystal. Jpn. J. Appl. Phys. *8*, 1166 (1969)
7.54 M. Fujimoto, O. Terasaki, D. Watanabe: Determination of atomic scattering factors of V and Cr by means of vanishing Kikuchi line method. Phys. Lett. A*41*, 159 (1972)
7.55 A. Rocher, B. Jouffrey: Contribution à l'étude des tensions critiques dans le Cu et Al. C. R. Acad. Sci. Paris B*275*, 133 (1972)
7.56 D. Watanabe, R. Uyeda, A. Fukuhara: Determination of the atom form factor by high voltage electron diffraction. Acta Cryst. A*25*, 138 (1969)
7.57 E.A. Hewat, C.J. Humphreys: "Si(111) and Ge(111) and (220) Scattering Factors Determined from Critical Voltage Measurements," in *High Voltage Electron Microscopy*, ed. by P.R. Swann, C.J. Humphreys, M.J. Goringe (Academic, London 1974) p.52
7.58 E.P. Butler: Application of the critical voltage effect to the study of compositional changes in Ni-Au alloys. Philos. Mag. *26*, 33 (1972)
7.59 I.P. Jones, E.G. Tapetado: "The Dependence of Electron Distribution and Atom Vibration in hcp Metals on the c/a Ratio: An Investigation Using the Critical Voltage Technique," in *High Voltage Electron Microscopy*, ed. by P.R. Swann, C.J. Humphreys, M.J. Goringe (Academic, London 1974) p.48
7.60 K. Kuroda, Y. Tomokiyo, T. Eguchi: "Temperature Dependence of Critical Voltages in Cu-Based Alloys," in *Electron Microscopy 1980*, Vol.4, ed. by P. Brederoo, J. Van Landuyt (Seventh European Congr. Electron Microscopy Foundation, Leiden 1980) p.112
7.61 C.G. Shirley, R.M. Fisher: "Application of the Critical Voltage Effect to Alloy Studies," in *Electron Microscopy 1980*, Vol.4, ed. by P. Brederoo, J. Van Landuyt (Seventh European Congr. Electron Microscopy Foundation, Leiden 1980) p.88
7.62 R. Leonhardt, H. Richter, W. Rossteutscher: Elektronenbeugungsuntersuchungen zur Struktur dünner nichtkristalliner Schichten. Z. Phys. *165*, 121 (1961)
7.63 M. von Laue: *Materiewellen und ihre Interferenzen* (Akademische Verlagsgesellschaft, Leipzig 1944)
7.64 M. Horstmann, G. Meyer: Messung der elastischen Elektronenbeugungsintensitäten polykristalliner Al-Schichten. Acta Cryst. *15*, 271 (1962)
7.65 M. Blackman: On the intensities of electron diffraction rings. Proc. Roy. Soc A*173*, 68 (1939)

7.66 C.J. Humphreys, P.B. Hirsch: Absorption parameters in electron diffraction theory. Philos. Mag. *18*, 115 (1968)
7.67 C.R. Hall: The scattering of high energy electrons by the thermal vibrations of crystals. Philos. Mag. *12*, 815 (1965)
7.68 K. Komatsu, K. Teramoto: Diffuse streak patterns from various crystals in x-ray and electron diffraction. J. Phys. Soc. Jpn. *21*, 1152 (1966)
7.69 N. Kitamura: Temperature dependence of diffuse streaks in single crystal Si electron diffraction patterns. J. Appl. Phys. *37*, 2187 (1966)
7.70 H.P. Herbst, G. Jeschke: "Diffuse Streak-Patterns from PbJ_2- and Bi-Single Crystals and Their Temperature Dependence," in *Electron Microscopy 1968*, Vol.1, ed. by D.S. Bocciarelli (Tipografia Poliglotta Vaticana, Rome 1968) p.293
7.71 E.M. Hörl: Thermisch-diffuse Elektronenstreuung in As-, Sb- und Bi-Kristallen. Optik *27*, 99 (1968)
7.72 M. Horstmann: Einfluß der Kristalltemperatur auf die Intensitäten dynamischer Elektroneninterferenzen. Z. Phys. *183*, 375 (1965)
7.73 M. Horstmann, G. Meyer: Messung der Elektronenbeugungs-Intensitäten polykristalliner Al-Schichten bei tiefer Temperatur und Vergleich mit der dynamischen Theorie. Z. Phys. *182*, 380 (1965)
7.74 M. Horstmann: Messung der thermisch diffusen Elektronenstreuung in polykristallinen Al-Schichten. Z. Phys. *188*, 412 (1965)
7.75 W. Zechnall: Temperaturabhängigkeit des Streuuntergrundes im Elektroneninterferenzdiagramm polykristalliner Ag-Schichten. Z. Phys. *229*, 62 (1969)
7.76 J. Hansen-Schmidt, M. Horstmann: Temperaturabhängigkeit der Streuabsorption schneller Elektronen in polykristallinen Au-Schichten. Z. Naturforsch. A*20*, 1239 (1965)
7.77 H. Boersch, O. Bostanjoglo, H. Niedrig: Temperaturabhängigkeit der Transparenz dünner Schichten für schnelle Elektronen. Z. Phys. *180*, 407 (1964)
7.78 W. Glaeser, H. Niedrig: Temperature dependence of dynamical electron diffraction intensities of polycrystalline foils. J. Appl. Phys. *37*, 4303 (1966)
7.79 W.W. Albrecht, H. Niedrig: Temperature dependence of dynamical electron diffraction intensities of polycrystalline foils. J. Appl. Phys. *39*, 3166 (1968)
7.80 G. Jeschke, D. Willasch: Temperaturabhängigkeit der anomalen Elektronenabsorption von Bi-Einkristallen. Z. Phys. *238*, 421 (1970)
7.81 C.R. Hall: On the thickness dependence of Kikuchi band contrast. Philos. Mag. *22*, 63 (1970)
7.82 H. Boersch: Über Bänder bei Elektronenbeugung. Phys. Z. *38*, 1000 (1937)
7.83 H. Pfister: Elektroneninterferenzen an Bleijodid bei Durchstrahlung im konvergenten Bündel. Ann. Phys. *11*, 239 (1953)
7.84 M. Komura, S. Kojima, T. Ichinokawa: Contrast reversal of Kikuchi bands in transmission electron diffraction. J. Phys. Soc. Jpn. *33*, 1415 (1972)
7.85 S. Takagi: On the temperature diffuse scattering of electrons. J. Phys. Soc. Jpn. *13*, 287 (1958)
7.86 J. Gjønnes: The influence of Bragg scattering on inelastic and other forms of diffuse scattering of electrons. Acta Cryst. *20*, 240 (1966)
7.87 K. Ishida: Inelastic scattering of fast electrons by crystals. J. Phys. Soc. Jpn. *28*, 450 (1970); *30*, 1439 (1971)
7.88 K. Okamato, T. Ichinokawa, Y.H. Ohtsuki: Kikuchi patterns and inelastic scattering. J. Phys. Soc. Jpn. *30*, 1690 (1971)
7.89 R. Høier: Multiple scattering and dynamical effects in diffuse electron scattering. Acta Cryst. A*29*, 663 (1973)

Chapter 8

8.1 A.J.F. Metherell, M.J. Whelan: Measurement of absorption of fast electrons in single crystal films of Al. Philos. Mag. *15*, 755 (1967)
8.2 A. Iijima: Intensity of fast electron transmitted through thick single crystals. J. Phys. Soc. Jpn. *35*, 213 (1973)
8.3 L. Reimer: Contrast in amorphous and crystalline objects. Lab. Invest. *14*, 939 (1965)
8.4 L. Reimer: Deutung der Kontrastunterschiede von amorphen und kristallinen Objekten in der Elektronenmikroskopie. Z. Angew. Phys. *22*, 287 (1967)
8.5 G. Dupouy, F. Perrier, R. Uyeda, R. Ayroles, A. Mazel: Mesure du coefficient

d'absorption des électrons accélérés sons des tensions comprises entre 100 et 1200 kV. J. Microscopie *4*, 429 (1965)
8.6 A. Mazel, R. Ayroles: "Etude dans des cristaux d'oxyde de magnesium des distances d'extinction correspondant a diverses reflexions systematiques," in *Microscopie Electronique 1970*, Vol.1, ed. by P. Favard (Société Francaise de Microscopie Electronique, Paris 1970) p.99
8.7 G. Möllenstedt: Elektronenmikroskopische Sichtbarmachung von Hohlstellen in Einkristall-Lamellen. Optik *10*, 72 (1953)
8.8 K. Shirota, T. Yamamoto, T. Yanaka, O. Vingsbo: On dark field techniques in transmission electron microscopy. Ultramicroscopy *1*, 67 (1975)
8.9 L. Reimer: Elektronenoptische Untersuchung zur Zwillingsbildung in Silber-Aufdampfschichten. Optik *16*, 30 (1959)
8.10 P. Rao: Separation and identification of phases with through-focus dark-field electron microscopy. Philos. Mag. *32*, 755 (1975)
8.11 G.M. Michal, R. Sinclaire: A quantitative assessment of the capabilities of 2 1/2 D microscopy for analysing crystalline solids. Philos. Mag. A*42*, 691 (1980)
8.12 L. Reimer: "Contrast in the Different Modes of SEM," in *Scanning Electron Microscopy: Systems and Applications 1973* (The Institute of Physics, London 1973) p.120
8.13 G.R. Booker, D.C. Joy, J.P. Spencer, H. von Harrach: "Contrast Effects from Crystalline Material Using STEM," in *Scanning Electron Microscopy 1974*, ed. by O. Johari (IIT Research Ins., Chicago 1974) p.225
8.14 D.M. Maher, D.C. Joy: The formation and interpretation of defect images from crystalline materials in a scanning transmission electron microscope. Ultramicroscopy *1*, 239 (1976)
8.15 L. Reimer, P. Hagemann: "Scanning Transmission Electron Microscopy of Crystalline Specimens," in *Scanning Electron Microscopy 1976/I*, ed. by O. Johari (IIT Research Inst., Chicago 1976) p.321
8.16 L. Reimer, P. Hagemann: Anwendung eines Rasterzusatzes zu einem Transmissionselektronenmikroskop. II. Abbildung kristalliner Objekte. Optik *47*, 325 (1977)
8.17 T. Yamamoto, H. Nishizawa: Imaging of crystalline substances in STEM. Phys. Status Solidi A*28*, 237 (1975)
8.18 H. Hashimoto: "High Voltage TEM - Contrast Theory," in *High Voltage Electron Microscopy*, ed. by P.R. Swann, C.J. Humphreys, M.J. Goringe (Academic, London 1974) p.9
8.19 C.J. Humphreys, L.E. Thomas, J.S. Lally, R.M. Fisher: Maximising the penetration in HVEM. Philos. Mag. *23*, 87 (1971)
8.20 A. Rocher, R. Ayroles, A. Mazel, C. Mory, B. Jouffrey: "Electron Penetration in Al, Cu, and MgO at High Voltages up to 3 MV," in *High Voltage Electron Microscopy*, ed. by P.R. Swann, C.J. Humphreys, M.J. Goringe (Academic, London 1974) p.436
8.21 C.J. Humphreys, J.S. Lally: Aspects of Bloch-wave channelling in high-voltage electron microscopy. J. Appl. Phys. *41*, 232 (1970)
8.22 J.W. Steeds: Many-beam diffraction effects in gold and measurement of absorption parameters by fitting computer-graphs. Phys. Status Solidi *38*, 203 (1970)
8.23 M.S. Spring: Electron channelling at high energies. Phys. Lett. A*31*, 421 (197
8.24 R. Uyeda, M. Nonoyama: The observation of thick specimens by high voltage electron microscopy. Jpn. J. Appl. Phys. *6*, 557 (1967)
8.25 G. Thomas: Electron microscopy at high voltages. Philos. Mag. *17*, 1097 (1968)
8.26 G. Thomas, J.C. Lacaze: Transmission electron microscopy at 2.5 MeV. J. Microscopie *97*, 301 (1973)
8.27 H. Fujita, T. Tabata: Voltage dependence of the maximum observable thickness by electron microscopy up to 3 MV. Jpn. J. Appl. Phys. *12*, 471 (1973)
8.28 H. Fujita, T. Tabata, K. Yoshida, N. Sumida, S. Katagiri: Some applications of an ultra-high voltage electron microscope on materials science. Jpn. J. Appl. Phys. *11*, 1522 (1972)
8.29 S. Mader: Elektronenmikroskopische Untersuchung der Gleitlinienbildung auf Cu-Einkristallen. Z. Phys. *149*, 73 (1957)
8.30 L. Reimer, C. Schulte: Elektronenmikroskopische Oberflächenabdrücke und ihr Auflösungsvermögen. Naturwissenschaften *53*, 489 (1966)

8.31 G.A. Bassett: A new technique for decoration of cleavage and slip steps on ionic crystal surfaces. Philos. Mag. *3*, 1042 (1958)
8.32 H. Bethge, K.W. Keller: Über die Abbildung von Versetzungen durch Abdampfstrukturen auf NaCl-Kristallen. Z. Naturforsch. A*15*, 271 (1960)
8.33 H. Bethge, K.W. Keller, N. Stenzel: Zur elektronenmikroskopischen Sichtbarmachung unterschiedlicher Bindungsenergien und Adsorptionseigenschaften an Lamellenstufen auf NaCl-Kristallen. Naturwissenschaften *49*, 152 (1962)
8.34 K. Kambe, G. Lehmpfuhl: Weak-beam technique for electron microscopic observation of atomic steps on thin single-crystal surfaces. Optik *42*, 187 (1975)
8.35 D. Cherns: Direct resolution of surface atomic steps by transmission electron microscopy. Philos. Mag. *30*, 549 (1974)
8.36 Y. Uchida, G. Lehmpfuhl, F. Fujimoto: "Dark Field Technique for a Direct Electron Microscopic Observation of the Surface Structure on Single Crystals," in *Microscopie Electronique à Haute Tension*, ed. by J. Jouffrey, P. Favard (Société Francaise de Microscopie Electronique, Paris 1976) p.113
8.37 S. Iijima: Observation of atomic steps of (111) surface of a silicon crystal using bright field electron microscopy. Ultramicroscopy *6*, 41 (1981)
8.38 G. Lehmpfuhl, K. Takayanagi: Electron microscopic contrast of atomic steps on fcc metal crystal surfaces. Ultramicroscopy *6*, 195 (1981)
8.39 J.W. Menter: The direct study by electron microscopy of crystal lattices and their imperfections. Proc. Roy. Soc. *236*, 119 (1956)
8.40 R. Scholz, H. Bethge: "High Resolution Study of 20° [001] Tilt Boundaries in Gold," in *Electron Microscopy 1980*, Vol.1, ed. by J. Brederoo, G. Boom (Seventh European Congr. Electron Microscopy Foundation, Leiden 1980) p.238
8.41 T. Komoda: On the resolution of the lattice imaging in the electron microscope. Optik *21*, 93 (1964)
8.42 H. Hashimoto, M. Mannani, T. Naiki: Dynamical theory of electron diffraction for the electron microscopical image of crystal lattices. Philos. Trans. Roy Soc. London A*253*, 459 (1961)
8.43 R. Sinclair: "Microanalysis by Lattice Imaging," in *Introduction to Analytical Electron Microscopy*, ed. by J.J. Hren, J.I. Goldstein, D.C. Joy (Plenum, New York 1978) p.507
8.44 W.C.T. Dowell: Das elektronenmikroskopische Bild von Netzebenenscharen und sein Kontrast. Optik *20*, 535 (1963)
8.45 R. Sinclair, R. Gronsky, G. Thomas: Optical diffraction from lattice images of alloys. Acta Metall. *24*, 789 (1976)
8.46 C.K. Wu, R. Sinclair, G. Thomas: Lattice imaging and optical microanalysis of Cu-Ni-Cr spinoidal alloy. Metall. Trans. A*9*, 381 (1978)
8.47 R. Sinclair, J. Dutkiewicz: Lattice imaging of the B19 ordering transformation and interfacial structure in Mg_3Cd. Acta Metall. *25*, 235 (1977)
8.48 V.A. Phillips: Lattice resolution measurements of strain fields at Guinier-Preston zones in Al-3.0% Cu. Acta Metall. *21*, 219 (1973)
8.49 R.G. Gronsky, G. Thomas: "Lattice Imaging of Grain Boundary Precipitation Reactions," in *Proc. 35th Annual Meeting of EMSA* (Claitor's Publ. Div., Baton Rouge, LO 1977) p.116
8.50 D.R. Clarke: "Determination of Grain Boundary Segregation by Combined X-Ray Microanalysis and Lattice Fringe Imaging," in *Scanning Electron Microscopy 1978/I*, ed. by O. Johari (SEM Inc., AMF O'Hare 1978) p.77
8.51 T. Komoda: Electron microscopic observation of crystal lattices on the level with atomic dimensions. Jpn. J. Appl. Phys. *5*, 603 (1966)
8.52 J.G. Allpress, J.V. Sanders: The direct observation of the structure of real crystals by lattice imaging. J. Appl. Cryst. *6*, 165 (1973)
8.53 J.M. Cowley, S. Iijima: "The Direct Imaging of Crystal Structures," in *Electron Microscopy in Mineralogy*, ed. by H.R. Wenk (Springer, Berlin, Heidelberg, New York 1976) p.123
8.54 J.L. Hutchison: "Lattice Images," in *Development in Electron Microscopy and Analysis*, ed. by J.A. Venables (Academic, London 1976) p.241
8.55 S. Iijima, S. Kimura, M. Goto: High-resolution microscopy of nonstochiometric $Nb_{22}O_{54}$ crystals: point defects and structural defects. Acta Cryst. A*30*, 251 (1974)
8.56 L.A. Bursill, A.R. Wilson: Electron-optical imaging of the Hollandite structure at 3 Å resolution. Acta Cryst. A*33*, 672 (1977)

8.57 M. Tanaka, B. Jouffrey: Many-beam lattice images calculated at 100 kV and 1000 kV. Acta Cryst. A*36*, 1033 (1980)
8.58 P.L. Fejes, S. Iijima, J.M. Cowley: Periodicity in thickness of electron-microscope crystal-lattice images. Acta Cryst. A*29*, 710 (1973)
8.59 D.F. Lynch, A.F. Moodie, M. A. O'Keefe: n-beam lattice images. V. The use of the charge-density approximation in the interpretation of lattice images Acta Cryst. A*31*, 300 (1975)
8.60 M. A. O'Keefe, J.W. Sanders: n-beam lattice images. VI. Degradation of image resolution by a combination of incident-beam divergence and spherical aberration. Acta Cryst. A*31*, 307 (1975)
8.61 T. Mitsuishi, H. Nagasaki, R. Uyeda: A new type of interference fringes observed in electron microscopy of crystalline substances. Proc. Imp. Acad. Jpn. *27*, 86 (1951)
8.62 G.A. Bassett, J.W. Menter, D.W. Pashley: Moiré patterns of electron micrographs and their application to the study of dislocations in metals. Proc. Roy. Soc. A*246*, 345 (1958)
8.63 O. Rang: Zur geometrischen Theorie der Moiré-Muster auf Elektronenbildern übereinander liegender Einkristalle. Z. Krist. *114*, 98 (1960)
8.64 R. Gevers: Dynamical theory of moiré fringe patterns. Philos. Mag. *7*, 1681 (1962)
8.65 J. Demny: Aussagen des Verdrehungsmoirés über Gitterfehler. Z. Naturforsch. A*15*, 194 (1960)
8.66 J.W. Matthews, W.M. Stobbs: Measurement of the lattice displacement across a coincidence grain boundary. Philos. Mag. *36*, 373 (1977)
8.67 L.A. Bruce, H. Jaeger: Geometric factors in fcc and bcc metal-on-metal epitaxy. Philos. Mag. *36*, 1331 (1977)
8.68 M.J. Whelan: An outline of the theory of diffraction contrast observed at dislocations and other defects in thin crystals examinated by TEM. J. Inst. Met. *87*, 392 (1959)
8.69 P.B. Hirsch, A. Howie, M.J. Whelan: A kinematical theory of diffraction contrast of electron transmission microscope images of dislocations and other effects. Philos. Trans. Roy. Soc. London A*252*, 499 (1960)
8.70 R. Gevers: On the dynamical theory of electron transmission microscope images of dislocations and stacking faults. Phys. Status Solidi *3*, 415 (1963
8.71 R. Gevers: On the dynamical theory of different types of electron microscopi transmission fringe patterns. Phys. Status Solidi *3*, 1672 (1963)
8.72 A. Howie, M.J. Whelan: Diffraction contrast of electron microscope images of crystal lattice defects. Proc. Roy. Soc. A*263*, 217 (1961); *267*, 206 (1962)
8.73 C.J. Ball: A relation between dark field electron micrographs of lattice defects. Philos. Mag. *9*, 541 (1964)
8.74 A. Howie: Inelastic scattering of electrons by crystals. Proc. Roy. Soc. A*271*, 268 (1963)
8.75 M. Wilkens: Zur Theorie des Kontrastes von elektronenmikroskopisch abgebildeten Gitterfehlern. Phys. Status Solidi *5*, 175 (1964)
8.76 M. Wilkens: Streuung von Blochwellen schneller Elektronen in Kristallen mit Gitterbaufehlern. Phys. Status Solidi *6*, 939 (1964)
8.77 M. Wilkens, M. Rühle: Black-white contrast figures from small dislocation loops. Phys. Status Solidi B*49*, 749 (1972)
8.78 J. van Landuyt, R. Gevers, S. Amelinckx: Fringe patterns at anti-phase boundaries with $\alpha = \pi$ observed in the electron microscope. Phys. Status Solidi *7*, 519 (1964)
8.79 C.M. Drum, M.J. Whelan: Diffraction contrast effects from stacking faults with phase-angle π. Philos. Mag. *11*, 205 (1965)
8.80 M.J. Whelan, P.B. Hirsch: Electron diffraction from crystals containing stacking faults. Philos. Mag. *2*, 1121 and 1303 (1957)
8.81 H. Hashimoto, A. Howie, M.J. Whelan: Anomalous electron absorption effects i metal foils. Philos. Mag. *5*, 967 (1960); Proc. Roy. Soc. A*269*, 80 (1962)
8.82 A. Art, R. Gevers, S. Amelinckx: The determination of the type of stacking faults in face centered cubic alloys by means of contrast effects in the ele tron microscope. Phys. Status Solidi *3*, 697 (1963)
8.83 R. Gevers, A. Art, S. Amelinckx: Electron microscopic images of single and intersecting stacking-faults in thick foils. Phys. Status Solidi *3*, 1563 (1963)

8.84 M.J. Marcinkowski: "Theory and Direct Observation of Antiphase Boundaries and Dislocations in Superlattices," in *Electron Microscopy and Strength of Crystals*, ed. by G. Thomas, J. Washburn (Interscience, New York 1963) p.333
8.85 S. Amelinckx: "The Study of Planar Interfaces by Means of Electron Microscopy," in *Modern Diffraction and Imaging Techniques in Material Science*, ed. by S. Amelinckx et al. (North-Holland, Amsterdam 1970) p.257
8.86 S. Amelinckx, J. Van Landuyt: "Contrast Effects at Planar Interfaces," in *Electron Microscopy in Mineralogy* (Springer, Berlin, Heidelberg, New York 1976) p.68
8.87 R. Serneels, M. Snykers, P. Delavignette, R. Gevers, S. Amelinckx: Friedel's law in electron diffraction as applied to the study of domain structures in non-centrosymmetrical crystals. Phys. Status Solidi B*58*, 277 (1973)
8.88 O. van der Biest, G. Thomas: Identification of enantiomorphism in crystals by electron microscopy. Acta Cryst. A*31*, 70 (1975)
8.89 A.J. Morton: Inversion anti-phase domains in Cu-rich γ-brasses. Phys. Status Solidi A*31*, 661 (1975)
8.90 R. Portier, D. Gratias, M. Fayard: Electron microscopy study of enantiomorphic ordered structures. Philos. Mag. *36*, 421 (1977)
8.91 R. Gevers, P. Delavignette, H. Blank, S. Amelinckx: Electron microscope transmission images of coherent domain boundaries. Phys. Status Solidi *4*, 383 (1964)
8.92 R. Gevers, P. Delavignette, H. Blank, J. van Landuyt, S. Amelinckx: Electron microscope transmission images of coherent domain boundaries. Phys. Status Solidi *5*, 595 (1964)
8.93 R. Gevers, J. van Landuyt, S. Amelinckx: Intensity profiles for fringe patterns due to planar interfaces as observed by electron microscopy. Phys. Status Solidi *11*, 689 (1965)
8.94 J. van Landuyt, R. Gevers, S. Amelinckx: Dynamical theory of the images of microtwins as observed in the electron microscope. Phys. Status Solidi *9*, 135 (1965)
8.95 H. Blank, S. Amelinckx: Direct observation of ferroelectric domains in barium titanate by means of electron microscopy. Appl. Phys. Lett. *2*, 140 (1963)
8.96 P. Delavignette, S. Amelinckx: Electron microscopic observation of antiferromagnetic domain walls in NiO. Appl. Phys. Lett. *2*, 236 (1963)
8.97 A.J. Ardell: Diffraction contrast at planar interfaces of large coherent precipitates. Philos. Mag. *16*, 147 (1967)
8.98 S.S. Sheinin, J.M. Corbett: Application of the multi-beam dynamical theory to crystals containing twins. Phys. Status Solidi A*38*, 675 (1976)
8.99 J. van Landuyt, R. Gevers, S. Amelinckx: Diffraction contrast from small voids as observed by electron microscopy. Phys. Status Solidi *10*, 319 (1965)
8.100 H.R. Wenk (ed.): *Electron Microscopy in Mineralogy* (Berlin, Heidelberg, New York 1976)
8.101 C.M. Wayman: "Martensitic Transformations," in *Modern Diffraction and Imaging Techniques in Material Science*, ed. by S. Amelinckx et al. (North-Holland, Amsterdam 1970) p.187
8.102 R. Gevers: On the kinematical theory of diffraction contrast of electron transmission microscope images of perfect dislocations of mixed type. Philos. Mag. *7*, 651 (1962)
8.103 W.J. Turnstall, P.B. Hirsch, J. Steeds: Effects of surface stress relaxations on the electron microscope images of dislocations normal to thin metal foils. Philos. Mag. *9*, 99 (1964)
8.104 M. Wilkens, M. Rühle, F. Häussermann: On the nature of the long-range dislocation contrast in electron transmission micrographs. Phys. Status Solidi *22*, 689 (1967)
8.105 P. Delavignette, S. Amelinckx: Dislocation nets in bismuth and antimony tellurides. Philos. Mag. *5*, 729 (1960)
8.106 A.K. Head: The computer generation of electron microscope pictures of dislocations. Aust. J. Phys. *20*, 557 (1967)
8.107 P. Humble: Computed electron micrographs for tilted foils containing dislocations and stacking faults. Aust. J. Phys. *21*, 325 (1968)
8.108 P. Humble: "Computed Electron Micrographs and Their Use in Defect Identifications," in *Modern Diffraction and Imaging Techniques in Material Science*, ed. by S. Amelinckx et al. (North-Holland, Amsterdam 1970) p.99

8.109 A.R. Thölén: A rapid method for obtaining electron microscope contrast maps of various lattice defects. Philos. Mag. *22*, 175 (1970)
8.110 D.J.H. Cockayne, I.L.F. Ray, M.J. Whelan: Investigations of dislocation strain fields using weak beams. Philos. Mag. *20*, 1265 (1969)
8.111 D.J.H. Cockayne: A theoretical analysis of the weak-beam method of electron microscopy. Z. Naturforsch. A*27*, 452 (1972)
8.112 D.J.H. Cockayne: The principles and practice of the weak-beam method of electron microscopy. J. Micr. *98*, 116 (1973)
8.113 I.L.F. Ray, D.J.H. Cockayne: The dissociation of dislocations in silicon. Proc. Roy. Soc. A*325*, 543 (1971)
8.114 A. Howie, Z.S. Basinski: Approximation of the dynamical theory of diffraction contrast. Philos. Mag. *17*, 1039 (1968)
8.115 C.J. Humphreys, R.A. Drummond: "The Column Approximation and High-Resolution Imaging of Defects," in *Electron Microscopy 1976*, Vol.1, ed. by D.G. Brandon (Tal International, Jerusalem 1976) p.142
8.116 D.J.H. Cockayne, M.L. Jenkins, I.L.F. Ray: The measurement of stacking-fault energies of pure face-centred cubic metals. Philos. Mag. *24*, 1383 (1971)
8.117 M.L. Jenkins: Measurement of the stacking-fault energy of gold using the weak-beam technique of electron microscopy. Philos. Mag. *26*, 747 (1972)
8.118 C.B. Carter, S.M. Holmes: The stacking-fault energy of nickel. Philos. Mag. *35*, 1161 (1977)
8.119 C.G. Rhodes, A.W. Thomson: The composition dependence of stacking fault energy in austenitic stainless steel. Metall. Trans. A*8*, 1901 (1977)
8.120 A. Gomez, D.J.H. Cockayne, P.B. Hirsch, V. Vitek: Dissociation of near-screw dislocations in Ge and Si. Philos. Mag. *31*, 105 (1975)
8.121 G.W. Groves, M.J. Whelan: The determination of the sense of the Burgers vector of a dislocation from its electron microscope images. Philos. Mag. *7*, 1603 (1962)
8.122 R. Siems, P. Delavignette, S. Amelinckx: Die direkte Messung von Stapelfehlerenergien. Z. Phys. *165*, 502 (1961)
8.123 M.H. Loretto, L.K. France: The influence of the degree of the deviation from the Bragg condition on the visibility of dislocations in copper. Philos. Mag. *19*, 141 (1969)
8.124 K. Marukawa: A new method of Burgers vector identification from electron microscope images. Philos. Mag. A*40*, 303 (1979)
8.125 M.F. Ashby, L.M. Brown: On diffraction contrast from inclusions. Philos. Mag. *8*, 1649 (1963)
8.126 M.F. Ashby, L.M. Brown: Diffraction contrast from spherically symmetrical coherency strains. Philos. Mag. *8*, 1083 (1963)
8.127 M.J. Makin, A.D. Whapham, F.J. Minter: The formation of dislocation loops in copper during neutron irradiation. Philos. Mag. *7*, 285 (1962)
8.128 U. Essmann, M. Wilkens: Elektronenmikroskopische Kontrastexperimente an Fehlstellenagglomeraten in neutronen-bestrahltem Kupfer. Phys. Status Solidi *4*, K53 (1964)
8.129 M. Wilkens: "Identification of Small Defect Clusters in Particle-Irradiated Crystals by Means of TEM," in *Modern Diffraction and Imaging Techniques in Material Science*, ed. by S. Amelinckx et al. (North-Holland, Amsterdam 1970) p.233
8.130 M. Rühle, M. Wilkens, U. Essmann: Zur Deutung der elektronenmikroskopischen Kontrasterscheinungen an Fehlstellenagglomeraten in neutronenbestrahltem Kupfer. Phys. Status Solidi *11*, 819 (1965)
8.131 M. Rühle: Elektronenmikroskopie kleiner Fehlstellenagglomerate in bestrahlten Metallen. Phys. Status Solidi *19*, 263 and 279 (1967)
8.132 M. Rühle, M. Wilkens: Small vacancy dislocation loops in neutron-irradiated copper. Philos. Mag. *15*, 1075 (1967)
8.133 K.H. Katerbau: The contrast of dynamical images of small lattice defects in the electron microscope. Phys. Status Solidi A*38*, 463 (1976)
8.134 B.L. Eyre, D.M. Maher, R.C. Perrin: Electron microscope image contrast from small dislocation loops. J. Phys. F*7*, 1359 and 1371 (1978)
8.135 M.L. Jenkins, K.H. Katerbau, M. Wilkens: TEM studies of displacement cascades in Cu_3Au. Philos. Mag. *34*, 1141 (1976)
8.136 M. Wilkens, M.L. Jenkin, K.H. Katerbau: TEM diffraction contrast of lattice defects causing strain contrast and structure factor contrast simultaneously. Phys. Status Solidi A*39*, 103 (1977)

8.137 M. Rühle, M. Wilkens: Defocusing contrast of cavities. Cryst. Lattice Defects *6*, 129 (1975)

Chapter 9

9.1 C.J. Cooke, P. Duncumb: "Performance Analysis of a Combined Electron Microscope and Electron Probe Microanalyser 'EMMA'," in *Fifth Intern. Congr. on X-Ray Optics and Microanalysis*, ed. by G. Möllenstedt, K.H. Gaukler (Springer, Berlin, Heidelberg, New York 1969) p.245

9.2 C.J. Cooke, I.K. Openshaw: "Combined High Resolution Electron Microscopy and X-Ray Microanalysis," in *Microscopie Electronique 1970*, Vol.1, ed. by P. Favard (Société Francaise de Microscopie Electronique, Paris 1970) p.175

9.3 J.B. LePoole: "Miniature Lens," in *Electron Microscopy 1964*, Vol.A, ed. by M. Titlbach (Czechoslovak Acad. Sci., Prague 1964) p.439

9.4 P.F. Chapman: "A Microanalysis Attachment for the Elmiskop I," in *Fifth Intern. Congr. on X-Ray Optics and Microanalysis*, ed. by G. Möllenstedt, K.H. Gaukler (Springer, Berlin, Heidelberg, New York 1969) p.241

9.5 H. Neff: Über die Röntgen-Emissionsanalyse von elektronenmikroskopischen Präparaten. Z. Instrumentenkd. *72*, 125 (1964)

9.6 E. Fuchs: X-ray spectrometer attachment for Elmiskop I electron microscope. Rev. Sci. Instrum. *37*, 623 (1966)

9.7 D.A. Gedcke: "The Si(Li) X-Ray Spectrometer for X-Ray Microanalysis," in *Quantitative Scanning Electron Microscopy*, ed. by D.B. Holt et al. (Academic, London 1974) p.403

9.8 T.A. Hall: Reduction of background due to backscattered electrons in energy-dispersive x-ray microanalysis. J. Micr. *110*, 103 (1977)

9.9 B. Neumann, L. Reimer: A permanent magnet system for electron deflection in front of an energy dispersive x-ray spectrometer. Scanning *1*, 130 (1978)

9.10 N.C. Barbi, A.O. Sandborg, J.C. Russ, C.E. Soderquist: "Light Element Analysis on the SEM Using a Windowless Energy Dispersive X-Ray Spectrometer," in *Scanning Electron Microscopy 1974*, ed. by O. Johari (IIT Research Inst., Chicago 1974) p.289

9.11 J.C. Russ: "Procedures for Quantitative Ultralight Element Energy Dispersive X-Ray Analysis," in *Scanning Electron Microscopy 1977/I*, ed. by O. Johari (IIT Research Inst., Chicago 1977) p.289

9.12 S.J.B. Reed: *Electron Microprobe Analysis* (Cambridge University Press, London 1975)

9.13 J.I. Goldstein, D.E. Newbury, P. Echlin, D.C. Joy, C. Fiori, E. Lifshin: *Scanning Electron Microscopy and X-Ray Microanalysis* (Plenum, New York 1981)

9.14 K.F.J. Heinrich: *Electron Beam X-Ray Microanalysis* (Van-Nostrand, New York 1981)

9.15 M.H. Jacobs, J. Baborovska: "Quantitative Microanalysis of Thin Foils with a Combined Electron Microscope-Microanalyser (EMMA-3)," in *Electron Microscopy 1972* (The Institute of Physics, London 1972) p.136

9.16 G.W. Lorimer, G. Cliff, J.N. Clark: "Determination of the Thickness and Spatial Resolution for the Quantitative Analysis of Thin Foils," in *Developments in Electron Microscopy and Analysis*, ed. by J.A. Venables (Academic, London 1976) p.153

9.17 R. König: "Quantitative X-Ray Microanalysis of Thin Foils," in *Electron Microscopy in Mineralogy*, ed. by H.R. Wenk (Springer, Berlin, Heidelberg, New York 1976) p.526

9.18 J.I. Goldstein, J.L. Costley, G.W. Lorimer, S.J.B. Reed: "Quantitative X-Ray Analysis in the Electron Microscope," in *Scanning Electron Microscopy 1977/I* ed. by O. Johari (IIT Research Inst., Chicago 1977) p.315

9.19 J. Philibert, R. Tixier: "Electron Probe Microanalysis of TEM Specimens," in *Physical Aspects of Electron Microscopy and Analysis*, ed. by B.M. Siegel, D.R. Beaman (Wiley, New York 1975) p.333

9.20 G.W. Lorimer, S.A. Al-Salman, G. Cliff: "The Quantitative Analysis of Thin Specimens: Effects of Absorption, Fluorescence and Beam Spreading," in *Development in Electron Microscopy and Analysis 1977*, ed. by D.L. Misell (The Institute of Physics, London 1977) p.369

9.21 C.R. Hall: On the production of characteristic x-rays in thin metal crystals. Proc. Roy. Soc. A*295*, 140 (1966)
9.22 D. Cherns, A. Howie, M.H. Jacobs: Characteristic x-ray production in thin crystals. Z. Naturforsch. A*28*, 565 (1973)
9.23 B. Neumann, L. Reimer: Anisotropic x-ray generation in thin and bulk single crystals. J. Phys. D*13*, 1737 (1980)
9.24 J. Bentley, N.J. Zaluzec, E.A. Kenik, R.W. Carpenter: "Optimization of an Analytical Electron Microscope for X-Ray Microanalysis," in *Scanning Electron Microscopy 1979/II*, ed. by O. Johari (SEM Inc. AMF O'Hare 1979) p.58
9.25 J. Philibert, R. Tixier: Electron penetration and the atomic number correction in electron probe microanalysis. J. Phys. D*1*, 685 (1968)
9.26 M.J. Nasir: "Quantitative Analysis on Thin Films in EMMA-4 Using Block Standards," in *Electron Microscopy 1972* (The Institute of Physics, London 1972) p.142
9.27 G. Cliff, G.W. Lorimer: "Quantitative Analysis of Thin Metal Foils Using EMMA-4 - the Ratio Technique," in *Electron Microscopy 1972* (The Institute of Physics, London 1972) p.140
9.28 G. Cliff, G.W. Lorimer: The quantitative analysis of thin specimens. J. Micr. *103*, 203 (1975)
9.29 M.N. Thompson, P. Doig, J.W. Edington, P.E.J. Flewitt: The influence of specimen thickness on x-ray count rates in STEM microanalysis. Philos. Mag. *35*, 1537 (1977)
9.30 T.P. Schreiber, A.M. Wims: A quantitative x-ray microanalysis thin film method using K-, L- and M-lines. Ultramicroscopy *6*, 323 (1981)
9.31 C.E. Lyman, P.E. Manning, D.J. Duquette, E. Hall: "STEM Microanalysis of Duplex Stainless Steel Weld Metal," in *Scanning Electron Microscopy 1978/I* ed. by O. Johari (SEM Inc., OMF O'Hare 1978) p.213
9.32 D.B. Williams, J.I. Goldstein: "STEM/X-Ray Microanalysis Across α/γ Interfaces in Fe-Ni meteorites," in *Electron Microscopy 1978*, Vol.1, ed. by J.M. Sturgess (Microscopical Soc. Canada, Toronto 1978) p.416
9.33 A.M. Ritter, W.G. Morris, M.F. Henry: "Factors Affecting the Measurement of Composition Profiles in STEM," in *Scanning Electron Microscopy 1979/I*, ed. by O. Johari (SEM Inc., AMF O'Hare 1979) p.121
9.34 T.A. Hall: "The Microprobe Assay of Chemical Elements," in *Physical Techniques in Biological Research*, Vol.1, Part A, ed. by G. Oster (Academic, New York 1971) p.157
9.35 T.A. Hall, H. Clarke Anderson, T. Appleton: The use of thin specimens for x-ray microanalysis in biology. J. Micr. *99*, 177 (1973)
9.36 T.A. Hall, B.L. Gupta: "EDS Quantitation and Application to Biology," in *Introduction to Analytical Electron Microscopy*, ed. by J.J. Hren, J.F. Goldstein, D.C. Joy (Plenum, New York 1979) p.169
9.37 A.R. Spurr: Choice and preparation of standards for x-ray microanalysis of biological materials with special reference to macrocyclic polyether complexes. J. Microscopie Biol. Cell *22*, 237 (1975)
9.38 G.M. Roomans, H.L.M. van Gaal: Organometallic and organometalloid compounds as standards for microprobe analysis of epoxy resin embedded tissue. J. Micr. *109*, 235 (1977)
9.39 H. Shuman, A.V. Somlyo, A.P. Somlyo: Quantitative electron probe microanalysis of biological thin sections: methods and validity. Ultramicroscopy *1*, 317 (1976)
9.40 T.O. Ziebold: Precision and sensitivity in electron microprobe analysis. Anal. Chem. *39*, 858 (1967)
9.41 D.C. Joy, D.M. Maher: "Sensitivity Limits for Thin Specimens X-Ray Analysis," in *Scanning Electron Microscopy 1977/I*, ed. by O. Johari (IIT Research Inst. Chicago 1977) p.325
9.42 A.J.F. Metherell: "Energy Analysing and Energy Selecting Microscopes," in *Advances in Optical and Electron Microscopy*, Vol.4, ed. by R. Barer, V.E. Cosslett (Academic, London 1971) p.263
9.43 W. Steckelmacher: Energy analysers for charged particle beams. J. Phys. E*6*, 1061 (1973)
9.44 H.T. Pearce-Percy: "The Design of Spectrometers for Energy Loss Spectroscopy" in *Scanning Electron Microscopy 1978/I*, ed. by O. Johari (SEM Inc., AMF O'Hare 1978) p.41

9.45 D.B. Wittry: An electron spectrometer for use with the TEM. J. Phys. D*2*, 1757 (1969)
9.46 H. Hintenberger: Improved magnetic focusing of charged particles. Rev. Sci. Instrum. *20*, 748 (1949)
9.47 S. Penner: Calculations of properties of magnetic deflection systems. Rev. Sci. Instrum. *32*, 150 (1961)
9.48 A.V. Crewe, M. Isaacson, D. Johnson: A high resolution electron spectrometer for use in transmission electron microscopy. Rev. Sci. Instrum. *42*, 411 (1971)
9.49 R.F. Egerton: A simple electron spectrometer for energy analysis in the transmission microscope. Ultramicroscopy *3*, 39 (1978)
9.50 R.F. Egerton: Design of an aberration-corrected electron spectrometer for the TEM. Optik *57*, 229 (1980)
9.51 H. Shuman: Correction of the second-order aberrations of uniform field magnetic sectors. Ultramicroscopy *5*, 45 (1980)
9.52 R.F. Egerton: The use of electron lenses between a TEM specimen and an electron spectrometer. Optik *56*, 363 (1980)
9.53 D.E. Johnson: Pre-spectrometer optics in CTEM/STEM. Ultramicroscopy *5*, 163 (1980)
9.54 A.W. Blackstock, R.D. Birkhoff, M. Slater: Electron accelerator and high resolution analyser. Rev. Sci. Instrum. *26*, 274 (1955)
9.55 J. Lohff: Charakteristische Energieverluste bei der Streuung mittelschneller Elektronen an Aluminium-Oberflächen. Z. Phys. *171*, 442 (1963)
9.56 Y. Kokubo, H. Koike, T. Someya: "Development of Energy Analyzer for Scanning and Transmission Microscope," in *Electron Microscopy 1974*, Vol.1, ed. by J.V. Sanders, D.J. Goodchild (Australian Acad. Sci., Canberra 1974) p.374
9.57 W. Kraus, P. Fazekas: Electron energy-loss spectrometry using an electron microscope in combination with an electrostatic cylindrical mirror. Siemens Forsch. Entwicklungsber. *6*, 172 (1977)
9.58 A.V. Crewe, J. Wall, L.M. Welter: A high resolution scanning transmission electron microscope. J. Appl. Phys. *39*, 5861 (1968)
5.59 H. Boersch: Experimentelle Bestimmung der Energieverteilung in thermisch ausgelösten Elektronenstrahlen. Z. Phys. *139*, 115 (1954)
5.60 H. Boersch, H. Miessner: Ein hochempfindlicher Gegenfeld-Energieanalysator für Elektronen. Z. Phys. *168*, 298 (1962)
9.61 H. Boersch, S. Schweda: Eine inverse Gegenfeldmethode zur Energieanalyse von Elektronen und Ionenstrahlen. Z. Phys. *167*, 1 (1962)
9.62 H. Brack: Über eine Anordnung zur Filterung von Elektroneninterferenzen. Z. Naturforsch. A*17*, 1066 (1962)
9.63 H. Boersch, R. Wolter, H. Schoenebeck: Elastische Energieverluste kristallgestreuter Elektronen: Z. Phys. *199*, 124 (1967)
9.64 M.T. Browne, S. Lockovic, R.E. Burge: "Instrumentation and Recording for the Vacuum Generators HB5 STEM Instrument," in *Developments in Electron Microscopy and Analysis*, ed. by J.A. Venables (Academic, London 1976) p.27
9.65 H. Boersch, J. Geiger, W. Stickel: Das Auflösungsvermögen des elektrostatisch-magnetischen Energieanalysators für schnelle Elektronen. Z. Phys. *180*, 415 (1964)
9.66 J. Geiger, M. Nolting, B. Schröder: "How to Obtain High Resolution with a Wien Filter Spectrometer," in *Microscopie Electronique 1970*, Vol.2, ed. by P. Favard (Société Francaise de Microscopie Electronique, Paris 1970) p.111
9.67 W.H.J. Anderson, J.B. LePoole: A double wienfilter as a high resolution, high transmission electron energy analyser. J. Phys. E*3*, 121 (1970)
9.68 W.H.J. Andersen, J. Kramer: "A Double-Focusing Wien Filter as a Full-Image Energy Analyser for the Electron Microscope," in *Electron Microscopy 1972*, (The Institute of Physics, London 1972) p.146
9.69 G.H. Curtis, J. Silcox: A Wien filter for use as an energy analyzer with an electron microscope. Rev. Sci. Instrum. *42*, 630 (1971)
9.70 G. Möllenstedt: Die elektrostatische Linse als hochauflösender Geschwindigkeitsanalysator. Optik *5*, 499 (1949)
9.71 G. Möllenstedt, W. Dietrich: Verbesserung der Optik des hochauflösenden elektrostatischen Geschwindigkeitsanalysators. Optik *12*, 246 (1955)
9.72 K. Keck, H. Deichsel: Die Verwendung der Elektronen-Einzellinse als "lichtstarkes" Energiefilter für Elektronenstrahlen. Optik *17*, 401 (1960)

9.73 A.J.F. Metherell, R.F. Cook: Resolution and dispersion of the four-classes of Möllenstedt electron energy analysers. Optik *34*, 535 (1972)
9.74 S. Kuwabara, T. Uefuji, Y. Takamatsu: A simple electrostatic energy filter for electron diffraction and electron microscopy. Jpn. J. Appl. Phys. *13*, 1495 (1974)
9.75 F. Lenz: Über das chromatische Auflösungsvermögen von Elektronenlinsen bei der Geschwindigkeitsanalyse. Optik *10*, 439 (1953)
9.76 R. Shirota, T. Yanaka: "An Energy Analyser with Rotation Symmetrical Lenses," in *Electron Microscopy 1974*, Vol.1, ed. by J.V. Sanders, D.J. Goodchild (Australian Acad. Sci., Canberra 1974) p.368
9.77 L. Reimer, U. Riediger: Energieverlustspektroskopie mit einer modifizierten Kaustikmethode in einem 100 keV-Transmissionselektronenmikroskop. Optik *46*, 67 (1976)
9.78 T. Ichinokawa: Electron energy analysis by a cylindrical magnetic lens. Jpn. J. Appl. Phys. *7*, 799 (1968)
9.79 K.Z. Considine, K.C.A. Smith: "An Energy Analyser for High Voltage Microscopy," in *Electron Microscopy 1968*, Vol.1, ed. by D.S. Bocciarelli (Tipografia Poliglotta Vaticana, Rome 1968) p.329
9.80 Y. Kamiya, K. Shimizu, T. Suzuki: The velocity analyser for high energy electrons. Optik *41*, 421 (1974)
9.81 R. Castaing: Quelques application du filtrage magnetique des vitesses en microscopie electronique. Z. Angew. Phys. *27*, 171 (1969)
9.82 R. Castaing, L. Henry: Filtrage magnêtique des vitesses en microscopie électronique. C. R. Acad. Sci. Paris *255*, 76 (1962)
9.83 H. Rose, E. Plies: Entwurf eines fehlerarmen magnetischen Energie-Analysators. Optik *40*, 336 (1974)
9.84 H.T. Pearce-Percy, D. Krahl, J. Jaeger: "A 4-Magnet Imaging Spectrometer for a Fixed-Beam Transmission Microscope," in *Electron Microscopy 1976*, Vol.1, ed. by D.G. Brandon (Tal International, Jerusalem 1976) p.348
9.85 G. Zanchi, J.Ph. Perez, J. Sevely: Adaption of a magnetic filtering device in a one megavolt electron microscope. Optik *43*, 495 (1945)
9.86 G. Zanchi, J. Sevely, B. Jouffrey: An energy filter for high voltage electron microscopy. J. Micr. Spectr. Electr. *2*, 95 (1977)
9.87 B.L. Jones, D.G. Jenkins, G.R. Booker: "Use of Silicon Linear Photodiode Arrays for Detection of High-Energy Electrons," in *Developments in Electron Microscopy and Analysis 1977*, ed. by D.L. Misell (The Institute of Physics, London 1977) p.73
9.88 B.L. Jones, D.M. Walton, G.R. Booker: "Developments in the Use of One- and Two-Dimensional Self-Scanned Silicon Photodiode Arrays as Imaging Devices in Electron Microscopy," in *Developments in Electron Microscopy and Anlaysis 1981*, ed. by M.J. Goringe (The Institute of Physics, London 1981) p.135
9.89 H. Shuman: Parallel recording of electron energy loss spectra. Ultramicroscopy *6*, 163 (1981)
9.90 P.E. Batson: Digital data acquisition of electron energy loss intensities. Ultramicroscopy *3*, 367 (1979)
9.91 R.F. Egerton, D. Kenway: An acquisition, storage, display and processing system for electron energy-loss spectra. Ultramicroscopy *4*, 221 (1979)
9.92 D.C. Misell, A.F. Jones: The determination of the single-scattering line profile from the observed spectrum. J. Phys. A*2*, 540 (1969)
9.93 D.W. Johnson, J.C.H. Spence: "Determination of the Single Scattering Electron Energy Loss Distribution from Plural Scattering Data," in *Electron Microscopy 1974*, Vol.1, ed. by J.V. Sanders, D.J. Goodchild (Australian Acad. Sci., Canberra 1974) p.386
9.94 J. Daniels, C. von Festenberg, H. Raether, K. Zeppenfeld: "Optical Constants of Solids by Electron Spectroscopy," in *Springer Tracts Mod. Phys.*, Vol.54 (Springer, Berlin, Heidelberg, New York 1970) p.77
9.95 R.W. Ditchfield, A.G. Cullis: "Identification of Impurity Particles in Epitaxially Grown Si Films Using Combined Electron Microscopy and Energy Analysis," in *Microscopie Electronique 1970*, Vol.2, ed. by P. Favard (Société Francaise de Microscopie Electronique, Paris 1970) p.125
9.96 R.F. Cook: "Electron Energy Loss Spectroscopy of Glass," in *Microscopie Electronique*, Vol.2, ed. by P. Favard (Société Francaise de Microscopie Electronique, Paris 1970) p.127

9.97 M. Isaacson: Interaction of 25 keV electrons with the nucleic acid bases, adenine, thymine, and uracil. J. Chem. Phys. *56*, 1803 and 1813 (1972)
9.98 J. Hainfeld, M. Isaacson: The use of electron energy loss spectroscopy for studying membrane architecture. Ultramicroscopy *3*, 87 (1978)
9.99 D.R. Spalding, A.J.F. Metherell: Plasmons losses in Al-Mg alloys. Philos. Mag. *18*, 41 (1968)
9.100 S.L. Cundy, A.J.F. Metherell, M.J. Whelan, P.N.T. Unwin, R.B. Nicholson: Studies of segregation and the initial stages of precipitation at grain boundaries in an Al-7wt% Mg alloy with an energy analysing electron microscope. Proc. Roy. Soc. A*307*, 267 (1968)
9.101 D.R. Spalding, R.E. Villagrana, G.A. Chadwick: A study of copper distribution in lamellar Al-$CuAl_2$ eutectics using an energy analysing microscope. Philos. Mag. *20*, 471 (1969)
9.102 R.F. Cook, S.L. Cundy: Plasmon energy losses in Al-Zn alloys. Philos. Mag. *20*, 665 (1969)
9.103 G. Hibbert, J.W. Eddington: Experimental errors in combined electron microscopy and energy analysis. J. Phys. D*5*, 1780 (1972)
9.104 G. Hibbert, J.W. Edington: Superposition effects in the energy analysing electron microscope. Philos. Mag. *26*, 1071 (1972)
9.105 R.F. Cook, A. Howie: Effect of elastic constraints on electron energy loss measurements in inhomogeneous alloy. Philos. Mag. *20*, 641 (1969)
9.106 D.R. Spalding: Electron microscopy evidence of plasmon-dislocation interactions. Philos. Mag. *34*, 1073 (1976)
9.107 R.F. Egerton: Measurement of inelastic/elastic scattering ratio for fast electrons and its use in the study of radiation damage. Phys. Status Solidi A*37*, 663 (1976)
9.108 D.B. Wittry, R.P. Ferrier, V.E. Cosslett: Selected-area electron spectrometry in the transmission electron microscope. J. Phys. D*2*, 1767 (1969)
9.109 C. Colliex, B. Jouffrey: Diffusion inelastique des electrons dans un solide par excitation de niveaux atomiques profonds. Philos. Mag. *25*, 491 (1972)
9.110 R.F. Egerton, M.J. Whelan: "High Resolution Microanalysis of Light Elements by Electron Energy Loss Spectrometry," in *Electron Microscopy 1974*, Vol.1, ed. by J.V. Sanders, D.J. Goodchild (Australian Acad. Sci., Canberra 1974) p.384
9.111 R.D. Leapman, V.E. Cosslett: Electron energy loss spectrometry: mean free paths for some characteristic x-ray excitations. Philos. Mag. *33*, 1 (1976)
9.112 M. Isaacson, D. Johnson: The microanalysis of light elements using transmitted energy loss electrons. Ultramicroscopy *1*, 33 (1975)
9.113 C. Colliex, V.E. Cosslett, R.D. Leapman, P. Trebbia: Contribution of electron energy loss spectroscopy to the development of analytical electron microscopy. Ultramicroscopy *1*, 301 (1976)
9.114 J. Sevely. J.Ph. Perez, B. Jouffrey: "Energy Losses of Electrons Through Al and Carbon Films from 300 keV up to 1200 keV," in *High Voltage Electron Microscopy*, ed. by P.R. Swann, C.J. Humphreys, M.J. Goringe (Academic, London 1974) p.32
9.115 R.D. Leapman, V.E. Cosslett: Electron spectrometry of inner shell excitation. Vacuum *26*, 423 (1977)
9.116 R.F. Egerton: Formulae for light-element microanalysis by electron energy-loss spectrometry. Ultramicroscopy *3*, 243 (1978)
9.117 G. Lehmpfuhl, J. Taftø: "The Channelling Effect in Electron Energy Loss Spectroscopy," in *Electron Microscopy 1980*, Vol.3, ed. by P. Brederoo, V.E. Cosslett (Seventh European Congr. Electron Microscopy Foundation, Leiden 1980) p.62
9.118 R.F. Egerton, C.J. Rossouw, M.J. Whelan: "Progress Towards a Method for the Quantitative Microanalysis of Light Elements by Electron Energy-Loss Spectrometry," in *Developments in Electron Microscopy and Analysis*, ed. by J.A. Venables (Academic, London 1976) p.129
9.119 D.C. Joy, D.M. Maher: Electron energy loss spectroscopy: detectable limits for elemental analysis. Ultramicroscopy *5*, 333 (1980)
9.120 H. Boersch: Gegenfeldfilter für Elektronenbeugung und Elektronenmikroskopie. Z. Phys. *134*, 156 (1953)
9.121 H. Watanabe: Energy selecting microscope. Jpn. J. Appl. Phys. *3*, 480 (1964)

9.122 S.L. Cundy, A.J.F. Metherell, M.J. Whelan: An energy analysing electron microscope. J. Sci. Instrum. *43*, 712 (1966)
9.123 S.L. Cundy, A. Howie, U. Valdrè: Preservation of electron microscopic image contrast after inelastic scattering. Philos. Mag. *20*, 147 (1969)
9.124 Y. Kihn, G. Zanchi, J. Sevely, B. Jouffrey: Application du filtrage en energie des electrons a l'observation des objets epais en microscopie electronique. J. Micr. Spectr. Electr. *1*, 363 (1976)
9.125 H.T. Pearce-Percy, J.M. Cowley: On the use of energy filtering to increase the contrast of STEM images of thick biological materials. Optik *44*, 273 (1976)
9.126 M. Isaacson, J.P. Langmore, H. Rose: Determination of the non-localization of the inelastic scattering of electrons by electron microscopy. Optik *41*, 92 (1974)
9.127 A.V. Crewe, J. Wall: Contrast in high resolution STEM. Optik *30*, 461 (1970)
9.128 A.El Hili: Analyse quantitative â haute résolution par images electroniques filtrées. J. Microscopie *5*, 669 (1966)
9.129 B. Jouffrey: "Electron Energy Loss Spectroscopy," in *Developments in Electron Microscopy and Analysis 1977*, ed. by D.L. Misell (The Institute of Physics, London 1977) p.351
9.130 C.J. Wilson, P.E. Batson, A.J. Craven, L.M. Brown: "Differentiated Energy Loss Spectroscopy in STEM," in *Developments in Electron Microscopy and Analysis 1977*, ed. by D.L. Misell (The Institute of Physics, London 1977) p.365
9.131 K.M. Adamson-Sharpe, F.P. Ottensmeyer: Spatial resolution and detection sensitivity in microanalysis by electron energy loss selected imaging. J. Micr. *122*, 302 (1981)
9.132 Y. Kamiya, R. Uyeda: Effect of incoherent waves on the electron microscopic image of crystals. J. Phys. Soc. Jpn. *16*, 1361 (1961)
9.133 Y. Kamiya, Y. Nakai: Diffraction contrast effect of electrons scattered inelastically through large angles. J. Phys. Soc. Jpn. *31*, 195 (1971)
9.134 S.L. Cundy, A.J.F. Metherell, M.J. Whelan: Contrast preserved by elastic and quasi-elastic scattering of fast electrons near Bragg beams. Philos. Mag. *15*, 623 (1967)
9.135 R. Castaing, P. Hénoc, L. Henry, M. Natta: Degré de cohérence de la diffusion électronique par interaction électron-phonon. C. R. Acad. Sci. Paris *265*, 1293 (1967)
9.136 S. Kuwabara, T. Uefuji: Variation of electron microscopic thickness fringes of Al single crystals with energy loss. J. Phys. Soc. Jpn. *38*, 1090 (1975)
9.137 J.B. LePoole: Ein neues Elektronenmikroskop mit stetig regelbarer Vergrößerung. Philips Tech. Rundsch. *9*, 33 (1947)
9.138 M.E. Haine, R.S. Page, R.G. Garfitt: A three-stage electron microscope with stereographic dark field and electron diffraction capabilities. J. Appl. Phys. *21*, 173 (1950)
9.139 W.D. Riecke, E. Ruska: Über ein Elektronenmikroskop mit Einrichtungen für Feinbereichsbeugung und Dunkelfeldabbildung durch Einzelreflex. Z. Wiss. Mikrosk. *63*, 288 (1957)
9.140 A.W. Agar: Accuracy of selected-area microdiffraction in the electron microscope. Br. J. Appl. Phys. *11*, 185 (1960)
9.141 W. Riecke: Über die Genauigkeit der Übereinstimmung von ausgewähltem und beugendem Bereich bei der Feinbereichs-Elektronenbeugung im LePoolschen Strahlengang. Optik *18*, 278 (1961)
9.142 W. Riecke: Verzeichnung und Auflösung der im LePoolschen Strahlengang aufgenommenen Beugungsdiagramme. Optik *18*, 373 (1961)
9.143 W.C.T. Dowell: Fehler von Beugungsdiagrammen, die mittels Elektronenlinsen erzeugt und abgebildet sind. Optik *20*, 581 (1963)
9.144 J.C. Lodder, K.G. van der Berg: A method for accurately determining lattice parameters using electron diffraction in a commercial electron microscope. J. Micr. *100*, 93 (1974)
9.145 F. Fujimoto, K. Komaki, S. Takagi, H. Koike: Diffraction patterns obtained by scanning electron microscopy. Z. Naturforsch. A*27*, 441 (1972)
9.146 A.P. Pogany, P.S. Turner: Reciprocity in electron diffraction and microscopy. Acta Cryst. A*24*, 103 (1968)

9.147 M.N. Thompson: "A Scanning Transmission Microscope: Some Techniques and Applications," in *Scanning Electron Microscopy: Systems and Applications*, ed. by W.C. Nixon (The Institute of Physics, London 1973) p.176
9.148 D.M. Maher: "Scanning Electron Diffraction in TEM and SEM Operating in the Transmission Mode," in *Scanning Electron Microscopy 1974*, ed. by O. Johari (IIT Research Inst., Chicago 1974) p.215
9.149 K.J. van Oostrum, A. Leenhouts, A. Jore: A new scanning micro-diffraction technique. Appl. Phys. Lett. *23*, 283 (1973)
9.150 R.H. Geiss: Electron diffraction from areas less than 3 nm in diameter. Appl. Phys. Lett. *27*, 174 (1975)
9.151 R.H. Geiss: "STEM Electron Diffraction from 30 Å Diameter Areas," in *Developments in Electron Microscopy and Analysis 1975*, ed. by J.A. Venables (Academic, London 1976) p.61
9.152 J.P. Chevalier, A.J. Craven: Microdiffraction, application to short range order in a quenched copper-platinum alloy. Philos. Mag. *36*, 67 (1977)
9.153 W.D. Riecke: Beugungsexperimente mit sehr feinen Elektronenstrahlen. Z. Angew. Phys. *27*, 155 (1969)
9.154 B. Bengtsson, B. Loberg, D.A. Porter, K.E. Easterling: "The Performance of a 200 kV STEM," in *Electron Microscopy 1976*, Vol.1, ed. by D.G. Brandon (Tal International, Jerusalem 1976) p.450
9.155 L.M. Brown, A.J. Craven, L.G.P. Jones, A. Griffith, W.M. Stobbs, C.J. Wilson: "Application of a High Resolution STEM to Material Science," in *Scanning Electron Microscopy 1976/I*, ed. by O. Johari (IIT Research Inst., Chicago 1976) p.353
9.156 H. von Harrach, C.E. Lyman, G.E. Verney, D.C. Joy, G.R. Booker: "Performance of the Oxford Field-Emission Scanning Transmission Electron Microscope," in *Developments in Electron Microscopy and Analysis 1975*, ed. by J.A Venables (Academic, London 1976) p.7
9.157 L. Reimer: Electron diffraction methods in TEM, STEM and SEM. Scanning *2*, 3 (1979)
9.158 W. Kossel, G. Möllenstedt: Elektroneninterferenzen im konvergenten Bündel. Naturwissenschaften *26*, 660 (1938)
9.159 W. Kossel, G. Möllenstedt: Dynamische Anomalie von Elektroneninterferenzen. Ann. Phys. *42*, 287 (1942)
9.160 P. Goodman, G. Lehmpfuhl: Elektronenbeugungsuntersuchungen im konvergenten Bündel mit dem Siemens Elmiskop I. Z. Naturforsch. A*20*, 110 (1965)
9.161 H. Raith: Elektronenbeugung im konvergenten Bündel an gekühlten Präparaten mit dem Siemens-Elmiskop I. Z. Naturforsch. A*20*, 855 (1965)
9.162 D.J.H. Cockayne, P. Goodman, J.C. Mills, A.F. Moodie: Design and generation of an electron diffraction camera for the study of small crystalline regions. Rev. Sci. Instrum. *38*, 1097 (1967)
9.163 J.M. Cowley, D.J. Smith, G.A. Sussex: "Application of a High Voltage STEM," in *Scanning Electron Microscopy 1970*, ed. by O. Johari (IIT Research Inst., Chicago 1970) p.11
9.164 P. Goodman: Observation of background contrast in convergent beam patterns. Acta Cryst. A*28*, 92 (1972)
9.165 C. van Essen: "SEM Channelling Patterns from 2 µm Selected Areas," in *Microscopie Electronique 1970*, Vol.1, ed. by P. Favard (Sociètè Francaise de Microscopie Electronique, Paris 1970) p.237
9.166 L. Reimer, P. Hagemann: "The Use of Transmitted and Backscattered Electrons in the Scanning Mode of a TEM," in *Developments in Electron Microscopy and Analysis 1977*, ed. by D.L. Misell (The Institute of Physics, London 1977) p.135
9.167 R.J. Woolf, D.C. Joy, J.M. Titchmarsh: "Scanning Transmission Electron Diffraction in the SEM," in *Electron Microscopy 1972* (The Institute of Physics, London 1972) p.498
9.168 A.J. Craven: Specimen Orientation in STEM," in *Developments in Electron Microscopy and Analysis 1977*, ed. by D.L. Misell (The Institute of Physics, London 1977) p.311
9.169 G. Möllenstedt, H.R. Meyer: Strahlengang zur Strukturanalyse von Einkristallen durch Elektronen-Transmissions-Doppelwinkelabrasterung. Optik *42*, 487 (1975)

9.170 J.A. Fades: "Another Way to Form Zone-Axis Patterns," in *Electron Microscopy and Analysis 1979*, ed. by T. Mulvey (The Institute of Physics, London 1979) p.9
9.171 H. Mahl, W. Weitsch: Kleinwinkelbeugung mit Elektronenstrahlen. Naturwissenschaften *47*, 301 (1960); Z. Naturforsch. A*15*, 1051 (1960)
9.172 R.P. Ferrier: "Small Angle Electron Diffraction in the Electron Microscope," in *Advances in Optical and Electron Microscopy*, Vol.3, ed. by R. Barer, V.E. Cosslett (Academic, London 1969) p.155
9.173 R.H. Wade, J. Silcox: Small angle electron scattering from vacuum condensed metallic films. Phys. Status Solidi *19*, 57 and 63 (1967)
9.174 J. Smart, R.E. Burge: Small-angle electron diffraction patterns of assemblies of spheres and viruses. Nature *205*, 1296 (1965)
9.175 V. Drahoš, A. Delong: "Low-Angle Electron Diffraction from Defined Specimen Area," in *Microscopie Electronique 1970*, Vol.2, ed. by P. Favard (Société Francaise de Microscopie Electronique, Paris 1970) p.147
9.176 R.T. Murray, R.P. Ferrier: Biological applications of electron diffraction. J. Ultrastruct. Res. *21*, 361 (1967)
9.177 G.A. Bassett, A. Keller: Low-angle scattering in an electron microscope applied to polymers. Philos. Mag. *9*, 817 (1964)
9.178 P.H. Denbigh, C.W.B. Grigson: Scanning electron diffraction with energy analysis. J. Sci. Instrum. *42*, 305 (1965)
9.179 L. Reimer, K. Freking: Versuch einer quantitativen Erfassung der Textur von Au-Aufdampfschichten. Z. Phys. *184*, 119 (1965)
9.180 M.F. Tompsett: Review: Scanning high-energy electron diffraction in materia science. J. Mat. Sci. *7*, 1069 (1972)
9.181 C.W. Grigson: Improved scanning electron diffraction system. Rev. Sci. Instrum. *36*, 1587 (1965)
9.182 F.C.S.M. Totthill, W.C. Nixon, C.W.B. Grigson: "Ultra-High Vacuum Modification of an AEI EM6 Electron Microscope for Studies of Nucleation in Evaporated Films," in *Electron Microscopy 1968*, Vol.1, ed. by D.S. Bocciarelli (Tipografia Poliglotta Vaticana, Rome 1968) p.229
9.183 A.M. MacLeod, J.N. Chapman: A digital scanning and recording system for spo electron diffraction patterns. J. Phys. E*10*, 37 (1977)
9.184 R.C. Newman, D.W. Pashley: The sensitivity of electron diffraction as a means of detecting thin surface films. Philos. Mag. *46*, 927 (1955)
9.185 F. Heise: Ein Zusatzgerät für Elektronenbeugung mit streifendem Einfall. Optik *9*, 139 (1952)
9.186 W. Riecke, F. Stöcklein: Eine Objektkammer mit universell beweglichem Präparattisch für Elektronenbeugungsuntersuchungen. Z. Phys. *156*, 163 (1959)
9.187 M. Eisfeldt, K.H. Herrmann, F. Thon: "Ein Elektronenbeugungsgerät als Zusatzeinrichtung zum Elmiskop I," in *Proc. European Regional Conf. on Electron Microscopy*, Vol.1, ed. by A.L. Houwink, B.J. Spit (Nederlandse Vereniging voor Electronenmicroscopie, Delft 1960) p.139
9.188 J.A. Venables, C.J. Harland: Electron back-scattering patterns - a new technique for obtaining crystal information in the SEM. Philos. Mag. *27*, 1193 (1973)
9.189 M.N. Alam, M. Blackman, D.W. Pashley: High-angle Kikuchi patterns. Proc. Roy Soc. A*221*, 224 (1954)
9.190 L. Reimer, W. Pöpper, B. Volbert: "Contrast Reversals in the Kikuchi Bands of Backscattered and Transmitted Electron Diffraction Patterns," in *Developments in Electron Microscopy and Analysis 1977*, ed. by D.L. Misell (The Institute of Physics, London 1977) p.259
9.191 D.G. Coates: Kikuchi-like reflection patterns obtained with the SEM. Philos Mag. *16*, 1179 (1967)
9.192 G.R. Booker: "Scanning Electron Microscopy: Electron Channelling Effects," in *Modern Diffraction and Imaging Techniques in Material Science*, ed. by S. Amelinckx (North-Holland, Amsterdam 1970) p.613
9.193 L. Reimer: "Electron Specimen Interactions in SEM," in *Developments in Electron Microscopy and Analysis*, ed. by J.A. Venables (Academic, London 1976) p.83
9.194 J.W. Steeds, G.J. Tatlock, J. Hampson: Real space crystallography. Nature *241*, 435 (1973)

9.195 G.J. Tatlocks, J.W. Steeds: Real space crystallography in molybdenite. Nature Phys. Sci. *246*, 126 (1973)

9.196 J.W. Steeds, P.M. Jones, G.M. Rackham, M.D. Shannon: "Crystallographic Information from Zone Axis Patterns," in *Developments in Electron Microscopy and Analysis*, ed. by J.A. Venables (Academic, London 1976) p.351

9.197 J.W. Steeds, P.M. Jones, J.E. Loveluck , K.E. Cooke: The dependence of zone axis patterns on string integrals or the number of bound states in high energy electron diffraction. Philos. Mag. *36*, 309 (1977)

9.198 M.D. Shannon, J.W. Steeds: On the relationship between projected crystal potential and the form of certain zone axis patterns in high energy electron diffraction. Philos. Mag. *36*, 279 (1977)

9.199 W. Witt: Zur absoluten Präzisionsbestimmung von Gitterkonstanten mit Elektroneninterferenzen am Beispiel von Thallium-(I)-Chlorid. Z. Naturforsch. A*19*, 1363 (1964)

9.200 J.M. Corbett, F.W. Boswell: Use of thin single crystals as reference standards for precision electron diffraction. J. Appl. Phys. *37*, 2016 (1966)

9.201 A.L. MacKay. Calibration of diffraction patterns taken in the electron microscope. J. Phys. E*3*, 248 (1970)

9.202 J.T. Jubb, E.E. Laufer: The beam-tilt device of an electron microscope as an internal diffraction standard. J. Phys. E*9*, 871 (1976)

9.203 E.E. Laufer, J.T. Jubb, K.S. Milliken: The use of the beam tilt circuitry of an electron microscope for rapid determination of lattice constants. J. Phys. E*8*, 671 (1975)

9.204 H. König: Gitterkonstantenbestimmung im Elektronenmikroskop. Naturwissenschaften *33*, 343 (1946)

9.205 F.W.C. Bosswell: A standard substance for precise electron diffraction measurements. Phys. Rev. *80*, 91 (1950)

9.206 C. Lu, E.W. Malmberg: ZnO smoke as a reference standard in electron wavelength calibration. Rev. Sci. Instrum. *14*, 271 (1943)

9.207 R. Rühle: Über Gesetzmäßigkeiten in Texturaufnahmen von Elektronenbeugungsbildern. Optik *7*, 279 (1950)

9.208 Z.G. Pinsker: *Electron Diffraction*, translated by J.A. Spink, E. Feigl (Butterworths, London 1953)

9.209 B.K. Vainshtein: *Structure Analysis by Electron Diffraction*, translated by E. Feigl, J.A. Spink (Pergamon, Oxford 1964)

9.210 J.A. Gard: Interpretation of electron micrographs and diffraction patterns: the electron optical investigation of clays. Mineralogical Soc. London (1971)

9.211 J.M. Cowley: Crystal structure determination by electron diffraction. Prog. Mater. Sci. *13*, 267 (1966)

9.212 S. Nagakura: A method for correcting the primary extinction effect in electron diffraction. Acta Cryst. *10*, 601 (1957)

9.213 B.K. Vainshtein, A.N. Lobacher: Dynamic scattering and its use in structural electron diffraction studies. Sov. Phys. Cryst. *6*, 609 (1961)

9.214 J.M. Cowley: Structure analysis of single crystals by electron diffraction. Acta Cryst. *6*, 516, 522, and 846 (1953)

9.215 J.M. Cowley: "The Theoretical Basis for Electron Diffraction Structure Analysis," in *Electron Microscopy 1962*, Vol.1, ed. by S.S. Breese (Academic, New York 1962) p.JJ-1

9.216 S. Fujime, D. Watanabe, S. Ogawa: On forbidden reflection spots and unexpected streaks appearing in electron diffraction patterns from hexagonal Co. J. Phys. Soc. Jpn. *19*, 711 (1964)

9.217 J.F. Brown, D. Clark: The use of the three-stage electron microscope in crystal-structure analysis. Acta Cryst. *5*, 615 (1952)

9.218 J.A. Gard: The use of the stereoscopic tilt device of the electron microscope in unit-cell determinations. Br. J. Appl. Phys. *7*, 361 (1956)

9.219 J.A. Gard: "Interpretation of Electron Diffraction Patterns," in *Electron Microscopy in Mineralogy*, ed. by H.W. Wenk (Springer, Berlin, Heidelberg, New York 1976) p.52

9.220 R.R. Dayal, J.A. Gard, F.P. Glasser: Crystal data on $FeAlO_3$. Acta Cryst. *18*, 574 (1965)

9.221 J.A. Gard, J.M. Bennet: "A Goniometric Specimen Stage, and Its Use in Crystallography," in *Electron Microscopy 1966*, Vol.1, ed. by R. Uyeda (Maruzen, Tokyo 1966) p.593
9.222 G. Cliff, J.A. Gard, G.W. Lorimer, H.F.W. Taylor: Tacharanite. Mineral. Mag. *40*, 113 (1975)
9.223 S. Kuwabara: Accurate determination of hydrogen positions in NH_4Cl by electron diffraction. J. Phys. Soc. Jpn. *14*, 1205 (1959)
9.224 V.V. Udalova, Z.G. Pinsker: Electron diffraction study of the structure of ammonium sulfate. Sov. Phys. Cryst. *8*, 433 (1963)
9.225 J.A. Gard, H.F.W. Taylor, L.W. Staples: "Studies in Crystal Structure Using Electron Diffraction of Single Crystals," in *Vierter Internationaler Kongreß für Elektronenmikroskopie Berlin 1958*, Vol.1, ed. by W. Bargmann et al. (Springer, Berlin, Göttingen, Heidelberg 1960) p.449
9.226 H.M. Otte, J. Dash, H.F. Schaake: Electron microscopy and diffraction of thin films. Interpretation and correlation of images and diffraction patterns. Phys. Status Solidi *5*, 527 (1964)
9.227 C. Laird, E. Eichen, W.R. Bitler: Accuracy in the use of electron diffraction spot patterns for determining crystal orientations. J. Appl. Phys. *37*, 2225 (1966)
9.228 K. Lücke, H. Perlwitz, W. Pitsch: Elektronenmikroskopische Bestimmung der Orientierungsverteilung der Kristallite in gewalztem Kupfer. Phys. Status Solidi *7*, 733 (1964)
9.229 F. Haessner, U. Jakubowksi, M. Wilkens: Anwendung elektronenmikroskopischer Feinbereichsbeugung zur Ermittlung der Walztextur von Kupfer. Phys. Status Solidi *7*, 701 (1964)
9.230 P.L. Ryder, W. Pitsch: The uniqueness of orientation determination by selected area electron diffraction. Philos. Mag. *15*, 437 (1967)
9.231 P.L. Ryder, W. Pitsch: On the accuracy of orientation determination by selected area diffraction. Philos. Mag. *18*, 807 (1968)
9.232 D.J. Mazey, R.S. Barnes, A. Howie: On interstitial dislocation loops in aluminium bombarded with alpha-particles. Philos. Mag. *7*, 1861 (1962)
9.233 M.H. Loretto, L.M. Clarebrough, P. Humble: Nature of dislocation loops in quenched Al. Philos. Mag. *13*, 953 (1966)
9.234 M. von Heimendahl: Determination of metal foil thickness and orientation in electron microscopy. J. Appl. Phys. *35*, 457 (1964)
9.235 S.S. Sheinin, C.D. Cann: The determination of orientation from Kikuchi patterns. Phys. Status Solidi *11*, K1 (1965)
9.236 R. Bonnet, F. Durand: Precise determination of the relative orientation of two crystals from the analysis of two Kikuchi patterns. Phys. Status Solidi A*27*, 543 (1975)
9.237 W. Griem, P. Schwaab, U. Stockhofe: Behandlung von Epitaxie-Fragen bei der Elektronenbeugung mit Hilfe der Datenverarbeitung. Arch. Eisenhüttenwesen *43*, 509 (1972)
9.238 W. Griem, P. Schwaab: Behandlung von gesetzmäßigen Verwachsungen nichtkubischer und teilkohärenter Phasen bei der Elektronenbeugung. Arch. Eisenhüttenwesen *44*, 677 (1973)
9.239 R. Bonnet, E.E. Laufer: Precise determination of the relative orientation of two crystals from the analysis of spot diffraction patterns. Phys. Status Solidi A*40*, 599 (1977)
9.240 M.D. Drazin, M.H. Otte: The systematic determination of crystallographic orientations from three octahedral traces on a plane surface. Phys. Status Solidi *3*, 824 (1963)
9.241 A.G. Crocker, M. Bevis: The determination of the orientation and thickness of thin foils from transmission electron micrographs. Phys. Status Solidi *6*, 151 (1964)
9.242 G. Thomas: *Transmission Electron Microscopy of Metals* (Wiley, New York 1962)
9.243 A. Baltz: Rotation of image and selected area diffraction patterns in the RCA-EMU3 electron microscope. Rev. Sci. Instrum. *33*, 246 (1962)
9.244 P. Delavignette: Determination of some instrumental constants of the electron microscope Philips EM 200. J. Sci. Instrum. *40*, 461 (1963)
9.245 H. Raether: Reflexion von schnellen Elektronen an Einkristallen. Z. Phys. *78*, 527 (1932)

9.246 R.D. Heidenreich: Theory of the 'forbidden' (222) electron reflection in the diamond structure. Phys. Rev. *77*, 271 (1950)
9.247 M. Takagi, S. Morimoto: The forbidden 222 electron reflection from Ge. J. Phys. Soc. Jpn. *18*, 819 (1963)
9.248 H. Göttsche: Zur Struktur dünner Ag-Schichten. Z. Phys. *134*, 517 (1953)
9.249 W. Pitsch: Kristallographische Eigenschaften von Eisennitrid-Ausscheidungen im Ferrit. Arch. Eisenhüttenwesen *32*, 493 and 573 (1961)
9.250 S. Ogawa, D. Watanabe, H. Watanabe, T. Komoda: "The Direct Observation of the Long Period of the Ordered Alloy CuAu (II) by Means of Electron Microscope," in *Vierter Internationaler Kongreß für Elektronenmikroskopie Berlin 1958*, Vol.1, ed. by W. Bargmann et al. (Springer, Berlin, Göttingen, Heidelberg 1960) p.334
9.251 D.W. Pashley, A.E.B. Presland: The observation of antiphase boundaries during the transition from CuAu I to CuAu II. J. Inst. Met. *87*, 419 (1959)
9.252 S. Ogawa: On the antiphase domain structures in ordered alloys. J. Phys. Soc. Jpn. *17*, Suppl.B-II, 253 (1962)
9.253 I. Ackermann: Beobachtungen an dynamischen Interferenzerscheinungen im konvergenten Elektronenbündel. Ann. Phys. *2*, 19 and 41 (1948)
9.254 P.M. Kelly, A. Jostsons, R.G. Blake, J.G. Napier: The determination of foil thickness by STEM. Phys. Status Solidi A*31*, 771 (1975)
9.255 R.G. Blake, A. Jostsons, P.M. Kelly, J.G. Napier: The determination of extinction distances and anomalous absorption coefficients by STEM. Philos. Mag. A*37*, 1 (1978)
9.256 J.W. Steeds, K.K. Fung: "Application of Convergent Beam Electron Microscopy in Materials Science," in *Electron Microscopy 1978*, Vol.1, ed. by J.M. Sturgess (Microscopical Soc. Canada, Toronto 1978) p.620
9.257 J.W. Steeds: "Convergent Beam Electron Diffraction," in *Analytical Electron Microscopy*, ed. by J.J. Hren, J.I. Goldstein, D.C. Joy (Plenum, New York 1979) p.387
9.258 P.M. Jones, G.M. Rackham, J.W. Steeds: Higher order Laue zone effects in electron diffraction and their use in lattice parameter determination. Proc. Roy. Soc. A*354*, 197 (1977)
9.259 B.F. Buxton: Bloch waves and higher order Laue zone effects in high energy electron diffraction. Proc. Roy. Soc. A*350*, 335 (1976)
9.260 J.W. Steeds: "Information About the Crystal Potential from Zone Axis Patterns," in *Electron Microscopy 1980*, Vol.4, ed. by P. Brederoo, J. Van Landuyt (Seventh European Congr. Electron Microscopy Foundation, Leiden 1980) p.96
9.261 J.R. Baker, S. McKernan: "Structure Factor Information from HOLZ Beam Intensities in Convergent-Beam HEED," in *Electron Microscopy and Analysis 1981*, ed. by M.J. Goringe (The Institute of Physics, London 1982) p.283
9.262 G.M. Rackham, P.M. Jones, J.W. Steeds: "Upper Layer Diffraction Effects in Zone Axis Patterns," in *Electron Microscopy 1974*, Vol.1, ed. by J.V. Sanders, D.J. Goodchild (Australian Acad. Sci., Canberra 1974) p.336 and 355
9.263 J.E. Loveluck, J.W. Steeds: "Crystallography of Lithium Tantalate and Quartz," in *Developments in Electron Microscopy and Analysis 1977*, ed. by D.L. Misell (The Institute of Physics, London 1977) p.293
9.264 G.M. Rackham, J.W. Steeds: "Convergent Beam Observation near Boundaries and Interfaces," in *Development in Electron Microscopy and Analysis*, ed. by J.A. Venables (Academic, London 1976) p.457
9.265 P. Goodman: A practical method for three-dimensional space-group analysis using convergent beam electron diffraction. Acta Cryst. A*31*, 804 (1975)
9.266 B.F. Buxton, J.A. Eades, J.W. Steeds, G.M. Rackham: The symmetry of electron diffraction zone axis patterns. Philos. Trans. Roy. Soc. A*281*, 171 (1976)
9.267 S.J. Pennycook, L.M. Brown, A.J. Craven: Observation of cathodoluminescence at single dislocations by STEM. Philos. Mag. A*41*, 589 (1980)
9.268 P.M. Petroff, D.V. Lang, J.L. Strudel, R.A. Logan: "Scanning Transmission Electron Microscopy Techniques for Simultaneous Electronic Analysis and Observation of Defects in Semiconductors," in *Scanning Electron Microscopy 1978/I*, ed. by O. Johari (SEM Inc., AMF O'Hare 1978) p.325
9.269 S.J. Pennycook, A. Howie: Study of single electron excitations by electron microscopy. Philos. Mag A*41*, 809 (1980)

9.270 A. Ourmazd, G.R. Booker: The electrical recombination efficiency of individual edge dislocations and stacking fault defects in n-type silicon. Phys. Status Solidi A*55*, 771 (1979)
9.271 H. Blumtritt, R. Gleichmann, J. Heydenreich, J. Johansen: Combined scanning (EBIC) and transmission electron microscopic investigations of dislocations in semiconductors. Phys. Status Solidi A*55*, 611 (1979)
9.272 T.G. Sparrow, U. Valdrè: Application of scanning transmission electron microscopy to semiconductor devices. Philos. Mag. *36*, 1517 (1977)
9.273 P.M. Petroff, D.V. Lang: A new spectroscopic technique for imaging the spatial distribution of nonradiative defects in a scanning transmission electron microscope. Appl. Phys. Lett. *31*, 60 (1977)
9.274 M.J. Leamy: Charge collection scanning electron microscopy. J. Appl. Phys. Phys. *53*, R51 (1982)

Chapter 10

10.1 I.G. Stojanowa, E.M. Belawzewa: "Experimentelle Untersuchung der thermischen Einwirkung des Elektronenstrahls auf das Objekt im Elektronenmikroskop," in *Vierter Internationaler Kongreß für Elektronenmikroskopie Berlin 1958*, Vol.1, ed. by W. Bargmann et al. (Springer, Berlin, Göttingen, Heidelberg 1960) p.100
10.2 M. Watanabe, T. Someya, Y. Nagahama: "Temperature Rise of Specimen Due to Electron Irradiation," in *Electron Microscopy 1962*, Vol.1, ed. by S.S. Breese (Academic, New York 1962) p.A-8
10.3 D.D. Thornburg, C.M. Wayman: Specimen temperature increases during transmission electron microscopy. Phys. Status Solidi A*15*, 449 (1973)
10.4 L. Reimer, R. Christenhusz, J. Ficker: Messung der Objekttemperatur im Elektronenmikroskop mittels Elektronenbeugung. Naturwissenschaften *47*, 464 (1960)
10.5 M. Fukamachi, T. Kikuchi: Application of the critical voltage effect to the measurement of temperature increase of metal foils during the observation with HVEM. Jpn. J. Appl. Phys. *14*, 587 (1975)
10.6 A. Winkelmann: Messung der Temperaturerhöhung der Objekte bei Elektronen-Interferenzen. Z. Angew. Phys. *8*, 218 (1956)
10.7 E. Gütter, H. Mahl: Einfluß einer periodischen Objektbeleuchtung auf die elektronenmikroskopische Abbildung. Optik *17*, 233 (1960)
10.8 G. Honjo, N. Kitamura, K. Shimaoka, K. Mihama: Low temperature specimen method for electron diffraction and electron microscopy. J. Phys. Soc. Jpn. *11*, 527 (1956)
10.9 L. Reimer, R. Christenhusz: Reversible Temperaturindikatoren in Form von Aufdampfschichten zur Ermittlung der Objekttemperatur im Elektronenmikroskop. Naturwissenschaften *48*, 619 (1961)
10.10 L. Reimer, R. Christenhusz: Experimenteller Beitrag zur Objekterwärmung im Elektronenmikroskop. Z. Angew. Phys. *14*, 601 (1962)
10.11 S. Yamaguchi: Über die Temperatur-Erhöhung der Objekte im Elektronenstrahl. Z. Angew. Phys. *8*, 221 (1956)
10.12 S. Leisegang: "Zur Erwärmung elektronenmikroskopischer Objekte bei kleinem Strahlquerschnitt," in *Proc. 3rd Intern. Conf. Electron Microscopy, London 1954*, ed. by R. Ross (Royal Microscopical Society, London 1956) p.176
10.13 B. Gale, K.F. Hale: Heating of metallic foils in an electron microscope. Br. J. Appl. Phys. *12*, 115 (1961)
10.14 K. Kanaya: The temperature distribution of along a rod-specimen in the electron microscope. J. Electr. Micr. Jpn. *4*, 1 (1956)
10.15 P. Balk, J. Ross Colvin: Note on an indirect measurement of object temperature in electron microscopy. Kolloid. Z. *176*, 141 (1961)
10.16 L. Reimer: Zur Zersetzung anorganischer Kristalle im Elektronenmikroskop. Z. Naturforsch. A*14*, 759 (1959)
10.17 L. Reimer: Ein experimenteller Beitrag zur Thermokraft dünner Aufdampfschichten. Z. Naturforsch. A*12*, 525 (1957)

10.18 D. Thornburg, C.M. Wayman: Thermoelectric power of vacuum evaporated Au-Ni thin film thermocouples. J. Appl. Phys. *40*, 3007 (1969)
10.19 G.R. Piercy, R.W. Gilbert, L.M. Howe: A liquid helium cooled finger for the Siemens electron microscope. J. Sci. Instrum. *40*, 487 (1963)
10.20 G.M. Parkinson, W. Jones, J.M. Thomas: "Electron Microscopy at Liquid Helium Temperatures," in *Electron Microscopy at Molecular Dimensions*, ed. by W. Baumeister, W. Vogell (Springer, Berlin, Heidelberg, New York 1980) p.208
10.21 S. Kritzinger, E. Ronander: Local beam heating in metallic electron microscope specimens. J. Micr. *102*, 117 (1974)
10.22 K. Kanaya: The temperature distribution of specimens on thin substrates supported on a circular opening in the electron microscope. J. Electr. Micr. Jpn. *3*, 1 (1955)
10.23 L. Reimer, R. Christenhusz: Determination of specimen temperature. Lab. Invest. *14*, 1158 (1965)
10.24 R. Christenhusz, L. Reimer: Schichtdickenabhängigkeit der Wärmeerzeugung durch Elektronenbestrahlung im Energiebereich zwischen 9 und 100 keV. Z. Angew. Phys. *23*, 397 (1967)
10.25 S. Leisegang: "Elektronenmikroskope", in *Handbuch der Physik*, Vol.33 (Springer, Berlin, Göttingen, Heidelberg 1956) p.396
10.26 V.E. Cosslett, R.N. Thomas: Multiple scattering of 5-30 keV electrons in evaporated metal films. II. Range-energy relations. Br. J. Appl. Phys. *15*, 1283 (1964)
10.27 L. Reimer: Monte-Carlo-Rechnungen zur Elektronendiffusion. Optik *27*, 86 (1968)
10.28 B. von Borries, W. Glaser: Über die Temperaturerhöhung der Objekte im Elektronenmikroskop. Kolloid. Z. *106*, 123 (1944)
10.29 J. Ling: A calculation of the temperature distribution in electron microscope specimens. Br. J. Appl. Phys. *18*, 991 (1967)
10.30 A. Brockes: Zur Objekterwärmung im Elektronenmikroskop. Kolloid. Z. *158*, 1 (1958)
10.31 R. Christenhusz, L. Reimer: Wärmeleitfähigkeit elektronenmikroskopischer Trägerfolien. Naturwissenschaften *55*, 439 (1968)
10.32 E. Knapek, J. Dubochet: Beam damage to organic material is considerably reduced in cryo-electron microscopy. J. Mol. Biol. *141*, 147 (1980)
10.33 I. Dietrich, F. Fox, H.G. Heide, E. Knapek, R. Weyl: Radiation damage due to knock-on processes on carbon foils cooled to liquid helium temperature. Ultramicroscopy *3*, 185 (1978)
10.34 Y. Talmon, E.L. Talmon: "Temperature Rise and Sublimation of Water from Thin Frozen Hydrated Specimens in Cold Stage Microscopy," in *Scanning Electron Microscopy 1977/I*, ed. by O. Johari (IIT Research Inst., Chicago 1977) p.265
10.35 Y. Talmon, E.L. Thomas: Beam heating of a moderately thick cold stage specimen in the SEM/STEM. J. Micr. *111*, 151 (1977)
10.36 L.G. Pittaway: The temperature distributions in the foil and semi-infinite targets bombarded by an electron beam. Br. J. Appl. Phys. *15*, 967 (1964)
10.37 H. Kohl, H. Rose, H. Schnabl: Dose-rate effect at low temperatures in FBEM and STEM due to object-heating. Optik *58*, 11 (1981)
10.38 L. Reimer: Irradiation changes in organic and inorganic objects. Lab. Invest. *14*, 1082 (1965)
10.39 K. Stenn, G.F. Bahr: Specimen damage caused by the beam of the transmission electron microscope, a correlative consideration. J. Ultrastruct. Res. *31*, 526 (1970)
10.40 D.T. Grubb, A. Keller: "Beam-Induced Radiation Damage in Polymers and Its Effect on the Image Formed in the Electron Microscope," in *Electron Microscopy 1972* (The Institute of Physics, London 1972) p.554
10.41 R.M. Glaeser: "Radiation Damage and Biological Electron Microscopy," in *Physical Aspects of Electron Microscopy and Microbeam Analysis*, ed. by B.M. Siegel, D.R. Beaman (Wiley, New York 1975) p.205
10.42 L. Reimer: "Review of the Radiation Damage Problem of Organic Specimens in Electron Microscopy," in [Ref.10.41], p.231
10.43 M.S. Isaacson: "Inelastic Scattering and Beam Damage of Biological Molecules," in [Ref.10.41], p.247

10.44 D.F. Parsons: "Radiation Damage in Biological Materials," in [Ref.10.41], p.259]
10.45 M.S. Isaacson: "Specimen Damage in the Electron Microscope," in *Principles and Techniques of Electron Microscopy*, Vol.7, ed. by M.A. Hayat (Van Nostrand-Reinhold, New York 1977) p.1
10.46 R.M. Glaeser, K.A. Taylor: Radiation damage relative to transmission electron microscopy of biological specimens at low temperature: a review. J. Micr. *112*, 127 (1978)
10.47 V.E. Cosslett: Radiation damage in the high resolution electron microscopy of biological materials: a review. J. Micr. *113*, 113 (1978)
10.48 W. Baumeister, W. Vogell (eds.): *Electron Microscopy at Molecular Dimensions* (Springer, Berlin, Heidelberg, New York 1980)
10.49 Z.M. Bacq, P. Alexander: *Fundamentals of Radiobiology* (Pergamon, Oxford 19
10.50 R.D. Bolt, J.G. Carroll (eds.): *Radiation Effects on Organic Materials* (Academic, New York 1963)
10.51 A. Charlesby: *Atomic Radiation and Polymers* (Pergamon, Oxford 1960)
10.52 A.J. Swallow: *Radiation Chemistry of Organic Compounds* (Pergamon, Oxford 1960)
10.53 H. Dertinger, H. Jung: *Molekulare Strahlenbiologie* (Springer, Berlin, Heidelberg, New York 1968)
10.54 H.C. Box: "Cryoprotection of Irradiated Specimens," in *Physical Aspects of Electron Microscopy and Microbeam Analysis*, ed. by B.M. Siegel, D.R. Beaman (Wiley, New York 1975) p.279
10.55 R. Spehr, H. Schnabl: Zur Deutung der unterschiedlichen Strahlen-Empfindlichkeit organischer Moleküle. Z. Naturforsch. A*28*, 1729 (1973)
10.56 H. Schnabl: Does removal of hydrogen change the electron energy-loss spect of DNA bases? Ultramicroscopy *5*, 147 (1980)
10.57 G.M. Parkinson, M.J. Goringe, W. Jones, W. Rees, J.M. Thomas, J.O. Willian "Electron Induced Damage in Organic Molecular Crystals: Some Observations and Theoretical Considerations," in *Developments in Electron Microscopy and Analysis*, ed. by J.A. Venables (Academic, London 1976) p.315
10.58 T. Gejvall, G. Löfroth: Radiation induced degradation of some crytstallin amino acids. Radiat. Eff. *25*, 187 (1975)
10.59 L. Reimer: Quantitative Untersuchung zu Massenabnahme von Einbettungsmitteln (Methacrylat, Vestopal und Araldit) unter Elektronenbeschuß. Z. Naturforsch. B*14*, 566 (1959)
10.60 W. Lippert: Über thermisch bedingte Veränderungen an dünnen Folien im Elektronenmikroskop. Z. Naturforsch. A*15*, 612 (1960)
10.61 G.F. Bahr, F.B. Johnson, E. Zeitler: The elementary composition of organi objects after electron irradiation. Lab. Invest. *14*, 1115 (1965)
10.62 K. Ramamurti, A.V. Crewe, M.S. Isaacson: Low temperature mass loss of thi films of l-phenylalanine and l-tryptophan upon electron irradiation. Ultramicroscopy *1*, 156 (1975)
10.63 R. Freeman, K.R. Leonard: Comparative mass measurement of biological mac molecules by scanning transmission electron microscopy. J. Micr. *122*, 275 (1981)
10.64 W. Lippert: Über Massendickeveränderungen bei Kunststoffen im Elektronenmikroskop. Optik *19*, 145 (1962)
10.65 G. Siegel: Der Einfluß tiefer Temperaturen auf die Strahlenschädigung vor organischen Kristallen durch 100 keV-Elektronen. Z. Naturforsch. A*27*, 325 (1972)
10.66 A. Brockes. Über Veränderungen des Aufbaus organischer Folien durch Elektronen-Bestrahlung. Z. Phys. *149*, 353 (1957)
10.67 A. Cosslett: "The Effect of the Electron Beam on Thin Sections," in *Proc. Europ. Reg. Conf. on Electron Microscopy*, Vol.2, ed. by A.L. Houwink, B.J. Spit (Nederlandse Vereniging voor Electronenmicroscopie, Delft 1960 p.678
10.68 L. Reimer: Interferenzfarben von Methacrylatschnitten und ihre Veränderu unter Elektronenbeschuß. Photogr. Wiss. *9*, 25 (1960)
10.69 L. Reimer: Veränderungen organischer Kristalle unter Beschuß mit 60 keV Elektronen im Elektronenmikroskop. Z. Naturforsch. A*15*, 405 (1960)

10.70 K. Kobayashi, K. Sakaoku: Irradiation changes in organic polymers at various accelerating voltages. Lab. Invest. *14*, 1097 (1965)

10.71 H. Orth, E.W. Fischer: Änderungen der Gitterstruktur hochpolymerer Einkristalle durch Bestrahlung im Elektronenmikroskop. Makromol. Chem. *88*. 188 (1965)

10.72 L. Reimer, J. Spruth: Information about radiation damage of organic molecules by electron diffraction. J. Micr. Spectr. Electron. *3*, 579 (1978)

10.73 W.R.K. Clark, J.N. Chapman, A.M. MacLeod, R.P. Ferrier: Radiation damage mechanism in copper phthalocyanine and its chlorinated derivatives. Ultramicroscopy *5*, 195 (1980)

10.74 N. Uyeda, T. Kobayashi, E. Suito, Y. Harada, M. Watanabe: Molecular image resolution in electron microscopy. J. Appl. Phys. *43*, 5181 (1972)

10.75 T. Kobayashi, Y. Fujiyoshi, K. Ishizuka, N. Uyeda: "Structure Determination and Atom Identification on Polyhalogenated Molecule," in *Electron Microscopy 1980*, Vol.4, ed. by P. Brederoo, J. Van Landuyt (Seventh European Congr. on Electron Microscopy Foundation, Leiden 1980) p.158

10.76 Y. Murata: "Studies of Radiation Damage Mechanisms by Optical Diffraction Analysis and High Resolution Image," in *Electron Microscopy 1980*, Vol.3, ed. by J.M. Sturgess (Microscopical Soc. Canada, Toronto 1980) p.49

10.77 R.M. Glaeser: Limitations to significant information in biological electron microscopy as a result of radiation damage. J. Ultrastruct. Res. *36*, 466 (1971)

10.78 D.T. Grubb, G.W. Groves: Rate of damage of polymer crystals in the electron microscope: dependence on temperature and beam voltage. Philos. Mag. *24*, 815 (1971)

10.79 L.E. Thomas, C.J. Humphreys, W.R. Duff, D.T. Grubb: Radiation damage of polymers in the million volt electron microscope. Radiat. Eff. *3*, 89 (1970)

10.80 A.V. Crewe, M. Isaacson, D. Johnson: "Electron Beam Damage in Biological Molecules," in *Proc. 28th Annual Meeting of EMSA* (Claitor's Publ. Div., Baton Rouge, LO 1970) p.264

10.81 L. Reimer: Veränderungen organischer Farbstoffe im Elektronenmikroskop. Z. Naturforsch. B*16*, 166 (1961)

10.82 N. Uyeda, T. Kobayashi, M. Ohara, M. Watanabe, T. Taoka, Y. Harada: "Reduced Radiation Damage of Halogenated Copper-Phthalocyanine," in *Electron Microscopy 1972* (The Institute of Physics, London 1972) p.566

10.83 W. Baumeister, U.P. Fringeli, M. Hahn, F. Kopp, J. Seredynski: Radiation damage in tripalmitin layers studied by mans of infrared spectroscopy and electron microscopy. Biophys. J. *16*, 791 (1976)

10.84 W. Baumeister, J. Seredynski: "Radiation Damage to Proteins: Changes on the Primary and Secondary Structure Level," in *Electron Micrsocopy 1980*, Vol.3, ed. by J.M. Sturgess (Microscopical Soc. Canada, Toronto 1980) p.40

10.85 W. Baumeister, M. Hahn, J. Seredynski, L.M. Herbertz: Radiation damage of proteins in the solid state: changes of amino acid composition in catalase. Ultramicroscopy *1*, 377 (1976)

10.86 M. Isaacson: Electron beam induced damage of organic solids: implications for analytical electron microscopy. Ultramicroscopy *4*, 193 (1979)

10.87 R.F. Egerton: Chemical measurements of radiation damage in organic samples at and below room temperature. Ultramicroscopy *5*, 521 (1980)

10.88 R.F. Egerton: Organic mass loss at 100 K and 300 K. J. Micr. *126*, 95 (1982)

10.89 H. Shuman, A.V. Somlyo, P. Somlyo: Quantitative electron probe microanalysis of biological thin sections: methods and validity. Ultramicroscopy *1*, 317 (1976)

10.90 T.A. Hall, B.L. Gupta: Beam-induced loss of organic mass under electron-microscope conditions. J. Micr. *100*, 177 (1974)

10.91 P. Bernsen, L. Reimer, P.F. Schmidt: Investigation of electron irradiation damage of evaporated organic films by laser microprobe mass analysis. Ultramicroscopy *7*, 197 (1981)

10.92 S.H. Faraj, S.M. Salih: Spectroscopy of electron irradiated polymers in electron microscope. Radiat. Eff. *55*, 149 (1981)

10.93a P.K. Haasma, M. Parikh: A tunneling spectroscope study of molecular degradation due to electron irradiation. Science *188*, 1304 (1975)

10.93b T. Wolfram (ed.): *Inelastic Electron Tunneling Spectroscopy*, Springer Ser. Solid-State Sci., Vol.4 (Springer, Berlin, Heidelberg, New York 1978)
10.94 M.J. Richardson, K. Thomas: "Aspects of HVEM of Polymers," in *Electron Microscopy 1972* (The Institute of Physics, London 1972) p.562
10.95 S.M. Salih, V.E. Cosslett: "Some Factors Influencing Radiation Damage in Organic Substances," in *Electron Microscopy 1974*, Vol.2, ed. by J.V. Sanders, D.J. Goodchild (Australian Acad. Sci., Canberra 1974) p.670
10.96 V.E. Cosslett, G.L. Jones, R.A. Camps: "Image Viewing and Recording in High Voltage Electron Microscopy," in *High Voltage Electron Microscopy*, ed. by P.R. Swann, C.J. Humphreys, M.J. Goringe (Academic, London 1974) p.147
10.97 M.V. King, D.F. Parsons: Design features of a photographic film optimized for the high-voltage electron microscope. Ultramicroscopy *2*, 371 (1977)
10.98 M. Fotino: "Improved Response of Photographic Emulsions for Electron Micrographs at Higher Voltages," in *Electron Microscopy 1974*, Vol.1, ed. by J.V. Sanders, D.J. Goodchild (Australian Acad. Sci., Canberra 1974) p.104
10.99 M.V. King, D.V. Parsons: Recording of electron-diffraction patterns of radiation-sensitive materials in the high-voltage electron microscope with luminescent radiographic screens. J. Appl. Cryst. *10*, 62 (1977)
10.100 S.M. Salih, V.E. Cosslett: "Studies on Beam Sensitive Substances," in *Developments in Electron Microscopy and Analysis*, ed. by J.A. Venables (Academic, London 1976) p.311
10.101 J. Dubochet: Carbon loss during irradiation of T4 bacteriophages and E.coli bacteria in electron microscopes. J. Ultrastruct. Res. *52*, 276 (1975)
10.102 W. Chiu, E. Knapek, T.W. Jeng, I. Dietrich: Electron radiation damage of a thin protein crystal at 4 K. Ultramicroscopy *6*, 291 (1981)
10.103 J. Dubochet, E. Knapek, I. Dietrich: Reduction of beam damage by cryoprotection at 4 K. Ultramicroscopy *6*, 77 (1981)
10.104 I. Dietrich, J. Dubochet, F. Fox, E. Knapek, R. Weyl: "Reduction of Radiation Damage by Imaging with a Superconducting Lens System," in *Electron Microscopy at Molecular Dimensions*, ed. by W. Baumeister, W. Vogell (Springer, Berlin, Heidelberg, New York 1980) p.234
10.105 I. Dietrich, F. Fox, H.G. Heide, E. Knapek, R. Weyl: Radiation damage due to knock-on processes on carbon foils cooled to liquid helium temperature. Ultramicroscopy *3*, 185 (1978)
10.106 D.F. Parsons, V.R. Matricardi, R.C. Moretz, J.N. Turner: Electron microscopy and diffraction of wet unstained and unfixed biological objects. Adv. Biol. Med. Phys. *15*, 161 (1974)
10.107 H.G. Heide, S. Grund: Eine Tiefkühlkette zum Überführen von wasserhaltigen biologischen Objekten ins Elektronenmikroskop. J. Ultrastruct. Res. *48*, 259 (1974)
10.108 K.A. Taylor, R.M. Glaeser: Electron microscopy of frozen hydrated biological specimens. J. Ultrastruct. Res. *55*, 448 (1976)
10.109 T.E. Hutchinson, D.E. Johnson, A.P. Mackenzie: Instrumentation for direct observation of frozen hydrated specimens in the electron microscope. Ultramicroscopy *3*, 315 (1978)
10.110 S.M. Salih, V.E. Cosslett: Reduction in electron irradiation damage to organic compounds by conducting coatings. Philos. Mag. *30*, 225 (1974)
10.111 R.S. Hartman, R.E. Hartman, H. Alsberg, R. Nathan: "The Improved Stability of an Organic Crystal in the Hitachi HV-1 High Vacuum Electron Microscope," in *Electron Microscopy 1974*, Vol.2, ed. by J.V. Sanders, D.J. Goodchild (Australian Acad. Sci., Canberra 1974) p.674
10.112 A. Rose: Television pickup tubes and the problem of noise. Adv. Electron. *1*, 131 (1948)
10.113 R.C. Williams, H.W. Fischer: Electron microscopy of tobacco mosaic virus under conditions of minimal beam exposure. J. Mol. Biol. *52*, 121 (1970)
10.114 M. Ohtsuki, E. Zeitler: Minimal beam exposure with a field emission source. Ultramicroscopy *1*, 163 (1975)
10.115 K.H. Herrmann, J. Menadue, H.T. Pearce-Percy: "The Design of Compact Deflection Coils and Their Application to a Minimum Exposure System," in *Electron Microscopy 1976*, Vol.1, ed. by D.G. Brandon (Tal International, Jerusalem 1976) p.342

10.116 Y. Fujiyoshi, T. Kobayashi, K. Ishizuka, N. Uyeda, Y. Ishida, Y. Harada: A new method for optimal-resolution electron microscopy of radiation-sensitive specimens. Ultramicroscopy *5*, 459 (1980)
10.117 I.A.M. Kuo, R.M. Glaeser: Development of methodology for low exposure, high resolution electron microscopy of biological specimens. Ultramicroscopy *1*, 53 (1975)
10.118 S.B. Hayward, R.M. Glaeser: Radiation damage of purple membrane at low temperature. Ultramicroscopy *4*, 201 (1979)
10.119 W. Chiu, R.M. Glaeser: "Evaluation of Photographic Emulsions for Low-Exposure Imaging," in *Electron Microscopy at Molecular Dimensions*, ed. by W. Baumeister, W. Vogell (Springer, Berlin, Heidelberg, New York 1980) p.194
10.120 M. Kessel, J. Frank, W. Goldfarb: "Low-Dose Microscopy of Individual Biological Macromolecules," in *Electron Microscopy at Molecular Dimensions*, ed. by W. Baumeister, W. Vogell (Springer, Berlin, Heidelberg, New York 1980) p.154
10.121 M.N. Kabler, R.T. Williams: Vacancy-interstitial pairs production via electron-hole recombination in halide crystals. Phys. Rev. B*18*, 1948 (1978)
10.122 H. Strunk: "High Voltage Transmission Electron Microscopy of the Dislocation Arrangement in Plastically Deformed NaCl Crystals," in *High Voltage Electron Microscopy*, ed. by P.R. Swann, C.J. Humphreys, M.J. Goringe (Academic, London 1976) p.285
10.123 L.W. Hobbs, A.E. Hughes, D. Pooley: A study of interstitial clusters in irradiated alkali halides using direct electron microscopy. Proc. Roy. Soc. A*332*, 167 (1973)
10.124 L.W. Hobbs: "Radiation Effects in the Electron Microscopy of Beam-Sensitive Inorganic Solids," in *Developments in Electron Microscopy and Analysis*, ed. by J.A. Venables (Academic, London 1976) p.287
10.125 L.W. Hobbs: Radiation damage in electron microscopy of inorganic solids. Ultramicroscopy *3*, 381 (1979)
10.126 T. Evans: Decomposition of calcium fluoride and strontium fluoride in the electron microscope. Philos. Mag. *8*, 1235 (1963)
10.127 L.E. Murr: Transmission electron microscope study of crystal defects in natural fluorite. Phys. Status Solidi A*22*, 239 (1974)
10.128 L.T. Chadderton, E. Johnson, T. Wohlenberg: "Transmission Electron Microscopy of 100 keV Electron Damage in Fluorite," in *Developments in Electron Microscopy and Analysis*, ed. by J.A. Venables (Academic, London 1976) p.299
10.129 R.D. Baeta, K.H.G. Ashbee: "Electron Irradiation Damage in Synthetic Quartz," in *Developments in Electron Microscopy and Analysis* (Academic, London 1976) p.307
10.130 D.E. McLennan: Study of ionic crystals under electron bombardment. Canad. J. Phys. *29*, 122 (1951)
10.131 O. Glemser, G. Butenuth: Veränderungen von $KMnO_4$ im Elektronenstrahl im Vergleich zur thermischen Zersetzung. Optik *10*, 42 (1953)
10.132 R.B. Fischer: Decompositions of inorganic specimens during observation in the electron microscope. J. Appl. Phys. *25*, 894 (1954)
10.133 J.H. Talbot: Decomposition of $CaSO_4.2H_2O$ in the electron microscope. Br. J. Appl. Phys. *7*, 110 (1956)
10.134 H.R. Oswald, W. Feitknecht: "The Oxydation of Manganous Hydroxide with Molecular Oxygen and the Transformation of the Products in the Electron Beam," in *Electron Microscopy 1962*, Vol.1, ed. by S.S. Breese (Academic, New York 1962) p.H-9
10.135 J. Sawkill: Nucleation in silver azide, an investigation by electron microscopy. Proc. Roy. Soc. A*229*, 135 (1955)
10.136 M.J. Makin: "Atom Displacement Radiation Damage in Electron Microscopes," in *Electron Microscopy 1978*, Vol.3, ed. by J.M. Sturgess (Microscopical Soc. Canada, Toronto 1978) p.330
10.137 M. Wilkens, K. Urban: "Studies of Radiation Damage in Crystalline Materials by Means of High Voltage Electron Microscopy," in *High Voltage Electron Microscopy*, ed. by P.R. Swann, C.J. Humphreys, M.J. Goringe (Academic, London 1974) p.332
10.138 K. Urban: "Radiation Damage in Inorganic Materials in the Electron Microscope," in *Electron Microscopy 1980*, Vol.4, ed. by P. Brederoo, J. Van Landuyt (Seventh European Congr. Electron Microscopy Foundation, Leiden 1980) p.188

10.139 V.E. Cosslett: "Radiation Damage by Electrons, with Special Reference to the Knock-On Process," in *Electron Microscopy and Analysis 1979*, ed. by T. Mulvey (The Institute of Physics, London 1979) p.277
10.140 M. Wilkens: "Radiation Damage in Crystalline Materials, Displacement Cross Sections and Threshold Energy Surfaces," in *High Voltage Electron Microscopy*, ed. by T. Imura, H. Hashimoto (Japanese Soc. Electron Microscopy, Kyoto 1977) p.475
10.141 N. Yoshida, K. Urban: "A Study of the Anisotropy of the Displacement Threshold Energy on Copper by Means of a New High-Resolution Technique," in *High Voltage Electron Microscopy*, ed. by T. Imura, H. Hashimoto (Japanese Soc. Electron Microscopy, Kyoto 1977) p.493
10.142 W.E. King, K.L. Merkle, M. Meshii: "Study of the Anisotropy of the Threshold Energy in Copper Using In-Situ Electrical Resistivity Measurements in the HVEM," in *Electron Microscopy 1980*, Vol.4, ed. by P. Brederoo, J. Van Landuyt (Seventh European Congr. Electron Microscopy Foundation, Leiden 1980) p.212
10.143 M.O. Ruault: In situ study of radiation damage in thin foils of gold by high voltage electron microscopy. Philos. Mag. *36*, 835 (1977)
10.144 L.E. Thomas: The diffraction dependence of electron damage in a high voltage electron microscope. Radiat. Eff. *5*, 183 (1970)
10.145 N. Yoshida, K. Urban: "Electron Diffraction Channelling and Its Effect on Displacement Damage Formation," in *High Voltage Electron Microscopy*, ed. by T. Imura, H. Hashimoto (Japanese Soc. Electron Microscopy, Kyoto 1977) p.485
10.146 D.W. Pashley, A.E.B. Presland: Ion damage to metal films inside an electron microscope. Philos. Mag. *6*, 1003 (1961)
10.147 J.J. Hren: "Barriers of AEM: Contamination and Etching," in *Introduction to Analytical Electron Microscopy*, ed. by J.J. Hren, J.I. Goldstein, D.C.Jo (Plenum, New York 1979) p.481
10.148 L. Reimer, M. Wächter: Contribution to the contamination problem in transmission electron microscopy. Ultramicroscopy *3*, 169 (1978)
10.149 H.G. Heide: Zur Vorevakuierung von Photomaterial für Elektronenmikroskope. Z. Angew. Phys. *19*, 348 (1965)
10.150 A.E. Ennos: The sources of electron-induced contamination in the electron microscope. Br. J. Appl. Phys. *5*, 27 (1954)
10.151 S. Leisegang: "Über Versuche in einer stark gekühlten Objektpatrone," in *Proc. 3rd Intern. Congr. on Electron Microscopy*, ed. by R. Ross (Royal Microscopical Soc., London 1954) p.184
10.152 H.G. Heide: Die Objektverschmutzung im Elektronenmikroskop und das Problem der Strahlenschädigung durch Kohlenstoffabbau. Z. Angew. Phys. *15*, 116 (1963)
10.153 H.G. Heide: Die Objektraumkühlung im Elektronenmikroskop. Z. Angew. Phys. *17*, 73 (1964)
10.154 J.T. Fourie: The controlling parameter in contamination of specimens in electron microscopes. Optik *44*, 111 (1975)
10.155 K.H. Müller: Elektronen-Mikroschreiber mit geschwindigkeitsgesteuerter Strahlführung. Optik *33*, 296 (1971)
10.156 G. Love, V.D. Scott, N.M.T. Dennis, L. Laurenson: Sources of contamination in electron optical equipment. Scanning *4*, 32 (1981)
10.157 M.T. Browne, P. Charalambous, R.E. Burge: "Uses of Contamination in STEM: Projection Electron Lithography," in *Developments in Electron Microscopy and Analysis 1981*, ed. by M.J. Goringe (The Institute of Physics, London 1981) p.47

Subject Index